Basics of Horticulture

3rd Revised Edition

Basics of Horticulture

3rd Revised Edition

K.V. Peter
Former Vice Chancellor
Kerala Agricultural University
Kerala

Former Director
ICAR-Indian Institute of Spices Research
Kozhikode-673012, Kerala
Email:kvptr@yahoo.com

NEW INDIA PUBLISHING AGENCY
New Delhi – 110 034

NEW INDIA PUBLISHING AGENCY

101, Vikas Surya Plaza, CU Block, LSC Market
Pitam Pura, New Delhi 110 034, India
Phone: + 91 (11)27 34 17 17 Fax: + 91(11) 27 34 16 16
Email: info@nipabooks.com
Web: www.nipabooks.com

Feedback at: feedbacks@nipabooks.com

ISBN: 978-93-86546-52-4

Composed and Designed by NIPA

Devotion

The book "Basics of Horticulture" is devoted to Prof. M S Swaminathan, a Great Teacher, Scientist and above all a great source of inspiration, encouragement and hope to Farmers of the World, Agricultural Scientists and the down trodden. Prof. M.S. Swaminathan is an Institution for the Poor, Weak and Toiling masses of the World. The movement for empowering Farm Women was initiated by him. A great visionary, he provided the research support to the Great Green Revolution, which transformed India from ship to mouth existance to a food surplus nation.

I also devote the book to Divine powers which gifted me grandson Antony Ajay Peter Kuruppacharil born on Tuesday 10th December 2013 at 11.50 am (Pooruttathi Nakshathram) and grand daughters Anna Vimala Anvar born on Sunday 31st August 2014 at 9.52 am (Chothi Nakshathram) and Annie Theresa' Anvar Born on Monday 16th January 2017 at 10.33 pm (Uthram Nakshathram) to keep my family happy and joyful.

प्रो. वी. एल. चौपड़ा
Prof. V.L. Chopra

सदस्य, योजना आयोग
योजना भवन, नई दिल्ली–110 001
Member, Planning Commission
Yojna Bhawan, New Delhi - 110 001

Foreword

In India, Horticulture encompasses 8.5% of arable land and contributes 24.5% of the GDP from Agriculture. Plantation crops alone occupying 0.95% of cropped area have a stake of 15.1% of the total agricultural export earnings. Horticulture provides a dynamic tool for enhancing economic returns, attaining nutritional security, creating employment avenues and ensuring ecological sustainability. The sector covers a broad spectrum of activities like production, post-harvest management, marketing, processing and export of fruits, vegetables, flowers, medicinal and aromatic plants, plantation crops, spices, bamboo, mushroom, honey and silk.

India is the largest producer of fruits (74.87 million tones) and vegetables (141.00 million tones) next to China. India's share in production is 9.9% in vegetables, 9.9% in fruits, 7.8% in Potatoes and 11.5% in dry Onions. In total world production, India ranks third in Potato (23.16 million tones), second in Onion (4.2 million tonnes), Brinjal (8.0 million tonnes), Cabbage (5.39 million tones), Cauliflower (4.4 million tones) and Peas (1.71 million tones). In fruits, India ranks first in Banana (13.30 million tonnes), Grapes (7.39 million tones), Mandarin Orange (18.53 million tones), Mango (12.73 million tones), 10th in Apple (1.35 million tones) and Papaya (0.15 million tones) and 4th in Pineapple (1.17 million tonnes). Despite the above positive indicators, performance of the sector in value addition, processing and export are not commensurate with its progress on the production front. In spite of having a 9.9% share of the global production of both fruits and vegetables, just 1.8% is processed and India's stake in export trade is a meagre 0.4%. The per capita availability of fruits and vegetables is also very low leading to nutritional insecurity, ill health, and loss in working hours. India has to increase the present level of 215.87 million tones of fruits and vegetables to 300 million tones by the end of XI plan. Use of high yielding planting materials, minimization of water use, living and productive soils, use of technologies like stress free nursery, precision farming, clean vegetable, fruit and ornamental production, vegetable and fruit forcing, improved packing houses, value added products, benefits through GI and IPR regimes etc. are to be explored. Uses of biotechnology, synthetic biology, information technology, functional genomics and modern marketing tools would benefit the farmers ultimately.

The present book "Basics of Horticulture" emphasizing Science in Horticulture would go a long way in upgrading and updating the present information. The post-graduate syllabus in Horticulture at ICAR level is getting revised under the convenership of Prof. K.V. Peter who is also the editor of this book. The editor has taken the information requirement of all stake holders-farmers, students, teachers, industry, seedsmen- for compiling the book. A large number of Deans, Professors, Faculty Members, Students and Progressive Farmers have been consulted.

I compliment all those Professors and also Prof. K.V. Peter for this important academic activity.

(V.L.Chopra)

Preface to 3rd Revised Edition

Science and practice of Horticulture have witnessed significant improvement in content, technological inputs, crops covered, diversified products development, marketing and above all awareness in healthcare and wellness. Globalization of horticulture and trade in products have made visible changes in the profession of marketing and trade. Modern nurseries with updated devices for control of water, nutrients, light and energy have resulted in the development of quality planting materials –coated seeds, hardened seedlings, seedling grafts (stions), rooted cuttings, Tissue Cultured (TC) plants. Biofortification of propagules using biocontrol agents has boosted seedling stand and survival in stressed soils. The limiting factors of production like reduced area (space), water, nutrients and energy are addressed by techniques of protected cultivation. Polytunnels, rain shelters, polyhouses,solar powered cold frames and modern glass houses have made cultivation of horticultural crops a reality throughout the year. Crops forcing is now a reality to grow crops during offseason and for periods of high demand. Fertigation (application of fertilizers dissolved in water) has minimized loss of nutrients and enhanced uptake considerably.Use of pheromones and colour traps inside the growing structures have reduced damages due to insect attack. Many biocontrol agents are identified which are eco-friendly and leaving no residues to crops as detected after application of pesticides especially systemic insecticides. High density planting is now practiced in mango, guava, avocado, pear, peach, plum and apple. Training and pruning are tailoring techniques to modify plant structure to utilize optimum space and harvesting within reach of hand both manual and mechanical. Peri-urban and urban horticulture are gaining attention for nutritional security, greening of cities and towns and ultimately to make earth green enough for humanity to live. Floriculture has emerged from the status of potted plants to commercial ventures. Harvest and post harvest technology are further strengthened by innovations much cherished by consumers and processors as well. Nano technology is emerging in products development. Vegetables, fruits, plantation crops, spices, ornamentals, tubers, medicinal and aromatic plants, bamboos, mushrooms and beverage crops are keys to human welfare and wellness. Information on above crops are updated. The present revised edition has 21 chapters including 10 appendices. Forty six scientists from eight Institutes, States Agricultural Universities and two organizations have contributed to the revised book. Since publication of first edition of BASICS OF HORTICULTURE in 2009, there is increase in number of Universities of Horticulture, Research Institutes and trade in products. A village market has now all kinds of vegetables, fruits, tubers and ornamentals which vouch for progress in the science and art of horticulture. Many educated youth are taking up Horticulture as a profession. Basic sciences like physiology, biochemistry,molecular biology and biotechnology, bioinformatics and economics are adding to the understanding of horticultural crops. Horticulture based science societies and academies are on rise. Publications like journals, magazines and proceedings of seminars/symposia are on the increase. The e-journals and e-books on the subject of Horticulture have made the subject closer to users.

I present the revised 3rd Edition -2018 for use of teachers, scientists, students, farmers, housewives and trade as a beacon of light in their search for updated information.

KV Peter

Acknowledgement

I acknowledge the contributors of 20 chapters and 10 appendices of the book **Basics of Horticulture 3rd Revised Edition** for their knowledge and willingness to share the same to the present and future generations of Horticulturists, students, scholars and traders. Prof. V L Chopra, my most respected Professor was kind enough to write Foreword to the book. My wife Vimala,sons Anvar and Ajay; daughters-in-law Anu and Cynara, grand son Antony Ajay Peter and grand daughters Anna Vimala Anvar and Annie Theresa Anvar were supportive in my struggle for good literature and books in Horticulture. Prof. P I Peter Chairman NoniBiotech Chennai and Dr Kirti Singh Chairperson World Noni Research Foundation Chennai granted me full freedom and facilities. My late parents Shri. Kuruppacharil Devassey Varkey and Mrs Rosa Varkey showered spiritual strength in this academic activity.

I acknowledge Mr Sumit Jain, Publisher, New India Publishing Agency, New Delhi for excellent printing of the revised edition.

Contents

Contributors

A. Chattopadhyay
Department of Vegetable Crops, BCKV
Mohanpur – 741 252, West Bengal

A.M. Ranjith
Department of Entomology
College of Horticulture
Kerala Agricultural University, P.O. KAU
Thrissur – 680 656, Kerala

Ancy joseph
Aromatic and Medicinal Plants Research Station
Odakkali, Ernakulam – 683 549

Asha Sankar
Department of Plantation Crops and Spices, College of Horticulture, P.O. KAU – 680 656
Thrissur, Kerala

Binoi K. Menon
Vegetable and Fruit Promotion Council of Keralam, Kochi
Kerala Agricultural University
P.O. KAU, Thrissur – 680 656, Kerala

Binoo Bonny
Department of Extension
Karela Agricultural University
P.O. Mannuthy, Thrissur – 680 656, Kerala

C.S. Anisha
Indian Cardamom Research Institute
Spices Board of India
Cochin – 682 025, Kerala

G.S.L.H.V. Prasada Rao
Kerala University of Veterinary and Animal Sciences
P.O. Mannuthy, Thrissur – 680 656, Kerala

G. Suja
Department of Entomology
College of Agriculture, P.O. Vellayani
Trivandrum – 695 522, Kerala

Hari Har Ram
U.P. Council of Agricultural Research
8th Floor Mandi Bhavan Vibhuti Khand,
Gompti Nagar, Lucknow - 226010

K. Dhinesh Babu
ICAR-NRC on Pomegranate, NH-9
Bypass Road, Shelgi (P.O)
Sholapur – 413 006, Maharashtra

K. Nirmal Babu
ICAR-IISR, Kozhikkode, Kerala – 673 012

K.V. Peter
World Noni Research Foundation
12, Srinivasa Nagar 2nd Street
Perungudi, Chennai – 600 096

K. Pradeep Kumar
Indian Cardamom Research Institute
Spices Board of India, Cochin – 682 025, Kerala

K. Mary Mathew
Indian Cardamom Research Institute
Spices Board of India, Cochin – 682 025
Kerala

K.S. Meenakumari
Department of Entomology
Onattukara Regional Agricultural Research Station, Kayamkulam, Alleppey, Kerala

K.R. Remya
College of Agriculture
Kerala Agricultural University
P.O.Vellayani, Trivandrum – 695 522, Kerala

K.B. Sheela
Department of Processing and Post Harvest Technology
College of Horticulture
Kerala Agricultural University
P.O. KAU, Thrissur – 680 656, Kerala

Lila Mathew
Department of Pomology and Floriculture
Kerala Agricultural University
P.O. KAU, Thrissur – 680 656, Kerala

M.N. Badukla
Krishidhan Vegetable Seeds India Pvt. Ltd.
Pune, Maharashtra

M.K. Pandit
Department of Vegetable Crops, BCKV
Mohanpur – 741 252
West Bengal

M.S. Sheela
College of Agriculture
Kerala Agricultural University
P.O.Vellayani, Trivandrum – 695 522, Kerala

N.K. Parameswaran
Kerala Agricultural University
P.O. KAU, Thrissur – 680 656, Kerala

P. Skaria Baby
Aromatic and Medicinal Plants Research Station
KAU, Odakkali, Ernakulam – 683 549

P. Maheswarappa
Co-ordinator, AICRP on Palms
ICAR-CPCRI, Kudlu P.O., Kasargold – 671124

P.B. Pushpalatha
Department of Processing and Post Harvest
Technology, College of Horticulture
Kerala Agricultural University
P.O. KAU, Thrissur – 680 656, Kerala

P.G. Sadhankumar
Department of Olericulture
Kerala Agricultural University
P.O. KAU, Thrissur – 680 656, Kerala

P.K. Valsala Kumari
College of Horticulture, KAU, P.O. KAU
Thrissur – 680 656, Kerala

P. Hazra
Department of Vegetable Crops, BCKV
Mohanpur – 741 252
West Bengal

P.P. Joy
Aromatic and Medicinal Plants Research Station
Odakkali, Ernakulam – 683549

Ramya Lakshmi
Indian Cardamom Research Institute
Spices Board of India
Cochin – 682 025, Kerala

Ragina Joseph
Aromatic and Medicinal Plants Research Station
KAU, Odakkali, Ernakulam – 683549

Sabeena Jose
College of Agriculture
Kerala Agricultural University
P.O.Vellayani, Trivandrum – 695 522, Kerala

Sapna Panwar
Division of Floriculture and Landscaping
I.C.A.R., IARI
New Delhi- 110012

S. Dutta
Department of Vegetable Crops, BCKVV
Mohanpur – 741 252, West Bengal

S.K. Sharma
ICAR-CIAH, Bikaner, Rajasthan – 334 006

Sherin Jose
Indian Cardamom Research Institute
Spices Board of India, Cochin – 682 025
Kerala

Swapna Sasidharan
Indian Cardamom Research Institute
Spices Board of India
Cochin – 682 025
Kerala

Shinoj Subramanian
KVK, ICAR-CMFRI, PB No 1603
Ernakulam North, O.P, Kerala

S. Nirmala Devi
Department of Olericulture
Kerala Agricultural University
P.O. KAU, Thrissur – 680 656, Kerala

S.S. Sindhu
Division of Floriculture and Landscaping
I.C.A.R., IARI
New Delhi- 110012

Samuel Mathew
Aromatic and Medicinal Plants Research Station
KAU, Odakkali, Ernakulam – 683549

T.R. Gopalakrishnan
Kerala Agricultural University
P.O. KAU, Thrissur – 680 656, Kerala

T. Bharati Meena
College of Agriculture
Kerala Agricultural University
P.O.Vellayani, Trivandrum – 695 522, Kerala

T. Jiji
Department of Entomology
College of Agriculture,
Kerala Agricultural University
P.O.Vellayani, Trivandrum – 695 522
Kerala

T. Pradeepkumar
Kerala Agricultural University, P.O. KAU
Thrissur – 680 656, Kerala

T. Radha
Department of Pomology and Floriculture
Kerala Agricultural University
P.O. KAU, Thrissur – 680 656, Kerala

U. Thapa
Department of Vegetable Crops, BCKVV
Mohanpur – 741 252, West Bengal

V.L. Sheela
College of Agriculture
Kerala Agricultural University
P.O.Vellayani, Trivandrum – 695 522, Kerala

Y.S. Rao
Indian Cardamom Research Institute
Spices Board of India
Cochin – 682 025
Kerala

Introduction

The book "Basics of Horticulture" published in 2009 is revised with updated information on production, protection, harvest and post harvest processing, value addition, products development and marketing.

Basics of Horticulture deals with principles and fundamentals of Fruit crops, Vegetables, Ornamentals, Spices, Aromatic crops, Medicinal plants, Plantation crops, tubers and Bamboo. Plants, nutrients, water, soil and climate are the pivotal points of any successful crop growing. Product development, value addition and marketing in a competitive global market are assuming importance. Green products and green marketing are emphasized in a clean human habitat. Horticulture as a Science and profession is to create employment and income in addition to nutrition security. Science of apomixis and polyembryony (PE) is very special and relevant in Horticultural Crops. Large scale multiplication of mangoes and citrus is possible by making use of apomixis. Now the genes for apomixis are transferred to self pollinated crops. Polyembryony is again a vital phenomenon worth elucidating. Climate plays a major role in crop distribution while weather in crop productivity. Climate decides the crop habit which depends on genotypes and phenotypes. Majority of plantation crops show seasonal behaviour in phonological events though they are perennials except crops like coconut and arecanut, having flowering phase round the year. Within the tropics, crops like Coffee, Cardamom and Tea prefer cool temperate climate as they grow along the high ranges where mean annual temperature varies 14 to 20°C while Coconut, Cashew, Cocoa and Rubber are seen cultivated in mid and high lands, where the annual mean temperature revolves 25 to 29°C. Black pepper is grown in the buffer zone as an intercrop between warm and cool climate, stretching between tropical and temperate climates within the tropics due to its wide adaptability. The Cashew growing regions across the world are located commercially between 15°N and 15°S from the equator. Small Cardamom- Queen of Spices is indigenous to evergreen rain forests of Western Ghats. The natural habit at of Cardamom is the high ranges between 1000 and 1500m amsl across the Western Ghats and characterized by cool, humid micro climate, which provides ideal conditions for Cardamom cultivation. Coffee is grown at high elevations across Western Ghats between 750 and 2000 m amsl. Robusta and Arabica are the two traditional cultivars grown over the high ranges along Western Ghats, covering three states of Karnataka, Kerala and Tamil Nadu. The "Monsooned Malabar Coffee" is processed at Mangalore, in coastal south Karnataka. High atmospheric humidity together with moderate temperature and gentle marine breeze with low vapour pressure deficit accompanied by continuous heavy rains during monsoon are characteristic features of Malabar Region. These peculiar climatic features along West coast (Malabar Region) in the humid tropics may be conducive for coffee beans processing. Night temperature and relative humidity are the most important factors which determine productivity of Tea in addition to rainfall. Irrigation during hot and dry periods that occur frequently in Nilgiris will be beneficial in crop production. Rainfall preferably between 1500 and 2000 mm annually with a dry season of not more than three months with less than 100 mm/month and temperature varying between 30-32°C as mean maximum are the climatic boundaries desirable for growing Cocoa, another beverage crop. Coconut is a tropical tree as its distribution is confined almost entirely to tropical zones. The effect of climate factors on perennial crops like coconut is different from that of annual crops.

Insect pests play havoc on crops productivity and cause considerable damage. Pests mean insect and non-insect pests of animal kingdom, diseases and weeds. Pest management is a strategy to reduce the conditions for multiplication of pests and to provide situations that minimize growth of crops. In modern Horticulture, pest management means environmentally safe, ecologically sound and economically sustainable crop production. The Agro-ecosystem encompassing the environment, the crop and pest complex is a total unit for pest management. Integrated Pest Management (IPM) is the total strategy or totality of actions in crops against individual or a group of pests that involve tactics based on cost: benefit analysis involving impact on crops, pests and environment and ultimately consumers. Integrated pest management strategies in Potato, tropical tuber crops like Aroids and Yams, Cassava, Mushroom, Chinese Potato, Sweet Potato, Ornamentals like Anthurium, Carnation, Chrysanthemum etc; Aromatic plants like Ambretta, Davana, Fennel, Lemon grass, Mentha and Patchouli; Plantation crops like Arecanut, Cashew, Cocoa and Oil palm, Coconut, Coffee and Tea; Spices like Black Pepper, Large Cardamom, Cardamom, Cinnamon, Clove and seed spices; fruit crops like Almond, Amla, Apple, Apricot, Citrus, Custard apple, Fig, Grape, Jackfruit, etc; Vegetables like Amaranth, Spinach, Beet root, Brinjal etc are elaborated. Pesticide free crop produces are the desired green products.

Nematodes constitute a group of primitive microscopic invertebrate animals which occur in all conceivable habitats in amazingly large numbers. Besides bacteria, virus, fungi, insects and nematodes are important pests of horticultural crops. The cost of crops loss is estimated more than Rs.242 billion annually. Continuous cropping of any single crop leads to a gradual increase in nematode density. Control of nematode is far more difficult than any other kind of pests because they usually inhabit soil and usually attack underground parts of plants. In principle, nematode management is an applied ecology. Nematode parasites are serious problems in most of the horticultural crops like fruits, spices, vegetables, tubers, plantations, medicinal and aromatic plants and ornamentals. Cultural methods like crop rotation; mulching, fallowing and organic amendments are effective. Biological methods, plant resistance and their integration are eco-friendly and are effective to manage nematodes.

Basics of Vegetable Science are oriented on nutritional and livelihood security, status of production, classification, influence of temperature, light and soil on production, management of seedlings in the nursery, raising of grafted seedlings, vegetable based cropping system, water management, plant nutrition and integrated nutrient management (INM), micro irrigation, pollination management, integrated pest management (IPM), biological control of diseases, weed management, poly house cultivation, post harvest handling and storage, marketing and export of vegetables and seed technology. Vegetables provide nutritional and health security. They are rich in vitamins (Vit.A; Vit.B1; Vit.B2; Niacin; Vit.B6;Folic acid; Vit.E; Vit.K; Vit.C), minerals (Calcium, Iron, Phosphorous, Iodine, Sodium), dietary fibre, anti oxidants (Carotenoids) and flavonoids. Medicinal values of vegetables (Tomato, Brinjal, Chilli, Sweet pepper, Cauliflower, Cabbage, Onion, Garlic, Pointed Gourd, Cucumber, Bottle gourd, Bitter gourd, Pumpkin, Okra, Cowpea, Radish) are explained for their alkaline properties, fiber content and non-calorific values. Vegetable productivity in India rose from 13.42 t/ha in 1995-96 to 17.38t/ha in 2011-12 due to use of high yielding varieties and practice of better crop management techniques. The potential productivity is still very high (Tomato-60-80t/ha; Brinjal-40-50 t/ha; Chilli-30-40 t/ha). Demand for Vegetables in India is projected 160 million tones in 2020-21. Hi-tech vegetable production involving hi-tech nursery, protected cultivation, efficient water and nutrient management and post harvest handling would lead to better price realization. Classification (botanical, hardiness, growing season, tolerance to soil reaction, plant parts used, methods of culture) of vegetables is elucidated for a better understanding of crops. Vegetable production is influenced by environmental parameters- temperature, light and soil characteristics. Grafting as a technology to raise healthy seedlings is a recent innovation. Plant nutrients and integrated nutrient management (INM) are elaborated as productivity enhancing factors. Nutrient requirement of vegetable crops, common diseases due to imbalanced nutrition and components of INM are discussed. Micro-irrigation and fertigation in vegetable crops manage water and fertilizer use to optimum levels. Pollination is a biological necessity for fruit and seed development in majority of vegetables. Bees are pollinators in Mustard, Onion, Cabbage, Cauliflower and Brussels sprouts. Pests and diseases devastate vegetable crops. Pesticide application leaves residues, undesirable for human health. Integrated Pest and Disease Management (IPDM) is an eco-friendly strategy to minimize crop losses leaving no residues in edible plant parts. Biological management of diseases and weeds is discussed elaborately. Weeds in vegetable gardens are classified, their spread studied and methods of control elaborated. Different classes of herbicides are dealt with. Vegetable forcing is growing the crops during off season using poly houses, green houses and climate controlled growing structures. Post harvest handling and storage of vegetables are dealt to illustrate extend of crop losses and methods to minimize the losses. Controlled Atmospheric Storage (CAS) and hypobaric storage are explained as possible methods to prolong storage life. Export trade of vegetables, main constraints of export and high value vegetables for export are brought to lime light. Seed technology in terms of information on seed moisture content, optimum moisture content, recalcitrancy and dormancy are highlighted.

Hybrid breeding technology is used widely by vegetable breeders. It was in 1916, hybrid vigour was first reported in cucumber. Hybrid vegetables are uniform, vigorous, early, high yielding and resistant to specific pests and pathogens. Several breeding systems are available making use of genetic male sterility, cytoplasmic-genetic male sterility, self incompatability, gynoecious lines, dioecy, monoecy, chemical suppressants, chemical hybridizing agents and manual methods. Reports on genetic male sterility are available in Tomato, Brinjal, Chilli, Cauliflower, Cabbage, Broccoli, Watermelon, Muskmelon, Cucumber, Summer Squash and Pumpkin. Genetic male sterility is used in hybrid seed production in Chilli and Muskmelon on a commercial scale. Cytoplasmic male sterility is used in Hybrid seed production in Tomato, Chilli, Cole crops, Onion, bunching Onion, Carrot and Radish. Production of Onion F_1 hybrids is elaborately discussed in the context of A, B and C or R lines. Self–incompatibility is another system used for hybrid seed production as in cole crops. Current policy on seed supply of GMS lines A, B and C/R lines is explained.

Biocontrol of pests and diseases is assuming prominence in the context of clean and zero residue vegetable production. It is the biological destruction of harmful micro-organisms causing plant diseases without disturbing the biological balance. Parasites and predators are used in an eco-friendly pest management strategy. Biological control programme is a three prolonged approach; importation and colonization of exotic natural enemies, conservation of existing natural enemies by modifying the conditions which favour and promote natural enemies and by making periodic releases of natural enemies for inoculation and inundation. As per ICAR-National Research Centre for Integrated Pest Management (2003), 165 pests were permanently controlled up to 1988 and there are 384 cases of successful control worldwide. The first deliberated introduction of an exotic natural enemy for biological control of an insect pest in India was the Australian lady bird beetle from Australia in 1898 to control the green scale *Coccus viridis* in Southern India. Conservation, augumentation, inoculative releases and inundative releases of predators and parasites are requisites for sustainable pest management. Majority of predators belong to Coccinellids, Syrphids, Acarine predators and Spiders. Parasites belong to mostly *Trichogramma*. Mass multiplication of predators and parasites are dealt in detail. Fungal pathogens with insectivorous properties are *Beaveria* spp; *Metarhizum anisopliae*; *Paecilomyces* spp; *Verticillium lecanii*; *Fusarium* spp; and *Rhizopus* spp. Nucleo polyhedro virus (NPV) is used effectively against *Agrotis ipsilon*, *Amsacta albistriga*, *Anadevidia peponis*, *Anomis flava*, *Helicoverpa armigera*, *Hellula undalis*, *Mythimma separata* etc. Other bio-control agents are selected protozoa, nematodes and novel microbes. Avermectin isolated from fermentation broth of *Streptomyces avermectlis* is effective against plant parasitic nematodes, anthrapods and leaf miners. Biological control of plant diseases is another challenging area of research.

Fruits play a significant role in the economic development, nutritional security, employment generation and overall economic growth of a country. The tropical and sub-tropical fruits consist of major fruits like Banana and Plantain, Citrus, Mango, Grapes and Pineapple and minor fruits like Avocado, Guava, Jackfruit, Litchi, Longan, Mangosteen, Papaya, Passion fruit, Rambutan, Sapota, Annonas, Bread fruit, Carambola, Langsat, Macadamia and White Sapota. The per capita availability of fruits in India is 62g/day while FAO recommends a consumption of 85g/day. The targeted fruit production in India by 2020 is 200 million tonnes. To achieve the above goal, the identified research priorities include collection and conservation of germplasm, measures to overcome chronic production problems, developing suitable horticulture based farming systems, developing high yielding varieties, faster multiplication of quality planting materials, introducing and popularizing suitable high quality fruit species from elsewhere and reducing post harvest losses. Botany and classification of fruit crops are relevant for designing agronomy and tree spacing. Classification is based on climate requirement, photo periodic responses, tolerance to shade, photo thermal requirements, drought tolerance, salt tolerance, soil reaction, specificity to soil texture, soil topography and based on bearing habit, longevity, rate of respiration during ripening, fruit size and nutritional aspects. Nutritive value of fruits is elaborated in terms of content of vitamins, minerals, anti oxidants, flavourants and calorific values. The fruits whose nutritive and medicinal values mentioned are Banana, Mango, Pineapple, Avocado, Citrus, Guava, Mangosteen, Papaya, Amla, Jamun, Jackfruit, Grapes, Olive, Pomegranate, Dates, Bread fruit, Carombola and Mango. There are 61 fruit crops in the World. The propagation methods of above crops like seeds, cuttage, graftage, budding, inarching, top working, layers and micropropagation are explained. Crop specific root stocks induce early fruit bearing, dwarfness and resistance to biotic and abiotic stresses. Design and Orchard management, precision farming and bio-regulator in fruit production are discussed in the context of enhancing productivity per unit land. Roles of Gibberellins, Cytokinins, Ethylene and Abscisic acid are dealt in detail for growth and development. Physiology of flowering, fruit set, fruit development and ripening are explained for a basic understanding of morphogenesis. Factors associated with growth and flowering like light, photoperiodism, temperature, altitude and soil are dealt in detail.

Post harvest losses are estimated 25-40% of fruit production. Fruits and vegetables are highly perishable. The losses are classified as physical, physiological and biological. Pre-harvest and post-harvest practices can minimize losses during bearing, harvest, processing, transport and marketing. Selection of varieties, manual harvesting/mechanical harvesting, sorting/grading, washing, trimming, curing, delatexing or desapping, waxing, pre-cooling, hot water treatment, vapour treatment, post harvest disease control, regulation of ripening, sprout inhibition, packaging, transportation, storage, refrigerated storage, controlled/modified atmosphere storage, hypobaric storage and zero-energy cool chamber are a few innovations to minimize pre and post harvest losses and to maximize returns for unit of marketable fruits harvested.

Preservation of fruits and vegetables is dealt in a separate chapter. Factors causing spoilage of foods, methods of preservation-asepsis, by filtration, low temperature, removal of water, use of chemicals, irradiation, hurdle technology of stabilization of foods etc are explained. Minimally processed (fresh cut) fruits and vegetables are gaining consumer acceptance. Selection of appropriate cultivars, good sanitation, pre-storage treatments with anti-oxidants, preservatives, edible coating, irradiation, low temperature and modified atmospheric storage are used to enhance shelf life of fresh cut products. Preservation of foods is by use of preservatives, classified as Class I preservatives (natural) and Class II preservatives (chemical). Commonly used natural preservatives are Common salt, Sugar, Honey, Dextrose, Vinegar, Vegetable Oil, Organic acids, Spices and Bacteriosins. Mode of action of above preservatives is mostly through plasmolysis of microbial cells and anti-microbial properties. Class II presevatives are sulphurous acids and their salts and benzoic acids and their salts. They interfere with genetic mechanism of cell division of microbes, enzyme systems, biosynthesis of vitamins and formation of complexes with their heavy metal ions. Value addition of fruits and vegetables, advantages of processing, value added products - fruit juice concentrates, intermediate moisture foods, minimally processed products, HT stabilized products and a large number of products from fruits and vegetables are nerrated.

Marketing of fruits and vegetables assume economic importance. Marketing scenario including systems, regulated markets, direct marketing, farmer markets, e-marketing, supply chain management and post harvest management are explained.

Biotechnology of horticultural crops is explained taking spices as an emerging sector. Micropropagation using tissue culture has been standardized. Protocols are available for micropropagation of 32 herbs and spices, the same being elaborately described for Vanilla, Large Cardamom, Thyme, Oreganum, Salvia and Mint.

Benefits of scale can be tapped only by mass production and minimizing cost of unit of production. Mechanization of farm operations, like land preparation, planting, field maintenance, harvesting and transporting is imperative *a priori* to low cost of production. Specific equipments are available for cleaning of land, leveling and grading, tillage, planting, field monitoring, vegetation control, animal/bird scaring, insect and disease management, disaster management, irrigation, manuring, harvesting, collection and transportation and integrated mechanization.

A separate chapter on Glossary of Selected Horticultural terms is given for ready reference. Eight appendices narrate varieties released in Aromatic crops, Fruit crops, Spices, Plantation crops, Ornamentals, and Vegetables. Appendices VII and VIII provide information on Research and Education Infrastructure in Horticulture in India and important web sites in Horticulture. List of released varieties and hybrids in aromatic crops from KAU, varieties of tuber crops released in India, released varieties of arid crops varieties of ornamental plants released for growing, released varieties of coconut, cocoa and arecanut; newly released varieties of spices from KAU and varieties of spice crops released through AICRP on spices are appended.

Chapter – 1

Science and Profession of Horticulture

K.V. Peter, Binoo Bonny and T. Pradeepkumar

The word horticulture was first used in 1600's and comes from two Latin words, hortus and cultûra which mean the culture of garden plants. Relf (1992) provided a comprehensive definition of horticulture as an applied science and art of growing fruits, vegetables, herbs and spices, nuts, tubers, aromatic and medicinal plants, plantation crops, mushrooms and ornamental plants which include shrubs, flowering plants and turfs stimulating minds, spirits and emotions of individuals, enriching health and wellness of communities, and linking human beings to nature and its landscaped exteriors for wellness and wellbeing. It is a science as it applies the basics of Biology like botany, morphology, anatomy, cytology, cytogenetics, biochemistry, ethnopharmacology and physiology and applies them to growing of crops. It also uses related sciences like Material science, Chemistry, Physics, Mathematics and Statistics. It is also an art requiring creative faculties for vegetative propagation like grafting, budding, layering, topworking, micro-propagation and conceiving floral and landscape architecture and designs. Horticulture forms a branch of agriculture concerned with intensively cultivated plants directly used by people for food, for medicinal purposes or for aesthetic gratification (Janick, 1979).

History and Evolution of Horticulture as a branch of science are intimately connected to the history of Agriculture and as such it forms a part of 10000 years of human history from hunter-gatherer existence to settled civilizations. Both continued to evolve together till the middle ages when Horticulture separated as a distinct branch from agriculture, though both share common practices like weeding, fertilizer application, irrigation etc,. Horticulture is distinguished by vegetative propagation, parthenogenesis and apomixis. It requires more intensive management and higher labor inputs and offers higher gross return per unit area per unit time compared to agriculture. Many historians believe that Horticulture formally got evolved around 3000 B.C. from the Egyptian temple gardens where fruit trees, palms and grape vines were cultivated and land preparation, irrigation and pruning were practiced. Archeological evidences showed that the Egyptian gardens contained many horticultural components like symmetry, water features, pargolas, edges, hedges and animations. Meanwhile Mesopotamia, Babylonia and Assyria established pre-modern gardens like hanging gardens of Babylonia built by Nebuchadnezzar that cultivated roses, figs, dates, grapes, and olives. Theophrastus who was a student of Plato and Aristotle is considered as the first scientific horticulturist and is credited with the authorship of *History of Plants* and *The Causes of Plants*. Later as the Renaissance evolved, creation of formal Gardens inspired by the romanticized garden of paradise was reported from varied places across the world. Gardens of Versailles, Belvedere in Vienna and Mughal gardens of India were examples of ancient formal gardens. The use of new plants and plant products eventually led to the advent of new operational branches within Horticulture which can be grouped into two broad sections – ornamentals and edibles (Table 1).

1. Major Branches of Horticulture Science

Table 1: Major Branches of Horticulture Science

Sl. No.	Name of Branch	Definition
1.	Olericulture (Vegetable Science)	Study of production and marketing of vegetables including tubers
2.	Pomology(Fruit Science)	Study of production and marketing of fruits and nuts
3.	Floriculture/ ornamental Horticulture	Study of production and marketing of flowers
4.	Spices Medicinal and Aromatic Plants	Study of spicies medicinal
5.	Plantation Crops	Study of plantation crops and aromatic plants
6.	Greenhouse Management	Growing and study of plants in greenhouses
7.	Viticulture	Study of production and marketing of grapes
8.	Turf Management	Study of production and maintenance of turf grass for sports, leisure use or amenity use
9.	Arboriculture(Silviculture in forestry)	Study of, and the selection, planting, care and removal of, individual trees, shrubs, vines and other perennial woody plants including bamboo
10.	Landscape Horticulture	Application of design and horticultural principles to placement and care of plants in the landscape
11.	Interiorscaping	Application of design and horticultural principles to placement and care of plants in indoor environments
12.	Horticultural Therapy	Use of horticultural plants and methods as therapeutic tools with disabled and disadvantaged people
13.	Oenology	Includes all aspects of study of wine and winemaking
14.	Post harvest Technology	Study of maintaining the quality of and preventing the spoilage of horticultural crops

Influence of discovery, cultivation and trade of horticultural crops are quite significant in human history. Explorations by Europeans in the middle ages were mostly related to the search of these high value crops like spices,aromatic and medicinal plants and for developing trade routes for rapid transportation. This eventually led to colonalization of major Asian and African countries which produced rich and diverse Horticultural crops like Coffee, Tea, Cocoa and Spices-black pepper, cardamom, cinnamon- by European countries. Importance of Horticultural crops in international Commerce and Trade was quite significant from time immemorial and led to its emergence as one of the fastest growing sectors in agriculture. The growth was quite radical in countries like India where fruits, vegetables, tuber crops, flowers, medicinal and aromatic plants,mushrooms, spices and plantation crops are grown in their diverse agro-climatic regions (Table 2).

2. Potential Production Zones of Various Horticultural Crops in India

Table 2: Potential Production Zones of Various Horticultural Crops in India

Name of the crop	Key production zones
Fruits	
Mango	Maharashtra (Alphonsa), Gujarat(Kesari), A.P. (Banganapalli), U.P, (Chausa and Dasheri), West Bangal (Himasagar)
Grapes	Maharashtra, A.P. and Karnataka
Litchi	W.B., Bihar, Uttarakhand, Parts of U.P, Parts of A.P
Cashew	Kerala, Maharashtra, Karnataka, NE States
Walnut	J&K, High hills
Mandarins including Kinnow	Maharashtra, Darjeeling, Punjab and Rajasthan for Kinnow
Pomegranate	Maharashtra, Karnataka, Rajasthan.
Vegetables	
Potato	W.B., U.P and Punjab
Onion	*Kharif* onion. Eastern states (Bihar, Odisha and W.B) Punjab, Maharashtra and Karnataka.
Chilli	Maharashtra, Rajasthan and A.P.
Bitter gourd and okra	Kerala, Maharashtra, W.B., Bihar
Spices	
Black pepper	Kerala, Maharashtra, and Karnataka
Ginger	Kerala, NEH including Sikkim and Darjeeling hills
Turmeric	Kerala, A.P., Meghalaya and Odisha
Cumin	Gujarat and Rajasthan
Large cardamom	Sikkim and NE States
Floriculture	
Rose	Maharashtra, and Karnataka
Orchid (Cymbidium)	Sikkim, Arunachal Pradesh and Darjeeling (West Bengal)
Anthurium	Kerala, Karnataka and NE States
Cut greens	Kerala, Karnataka, A and N Islands and NE States
Dry flowers	Tamil Nadu, West Bengal

Source: Pradeepkumar *et al.*, 2008

3. Indian Horticulture

Modern and scientific horticulture was borne in 1928 when Imperial Council of Agricultural Research later named as Indian Council of Agricultural Research was constituted in India. It gained focused attention only from the third Five year plan, though systematic research on fruits, vegetables and ornamental crops began in 1954. Establishment of the ICAR Indian Institute of Horticultural Research (IIHR) at Bangalore and starting of eight All India Coordinated Research Projects (AICRP) to cover different horticultural crops in 4th Five Year Plan (1969-74) gave impetus to horticulture development in India. Horticulture education is now offered in 67 colleges (49 in State Agricultural Universities (SAUs), 2 in deemed Universties, 3 Central Agricultural Universities and 4 Universities with Agricultural Faculties 8 affiliated to SAUs and 3 others). Fourteen SAUs have facilities for education in this field at graduation level and 32 at post –graduation level. The intake capacity has sharply increased from 1,100 in 2000-01 to 3,000 in 2009-10 with expansion of intake in SAUs, including entry of private colleges affiliated to SAUs and other general universities. Against actual admitted strength of about 2,282 during 2009-10, 1,465 graduates passed out.Diploma in Horticulture is offered as a part of skill development by School of Agriculture, IGNOU. The Regional Centre ,IGNOU Cochin offers a Post Graduate Diploma in Plantation Management. TNAU Coimbatore offers B.Tech (Horticulture) emphasizing engineering aspect of Horticulture.The Krishi Vijyan Kendras offer training on nursery management,crop canopy management, protected cultivation, tissue culture, biocontrol and biofertilizers,water management, packaging and recently retailing to unemployed youth. Training on coconut climbing using a climber has empowered men and women.There is considerable employbility in urban, peri-urban and even abroad especially middle east and North East Asian countries.

Horticultural research in the country is being carried out at 11 ICAR institutes (with 26 regional stations), 5 National Research Centres (on major crops),6 network projects/ outreach programmes and 5 Project Directorates. Area specific multi-disciplinary research is also being conducted under 13 ICAR All India Coordinated Research Projects each on Tropical, Subtropical and Arid Fruits; Vegetables, Potato, Tuber Crops, Mushrooms, Floriculture, Medicinal and Aromatic crops, Palms, Cashew, Spices and Betel vine; and Post Harvest Technology at 251 research centres located at various research Institutes and universities.There are five full fledged Universities of Horticulture and Forestry; 56 State Agricultural Universities and two Deemed Universities ICAR(IARI) and SHIATS where teaching, research and extension on horticultural crops are in progress. In addition, a few R&D establishments of crop/commodity boards (Coffee Board, Tea Board, Spices Board, Coir Board, Rubber Board, Coconut Development Board, Directorate of Cashew and Cocoa Development, Directorate of Spices, Aromatic and Medicinal Plants) and private sectors are providing research and development support to Indian horticulture. Research is being funded by DBT, DST, DRDO, CSIR, AERC and DOS in basic and strategic areas.

The period 1991-2003 saw planned investment in rapid expansion of infrastructure which helped emergence of horticulture as a sustainable livelihood option in India and is heralded as the period of golden revolution. This led to emergence of India as the second largest producer of fruits and vegetables,largest producer of mango, banana, coconut, cashew, papaya and pome granate, largest producer and exporter of spices and ranks first in productivity of grapes, banana, cassava, peas and papaya. It is estimated that all the horticulture crops put together cover nearly 21 million hectares with an annual production of 240.5 million tones. Horticulture crops occupy 11% of the total cropped area in India and constitute nearly 30.4% of agricultural GDP. Its contribution to national economy is about four % and is one of the fastest growing sectors in agriculture. It is a thrust area to stimulate agricultural growth, spurred by the changing domestic food habits towards more nutritious food and increasing overseas demand (Table 3).

4. Current Production and Projected Demand of Horticulture Products

Table 3: Current Production and Projected Demand of Horticulture Products*

Commodity	Production in million tones			Annual Growth Rate required (%)	
	Production in 2014-15	Production in 2015-16	Demand in 2020-21	2008-09 to 2011-12	2008-09 to 2020-21
Fruits	89.51	91.44	98.00	5.7	3.0
Vegetables	129.08	166.57	220.00	12.7	4.5
Spices	4.15	6.11	6.50	10.3	3.9
Others	13.02	16.87	35.50	29.9	8.7
All commodities	214.72	280.99	360.00	11.7	4.4

*NHB Data base 2015

Though its performance in production is laudable, in value addition, processing and export segments, India's contribution is not as expected. The major horticulture crops exported from India include spices, mango, grapes, orange, apple, banana, potato and vegetables. Major importers of Indian horticulture products are Bangladesh, Nepal, UAE, UK and Malaysia. Today horticultural crops cover about 25 % of total agricultural exports of the country. Apart from bringing in revenue from exports, horticulture plays a significant role in improving the livelihood of rural population. Flower harvesting, nursery maintenance, seed production and tissue culture, propagation of fruits and flowers and food processing are highly remunerative employment options for women in rural areas. This sector provides employment to around 19 % of the total labour force in rural India by generating direct and indirect employment opportunities.

Horticultural products are referred to as protective foods and assume great importance in nutritional security of people. According to estimates, the per capita consumption of fruits and vegetables in India is around 46g and 130g against the Indian Council of Medical Research (ICMR) recommendation of 92g and 300g respectively. This is aggravated by the fact that more than 20-22% of the total production of fruits and vegetables are lost due to spoilage at various post harvest stages. In India ,only less than 2% of total vegetables produced are commercially processed as compared to 70% in Brazil and 65% in USA (Table 4). This leaves great scope for technology adoption both in traditional horticultural enterprises and in commercial horticulture sectors. Moreover, a major shift in consumption pattern of fresh and processed fruits and vegetables along with greater interest shown by corporate sector, horticulture holds great promise for huge influx of investment with diversification and value addition holding the key. The 11th Plan initiated a massive National Horticulture Mission (NHM) with an outlay of Rs. 20,000 crores and is being continued in 12th plan(2012-2017) with additional investment. The NHM was launched in 2005-06 with a mandate to promote integrated development in horticulture, to help in coordinating, stimulating and sustaining the production and processing of fruits and vegetables and to establish a sound infrastructure in the field of production, processing and marketing with a focus on post harvest management to reduce losses.

5. Extent of Value Addition in Fruits and Vegetables

Table 4: Extent of Value Addition in Fruits and Vegetables

Country	Extent of Value Addition (%)
Brazil	70
France	50
USA	60-70
Malaysia	83
Thailand	50
Israel	50
New Zealand	40
China	10
India	<2

Source: Pradeepkumar, *et al.*, 2008

6. Horticulture as Profession

Horticulture, being one of the fastest growing sectors in agriculture, opened a wide range of career opportunities and jobs requiring knowledge, skill and continued training in horticulture. Direct jobs in Horticulture require a basic level of training that could be vocational certificate/diploma at Polytechnic/Rural Institute/Distance education mode or degree in horticulture or agriculture at college level. The work may be indoors or outdoors or both. Supply: demand trends in Horticulture with 2010 as base year are given as Fig.1. College education provides more in depth knowledge of the field and offers job opportunities at extension,consultancy, supervisory or managerial levels and to conduct research-basic and applied. Areas of Employment of Horticulture Graduates in government/private sector and the required qualifications are given in Table 5.

7. Projected Demand for Horticulture Professionals in 2020

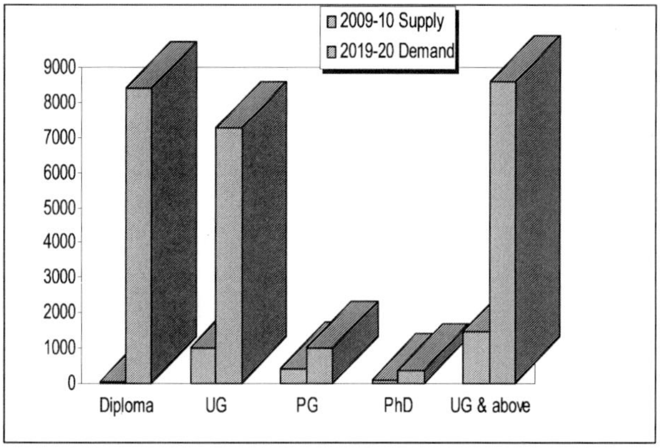

Fig. 1: Projected Demand for Horticulture Professionals in 2020

8. Major Career Opportunities in Horticulture

Table 5: Major Career Opportunities in Horticulture

Sl. No.	Minimum Qualification	Job description	Posts	Recruitment Agency
Public Sector				
1.	M.Sc Ag. /Hort and PhD in Horticulture	Research	Scientist	Agriculture Scientist Recruitment Board, (ICAR)
		Education	Lecturer, Reader, Assistant Professor, Associate Professor etc.	Agricultural Universities
		Transfer of Technology	Training Organizer & Training Associate	Krishi Vigyan Kendra
2.	Graduate in Horticulture/ Agriculture	Administration	Civil Services	Union Public Service Commission (U.P.S.C)
		Development/ TOT Research/ Training	Horticulture Officer/ Agriculture Officer Training Assistant/ Technical Officers/ Farm Supervisor	State Public Service Commission Agricultural Universities, ICAR, DRDO, CSIR Banking Sector
		Quality Control	Horticulture Inspector / Marketing Inspector	State Departments
Private Sector				
1.	Graduate in Horticulture/ Agriculture	Technical Expert	As Horticulturist /Horticulture Officer or Supervisor (Landscape)	Industries, Farm Houses, Hotels, Golf Courts & Construction Companies
		Technical Expert	Horticulture officer	Processing companies
	Marketing	Field marketing Officer		Pesticides, seeds and insecticide companies

Professionals with the skill for Total Quality Management (TQM) are required in areas of Fruit Science, Vegetable Science, Floriculture and Land Scape Architecture, seed production and processing industry, spices, medicinal and aromatic crops. Major professional areas in floriculture industry cover trade of i) cut flowers ii) nursery and potted plants, seeds and bulbs iii) micropropagated plants and iv) essential oils from flowers for perfumes. It is estimated that annual growth rate of domestic trade of floriculture products is about 25-30 %. The exports of medicinal and aromatic plants and their products have significant potential particularly to advanced countries like U.S.A., Japan and Europe. Herbal medicine is explored as alternatives to antibiotics, steroids and hormonal drugs for avoiding side effects on body functions. The international market of medicinal plants related trade is estimated US $ 60 billion per year having a growth rate of 7 %. Another promising horticulture sector is spices and India commands 46% in global trade in terms of quantity and 28% in terms of value/ annum.

Vegetable seed production is a remunerative enterprise compared to fresh vegetable production. Vegetable seeds, which are low in volume but high in value, have attracted private investments. Production of genetically pure seed requires high technical skill and expertise. India ranks 10th position in the world regarding production of seeds. In vegetables, major crops in which seed production are being carried out are tomato, cabbage, gourds, melons, brinjal, cauliflower and chilli. India is the largest producer of gourd's seeds. Twenty four national and multinational companies are involved in the production and distribution of gourd seeds. Thirty % of requirements of OP varieties are met by private companies, 15 % by public sector and remaining from farmer's own seeds. F_1 hybrid seed production is a monopoly of private seed industry in India. Graduates specialized in vegetable seed production and breeding have greater employability as breeders in public, corporate organizations and private companies. The employment opportunities in seed industry range from technicians, farm supervisors, farm/estate managers, seed experts and supervisors and project coordinators. Research and teaching are other avenues of employment both in India and abroad.

The food processing industry ranks fifth in its contribution to value addition but tops the list in terms of employment opportunities with approximately 15 lakhs people employed. It contributes 19 % of total Gross Domestic Product in the industrial sector. Every 10 million rupees invested in fruit and vegetable processing in the organized sector generates employment to the tune of 140 persons per year. The same level of investment in Small Scale Investment (SSI) units creates employment to 1050 persons per year. The SSI unit in food industry employs 4,80,000 persons, which account for 13% of all SSI units employed. The graduates can work as consultants, plant managers and sourcing specialists after proper training. One can also be an entrepreneur and offer employment to others. Avenues exist in teaching, research and training. R&D area is the most attractive as product development leads to immediate market application (Table 6).

9. Level of Employment (%) in Horticulture Processing Sector

Table 6: Level of Employment (%) in Horticulture Processing Sector

Level of Agriculture Education	Vegetable and Fruit processing units	Alcoholic beverages units
No qualification in agricultural subjects	89.04	59.44
Certificate/training	2.25	33.57
Diploma	1.74	1.40
Graduates	5.82	5.58
Post Graduates	0.78	0.00
Doctorates	0.37	0.00
All levels	10.96	40.56

For persons with formal qualification in Horticulture, self employment opportunities as Horticulture Consultant who provides advice, design, evaluation, supervision of garden or orchards is in vogue. Government support to the tune of Rs.10 lakhs is available for qualified professionals to start Agri-clinics which provide technical advice and inputs to farmers. Commercial nurseries, seed production and marketing, fruit and vegetable farms, florist shop, landscape architect, cold chain proprietor, processing enterprises and mushroom production are other vocations popular among horticulturists. The horticultural industry has also spawned a number of supporting or service industries where indirect jobs are created.Outsourcing for quality seeds, chemical fertilizers, pesticides, growth hormones, biofertilizers, biocontrol agents, implements, potting media, mulches, rain shelters, protected cultivation structures, vegetable forcing, hydroponics, aeroponics, packaging and marketing have spurred job opportunities in this sector. Similarly perishable nature of horticulture produces and products warrant special handling on transportation to retain their quality for a long time. This led to the development of cold chains in storing, handling and transportation from the areas of production to marketing outlets creating job opportunities. Secondary and tertiary industries related to processing of horticulture products, development of roads and other infrastructural facilities have also gained prominence leading to generation of skilled and trained employment.Retailing in the context of Foreign Direct Investment is assuming further employability.

10. Conclusion

Horticultural crops are essentially high value crops ensuring maximum returns to growers with multiple scope for value addition. Industry prefers these crops due to their consumer appeal. This sector is providing diversification of agriculture owing to better land use, creation of employment opportunities and overall nutritional security. Manpower ratios in the sector show a large demand for professionals and skilled manpower. It is estimated that for every graduate, three para-professionals are in demand. Even though there is need for 50% more graduates, quality concerns are to be addressed stringently. Many non-government jobs need a degree in agriculture/horticulture but does not specify any specialization. Of the emerging sciences, Food Technology, Biotechnology and Agribusiness Management are in demand, which are at present produced more in traditional universities. Horticultural Education in Private Universities is emerging but need legal and statutory regulation for quality and standards. Central Agricultural Universities offering broad based courses in Horticulture and it's related disciplines are in anvil. .

References

Janick, J. (1979). Horticultural Science. 3rd ed. San Francisco, Freeman and Company. 327 p.

Kumar, P and Kumar,K.(2003). Demand, Supply and Trade Perspective of Vegetables and Fruits in India. *Indian Journal of Agricultural Marketing* 17: 121-130.

Peter, K.V. (2009) Plantation Crops. National Book Trust, New Delhi – 110070,332p

Peter, K.V. (2007) Tuber Crops. National Book Trust, New Delhi-110070, 222p

Peter, K.V. (2015) Basics of Horticulture 2nd Revised Edition, New India Publishing Agency, New Delhi, 703 p.

Pradeepkumar,T, Suma, B., Jyothibhaskar and Satheesan,K.M. (2008). Management of Horticultural Crops, : Vol. 11:Horticulture Science Series: In 2 Parts , Peter, K.V. (Series Editor). New India Publishing Agency, New Delhi (www.nipabooks.com).

Relf, P.D.(1992). Human issues in Horticulture. *HortTechnology*. 2(2): 159–287.

Chapter – 2

Horticulture and Climate Change

G.S.L.H.V. Prasada Rao

Climate change and variability are concerns of humankind. The recurrent drought and desertification threaten seriously the livelihood of over 1.2 billion people who depend on land for most of their needs. The global economy has adversely been influenced due to droughts and floods, cold and heat waves, forest fires, landslips and mudslips, icestorms, duststorms, hailstorms, thunder clouds associated with lightning and the sea level rise (Fig. 1). The year 1998 was the warmest and declared as the weather related disaster year, which caused hurricane havoc in Central America and floods in China, India and Bangladesh. Canada and New England in the U.S. suffered heavily due to icestorm in January while Turkey, Argentina and Paraguay with floods in June 1998. In contrast, huge crop losses were noticed in Maharashtra (India) due to unseasonal and poor distribution of rainfall during 1997-98. The 1997/1998 El Nino event, the strongest of the last century affected 110 million people and costed the global economy nearly US$ 100 billion. A string of 16 consecutive months saw record high global mean temperature in 1997-98. Statistics compiled from insurance companies for 1950-1999 showed that major natural catastrophes, which were weather related, caused estimated economic losses of US$960 billion. Most of the losses were recorded in recent years since 1995 onwards as the top ten warmest years occurred during the first thirteen years of the current century of 2100 A.D.

The year 2005 was another historic second worst warmest year on record for hurricanes. The hurricane Katrina over new Orleans (USA) in August; the hurricane Rita in Texas, Central and Western Cuba and Southern Florida and typhoon over Hainan Province in South China and Vietnam during the last week of September, while early October over Mexican's Gulf coast; heavy downpour over Mumbai on 26th July, 2005 (Single-day the highest record rainfall of 944 mm) and 3rd September, 2005 over Bangalore; severe tropical storms in Andhra Pradesh in September; floods in Kerala, Karnataka, Maharashtra, Gujarat, Orissa and Himachal Pradesh during the Southwest monsoon (June-September), 2005 in India devastated cropped area to a large extent in addition to losses of thousands of human lives. In contrast, it was declared as a famine year in 24 sub-Sahara African countries due to drought and attack of locusts in 2005. Similarly, Australia experienced severe droughts in 2002 and 2006. The wheat crop was the worst affected in Australia due to unprecedented drought in 2006. It led to price rise globally since the wheat growing countries experienced severe drought in 2006. Occurrence of droughts and floods devastated rice and other crops in Andhra Pradesh and 40% cereal production was affected in Karnataka due to drought in 2006. Similar was the case during monsoon 2007, causing floods across several continents (Hurricane Dean in August slammed into Mexico) including India and Bangladesh. Torrential downpour in June, 2007 over Kerala, Karnataka, Andhra Pradesh and Maharashtra while in July and August over Gujarat, West Bengal, Odisha, Bihar, Uttar Pradesh and Assam, led to floods. Heavy rains again in September in Andhra Pradesh, Karnataka and Kerala led to floods and thus the year 2007 can be declared as the flood year in India. A huge crop loss was noticed in several states of the Country due to floods in *kharif*, 2007. A major food shortage is expected in majority of African countries due to heavy floods, which devastated several crops in the region. Mali, a West African country more often plagued by droughts, received unprecedented rains during 2007. Similar was the case in Algeria, Uganda, Sudan, Ethiopia and Kenya. The year 2007 was also recorded as the warmest winter globally. The year 2010 was recorded as the global wettest and one of the warmest years. It was the warmest year in India, followed by 2009. The years 2002, 2009

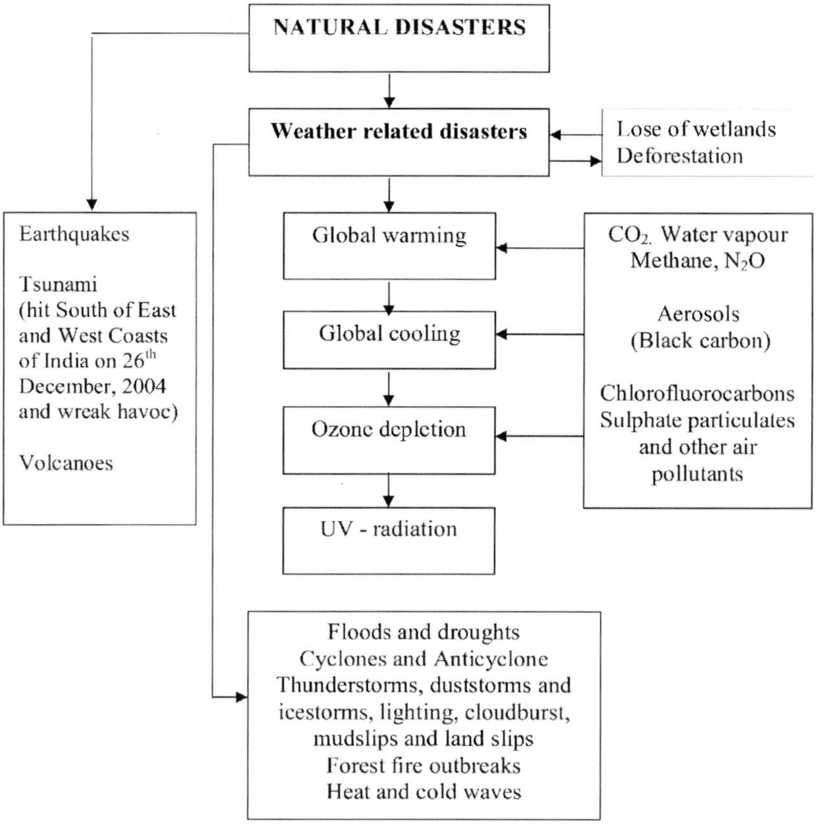

Fig. 1: Natural and weather related disasters

and 2012 were declared as the severe droughts during *kharif* in India and the *kharif* foodgrains production was adversely affected.

The Indian economy is mostly agrarian based and depends on onset of monsoon and its further behaviour. The year 2002 was a classical example to show how Indian foodgrains production depends on rainfall of July and it was declared as the all-India drought, as the rainfall deficiency was 19% against the long period average of the country and 29% of area was affected due to drought. The *kharif* foodgrains production was adversely affected by a whopping fall of 19.1%. Similar was the case during all-India droughts in 2009 and 2012 during which *kharif* foodgrains production was adversely affected. Occurrence of droughts and floods during South West monsoon across the Country adversely affects foodgrains production to a greater extent as evident from Fig. 2. It is one of the reasons that the foodgrain production is not in tune with plan estimates and the foodgrains production is likely to touch only a maximum of 260 million tonnes by 2020 at the present rate though it is projected as 400 million tonnes to declare India as one of the developed countries.

Increase in aerosols (atmospheric pollutants) due to emission of greenhouse gases including black carbon and chlorofluorocarbons (CFCs), hydrochlorofluorocarbons (HCFCs), hydrofluorocarbons (HFCs) and perfluorocarbons (PFCs), Ozone depletion and UV-B filtered radiation, eruption of volcanoes, the "human hand" in deforestation in the form of forest fires and loss of wetlands in the process of imbalanced development are causal factors for climate variability and change. The loss of forest cover and wetlands, which normally intercept rainfall and allow it to be absorbed by the soil, causes precipitation to reach across the land, eroding top soil, causing floods and droughts. Paradoxically, the lack of trees also exacerbates drought in dry years by attaining soil to dry out more quickly.

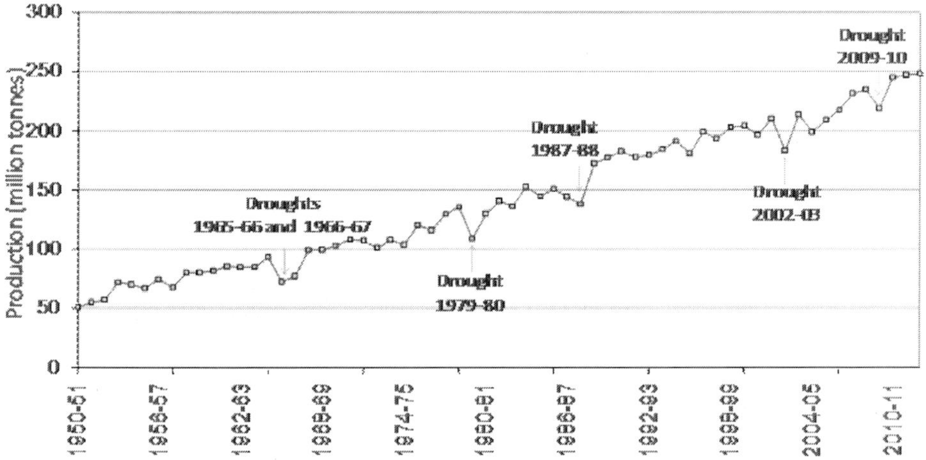

Fig. 2: Impact of droughts on Indian foodgrains production from 1950-51 to 2012-13

1. Greenhouse Effect

Carbon dioxide, water vapour, methane, carbon monoxide, sulphur and nitrous oxides, chlorofluorocarbons and chlorofluoromethanes are the atmospheric constituents of major importance. Incoming shortwave radiation is not absorbed by the above atmospheric constituents and the outgoing longwave radiation from the earth is absorbed by them and reradiate back to earth surface. In the process, earth warms up and its temperature raises. It is also known as the glasshouse effect as property of glass is that it allows solar radiation through it and disallows longwave radiation from earth, thereby warming takes place. The atmospheric constituents which have the property of absorbing longwave radiation and transparent to shortwave radiation are known as the greenhouse gases (GHG). Human activities like fossil fuel combustion, production of synthetic chemicals, biomass burning, deforestation, excess use of chemical fertilizers and pesticides change the chemical composition of atmosphere, thereby enhancing greenhouse effect. The frequent eruption of volcanoes in different parts of the globe in recent times is one of the predominant factors to change the chemical composition of atmosphere on which our knowledge is limited. The present growth rate of GHG is 1% in CH_4, 0.4-0.5% in CO_2 and 0.2-0.3% in N_2O as per the Intergovernmental Panel on Climatic Change (IPCC, 2007). In addition to the above, chlorofluorocarbons and chlorofluoromethanes contribute significantly to greenhouse effect. Is the greenhouse effect mitigated by regional cooling effect from aerosols like black carbon? New research showed that the aerosol effect could dominate the greenhouse effect over the continents and by and large the effect of greenhouse will be slowed down. In contrast, the ozone - friendly substitutes for chlorofluorocarbons (CFCs) like hydrofluorocarbon (HFCs), hydrochlorofluorocarbons (HCFCs) and perfluorocarbons (PFCs) are also powerful greenhouse gases that contribute to global warming. There is also evidence that faster rate of global warming since 1976 is human-induced.

2. Global Warming

Global warming is the biggest long term threat to life on earth. Rise in temperature may drive thousands of species to extinction, trigger more frequent floods and droughts and sink low lying islands and coastal areas by rising sea levels. It is the result of rising atmospheric content of CO_2 mainly owing to burning of hydrocarbons or fossil fuels like as petrol and diesel. Destruction of forests and their degradation too contribute to rise in carbon dioxide levels. The IPCC (2006) projected the rate of warming for the 21st century to be between 0.8 and 4.4 °C at various stabilized CO_2 levels in atmosphere and it is most likely to be 3°C by end of this century. It could cost global economy almost $7 trillion by 2050, is equivalent to a 20% fall in growth if no action is taken on greenhouse gas emissions (Fig. 3). If action is taken, it will cost only $350 billion due to climate change already taken place, just 1% of the global GDP. The winter 2007 was the warmest and recorded 0.85 °C above average of 12 °C and the previous highest was 0.71 °C, which occurred in 2002 in Northern Hemisphere. The entire Europe Union recorded the warm winter, having more than 2 °C above average. New York experienced the highest temperature of 21.7 °C on a day in January, 2007 and the second highest was recorded as 17.2 °C in 1950. Night temperatures are in increasing trend in certain parts of the Country. It is detrimental to fruit crops as well as temperate crops like wheat.

Economics of climate change: It could cost the global economy almost $7 trillion by 2050-equal to a 20 per cent fall in growth – if no action is taken on greenhouse gas emissions. Taking action now could cost just one per cent - $350 billion – of global GDP (*Source*: Stern report, IPCC, DoE, 2006)

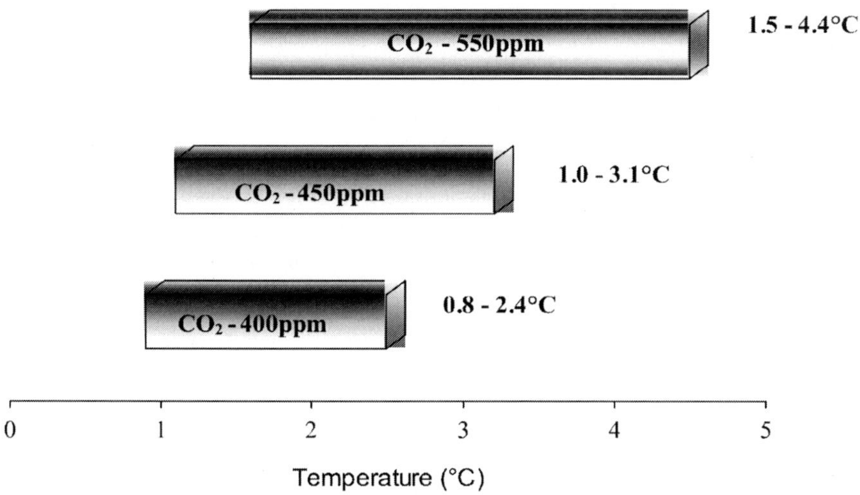

Fig. 3: Projected CO_2 versus Temperature rise and its economics

The increase in all-India mean temperatures is almost solely contributed by increase in maximum temperature (0.6°C/100 years) with minimum temperature remaining practically trendless. Consequently, there is a general increase in diurnal range of temperature. In rainfall, there was a decrease since last 50 years. A marked increase in rainfall and temperature is projected in India during the current century. The maximum expected increase in rainfall is likely to be 10-30% over central India. Temperatures are likely to increase by 3-4 °C towards end of the Century. It is more pronounced over Northern parts of India. The mean sea level rise is likely to be slightly less than 1mm/ year along the Indian coast. Greater number of high surges and increased occurrences of cyclones in post-monsoon period along with increased maximum wind speed are also expected as per Ministry of Environment and Forests (MoEF), Govt. of India and Department of Environment, Food and Rural Affairs (DEFRA), U.K. This phenomenon of climate change threatens the area of land availability for farming. As per the United Nations Report of FAO, India stands to lose 125 million tonnes, equivalent to 18% of its rainfed cereal production from climate change by 2015. China's rainfed cereal production potential of 360 million tones is expected to increase by 15% during the same period. It would also cause a worldwide drop in cereal crops between 20 and 400 million tonnes, put 400 million more people at risk of hunger, and put up to 3 billon people at risk of flooding and without access to fresh water supplies. The crop production losses due to climate change may also drastically increase number of undernourished people, severely hindering progress in combating poverty and food security. The severest impact is likely to be in sub-Saharan African countries, which are the least able to adapt climate change or to compensate for it through increase in food imports. In 2004 and 2005, 24 sub-Saharan African countries faced food emergencies, caused by a lethal combination of locusts and drought. In addition, adverse hot and dry weather in United States and drought conditions in parts of the European Union lowered cereal output during 2005 when compared to that of 2004. The simulation models indicate that the global warming leads to reduction in rice and wheat production in Northern India.

3. Atmospheric Carbon Dioxide (CO_2)

CO_2 being an efficient absorber of long wave radiation from earth, it is of great climatic significance. Carbon dioxide has important absorption peaks at about 2, 3 and 4μ and a major absorption peak at about 15μ. This gas emits half of the absorbed heat back to earth, influencing flow of energy between earth and atmosphere. CO_2 is a very important atmospheric constituent in the heat budget of earth and atmosphere. There is a sharp increase in carbon dioxide in recent decades due to human-made interventions in terms of burning fossil fuels like coal, oil and natural gases. The projected concentration of CO_2 in the atmosphere by 2050 is as follows:

Projected concentration of CO_2 in the atmosphere

Periods	CO_2 concentration
Before pre-industrial revolution	250-290 ppm
1958	315 ppm
1984	354 ppm
2013	389 ppm
Projection by 2050	>450 ppm (?)

The concentration of carbon dioxide in atmosphere was in a steady state at 280 ppm till the pre - industrial period, 1950. It is rising since then at the rate of 1.5 to 1.8 ppm/ year. The concentration of carbon dioxide in atmosphere increased from 280 before 1950 to 370 ppm in 2000. It is likely to be doubled by the end of 21st century if preventive steps are not taken. Measurements at IARI, New Delhi showed that the concentration of carbon dioxide increased from 330 to more than 370 ppm, indicating similar trend of global phenomena. Over the same period, atmospheric concentrations of methane and nitrous oxide increased by 151 and 17%, respectively. This resulted in an increase of global temperature by 0.6 °C, causing global warming. As a result of warming, the global mean sea level rose by between 10 and 20 cm. The rise is as much as 10 times the average increase in the last 3000 years. An increase of 2.5mm/year is noticed in sea level of the Indian Ocean since 1950s. The trend appears to be higher on the East coast when compared to that of the West Coast. It is likely that the rate and duration of warming of 20th century were larger than at any other time during the last 1000 years.

In temperate soils, unexpected loss of carbon is taking place due to global warming. It leads to more carbon dioxide in the atmosphere, which means even more global warming. The findings showed that the carbon was being lost from soil at an average of 0.6% a year and richer the soils, higher the rate of carbon loss. There was no single factor other than global warming that could explain such losses in non-agricultural soils.

4. Aerosols and Global Cooling

Climate change is usually associated with increasing levels of greenhouse gases, a large amount of soot and other pollutants into the atmosphere in the form of tiny particles, known as aerosols. Aerosols can, however, be produced naturally too. There is great variation in the amount of aerosols around the globe and are unevenly distributed. They change from time to time depending on type of aerosols and quantities in which they are present. Many aerosols are invisible to naked eye and are microscopic. They include sea salts from breaking sea waves, pollen, fine seeds of plants, spores, bacteria and various organisms lifted by wind, smoke and black carbon (soot) from fires, tiny sand particles, volcanic ashes and meteoric dust. Likewise, sulphur-containing compounds known as sulphates arise by natural processes and plants release organic materials. Aerosols in the process of scattering, contribute to varied colours of red and orange at sunset and sunrise. Blue of the sky is also due to selective scattering by microscopic dust particles.

From global warming point of view, these tiny soot particles are very important. They absorb a part of incoming shortwave solar energy, thereby heating up the lower atmosphere and totally burning off clouds that might have formed. The combined effect of increasing greenhouse gases and human-induced soot in the warming of lower atmosphere could be as much as 0.25 degrees Celsius/ decade (positive effect). Aerosols are tiny particles, of about one micron, which scatter sunlight back to space and then cause a regional cooling effect (negative effect). Such aerosols are sulphates, soot, organic carbon and mineral dust. Black carbon absorbs solar radiation in visible spectrum and reduces amount of radiation reaching earth surface. Increase in black carbon alters large scale atmospheric circulation and hydrological cycle. Black carbon is a product of incomplete combustion of coal, diesel fuels, biofuels and outdoor biomass burning generated from industrial pollution, traffic, outdoor fires and household burning of coal and biomass fuels. The imbalance between positive (warming effect) and negative (cooling effect) aerosol forcing may slow down sometimes the global warming or may lead to global cooling. As the life span of human induced aerosols in the atmosphere is short- lived (one to three weeks) while the long-lived greenhouse gases may persist and thereby accumulate, leading to overall global warming. Many models indicate that global warming would increase rainfall over India while decrease in rainfall was noticed when the effect of aerosols was introduced. The gradual decrease in rainfall over India since last fifty years supports the theory of effect of aerosols on decline in rainfall. If that is the case, frequency of droughts in India could increase in coming decades if pollution continued unabated.

5. Clouds and Global Warming/Cooling

Hygroscopic aerosols act as nuclei for cloud formation. Clouds would not form if there were no fine particles in the atmosphere around which water vapour could condense into droplets. Brown clouds contain a variety of aerosols, including sulphates from coal combustion, nitrates produced by vehicular emissions, and soot resulting from fossil fuel combustion and the burning of biomasses. They enhance warming quite a bit in lower atmosphere. Clouds play a dual role in cooling and warming of earth. If the sky is totally covered with clouds, clouds act as an active surface instead of earth surface. The incoming solar radiation is reflected back to a large extent from top of clouds and thus low solar radiation reaches earth surface. This keeps the planet cool. On the other hand, they absorb long wave radiation and emit back in all directions and transmit little long wave radiation to space. In the process, atmosphere traps huge amounts of heat emitted from earth's surface and clouds, leading to green house effect. Clouds are known as warm clouds in the tropics while cold clouds in temperate zones. Clouds over the oceans differ in character and heating effects to that of clouds over the continents. Their nature and effects also vary with season and time of day. Tropical clouds trap three times as much heat as clouds trap on a global average, but also reflect so much sunlight that the heating effect is cancelled out. Since the tropics account for 20% of earth's surface and effects of tropical cumulonimbus clouds are disproportionate, the delicate heat balance they maintain is critical. A warmer ocean would likely to cause more clouds to form, increasing both their heating and cooling effect. Analysis of satellite data also showed that the clouds like stratocumulus over temperate regions reflect enough sunlight to exert a net cooling effect locally, and this appears to account for clouds overall cooling effect worldwide. The behaviour of disproportionate cooling and heating may change itself under global warming, perhaps. If it is understood, climatologists may be able to forecast the timing and extent of expected global warming with far more confidence than today.

6. Ozone Depletion and UV Radiation

The greatest concentration of ozone is at an average in height of 25 km and above in the stratosphere. The ozone molecule is made up of three atoms of oxygen. It is the most efficient absorber of ultraviolet radiation from sun and thus protects all life forms in planet earth. Ozone depletion due to industrialization in recent decades is the concern of humankind and biological activities. Release of compounds like chlorofluorocarbons (CFCs), carbon tetrachloride and methyl chloroform could significantly deplete ozone layer that shielded the planet from ultraviolet radiation. The CFCs are used in a variety of industrial, commercial, and household applications. These substances are non-toxic, non-flammable and non-reactive. They are used as coolants in commercial and home refrigeration units, aerosol profellants and electronic cleaning solvents.

The global average thickness of ozone is 300 Dibson units, equivalent to 3mm. In contrast, it is about 100 Dibson units, which alarm us that we are in great danger if precautionary steps are not taken up. The ozone level fell to 90 Dibson Units, which was the lowest value recorded on 30-09-98 nearly equalled to the lowest value ever recorded 88 Dibson Units on 28-09-94 over Antarctica. The ozone losses are caused by chlorine and bromine compounds released by chlorofluorocarbons and halons. Year-to-year variation of size and depth of ozone hole depend on variations in meteorological conditions. The unusual cold temperature by 5-9°F over Antarctic zone enables greater activation of reactive chlorine that ultimately causes more ozone loss and lower ozone levels. Increase in chlorine levels should peak in the Antarctic stratosphere within a few years (Shashi *et al.*, 2002).

Among UV radiations, UV-B radiation in the range of 280-320 nm is more sensitive to ozone fluctuations and reaches earth surface. A global network is created to monitor UV-B filtered radiation in terms of Minimum Erythema Dose (MED). The human-made interventions in industrial development lead to ozone depletion, thereby filtered UV radiation reaches the ground, resulting in various human, animal and crop diseases. The ozone loss has the potential to increase incidence of skin cancer, cataracts and damage to people's immune system, harm some crops and interfere with marine life. However, little is known on impact of ozone depletion and increasing UV-B radiation on ontogeny of tropical plants and human and animal diseases since studies in this direction are lacking. Because CFCs remain in the atmosphere for100 years, continued accumulation of these chemicals pose ongoing threats, even after their use is discontinued.

Models on global warming indicate that the rise in temperature is likely to be around 2-2.5°C by end of this century. It is likely that the extreme weather events like droughts and floods, cold and heat waves increase in coming decades. The human and crop losses are likely to be heavy. The global economy will be adversely affected as mentioned in the latest report of IPCC. If the sea level increase as projected, the coastal areas which are thickly populated will be in peril and for the existing population, the safe drinking water will be a great problem. The whole climate change is associated with increasing greenhouse gases and human- induced aerosols and the imbalance between them may lead to uncertainty even in year-to-year monsoon behaviour over India. Therefore, there should be a determined effort from developed and developing countries to make industrialisation environment-friendly by reducing greenhouse gases pumping into atmosphere. In the same fashion, awareness programmes on climate change and its effects in various sectors viz., agriculture, health, infrastructure, water, forestry, biodiversity and sea level and the role played by human beings in climate change need to be taken up on priority. In the process, life style of people should be changed so as not to harm Earth-Atmosphere continuum by pumping CFCs into atmosphere. From the agriculture point of view, effects of extreme weather events on crops are to be documented so that it will be handy to planners in such reoccurrence events for mitigating the ill effects. Also, there is need to guide planners on projected future crop scenarios based on climate change events, which will be more realistic at field level as models always overestimate the impacts. Finally, we have to foresee these extreme events and prepare ahead to combat them so that the losses can be minimised.

7. Climate and Horticulture

Climate plays a major role in crop distribution while weather in crop productivity. Climate decides the crop habitat while crop habit depends on genotype and phenotype. By genotype it is meant the heredity received by the organism, which is relatively constant throughout life. Phenotype refers to the appearance of the organism-what it looks like-which is subject to change throughout life (Johannsen, 1903). The crop phenologys like flushing, flowering, fruit development and harvest time are sensitive to vagaries in weather and differs from crop to crop. The phenological events undergo several abiotic and biotic stresses or environmental stresses as seen in the case of cashew flowering (Fig. 4) during the process of crop growth and development. The effect of environmental stress on phenology could adversely affect final yield in the same year or the following years depending upon type of crop in the case of plantation crops. Of course, a majority of plantation crops show seasonal behaviour in phenological events though they are perennial except crops like coconut, having flowering phase round-the-year. The habitat of plantation crops is

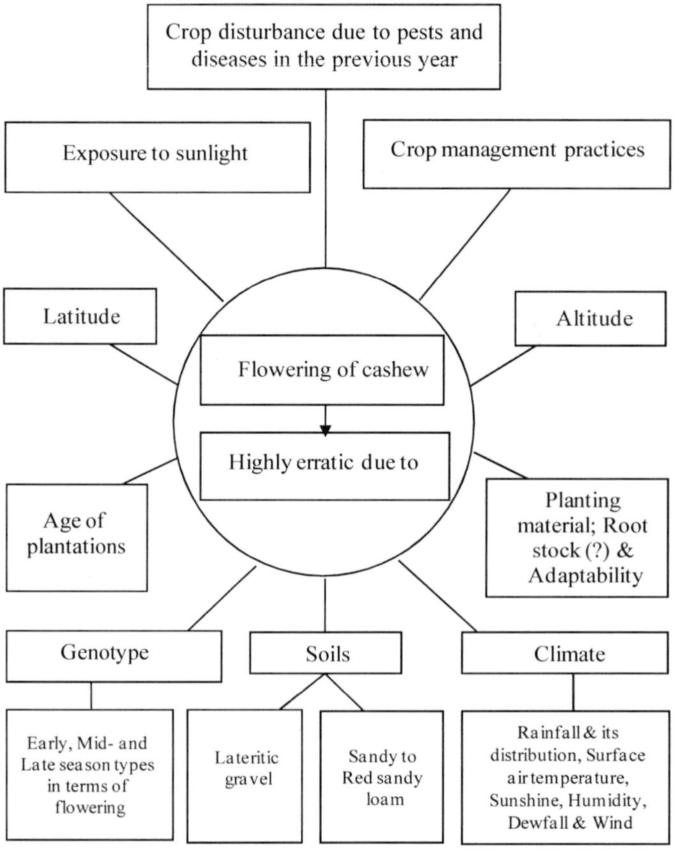

Fig. 4: Effect of abiotic and biotic factors on flowering of cashew

different as they are perennials traditionally grown in the humid tropics under the rainfed conditions. Within the tropics, crops like coffee, cardamom and tea prefer cool-temperate climate as they grow along the high ranges, where mean annual temperature varies between 14 and 20°C while coconut, cashew, cocoa and rubber are seen cultivated in mid - and - high lands, where the annual mean temperature revolves between 25 and 29°C. Interestingly, black pepper is in the buffer zone as an intercrop between warm and cool climates, stretching between tropical and temperate climates within the tropics due to its adaptability. According to thermal regime, black pepper is grown under "Meso therms and Micro therms-I and II" while cardamom, coffee (arabica) and tea under "Micro therms III". Black pepper likes moderate and low surface air temperature to some extent while cardamom low temperature only round-the-year, thus indicating different habitat under which they grow in rainfed conditions. Similarly, the flowering behaviour in fruit crops is influenced by geographical coordinates like latitude, longitude and altitude as per the Hopkins' Bioclimatic law (1938), which states that a biotic event in North America will, in general, show "a lag of four days for each degree of latitude, five degree of longitude and four hundred feet of altitude, northward, eastward and upward in spring and early summer". In mango, the India Meteorological Department also established that there is a delay in flowering from South to North of India, which generally followed Hopkins' Bioclimatic Law (IMD, 1957). Similar results were obtained in flowering of cashew.

Thermal regime and plantation crops distribution across the Western Ghats of Kerala

Classes	Regions	Temperature conditions	Altitude above MSL	Crops
Mega therms	Low land	High to moderate temperature throughout the year	0-10 m	Coconut, arecanut and cashew
Meso therms	Mid land	Moderate temperature throughout the year, winter temperature is relatively low	10-100 m	Coconut, cocoa arecanut, rubber, cashew and black pepper
Micro therms - I	High land	Moderate to low temperature throughout the year, winter temperature is low	100-500 m	Rubber, coconut, cashew, arecanut and black pepper
Micro therms – II	High land	Low temperature throughout the year, winter temperature is low	500-1000 m	Coffee (robusta), rubber, arecanut and black pepper
Micro therms – III	High ranges	Low temperature throughout the year, winter temperature occasionally goes below 0° C.	1000-2500 m	Tea, coffee (arabica) and cardamom

The response of plants to diurnal/seasonal fluctuation in temperature is known as thermoperiodism. A number of physiological processes, including germination, stem elongation, fruiting, floral development and increase in frost-hardiness proceed most satisfactorily under a rhythm of alternating temperatures. The tomato is sensitive to night temperatures. The optimum temperature may be 26-30°C in young seedlings. As the plant grows older, it favours lower temperature and the optimum night temperature may be a range of 13-20°C, depending on the variety. The fruit set is abundant with day light temperature of 26.5°C and night temperature of 15-20°C. A high optimum daytime temperature is effective because it increases photosynthetic rate. Tomatoes grow mainly at night and with temperatures above 18°C, translocation of sugars becomes a limiting factor. Actually, a little fruit set is observed in the field with night temperatures above 22°C or below 10°C. In Southern California, in winter and early spring, night temperatures are often below 10°C, but late afternoon temperature is between 15 and 20°C. But covering plants at 3 pm, an optimum dark (night) temperature exists for a few hours and fruit set is improved considerably. In India also, more tomatoes could be produced / plant and fruit size is high if harvested during winter when compared to that of other warm seasons during which temperature may be a limiting factor.

The growing degree days or heat unit concept or thermal unit concept was developed assuming that there is a direct and linear relationship between crop growth and temperature. It states with the assumption that growth of plants dependents on total amount of heat to which it is subjected during its lifetime. The heat unit or growing degree days are defined as follows:

$$GDD=[(Tmax + Tmin)/2]-Tt$$

where, Tmax is the maximum temperature in °C; Tmin is the minimum temperature in °C and Tt is minimum threshold or base temperature.

Minimum threshold temperature is the temperature below which no growth takes place. It varies from 4.5 to 12.5°C for different crops. It is 5°C for rice, 8°C for millets (jowar, ragi and bajra), 10°C for maize, 14-15°C for cotton and 13°C for coconut and cashew. The degree day concept can be used for forecasting duration of crop and its harvest dates. The crop yields also could be predicted based on crop-specific thermal units.

In okra, the fruit yield is directly related to number of heat units. The fruit yield is high if number of heat units is high irrespective of time of sowing and vice-versa (Fig.5). It clearly indicates that okra produces better fruit yield if sowing is taken up from February to April and from September to October. More studies are needed in this direction at different locations to draw such conclusions.

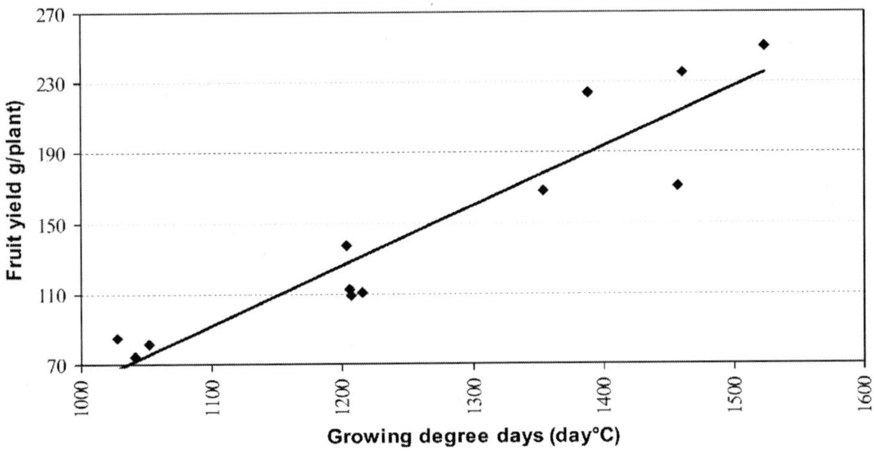

Fig. 5: Growing degree days versus fruit yield of okra at Vellanikkara (Thrissur)

8. Climate and Cashew

The cashew growing regions across the World are located commercially between 15°N and 15°S from the equator. In India, it is grown along the East and West Coasts and regarded as a coastal tree since confining the coastal area never exceeds 300 m contour line. Though it is known as a coastal tree, it is seen cultivated at higher altitudes too. The altitude of the region where cashew can be grown depends on latitude of the place. For example, cashew in Tanzania and Brazil is grown up to 1000m amsl whereas in Assam (India) cashew is not suitable for altitudes 170 m amsl. Cashew productivity is poor in inland plateau and at higher altitudes depending upon distribution of rainfall and temperature as it prefers only coastal climates. Cashew shows a tendency for late flowering and fruiting irrespective of latitude depending upon the altitude. Being a tropical crop, cashew has the ability to thrive in high temperatures. In its natural habitat, it extends into the semiarid regions like northern Mozambique where daily maximum temperatures exceed 40°C. It is similar in the East Coast of India also, where cashew is grown. It appears that cashew responds to phototropism as it produces better under sunlight. A delay of one week to 10 days in all biotic events of cashew like bud break and flowering may also be noticed towards north direction within a cashew tree (Fig.6) Cashew is not only photosensitive and thermo-sensitive as the biotic events respond to low minimum and high maximum surface air temperatures. It seems that a temperature range between 15 and 35°C may be the optimum for better growth and production of cashew. The surface air temperature is conducive across the West coast for better cashew yield while it may not be so across the East Coast under rainfed conditions.

A cashew plant, undergoes first a dry spell and waits for the second dry spell to occur for break of bud. It appears that both the dry spells at the time of bud break and thirty days prior to bud break with bright sunshine are critical as far as the cashew bud

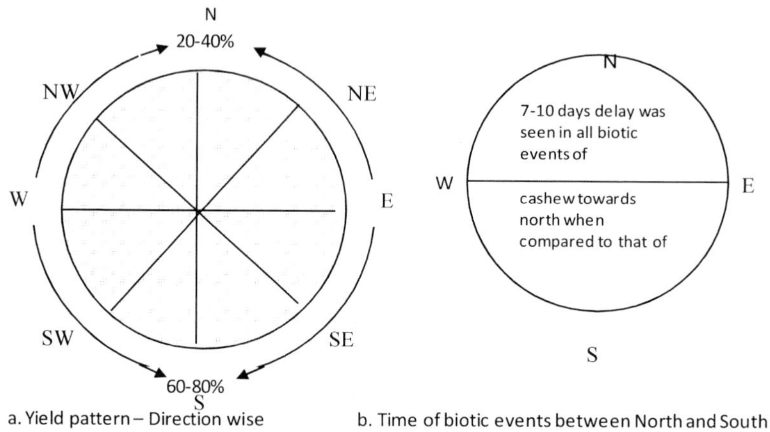

Fig. 6: Directional influence on biotic events and yield pattern of cashew

break is concerned. Influence of dry spell and bright sunshine in the mechanism of bud break of cashew are yet to be understood. The soil moisture stress has no relevance as the bud break of cashew begins much earlier before soil moisture stress starts. Instead, flowering of cashew requires relatively dry atmosphere with mild winter for better flowering. The mild winter may be defined as "Low minimum surface air temperature ranging between 15 and 20°C coupled with more dew nights having moderate dew". The continuous rains without critical dry spell may delay bud break of cashew. The unusual rains during November and December may lead to delay in reproductive phase of cashew along the West Coast (Fig. 7), missing the mild winter and thus yields low. Cashew is enamoured with dry spell/drought during its reproductive phase and provides relatively better yield as it is seen grown under uni-model distribution of rainfall (Fig. 8), having 650 to 3600 mm rainfall spread in 45 to 120 days. The abnormal drought situation affects the quality of cashew nuts very much under rainfed conditions. The delay in crop duration could be explained with geographical co-ordinates once the bud break in cashew is initiated immediately after the rain spell is over. The law stated by Hopkins (1938) holds good in all biotic events of cashew under better crop management in rainfed situations of tropical monsoon climates, provided the genotype and rainfall distribution are uniform. There was a difference in number of days delayed in cashew flowering at each degree of north latitude while the effect of altitude on time of cashew flowering is similar as stated by Hopkins (1938). In the humid tropical monsoon climates, there is a delay of six days in cashew flowering at every 1° of North latitude and for every 100 meters of altitude, the delay in cashew flowering is three days (Fig. 9). Of course, increase in night temperature during flowering season that is noticed in recent years in certain parts of the Country adversely affected flowering in mango (including the delay in flowering time). It is yet to be established in projected global warming and climate change scenario.

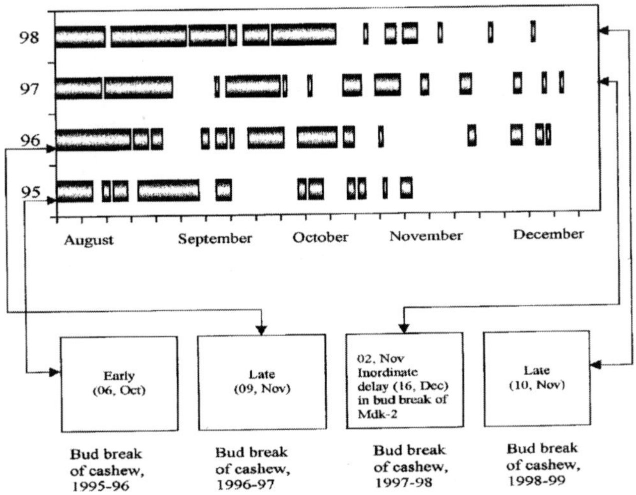

Fig. 7: Daily rainfall distribution during August-December from 1995 to 1998 at RARS, Pilicode

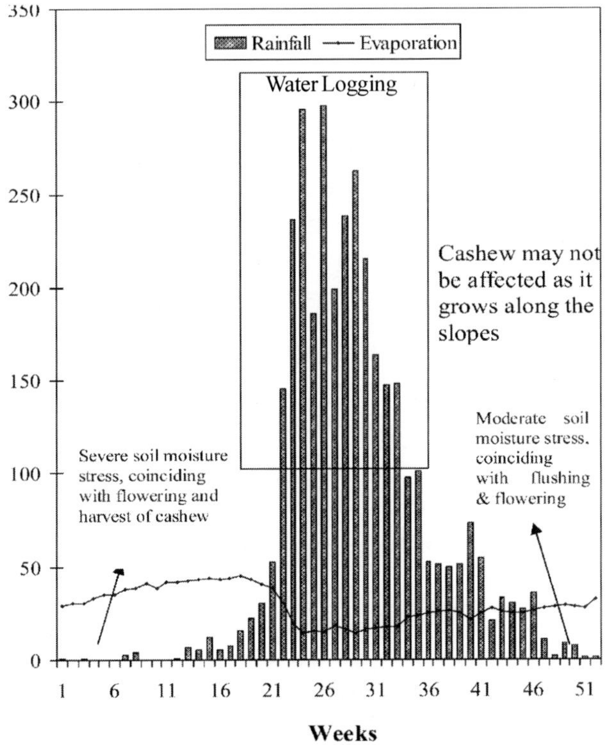

Fig 8: Mean weekly rainfall versus evaporation at RARS, Pilicode

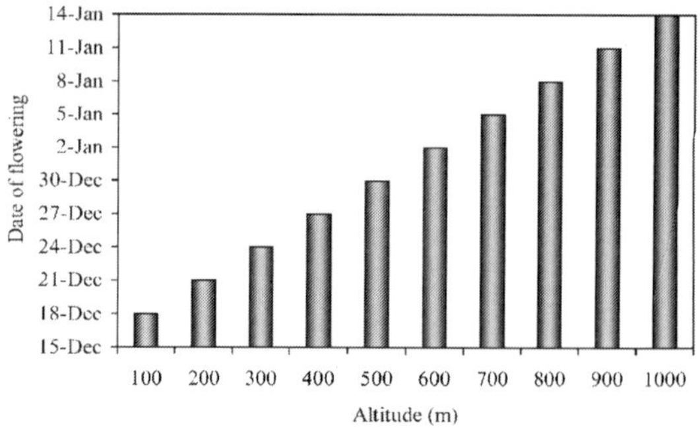

Fig. 9: Influence of altitude on cashew flowering

The kernel weight depends on availability of soil moisture, important during nut setting and development. It is one of the reasons why weight of cashew nuts is relatively more if a good shower is noticed during nut development phase. The shelling (%) also is high under above conditions. The nut and kernel weights may be better with altitude in early season types though the shelling (%) is less. It may not be the case with late season types. Depending upon genotype and its adaptability, the nut and kernel weights are superior in both West and East Coasts of India. The shelling (%) is low across the East Coast when compared to that of the West Coast. This could be explained due to high maximum temperature (35-40°C), which reduces nut development phase of cashew. The day maximum temperature shoots up to 42°C during nut development phase of cashew across the East coast while not so in the West Coast and hence the nut size is adversely affected along the East Coast under rainfed conditions. It could be some extent manipulated through summer irrigation. The cashew yield and kernel weight are influenced by number of heat units if other meteorological factors are not in hindrance with nut development and yield. The early varieties need less number of heat units while more in case of late varieties of cashew. The nut characters also respond to phototropism also. Further studies are needed in this direction.

Night temperature between 15 and 20° C, cloudiness and relatively dry weather with the afternoon relative humidity between 40 and 60% may be optimum for triggering pest population of tea mosquito bug in cashew. Tea mosquito bug is a menace over the West Coast and coincides with flushing and flowering time of cashew. The cold (< 12°C of minimum temperature), hot (>35°C of maximum temperature) and wet (Continuous heavy rains) weather conditions may not be conducive for triggering pest population. Continuous cloudiness (< 2h/day) consecutively for five days during second week of December, immediately followed by continuous dewfall under ideal thermal and humid conditions may not only trigger the growth of the fungi but help in the sporulation, multiplication and spread of the Inflorescence blight in cashew during 1998-1999. Interestingly, there is a lead - time of one month between incidence of disease and crop damage. Such studies will be of immense use in integrated pest management to overcome menace of tea mosquito bug in cashew.

9. Climate and Cardamom

Small cardamom (*Elettaria cardamomum* Maton) - popularly known as the Queen of Spices - is indigenous to the evergreen rain forests of the Western Ghats. It is confined to the States of Kerala, Karnataka and Tamil Nadu, accounting for an area of 41,288 ha,

25,947 ha and 5,085 ha, respectively. Among these States, Kerala accounts for major production (72.4 %), followed by Karnataka (20 %) and Tamil Nadu (7.6 %). The natural habitat of cardamom is the high ranges between 1000 and 1500 m amsl across the Western Ghats and characterized by cool-humid- microclimate, which provides ideal conditions for cardamom cultivation. The annual rainfall and rainfall during the South West monsoon (June-Sep) increase from South to North across the cardamom tract and the West (Kerala and Karnataka) of the Western Ghats receive more rainfall when compared to that of East (Tamil Nadu). The reverse trend was noticed during post monsoon (Oct-Nov) and winter (Dec-Feb) and the East receives more rainfall than West of Western Ghats. In summer (Mar-May), Kerala and Tamil Nadu receive relatively more rainfall than Karnataka. The mean annual rainfall of cardamom growing areas vary from 1400 mm over Tamil Nadu to 2600 mm over Karnataka (Fig.10). The number of rainy days varies from 80 to 130 days. A declining trend was noticed in annual rainfall at all the locations except Pampadumpara and Madikeri. Similar was the case during the South West monsoon except Madikeri. In other seasons, no such uniform increasing or decreasing trend was noticed across the cardamom tract except Thandikudi, where rainfall was declining in all the seasons, The forests exert a domineering influence on soil, water resources and microclimate of cardamom. In the turn of twenty-first century, forests constituted only 24% when compared to 70% in the middle of nineteenth century in Kerala. The fast-dwindling forest cover and its consequence over climate are the concern across the cardamom tract of Western Ghats.

The surface air temperature and its range were high (5.3-15.9° C) over Karnataka, followed by Kerala (5.2 -11.7° C) and Tamil Nadu (6.9 -10.7° C) across the cardamom tract of Western Ghats. Majority of locations showed increasing trend in annual maximum temperature and it was true in South West monsoon and post monsoon periods while declining during winter and summer. Interestingly, an increasing trend was noticed in majority of locations in case of minimum temperature also. The mean annual maximum temperature across cardamom tract is the lowest (24.1°C) over Tamil Nadu and the highest (26.9 – 30.0°C) over Karnataka while intermediary over Kerala (24.6 – 27.4°C). The minimum temperature ranges from 15.4 to 17.8 °C, dropping down to 13.0°C over Tamil Nadu occasionally (Fig.11). Coffee and tea are grown under the identical thermal and moisture regimes of cardamom.

Production and productivity of small cardamom were in increasing trend though a sharp decline was noticed in cardamom area across Western Ghats. This could be attributed to technological interventions since last three decades. However, the inter-annual fluctuations in cardamom production were common due to weather aberrations. For example, the cardamom production was badly hit during 1983 due to unprecedented drought that occurred from November 1982 to May 1983 across the cardamom tract of Western Ghats. Similar was the drought during 2003-04 over Kerala. It is also noticed that the cardamom production over Kerala in recent years was also badly hit due to dry spells that occurred during monsoon of 2002 and 2003. There existed a strong relationship between dry spells and cardamom production and the climate risk is more over the Karnataka region when compared to other regions across Western Ghats. The area decline in cardamom could be attributed

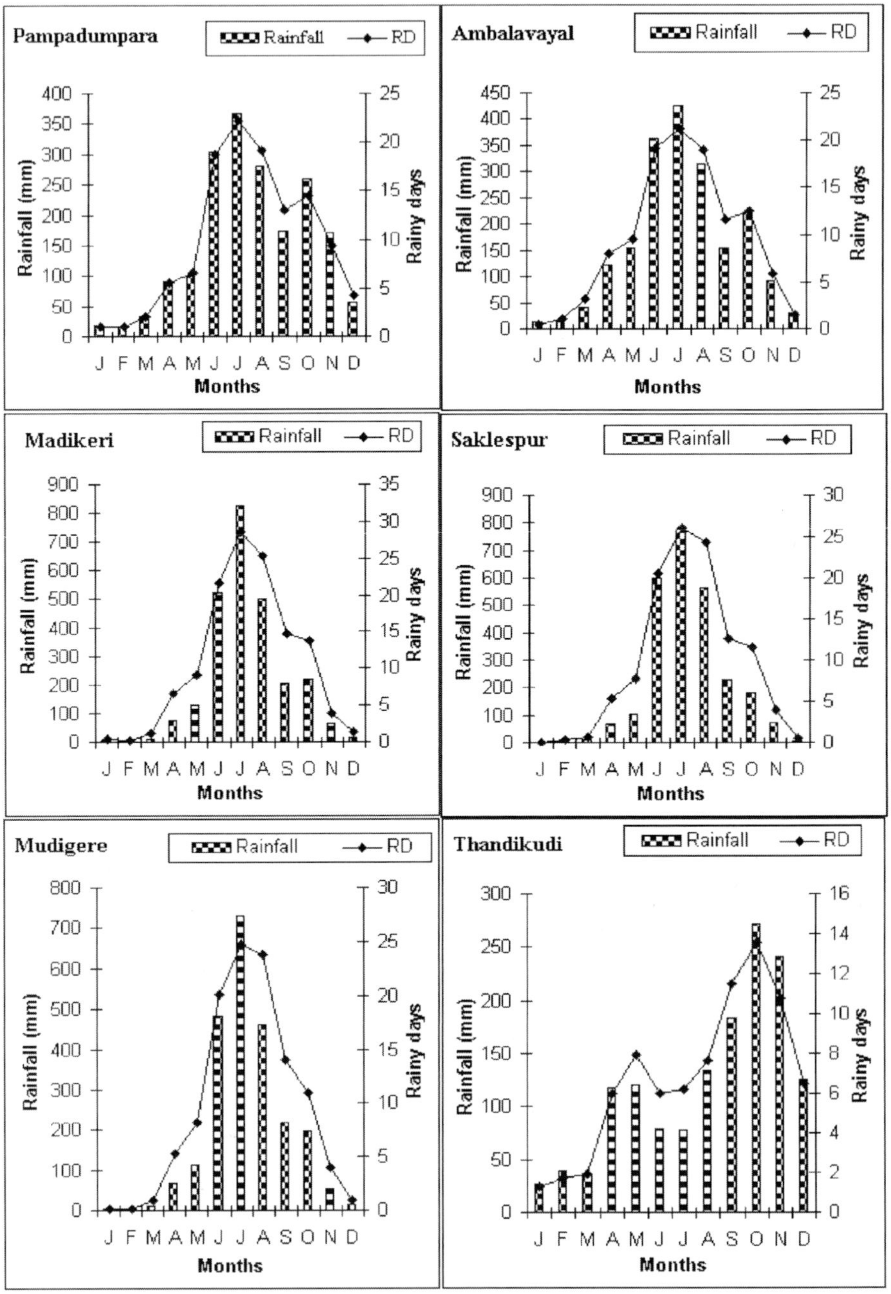

Fig. 10: Monthly rainfall distribution over the Western Ghats (Cardamom, Coffee and Tea growing areas)

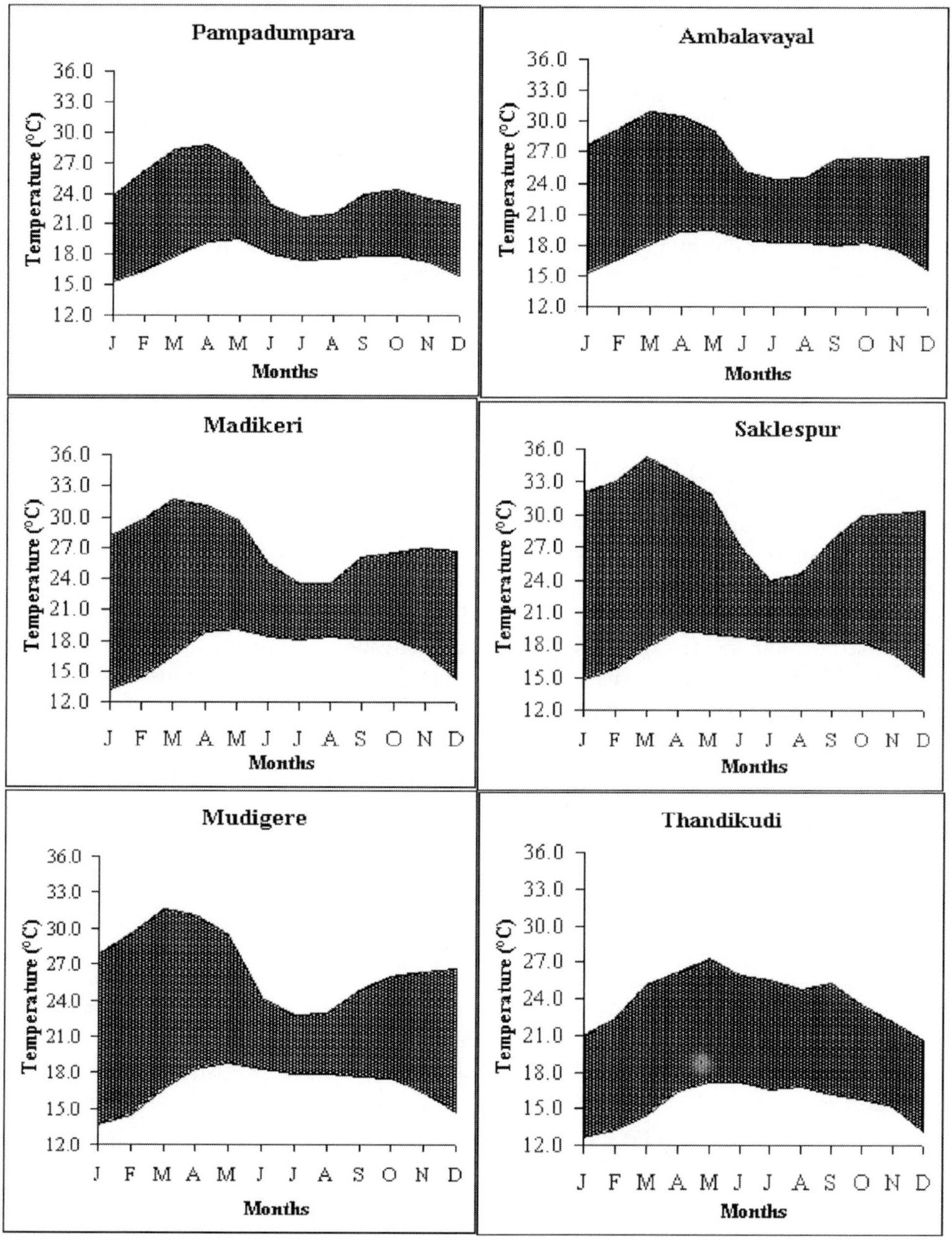

Fig.11: Temperature regime over the Western Ghats (Cardamom, Coffee and Tea growing areas)

to instability in prices and climate variability in terms of decrease in rainfall and temperature increase over cardamom tract of Western Ghats due to deforestation.

10. Climate and Coffee

Coffee is grown at higher elevation across Western Ghats between 750 m and 1200 m amsl. Robusta and Arabica are the two traditional cultivars grown over the high ranges along Western Ghats, covering three States *viz.*, Karnataka, Kerala and Tamil Nadu. The major factors that affect production of Arabica and Robusta are lack of blossom and backing showers during summer months as the coffee growing areas experience uni-model rainfall. The uni-model rainfall means heavy rains during summer monsoon (June-September), followed by prolonged dry spell from November to May. Heavy rainfall during monsoon leads to waterlogging while continuous cloudiness restricts uptake of nutrients as evapotranspiration demand is low. The Indian summer monsoon lands over coffee tract around 1st June and normally between 25th May and 8th June. Once monsoon commences, heavy downpour of rains are expected during June and July with cloudy weather and overcast sky. The normal annual rainfall across the coffee growing tract varies widely between 1400 and 4000 mm. About 80% is recorded during monsoon months of June, July, August and September. Number of rainy days varies between 20 and 27 in a month during peak rainy months and July is the rainiest month, followed by June and August. The mean monthly maximum temperature is always less than 30°C while the minimum temperature revolves around 24°C and thus the surface atmospheric temperature range is 6-7°C only. The crop is also sensitive to wind disturbances and light conditions. Arabica and Robusta, popularly grown across the West coast, respond differently to various weather elements. The two cultivars are also well demarcated with elevation of growing environment as Arabica prefers high altitudes when compared to that of Robusta, indicating that Arabica is relatively tolerant to low temperature and humidity when compared to that of Robusta. Growth of coffee across coffee tract appears to be affected when exposed under open conditions due to high temperature during summer and low temperature during winter. It requires a certain degree of shade for its better performance. Similar is the situation in cardamom also.

Weather effects on Arabica and Robusta across the Western Ghats

Climatic factors	Arabica	Robusta
Lack of blossom showers	Absence of rains in March – April	Absence of rains in March
Rain on opening day of blossom	Partial to complete failure	Partial failure
Hail storm	Injure the floral and Vegetative phase	Partial failure
Backing showers	Absent in May- poor crop set	Absent in April- poor crop set
Severe western exposures (with lowering of sub-soil-soil moisture)	Partial to complete	Partial to complete
Excessive wetness and waterlogging	Partial to complete	Partial to complete
Wind	Both are sensitive to wind disturbance	

The annual rainfall across the coffee tract varies between 1400 mm over Tamil Nadu and 2600 mm over Karnataka. It revolves around 2000 mm over Kerala. The annual number of rainy days varies between 80 and 130 over the coffee tract. Bi-model distribution of rainfall is noticed over some of the coffee growing areas at Pampadumpara, Ambalavayal and Thandikudi due to both The South West and North East monsoons. The coffee tract in Kerala receives about 60 % of annual rainfall during rainy months (June-August) while Karnataka receives about 75% and 20% only over Tamil Nadu. Interestingly, the monthly rainfall is high from August to December over Tamil Nadu due to influence of North East monsoon unlike other coffee growing locations. Uni-model rainfall distribution is noticed over Karnataka, having the highest monthly rainfall from June to August when compared to Kerala and Tamil Nadu across the coffee tract as seen in Fig.10. At the same time, Karnataka receives insignificant rainfall from December to March, indicating that the coffee plantations are under moderate to severe soil moisture stress if the pre-monsoon showers fail. It is not the case across coffee tract of Kerala and Tamil Nadu. It reveals that the coffee tract over the highranges across the West Coast is characterized by heavy rainfall during June, July and August except at Thandikudi (Tamil Nadu). The prolonged dry spell from December to March coupled with high rates of evapotranspiration results in the rapid depletion of soil moisture and surface water sources. Both these phenomena - high water surplus during June – August and severe water deficit during December - May affect adversely productivity of coffee plantations. It is more relevant over the Karnataka Coffee Plantations when compared to that of Kerala and Tamil Nadu.

The mean annual maximum temperature of coffee tract varies between 24 and 30°C while 15 and 18°C in minimum temperature. The coffee tract of highranges over Karnataka experiences the highest maximum temperature during March, varying between 31.6 and 35.3°C while the night temperature in January varies between 13.2 and 14.8°C. In contrast, the highest maximum temperature of 27.3°C is only observed in May across eastern parts of coffee tract (Tamil Nadu) and a low of 12.7°C in January as noticed in Fig. 11. The mean monthly maximum temperature is high (28.8°C) in April over Pampadumpara while at Ambalavayal during March (30.9°C). The minimum temperature at the above locations is in January and 15.3°C at both locations. It reveals that maximum temperature during summer is high over Karnataka and low over Tamil Nadu across the coffee tract. It is intermediary over Kerala. January experiences the lowest temperature at all the locations, varying between 12.7 and 15.3°C. The maximum temperature drops down significantly during summer monsoon, which is a climatic characteristic feature in the humid tropics due to heavy and continuous rainfall.

Relatively uniform temperature range (difference between maximum and minimum temperatures) is maintained (6.9 -10.7°C) over Tamil Nadu, followed by Kerala (5.2 - 11.7°C). It is high (5.3 -15.9°C) over Karnataka (Fig 12). The surface air temperature and its range are high over coffee tract of Karnataka and Kerala when compared to that of Tamil Nadu. It can be attributed to altitudinal difference across coffee growing tract. It also reveals that the climatic risk is high over Karnataka Coffee Plantations.

The "Monsooned Malabar Coffee" is processed at Mangalore, located in Coastal Karnataka. The processing location experiences heavy rainfall of more than 1000 mm in July, followed by 957 mm in June and 605 mm in August. Out of 3264 mm of annual rainfall, 2852 mm is received during the monsoon period. It accounts more than 80% of annual rainfall. Similar is the case over Kasaragod District, adjoining to Mangalore. The maximum temperature round-the-year varies between 29 and 33.5°C. April is the hottest month while January is the coldest (around 20°C). The night temperature round-the- year ranges from 20 to 26°C. Being located in the tropics along the West Coast, the crop processing environment in terms of surface air temperatures is relatively uniform throughout the year. However, the maximum temperature dips during monsoon months irrespective of coastal or high lands and the difference in day and night temperatures is relatively less (5-7°C) when compared to that of other months. The monthly temperature range appears to be high (around 12°C) in December. It reveals that continuous high rainfall (>2500 mm) together with moderate night and day temperatures (24-30°C) during monsoon season appears to be conducive for coffee bean processing as the processing units are located along coastal belt. It is the reason why the 'Malabar Region' is chosen, where coffee beans processed and marketed are characterized with unique monsoon features. It appears to be optimum for coffee beans processing during monsoon period. Water vapour in atmosphere during monsoon period is such that the vapour pressure deficit is naturally maintained at minimum level with gentle marine breeze of three to six km/hr. The relative humidity during the night and forenoon revolves more than 90% which drops down to 70-80% in the afternoon on many occasions during monsoon months. It reveals that high atmospheric humidity together with moderate temperature and gentle marine breeze with low vapour pressure deficit accompanied by continuous heavy rains during the monsoon season are the characteristic features of the Malabar Region. These peculiar climatic features along West Coast (Malabar Region) in the humid tropics may be conducive for coffee beans processing.

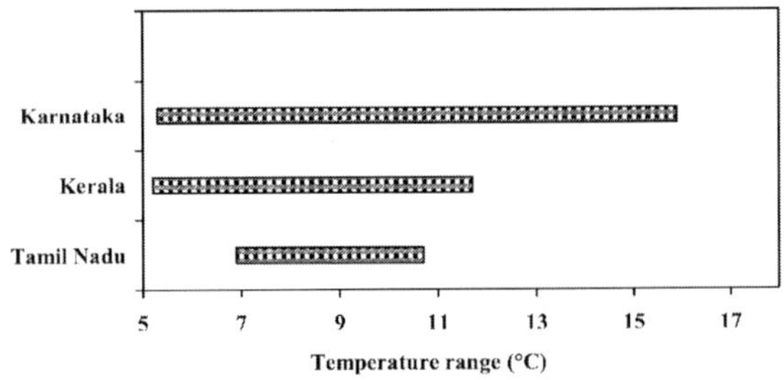

Fig.12: Temperature range (°C) over the Western Ghats (Cardamom, Coffee and Tea growing areas)

11. Climate and Tea

Two growth periods are seen in tea (mid-March to mid-June and mid-August to mid-November). Night temperature and relative humidity are the most important factors which determine productivity of tea in addition to rainfall. In Annamalai, night temperature occasionally drops below 5°C and humidity is below 10% when the dry atmosphere is prevalent. This situation leads to tea production at lower levels.

In tea, there is a time lag between the physiological function of tea bush and the vegetative harvest. It is about five weeks in the pruned year and four weeks in other years. As such, the weather conditions of previous month are generally correlated to current month's harvest. About 50% reduction in green tea leaves was noticed in April 2004 when compared to 2003 and 2005 in Himachal Pradesh due increase in maximum temperature of the order of 2.1 to 7.9°C in March (Prasad and Rana, 2006). In apple, flowering was advanced by 15 days and fruit set was very poor.

Effect of maximum temperature in March 2004 on Apple and Tea in Himachal Pradesh

Apple	Tea
Flowering was early by 15 days	About 50% reduction in green tea leaves in April when compared to 2003 and 2005
Large – scale flower drop due to acute moisture stress	The yield reduction was seen only after one month.
Heavy rainfall during second fortnight of April accompanied by sharp fall in temperature caused poor fruit set	Heavy losses in yield were noticed in Potato (matured ahead), vegetables and pea
Optimum temperature for fruit blossom and fruit set is 24°C while the region experienced above 26°C for 17 days	

Irrigation during cool dry season (mid-January to mid- March) will not produce desired effect on production as temperature is the limiting factor during the period. Irrigation during hot and dry period (March to September) that occurs frequently in Nilgiris will be beneficial in crop production. In recent years, tea production has come down across the Northeast region. It is attributed to the global warming.

12. Climate and Cocoa

Rainfall preferably between 1500 and 2000 mm annually with a dry season of not more than three months with less than 100 mm/month and temperature varying between 30-32° C as mean maximum and 18-21°C mean minimum with an absolute minimum of 10°C are the climatic boundaries desirable for growing cocoa. The loss of apical dominance in cocoa is seen constantly at a temperature of above 32°C. Loss of apical dominance means development of side shoots of plant. As in other fruit crops, duration of pod, its size and quality and pod development are controlled by surface air temperature. A decline of 39 % in annual cocoa yield was noticed in 2004 when compared to that of 2003 due to rise in maximum temperature of the order of 2-3° C from 14th January to 16th March, 2004 along with prolonged dry spell. Such trend was noticed whenever summer temperature shoots up by 2-3°C when compared to that of normal maximum temperatures of 33-36.5 °C. The adversity of weather aberrations on rhythm of normal growth of cocoa is reflected in yield after a lag period of 4-5 months.

High rainfall has a malevolent effect, where crop is grown under waterlogged conditions while summer rainfall has a benevolent effect with yield. On augmenting dry season's rainfall by irrigation, the results have not always been very marked; because the water lost from the leaves is so high with low humidity that the roots cannot match the loss. Crop yields are very low with heavy shade and increase with increasing light up to 50

% level. Cocoa yield is affected by presence or absence of fertilizer if the level of light is above 50%. With added fertilizer, yields increase almost up to full light, whereas in absence of fertilizer, yields fall off. The theory has therefore been advanced that the light regime for optimum yield of cocoa is a function of its mineral nutrition.

13. Climate and Coconut

The coconut is a tropical tree plant as its distribution in the world is confined almost entirely to tropical zone. In fact, it is highly adaptable to a variety of environmental conditions though it does exhibit some growth preferences. The West Coast of India, the major coconut belt of the country, rainfall varies from 1500 to 3500 mm or more/annum and the crop is grown under rainfed conditions. Within the State of Kerala, the per palm production is much better towards South due to uniform distribution of rainfall as the palm grows under rainfed conditions. High rainfall during heavy rain months (June-August) adversely affects coconut yield due to waterlogging and lack of aeration. Heavy button shedding is likely under waterlogged and severe soil moisture stress conditions. In non-traditional areas like Tamil Nadu, Andhra Pradesh and Karnataka, coconut palms are grown commercially under irrigated conditions only, otherwise crop does not come up due to climate extremes in terms of surface air temperature and poor annual rainfall of about 1000mm. Of all the climatic factors affecting the coconut, rainfall appears to be the most important under rain fed cultivation. There is a lag period between influence of meteorological variables and crop yield in coconut as initiation of primodium takes place 44 months before harvest. The yield of a particular year was influenced by January to April rains for two years prior to harvest together with rains during same period of the year of harvest (Patel and Anandan, 1936). Park (1934) found that severe drought experienced in puttalam (Srilanka) affected yield of nuts for a period of about two years with the maximum effect at about 13 months after conclusion of drought. Decline in monthly nut yield was noticed in the following year from February 1984 to January 1985 due to severe drought of summer 1983. The effect of drought on monthly nut yield was noticed in the eighth month after drought period was over with a maximum (64.1%) reduction in nut yield in July 1984 (ie, 13 months after the drought period was over) and the minimum (23.6%) in January 1985. Similar was the case during 1988-89 (Fig. 13). The recent summer drought during summer 2004 also adversely affected coconut yield to some extent over Kerala. In the semiarid tropics like Tamil Nadu, heavy crop loss was reported due to continuous failure of rains in 2001 and 2002 and lack of irrigation due to poor water recharge in wells. It was similar in 2012-13. Coconut palms show the following characteristics under severe soil moisture stress depending upon duration and intensity of drought (Rao, 2002).

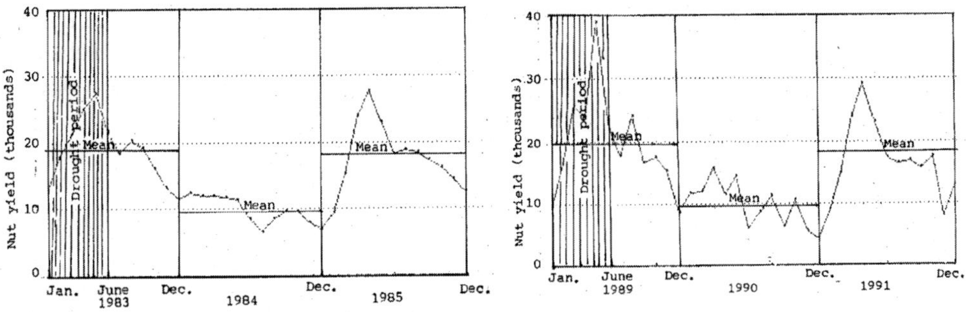

Fig. 13: Effect of drought on coconut yield at RARS, Pilicode (Kerala)

Withering and mortality in young seedlings under poor management

Drooping, wilting and drying of lower whorl of leaves

Breakage of leaves at petiole or just above it

Spindle leaf breaking which lead to mortality in senile palms under conditions of poor management

Abortion of spadices, starts from October/ November onwards

Button shedding and immature nut fall

Nut size decline and

Finally decline in nut yield in subsequent year up to 50% depending upon the type of management and genotype

In fact, it is temperature that sets limit to latitude and altitude up to which coconut can be successfully cultivated. The farther one goes from equator the more is the palm confined to lowlands. It is restricted to an altitude of 600 m amsl, beyond which it is not economic as crop comes up very late for bearing and nut development process takes too long a period. Hence, the coconut palm likes equable temperature neither very hot nor very cold. The optimum mean annual temperature for better growth and maximum yield is 25-27°C with a diurnal variation of 5-10°C. The coconut palm grows under profuse irrigated conditions where the range of variation in temperature is considerable as seen in Tamil Nadu, Karnataka and Andhra Pradesh. On the West Coast of India, the monthly mean minimum temperature does not fall below 20°C and the mean monthly maximum rarely goes above 33°C. In Malaya maximum temperature varies from 30 to 32°C on the coast and 34° to 37°C in the inland. The diurnal variation in temperature is similar to that of 5-10°C. High temperature causes drying up due to desiccation of inflorescence in primordial stage. These would normally open 16 months later in monsoon season and because of their abortion, the leaves, which subtend them, show up barren axils. Abortion of spadices is influenced by high temperature prevailing during hot weather season. The second phase of nut development in coconut is sensitive to heat units since it is the critical phase, which finally decides nut size in coconut. If the number of heat units during second phase of nut development (4-7 months old nut) is more than 2100 day °C, the final nut size may be diminished. Due to high maximum temperature and low humidity, oil content is relatively less in non traditional areas. It clearly indicates that number of heat units influence crop duration, nut size, oil content and nut yield to a large extent. The genetic coefficients are derived for each crop based on thermal units and used in crop growth simulation models, which are widely operational for decision making. Strong winds are not desirable as they considerably damage coconut plantations. They not only uproot or break the stem but also twist the crown or break the leaves and destroy considerable part of the crop. Heavy and extensive damage to coconut palms due to cyclones are reported from Philippines, South India and Jamaica due to strong winds.

It is well known that climatic factors do greatly influence incidence and intensity of pests and diseases. On West Coast of India, incidence and severity of the pests such as rhinoceros beetle (*Oryctes rhinoceros*) and black-headed caterpillar (*Nephantis serinopa*) are more in summer. Again large scale emergence of cock chafer grubs (*Leucopholis coneophora*) on the West Coast is noted only about a fortnight after onset of South West Monsoon in sandy and sandyloam soils. Regarding diseases, the incidence of bud -rot and leaf – rot diseases of coconut palm is always noticed in a severe form only in rainy months when atmospheric humidity is high. The stem bleeding is prevalent in northern districs of Kerala, where heavy rains during monsoon followed by prolonged dry spell are noticed.

Climate change and coconut in Kerala

Crop growth simulation models indicate that the climate change is likely to affect the coconut production differently across the coconut growing areas of the country. The coconut productivity is likely to increase in Kerala as per the crop model studies. However, the studies conducted at the Academy of Climate Change Education and Research, Kerala Agricultural University indicated that increase in maximum temperature, minimum temperature, average temperature and temperature range from 1951 to 2009 reduced marginally the coconut productivity in Kerala (Fig. 14). The decline in coconut productivity over a long period of time can be attributed to various factors. One of them could be climate change/climate variability.

Climate variability and coconut price

A steep increase in coconut price was also seen in 1984 due to low coconut production in Kerala as a result of summer drought in 1983 and for the first time, the coconut growers got a high premium of more than Rs. 2200/ quintal of copra. Under open auction sale, on an average Rs. 3.07/ coconut was obtained at the Regional Agricultural Research Station, Pilicode in 1984 as against Rs. 1.70/ nut in 1983. Such trend was seen in lean crop years. A sudden price hike was also noticed from 2002 to middle of 2005 due to low coconut production over Kerala and a large number of coconut palms withered in Tamil Nadu due to all India drought 2002. The effects of summer droughts in 2004 and 2013, similar to that of summer drought 1983, reflect on price rise in coconut (Fig. 15). Of course, the coconut price in Kerala depends on several factors viz. import-export

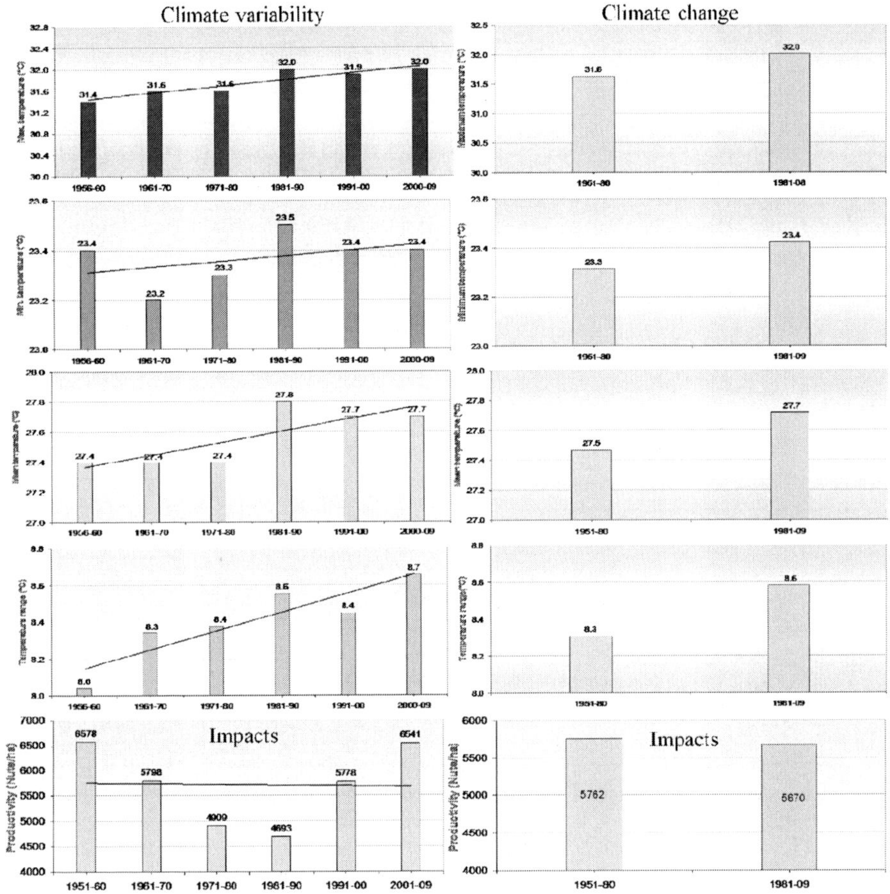

Fig. 14: Decadal and tri-decadal maximum, minimum, mean temperature, temperature range and productivity of coconut

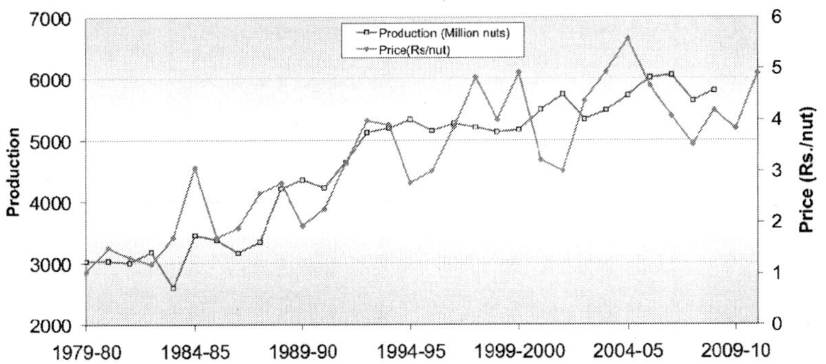

Fig. 15: Coconut production and nut prices over Kerala

policy, total coconut production outside Kerala, less edible oil consumption and coconut oil use in industries as several substitutes are used in place of coconut oil. Import of palmolene oil may be one of the major contributing factors in deciding coconut oil price. Of course, this is to highlight the role of climate variability on coconut price and as such the subject area of coconut price and production and vice versa is beyond scope of the study.

The effect of climatic factors on perennial crops like coconut is different from that of annual crops. In seasonal crops, the effect is limited to short period of crop season and is manifested immediately. In the case of perennial crops, the effect may be felt for longer periods, may be discernible only after some time lag. Thus what is claimed as the effect of a particular season may not in fact be due to that season alone but may also be partly due to delayed effects of a foregone season. These are inevitable on account of peculiar growth features exhibited by the crop. Extreme weather events like droughts and floods, cold and heat waves (Samra *et al.*, 2004) adversely affect crop production to a large extent. At the same time, effects of extreme weather events persist for more number of years in plantation crops depending upon crops' phenophase.

Occurrence of cold wave (frost and cold spell) during December-January, 2003 and its effects (Samra *et al.*, 2004)

States affected	Crops suffered	Percentage loss	How to reduce the impact (?)
Parts of Jammu, Punjab, Haryana, Himachal Pradesh, Bihar, Uttar Pradesh and North Eastern States	Mango, Litchi, Guava, Papaya, Ber Kinnow, Pine apple, Sapota, Amla, Assam lemon, Jack fruit and Peach	10-100% depending upon crop and variety within the crop (mango)	Proper selection of fruit species /varieties, wind breaks or shelter belts, frequent irrigation, smoking, covering young fruit plants with thatches or plastic shelter and air mixing
	Fruit size and quality were affected in horticultural crops	Damage is more in low - lying areas where cold air settled and remain for a longer time on ground	
		Temperate fruits such as apple, plum and cherry gave higher yield due to extended chilling	Weather forewarning

Though to be reduced due to increase in temperature in the case of coffee, tea and cocoa as well in other crops. Though the models indicate reduced crop yields due to long term climate changes (increase in temperature), it is in fact not seen such impacts as on today, but at the same time the adverse impacts of climate variability (occurrence of heat and cold waves and floods and droughts) on crop yields is significant. The GCM models indicate that the frequency of weather abnormalities is likely to increase in ensuing decades under the projected climate change scenario. Therefore, the adverse impact of climate change on horticultural/field crops is seen only through climate variability rather than the direct long term effects of climate change, which is doubtful. Therefore, short term crop strategies need to be worked out as a part of climate resilience horticulture to combat the ill effects of climate variability/climate change.

References

Cosgrove, B. (1991) Weather: Dorling Kindersley Eyewitness Guides. Dorling Kindersley Ltd., 9 Henrietta Street, London, 64p.

Carroll P. Wilsie (1974) Crop Adaptation and Distribution. Eurasia Publishing House (P) Ltd; New Delhi, India, 448p.

Hopkins, A. D. (1938) Bio climates, a science of life and climate relations. U.S. Department of Agric. Misc. Pub, 280p.

IMD (Indian Meteorological Department) (1957) Agricultural Meteorology in India Meteorological Department: A Review. Issued in commemoration of silver jubilee of Agricultural meteorology, Division of India Meteorological Department, 42p.

IPCC (Intergovernmental Panel on Climatic Change) (2006) The Economics of Climate Change: Stern Review. The Summary of conclusions. Survey of the Environment 2007, The Hindu, pp 141-145.

IPCC (Intergovernmental Panel on Climatic Change) (2007) Climate Change: The Physical Science Basis. Extracts from the IV Assessment Report. Survey of the Environment 2007, The Hindu, pp 147-155.

Johannsen, W. (1903) Adaptation through natural selection. (In) *Crop adaptation and distribution*. Eurasia Publishing House (P)Ltd, New Delhi, 448p.

Nybe, E.V. and Rao, G.S.L.H.V.P. (2005). Drought management in plantation crops. Kerala Agricultural University, Thrissur, Kerala, India, 176p.

Park, M. (1934) Some notes on the Effect of drought on the yield of coconut palm *Trop.Agriculturist*,83: 141-150.

Patel, J.S. and Anandan, A.P. (1936) Rainfall and yield in coconut. *Madras Agric*. J. 24: 5-15

Peter, K.V. (2018) (Ed) Climate Resilient Horticultural Crops of Future, New India Publishing Agency, New Delhi 431 p.

Prasad, R and Rana, R. (2006) A study on maximum temperature during march 2004 and its impact on rabi crops in Himachal Pradesh. *J. Agrometeorology*, 8(1): 91-99.

Ramakrishna, Y.S., Rao, G.G.S.N., Rao, S.G. and Vijayakumar, P. (2006) Impact of climate change in Agriculture. In: Environment and Agriculture (eds. Chadha, K.L. and Swaminathan, M.S.). Malhotra Publishing House, New Delhi, pp. 1-30.

Rao, G.S.L.H.V.P. (2002) Climate and Cashew. Kerala Agricultural University. Thrissur, Kerala, India, 100p.

Rao, G.S.L.H.V.P. (2002) Climate variability and coconut production in the humid tropics. Kerala Agricultural University, Thrissur, India. 38p.

Rao, G.S.L.H.V.P. (2008) Agricultural Meteorology. Prentice – Hall of India Private Limited, New Delhi, 366p.

Rao, G.S.L.H.V.P. (2011) Climate Change Adaptation Strategies in Agriculture and Allied Sectors. Scientific Publishers, Jodhpur (India), 330p.

Rao, G.S.L.H.V.P. and Nair, R.R (1988) Agrometeorology of Plantation Crops. Kerala Agricultural University, Thrissur, Kerala, India. 187p.

Rao, G.S.L.H.V.P., Rao, G.G.S.N. and Rao, V.U.M. (2010) Climate Change and Agriculture over India. PHI Learning Private Limited, New Delhi, 352p.

Samara, J.S., Singh, G and Ramakrishna, Y.S. (2004) Jan.10 cold wave during 2002-03 over north India and its effect on crops. *The Hindu*.p.6.

Shashi, K., Pathak and Mason, N.J. (2002) Our shrinking ozone layer: General perspectives and remedial measure. Resonance, December pp.71-80.

Shukla, P.R., Sharma, S.K and Ramana, V.P. (2002) Climate Change and India - Issues, Concerns and Opportunities. Tata – McGraw-Hill Publishing Company Ltd, New Delhi, 314p.

Chapter – 3

Vegetable Science

Pranab Hazra, A. Chattopadhyay, U. Thapa, M.K. Pandit and S. Dutta

Basics of Vegetable science are oriented on different aspects of Vegetable science viz., Vegetables for nutritional and health security, present status of vegetable production, Classification of vegetable crops, Influence of environment on vegetable production (Temperature, Light, Soil), Management of seedlings in the nursery, Raising of grafted seedlings in vegetable crops, Vegetable based cropping system, Water management, Plant nutrients and integrated nutrient management, Microirrigation and Fertigation, Pollination management, Integrated pest management in vegetable crops, Biological management of diseases of vegetable crops, Weed management in vegetable crops, Management of vegetable crops under polyhouse, Post-harvest handling and storage of vegetables, Marketing and Export of vegetables and Seed technology in vegetable crops.

1. Vegetables for Nutritional and Health Security

In the first joint workshop of WHO and FAO on Fruits and Vegetables for Health held at the WHO Centre for Health Development in Kobe, Japan on 1-3 September 2004, the observations recorded were: a) non-communicable diseases (NCDs), especially cardiovascular diseases (CVDs), cancer, obesity and type 2 diabetes mellitus, currently kill more people every year than any other cause of death. Four factors in the epidemiology of these diseases: poor diet, physical inactivity and tobacco and alcohol use are of overwhelming concerns to public health. Fruits and vegetables are an important component of a healthy diet and, if consumed daily in sufficient amounts, could prevent major diseases such as CVDs and certain cancers. According to The World Health Report 2002, low fruit and vegetable intake causes about 31% of ischemic heart disease and 11% of stroke worldwide and b) overall it is estimated that up to 2.7 million lives could potentially be saved each year if fruit and vegetable consumption was sufficiently increased. Recommendations in this direction tend to complement and reinforce other valid messages based on the long known health benefits of consuming vegetables and fruits as dietary sources of fibre, vegetable proteins and protective micronutrients. The Joint FAO/WHO Expert Consultation on diet, nutrition and the prevention of chronic diseases, recommended the intake of a minimum of 400g of fruits and vegetables per day (excluding potatoes and other starchy tubers) for prevention of chronic diseases such as heart disease, cancer, diabetes and obesity, as well as for prevention and alleviation of several micronutrient deficiencies, especially in less developed countries. WHO responded to the global rise in NCDs by giving increasing attention to their prevention and control, most recently through the Global Strategy on Diet, Physical Activity and Health endorsed at the Fifty-seventh World Health Assembly on 22 May 2004.

According to the studies of Indian Council of Medical Research (ICMR), New Delhi and National Institute of Nutrition (NIN), Hyderabad, meagre intake of vegetables and low cost protective foods, is largely responsible for malnutrition among the majority of our population. The dieticians advocate intake of 125 g leafy vegetables, 75 g other vegetables and 100 g root and tuber vegetables (total 300 g) everyday to make our diet balanced. Vegetables provide all the nutrient components like carbohydrates, protein, fat, vitamins, minerals and water along with dietary fibre which are the essential constituents of a balanced diet. Supply of nutrient components through balanced diet is much effective than supplementation through synthetic vitamins and mineral tablets.

Carbohydrates

Carbohydrates serve as the chief source of energy in the food. It is also called protein sparing food. The carbohydrates may be classified into groups according to their complexity viz., monosaccharides (e.g, glucose, fructose, mannose, etc., $C_6H_2O_6$), union of two or more monosaccharide units results in the formation of oligosaccharides (e.g, sucrose, lactose, maltose, $C_{12}H_{22}O_{11}$ etc.) and polysaccharides (e.g., starch, cellulose, inulin, etc., ($C_6H_{10}O_5$). Non-digestible carbohydrates like, cellulose, hemicellulose, gums, pectin and lignin provide roughage and insoluble dietary fibre. Upon burning 1 g glucose 4.0 calories of energy are liberated. Daily requirement of carbohydrates is 400-500 g and major sources of carbohydrates are potato, sweet potato, cassava, elephant-foot yam, taro, garlic, pea, onion, bitter gourd, Chinese potato, Brussels sprouts, carrot, etc.

Protein

Proteins are extremely complex nitrogen-containing organic compounds. They constitute a major part of the protoplasm. Function of proteins in diet is not primarily to supply energy on oxidation but to furnish certain essential components of the living tissue of the organism itself. Essential amino acid constituents of protein are arginine, histidine, lysine, tyrosine, phenylalanine, cystine, methionine, threonine, serine, leucine, isoleucine, valine, glutamic acid, aspartic acid, glycine, alanine, tryptophan and proline. Proteins in the diet are completely broken down in the gastrointestinal tract to individual amino acids which are absorbed into the blood and are passed to various parts of the body. Metabolism of 1 g protein liberates 4.0 κ calories of energy. Daily requirement of protein is 60-70 g and good sources of protein are pea, cowpea, broad bean, lima bean, leaves of fenugreek, mustard, pumpkin, pointed gourd, drumstick, agathi, celery, garlic, Brussels sprouts etc.

Fat

Fats and oils are lipids (esters of fatty acids or substances capable of forming such esters) in which excess nutrients and heat energy are stored in human body. The fats and oils in vegetables and other foods are largely mixtures of palmitin, stearin and olein. Fats are hydrolysed with liberation of glycerol ($C_3H_5(OH)_2$) and different fatty acids. Fats have a high fuel value and upon oxidation, 1 g of fat liberates 9 calories of energy. Daily requirement of fats and oils is 50 g and most of this requirement is met through cooking oil and animal fats. Vegetables contain very low fat ranging mostly between 0.1 and 0.2% and the vegetable crops containing 0.3% and more fat are Bengal gram leaves, small bitter gourd, chilli, sweet pepper, brinjal, Brussels sprouts, snake gourd, pointed gourd, lettuce, pink radish, sweet corn, hyacinth bean, cluster bean, globe artichoke, palak, spinach, fenugreek leaves, mustard leaves, coriander leaves, radish top and mint.

Vitamins

Vitamins are biologically active compounds, occurring in varying and minute proportions in foods, which must be available to the human being and other animals from exogenous sources in order to maintain normal physiological process.

Vitamin A

This fat soluble vitamin is obtained from food in two forms: as retinol or pre-formed vitamin A and as β -carotene or provitamin A, which gets converted into vitamin A in the liver and intestine. Vitamin A is essential for clear vision in dim light and at night. It maintains integrity of epithelial tissues, maintains healthy mucous membranes of eyes, mouth, gastro intestinal, respiratory and genito-urinary systems. Daily requirement is 5000 IU (0.6 ìg â carotene = 1 IU vitamin A). Abundant vitamin A (more than 2500 IU per 100 g) is present in carrot, amaranth, palak, spinach, Indian spinach, water spinach, fenugreek leaves, mustard leaves, pumpkin leaves, drumstick leaves, portulaca, cowpea leaves, beet top, coriander leaves, broccoli, kale, giant spine gourd, muskmelon, winter squash, etc.

Thiamine or vitamin B_1

This water soluble vitamin is essential for the proper utilization of carbohydrates in the body. Daily requirement is 1.2 mg. Most of the vegetables contain 0.05 - 0.06 mg thiamine per 100 g and the vegetable crops containing 0.1 mg or more thiamine per 100 g are palak, pea, ripe tomato, chilli, muskmelon, garlic, leek, pea, hyacinth bean, giant taro, tannia, asparagus, globe artichoke, palak and agathi.

Riboflavin or vitamin B_2

This vitamin is essential as a part of coenzymes for several oxidation processes inside the cell and is concerned with energy and protein metabolism. Daily requirement is 1.7 mg. Range of riboflavin content in most of the vegetables is 0.01 - 0.08 mg per 100 g of edible part and the vegetables containing 0.1 mg or more riboflavin are palak, chilli, sweet pepper, cauliflower, knol khol, broccoli, Brussels sprouts, giant spine gourd, garlic, lettuce, celery, okra, winged bean, asparagus, portulaca, pigweed, radish top, beet top, mint, cowpea, leaves, chicory, watercress, endive and parsley.

Niacin or nicotinic acid

It is involved as component of a coenzyme in oxidative reactions and is concerned with the metabolism of carbohydrate, fat and proteins. This vitamin can be synthesized in the body from trytophan. Daily requirement is 19 mg. Most of the vegetables contain niacin to the tune of 0.1 - 0.7 mg per 100 g and the vegetable crops containing 0.5 - 0.9 mg niacin/100 g are palak, amaranth, bitter gourd, giant spinegourd, pointed gourd, bottle gourd, pumpkin, chilli, radish, lettuce, carrot, pea, cowpea, hyacinth bean, okra, sweet potato, spinach and fenugreek leaves.

Pyridoxine or vitamin B_6

The coenzyme form of this vitamin, pyridoxal phosphate is required for the metabolism of amino acids and conversion of tryptophan to nicotinic acid and is also associated with metabolism of essential fatty acids. It is widely distributed in different vegetable crops.

Folic acid or folacin

It is associated with reduced risk of cardiovascular disease and colon cancer. Folic acid lowers homo cysteine that has an adverse effect on the lining of the arteries. It is also required for the multiplication and maturation of red cells and is particularly important for women. Actual requirement of folic acid is 50-100 mg depending on age and vegetable crops like spinach, palak, lettuce, cabbage, cowpea and French bean supply good amount of folic acid.

Vitamin E or a-tocopherol

It possesses anti-oxidant property and prevents oxidation of β carotene and vitamin A in the intestine. It also prevents oxidation of polyunsaturated fatty acids (PUFA) in cells and thereby maintains integrity of cell membranes. Daily requirement is 5.0 mg and different green leafy vegetables and sweet corn contain adequate vitamin E.

Vitamin K

It is required for the formation of pro-thrombin (blood clot) and for normal function of liver. Daily requirement is 0.015 mg and different green leafy vegetables contain vitamin K.

Vitamin C or ascorbic acid

It is a strong water soluble reducing agent which is easily oxidized to form semi-dehydro ascorbic acid that is relatively stable. It is involved in collagen synthesis, bone and teeth calcification. It helps in absorption of dietary iron by maintaining its reduced form. It maintains connective tissues in blood capillaries and hastens the process of healing wounds and prevents hemorrhages of small blood vessels. About 50 mg of vitamin C is required daily and the vegetable crops particularly chilli, sweet pepper, cabbage, broccoli, kale, leafy greens, drumstick, radish top, coriander leaves, bitter gourd, amaranth and Indian spinach are good sources of vitamin C.

Minerals

Minerals play major role in the functioning of the physiological activities and reproduction and they are the components of various vital body constituents.

Calcium

Normal functioning of heart and nervous system and development of teeth and bone depend on calcium. Daily requirement is 500-600 mg. Very high calcium containing (200-400 mg/100 g) vegetables are curry leaves, hyacinth bean, amaranth, palak, Indian spinach, fenugreek leaves, Bengal gram leaves, pumpkin leaves, pointed gourd leaves, radish top, agathi, carrot top, mint, cowpea leaves and drumstick leaves and the vegetables containing 100-200 mg calcium/100g are coriander leaves, pigweed, portulaca, beet top, Ceylon spinach, Swiss chard, young green onion, onion, chow-chow and parsley.

Iron

Haemoglobin in blood is decreased due to iron deficiency. Daily requirement is 10 mg and most of the vegetables other than leafy greens contain 0.4-1.7 mg iron/100 g. Green leafy vegetables are exceedingly rich in iron with the highest in amaranth (25.5 mg/100 g) followed by Bengal gram leaves (23.8 mg/100g). Different leafy vegetables containing more than 10 mg iron are palak, spinach, fenugreek leaves, mustard leaves, coriander leaves, portulaca, radish top, mint and cowpea leaves. Other vegetables containing appreciably high iron content are lettuce, green banana, hyacinth bean, pigweed, agathi, drumstick leaves, giant spine gourd, watermelon, sorrel and water spinach.

Phosphorus

Normal growth and development is retarded due to phosphorus deficiency. Daily requirement is 1000 mg. Vegetables are quite rich in phosphorus. Tree highest of 310 mg phosphorus/100 g is available in garlic. Other vegetables containing 70-100 mg phosphorus per 100 g are pea, lima bean, taro, pumpkin leaves, carrot top, mushrooms, chilli, cauliflower, broccoli, brussels sprouts, bitter gourd (large), parsnip, cowpea, hyacinth bean, winged bean, globe artichoke and drumstick leaves.

Iodine

Iodine is required for the synthesis of thyroxin hormone from thyroid gland. Daily requirement is 0.1 - 0.2 mg and different vegetables like, tomato, sweet pepper, carrot, onion, garlic, beet, agathi leaves, etc. contain appreciable quantity of iodine.

Sodium

Sodium deficiency is generally manifested by weight loss and nerves breakdown. Daily requirement is 4000-6000 mg and different vegetables particularly celery, green onion, Chinese cabbage and radish are rich in sodium. Sodium requirement is mostly met through intake of common salt.

Dietary Fibre

Dietary fibre is defined as the sum of lignin and polysaccharides that are not digested by the endogenous secretions of the digestive tract. Plant cell walls comprising of cellulose, pectin, xyloglucan, xylan, mannan, free arabinan, etc. are main sources of dietary fibre. Of the total dietary fibre, lignin, cellulose and some components of hemicelluloses form insoluble dietary fibre and gums, pectin and mucilaginous substances constitute soluble dietary fibre. Dietary fibre renders to several health benefits through protection against certain types of cancer, lowering blood cholesterol, etc. It binds to bile salts and thus prevents their re-absorption which ultimately leads to reduction in cholesterol level in blood resulting reduced incidence of coronary heart disease. The gums and pectins reportedly reduce the post-prandial levels of glucose in blood. It is recommended to take 40 g dietary fibre in our daily diet. Vegetables, the rich source of dietary fibre contribute over 50% of the total fibre intake. Dietary fibre content in vegetables ranges widely from 0.2 g/100 g edible portion in watermelon to 3 g/100 g in pointed gourd.

Antioxidants Value of Vegetable Crops

Aerobic organisms are constantly exposed to systems that generate reactive oxygen species (ROS), including the superoxide radical anion (O."), hydroxyl radical (.-OH), hydrogen peroxide (H2O2), various peroxyl radicals (ROO.), and singlet oxygen ($^1O_2$). These ROS are highly reactive and can damage living cells if formed in significant amounts. To avoid cellular damage by ROS, the most biological systems have developed defense mechanisms, antioxidants that convert ROS to unreactive derivatives. Antioxidants are broadly defined as molecules that, when present at low concentrations, compared to an oxidizable substrate, significantly delay or prevent oxidation of that substrate. These include enzymes, frequently known as preventive antioxidants, that is, superoxide dismutase (SOD), catalase (CAT) and peroxidase (POD), which prevent or inhibit formation of ROS, and chain-breaking antioxidants, that is, vitamin C, vitamin E, and carotenoids; and phenolics scavenge ROS and break radical chain sequences. The antioxidant defense system of scavenging the free radicals and reactive oxygen species is divided into two parts: the preventive antioxidants and chain breaking antioxidants. There are numerous compounds in vegetables that function as antioxidants and of these, the followings are the most potent in action.

Carotenoids: These are potent singlet oxygen quenchers in biological system and lycopene appears to have the highest antioxidant activity followed by β-carotene, lutein and cryptoxanthin. Carotenoids which provide colour pigments are present abundantly in tomato, carrot, spinach, amaranth and other leafy greens, pumpkin, watermelon and muskmelon.

Ascorbic acid: The antioxidant activity of ascorbic acid is due to the ease with which it loses electrons, making it very effective in biological systems. It serves as a reducing agent for many reactive oxygen species because it is an electron donor. It reduces tocopherol radicals back to their active form at the cellular membrane. Ascorbic acid is easily oxidized to form relatively stable semi-dehydroascorbic acid and further oxidation generates diketogluconic acid having no biological function.

Vitamin E: It is the major lipid soluble antioxidant responsible for protecting the polyunsaturated fatty acids in membranes against lipid peroxidation. Tocopherols scavange the free radicals by reacting with lipid peroxyl radicals to form tocopheroxyl radicals which react with ascorbate to regenerate tocopherol and produce an ascorbate radical. The antioxidant vitamin (β carotene, tocopherol and ascorbic acid) contents is the highest in the cole crops, kale containing the highest levels followed by broccoli, brussels sprouts, cabbage and cauliflower.

Flavonoids: The flavonoids are a large family of low molecular weight poly-phenolic compounds ubiquitous in plant tissues which include flavones, flavonols, flavonones, isoflavones, flavan-3-ols, anthocyanins, anthocyanidins, catechins, biflavans and isoflavonoids. Most of the flavonoids are structurally polyphenols having antioxidant activity. Different vegetables like kale, spinach, broccoli, brussels sprouts, beets, red bell pepper, onion, garlic, corn, brinjal, cauliflower, cucumber, garlic, beans and mushroom are rich sources of flavonoids hence, have potent antioxidant activity. Potential flavonols in the vegetables are quercetin, kaempferol, myricetin and luteolin. High levels of quercetin are found in onion, kale, tomato and certain varieties of lettuce and the level of kaempferol is high in kale, broccoli and endive.

Medicinal Value of Vegetable Crops

Consumption of different vegetable crops is documented in different ancient scriptures and epics of India and it is believed that these crops are consumed for their different medicinal values. Prominent medicinal values documented in some important vegegetables through different studies are hereunder presented.

Tomato

- Several epidemiological studies show that consumption of tomato protects against cancers of digestive tract, stomach, colon, rectum, gastrointestinal neoplasm and prostate, and this beneficial effect is due to presence of antioxidant compounds like lycopene, carotenoids and vitamin C.
- Among all the antioxidants lycopene- a major carotenoid without pro-vitamin A activity is responsible for their beneficial effects, and is the most efficient quencher of singlet oxygen.

- Tomato is useful in curing cankers and sores in the mouth.
- The secondary metabolites, i.e., colorless glycoalkaloids a-tomatine and dehydrotomatine, significantly reduce plasma cholesterol that feeding high tomatine green tomatoes induce a greater reduction in plasma LDL (bad) cholesterol and triglyceride levels than feeding low tomatine red tomatoes.
- The tomato pulp and juice are mild aperients (laxative), a promoter of gastric secretion and blood purifier and is considered the intestinal antiseptic.
- It is a very good appetizer and remedy for patients suffering from constipation if used as soup.
- It stimulates torpid (sluggish, dull, inactivated) liver and is good in chronic dyspepsia.
- Tomatoes poultice is used in the treatment of edema during pregnancy.

Brinjal

- Fruits are excellent remedy for those suffering from liver troubles and high blood cholesterol problem.
- The de-cholesterolizing action is attributed to presence of higher amount of (65%) ploy-unsaturated fatty acids (linoleic and lenolenic acids) in the flesh and seeds of the fruit.
- Magnesium and potassium contents in the fruits further help in reduction of cholesterol.
- It is rich in bioflavonoids which can prevent stomach cancer.
- Piercing brinjal fruits with a needle and fried in sesame oil is a remedy for toothache.
- White brinjals are said to be a remedy for diabetic patients.
- Aqueous extracts of fruit juice inhibit choline esterase activity of human plasma and dry fruit contains goitrogenic principles.

Chilli

- Stimulation of the hot chilli causes the heart to beat faster thus causing the body to sweat and act as body's natural air conditioner.
- The warmth of chillies not only stimulates the taste buds but also the appetite and digestion through mouth watering effects of increasing saliva flow.
- Chilli has the protective effect against peptic ulcer disease.
- Chilli peppers get rid the body of enough fats to lower the blood cholesterol level and reduce the chances of heart attack.
- Capsaicin, the pungent principle has substantial antigenotoxic and anticarcinogenic effects.

- The carotenoides, predominantly, capsanthin and beta carotene have anti oxidant property.
- The flavonoides in red pepper particularly, quercetin, kaempferol, myricetin and luteolin are responsible for induction of enzymes that detoxify carcinogens.
- Capsaicin, once ingested, causes the brain to release endorphins into the blood stream which can induce a natural feeling of well being similar to that achieved by long distance runners.
- Capsaicin is also used in many pharmaceutical preparations like pain balms, vapour linements, skin ointments for cold, sore throat, chest congestion, etc.
- Capsaicin is also used in the preparation of prickly heat powder.
- The oil content of the capsicum, if dissolved in ether and applied with cotton wool is very useful in relieving rheumatic pains.

Sweet pepper

- The delicate flavour of capsicum stimulates the taste buds and also the appetite.
- Appreciably high Vitamin A and C contents in capsicum act as potential antioxidants.
- The carotenoides, predominantly, capsanthin and beta carotene have anti oxidant property.
- The flavonoides in red pepper particularly, quercetin, kaempferol, myricetin and luteolin are responsible for induction of enzymes that detoxify carcinogens.
- Capsicum preparations are used to counter irritants, in lumbago, neuralgia and rheumatic disorders.
- It has tonic and carminative action especially in dyspepsia.

Cauliflower

- Cauliflower is fairly a high source of glucosinolates (40-80 mg/100g fresh), predominantly sinigrin and glucobrassicin which have prominent anti-carcinogenic property.
- The products of hydrolysis of these glucosinolates formed by enzymatic or non-enzymatic action are biologically active compounds with diverse effect on human health.
- The products of glucosinolate breakdown, particularly the isothiocyanates act as anti-carcinogens.
- Cauliflower also contains S-methylcysteine sulfoxide (14 mg/100 g fresh) which also show anti-carcinogenic property.
- Beta carotene in cauliflower acts as potential anti-oxidant.

Cabbage

- Cabbage is fairly a high source of glucosinolates (40-90 mg/100g fresh), predominantly sinigrin and glucobrassicin which have prominent anti-carcinogenic property.
- The products of hydrolysis of these glucosinolates formed by enzymatic or non-enzymatic action are biologically active compounds with diverse effect on human health.
- The product of glucosinolate breakdown, particularly the isothiocyanates act as anti-carcinogens.
- Cabbage also contains S-methyl cysteine sulfoxide (18 mg/100 g fresh) which also shows anti-carcinogenic property.
- Cabbage protects against bowel cancer due to presence of indole-3 carbinol.
- Beta carotene in cabbage acts as potential anti-oxidant.
- Rich in vitamin C and fairly high fibre and calcium contents reduce the risk of colon cancer.
- Cabbage used against ailments like gout, diarrhoea, stomach and coeliac troubles.
- Cabbage juice is a remedy against poisonous mushrooms and as a gargle against hoarseness.
- The leaves are traditionally used to cover ulcers and wounds.
- Fermented product of cabbage-shredded leaves (sauerkraut) has curative effect on scurvy disease.

Onion

- Onion bulbs contain different flavonoids, the plant secondary metabolites like, quercetin (28.4-48.6 mg), kaempferol (0.2 mg), myricetin (0.1 mg), luteolin (0.1 mg), etc. per 100 g fresh which are potential antioxidants.
- The highly biologically reactive organosulfur compounds responsible for the characteristic odour and taste have potential influence on phase-I (carcinogen activation) and phase-II (carcinogen detoxification) enzymes.
- Consumption of onion results in a fewer tendencies to form blood clots and lower the levels of cholesterol and lipoproteins associated with heart disease.
- Onions are diuretic, relieve heat sensation, insect bite and soar throat and the juice is used as smelling on hysterial fits in faintness.
- It is applied on bruises, boils and wounds.

Garlic

- The highly biologically reactive organosulfur compounds responsible for the characteristic odour and taste have potential influence on both phase-I (carcinogen activation) and phase-II (carcinogen detoxification) enzymes.
- Garlic cloves contain different flavonoids, the plant secondary metabolites like, quercetin, kaempferol, myricetin , luteolin etc. which are potential antioxidants.
- Allicin (diallyl thiosulfinate), which has hypocholesterolemic action, reduces the cholesterol concentration in human blood.
- Garlic consumption also reduces serum lipids perhaps by decreasing their synthesis and/or increasing their excretion through the intestinal tract.
- Garlic oil also reduces blood sugar and this hypoglycemic effect may be due to increased glycogenesis in the liver and better utilization of glucose in the peripheral tissue.
- Garlic is described as hot stimulant, carminative and anti-rheumatic.
- Garlic oil is a powerful anticeptic mainly due to presence of allicin.It is used as vermifuge for expelling round worms and has long been recommended for cure of a number of ailments viz., wounds, ulcers, pneumonia, bronchitis, atopic dyspepsia and gastro-intestinal disorders.
- Inhalation of garlic oil or garlic juice has generally been recommended by doctors in pulmonary tuberculosis, rheumatism, sterility, impotency, cough and red eyes.
- Garlic is a gastric stimulant and helps in digestion of food and absorption of nutrients from it.

Pointed Gourd

- It is easily digestible, diuretic, laxative and cardiatonic.
- The fruit is particularly recommended during convalescence.
- It is recommended also for bronchitis, biliousness, and high fever and nervousness.
- It also invigorates the heart and brain and is useful in the disorders of the circulatory system.
- The fruits show prospects in the control of certain cancer like conditions.
- It is particularly beneficial for colon problem due to very high dietary fibre content.

Cucumber

- The fruits and seeds have cooling effect and prevent constipation.
- Fruits are good for people suffering from constipation, jaundice and indigestion.

- Fruits are demulcent and diuretic.
- Fruits contain photolytic enzymes, ascorbic acid oxidase, succinic and malic dehydrogenases.
- Fruits are also used as an astringent and antipyretic.
- Seed oil is used as antipyretic.

Bottle Gourd

- Fruits are easily digestible and have cooling effect.
- It is good for people suffering from biliousness and indigestion.
- Fruits are cardiatonic and diuretic.
- The pulp is good for overcoming constipation, cough and night blindness.
- The pulp also acts as an antidote against certain poisons.
- The plant extract is used as a cathartic.
- The seeds are used for treating dropsy.
- Decoction of leaves mixed with sugar is given for treating jaundice.

Bitter Gourd

- Fruits are antidotal, antipyretic tonic, appetizing stomachic, antibilius and laxative.
- Several researches showed that it contains a hypoglycaemic or insulin like principle, designated as 'plant –insulin' which bas been found highly beneficial in lowering the blood and urine sugar levels.
- It is highly beneficial in the treatment of blood disorders like blood boil, scabies, itching, psoriasis, ring worm and other fungal diseases.
- It is also good for curing rheumatism, diabetes and asthma.
- Fruits and leaves are anthelmentic and vermifuge and useful in piles, leprosy and jaundice.
- Fresh leaf juice is an effective medicine in early stages of cholera and other types of diarrhoea during summer.
- Leaf juice is used as emetic, purgative, given in bilious affections and rubbed in burning of soles of the feet.
- Roots are astringent and useful in haemorrhoids.

Pumpkin

- Pumpkin is particularly important for the supply of antioxidants and especially β-carotene in the foods.
- It alleviates cardiovascular diseases and constipation.
- It is also a good diuretic food because of having high K : Na ratio.
- Its combination with sugar and honey has soothing effect to ulcers and blisters.
- It is used in the treatment of diabetis, rheumatism, eczema and burns and against worms and other parasites.
- The fruit pulp is also used for poultice for boils, carbuncles and ulcers etc.
- Its seeds along with sugar are taken to kill tape worms and are useful in urinary diseases.

Okra

- Okra contains high amount of dietary fibre (1.2%) which includes a large spectrum of molecules viz., cellulose, non-starch polysaccharides and lignin which give protection against certain types of cancer particularly colon cancer, regulate transit and lower blood cholesterol.
- It is very useful against genito-urinary disorders, spermatorrhoea and chronic dysentery.
- Its use in regular diet increases is sperm count in men.

Cowpea and Hyacinth Bean

- Among the food legumes, cowpea and hyacinth bean are particularly important for supply of antioxidants and especially b-carotene in foods.
- Cowpea is particularly important for the supply of vitamin B in foods.
- Cowpea contains high amount of dietary fibre which gives protection against certain types of cancer particularly colon cancer, regulates transit and lowers blood cholesterol.

Radish

- In the Sutra period (800 B.C to 300 B.C) the mention of radish root was made as aiding digestion when munched after meals.
- Radish roots are good for liver and gallbladder trouble.
- In homeopathy, they are used for neuralgic headaches, sleeplessness and chronic diarrhoea.

- The roots are useful in urinary complaints, piles and problems of spleen.
- The seeds are peptic, expectorant, diuretic and carminative.
- The juice of fresh leaves is used as diuretic and laxative.

Awareness Programmes on Health Security

1. Extensive training programmes to the vegetable farmers regarding the use of Good Agricultural Practices (GAP) in different vegetable crops and appropriate pre and post-harvest technologies for vegetable crops to extend availability of safe produce.
2. Television and radio programmes need to be broadcast to the public on the benefits of fruit and vegetables, their proper washing, storage, healthy preparation, and cooking methods.
3. Educational activities need to be organized for both pre-school and school-going children through puzzles, posters, songs and comics to create awareness about fruit and vegetables and what constitutes a healthy diet, to educate how to select fruit and vegetables, how to prepare foods and to learn how to grow vegetables in home garden.
4. Consorted programmes on the development of home vegetable gardens combined with community nutrition education campaigns particularly in rural areas which will offer a proven means of enhancing household food security, improving nutritional status and providing a much-needed supplement to the household income.

2. Status of Vegetable Production

Horticultural crops are the best suitable alternative to the urgently needed balanced diversification of Indian Agriculture and vegetable crops emerge as the key horticultural crops both in domestic and export market. Presently India is the second largest producer of vegetables sharing 11.5 % world production of vegetables and other important vegetable producing countries in the world being China, USA and Turkey with the production share of 46.1, 4.7 and 3.1 %, respectively. Total vegetable production in India before independence was 15 million tones and since independence for decades the growth rate was stabilized near 0.5 %. The impetus on vegetable research and policy intervention to promote vegetable crops witnessed a sudden shot up of growth rate of 2.5 %, a hike of five times in the last decade giving total annual production of over 156 million tones during 2011-12. The overall productivity of 17.38 tones / hectare is far below the potential productivity of different vegetable crops.

Area, production and productivity of vegetable crops in India over last decade

Year	Area (000'ha)	Production (000't)	Productivity (t/ha)
1994-95	5013	67286	13.42
1995 - 96	5335	71594	13.42
1996-97	5515	75074	13.37
1997-98	5607	72683	12.96
1998-99	5866	87536	14.92
1999-2000	5993	90831	15.15
2000-01	6250	93850	15.01
2001-02	6155	88622	14.40
2002-03	6092	84815	13.92
2003-04	6309	93165	14.76
2004-05	6744	101246	15.01
2005-06	7213	111399	15.44
2006-07	7581	114993	15.16
2007-08	7848	128449	16.36
2008-09	7981	129077	16.17
2010-11	8495	146554	17.25
2011-12	8990	156325	17.38
2012-13	9205	162187	17.62
2013-14	9396	162897	17.34
2014-15	6358	88819	13.97
2015-16	9575.38	166608	17.40

(*Source:* NHB, Indian Horticulture Database-2012)

Statewise area, production and productivity of vegetables (2015-16)

States/UT's	Area 000'ha)		Production (000't)		Productivity (t/ha)	
	2011-12	2015-16	2011-12	2015-16	2011-12	2015-16
Andaman & Nicobar	6.31	15.99	43.21	45.54	6.84	7.60
Andhra Pradesh	660.97	234.95	12025.28	5409.07	18.19	23.04
Arunachal Pradesh	6.34	0.87	83.50	12.55	13.17	14.45
Assam	266.00	306.55	3045.56	3776.16	11.44	12.32
Bihar	857.01	3838.51	15552.38	14400.12	18.14	5.07
Chhattisgarh	351.55	475.23	4582.63	6318.00	13.03	13.30
D & N Haveli	1.10	0.00	5.50	0.00	5.00	15.72
Delhi	27.89	23.0	466.68	362.82	16.73	11.41
Goa	6.50	7.24	78.20	82.59	12.03	20.25
Gujarat	517.63	626.23	10049.81	12682.26	19.41	14.99
Haryana	356.77	410.74	5068.42	6156.88	14. 21	19.22
Himachal Pradesh	85.68	88.64	1561.51	1704.35	18. 22	22.14
Jammu & Kashmir	63.06	62.69	1395.47	1386.37	22.12	13.21
Jharkhand	261.24	193.92	3902.63	25.61	14.94	17.82
Karnataka	454.70	509.54	7662.50	9080.80	16.85	14.36
Kerala	149.05	141.91	3626.00	2039.33	24.32	2.00
Lakshadweep	0.25	0.33	0.31	0.66	1. 24	21.04
Madhya Pradesh	506.99	7.3.39	10084.01	14797.74	19. 88	105.14
Maharashtra	591.00	619.14	8778.00	9376.29	14.85	8.93

Contd.

States/UT's	Area 000'ha)		Production (000't)		Productivity (t/ha)	
	2011-12	2015-16	2011-12	2015-16	2011-12	2015-16
Manipur	20.83	32.81	200.32	293.15	9.61	10.63
Meghalaya	39.46	48.62	385.01	517.22	9.75	10.63
Mizoram	37.42	54.14	221.10	281.82	5.91	5.20
Nagaland	33.04	45.95	222.63	29.70	6.74	11.55
Odisha	690.07	578.00	9520.56	8193.20	13.79	14.18
Puducherry	0.55	0.67	7.49	10.66	13.62	15.91
Punjab	178.22	217.01	3674.53	4368.43	20. 61	20.13
Rajasthan	181.71	146.75	1287.41	1150.21	7.08	7.84
Sikkim	25.03	25.78	127.65	129.08	5.09	5.01
Tamilnadu	306.66	290.69	9068.49	7771.58	29.57	26.73
Tripura	34.20	49.58	552.55	830.16	16.15	16.74
Uttar Pradesh	852.09	1202.39	18563.75	25688.95	21.78	21.36
Uttarakhand	89.29	100.92	1066.71	1110.05	11. 94	10.99
West Bengal	1330.94	1391.43	23415.69	22825.57	17.59	16.40

(*Source:* NHB, Indian Horticulture Database-2012-2015)

Average and the highest productivity of major vegetables

Vegetables	India (tones/ha)	World (tones/ha)	Potential productivity (tones/ha)	Maximum productivity (tones/ha)
Tomato	19.5 0	32.80	60-80	70.45 (USA)
Brinjal	16.08	17.48	40-50	34.7 (Japan)
Chilli	9.18	14.4	30-40	44.5 (Spain)
Okra	11. 60	6.90	15-20	17.78 (Jordan)
Pea	9 .5 0	8.35	18-20	20 (Lithuania)
Melons	20.48	20.95	30-40	45.83 (Cyprus)
Gourds	9.72	12.97	25-30	41.33 (Israel)
Cucumber	6.67	16.98	40-50	67.67 (Korea)
Watermelon	12.75	27.13	30-40	40.96 (Spain)
Cabbage	21.50	27.70	30-40	42.59 (Japan)
Cauliflower	18.30	16.90	35-40	45.25 (NewZealand)
Onion	14.20	19.10	40-50	60.33 (Korea)
Garlic	4.17	12.37	15-20	23.23 (Egypt)

(*Source:* NHB, Indian Horticulture Database-2012)

The potential technological intervention with improved gene pool and precise management can very well shoot up the growth rate to near 6%/annuum. This may lead to increased production of vegetable crops to 228 million tones by 2015, which may go in a long way to achieve the much-awaited target of making available 300g vegetables /capita / day or even more. To meet the availability of recommended 300 g/ caput/ day for balanced diet, India will need to produce 200/260 million tones of vegetables by 2016 and 2021 respectively.

Period	Production (million tones)	Demand (million tones)
2005-06	111.40	120.00
2011-12	156.33	196.00
2020-21	190.00	260.00

There is a vast gap between actual and potential productivity which indicates ample scope for development of technologies to harness the potential to meet domestic and export demand. Although commendable progress is made on research front and a number of technologies are developed, several target oriented research strategies are needed to get the answer of some of the unsolved questions.

Priorities for developing multiple resistant cultivars with premium attributes

Crops	Targeted biotic stress	Priortised trait (s) to be incorporated
Tomato	TLCV + early blight + bacterial wilt + RKN	Tolerant to high temperature, highTSS and lycopene
Brinjal	Phomopsis + bacterial wilt + fruit- and shoot-borer	Early maturing
Chilli	Leaf curl + thrips + mites + anthracnose	High oleoresin and less capsaicin
Capsicum	*Phytophthora* + thrips + mites	Adapted to tropical regions of India
Okra	YVMV + ELCV + fruit-borer	Dark green pods with five ridges
Onion	Stemphyllium + purple blotch + thrips	Brown bulb, white bulb with high TSS
Cucumber	DM + mosaic	Pickling type
Muskmelon	PM + DM + anthracnose + fusarium	High TSS
Watermelon	PM + DM + anthracnose	High TSS and seedless fruits
Cabbage	Black rot + diamondback moth	Tolerant to high temperature
Cauliflower	Black rot + diamondback moth	Tolerant to high temperature

Note: TLCV, Tomato leaf curl virus; RKN, root-knot nematode; YVMV, yellow-vein mosaic virus; ELCV= Enation leaf curl virus; DM, downy mildew; PM, powdery mildew

Poor adoption of technologies has always been a major handicap in increasing productivity in India. Therefore, promotion of developed technologies through various schemes is the key to harness available technologies ensuring increased productivity and profitability to the farmers. Likewise pro-active government policies for development of infrastructure and law enforcement to support and execution of such promotional activities are equally pertinent to achieve higher productivity and the major promotional activities which need due attention are mentioned below.

Area increase under hybrids/improved varieties: Promotion of hybrid vegetable technology/improved varieties is a major strategy to increase vegetable productivity. Out of the total area under vegetable cultivation, hardly 15-20 % area come under hybrid vegetable crops, moreover major share goes to tomato and cabbage. Concerted efforts need to be made to cover about 35-40 % area under hybrids of different important vegetable crops viz, okra, chilli, sweet pepper, cauliflower, bottle gourd, bitter gourd, cucumber, ridge gourd, pumpkin, watermelon, muskmelon, summer squash, etc.

Increased seed replacement: Promotion of seed production of improved varieties/ F_1 hybrids of vegetable crops in India capitalizing huge and low cost human resource should be the basic strategy for increasing productivity by ensuring quality seeds to the farmers and increasing profitability through export of vegetable seeds. Faster rate of seed replacement need to be acomplished through awareness among the growers about the new variety/hybrid and high level of quality seed production.

- **Front line demonstrations:** Large scale Front Line Demonstrations (FLDs) and distribution of minikit trials for hybrids/improved varieties and newly introduced future potential vegetable crops should be encouraged in selected village clusters identified for specific vegetable crops through Agricultural universities/Krishi Vigyan Kendras Government line department/NGOs, etc. to popularise the new and improved genetic materials among the vegetable growers.

- **Seed village clusters/cluster farming/organic farming/export zones:** Cluster farming/Seed village concept need to be encouraged to increase the production of seeds, planting materials and specified market driven product through identification of village clusters based on production potential of vegetables/vegetable seeds and pesticide free areas/zones need to be declared preferably in the north-eastern hill region for promoting organic cultivation of specialized vegetables. Such cluster village concept will also strengthen cause of vegetables suitable for export. Production of manures/vermicompost can also be encouraged and strengthened in cluster village concept to promote organic farming.

- **Seed multiplication through public–private tie up:** Adequate multiplication of the seeds of public sector open pollinated improved varieties and hybrids through strengthening public–privated tie up and periodical renewing of basic seed from the public sector.

Development of Bioresource centres for conservation of genetic resources: Genetic resources are the building blocks for the new varieties as well as hybrids. The Indian gene center is extremely rich. The Indian sub-continent is one of the centers of origin and diversity of several vegetable and spice crops. Wide and prolonged cultivation of different introduced vegetable crops, indigenous, tropical and tropicalized vegetable crops triggered the development of huge diversity in local cultivars of most of the vegetable crops in the country. These indigenous varieties evolved through both natural and human selection, and produced and used by the farmers of the developing countries world wide including India are called primitive cultivars by the corporate sectors of seed business and germplasm of these freely available resources are apparantly regarded valueless. The disappearance of ecological diversity and the intolerance of cultural diversity are processes that are intimately linked. The survival struggles of different communities are infact, struggles for the preservation of diversity. In the last two decades or so, in addition to the proximate causes leading to the loss of biological resources like, clearing of forests, over-harvesting of plants, indiscriminate use of pesticides and rapid urbanization, deliberate substitution of diversity by uniform improved varieties and hybrids has worsened the biodiversity crisis in different vegetable crops. It is urgently needed to establish "Bioresource centre" in different zones of the country with active participation of the farmers to conserve the indigenous genetic resources of different vegetable crops for future use in breeding.

Crop diversification: Most of the vegetables are short duration and are suitable for growing in mixed, relay and companion cropping systems. Taking this advantage, crop diversification among the vegetables, the underutilized and exotic vegetables rich in nutrient for both domestic and export markets should be promoted through adoption of different developmental schemes. Some of these vegetables which need to be promoted are sweet corn, baby corn, moringa, leek, bunching onion, welsh onion, broccoli, brussels sprouts, Chinese cabbage, pointed gourd, kakrol, kartoli, ivy gourd, sweet gourd, faba bean, summer bean, lima bean, winged bean, sword bean, velvet bean, etc.

Hi-tech vegetable production: To bridge the vast gap between actual and potential productivity, there is urgent need to adopt efficient and mechanized vegetable production technologies which will not only increase productivity but also enhance the product quality, thereby, ensuring better returns to farmers.

1. Development of Hi-tech nursery for production of healthy seedling.
2. Protected vegetable cultivation around peri-urban areas.
3. Efficient use of water and nutrient through drip/sprinkler irrigation and fertigation.
4. Raised bed cultivation and use of organic/degradable biomass for mulching which will economize irrigation water, retain moisture, stabilize soil temperature, minimize weed population and ensure high quality production.
5. Promotion of precision farming for maximizing input use efficiency to get higher and quality return.
6. Promotion of off-season production of vegetables to enhance availability of quality vegetables throughout the year and maximize the use of natural resources through use of improved genetic materials and protected cultivation.

Periurban vegetable farming: Periurban vegetable farming through different forms of market gardens has gained momentum around different cities and metropolies, but it need to be further developed and strengthened through imparting education to the practicing farmers for the use of "Good Agricultural Practices" for production of safe produce and appropriate pre and post harvest technologies to extend availability of produce. Codexes for pesticide residue level need to be created/developed and enforced through pesticide residue testing in the "Quality control laboratory" to ensure safe and fresh vegetables to the consumers.

Integrated plant nutrient management system (IPNMS): The integrated plant nutrient system (IPNS) includes balanced use of chemical fertilizers; organic manures and biofertilizers which not only increase the productivity in sustainable manner but also take care the soil health and productivity. The IPNS system is more result oriented in protected vegetable cultivation where crop rotation and soil replacement are highly difficult tasks. It will be more effective and sustainable if both IPNS and IPM systems are adopted together.

Integrated pest management (IPM): It is the eco-friendly system approach in both strategies and technologies to reduce the dependence of chemicals for controlling the pest and diseases. Nevertheless, multiple support and co-operation from public and private organizations is needed to harness full potential of available technologies and strategies.

Use of resistant/tolerant varieties

1. Promoting awareness among the farmers about the predisposing factors for the occurrence of diseases and pests and at the same time harmful effects of indiscriminate use of chemical pesticides.
2. Emphasis on the development of behavioural and cultural management practices, use of sex pheromone, trap crop, predators, parasitoids and entomo-pathogens based management tactics to control pests and diseases.
3. Incorporation of botanicals, microbials and pheromones in IPM schedule to reduce dependency on chemical insecticides.
4. Emphasis on the quality control for the production of bioagents and botanicals and development of bioagent release technique to improve performance of bioagents.
5. Policy frame work for proper monitoring of the quality of commercially available bio-control agents as well as products starting from production unit till its delivery to the farmers.

Developemnt of plant health clinic: With the shrinking land resources, more intensive crop husbandry and even more complex agri-horticultural production and marketing systems, the agro-biologists do face newer and intricate problems in combating different biotic stresses caused by fungi, bacteria, virus, insect pests, nematodes, rodents, weeds and abiotic stresses caused by nutrient dificiency, soil reaction, high temperature and high moisture condition. It is a must now to understand these stress factors, their mode of action and preventive measures with greater precision and vigour with strong laboratory support. Plant health clinic is district or specific production zone level laboratory based and service oriented project for the growers and other stake holders to elevate their skill and operative standards to produce high quality and safe product throgh Good Agricultural Practices.

Post harvest management: Governmental policy interventions are needed for adequate post harvest management of produce through development of production centres based on cluster village concept and establishment of processing units which may include input subsidies for establishment of cool chain, processing into value-added products, pricing policies, import/export tariffs and facilities for marketing.

Farmers' cooperative societies/Agri-business consortium: The pro-active government policy with liberal funding to support cooperative societies would play a key role in sustaintaining vegetable productivity, production and marketing ensuring profitability to the farmers. Vegetable farming as well as marketing of products under co-operatives will ensure:

- Assured availability of quality seed, other planting materials and agricultural inputs like fertilizers, micro-nutrient chemicals, biofertilizers, biopesticides, botanicals, pheromone trap, mulching material, machinery, etc. to sustain high productivity.
- High quality produces with higher productivity because of community level effectiveness of adoption of technologies and concepts like integrated plant nutrition management system and integrated insect pest and disease management.
- Assured and cost effective availability of vegetables to post harvest industries and for bulk marketing by big corporate sector.
- Better access to domestic and export markets to sell the produce at competitive prices.

Mechanization of vegetable farming: Precision farming and less labour intensive but efficient vegetable farming need mechanization in the different farming practices viz., direct seed sowing, transplanting, weeding, hoeing, spraying, harvesting etc. So, adequate and cost effective machinery and tools should be made available to the farmers with proper demonstration facilities.

Protection to farmers/Risk assistance: Crop insurance scheme should be strengthened for protecting/safeguarding the interest of vegetable growers in the situation of crop loss due to adverse weather conditions and poor quality seeds, fertilizers and plant protection inputs.

Improvement/Establishment of regular markets/strengthening of fresh vegetable marketing: More regular markets for vegetables need to be established at suberban areas, town, and metropolices and the existing markets need to be modernized to encourage not only efficient marketing but also assure good quality and hyginic produces to the consumer. The condition of the weekly or bi-weekly big markets in the rural areas, local markets in the subarban areas need to be be improved through providing adequate space, cleanliness inside the market and efficient transport facilities.

Grading standards: In India grading started as early as 1937 with the Agricultural Produce (Grading and Marketing) Act. At present only three vegetables viz., potato, onion and garlic comes under grading only for export. In this era of WTO, export oriented competitive vegetable farming is gaining momentum therefore, grading standards need to be developed to compete in the global market.At the same time, proper follow up programmes need to be taken to introduce grading for domestic market also which will ensure better return to the growers.

Increasing the purchasing power of rural population: Purchasing power of the population of the rural sector should be increased through increasing productivity of the crops and at the same time, by generating employment in the rural areas by encouraging them for specified farm and other related activities like, open pollinated and hybrid seed production, on farm processing of fresh vegetables, home scale vegetable processing, packaging, making organic manure from farm waste, etc.

Vocational courses for entrepreneurs/Master farmers: Development of knowledge base to the progressive vegetable growers through vocational courses in the Agricultural universities, Krishi Vigyan Kendras and Government establishments will play a pivotal role amongst a larger section of farmers to promote cluster farming in specified areas.

Policies related to awareness and human resource development: There is need to develop specialised human resource through trainings arranged nationally/internationally for knowledge empowerment of scientists, Govt. officials of line departments and the vegetable growers on various aspects of vegetable improvement, production and protection technologies, vegetable seed production, vegetable post harvest management and processing technologies. Different facets of such awareness and human resource development programmes may be out lined as-

- Development of specialized human resource at the scientist and Govt. official level through trainings arranged nationally/internationally on various aspects of vegetable breeding, production and protection technologies, vegetable seed production, vegetable post harvest management and processing.
- Extensive training programmes to vegetable farmers regarding use of GAP (Good Agricultural Practices) and production and protection management in different vegetable crops and appropriate pre and post-harvest technologies for vegetable crops to extend the availability of safe produces.
- Television and radio programmes on the benefits of fruits and vegetables, their proper washing, storage, healthy preparation and cooking methods.
- Educational activities for pre-school children including puzzles, posters and songs and such activity for the school children may encompass the followings-
    - Educating children and creating awareness about fruit and vegetables and what constitutes a healthy diet e.g. through comics, posters and visible reminders in classrooms.
    - Building children's food skills, e.g. how to select fruits and vegetables, how to prepare foods.
    - Producing and teaching about fruits and vegetables via school gardening programmes.

3. Classification of Vegetable Crops

Vegetables are defined as of herbaceous plant origin, material with no calorific value providing micronutrients, vitamins, antioxidants, fibre and alkaline reaction inside human body. The crops used for vegetable purpose in the world belong to 1200 species under 78 families and of them, more than 860 species under 59 families belong to dicotyledoneae and about 340 species under 19 families belong to monocotyledoneae. In the tropical and subtropical parts of the world, about 90 species of vegetable crops are cultivated, but hardly 20 of them are commercially important. Vegetable crops are not only botanically different but also their cultural and climatic adaptation also show wide variability. It is, therefore, necessary to classify the vegetable crops for their easy recognition. Vegetable crops can be classified according to the following ways: (1) botanical classification, (2) classification based on hardiness, (3) classification based on growing season, (4) classification based on tolerance to soil reaction, (5) classification based on salt tolerance, (6) classification based on the plant parts used as vegetable and (7) classification based on methods of culture.

Botanical Classification

This is by far the best method of classification especially for vegetable scientists and breeders as it is based on botanical and taxonomic relationships among different vegetable crops. However, from the standpoint of vegetable cultivation, this classification would not be of much use because cultural requirements for the vegetable crops belonging to the same group vary widely except in few cases.

In the present classification, over 140 different types of vegetable crops under 6 monocotyledoneae and 20 dicotyledoneae families are included. Different fruit crops in their unripe stage viz., green banana (*Musa paradisiaca* L, family: Musaceae), green papaya (*Carica papaya* L., family: Caricaceae) and immature jackfruit *(Artocarpus heterophyllus* Lamk. family: Moraceae) are popularly used as vegetables in Indian cuisine for the preparation of different dishes.

Vegetable crops, their botanical name and family and chromosome number

Names	Botanical name	Chromosome number (2n)
	Monocotyledoneae	
	1. Amarylidaceae (Alliaceae)	
Onion	*Allium cepa* L.	16
• Multiplier onion	*A. cepa* var. *aggregatum* L.	16
• Top onion	*A. cepa* var. *viviparum* (Metz.) Alef.	16
Garlic	*A. sativum* L.	16-32 (2x – 4x)
Leek	*A. ampeloprasum* L. var. *porrum*	32, 48 (4x, 6x)
Welsh onion	*A. fistulosum* L.	16
Shallot	*A. ascalonicum* L.	16
European chive	*A. schoenoprasum* L.	16
Chinese chive	*A. tuberosum* Rottler ex Spreng	32
Kurrat	*A. kurrat*	32 (4x)
	2. Araceae	
Taro	*Colocasia esculenta* (L.) Scott	-
• Eddoe type	*C. esculenta* var. *antiquorum*	28,42 (2x, 3x)
• Dasheen type	*C. esculenta* var. *esculenta*	42 (3x)
Giant taro	*Alocasia macrorrhiza* (L.) Scott, *A. indica* (Roxb.) Scott.	26,28
Swamp taro	*Colocasia esculenta* var. *Stolonifera* (L.) Scott	26,28

Contd.

Tannia	*Xanthosoma sagittifolium* (L.) Scott.	26
Elephant-foot yarn	*Amorphophallus paeoniifolius* (Dennst) Nicholson	26,28
	3. Bambusaceae	
Bamboo	*Bambusa vulgaris* Schrader ex Wendland	
	4. Dioscoreaceae	
Yam	*Dioscorea* spp.	30-80 (3x-8x)
• Greater yam	*D. alata* L.	40 (4x)
• Lesser yam	*D. esculenta* (Lour.) Burkill	40 (4x)
• White yam	*D. rotundata* (L). Poir.	40 (4x)
	5. Gramineae (Poaceae)	
Sweet corn	*Zea mays* L. var. *rugosa*	20
	6. Liliaceae	
Asparagus	*Asparagus officinalis* L.	20
	7. Nelumbonaceae	
Lotus	*Nelumbo nucifera* Gaertn.	
	Dicotyledoneae	
	1. Aizoaceae	
New Zealand spinach	*Tetragonia expansa* (Murr.)	
	2. Amaranthaceae	
Leafy amaranth	*Amaranthus tricolor* L., *A. blitum* L., *A. viridis* L., *A. tristis* L., etc.	32 or 34
	A. dubius L.	64 (4x)
	3. Basellaceae	
Indian spinach (Malabar night shade)	*Basella rubra* L. var. *alba* (green type)	24
	B. rubra L. var. *rubra* (red type)	24
	4. Chenopodiaceae	
Garden beet	*Beta vulgaris* L. subsp. *vulgaris*	18
Palak	*B. vulgaris* L. var. *bengalensis* Hort.	18
Chard	*B. vulgaris* L. var. *cicla*	18
Spinach	*Spinacea oleracea* L.	12
French spinach (orach)	*Artiplex hortensis* L.	12
Pigweed	*Chenopodium album* L.	36 (4x)
	5. Compositae (Asteraceae)	
Lettuce	*Lactuca sativa* L.	
• Head type	*L. sativa* var. *capitata* L.	18
• Leaf type	*L. sativa* var. *crispa* L.	18
• Cos type	*L. sativa* var. *longifolia* L.	18
• Asparagus type	*L. sativa* var. *asparagina* L.	18
Chicory	*Cichorium intybus* L.	18
Endive	*Cichorium endivia* L.	18
Globe artichoke	*Cynara scolymus* L.	34
Jerusalem artichoke	*Helianthus tuberosus*	102 (6x)

Terragon	*Artemisia dracunculus* L.	18
Salsify	*Tragapogon porrifolius* L.	-
Black salsify	*Scorzonera hispanica* L.	-
Spanish salsify	*Scolymus hispanica* L.	-
Dandelion	*Taraxacum officinale* Weber ex Wiggers	-
	6. Convolvulaceae	
Sweet potato	*Ipomea batatus* (L.) Lam.	90 (6x)
Water spinach	*Ipomea aquatica* Forsk	-
	7. Cruciferae (Brassicaceae)	
Cabbage	*Brassica oleracea* L. var. *capitata* L.*f. alba* DC	18
Red cabbage	*B. oleacea* L. var. *capitata f. rubra* (L.) Thell	18 D.C.
Savoy cabbage	*B. oleacea* var. *sabauda* L.	18
Cauliflower	*B. oleacea* L. var. *botrytis* L.	18
Brussels sprouts	*B. oleracea* L. var. *gemifera* DC.	18
Sprouting broccoli	*B. oleracea* L. var. *italica* Plenck	18
Knol khol (Kholrabi)	*B. oleracea* var. *gongylodes* L.	18
Kale/collard	*B. oleracea* var. *acephala*	18
Chinese cabbage	*B. chinensis* L. (Pak-choi), *B. chinensis* L. var. *pekinensis* (Pet-sai)	20
Chinese kale	*B. alboglabra* Bailey	-
Mibuna	*B. japonica*	-
Turnip	*B. rapa* L. var. *glabra* Kitamura	20
Rutabaga (swede)	*B. napus* var. *napobrassica* Peterm.	38 (amphidiploid)
Leaf mustard	*B. juncea* (L.) Czern. & Coss. var. *cuneifolia* Roxb.	36 (4x)
Curled mustard	*B. juncea* var. *crispifolia*	36 (4x)
White mustard	*Sinapis alba* L.	24
Radish	*Ranphanus sativus* L.	18
• European radish	*R. sativus* L. var. *radicula* DC	18
• Japanese & Chinese winter radish	*R. sativus* L. var. *raphanistroides* Makins	18
• Black or Spanish radish	*R. sativus* L var. *niger* (Mill.) Pers.	18
• Oilseed radish	*R. sativus* L var. *caudatus* (L) Hooker et Anderson	18
Horse radish	*Armoracia rusticana* Gaertn.Mey. & Schred.	32 (4x)
Water cress (brahmi sag)	*Rorippa nasturtium-aquaticum* (Syn. *Nasturtium officinale* R. Br.)	32
Garden cress	*Lepidium sativum* L.	16,32
Upland cress	*Barbarea vulgaris* R. Br.	-
Sea kale	*Crambe maritime*	-
Catran	*Crambe tatarica* Busch	-
Tuberous rooted Chinese mustard	*Sinapis juncea* var. *napiformis* Paill & Bois.	-
	8. Cucurbitaceae	
Cucumber	*Cucumis sativus* L.	14
Muskmelon	*Cucumis melo* L.	24
• Snap melon	*C. melo* L. var. *momordica*	24
• Long melon	*C. melo* L. var. *utilissimus* Duth.& Full.	24
• Snake melon	*C. melo* L. var. *flexuosus* Naud	24

• Netted melon	*C. melo* L. var. *reticutalus*	24
• Pickling melon	*C. melo* L. var. *conomon* Mak.	24
• Cantaloupe	*C. melo L.* var. *cantalupensis* Naud.	24
• Winter melon	*C. melo* L. var. *inodorus*	24
• Mango melon	*C. melo* L. var. *dudain*	24
• Smazll gourd	*C. melo* L. var. *agretis* Naud.	24
Gherkin	*Cucumis anguria* L.	24
Watermelon	*Citrullus lanatus* (Thumb.) Matsum & Nakai	22
Round melon	*Praecitrullus fistulosus (*Stocks) Pang.	24
Pumpkin	*Cucurbita moschata* (Duch.) Poir.	40
Summer squash	*C. pepo* L.	40
Winter squash	*C. maxima* Duch.	40
Buffalow gourd	*C. ficifolia* Bouche	40
Bottle gourd	*Lagenaria siceraria* (Molina)Stadl.	22
Bitter gourd	*Momordica charantia*	22
Balsam apple	*M. balsamina* L.	22
Giant Spine gourd or sweet gourd	*M. cochinchinensis* (Lour.)Spreng	28
Teasle gourd	*M. subangulata* Blume subsp. *renigera* (G. Don) W.J.J. de Wilde	28
Ridge gourd	*Luffa acutangula* (L.) Roxb.	26
Sponge gourd	*L. cylindrica* Roem.	26
Pointed gourd	*Trichosanthes dioica* Roxb.	22
Snake gourd	*T. anguina* L.	22
Wax gourd	*Benincasa hispida* (Thunb.) Cogn.	24
Ivy gourd	*Coccina grandis* (L.) Voigt.	24
Chow Chow	*Sechium edule* (Jacq.) Swartz.	28
Mitha karela	*Cyclanthera pedata* (L.) Schrader	32
	9. Euphorbiaceae	
Cassava	*Manihot esculenta* Crantz.	36
Chekurmanis	*Sauropus androgynus* (L.) Merr.	-
	10. Labitae	
Chinese potato	*Solenostemon rotundifolius (Poir) Morton* (Syn. *Coleus parviflorus* Benth.)	-
Chinese artichoke	*Stachys siebeldi* Miq.	-
Hoary basil	*Ocimum americanum* L.	-
	11. Leguminoceae (Fabaceae)	
Garden pea & snow pea	*Pisum sativum* L. sub sp. *hortense*	14
French bean	*Phaseolus vulgaris* L.	22
Lima bean	*P. lunatus* L.	22
Runner bean	*P. coccineus* L.	22
Tepary bean	*P. acutifolius* Grey	22
Thicket bean	*P. polystachyus*	22
Adzuki bean	*P. angularis* (Wild.) Wright	22
Sword bean	*Canavalia gladiata* (Gucav.) DC.	22,24
Gotani bean	*C. plagosperma*	22
Jack bean	*C. ensiformis* (L.) D.C.	22
Valvet bean	*Mucuna deeringiana* (Bor.) Merr.	22

Cluster bean	*Cyamopsis tetragonolobus* (L.) Taub.	14
Winged bean	*Psophocarpus tetragonolobus* (L.) Taub.	18,22
Broad bean	*Vicia faba* L.	12
Hyacinth bean	*Lablab perpureus* (L.) Sweet.	22,24
• Vegetable type	*Lablab perpureus* (L.) var. *typicus*	22,24
• Pulse type	*Lablab perpureus* (L.) var. *lignosus*	22,24
Horse gram	*Dolichos uyniflorus* Lam.	22,24
Cowpea	*Vigna unguiculata*	22
• Small podded vegetable and pulse type	*Vigna unguiculata* sub sp. *unguiculata*	22
• Long podded vegetable type	*V. unguiculata* sub sp. *sesquipedalis*	22
Soybean	*Glycine max* (L.) Merrill	40
Yam bean	*Pachyrrhizus erosus* Urbn.	22
African yam bean	*Sphenotylis stenocarpa*	-
African-locust bean	*Parkia clappertonia*	-
Ground bean	*Voandzeia subterranea* (L.) Thou.	-
Marama bean	*Tylosema esculentum* (Burchell) Schreiber	-
Agathi	*Sesbania grandiflora* Poit.	24
Fenugreek	*Trigonella foenum-graecum* L.	16
	12. Malvaceae	
Okra	*Abelmoschus esculentus* (L. Moench)	126,130,132 (Allopolyploid)
	13. Martynaceae	
Martynia	*Proboscidea jussieeui* (Keller)	-
	14. Moringaceae	
Drumstick	*Moringa oleifera* Lamk.	28
	15. Polygonaceae	
Rhubarb	*Rheum rhaponticum* L., *Rheum rhabarbarum* L.	22, 44 (4x)
Sorrel	*Rumex acetosa* L.	-
Patience dock	*R. patientia* L.	-
Bladder dock	*R. vesicarius* L.	-
Buck wheat	*Fagopyrum tatariocum* (L.) Gaertn.	16
	16. Portulaceceae	
Pusley	*Portulaca oleracea* L.	54
Ceylon spinach	*Talinum triagulare* Willd.	-
	17. Rutaceae	
Curry leaf	*Murraya koenigii* (L.) Spreng	18
	18. Solanaceae	
Potato	*Solanum tuberosum* L.	48 (auto 4x)
Brinjal	*Solanum melongena* L.	24
• Fruit round or egg-shaped	*S. melongena* var. *esculentum*	24
• Fruit long & slender	*S. melongena* var. *serpentinum*	24
• Plant dwarf	*S. melongena* var. *depressum*	24
Tomato	*Lycopersicon esculentum* Mill. (*Solanum lycopersicum* L.)	24

• Common	*L. esculentum* var. *commune Bailey*	24
• Large-leafed	*L. esculentum* var. *grandifolium* Bailey	24
• Upright	*L. esculentum* var. *validum* Bailey	24
• Cherry	*L. esculentum* var. *cerasiforme* Bailey	24
• Pear	*L. esculentum* var. *pyriforme* bailey	24
Currant tomato	*L. pimpinellifolium* (Juslen) Mill.	24
Sweet pepper & Chilli	*Capsicum annuum* L.	24
Tabasco pepper	*C. frutescens* L.	24
Husk tomato	*Physalis pubescens* L.	24
Tomatillo	*P. ixocarpa* Brot.	24
Tree tomato	*Cyphomandra betacea*	24
	19. Tiliaceae	
Jute	*Corchorus olitorius* L.	14
	20. Umbelliferae (Apiaceae)	
Carrot	*Daucus carota* L.	18
Celery	*Apium graveolens* L. var. *dulce* (Mill.) Pers.	22
• Celeriac	*A. graveolens* L. var. *rapaceum* (Mill.) Gaud.	22
• Leaf celery	*A. graveolens* L. var. *secalinum*	22
Parsley	*Petroselinum crispum* (Mill.) Nym. Ex. Hill.	22
• Turnip-rooted parsley	*P. crispum* var. *tuberosum*	22
Parsnip	*Pastinaca sativa* L.	22
Turnip-rooted chervil	*Chaerophyllum bulbosum* L.	-
Skirret	*Sium sisarum* L.	-
Chervil	*Anthriscus cerefolium*	-
Dill	*Anethum graveolens* L.	22
Coriander	*Coriandrum sativum* L.	22

Classification based on hardiness

This classification is based on the endurance of vegetable crops against frosts and accordingly, the crops may be classified as hardy, semi-hardy and tender. Hardy vegetable crops can withstand frosts without injury whereas tender and very tender vegetable crops cannot, and are even killed by light frost. Tender vegetable crops are comparatively more tolerant to cold temperature than the very tender vegetable crops. The semi-hardy vegetable crops are not generally injured by light frost.

Hardy vegetable crops: Asparagus, broccoli, Brussels sprouts, cabbage, chive, collards, garlic, kale, knol khol, leek, onion, pea, parsley, radish, horse radish, mustard, sorrel, rhubarb, spinach, turnip, broad bean, etc.

Semi-hardy vegetable crops: Beet, carrot, cauliflower, celery, lettuce, globe artichoke, palak, parsnip, potato, salsify, chard, chicory, Chinese cabbage, endive, rutabaga,etc.

Tender vegetable crops: Tomato, sweet pepper, French bean, New Zealand spinach, soybean, sweet corn, pumpkin, summer squash, cowpea, sweet potato, etc.

Very tender vegetable crops: Cucumber, brinjal, chilli, okra, muskmelon, watermelon, bottlegourd, bittergourd, ridgegourd, snakegourd, spongegourd, pointedgourd, ashgourd, amaranth, cassava, yam, elephant-foot yam, cluster bean, lima bean, winged bean, hyacinth bean, Indian spinach, taro, etc.

Classification based on growing season

Classification based on hardiness for which tolerance to frost is the main criteria cannot explain the conditions under which the crops should grow at its best. For example, some hardy crops, like asparagus, pea, turnip, cabbage, etc., cannot grow under hot environment while other hardy crops like, radish, onion, etc., can tolerate hot environment much better. Similarly, some tender vegetable crops, like brinjal, muskmelon, cowpea, etc., cannot grow well in cool weather even if there is no frost. From this point of view, classification based on growing season may be more useful for the farmers. Based on the distinct growing seasons in the plains of India, vegetable crops may be classified as summer or spring-summer season, rainy season and winter season vegetable crops. Sumner and rainy season vegetable crops are tropical, *i.e.,* heat-loving and heat-resistant vegetable crops and winter season vegetable crops are temperate *i.e.,* cold-resistant and/or frost resistant vegetable crops. In the plains of India, generally summer or spring-summer season prevails from February to June, rainy season from June to September and winter season from October to January.

Summer or spring-summer season vegetable crops: Brinjal, chilli, muskmelon, watermelon, long melon, snap melon, round melon, cucumber, bottle gourd, bitter gourd, snake gourd, ash gourd, ridge gourd, sponge gourd, pumpkin, summer squash, winter squash, cucumber, okra, tomato, chilli, cowpea, cluster bean, amaranth, Indian spinach, pointed gourd, water spinach, etc.

Rainy season vegetable Crops: Okra, brinjal, chilli, bottle gourd, bitter gourd, snake gourd, ash gourd, ridge gourd, sponge gourd, pointed gourd, ivy gourd, cowpea, cluster bean, etc.

Winter or autumn-winter season vegetable crops: Tomato, potato, sweet pepper, cabbage, cauliflower, knol khol, Brussels sprouts, sprouting broccoli, radish, carrot, turnip, beet, spinach, palak, onion, garlic, French bean, pea, hyacinth bean, fenugreek, lettuce, etc.

Classification Based on tolerance to soil reaction

Vegetable crops can be classified into three groups according to their relative tolerance to soil acidity viz., less tolerant, moderately tolerant and very tolerant although most of the vegetable crops require slightly acidic to neutral pH values (6.5-7.0) of the soil solution for their normal growth and development.

Less tolerant (pH 6.8-6.0)	Moderately tolerant (pH 6.8-5.5)	Very tolerant(pH 6.8-5.0)
Asparagus	French bean	Chicory
Beet	Lima bean	Dandelion
Broccoli	Brussels sprouts	Endive
Cabbage	Carrot	Fennel
Cauliflower	Collard	Rhubarb
Celery	Cucumber	Shallot
Palak	Brinjal	Sorrel
Swiss chard	Garlic	Sweet potato
Cress	Horse radish	Watermelon
Chinese cabbage	Kale	
Leek	Knol khol	
Onion	Parsley	
Lettuce	Pea	
Muskmelon	Chilli	
New Zealand spinach	Pumpkin	
Okra	Radish	
Onion	Rutabaga	
Orach	Summer squash	
Parsnip	Winter squash	
Salsify	Tomato	
Spinach	Turnip	
Water cress	Potato	

Source: *Knott's Handbook for Vegetable Growers*

Classification based on salt tolerance

Vegetable crops may be categorized into three groups: sensitive, moderately resistant and resistant based on their relative tolerance to soil salinity.

Less tolerant	Moderately tolerant	Tolerant
Celery	Tomato	Beet
Bean crops	Broccoli	Kale
Radish	Watermelon	Turnip
Potato	Chilli	Palak
Snake gourd	Cucumber	Asparagus
Sweet pepper	Amaranth	Bitter gourd
Pea	Pumpkin	Ash gourd
Brinjal	Summer squash	Lettuce
Artichoke	Cauliflower	
Sweet potato	Cabbage, Muskmelon	
	Bottle gourd	
	Carrot, Spinach	
	Corn, Onion	

Source: Knott's Handbook for Vegetable Growers, An introduction to Soils and Plant Growth (Donahue *et al)* and *Irrigation Principles* and *Practices* (Hansen *et al.)*

Classification based on plant parts used as vegetable

In this approach, vegetable crops can be grouped according to their edible plant parts.

Leaves: Leaf lettuce, spinach, palak, amaranth, coriander, fenugreek, Indian spinach, water spinach, etc.

Growing point surrounded by overlapping leaves: Cabbage, head lettuce, etc.

Modified above ground stem: Knol khol, asparagus, etc.

Flower head: Broccoli, globe artichoke, etc.

Arrested inflorescence: Cauliflower

Flower: Agathi, male flowers of pumpkin

Underground plant parts

Modified stem: Potato, Jerusalem artichoke, yam (tuber), elephant foot yam, taro (corm), onion, garlic (bulb).

Modified root: Radish, carrot, beet, turnip, cassava, Chinese artichoke (modified tap root), sweet potato, and Chinese potato (modified adventitious roots).

Fruits: Brinjal, tomato, chilli, sweet pepper, peas, French bean, hyacinth bean, cluster bean, muskmelon, watermelon, cucumber, pumpkin, bottlegourd, bittergourd, ridgegourd, pointedgourd, snakegourd, spongegourd, ashgourd, chowchow, okra, etc.

Classification based on method of culture

This classification is the best for the vegetable growers because the vegetable crops requiring the same cultural practices (but not same season for cultivation) are grouped together though they maybe divergent botanically. However, the crop species in the groups like cole crops, bulb crops, cucurbits, peas and beans belong to the same family.

Group 1 : **Potato**

Group 2 : **Solaneceous fruits** (Tomato, brinjal, chilli, sweet pepper)

Group 3 : **Cole crops** (Cabbage, cauliflower, sprouting broccoli, Brussels sprouts, knol khol, Chinese cabbage).

Group 4 : **Cucurbits** Pumpkin, summer squash, winter squash, muskmelon, watermelon, cucumber, bottle gourd, ash gourd, ridge gourd, snake gourd, sponge gourd, bitter gourd.

Group 5 : **Root crops** Radish, carrot, beet, turnip, horse radish, rutabaga, salsify, turnip-rooted chervil, Spanish salsify, black salsify

Group 6 : **Bulb crops** Onion, garlic, leek, shallot, welsh onion, chive.

Group 7	:	**Salad crops** Lettuce, celery, endive, chicory, parsley, sorrel, water cress, garden cress, and upland cress.
Group 8	:	**Greens and pot herbs** Amaranth, palak, spinach, New Zealand spinach, coriander, Indian spinach, French spinach, dandelion, pigweed, tarragon, parsley, Ceylon spinach, fenugreek.
Group 9	:	**Peas and beans** Pea, French bean, cowpea, hyacinth bean, asparagus bean, cluster bean, winged bean, lima bean, velvet bean.
Group 10	:	**Tuber crops other than potato** Taro, giant taro, tannia, yam, cassava, elephant-foot yam, Chinese potato.
Group 11	:	**Sweet potato**
Group 12		**Okra**
Group 13	:	**Pointed gourd**
Group 14	:	**Chayote (chow chow)**
Group 15	:	**Tropical perennial vegetable crops** Drumstick, agathi, chekurmanis, curry leaf
Group 16	:	**Temperate perennial vegetable crops** Asparagus, rhubarb, seakale, globe artichoke

4. Influence of Environment on Vegetable Production

Agriculture is actually an interaction of a biological system (i.e. crop plants) and the physical environment that surrounds them. The most important environmental parameters that influence growth and development, are temperature, photo conditions (light), relative humidity, water, soil factors, air, wind and air pollutants, among them some are of greater importance, namely, temperature, light, water and soil factors, which continue to interact in a complex manner to influence the entire life cycle of crop plants. These environmental parameters are essential for normal growth and development of vegetable crops. If one becomes short in supply, it is called a limiting factor. The special mechanisms and properties of the plant biological system and environmental dynamics, through various means and processes, make total growth and production of biomass happen. In organized crop culture, growth (irreversible increase in shape, size and dry matter) and development (switching over from vegetative phase to reproductive phase) together with yield are physiological expressions of a crop modified by cultural manipulations like transplanting, manuring, training & pruning, grafting, applying growth substances and some stress (both biotic and abiotic) regulatory steps.

In the dynamic and living environment on the surface of the earth, vegetables, like all other green plants, fix the kinetic energy from the sun and change it to the potential form of food and other compounds; this process of producing optimum yield becomes possible through a balanced supply of nutrient, water and growth regulators in an optimum range of temperature, light, relative humidity and CO_2 status. As a matter of fact, the essential factors act in an interrelated and combined manner to result in a particular effect. The interaction between these environmental factors and genetic architecture determine the timing and rate of growth and progress of crop plants through the cycle of life. These environmental factors directly influence photosynthesis— the basic source of all energy requirements to the human species. Along with these factors, the role of protection measures against both biotic and abiotic stresses cannot be ignored.

Temperature

Plant growth and development rates do vary at different temperature regimes. The thermal energy from sun is required in all the biological, chemical and physiological processes that occur in plants. The influence of temperature on the plant growth and development (i.e. thermomorphogenesis) is of immense importance and this aspect is taken care of by a branch of agro-meteorological science called "Phenology", which deals with the relationship between climate or weather and periodic biological developmental phases of plants. From germination of seeds to formation of new seeds, plant show several visible external changes caused due mainly to the influence of different environmental parameters; these external changes during different stages are called phonological phases or phenophases. Every plant needs a particular duration of warmth for a specific growth phase to occur.

Temperature affects growth and development of plants primarily by controlling the manner and speed of biochemical processes (following the Q_{10} rule proposed by Dutch chemist Vant Hoff that rate of chemical reaction is doubled for each 10°C rise in temperature, obviously within a certain range). Temperature emerges as the major growth factor through its control on plant development including morphogenesis and product quality and induction of flowering stimuli such as vernalization. Temperature also determines climatic zones and controls plant distribution, growth cycles, growth rates and ultimately yield.

Biological activity of plants is limited on the lower side by freezing point of water (0°C) and on the upper side by denaturation point of protein (50°C). Plants can only grow and survive within these temperature limits, if during the seasonal cycle sufficient time is available for efficient growth. Plants endure temperature extremes due to combined effect of their morphology and physiology. In high temperature regime, small leaf, very deep root system, minimum water loss mechanism, orientation of leaf plane, etc. are some of the heat avoiding adaptations. On the other hand, plant can avoid low temperature extreme most commonly through dormancy as in the case of biennial crops like cabbage, long day onion, European radish and carrot, beet, etc. For this reason, it is difficult to specify the range of survival temperature because of the influence of certain factors including adaptation, developmental stage, duration of cold stress, level of dormancy arid cultivar differences

Influence of temperature on plant physiology

Molecules are in continuous motion at any temperature above absolute zero and this motion is accentuated in proportion to the amount of heat supplied. The energetic of the molecular or atomic units is manifested in plant temperature. Hence, temperature is the most important environmental factor for plant growth and development because of its unique regulatory influences on plant physiology. Temperature controls all physiological activities like photosynthesis, respiration, enzymatic activity, water status, germination, development of economic parts, flowering, sex expression, pollen viability, pollination, fruit set, fruit quality, maturity and senescence, seed production, seed storage, shelf life, vernalisation of biennials and ultimate yield. Microbial growth and development and pests and disease occurrence are also influenced by temperature. At lower and upper extremes of temperatures, tropical vegetable crops are damaged by frost (at or below 0°C) and chilling (0-2°C) and wilting and scorching respectively. Freeing and sub-zero temperature cause injury to plants by extra-cellular ice formation with the geographic location and season and as the daily temperature fluctuates.

Plants display remarkable thermal resilience despite the fact that biologically important reactions are limited to a small range of temperature. Plants' temperature response curve for growth rises rapidly from 0°C to 15°C followed by a steady increase above 15°C to an optimum value between 20° and 30°C. Optimum temperature range depends on the particular crop. In most of the vegetable crops, growth and reproductive processes declines above 35°C and thermal death point for most of the crops including the vegetable crops reaches at temperature nearing 50°C. The evidence of global warming has increased the interest in the cause of yield declines at temperatures only slightly above optimal in many vegetable crops. Above optimal temperatures can shorten the crop life period and markedly hasten flowering and ripening of fruits in many vegetable crops. Temperature requirement for successful cultivation of different vegetable crops are given below.

Monthly mean temperature (°C)(day and night)			Vegetable Crops
Minimum	Maximum	Optimum	
Group: Hot18.0	35.0	25.0-27.0	Okra, watermelon, muskmelon, chilli, sweet potato, martynia, yam, cassava, amaranth, winged bean, cluster bean.
Group: Warm10.0	35.0	20.0-25.0	Tomato, sweet pepper, brinjal, lima bean, French bean, hyacinth bean, cowpea, sweet com, palak, Newzealand spinach, cucumber, pumpkin, pointed gourd, bottle gourd, bittergourd, ridge gourd, snake gourd, summer squash, drumstick, Indian spinach, portulaca, taro, elephant foot yam.
Group : Cool-hot 5.00	30.0	20.0-25.0	Onion, leek, garlic, shallot, chicory, chives, salsify, scorzonera, pak-choi, globe artichoke, Indian cauliflower, Asiatic radish
Group: Cool-warm 5.0	25.00	18.0-25.0	Pea, broad bean, late cauliflower, Chinese cabbage, cabbage, broccoli, Brussels sprouts, kale, knol khol, turnip, European radish, carrot, beet, rutabaga, horse radish, parsnip, potato, parsley, fennel, dil, Swiss chard, spinach, lettuce, celery, celeriac, asparagus, garden cress, mustard, cardoon.

Source: *The Physiology of Vegetable Crops* (ed. H. C. Wein)

It is also not true that crops yield the highest at a constant temperature within the optimum range of thermo sensitivity. Optimum temperature range is fluctuating thermo-conditioning of plants within which maximum photosynthesis and normal respiration occur to realize highest marketable yield due to storing of more carbohydrates. Total dry matter production is the outcome of the efficient interception (by the crop canopy) of the available insolation and its effective utilization during the growing season. In general, at a constant temperature condition lower than the required for a particular crop, low rate of photosynthesis and consequently low rates of growth and yield would result. Conversely, at constant temperature higher than the optimum for the crop, high rate of photosynthesis coupled with excessively high rate of respiration which would ultimately result in low yield.

For the same mean daily temperature, temperature does vary in temperate and tropical latitudes to an extent of 12-15°C and 2-6°C respectively. This alternating day and night temperature within any given 24-hour period is called diurnal variation of temperature or Thermoperiodism. In general due to wider diurnal variations and longer day lengths, expected yield may be higher in temperate locations but the rate of biomass formation is almost equal (or barely higher) to that in the tropics. Responses of growth and development of vegetable crops to diurnal temperature amplitudes have not been sufficiently investigated. In some crops conspicuous responses have been observed such as, low night temperature between 15° and 21°C for fruitset in tomato, higher day temperature for successful growth in lettuce. On the other hand, some crops like cucumber, radish, etc. do not show specific response to day and night temperatures. In general, diurnal differential is beneficial for growth and development. At comparatively low night temperature, quality of growth, earliness, intensity of flowering, translocation of carbohydrates and fruit development are all benefited.

Every plant has a temperature requirement and there is a temperature, below which plants will not grow and develop, this is called the base temperature. Base temperature is specific for a particular crop. Growing degree-day (GDD) is another parameter of measurement of thermo-sensitive growth and is obtained by subtracting plant base temperature from daily mean temperature. Adding the GDD of each day up to the maturity, we get the measure of total sum of heat required by a crop. This is called day degrees or summation constant, irrespective of the of the crop locality; e.g. for pea day-degree is 700-1700 and for lettuce it is 1450-1580.

At the temperature range below optimum, both photosynthesis and respiration decrease but rate of decrease is more for photosynthesis. Very little carbohydrate will be available for growth and development of the crop. Further, the rate of protein synthesis in the development of new cells is also low. Hence, yield is reduced markedly. Immature foliage tissues of the vegetable crops are liable to freezing injury if they are grown persistently in very low night temperature condition. In the upper range of optimum night temperature, vigorous vegetative growth takes place provided there is high rate of photosynthesis and normal rate of respiration during daytime and other environmental factors become favourable. So, little carbohydrate is stored at this range of high night temperature condition. On the contrary, at lower range of night temperature, moderately vigorous vegetative growth is induced with consequent storing of more carbohydrate. This situation is ideally meet with in potato, sweet potato and yam in which high bulking rate occurs at comparatively low night temperature condition. Ideal crop growth and yield may be achieved if upper range of the optimum night temperature in the early vegetative phase and lower range of the optimum night temperature in the late vegetative phase (bulking phase) of these crops are maintained.

Influence of temperature on growth and development

Temperatures both of soil and atmosphere, influence all aspects of growth and development of vegetable crops, *viz.*, seed germination, survivability and growth, development of economic plant parts, flowering and seed production, pollination, fruitset, quality of the produce, seed storage and occurrence of diseases and pests.

Soil temperature mainly influences seed germination and water uptake by the plants. Plants extract water from warm soils more quickly and easily than from cold soils. Root growth is also inhibited in cold soil. On the other hand, atmospheric temperature influences most of the other aspects of growth and development of the plants.

Seed germination

Minimum soil temperature for seed germination in majority of vegetable crops ranges 2°-15°C. Seeds of some cool season vegetable crops, like onion, parsnip, spinach, beet, cabbage, carrot, cauliflower, Swiss chard, parsley, pea, radish, turnip, etc., can germinate at the minimum temperature range of 1.7° to 4.4°C. For most of the warm season vegetable crops, like cucurbits, beans, brinjal, chilli, okra, etc., a minimum of 15°C soil temperature is required for seed germination. Seeds of some vegetable crops, like tomato, asparagus, etc., require a minimum of 10°C for seed germination. Optimum soil temperature for seed germination in most of the vegetable crops lies between 20° and 30°C. Maximum temperature range for seed germination in the warm season vegetable crops is 35° to 40°C. Seeds of some cool season vegetable crops also can germinate at the maximum range of 32° to 35°C, yet impaired germination of these crops mostly occur at soil temperature beyond 30°C.

Survivability and growth

Base temperature is the threshold atmospheric temperature level below which the plant can not develop, and each plant has its own base temperature, *e.g.*, pea (4.4°C). French bean (10°C), asparagus (5.5°C), spinach (2°C), lettuce (4.4°C), pumpkin (13°C), tomato (15°C), etc. The concept of heat unit, expressed as day-degree, is based on the proportional growth of the crop to the ambient temperature and such correlation has been used to predict a probable harvest time for some vegetable crops like peas, muskmelon, etc. However, the system of heat unit works when other environmental factors that have a direct influence on plant growth vary in the same direction as temperature. The structural and physiological development of the crop plants vary greatly with the temperature of the environment. The minimum and maximum temperatures for growth and development of most of the vegetable crops lie within the range of 7° to 35°C. According to the temperature requirement for proper growth and development, vegetable crops are broadly categorized into two groups: cool and .warm season crops. Within a crop, varieties differ in temperature requirement. As for example, varieties of group-l cauliflower (very early) can produce curd at 22-27°C temperature condition whereas group-IV cauliflower (late) produces curd at relatively much lower temperature (l0-12°C) condition. Similarly, cabbage itself is a cool season vegetable crop, but some high-temperature tolerant hybrids can produce heads even at 30°C temperature. So, this categorization should be treated in a general manner.

Cool season vegetable crops: These vegetable crops are generally tolerant to low temperature and sometimes even frost. Among these crops, rhubarb, horse radish, sorrel, etc. are frost-resistant and can stand subzero temperature. Crops like cabbage, onion, garlic, Brussels sprout, knol khol, pea, turnip and radish are more tolerant to cold temperature than cauliflower, carrot, celery, beet, palak, etc. According to the relative insensitivity to weak frost, the cool season vegetable crops can be classified into hardy (more tolerant to cold temperature) and semi-hardy (comparatively less tolerant to cold temperature). Chilling causes bolting in some of these crops.

Warm season vegetable crops: These tropical and tropicalized vegetable crops cannot withstand prolonged cold weather, and are the best suited in relatively high temperature condition. Among these crops, tomato, brinjal, sweet pepper, pumpkin, cucumber, etc. are heat-loving crops while, muskmelon, watermelon, sweet potato, winged bean, cluster bean, etc. are heat-resistant crops. Heat-resistant crops can be grown even in very high temperature (40°C) condition. These warm season vegetable crops are called tender crops which are sensitive to weak frost although, relative susceptibility to cold temperature varies with the crops and accordingly these vegetable crops can be classified as tender and very tender as per their sensitivity to cold temperature. Some tender crops like, tomato, summer squash, French bean, etc. are fairly cold-tolerant. These habitually warmth preferring plants are injured at low but nonfreezing temperatures and show signs of injury below 10^{0}C. Outward signs of injury can vary as per the plant types, age of the plant and length of stress condition and include wilting, chlorosis (chloroplast breakdown), sterility, frost banding, reduced leaf expansion, necrosis, malformations, browning and even death.

Development of vegetative plant parts of economic use

Development of the different modified vegetative plant parts which are economic product of several vegetable crops is greatly influenced by the prevailing atmospheric temperature.

Cauliflower: In cauliflower, development of curd marks the onset of reproductive stage of the plant. Curding is initiated in the plant under wide range of' temperature condition from 5-28°C depending on the cultivars and optimum temperature ranges for curd initiation and development of different cauliflower cultivars grown in India are 20-27°C for group

1A(very early); 20-25°C for group 1B(early); 16-19°C for group II (mid-early); 12-16°C for Group III (mid season) and 10-12°C for group IV(late).

Cabbage: The open pollinated European cultivars grown in India require 15-18°C temperature for proper growth and heading. Most of the traditional European cultivars under mild and short span of winter condition of particularly eastern and western India do not form tight heads. In this situation, early short duration cultivars and tropical hybrids perform well. Some high temperature tolerant hybrids developed in Japan, Korea and Taiwan are capable of forming tight heads even at an average temperature of 25°C-30°C and these hybrids have become popular in India.

Knol khol: The common European cultivars that are grown in India require 15° to 18°C for good growth and formation of stem knob, although quality of knob becomes better at comparatively low temperature of 10-15°C.

Brussels sprout: It requires a temperature range of 10-15°C for the successful development of miniature heads in the leaf axil.

Sprouting broccoli: Development of flower heads require 10-12°C temperature however, for proper development of flower head, night/day temperature regime of 10/15°C is optimum. Warm weather condition makes the bud clusters loose.

Chinese cabbage: High quality head can be obtained under moderate to cool temperature condition of 15.5°-21°C and temperature below 15°C induce the formation of seed stalks before good heads are produced.

Onion: Photoperiod seems to have major role in bulbing, but influence of temperature to bulbing and bulb development cannot be ignored. In general, moderately low temperature condition of 10-15°C tend to prolong the vegetative phase and thereby, increase the size and yield of bulbs and in contrast, higher temperature hasten the onset of reproductive phase. In the cultivars of rabi onion (12-13 hours day length for bulb formation and development) that are widely grown in India during December-January to April-May, temperature above 15°C is necessary for bulbing and bulb development, 19-20°C being the optimum. In the cultivars of kharif onion (10-11 hours day length for bulb formation and development) that are mostly grown in Maharashtra during mid August to November-December, 20-22°C is optimum for bulbing. Rapid bulb growth takes place at 25-30°C temperature condition, however, bulb development is retarded significantly at high day temperature of 40°C.

Garlic: Low temperature is the pre-requisite for higher yield in garlic. Cool and moist condition is required for proper vegetative growth and cool and dry period favour the production of best quality bulbs. During bulbing and bulb development, cool temperature of 10-15°C helps in producing larger bulbs.

Potato: Relatively high temperature of 20-25°C is optimum for stem and leaf growth. Optimum night temperature for tuberization is 18-20°C which is inhibited above 21°C night temperature. This optimum night temperature for tuberization, however, varies with the prevailing photoperiod. At 12-hour photoperiod, night temperature of 20°C is optimum but with the lengthening of photoperiod, optimum night temperature requirement for tuberization decreases up to 16°C at 24-hour photoperiod. High night temperature decreases tuber yield more than high day temperature.

Radish: Radish, in general, is best suited to cool or moderate climate. The Asiatic or tropical cultivars are more adapted to high temperature condition than European or temperate cultivars. Relatively high temperature of 20-25°C for early root growth and comparatively low temperature of 10-18°C in the later part are required for proper root development. These temperature requirements will be at higher side for Asiatic cultivars and lower side for European cultivars. The roots become tough during hot weather.

Carrot: Very high or low temperature is not conducive to proper growth and colouration of the roots. So, it requires moderate climate. However, the Asiatic or oriental types are well adapted to comparatively high temperature condition than the European types. A temperature range of 16-21°C is ideal for root development and good colouration. In the temperature regime below 15°C, roots become slender with inadequate colour development. At higher temperature condition around 24-25°C, roots become shorter, thicker and less coloured. This temperature sensitivity is more pronounced in European cultivars

Beet: Beet is fairly hardy and can tolerate freezing temperature for some time. It can be grown in warm weather but best quality roots are produced at the temperature regime of 15-21°C. In warm weather (above 25°C), zoning of beet marked by alternate white and coloured circles in the root tissue is pronounced. At temperature between 10-15°C prevailing for some period, seed stalks are produced prematurely before completion of vegetative phase

Turnip: It is best suited to cool or moderate temperature condition of 15-20°C. At temperature below 10°C, flowers stalks may emerge. Though it can be grown in warm weather (25°-26°C), yet roots become tough and bitter in taste in high temperature condition.

Sweet Potato: The plant is tender in nature and requires long, warm growing season. A temperature range of 20-27°C is best suited to vine growth. A slightly lower temperature regime of 20-25°C is ideal for tuberization in the adventitious roots. Tuberization is hampered at very high temperature (35-40°C) and low temperature (15°C) although tuberization is greatly influenced by short-day condition.

Cassava: Cassava is only suited in places where average annual temperature lies above 20°C. Ideal temperature for the crop growth and root tuberization ranges between 25-32°C.

Taro: Corm, the economic plant part is the swollen and solid base of a stem with distinct nodes and internodes and apical as well as axillary buds. Plants growth and corm development are best suited in swampy tropical climate with a temperature range of 25° to 30°C however, in moist environment, it can be grown at a temperature of 20-21°C.

Elephant-foot Yam: The plant requires fairly high temperature of about 30°C with adequate moisture for shoot emergence and subsequent vegetative growth. Wide fluctuation in day and night temperatures during growth stage is not desirable. Bulking of corm requires comparatively less temperature of 20-25°C and dry weather condition.

Yam: Warm climate (25-30°C) and adequate moisture are ideal for vegetative growth and bulking of stem tuber. Growth is adversely affected below 20°C temperature condition.

Yam bean: Warm (25-30°C) and humid climate is needed for initial vegetative growth, but cool night temperature of 18-20°C along with dry sunny days are required for bulking and development of root tuber.

Asparagus: It requires a temperature range of 15-24°C, 16-18°C being the optimum throughout the growing season for good spear production. During dormancy, low temperature below 10°C is required for 2-3 months.

Artichoke: The plants are killed by freezing temperature. On the other hand, hot weather causes the buds to open thereby reducing the tenderness of the edible portion. So, mild winter and cool summer are ideally suited for its growth and production of flower heads.

Palak: It can be grown in a wide range of temperature condition. However, mild temperature condition of 16-21°C is the best suited good and quality leaf growth. High temperature condition of 25°C and above under long day regime leads to bolting, causing reduction in leaf yield.

Spinach: It thrives best during relatively cool weather and can withstand frost. It cannot do well in hot weather. Favourable temperature for best yield ranges between 16° and 21°C under short-day condition.

Lettuce: It is best grown in a relatively cool growing season with a temperature range of 13-18°C. More precisely, day temperature of 20°C with 18°C night temperature is the best for growth and development.

Melons: Lower night temperature increases sugar content in the fruit which may be due to higher digestion (hydrolysis) of carbohydrates *i.e.*, starch $\xrightarrow{\text{Diastase}}$ soluble starch, various dextrins and maltose ’! glucose, sucrose ($C_{12}H_{22}O_{11}$) + H_2O $\xrightarrow{\text{Invertase}}$ glucose ($C_6H_{12}O_6$) + Fructose ($C_6H_{12}O_6$) or higher mobilization of sugar to the fruits. For this reason, sweetest melons are produced in the arid climate where very hot day is followed by cool night.

Celery: Optimum temperature condition for successful growth is 15° to 18°C and low temperature range of 10-15°C causes bolting while high temperature causes bitterness in the leaves.

Flowering, Pollination and Fruit Set

Photoperiod and temperature are the most important environmental factors for the induction of flowering. Photo-thermal flowering, though common, yet influence of photoperiodism is stronger on flowering responses. The photo-thermal control of the timing of flowering is especially important with regard to adaptation of the crop.

Monoecious cucurbits: In cucumber, pumpkin, bitter gourd, ridge gourd, bottle gourd, etc., yield is increased under moderate temperature condition of 20-25°C coupled with short day condition because of emergence of more pistilate flowers under this condition.

Pea: It thrives best in cool weather, optimum monthly average temperature being 10-18°C. High temperature of above 25°C hastens the maturity, lowers pod set and adversely affects the quality of peas.

French bean: Short-day condition and a moderate temperature range of 18-25°C are ideal for its cultivation. High temperature of 27-32°C though increases flower number per plant, yet markedly decreases pod set.

Lima bean: High temperature of 26-30°C drastically affects fruit set.

Other beans: All other beans excepting broad bean, *viz.,* cowpea, hyacinth bean, cluster bean, winged bean, etc. are typical warm season crops however, they are also adapted to subtropical conditions. Ideal temperature for most of the cultivars of these beans ranges between 18- 25°C. Very high temperature above 35°C adversely affects fruit set, particularly in reduced relative humidity condition.

Tomato: Daily mean temperature optimum being 22-26°C is more critical than night-time temperature alone for optimum fruit set and yield. Earlier studies on thermoperiodism suggested the critical factor in setting of fruits is the night temperature, the optimum range being 15-21°C. In general, the fruits fail to set at low night temperature of 13°C or below and at high 32.0-35.0°C/20-22°C day/night temperature stress mainly because of distortion in the development of male gametophyte. Some cultivars which can set fruit above and below the critical temperature are called high-temperature tolerant and coldset cultivars, respectively.

Brinjal: It requires long and warm growing season with a temperature range of 17-25°C for successful growth, fruit set and yield and high day temperature above 35°C adversely affects both growth and . Fruit set is mostly influenced by the genotype. In low temperature condition of 10°C and below, sometimes ovary in the bud stage splits resulting development of deformed fruits.

Chilli: Chilli is a tropical crop however, somewhat moderate night temperature of 15-20°C is ideal for fruit set and fruit development. High night temperature of 24°C and above causes flower drop. High day temperature of 30°C and above coupled with low light intensity leads to flower drop.

Sweet pepper: Sweet pepper though basically is a tropical crop yet requires somewhat moderate temperature for proper growth and fruit set and day /night temperature

20-25°C/ 15-16°C being optimum for fruit set and development. High day temperature of 30°C and above causes flower drop. Low night temperature of 10° to 15°C reduces pollen viability which some times lead to parthenocarpic development of fruits (seedlessness) in both chilli and sweet pepper.

Potato: Flowering and subsequent fruit set in potato is the result of interaction between photoperiod and temperature. Long day length (longer than 12 hours) coupled with cool temperature of 10-12°C induces the emergence of inflorescence and subsequent fruitset.

Indian or tropical cauliflower: Indian cauliflowers (Group I & II) are tropical annual in nature and do not require vernalization (specific cold temperature treatment) for the emergence of flower stalks. During curd formation and development stage, the temperature should be uniformly cool with moderate humidity to avoid development of different disorders like riceyness, fuzziness, etc.

Biennial and temperate annual crops: In the crops, like cabbage, Group III and IV cauliflower, knol khol, Brussels sprouts, sprouting broccoli, kale, Chinese cabbage, long-day onion, leek, beet, turnip, European radish and carrot, parsley, parsnip, celery, Swiss chard, rutabaga, salsify, etc. exposure of the plants to low temperature of 4-7°C (may be more in some crops) for 6-8 weeks (vernalization) is required after they have passed the juvenile phase for the production of flower stalks. Such promotive effect of low temperature on flowering is called vernalization. Vernalization stimulus is received by the apical meristem and expanding leaves. Some of the vernalization stimulus is related to the production of natural gibberellin substances in the plants. This proposition is based on two main observations: (a) gibberellin plays a role in the flowering of much long-day plant, (b) vernalization effect could be obtained with gibberellin application in some crops, like cabbage, carrot, etc. However, the situation is not straight forward in all the cases. So, it is suggested that control of flowering by vernalization occurs through multiple interactions between indigenous plant hormones. This vernalization has some crop specific considerations, like (a) juvenile stage for low temperature exposure varies with the crops (in cabbage and cauliflower it is after 10-12 weeks of planting while in Brussels sprouts, it is after 12-16 weeks), (b) sometimes very low temperature shortens the vernalization period, (c) low temperature requirement varies with the crop and cultivars of the particular crop (e.g.,tropical cabbage requires 10-13°C for 6-8 weeks, rabi onion varieties of India generally need 10-12°C for 5-6 weeks and beet, turnip, etc., require 8-10°C for 7-8 weeks).

Cucurbits: In cucumber, pumpkin, summer squash, muskmelon, watermelon and bitter gourd, pollination and fruitset take place in the morning hours at the optimum temperature range of 12.8-18.3°C. On the other hand, fruitset in bottle gourd and ridge gourd occurs at comparatively high temperature condition of 25°C and above during midday to afternoon hours. In snake gourd and pointed gourd, anthesis occurs at night which is hastened with the rise in night temperature and fruitset occurs in moderate temperature condition of the morning hours.

Palak: Temperature above 25°C coupled with long-day condition leads to bolting in the plants.

Spinach: Normally, long days in combination with warm weather (above 25°C) are favourable for seed stalk formation. Photoperiodic response in spinach can be altered by low temperature treatment. Exposure of the plants to 10-15°C for a month can induce flowering even in short-day condition.

Lettuce: Temperature above 22°C accelerates seed stalk development.

Product quality

Quality of the produce is affected by temperature and some of such influences on different vegetable crops are mentioned below:

Tomato: Of the five climatic factors recorded, temperature was predominantly implicated in affecting tomato fruit quality. Colour development is one of the important quality criteria of tomato and temperature markedly regulates colour development in fruits. Ideal temperature range for the development of both red and yellow colours is 10-25°C. High temperature inhibits fruit ripening and fruit colour deteriorates by inhibiting the accumulation of ripening-related mRNAs and ripening processes that depend on continuous protein synthesis, ethylene production, high respiration rate, reduction in carotene synthesis, lycopene accumulation, and cell-wall dissolution, which are all diminished under high temperatures of 32°C and above At temperature below 10°C, both red and yellow colours do not develop. The fruits that matured during summer under high temperatures showed a higher accumulation of organic acid other than ascorbic acid with the shortening of crop duration while the fruits that matured during winter to spring under optimum temperature conditions were firmer, had higher soluble sugars, and a longer growing period for consumption. High temperature also causes scalding in fruits.

Chilli: High night temperature influences the accumulation of higher capsaicin content in the fruit.

Cauliflower: Occurrence of riceyness (development of small flower buds over curd) is often related to fluctuation in temperature both high and low than the optimum required for curding for a particular cultivar. High temperature is mostly responsible for riceyness of curd. High temperature exposure after curd formation can cause loose, leafy or yellowing of the curds. All these .temperature related disorders markedly lower curd quality.

Cabbage: Traditional European cultivars do not form tight heads in the temperature above 18-20°C. Tropical hybrids are capable of forming tight heads at temperature regime of 25°-30°C.

Carrot: In the European or temperate cultivars, proper orange colouration in the root is the prime quality consideration. A temperature range of 16-21°C is the best for colour development. Synthesis of carotene (proportional to orange colouration) is decreased at temperature below 15°C and above 24°C although, intensity of colour increases with the age of the root.

Beet: Good quality roots having high sugar content and dark internal colour throughout the root are produced under cool weather condition (15-18°C). In warm weather (above 25°C) during growth period, the roots show alternate white and coloured rings when sliced which is called zoning of beet which reduces the quality.

Radish: Hot weather condition makes the root tough and pungent before reaching edible size particularly, in temperate radish cultivars (Japanese radish and summer radish of temperate region).

Turnip: Best quality root is developed in moderate climatic condition. In temperature above 25°C, the roots become tough and bitter in taste and in low temperature condition (below 10°C), bolting occurs before good roots are formed.

Pea: Quality of pea is associated with tenderness (less starch content) and high sugar content. High temperature of 30°C and above sharply reduces the quality by toughening the peas. At high temperature, enzymatic conversion of sugar to starch and at the same time comparatively high respiration resulting evolution of $CO_2 + H_2O$ + heat at the expense of sugar is more which cause less sugar content in the peas.

Palak: Hot weather condition of 25°C and above reduces leaf quality by triggering early bolting and by decreasing leaf succulence quickly.

Spinach: In moderate temperature of 16-21°C, the leaves become lush green in colour, containing more sugar and dry matter content which ensures good quality. In temperature condition of 25°C and above, leaves become less succulent, yellow and contain less sugar and dry matter due to bolting tendency.

Lettuce: It requires cool temperature condition of 20/18°C day and night for best quality produce, like solid head, leaf crispiness, etc. Temperature above 22°C reduces the quality of the head and leaves.

Melons: High day temperature followed by cool night enhances sugar content in the fruits.

Potato: Plant tissue commonly shows increase in sugar concentration at temperature below 10^{0}C. For this reason, potatoes become undesirably sweeter when stored in cold storage at 4-5°C temperature condition. Such development of sweetness of tuber in low temperature condition happens because enzymatic conversion of sugar to starch and at the same time respiration at the expense of sugar is reduced drastically in comparison to enzymatic conversion of starch to sugar.

Incidence of diseases and pests

Incidence of some diseases and insect pests is highly influenced by the prevailing atmospheric temperature and relative humidity condition and some of such incidences are mentioned hereunder.

Late blight of potato and tomato: The fungus *(Phytopthora infestans)* rapidly multiplies under cool temperature (10-20°C) and high humidity (above 80% relative humidity) condition.

Early blight of potato and tomato: Incidence of this disease caused by the fungus *Alternaria solani* is severe in comparatively high temperature (20-25°C) and under intermittent rain condition.

Black rot of cole crops: This bacterial disease caused by *Xanthomonas campestris* is particularly destructive in warm and humid climatic condition but is not pronounced in low temperature condition.

Fusarium wilts of vegetable crops: Incidence of these wilt diseases incited by different f.sp. of the fungus *Fusarium oxysporum* is severe under the optimum temperature of 18- 24°C and growth of the fungus is inhibited at 35°C.

White fly: Population of this insect pest and virus vector *(Bemisia tabaci)* increases with the rise in temperature. For this reason, incidence of leaf curl virus in tomato, yellow vein mosaic virus in okra, etc. is low when they are grown in comparatively low temperature condition.

Aphid: Population of this insect pest virus vector *(Myzus persicae)* increases with the rise in temperature at the onset of spring.

Light

Solar energy, a form of kinetic energy, is produced in the sun due to thermo-nuclear reactions at the expense of matter. The incident energy from the sun in the form of light exhibits the properties of electromagnetic radiation like wavelength and frequency. The visible spectrum of light (400-700 nm wavelengths) is the effective spectrum as light spectrum of this wavelength range is utilized for vision by human beings and animals and for biochemical reactions, like photosynthesis by plant community. Ultraviolet (15-390 nm wavelengths) and infrared rays (760-2500 nm wavelengths) are not at all beneficial to crop growth, rather these are damaging to the plants. Energy level per quantum of light (according to quantum concept, light comes from the sun in the form of tiny particles having certain weight called quanta or photons) increases with the decrease in wavelength. Hence, the ultraviolet rays dump huge packages of energy into cellular components resulting in molecular disorganization. Again long wavelengths of infrared rays are absorbed by cellular moisture resulting in damage to the living cells. For this reason, excellent growth and high yield are generally realized in the vegetable crops grown under glasshouse, polyhouse or polytunnel because glass or polyethylene used as cladding material in these structures have the property of absorbing ultraviolet and infrared rays.

Utilization of light

The visible spectrum of light between two potentially harmful categories of invisible light (ultraviolet and infrared) represent about 50 per cent of the solar radiation. Wavelengths of the visible spectrum (400-700 nm) are selectively absorbed by numerous carbonaceous molecules like chlorophyll and various pigments mainly, phytochrome, carotenoids and flavins. Light is utilized by the plants through its absorption and simultaneous selective translation of its energy into organized biochemical systems. The light-regulated organized biochemical systems include photosynthesis, seed germination, photoperiodism, pigmentation responses, etc.

Photosynthesis is by far the most important light mediated biochemical reaction in the earth. We all know that the first manufactured food of the plant is produced through assimilation of CO_2 followed by its combination with water upon the mediation of light energy that is trapped in chlorophyll. In photosynthesis, light energy is converted to chemical energy through a series of biochemical reactions. Chloroplast is the photosynthetic apparatus of the plant. There are two locales in the chloroplasts which are photosynthetically important, namely, i) the lamellae or membranes – an area of concentrated photosynthetic pigments, where transformation of light energy to chemical energy (Photophosphorylation) occurs ultimately by producing Adenosine Tri Phosphate (ATP). ATP is the energy currency available in a histological system and, ii) the stroma, a less dense, fluid area where CO_2 is reduced (dark reaction). As with temperature, light has also an optimum range for growth particular to each crop and within which, other factors being not limiting, the rate of photosynthesis is high and there are abundant carbohydrate food for growth. But at compensation point at which the rate of respiration equals the gross photosynthetic rate, the net amount of food available is nil.

At a normal rate of respiration, the higher the light intensity, the higher is the net photosynthesis. Hence, photosynthesis is limited by absorbed light, i.e., light reactions and not by CO_2 fixing capacity, *i.e.,* dark reactions. In the plant community as a whole, efficiency of light energy utilization is very low. High light quantum requirement is the single cause of this inefficiency. For this reason, increased photosynthetic efficiency is expected in place of high light intensity; obviously other factors should be favourable. In fact, when crop plants are adequately supplied with water and nutrients, both water use and dry matter production depend upon the receipt and utilization of solar radiation. Optimum light intensity is the intensity range at which rate of gross photosynthesis is high and rate of respiration is normal resulting in high net photosynthesis in the particular crop. Although most the vegetable crops require high light intensity yet they may be categorized under two categories according to the range of light intensity required for proper growth and development.

Sun plants: These plants require high light intensity of 3000-8000 ft-c. e.g., most of the vegetable crops like tomato, brinjal, chilli, all cucurbits, peas and beans, sweet potato, all root crops, etc.

Slight shade and direct sun tolerant plants: These plants thrive well over wide range of light intensity of 2000-8000 ft-c. e.g., vegetable crops like cabbage, potato, etc.

High light intensity requirement of the vegetable crops suggests the probable inconvenient consequences of growing them under the light intensity below the optimum range.

- Reduction in chlorophyll content of the leaves.
- Chloroplast of the mesophyll cells does not get filled with dense granular structure.
- Reduction in the number of palisade cells in the leaves.
- Reduction in the rate of photosynthesis.
- Reduction in growth and yield.

Similarly, there are some harmful consequences of growing vegetable crops under very high light intensity condition above the optimum range.

- Light harvesting system captures light energy much faster than its utilization for transformation to chemical energy hence, in most of the cases strong irradiance is of little use for photosynthesis.
- Increased heat load due to high light intensity increases the rate of respiration resulting in less net photosynthesis.
- Increased leaf temperature induces high rate of transpiration resulting in high water demand.
- In reduced water supply situation, lack of turgidity in the stomatal guard cells induces partial closure of stomata resulting in slow diffusion of CO_2 which reduce photosynthesis.
- High heat due to strong irradiance may inactivate some enzyme system.
- Reduction of chlorophyll content in the leaves which may reduce photosynthesis.

Influence of photoperiod on growth and development

Photoperiod refers to the length of light and dark period for any 24-hour day in a specified location. The light period of the day in the equatorial region is almost 12 hours throughout the year, but in polar region it may be 24 hours in summer.Day and night is the result of diurnal motion of the Earth but difference in the length of day and night occurs due to annual motion of the Earth in its elliptical orbit around the sun. Because the Earth's axis is inclined at 23.5° towards the plane of the Earth's orbit around the Sun, there is a regular annual variation in day length in any part of the world, although the extent of variation depends on latitude (the angular distance north or south from the equator of a point on the earth surface, measured on the meridian of the point). In zero latitude (Entebbe, Uganda), both longest and shortest day length are 12.50 hour; in 20° latitude (Hawaii), longest and shortest day length are 14.11 and 11.44 hour, respectively; in 40° latitude (Philadelphia, USA), these are 16.01 and 10.22 hour, respectively; in 60° latitude (Shetland Isles), these are 22.25 and 7.48 hour, respectively; and in 70° latitude (North Alaska), 24 hour is the light period. In 21 March and 23 September, the Sun remains directly over the equator resulting in equal length of light (12-hour day) and dark (12-hour night) period all over the world. On 21 December, the Sun remains in the farthest south of the equator, revealing the shortest day and the longest night (10-hour day and 14-hour night) in the regions of northern hemisphere (northern half of terrestrial globe or celestial sphere) including India, and the reverse situation occurs in southern hemisphere. From 22 December, day length starts increasing with simultaneous shortening of night length in the northern hemisphere. In 21 June, the Sun remains in the farthest north of the equator revealing the longest day and the shortest night (14-hour day and 10-hour night) in India and other regions of northern hemisphere. The opposite will happen in southern hemisphere. From 21 June, day length starts shortening till 21 December when the day length is minimum in the northern hemisphere. Within the hemisphere, the longest light and the shortest dark periods vary with the locations according to their distance from the equator. For this reason, length of the longest day on 21 June in a place in North India will be different from a place in South India. Whatever may be the location wise variations, pattern of change in the photoperiod remains the same in the particular hemisphere.

Photoperiodism has a great influence on the evolution and distribution of plant species on earth. The term "Photoperiodism" was coined by W.W.Garner and H.A. Allard (1920) working with a mutant cultivar of tobacco (*Nicotiana tabacum*), named 'Maryland Mammoth'. Photoperiod influences growth and development of the crop in various ways like induction of flowering, carbohydrate production, development of storage organs and sex expression in cucurbits.

Induction of flowering

Garner and Allard (1920) first demonstrated that flowering was neither a matter of age maturity nor just switching over from vegetative to reproductive stage. The length of required photoperiod for induction of flowering is not a continuous requirement but to the least, during the development of floral primordial is a must. Such photo-induction is perceived in the plant leaves; some crops may be photo-induced to flower with a single inductive cycle and some may require multiple cycles. The minimum or maximum photo duration, as the case may be, required for differentiation of flower bud and consequent flowering is called critical day length.

Relative length of light and dark period determines the time of flowering in some vegetable crops while in others, photoperiod has no effect on flowering. Accordingly, the vegetable crops can be classified into three groups: Long-day, short-day and day-neutral plants. All the crops of tropical or subtropical origin are essentially short-day plants (SDPs), whereas the vast majority of the crops of Mediterranean or temperate origin are long-day plants (LDPs). Day-neutral plants (insensitive to photoperiod) are often the product of special selection or the consequence of adaptation to regions outside the latitude of origin. For example, tomato and French bean, the DNPs are essentially SDPs according to place of origin but tend to relatively insensitive to day length as a result of their intensive breeding in temperate latitudes.

Long-day crop: These crops require long light and short night (generally 8-10 hours of continuous dark) for induction of flower buds, *e.g.*, potato, onion, lettuce, cabbage, Chinese cabbage, cauliflower, knol khol, radish, spinach, palak, beet, turnip, carrot, etc.

In general, critical photoperiod for these long day plants varies from 12-14 hours, which is the minimum length of light period required for differentiation of flower buds in these crops. The plants will not flower below this critical photoperiod.

Short-day crop: These crops require short light and long dark period (10-14 hours of continuous dark) for the formation of flower buds, *e.g.* sweet potato, Indian spinach, hyacinth bean, cluster bean, winged bean, etc. In general, critical photoperiod varies from 11-14 hours *i.e.,* it is the maximum length of the light period at which flower buds are developed. So, the plants will not flower when the light period exceeds the critical photoperiod.

Day-night neutral crop: These crops are insensitive to photoperiod *e.g.,* tomato, brinjal, chilli, sweet pepper, pumpkin, summer squash, winter squash, muskmelon, watermelon, cucumber, bottle gourd, ash gourd, ridge gourd, snake gourd, sponge gourd, bitter gourd., cowpea, okra, French bean, amaranth, etc.

Effect of phytochrome in flowering response

Photoperiodism indicates the response of a plant to sense the length of a dark period between two light periods. The action spectra of the light break imply that Photoperiodism is driven by phytochrome pigment system. Phytochrome is a pale blue chromoprotein with a prosthetic group, which is possibly an open-chain pyrrole. This photoreceptive pigment is present in the cytoplasm of the green plants widely in minute amounts (ppb) and can exist in revertible forms called red light (P_R) and far-red light (P_{FR}) absorbing phytochromes. Its acts as a time censor inside the plants biological system to help ensure them to flower in their temporal niche. The P_R is stable, inactive, storage form and the P_{FR} is the unstable and bio-chemically active form. In a 24 hour cycle, during light period, P_R changes to P_{FR} and during the dark period P_{FR} gradually changes to P_R. The enigma of synchronization among timing, season and minimum physiological maturity of green plants for flowering is still unsolved with the phytochromes present in green parts of plants and leaves being the photoreceptors, it is the buds that flower. Phytochromes do not probably directly take part in promoting flowering but stimulates messenger in the form of growth hormones to travel from leaf to bud; the presence of such a postulated flowering substance has been given the name "florigen" by Chailakhyan in 1936; it possibly reduces the flowering inhibiting substances inside the plants.

Development of economic plant parts

Different plant organs modify and enlarge due to powerful mobilizing actions of stored foods in response to induction phenomenon incited by environmental factors like photoperiod, temperature, light, etc. These storage organs, *i.e.,* repository for photosynthates are the economic plant parts of many vegetable crops *e.g.,* hypocotyl combined with taproot (carrot, beet, radish, turnip, parsnip), stem tuber (potato, yam, Jerusalem artichoke), root tuber (sweet potato, cassava), fleshy scale leaves of the bulb (onion, garlic, leek), corm (elephant foot yam, taro), stem knob (knol khol), petioles

(Swiss chard, celery), leaves (cabbage, lettuce), modified buds (Brussels sprouts), arrested inflorescence (cauliflower). In all these cases, the storage polysaccharides become the central part of storage organ formation, although fats and proteins are also stored according to the kind of storage organ.

Sucrose is the principal carbohydrate translocated to the storage organ due to the stimulus of photoperiod. The stimulus is perceived by the leaves and downwardly translocated to the responding organs. Biosynthesis of starch from sucrose or other carbohydrates, fats from fatty acids, proteins from amino acids proceed in the storage organs. Although the mechanism of mobilization activities is still not well understood, evidences is there that the stimulus of photoperiod for the development and regulation of storage organs are different plant hormones. In general, gibberellin appears to inhibit storage organ formation. In the tuber and stem-knob inducing short photoperiod, gibberellin content in potato and knol khol plants decline rapidly. In short-day induction of flowering response, auxin content of leaves also declines. during Implication of auxin is also there for the bulbing of onion.

Long day, in general, stimulates bulb formation and development of onion and garlic and taproot enlargement of radish and beet. On the other hand, short day stimulates tuber formation of potato, Jerusalem artichoke, sweet potato and yam. Photoperiodic stimulus is not so prominent in corm formation and development of elephant-foot yam and taro.

In onion, responses to the photoperiod for bulbing differ with the cultivars. The rabi onion cultivars of India which are cultivated widely during December to May in India are biennial and require 12-13 hours day length for bulb formation and development whereas the kharif onion cultivars which are cultivated during August to December, particularly in Maharashtra and Karnataka, are annual and require 10-11 hours day length for bulb formation and development. The long-day cultivars of the European countries require13-14 hours day length for bulb formation and development and they do not form bulb in short-day condition. In garlic, critical photoperiod for bulbing is 12 hours and short-day condition promotes secondary growth and suppresses bulbing and exposure to long-day condition not only accelerates bulbing but also hastens senescence.

In potato, there is a wide range of response of the cultivars to photoperiod although, in general, tuberization is accelerated by short days. Obviously, there are cultivars which produce high yields under long day condition. In most of the cases, tuberization is favoured by short days coupled with low night temperature. Effect of photoperiod is supplemented by the night temperature, in that optimum night temperature requirement decreases with the lengthening of photoperiod.

Soil

Soils are produced due to physical and chemical withering of the rocks to different small particles and simultaneous mixing of those inorganic particles with humus, the organic materials, and other inorganic constituents. Soils are unconsolidated, soft and changeable mass of the earth's crust developed through mixing of the soil particles with different inorganic and organic materials and are capable of supporting plant growth. Soils are the source of nutrients and water for the crops as well as the anchor of the roots. Productivity of the crops largely depends on the physical and chemical composition of the soil.

Physical properties of soil

Physical properties of soils include texture, structure, density, porosity, temperature, water content, etc. which all influence vegetable production. Soil texture which is judged by the proportions of sand, silt and clay particles is an important soil characteristic because it determines water holding capacity, the ease of tilling, soil aeration and soil fertility. High clay soils are heavy, rich in humus with high moisture retentive capacity, low heat conductivity, etc. and although contain considerable quantity of nutrients this type of soil is difficult to till, wet and drain. These soils are particularly unsuitable for potato, sweet potato, radish, carrot, beet, turnip, onion, garlic, leek, elephant foot yam, yam, etc. Heavy soils can be made suitable by adding organic matters, lime and sand. Loose textured soils like, sandy and sandy loam soils are not rich in humus and nutrients but easy to till, have plenty of aeration and are easily wetted but dry rapidly and lose nutrients easily. Vegetable crops yielding underground economic plant parts viz., potato, sweet potato, radish, carrot, beet, turnip, onion, garlic, leek, elephant foot yam, yam, etc. do particularly well in loose textured soil. These soils can be improved by incorporating adequate organic matter and fertilizers. Loose loamy soil which contains sufficient organic matters and nutrients is the most favourable soil for growing all the vegetable crops.

Soil temperature regulates seed germination, root growth and water uptake. Plants extract water from the warm soils more quickly and rapidly than from the cold soils. Root growth is also inhibited in the cold soils. The minimal soil temperature for the germination of the seeds of the majority of vegetable crops ranges between 2.0° and 15.0°C optimum being 20°- 30°C. Soil temperature above 30°C generally impairs seed germination. In a thumb rule, the average annual soil temperature can be approximated by adding about 1°C to the mean annual air temperature.

Soil fertility

It is the ability of the soil to supply nutrient elements in the amounts, forms and proportions required for maximum plant growth, and it depends on the physicochemical properties and organic matter content of the soil. Humus and other organic matters improve the hydro-physical, chemical, and biological properties of the soil and thereby increase soil fertility. Soil fertility can be increased by: (a) judicious soil management by incorporating organic matters, lime, etc., (b) tillage, drainage, etc. and (c) supplying nutrients to the soils through inorganic fertilizers and other chemicals. In general, vegetable crops have high nutrient requirements due to their high productivity. Most of the vegetable crops have short growing span (4-5 months), while some others have shorter even. Heavy production within such a short period of time also suggests the need of adequate soil fertility for successful vegetable production. Again, vegetative parts are important in some crops like leafy vegetables, cabbage, cauliflower, root crops, bulb crops, etc. while reproductive parts in the other crops like tomato, brinjal, okra, cucurbits, etc. Optimum vegetative growth is essential for all the vegetable crops. Excessive vegetative growth sometimes adversely affects yield in some crops like pumpkin, summer squash, tomato, okra, etc. Soil fertility needs to be handled judiciously according to the need of the crops.

Soil reaction

Each type of soil contains a certain amount of soluble matters and for this reason, it possesses a specific reaction which is manifested during interaction with water or salt solution. This reaction of soil *i.e.,* soil pH is determined by the proportion of hydrogen ion to bases absorbed on the clay and organic colloids in the soil. The pH scale has long been used as an index of acidity or alkalinity of soils. Soil pH is a key factor in regulating nutrient supply.

Soil acidity: Soils become acidic when water containing weak acids (carbonic and organic acids) flow through them and basic cations (Ca^{++}, Mg^{++}, Na^{++}, K^{+}) are replaced from the cation exchange sites with hydrogen (H^{+}) and aluminium (Al^{+++}) ions. Both humus and clay particles have large surface areas with proportionately higher number of cation exchange sites than silt and sand. Clayey and/or organic soils have the potential of being strongly acidic. Optimum availability of all the macro and micro-nutrient elements is realized in weak acidic to neutral soils (pH 6.5-7.0). Availability of phosphorus, calcium, magnesium and molybdenum sharply declines when soil pH falls below 6.5 while that of nitrogen, potassium and sulphur declines sharply below soil pH value 6.0. Availability of micronutrients like boron, copper, zinc and manganese falls sharply when soil pH falls below 5.0.

Soil acidity has become the major problem for vegetable cultivation in many humid tropical and subtropical areas. The adverse effect of acid soils on plant growth is mainly related to nutrient imbalances and slow microbial activity. Low yield of the crops grown on acid soil may be due to (i) toxic concentration of aluminium, (ii) toxic concentration of manganese, (iii) iron toxicity in some soils, (iv) deficiency of bases like calcium, magnesium, sodium and potassium, (v) molybdenum deficiency, (vi) nitrogen, phosphorus and sulphur deficiency, (vii) very slow microbial activity and (viii) very slow organic matter decomposition.

Most of the vegetable crops require slightly acidic to neutral pH values (6.5-7.0) of the soil solution for normal growth and development. However, they can also be grown successfully on soils that are on the alkaline side of neutrality (pH 7.0-7.5). They do well up to pH 7.6 if there is no deficiency of essential nutrients. Some vegetable crops viz., chicory, dandelion, endive, fennel, rhubarb, shallot, sorrel, sweet potato and watermelon show considerable tolerance to soil acidity and can be grown adequately at a soil pH as low as 5.0. Vegetable crops showing moderate tolerance to soil acidity are French bean, lima bean, Brussels sprouts, carrot, collard, cucumber, brinjal, garlic, horse radish, kale, knol khol, parsley, pea, pumpkin, radish, rutabaga, summer squash, winter squash, tomato, potato, etc. Crops grown on acidic soil low in organic matter are likely to be injured more by acidity than on the soil having the same acidity but higher in organic matter content. Soil acidity is neutralized by the application of lime and quantity of lime to be added depends on a) extent of soil acidity, b) soil texture and kind of lime. To change the soil pH from 5.5 to 6.5 up to the plough depth layer lime stone ($CaCO_3$) requirement will be 2.9 t/ha for sandy loam soil and 5.1 t/ha for clay loam soil however, for the correction of same soil acidity burnt lime (CaO) requirement will be 1.63 t/ha for sandy loam soil and 2.9 t/ha for clay loam soil.

Soil alkalinity: Alkali or sodic soils have generally high pH values, high exchangeable sodium percentage and is very much deficient in soluble calcium. A pH value of 7.5 to 8.0 usually indicates the presence of carbonates of calcium and magnesium, and a pH of 8.5 or above usually indicates appreciable exchangeable sodium. In non-saline alkaline soil, the exchangeable sodium percentage is greater than 15, and the electrical conductivity of the saturation extract is less than 4 mmhos/cm (at 25°C) and in saline-alkali soil, the exchangeable sodium percentage is greater than 15 and conductivity is greater than 4 mmhos/cm (at 25°C). Alkali soils are basically problematic because of its impervious nature owing to very high exchangeable sodium content and are generally reclaimed by adding gypsum ($CaSO_4$,$2H_2O$) although other materials such as, pyrites, phosphogypsum, elemental sulphur and sulphuric acid can also be used.

Soil salinity

It is an important soil factor influencing vegetable production. The soil is saline when it contains an excess amount of soluble salts. In saline soil, the electrical conductivity of the saturation extract is greater than 4 mmhos/cm (at 25°C) and the exchangeable sodium percentage is less than 15. The pH is usually less than 8.5. In soil science term, soluble salts are those inorganic salts that are more soluble than gypsum ($CaSO_{4,}$ $2IH_2O$) and soluble salts of soils are mostly chloride, sulphate and bicarbonates of sodium, calcium and magnesium. Concentration of soluble salts in the soil is expressed as electrical conductivity (EC). Electrical conductivity is expressed in reciprocal ohms (reciprocal of electrical sensitivity or resistivity) or mmhos per centimeter. Normal salt concentration in soil and water mixture of 1: 2 ratio expresses EC of not more than 1 mmhos/cm and EC beyond 3 mmhos/cm is harmful for most of the vegetable crops. Soil salinity is harmful for the plants because of its osmotic and toxic effects on the metabolic processes of the cell. Saline soil solution increases the concentration of water and makes it less accessible to plants. Greater the soil salinity the lesser is the water that the crop can extract from the soil and the crop suffers from water shortage.

Most of the vegetable crops are either less tolerant or moderately tolerant to soil salinity however, some vegetable crops viz., beet, kale, turnip, palak, asparagus, bitter gourd, ash gourd and lettuce show considerable tolerance to soil salinity. Of the vegetable crops viz., chilli, watermelon, pumpkin, bottle gourd, tomato, broccoli, cucumber, amaranth, summer squash, cauliflower, cabbage, muskmelon, carrot, spinach, corn and onion showing moderate tolerance to soil salinity, chilli, watermelon, pumpkin and bottle gourd perform comparatively better than the others in soil salinity condition.

5. Management of Seedlings in Nursery

Seedlings of many seed propagated vegetable crops like, tomato, brinjal, chillies, sweet pepper, cauliflower, cabbage knol khol, broccoli, Brussels sprout, kale, endive, chicory, celery, lettuce, parsley, onion, etc. are first raised in the nursery beds before transplanted in the main field although, many of these crops like, tomato, sweet pepper, omiom, cabbage, cauliflower, broccoli, etc. are also direct seeded using precision seed drills in the USA, Japan, the Netherlands, Italy and other European countries. Many seed propagated vegetable crops viz., okra, peas, beans, radish, carrot, beet, turnip and all the cucurbits (e.g. bitter gourd, bottle gourd, pumpkin, cucumber, watermelon, muskmelon, snake gourd, sponge gourd, ridge gourd, etc.) are conventionally grown through direct seeding in the main field.

Transplant groups

1. **Plants easy to transplant:** They form new roots rapidly, put forth shoots rather slowly and absorb water readily e.g. beet, broccoli, Brussels sprouts, cabbage, cauliflower, Swiss chard, lettuce and tomato.

2. **Plants transplanted with moderate ease:** This group replace roots and grow shoots a bit slower than the previous group and do not absorb water very efficiently e.g. celery, brinjal, onion, pepper, leek and cauliflower.

3. **Plants transplanted with difficulty, requiring special care:** They are very poor or slow to both root replacement and enhanced shoot growth. They take long time to cope up root injury and do not survive very ordinary transplanting. They require special care in handling e.g. cucumber, sweet corn, legumes, pumpkin, squash and melons.

A number of vegetable crops are also propagated and grown by different vegetative plant parts like, underground stem tuber (potato), underground root tuber and above ground vines (sweet potato), corm, corm pieces and cormel (elephant-foot yam), cormel (taro), bulb (onion), clove i.e., bulblet, bulbil (garlic), fleshy root and vine cuttings (pointed gourd), fleshy root (giant spine gourd or kakrol), stem cutting (cassava), limb cutting (drumstick), root tuber (Jerusalem artichoke), root suckers or offshoots and division of old crown (globe artichoke), pieces of stem tuber and vine cuttings (yam) and division of crowns (rhubarb). Followings are different advantages of raising seedlings in the nursery.

1. Uniform, vigorous and healthy planting materials can be raised for better growth, development and yield.

2. Economy of seed rate which cut down the cost of cultivation particularly for hybrids.

3. Seedlings can be raised in off-season vegetable forcing.

4. Favourable growing conditions can be provided to the seedlings in nursery bed through convenient protection against biotic and abiotic stresses.

5. Efficient time management for the preparation of main field.
6. Improvement in land use efficiency by increasing the cropping intensity.
7. Facilities for intercropping and relay cropping.
8. Easy adjustment of date of planting according to pest and disease vulnerability and market demand.
9. Maintainance of desired plant density in the field.
10. Efficient management of weed, pest and disease.

Different Features of a Nursery

Nursery site

The nursery site should be located away from shade at higher topography of land to facilitate proper drainage. The area should preferably be fenced or protected from pet and wild animals. Irrigation, transport, electricity, communication facilities, availability of basic material are some important considerations for a commercial nursery.

Nursery soil

Success of seedling nursery depends mainly upon the quality of soil. Before selecting a nursery, the soil analysis must be done to avoid any risk. The soil should be light to medium in texture with high porosity, good fertility, sufficient water holding capacity and good drainage. A soil pH of 6.5-7.5 is most suitable for vegetable nursery. Well drained sandy loam soil is the best for raising the seedlings. If the nursery soil is heavy, sand @ 2 kg/sqm and if the soil is light farm yard manure or compost @ 2.5 kg/sqm need to be added to raise good seedlings. Farm yard manure or compost also replenishes the macro and micronutrients in the growing medium. Well rotten and powdered FYM or compost is also used as organic mulch to improve the water holding capacity of soil.

Growing media for the seedlings

Seedling growing media are one of the most important factors in raising vegetable nursery because the growth and development of nursery plants and roots proliferation depends on it. Growing media are divided into soil, soil-less group or mixture of both. The growing media are soil, sand, coco peat, compost, vermicompost, peatmoss, perlite, pumice, sawdust, ashes, etc. Under open field condition, farmers generally use soil, sand and well rotten farm yard manure as they are cheap, easily available, retain sufficient water, air and allow sufficient drainage. Growing media should be easily available and must have the properties like, neutral in reaction, sufficient water and air retentive capacity and allow proper drainage, free from weed seeds, nematodes, fungi and bacterial inoculum and capable of being sterilized. Different growing media are hereunder mentioned.

Vermicompost: Vermicompost is a useful growing medium and it can be prepared easily by rearing of earthworms which provides all the required qualities like firmness, good water and air holding capacity and better nutrition and support to the seedlings.

Saw dust: This is a by-product of wood and timber industry and generally free from harmful substances. It is available at a very cheap rate everywhere. It has very good moisture retention capacity, free from microorganism inoculum and neutral in reaction.

Peat: Peat consists of the remains of aquatic, marsh, or swamp vegetation, formed as a result of partial decomposition under anaerobic or semi-anaerobic conditions. It is dark brown to black in colour, low in pH with a low moisture-holding capacity but contain 2.0 to 3.5 percent nitrogen.

Sphagnum moss: Commercial sphagnum moss is the dehydrated young residue or living portions of plants in the genus *Sphagnum,* such as *S. papillosum, S. capillaceum,* and *S. palustre.* It is relatively sterile, light in weight and has a very high water holding capacity, being able to absorb water 10 to 20 times of its weight. This material is generally shredded, either by hand or mechanically, before it is used in a propagation or growing medium. It has a pH value of about 3.5 to 4.0 and contains small quantity of minerals.

Perlite: Perlite is a grey white aluminium silicate mineral from lava flows. It is light in weight, highly expandable (4-20 times to its original volume) when heated approximately at 957^{0}C. The surface holds moisture and nutrients to make them available to the seedlings. Since it is sterile in nature, it is free from the diseases causing inoculums and nematodes. It has low cation exchange capacity and does not influence the pH of the media. High porosity of perlite helps to control water holding capacity and aeration of substrates. A course perlite is 1.2-2.0 mm in diameter with pH of 7.5.

Vermiculite: Vermiculite is a micaceous mineral that expands markedly when heated. Chemically it is hydrated magnesium aluminium/iron silicate. When expanded, vermiculite is very light in weight, 90 to 150 kg per cubic meter, neutral in reaction with good buffering properties, and insoluble in water. It is able to absorb large quantities of water (80 percent porosity). It has a relatively high cation exchange capacity (CEC 100-150 meq/100 g) and thus can hold nutrients in reserve and later on release them for the use of seedlings. It contains 9-12% magnesium and 5-8% potassium for the growing plants.

Pumice: Chemically pumice is mostly silicon dioxide and aluminium oxide, with small amounts of iron, calcium, magnesium and sodium in the oxide form. Pumice is screened to different size grades but is not heat treated. It increases aeration and drainage in rooting mixture and can be used alone or mixed with peat moss.

Rock wool: Rock wool is a man made mineral fibre. It has a high percentage of pore space, which holds both the nutrient solution and air for the growing plants. The rock wool contains 40-50% easily available water and 40-45% air capacity by volume. It acts as a support medium for the root system and as a means for holding nutrient solution and air in close proximity to the roots. It is sterile, easily managed and transported, completely consistent in performance and has a pH of about 7.0-7.5.

Coco peat: Coco peat is biodegradable and eco-friendly growing medium prepared from natural coir fibre. It is completely organic and soil free potting medium, a mixture of corky cellular material and short fibre which can replace soil and peatmoss. It has uniform texture and consistency, free from pathogen, weed seeds and toxins. It has a high water holding capacity, almost five times of its volume and retains and releases nutrients over a long period which provides better aeration and enhances strong and healthy root growth.

Soil treatment

For raising healthy seedlings, nursery soil must be treated by physical, chemical or biological means to kill the soil borne disease causing pathogens particularly, *Pythium, Rhizoctonia, Fusarium, Sclerotinia, Botrytis, Phytophthora, Phomopsis, Sclerotium, Alternaria* and *Pseudomonas solanacearum*, nematodes (particularly *Meloidogyne* spp.) and insect pests. Different species of *Pythium, Sclerotinia* and *Phytophthora* are more likely to cause damping off in cool wet soils, where as different species of *Rhizoctonia, Fusarium* and *Sclerotium* cause damping off in warmer and drier conditions which are most common disease in the nursery. Symptom manifestation of bacterial wilt in different vegetable crops particularly tomato and brinjal caused by *Pseudomonas solanacearum* mostly occur in the main field but very often the bacteria harbour the seedling in the nursery which is carried forward in the main field. Similarly infestation of nematode in the seedbed is carried forward to the main field. Besides, different fungi and bacteria cause some specific disease in the seedling stage viz., early blight of tomato (*Alternaria solani*), leaf blight of tomato (*Alternaria alternata* f.sp. *lycopersici*), collar rot of tomato, brinjal, chilli (*Sclerotium rolfsii*), bacterial spot of tomato (*Xanthomonas campestris* pv. *vesicatoria*), anthracnose of chilli and capsicum (*Colletotrichum orbaculare*), purple blotch of onion (*Alternaria porri*), etc. Nursery soil may be treated by different methods as mentioned below.

Soil solarization: The whole nursery area after being irrigated lightly is covered with transparent polyethylene sheet of 200 gauge for about 5-6 weeks during the summer months (from mid-April to June) when air temperature may raise upto 40-48^{0}C and soil temperature inside the polyethylene reaches upto 48-56^{0}C which is sufficient to kill most of the harmful pathogens, nematodes and weed seeds. After 5-6 weeks, the polyethylene sheet is removed and the area is used for the preparation of nursery beds for sowing the seeds.

Steam treatment: The nursery area is first covered with polyethylene sheet in such a manner that air can not pass in or out and then steam is passed inside for at least 4-6 hours continuously to kill the harmful pathogen, nematodes and insects pest. The soil is then ploughed and prepared for seed sowing.

Formalin treatment: Formalin solution is prepared by adding 250 ml commercial grade formaldehyde (40%) per 10 litre of water and soil is drenched @ 4.5 litres per square meter so as to saturate top soil surface upto a depth of 15-20 cm 15-20 days before seed sowing and the drenched area is covered with 200 gauge polyethylene sheet so that fumes of formalin penetrate inside the soil to kill the pathogens. The polyethylene cover is removed after 24 hours and the bed kept open for 15-20 days prior to seed sowing.

Soil treatment with fungicides: Fungicides like captan/thiram@ 5 g / square meter for dry treatment or 0.25-0.30% solution @ 4-5 litres / square meter or mancozeb@ 2g + carbendazim @ 1 g/ litre of water or chlorothalonil 2g + carbendazim @ 1 g / litre of water can also be used as soil drenching

Soil treatment with insecticides: Granular insecticides like, Furadon, Phorate 10 G, etc. are applied to the soil @ 4-5 gram per square meter and then thoroughly mixed to a depth of 15 to 20 cm to kill insects, their eggs, nematodes

Soil treatment with bio-agents: Different species of *Trichoderm* spp., *Pseudomonas floroscence* and *Aspergillus niger* are very much effective biological agents to controls majority of the soil borne pathogens and these bio-agents are added @ 10-25 gram per square meter and well mixed in the soil and seeds should be sown 2 days after the application of bio-agents. Success of biological control depends on the activity of the spores, spore counts in the formulation and sufficient organic matter in the beds for rapid multiplication of the bio-agents.

Seed treatment

Most of the vegetable seeds are treated with Captan or Thiram @ 2.5-3 g/kg seed to disinfect the seed surface. Hot water treatment of seeds of cabbage and cauliflower at 50oC for 30 minutes just before sowing prevents seed borne black rot disease caused by the bacteria *Xanthomonas campestris* pv. *campestris*. Biological control of the soil borne damping off diseases caused by *Pythium* sp., *Rhizoctonia* sp., *Fusarium* sp., etc can be done through seed treatment with different antagonistic fungi like, *Trichoderma* spp., *Gliocladum virens*, *Penicillium* spp., and bacteria like, *Bacillus subtilis* and *Pseudomonas fluoroscens*.

Use of coated or pelleted seeds has now become quite common in vegetable crops especially those having small seeds like tomato, brinjal, chilli, sweet pepper, lettuce, etc. The pelleted seeds are often coated with pesticides for better protection against pest and disease causing pathogens. In the developed countries enhanced seeds are also available which are developed by osmoconditioning or priming. In this process, seeds are allowed to imbibe water in a temperature controlled dilute aerated solution of an organic or inorganic osmoticum to ensure faster and uniform germination.

Preparation of Nursery Bed

Nursery bed should generally be raised 15-20 cm from ground level and the soils must be fine and have optimum moisture for uniform and high percentage of germination. The length of the bed may be kept 3 to 5 meter, and width is restricted to a maximum of 1 meter only to facilitate intercultural operations. The beds should preferably be prepared in the east and west direction and line should be made north to south direction on the beds to explore and capture maximum radiation energy especially during winter.

Seed sowing in the nursery

Proper seed sowing in the nursery beds is one of the important factors to get healthy and vigorous seedlings. After the seed bed preparation, seeds are sown in the nursery bed either by broadcasting or in lines depending upon nature and season of crop and availability of time and labour, although line sowing is the best. About 1 cm deep lines parallel to the width at a distance of 5.0 cm apart are opened and the seeds are sown or placed at a distance of about 1.0-2.0cm in the line. After sowing, the seed bed should be covered with 0.5 cm. thick mixture of farm yard manure: sand: soil in the ratio of 1:1:2 for better moisture conservation and germination.

First watering of seedbed with chloropyriphos solution@0.25% protects the seeds from termites and antsattack. Sprayingtheseedbed with mancozeb@ 2g + carbendazim @ 1 g/litre of water or chlorothalonil 2g + carbendazim @ 1 g/litre of water mixture protects seedlings from damping offdisease.

The seedlings in the nursery should be protected from different sucking insects particularly white fly which transmit different virus for 25 to 30 days by spraying Imidacloprid (0.2 ml/litre of water) at weekly interval. Mulching the seedbeds with paddy straw or sugar cane trash in hot weather with intermittent sprinkling of water and polyethylene mulch during winter helps in quick and uniform germination of seeds. The time to remove the mulch from the seedbed is extremely important. The seedbeds should be observed daily and as and when the white thread like structure is seen above the ground, the mulch should be removed carefully in the evening hours to avoid any damage to the emerging seedlings. During the seedling growth, if temperature reach above 30^{0}C then the beds should be covered by green or black coloured 50% or 60% shade nets about 1 meter above the ground with suitable support to protect the seedlings from direct scorching sunlight. The newly introduced translucent blue shade net gives blue light, which is excellent for plant growth. The nursery beds require light irrigation with the help of rose can from the seed sowing until the seedlings become ready to transplant.Young roots of seedlings absorb water primarily by osmotic forces. Anaerobic condition caused by excessive soil moisture retard seed germination. It is important to remove weak, diseased, insect pests damaged and dense plants from the nursery beds keeping 5 to 1.0cm distance from plant to plant for proper growth of the seedlings.

Pricking out of seedlings

It is a method of producing healthy and hardened seedlings by pricking out the young seedlings when first true leaves have come out and transplanting them to another seed bed or seed box or grown individually in small plastic thumb pots of about 5cm diameter. The seedlings should be uprooted carefully with the help of V-notched wooden stick for second transplanting and spaced about 5cm within a row which are 7-8cm apart, both in second seed beds or seed boxes. The soil mixture generally used to grow the pricked out seedlings is slightly richer than the previous one for first seed bed. Seedlings become ready for planting in the main field generally three weeks after first transplanting.

Seedling raising under polyhouse

The direct sown crops like, okra, peas, beans, radish, carrot, beet, turnip, bitter gourd, bottle gourd, pumpkin, cucumber, watermelon, muskmelon, snake gourd, sponge gourd, ridge gourd, etc. cannot be transplanted after uprooting because of the formation of an external layer of corky cell "suberine" in the existing cells of the roots. Since suberine is impervious to water which results in the death of the seedlings due to water deficit. Poor root regeneration also hinder plant establishment after transplanting in these crops. But these difficult to transplant vegetable crops can be transplanted without disturbing the roots in the growing media to minimize root damage. Generally nursery raised plants of these crops are economic in the adverse situation when there is too low temperature for seed germination or the field preparation is late due to some standing crop in the field. It is now a common practice to sow seeds of different cucurbits like, bottle gourd, bitter gourd, muskmelon, watermelon, cucumber etc in winter months (December – February) in small polybags (15 x 10cm size of 200-300 gauges) filled with soil, farm yard manure /compost and sand mixture (1:1:1) to get early crop. Seeds having hard or leathery seed coat such as bottle gourd, bitter gourd, pumpkin, watermelon, etc. may be soaked in water overnight before sowing for better germination. The polybags are kept in polyhouse where temperature is raised above 20°C during day time. Simple, improvised structure with bamboo poles or iron pipes with a covering of polyethylene sheet of 200 gauge can be used in raising the vegetable seedlings under unfavorable environmental condition. Micro sprinklers are the best source to irrigate such polyhouse grown vegetable seedlings. The micro sprinklers connected with water supply and sensors automatically sprinkle water at regular intervals. Seedlings of different cucurbits become ready within 28-30 days under the polyhouses especially during winter months (Dec-Feb) while under open condition it takes 50-60 days. As soon as field conditions are favourable for planting, the seedlings are transplanted in well prepared mounds either by removing the polyethylene or by cutting lower portion of the polyethylene so that roots may penetrate into the soil. In this way the crops can be advanced by 30-45 days compared to normal direct seeding in the main field.

Protection of seedlings

Different vegetable crops like, tomato, chilli, sweet pepper, etc are very susceptible to virus particularly leaf curl virus which are usually deposited inside the plant at the seedling stage by the insect vector primarily white fly (*Bemisia tabaci*). Fruit and shoot borer also attacks brinjal even in the seedling stage. Seedling of these vegetable crops should be raised inside the fine mosquito net coupled with spraying of insecticides outside the mosquito net to keep away the infestation of the insect vectors to produce virus free seedlings.It is better to spray the combination of fungicides (0.2% chlorothalonil or 0.25% mancozeb + 0.1% carbendazim), insecticides (metasystox 1.5 ml/litre or imidachloprid 0.2 ml/litre) and sticker (triton or sandovit @ 1 ml/litre of solution) to control the fungal pathogen and sucking insect pests like, white fly, jassids, mite, thrips, etc. Red pumpkin beetle which is the most prominent insect pest of seedling of cucurbits can effectively be controlled by spraying carbaryl 50 WP @ 2 g/litre at 2-3 leaf stage.

Hardening of seedlings

Hardening of seedlings is done for better field stand by different ways i) lowering the available soil moisture to 20% by withholding watering to the plant 5-7 days before transplanting, ii) lowering the temperature which retards the growth, iii) by application of 4000 ppm NaCI with irrigation water, iv) by spraying of 2000 ppm of cycocel + 0.25% $ZnSO_4$ and v) by spraying of 0.2% muriate of potash.

Raising grafted seedlings in vegetable crops

Grafting is a simple method of propagation in which desired rootstocks are utilized to induce vigour, precocity, enhanced yield and quality and better survival under biotic and abiotic stress conditions. In recent past, grafting of cucurbits, tomato, eggplant and pepper has gained much popularity in Japan, Korea and United States. However, raising of grafted plants of vegetables first started with watermelon onto gourd rootstock in Japan and Korea in the late 1920s. Later, eggplant was grafted onto scarlet eggplant *(Solanum integrifo/ ium* Poir.) in the 1950s. In Greece also, cucurbit grafting has a 50-year history and is much popular in southern areas. Since then, the area of grafted vegetables particularly watermelon, cucumber, melon, tomato and eggplants has increased multifold. The number of vegetable being grafted is estimated to be 600 million seedlings per year in Korea and 800 million in Japan which clearly suggest the benefits of growing vegetables using grafted seedlings.

Benefits of grafted seedlings in vegetable crops

Basic objective of using grafted seedlings is to reduce the dependency upon chemicals required to treat the soil to avoid the soil borne diseases which has opened up new vista in organic farming of vegetables. Development of mechanized and robotic grafting technology has given a fillip to this novel eco-friendly approach. Different benefits imparted by the grafted seedlings are documented.

Avoidance of disease and pest incidence: The main objective of grafting is to avoid soil borne diseases such as Fusarium wilt in different cucurbits particularly cucumber, muskmelon, watermelon and bittergourd and bacterial wilt in tomato and brinjal through the use of resistant rootstocks. Use of resistant rootstocks in grafting in combination with integrated pest management practice is the most viable option to reduce the need for soil fumigation with methyl bromide to kill these soil borne pathogens which also prove to be a boon in organic farming of vegetables.

Avoidance of nematode infestation: Another important objective of grafting is to avoid the infestation of plant parasitic nematodes particularly root knot nematodes (*Meloidogyne* spp.) which harbour different vegetable crops with the use of nematode resistant rootstock through interspecific or intergeneric grafting. Some of such nematode resistant rootstocks are *Sicyos angulatus* and pumpkin for cucumber, Nematex variety of tomato and *Solanum nigrum* for tomato, etc.

Minimizing the autotoxic effect: Recently autotoxic potential of some cucurbits crops have been pointed out which may pose serious problem of soil sickness in commercial production of cucurbits although there are differences in autotoxic potential among the cucurbits and watermelon, muskmelon and cucumber exhibits have more auto-toxic potential than the others. The autotoxins affect ion uptake, membrane permeability, photosynthesis and phytohormone balance. It seems possible to overcome autotoxicity in different cucurbits crops through grafting them on *Cucurbita ficifolia*.

Providing cold hardiness: Grafting is also useful to initiate the flowering and fruit set at low temperature which is useful in saving electricity energy of to maintain requisite day/night temperature regime in the polyhouse during winter particularly in the temperate countries. Early yield of cucumber could be obtained by grafting it on *Cucurbita ficifolia* because the grafted plants can survive at very low temperature of 20°C/12°C day/night.

Survival of grafted plants under excessive moisture condition: Flooding or excessive soil moisture condition generally reduces photosynthetic rate, stomatal conductance, transpiration, soluble protein and activity of ribulose-1, 5-biophosphate carboxylase/oxygenase (Rubisco). Intergeneric grafting imparts the attributes of flood tolerance particularly in different in cucurbits by upkeeping these physiological manifestations through sustained enzyme activity like Rubisco activity via Rubisco activase enzyme.

Improving quality traits: Improved quality in the grafted seedlings has not yet been achieved prominently. It was observed through different experiments that grafting did not affect the quality traits viz., taste, size and shape in cucumber however, more number of commercially acceptable and shiny fruits of cucumber could be obtained by grafting it on some hybrids of summer squash. Enhanced sugar content in Galia and Haon varieties of melon has been found by grafting them on the melon cultivar Suiker. In tomato, grafted seedlings generally produce increased number of marketable fruits with better colour and enhanced lycopene content and decreased number of malformed, under developed and gray mold infected fruits.

Manipulating the harvesting period: Altering in harvesting period either early or late often prove to be advantageous to both the growers and consumers. Cucumber grafted onto *C. ficifolia* generally grows faster and give high percentage of early yield.

Stionic effect on mineral nutrition: Grafting also influences growth and mineral content of the plant tissues. Interveinal chlorosis and marginal greening of cucumber leaves called ‘green ring’ is caused due to Mg deficiency which is induced mainly by excessive accumulation of ammonium ions. This disorder becomes more severe in cucumber grafted onto the summer squash cultivar Shintosa than in those grafted onto fig leaf gourd (*Cucurbita ficifolia*). Cucumber cultivars which produce fruit with a heavy surface bloom due to origination of trichomes, principal component of which being silica, on the fruit surface are undesirable for marketing. Development of bloom can be depressed by grafting the cucumber onto appropriate cultivars of summer squash called ‘bloomless rootstocks’.

Influence on sex expression: Sex expression and flowering pattern of the scion of different cucurbits changes with grafting because root may control floral formation by production of different flower inducing factors.

Prime objectives of grafting in different vegetable crops are hereunder mentioned

Vegetables	Objectives of grafting
Watermelon	Tolerance to Fusarium wilt (*F. oxysporum*), low temperatures, wilting due to physiological disorders, drought tolerance.
Cucumber	Tolerance to *Fusarium* wilt, *Phytophthora melonis,* low temperatures,
Melon	Tolerance to *Fusarium* wilt (*F. oxysporum),* low temperatures, wilting due to physiological disorders, Phytophthora disease.
Tomato	Tolerance to bacterial wilt (*Pseudomonas solanacearum*), *Fusarium oxysporum,* nematodes (*Meloidogyne* spp.), *Verticillium dahliae.*
Eggplant	Tolerance to bacterial wilt (*Pseudomonas solanacearum*) *Verticillium alboatrum, Fusarium oxysporum,* low temperatures, nematodes and induction of greater vigor.
Bittergourd	Tolerance to *Fusarium* wilt (*Fusarium oxysporum .f.*sp. *momordicae*)

Promising rootstocks for grafting different vegetable crops

Scion	Rootstocks
Cucumber	*Cucurbita moschata, Cucurbita ficifolia, Cucurbita maxima, Sicyos angulatus*
Melons (for open field)	*Cucurbita sp., C. moschata x C. maxima, Cucumis melo*
Melons (for green house)	*Cucumis melo, Benincasa hispida, Cucurbita* spp.*C. moschata x C. maxima*
Watermelon	*Citrulus lanatus, Cucurbita maxima, C. moschata* F_1 *C. moschata x C. maxima, Lagenaria siceraria*
Bitter gourd	*Cucurbita moschata, Lagenaria siceraria, Luffa aegyptica*
Tomato	*Lycopersicon pimpinellifolium, L. esculentum, Solanum nigrum*
Eggplant	*Solanum torvum, Solanum integrifolium, Solanum melongena, Solanum nigrum*

Infrastructure facilities for grafting on a commercial scale

Grafting is carried out in a shady place sheltered from the wind to avoid wilting of the grafted plants. When grafting is performed, it is important to increase the chances for vascular bundles of the scion and rootstock to come in close contact by maximizing the area of the cut surface and taking care that the cut surface does not dry out. After grafting it is necessary to keep the grafted plants at about 30°C and with more then 95% humidity for three days. Gradually, the relative humidity should be lowered with the increase in light intensity. Successful commercial grafting programme requires two types of facility *viz* screen house and grafting chamber.

Screen house: A screen house is used for growing seedlings prior to grafting and for hardening of grafted plants under controlled light condition prior to disposing the seedlings or transplanting in the main field and to protect the seedlings from rain. Screen house is also used to check the infestation of virus-transmitting insect vectors such as aphids and whiteflies.

Grafting chamber: Grafting chamber is basically a polished with mist generation facility through micro-sprinkler where the grafted seedlings are kept for about one week under high humidity and reduced light intensity condition. Immediately after grafting high humidity is required to minimize drying and shade is required to minimize shock and induce fast healing of the wounds in both stock and scion.

Methods of grafting

Several methods of grafting are in vogue, each involving the joining of the leaf bearing part (scion) of one plant with the rootstocks of another. Different grafting methods include cleft grafting, tube grafting. whip and tongue grafting, splice grafting, flat grafting, bud grafting, hole insertion grafting, tongue approach grafting etc and some of the widely used methods of grafting are briefly described.

Cleft grafting: Tomato plants are mainly grafted by this method where seeds of the rootstocks are sown 5-7 days earlier than those of the scion. The stem of the scion (at the four leaf stage) and the rootstock (at the 4-5 leaf stage) are cut at right angles, each with 2-3 leaves remaining on the stem. The stem of the scion is cut in a wedge and the tapered end fitted into a cleft cut in the end of the rootstock. The graft is held firm with a plastic clip.

Tube grafting: This method of grafting requiring tubes with a smaller inside diameter makes it possible to graft small plants grown in plug trays two or three times faster than the conventional method and is quite popular among Japanese seedling producers. The optimum growth stage for grafting varies according to the kind of plug used. First the rootstock is cut by a slant cut and the scion is cut also in the same way. Plastic tubes with side slit are placed onto the cut end of the rootstock. The cut ends of the scions are inserted into the tube, splicing the cut surfaces of the scions and rootstocks together. While practicing the tube grafting in eggplant the seeds of *S. torvum* must be sown a few days earlier than those of the other rootstock species.

Tongue approach grafting: Melons and other cucurbitaceous vegetable crops are generally grafted by this method. It gives higher survival ratio because the root of the scion remains until the formation of the graft union. In this method, seeds of cucumber are sown 10-13 days before grafting and pumpkin seeds 7-10 days before grafting to ensure uniformity in the diameter of the hypocotyl of the scion and rootstock. The shoot apex of the rootstock is removed so that the shoot cannot grow. The hypocotyl of the scion and rootstock are cut in such a way that they tongue into each other, and the graft is secured with a plastic clip. The hypocotyl of the scion is left to heal for 3-4 days and then crushed between the fingers and later the hypocotyl is cut off with a razor blade three or four days after being crushed.

Mechanized grafting

Grafting is arduous task and efforts are being made to reduce the labour required. Attempts have been made to mechanize grafting since 1987. There are several basic factors which govern the success of grafting by machine or robot such as seedling shape, location of cut, seedling gripping, cutting method, fixing materials and tools etc. Grafting robots for plug have been developed by combining the adhesive and grafting plates. This robot makes it possible for eight plugs of tomato, eggplant or pepper to be grafted simultaneously. Recently a fully automatic grafting system for cucurbitaceous vegetables has been designed through which seedling quality is estimated before being used for grafting.

6. Vegetables Based Cropping System

Cropping system is defined as the cropping pattern followed on a farm and its interactions with farm resources, other farm enterprises and production technology while the yearly sequence and spatial arrangement of crops and fallow on a given area is termed as cropping pattern.

In India, land resources are very limited which will further be decreased due to explosive increase in population, rapid urbanization and industrialization. At present, hardly 3 per cent of our total arable land of India is under vegetable cultivation. However, it is very difficult to get such horizontal expansion in vegetable cultivation. In this situation, increase in total production per unit area leaves the most viable option for enhancing the total production in the country and such target can be achieved through (a) replacing the local and unimproved varieties by improved high yielding varieties and hybrids, (b) following integrated plant nutrient and protection measures, (c) judicious management of irrigation water and (d) increasing cropping intensities following different cropping patterns.

Multiple cropping in vegetable crops

It is a cropping system representing the philosophy of maximum crop production per unit area of land within a calendar year or other relevant time span with minimum soil deterioration. In its simplest form, multiple cropping is a one-year cropping system in which two or more crops are grown in succession within a year. In multiple cropping systems, inherent plant nutrients are utilized exhaustively due to intensive cropping, particularly with vegetable crops. Any loss in soil productivity due to such depletion is temporary and can be corrected by balanced application of organic manure and inorganic fertilizers. In fact, crop rotation without applying fertilizers may maintain organic matter in the soil, but does not maintain high productivity levels. Both yield and soil fertility status need to be maintained in sustainable manner following integrated plant nutrient management. Judicious plant nutrient management in multiple cropping systems not only increases crop productivity but also ensures abundant crop residues and their roots which enrich soil organic matter. Soil microbial population is in a dynamic equilibrium with one another in a given environment. Any change with one another in this equilibrium due to intensive cropping is merely temporary and enhanced crop residues in various stages of decomposition can sustain diversified flora and fauna. Multiple cropping should necessarily be based on some important considerations (a) use of varieties and hybrids for specific climatic requirements or duration, (b) adequate soil management to sustain soil health, (c) adequate nursery management for timely supply of healthy seedlings, (d) management of crop residue for recycling of organic matter, (e) integrated plant nutrient and protection measures, (j) judicious management of irrigation water and (g) compatibility of the crop combination with capital, enterprise preference and farmers' skill.

Multiple cropping in vegetable crops has several merits like (i) generation of more income from the farm through increased production per unit area in the year, (ii) generation of more livelihood through intensive cropping throughout the calendar year, (iii) maintenance of soil fertility and health through balanced nutrition and scientifically based crop alterations, (iv) complete and economical use of natural resources, particularly nutrients and water, (v) rational use of capital and time, and (vi) management of pest and diseases and weeds.

Vegetable cultivation is a capital intensive venture hence, effective land utilization through intensive cropping pays high dividend particularly in market garden where vegetable farming is generally done on high-priced land. Various forms of multiple cropping in vegetable crops are crop rotation, succession cropping, intercropping and relay cropping.

Crop Rotation: It is a cropping system of growing different crops in a regular sequence on the same land repeatedly for the period of two, three or more years. Here the cycle of cropping sequence takes more than one year to complete.

Succession Cropping: It is the system of growing two or more crops in succession on the same land within a year. This cropping system is generally followed in most market gardens where the aim is to keep the high-priced land occupied with the high value crops for most part of the year.

Relay Cropping: It is also a system of growing different crops on the same land within a year but in this system, the succeeding crop is sown / planted before the preceding crop is ready for harvest. So, the growing spans of two crops overlap for a short period.

Intercropping: It is a cropping system where two or more crops are grown simultaneously in alternate rows or otherwise in the same land showing significant amount of intercrop competition. The crops may or may not be sown/planted and harvested at one time. It is mainly aimed at increasing the yield of the companion crop without reducing the yield of the main crop.

Principles for arranging the cropping sequence

The economic consequence of crop rotation, succession cropping and relay cropping depend on the set of crops. There should be definite planning in arranging the cropping sequence based on the following principles.

1. Repetition of the crops having common diseases and pests should be avoided. For example, solaneceous crops should be avoided in the rotation when *Fusarium* and bacterial wilt becomes a problem. Many serious diseases and pests can be controlled

by scientific crop rotation. Club root of cabbage and other crucifers caused by the fungus *Plasmodiophora brassicae* can effectively be controlled by keeping the land free from cruciferous crop for a period of at least three years. Root-knot nematodes (*Meloidogyne* spp.) can be controlled by including nematode non-host crops like marigold, corn, wheat, rice, etc. in the cropping sequence. Rice-sweet potato-cowpea rotation is effective for the control of sweet potato weevil.

2. The relation of individual crops to their predecessors must be taken into consideration. Generally, vegetable crops of different families grow well after majority of the crops. Growing different cucurbits after solanaceous fruit vegetable crops is in general not profitable. Cabbage has marked depressing effect on the yield of onion that follows. A number of vegetable crops, like onion, spinach, tomato, etc. are good forerunners for the other crops when the compatibility conditions are fulfilled.

3. The deep-rooted crops should be followed by shallow-rooted ones. Deep-rooted vegetables, like pumpkin, summer squash, tomato, peas and beans, etc. are able to use nutrients extracted from deep layers of the soil, and these crops should be alternated with the crops having weak and shallow root systems like, onion, cauliflower, cabbage, potato, leafy greens, lettuce, etc.

4. Heavy feeder crops like, cabbage, cauliflower, potato, brinjal, etc., should be followed by low-feeding crops like okra, cucumber, pumpkin, etc.

5. Specific nutritional requirements of various crops should be taken into consideration. Cabbage, cauliflower, tomato, leafy greens, etc. extract large amount of nitrogen from the soil, and these crops should be followed by leguminous vegetable crops like cowpea, French bean, hyacinth bean, etc. Symbiotic nitrogen fixation by the *Rhizobium* bacteria that grows in the root nodules of the leguminous vegetable crops provides benefit to the following crops.

6. The crops which efficiently utilize the organic manure residues should be grown after the crop which do not utilize fully. For that matter, immediately after the addition of organic matter to the soil, cucumber, pumpkin, sunmer squash, cabbage, leek, etc. should be grown and the root crops like, radish, carrot, beet, turnip which can utilize the organic residues fully should be followed.

7. Leguminous vegetable crops should be included in the cropping sequence which not only upgrade the protein status of the farm produce but also enhance soil fertility. It is to be mentioned in this context that legumes increase the susceptibility of the following crops to moisture stress because legumes are large users of water and normally leave the soil dry after harvest.

8. Leguminous green manure crops like cowpea, daincha (*Sesbania* sp.), rice bean, clover, etc. should be accommodated in the rotation in order to increase the organic matter status of the soil.

Examples of different cropping sequences

Succession cropping

1. Vegetable cowpea (June to Oct) - mid-season cauliflower(Oct to Jan) - onion (Jan to May).
2. Pumpkin (Sept to Feb) - okra (Feb to June) – cabbage (June to Sept).
3. Potato (Nov to Feb) - amaranth (March to May) - okra (May to Oct).
4. Early radish (Aug to Oct)-French bean. (Nov to Feb)-bitter gourd (Feb to June) - amaranth (June to Aug).
5. Cabbage (Aug to Nov)-tomato (Nov to March)-ridge gourd (March to July)- cowpea/ rice bean for green manure (July to Aug).
6. Early cauliflower (July to Oct)-brinjal (Oct to April) - amaranth (May to June).
7. Early tomato (August to Jan)-onion (Jan to May)-green manure crop (June to July).
8. Brinjal (May to Nov) - pea (Nov to March) - amaranth (March to May).
9. Sweet pepper (Sept to Feb) – bitter gourd (Feb to June)-vegetable cowpea (June to Sept).
10. Chilli (Oct to April) - okra (April to Aug) - palak (Aug to Oct).
11. Late tomato (late Dec to April) - okra (April to August) - hyacinth bean (Aug to Dec).
12. Radish (Oct to Dec) - watermelon (Dec to April) - brinjal (April to Oct).
13. Elephant-foot yam (March to Oct) - pumpkin (Oct to Feb).
14. Bottle gourd (Sept to Feb) - okra (Feb to July) - Radish (July to Sept).

Note: To make such multiple cropping successful, cucurbit seeds need to be raised in polyethylene packets. When these are to be started during winter, the seedlings must be raised under polyethylene shed to avoid cold injury.

Relay cropping

1. Potato-pumpkin (potato is harvested during February to March and pumpkin seeds are sown in last week. of January. After harvesting of potato, pumpkin is continued up to July).
2. Early cauliflower-pumpkin (cauliflower seedlings are transplanted in July and harvested during October-November. Pumpkin seeds are sown in September in the standing crop of cauliflower and continued up to February-March).
3. Brinjal - ridge gourd/bitter gourd/watermelon (brinjal is grown during July to February. Seedlings of the cucurbits grown in polyethylene packets are to be planted in December

and continued up to May).

4. Bitter gourd-okra (bitter gourd is grown during October to February. Pre-germinated seeds of okra are to be sown in January and continued up to May).
5. Bottle gourd/summer squash/cucumber - bush type vegetable cowpea (the cucurbits are grown during November to May following ridge and furrow method. Cowpea seeds are sown in the beds between two furrows in April and continued up to July).
6. Cabbage - watermelon/cucumber (cabbage is grown from October to January while the cucurbit is started with the seedlings raised in polyethylene packets from December and continued up to April).

Intercropping

1. Cabbage + radish (cabbage is the main crop. The companion crop radish is of comparatively short duration and harvested early).
2. Tomato + radish + lettuce (tomato is the main crop and to be grown at higher spacing of 75-90 cm. Of the companion crops, radish is harvested 50 days and lettuce 80-85 days after sowing. So, enough space is left after harvesting the companion crops to continue tomato successfully up to 150 days after planting. In this intercropping, tomato must he staked and trained).
3. Cucumber + cabbage/cauliflower (cucumber is the main crop and is sown in October with a view to capitalize the advantage of early harvest in spring. Cabbage/cauliflower is grown in the interspaces successfully for the initial 80- 85 day period).
4. Carrot + peas (carrot, the main crop is very slow growing for initial 45 days. Early pea varieties, like Arkel from which first picking of pods can be done 55-60 days after sowing can successfully be grown in between the rows for the period of 85 days. At the time of last harvesting of peas, the plants are to be uprooted with subsequent earthing up in carrot which is continued upto 135-140 days after sowing).
5. Tomato + palak (palak can be grown successfully in between tomato for 55 - 60 days duration. Harvesting of palak by uprooting the whole plant starts 35 days after sowing).
6. Pointed gourd + palak/radish/early cauliflower (vine or root cuttings of pointed gourd, the main crop is planted in October at the side of the raised bed of 2.5 to 3 m width. Initial growth of the regenerated vine is slow due to advent of comparatively low temperature. Short-duration crops (75 to 80 days) can be grown profitably as companion crop in the beds. Pointed gourd is continued up to August-September. In case of two successive crops, ratooning in pointed gourd is done in October enabling the companion crop to grow again).
7. Cassava + onion/cowpea (cassava is the main crop of long 200 days duration and onion can be grown successfully in between the rows of cassava).

8. Swamp taro + garlic (regenerated plants of swamp taro from runner are planted at 75 cm spacing in both the ways in October in depressed swampy land. With the advent of dry spell in winter, excessive moisture from the soil dries out leaving it fit for planting garlic in December. Garlic is harvested in March and swamp taro is continued up to last week of August with harvesting from June onwards)
9. Elephant-foot yam + okra/ sweet potato/cucurbits/cowpea (elephant-foot yam is planted in April with wide spacing of 75 to 90 cm in both ways and continued up to January. Successful intercrops can be raised up to September).

Different vegetable crops can be grown as intercrop in the early stages of orchard development (first 2-3 years). Afterwards, shade provided by the orchard trees hinders successful vegetable cultivation. However, some hardy crops like elephant-foot yam, taro, sweet potato and cowpea can be grown as intercrop for some more periods. Similarly, vegetable crops like cowpea, cluster bean, brinjal, tomato, sweet pepper, onion can be grown as intercrop with long-duration field crop like sugarcane.

7. Water Management

Water is the most vital component of the living cells and the basis of plant feeding and in fact, both literally and poetically is the blood of civilization. Water is also characterized with extraordinary physical properties: it has high specific heat, biggest latent heat vaporization, very high thermal conductivity and greatest surface tension. Large amounts of heat are tied up with water because of very high specific heat, which contribute to both climatic stability and to heat budgets of the plants and other organisms.

Integral relationship of plant and water

Water molecules are integral part of living systems being the solvent for metabolites and structural component of proteins and nucleic acids. In fact, biological macromolecules cannot exist in their complicated secondary folding or spiraled helixes without the support of hydrogen bonding within their water matrix. The water matrix is also critical for metabolic activity because its properties of hydration permit ready reactivity between molecules in solution as well as between enzymes and their substrates, and the biochemical processes are sensitive to tissue hydration. A reduction of only 20-25 per cent below maximum hydration is sufficient to cause metabolic malfunctioning of many enzymes. It is the substrate for photosynthesis and a product of respiration. Water also helps to buffer plant tissue against fluctuations in the external environment. Plant growth including leaf expansion, organ enlargement, reproductive development and stomatal function are all clearly associated with moisture status. Hormonal changes (abscisic acid and cytokinins) in plant body occur with brief reduction in water potential. Water is the also major force in shaping the climatic patterns.

Transpiration lowers water potential at evaporative sites within leaves, and this effect is immediately translated to the root system through creating tension in the plants' vascular system. This demand of water must be satisfied continuously to maintain leaf water potential. The internal water balance of plants depends on the relative rates of water absorption and water loss. Despite stomatal resistance, the steep gradient in vapour pressure from leaf to air favours the transpiration rate well in excess of CO_2 fixation. Transpiration rate commonly ranges between 500 to 2500 mg H_2O/dm^2-h whereas CO_2 fixation rates generally between 5 to 25 mg CO_2/dm^2-h. This physiological phenomenon amply justifies high water requirement to maintain high dry matter productivity. For this reason, actively growing plants as the case of all the vegetable crops need to maintain liquid phase continuity from soil water through its vascular system and all the way to evaporative sites in leaves. Vegetables contain large amount of water (more than 85%) and the product quality like, tenderness, succulence, crispness and flavour is very much related to water supply at proper stages. In fact, texture of the vegetables is determined by the combination of tissue structure, cell wall properties and turgor pressure.

Development of moisture stress in foliage results different physiological consequences *viz.* (1) decrease in stomatal opening, (2) reduction in photosynthesis and transpiration, (3) dehydration of protoplasm, (4) reduction in cell division and enlargement, (5) increase in respiration initially which later decrease due to less availability of respiratory substrate, (6) decrease in total dry matter production and growth, (7) hastening in maturity, (8) accumulation of sugar particularly during later part of growth, etc. Virtually every facets of the plant's physiology are dislocated to some degree during temporary moisture stress which is recovered subsequently upon adequate moisture supply later and under severe moisture stress such recovery of essential physiological functions is not operative causing permanent wilting of the plant.

Consumptive use and water use efficiency

Soil water is lost by evaporation from the soil surface, water surface and from the surfaces of leaves of the plant and transpiration from plant body. In the field, these two processes cannot be separated so far as plant water requirements are concerned. The water lost in these two ways is called evapotranspiration which is nearly equal in amount to the consumptive use or net water requirement of the crop. Of the total absorbed water, hardly 0.1 to 0.5 per cent water is actually incorporated into the plant tissues and the remaining water is ward off from the plant body by transpiration. So, consumptive use is the quantity of water lost by evapotranspiration plus the quantity that is actually incorporated in the plant tissues. Water loss from mature plants is about 50 to 90 per cent water which is lost from a free water surface on the same area. Evaporation from free water surface is estimated by using "Evaporation pan". Potential evapotranspiration is indirectly estimated from pan evaporation data after necessary adjustment.

Water requirement of a crop can be calculated from consumptive use or evapotranspiration of a crop which is influenced by temperature, irrigation practices, length of growing season, precipitation and other factors. The volume of water transpired by crop plants depends on availability of water, absorption of water by the roots, temperature and humidity of the atmosphere, wind movement, intensity and duration of sunlight, stage of development of plant, type of foliage, nature of the leaves and kind of crop. Variations in consumptive use rate occur from day to day because of changes in climate. It is increased in hot, dry and windy day and is decreased in cool day. Consumptive use also varies with relative growth of the crop. The phenology of the growing plant can be characterized by vegetative, flowering, fruiting, and other distinctive characteristics in vegetative stage like, curding in cauliflower, heading in cabbage and head lettuce, bulbing in onion and garlic, tuberization in potato and sweet potato, etc. During the vegetative stage consumptive use continues to increase till the end of this stage of growth and flowering occurs near and during the peak of consumptive use. The fruiting stage is accompanied by a decrease in consumptive use until the transpiration essentially ceases during the latter part of the development of seeds inside the fruit.

Stomata are the compulsory passage way for CO_2 and H_2O gas exchange between plant and the atmosphere. Therefore, in order to fix carbon, the plant loses water to the atmosphere, the ratio being variable depending on species and growing conditions. This ratio is called water use efficiency *i.e.,* the quantity of water transpired per unit carbon fixed in dry matter. Water use efficiency increases as the amount of water required per unit of dry matter production decreases. "Transpiration ratio" is used to express the water use efficiency. Transpiration ratio refers to the quantity of water (in gram) transpired by the plant to accumulate 1 g of dry matter. This ratio ranges from 200 to 1000 *(i.e.,* 200: 1 to 1000: 1) depending on the crop species, cultivation conditions and vegetation period. Vegetable crops can be categorized into four groups according to such index of water exchange. Leafy vegetables require much more water to produce 1 g dry matter than pumpkin. Vegetable crops showing high transpiration ratio cannot endure water stress, and any shortage of water during the period of growth and development severely affects yield. On the other hand, vegetable crops characterized by very low transpiration ratio can somewhat endure water stress condition and can give satisfactory yield even in moisture-deficient condition. Vegetable crops can be classified into four groups according to the water requirement which is expressed as the units of absorbed water required for the production of one unit of dry matter.

Classification of vegetable crops according to their water requirement

High	Moderate	Low	Very low
Palak	Onion	Pea	Watermelon
Amaranth	Cucumber	French bean	Muskmelon
Lettuce	Chilli	Cowpea	Pumpkin
Celery	Brinjal	Broad bean	Wax gourd
Other leafy greens	Tomato	Cluster bean	
Sweet pepper	Carrot	Winged bean	
Cabbage	Potato	Hyacinth bean	
Cauliflower			
Asparagus			
Broccoli			
Rhubarb			
Brussels sprouts			

Contd.

Contd. Table

High	Moderate	Low	Very low
Horse radish			
Radish			
Beet			
Turnip			
Ridge gourd			
Rutabaga			
Green onion			

Source: Vegetable Growing in Home gardens of Tropical and Subtropical Areas (Konokov and Vemulapalli)

Scheduling of irrigation

Frequency of irrigation and amount of water to be applied depend on a number of factors such as, depth of root system, water use efficiency, stages of growth, soil type, prevailing weather condition and actual consumptive use of the crop however, it is a rather complex consideration.

Irrigation requirement depends on the stages of growth of the particular vegetable crop. When the plants are young although they require and use less water, need highly moist soil because the root system of the young plants is weak, sparsely distributed and located in the upper 15 to 20 cm layer of the soil that gets quickly dried. The vegetable crops utilize and transpire more water in the later stages of growth with more area and size of the leaves. Although most plants have critical stages of growth during which water stress markedly reduce yield. However, all the leafy vegetable crops, cauliflower, radish, carrot, beet, celery, etc. require optimum soil moisture level constantly during the whole vegetation period for proper growth and yield.

Over-irrigation also causes other harms like, bursting of heads in cabbage, branching of roots in carrot, fruit cracking in tomato, fruit rot in pointed gourd, etc. Abundant irrigation during ripening of fruits causes an increase in the water content of fruits and subsequently reduces their sweetness and storability. It is thus suggested to stop irrigation before fruit ripening of watermelon, muskmelon, pumpkin, winter squash, etc. In potato, onion, elephant foot yam, sweet potato, etc. irrigation should also be stopped 15 days before the harvest which greatly extends their storability.

SMT: Soil Moisture Tension; ASM: Available Soil Moisture

The depth, extent and configuration of the root system are also important in judging water absorptive potential of the particular crop. Roots of the shallow-rooted vegetable crops are confined to the surface soil which remove water much more rapidly and dry the soil. So, shallow rooted vegetable crops require frequent irrigations. On the other hand, crops with extensive root system continue to extract water from greater soil depths. Thus, interval between irrigation for the deep-rooted vegetable crops can be extended.

Irrigation needs and critical soil moisture period and impact of water deficit of vegetable crops.

Crops	SMT(bars)	ASM	Critical stage of growth	Major defects apart from low yield caused by water deficit
Potato	-0.30	65%	Stolon formation, tuberization and tuber enlargement	Poor and mis-shaped tuber
Tomato	-0.35	60%	Flowering and fruit development and after each harvest	Blossom-end rot and fruit cracking
Brinjal	-0.45	50%	Flowering and fruit development and after each harvest	Blossom-end rot,misshapen fruit, poorfruit colour
Chilli	-0.45	50%	Tenth leaf to flowering and fruit development and after periodical harvests	Low fruit set, shriveledfruits
Sweet pepper	-0.35	60%	Flowering and fruit development and after each harvest	Poor fruit size and blossom-end rot
Cauliflower	-0.25	70%	Curd formation and enlargement	Ricey curd, buttoning
Cabbage	-0.35	60%	Head formation and enlargement	Growth cracks andtough leaf
Broccoli	-0.25	70%	Heads development	Strong flavour, ricey curd and premature bolting
Onion	-0.35	60%	Bulb formation and bulb enlargement	Poor-sized bulb and early bolting
Garlic	-0.35	60%	Bulb formation and bulb enlargement	Poor-sized cloves
Carrot, beet, turnip	-0.40	55%	Root enlargement	Growth cracks, misshapen roots
Cucumber	-0.45	50%	Flowering and fruit development	Pointed and cracked fruits with bitterness
Pumpkin and all gourds	-0.50	55%	Flower bud development and early fruit development	Misshaped, small and cracked fruits
Watermelon	-0.60	40%	Flowering and fruit development	Blossom-end rot
Muskmelon	-0.60	40%	Flowering and fruit development	Poor fruit growth and quality
Okra	-0.55	45%	Flowering and pod development	Toughness and fibredevelopment in pods
Peas	-0.60	40%	Flowering and pod filling	Poor-pod filling andshriveled grains
French bean, hyacinth bean	-0.45	50%	Pod setting and pod enlargement	Poor pod filling with stingy pods
Cowpea	-0.45	50%	Prior to flowering and after pod set	Poor pod filling with fibrous pods
Leafy vegetables	-0.30	70%	Entire crop duration	Toughness in leaves
Celery	-0.30	70%	Entire crop duration	Toughness in leaves
Head lettuce	-0.35	60%	Head formation	Poor head with tough leaves
Sweet potato	-0.70	35%	40-45 days after planting at tuber formation	Poor and misshaped tuber
Elephant foot yam	-0.60	40%	Shoot emergence	Low bulking of corm

Knowledge of rooting depths is, therefore, important for scheduling the time and quantity of irrigation. Rooting depths vary greatly among the vegetable crops. The vegetable crops can be classified into 5 groups according to their rooting depth, *viz.,* very shallow-rooted, shallow-rooted, moderately deep-rooted, deep-rooted and very deep-rooted.

Rooting depth of different vegetable crops

Very shallow rooted (15-30cm)	Shallow-rooted (30-60 cm)	Moderately deep-rooted(60-90 cm)	Deep-rooted (90-120 cm)	Very deep-rooted (120-180 cm)
Onion	Broccoli	French bean	Pea	Artichoke
Lettuce	Brussels sprouts	Beet	Chilli	Asparagus
Small radish	Cabbage	Carrot	Rutabaga	Lima bean
	Cauliflower	Chard	Summer squash	Parsnip
	Celery	Cucumber	Turnip	Pumpkin
	Chinese cabbage	Brinjal		Winter squash
	Endive	Muskmelon		Sweet potato
	Garlic			Tomato
	Leek			
	Parsley			
	Potato			
	Palak			
	Spinach			
	Radish			
	Cowpea			

The relationship between rooting depth and water requirement is not always direct. Some shallow-rooted crops, like cabbage, cauliflower, spinach and onion require relatively lesser quantity of water as compared to the crops with deep root system like, tomato, muskmelon, etc., which is quite obvious. On the other hand, water requirement of some shallow-rooted crops, like potato, celery, etc., is much higher than some very deep-rooted crops like sweet potato, artichoke, lima bean, etc. Similarly, water requirement of very shallow-rooted crops like onion and lettuce is same as that of the moderately deep to deep-rooted crops like pole bean, beet, carrot, cucumber, pea, chilli, etc. In most of the vegetable crops, a great proportion of the total root weight is frequently found to occur near the upper surface of the soil. Hence, the deep-rooted vegetable crops often extract water mostly from the upper layers of the soil. Water and nutrient absorption also decreases rapidly with the depth of roots. Moisture extraction depths of shallow- and deep-rooted vegetable crops are sometimes equated. In such situations, water requirement of the shallow- and deep rooted vegetable crops differs with the parameters other than the rooting depths, like water use efficiency, duration of the crop, consumptive use, etc. Generally, shallow-rooted crops grown in sandy soil require moderate and frequent irrigation while deep-rooted crops in clayey soils need heavy irrigation at longer intervals.

8. Plant Nutrients and Integrated Nutrients Management

Plants obtain from soil, besides water, different inorganic substances that are essential in the general metabolism of the plants. The supply and absorption of inorganic elements needed for growth and development is defined as nutrition, and these elements are called plant nutrients. An essential nutrient element may be regarded as one which is required for normal life processes and whose functions cannot be substituted by other elements.

Carbon (C), Hydrogen (H), Oxygen (O), Nitrogen (N), Phosphorus (P), Potassium (K), Calcium (Ca), Magnesium (Mg), Sulphur (S), Iron (Fe), Manganese (Mn), Copper (Cu), Zinc (Zn), Molybdenum (Mo), Boron (B) and Chlorine (Cl) are known to be essential for higher plants. Sodium (Na), Silicon (Si) and Cobalt (Co) have been found to be essential in some higher plants and hence, cannot be regarded as essential for all the higher plants. Growth and yield get restricted; any if of these elements falls short hence, balanced nutrition demands supply of all the essential elements. Contrast to the essential elements, there are some, like Aluminium (AI), Nickel (Ni), Selenium (Se) and Fluorine (Fl) which are toxic to the plants.

Classification of plant nutrients

1. Based on the quantity used by the plant

These classifications of plant nutrients are based on the nutrient element content in the plant body.

Macronutrients: Those nutrients which are used by the crop in higher quantities *e.g.,* N, P, K, Ca, Mg, and S (Na and Si may also be regarded as macronutrients for the plant which requires those nutrient elements). The elements carbon, hydrogen and oxygen are very much needed for plant body but these are not essentially supplied, rather crop plants get these essential elements from nature itself as C and O from CO_2 and O_2 from the atmosphere and H from H_2O in soil or precipitation.

- **Primary nutrients:** Those nutrients which are used in considerable quantities by all the crops e.g., N, P, K.
- **Secondary nutrients:** Those nutrients which are used in moderate quantities by the crops and influence soil pH. e.g., Ca, Mg, S.

Micronutrients: Those nutrients which are essential but used by the crop in relatively smaller quantities by the crops e.g., Fe, Mn, Cu, Zn, Mo, B and Cl (CO is also micronutrient for the plant which needs it).

However, these classifications of plant nutrients are somewhat arbitrary. For example, the Fe and Mn contents in plant tissues are sometimes found as high as that of S or Mg. Again, comparatively higher concentration of Cl is present in many plants but this nutrient element is needed in minute quantities for its specific role in photosynthesis.

2. Based on physiological point of view

Group 1: This group contains C, H, O, N and S which are the major constituents of the organic materials of the plant body. Carbon is taken up in the form of CO_2 from the atmosphere at the time of photosynthesis, and possibly in the form of HCO_3^- ion from the soil solution. Hydrogen is taken up in the form of water which is reduced to H+ ion (photolysis) at the time of photosynthesis. Nitrogen is taken up in the form of NO_3^- or NH_4^{++} ions from the soil solution. The leguminous plants may get elemental N_2 gas from the soil air through the symbiotic nitrogen fixation by *Rhizobium* bacteria. Certain free living bacteria *(Azotobacter; Clostridium butyricum),* blue-green algae and fungi are able to fix elemental N_2 from soil air into their body tissues which in turn enrich the soil with nitrogen which is called non-symbiotic nitrogen fixation. Sulphur is not only taken up in the form of $SO_4^=$ or SO_3^- ion from the soil solution but can also be absorbed as SO_2 gas from the atmosphere.

Group 2: This group contains P, B and Si. Phosphorus is taken up in the form of $H_2PO_4^-$, $HPO_4^=$ or $PO_4^=$ ions, boron in the form of boric acid (H_3BO_3) or borate ($B_4O_7^=$) ions, and silicon in the form of silicate ions from the soil solution.

Group 3: This group contains K, Na, Mg, Ca, Mn and Cl and these nutrient elements are taken up in the form of their ions from the soil solution.

Group 4: This group contains Fe, Cu, Zn and Mo. These elements, with the exception of Mo, are taken up as chelate complexes (an organic compound which holds tightly certain cation attached to it) from the soil solution in addition to their ionic forms. Molybdenum is taken up in the form of molybdate (MoO_4^-) ions from the soil solution.

Nutrient requirements of vegetable crops

Nutrient requirement is basically a genetic characteristic of the crop plant, and this requirement may vary with the cultivars of the particular crop. Nutrient removal by high yielding cultivars and hybrids is much higher than what are removed by the low yielding cultivars of the same crop. Again, nutrient removal by any crop depends on the nutrient availability and their absorption which is influenced by soil pH, moisture and soil temperature. In general, vegetable crops exert tremendous pressure on the soil for nutrients due to their high productivity. When the crop is harvested it marks the removal of plant nutrients from the soil. There is no effective way to reduce the nutrient demand because nutrient requirement at a given yield level is quite stable. Nutrient removal is perhaps the most critical factor when sustainability of a farming system is considered. If the nutrients removed from the field are not replaced, the fanning system will not remain sustainable due to decline in soil productivity. Some idea of the optimum fertility levels needed for vegetable production can be obtained from the quantities of nutrients removed by the crops.

Removal of primary nutrients from the soil by vegetable crops

Crops	Nutrient removal (Kg/100 kg yield)		
	Nitrogen	Phosphorus	Potassium
Potato	0.74	0.28	1.39
Tomato	0.33	0.11	0.44
Brinjal	0.29	0.08	0.50
Cabbage	0.53	0.16	0.50
Cauliflower	0.40	0.16	0.50
Knol khol	0.50	0.42	0.85
Carrot	0.42	0.18	0.67
Radish	0.60	0.30	0.60
Beet	0.24	0.16	0.45
Onion	0.30	0.13	0.40
Leek	0.34	0.13	0.80
Cucumber	0.18	0.13	0.30
Pumpkin	0.18	0.14	0.32
Muskmelon	0.36	0.11	0.64
Bottle gourd	0.15	0.16	0.16
Pea	1.25	0.45	0.90
Beans	0.87	0.27	1.07
Okra	0.30	0.13	0.45
Celery	0.67	0.27	1.01
Lettuce	0.30	0.12	0.53
Spinach	0.50	0.15	0.30
Asparagus	2.40	1.20	3.00
Cassava	0.38	0.18	0.88
Sweet potato	0.48	0.19	0.85
Elephant foot yam	0.34	0.08	0.49
Yam	0.57	0.14	0.61

Vegetable crops may be categorized into four groups based on the quantity of nutrient removal *viz.,* very high removal (asparagus, peas, beans, potato, celery, etc.), high removal (cabbage, cauliflower, knol khol, radish, sweet potato, yam, spinach, carrot, etc.), moderate removal (onion, garlic, tomato, brinjal, beet, leek, muskmelon, okra, cassava, elephant foot yam, lettuce, etc.) and low removal (cucumber, pumpkin, bottle gourd, etc).

From the pattern of primary nutrient removal by different vegetable crops, it appears that removal of phosphorus is comparatively low while that of potassium is luxuriantly high. This does not mean that application of very high potassium will give good yield in all the vegetable crops. Again, plants require more phosphorus than its removal for giving optimum yield. On the other hand, they require as much sulphur as phosphorus. Nutrient requirements depend on the specific demands of roots, shoots, leaves and fruit development of the crop, although all the vegetable crops are very responsive to nitrogen application. In general, the vegetable crops whose above ground vegetative parts are consumed (cabbage, cauliflower, knol khol, amaranth, palak, spinach and other leafy vegetables, lettuce, celery and other salad crops, etc.) require more nitrogen; the bulb, root and tuberous vegetable crops (carrot, radish, beet, turnip, onion, garlic, leek, shallot, potato, sweet potato, elephant-foot yam, yam, taro, etc.) require more potash; and fruit vegetable crops (tomato, brinjal, chilli, okra, pumpkin, bottle gourd, bitter gourd, ridge gourd, summer squash, cucumber, gherkin, watermelon, muskmelon, etc.) require more phosphorus. Peas and beans require high nitrogen but most of the required nitrogen is supplied through symbiotic nitrogen fixation by *Rhizobium* bacteria in their root nodules. In these crops, a low dose of nitrogen should be applied as basal to stimulate early growth. Cabbage, cauliflower, radish, onion, turnip require comparatively more sulphur as the compounds which impart characteristic odour to these vegetable crops contain sulphur. Removal of sulphur from soil per 100 kg yield by some vegetable viz., cabbage, cauliflower, tomato, onion, radish and cassava is 0.11 kg, 0.08 kg, 0.06 Kg, 0.09 kg, 0.20 kg and 0.03 kg, respectively.

Evaluation of soil fertility

Plant tissue nutrient analysis: The basic principle behind this technique is that the nutrient concentration of plants is related to the amount of nutrient element available in soil. General range of nutrient content in fully developed leaves of vegetable crops is given below. If the nutrient level in the tissues falls below the critical concentration, the soil may be deficient in that element for optimum plant growth and deficiency symptoms in plant may occur.

Sufficient and critical level of nutrients in the leaf tissue of vegetable crops (general level)

Nutrient		Per cent or ppm of dry weight of leaf	
		Sufficient level	Critical level
Nitrogen	(%)	1.50-4.0	1.00-1.50
Phosphorus	(%)	0.25-0.80	0.20-0.25
Potassium	(%)	2.00-9.00	1.50-2.00
Calcium	(%)	0.35-2.00	0.35-0.80
Magnesium	(%)	0.25-1.00	0.20-0.25
Sulphur	(%)	0.16-0.50	0.10-0.16
Manganese	(ppm)	30-200	30-50
Iron	(ppm)	50-250	50-80
Zinc	(ppm)	30-100	20-30
Boron	(ppm)	30-60	20-30
Copper	(ppm)	8-20	4-8
Molybdenum	(ppm)	0.5-5.0	0.2-0.5

Different plant parts of the same plant contain different concentrations of the same nutrient and nutrient concentration again varies with the stage of the crop. So, leaf samples for tissue analyses should be selected on the basis of specific physiological age *i.e.*, developmental stage. It is important that the samples must be free from diseases, insect damage and physical or chemical injury. Leaf near the fruit should not be sampled as the nutrients, it might have contained, are often translocated to the fruits.

Common disorders/problems of vegetables associated with imbalanced nutrition

Disorder/other problems	Due to	Remedy
Blossom end rot of tomato, sweet pepper and watermelon	Low pH and calcium deficiency in tomato, sweet pepper, watermelon and onion; In cauliflower and cabbage club root is incited by the fungus *Plasmodiophora brassicae* but aggravated by low pH and calcium deficiency.	Soil application of ground lime stone ($CaCO_3$) 10-14 days before planting
Club-root of cabbage and cauliflower		
Damping off of onion		
Black rot in cabbage and cauliflower	Incited by the bacteria *Xanthomonas campestris* pv. *campestris* but aggravated by disproportionately higher application of nitrogen	Application of P and K in proportionately higher quantity
Browning of cauliflower, hollow stem and head of cabbage and cauliflower, black heart of beet, internal browning of turnip, stem resetting of tomato, fruit cracking in tomato, cracked stem in celery	Deficiency of boron	Soil or foliar application of borax with basal NPK fertilizer
Root rot of cucumber	Deficiencies of manganese and zinc	Soil or foliar application of sulfates of manganese and zinc
Interveinal chlorosis and necrosis in older leaves in tomato, brinjal and chilli, interveinal chlorosis and puckering of older leaves in cabbage, interveinal yellow spots followed by chlorosis in leaves in okra, die back of leaves in onion	Deficiency of magnesium	Soil or foliar application of magnesium sulfate
Root splitting in radish	High ammonium in soil	Application of N through non ammonium fertilizer or organic matter.
Whiptail of cauliflower, upwardly rolled leaves with diffused marginal and interveinal yellow mottling in potato, purple young leaves in tomato, discoloured leaf and death of growing tip in radish, yellow or white veins in the leaves of pea, pale green leaves with interveinal mottling and scorching in bean, deep blue leaves with conspicuous yellow and green mottling in onion	Deficiency of molybdenum	Application of ammonium molybdate
Susceptibility of brinjal to leaf hopper of nitrogen	Disproportionately higher application in proportionately higher quantity	Application of P and K fertilizer

Soil analysis: It is a rapid chemical method and is important for predicting the fertility status of the soil. Fertilizer recommendations can be advocated judiciously if the results of soil tests are available. Soil tests and their interpretations are based on the soil samples. It is, therefore, important that the soil samples should be properly collected and should represent the area under soil test. After scrapping the surface litter, uniform core of a thin slice of soil from the surface to plough depth (15-22 cm) from 15 to 20 spots of each sampled field should be taken. When soil chemical analyses (soil tests) are employed to determine which nutrients are deficient and by how much, nutrient balances in the soil can be achieved by fertilizer application. Ranges of nutrient content commonly found in mineral soils are given below.

Range in nutrient content commonly found in soils

Nutrients	Normal range	
	Per cent	ppm
Nitrogen	0.02-0.50	200-5000
Phosphorus	0.01-0.20	100-2000
Potassium	0.17-3.30	1700-33000
Calcium	0.07-3.60	700-36000
Magnesium	0.12-1.50	1200-15000
Sulphur	0.01-0.20	100-2000
Iron	0.50-5.00	5000-50000
Manganese	0.02-1.00	200-10000
Zinc	0.001-0.025	10-250
Boron	0.0005-0.015	5-150
Copper	0.0005-0.015	5-150
Chlorine	0.001-0.10	10-1000
Cobalt	0.0001-0.005	1-50
Molybdenum	0.00002-0.0005	0.2-5

Source: The Nature and Properties of Soil (N C. Brady)

Percentage or ppm of nutrient element can be converted to kg of nutrient per hectare *e.g.,* the soil containing 0.1 per cent potassium means 1 g potassium per kg of soil. One hectare furrow slice, about 15 cm deep of upland mineral soil weighs about 2,240,000 kg. So, the conversion of nutrient is 2,240 kg of potassium per hectare.

Based on the soil tests, soils can be classified into three fertility level categories *viz.,* high, medium and low. Fertilizer recommendations should be based on this soil fertility level.

General rating chart of the soil test data for N, P and K

Fertility level	Organic C (%)	Available N (kg/ha)	Available P_2O_5 (kg/ha)	Available K_2O (kg/ha)
High	> 0.75	> 450	> 90	> 340
Medium	0.50-0.75	280-450	45-90	150-340
Low	< 0.50	< 280	< 45	< 150

Management of nutrients in soil

Organic matter

Soil organic matter consists of decomposed plant and animal residues and numerous dead microorganisms particularly which are nitrogen fixers (both free living and symbiotic) which are very active and important portion of soil. After active decomposition, the organic matter residues are collectively called humus. Beneficial effect of soil organic matter can be summarized as:

1. It is the nitrogen reservoir and source of 90-95 per cent nitrogen in unfertilized soil.
2. It supplies large portions of soil phosphorus and sulphur. It is also the source of other macro- and micronutrients.
3. It supplies cementing substances for desirable aggregate formation of soil.
4. It loosens up the soil to provide better aeration and water movement mainly due to soil aggregation which increases soil pores and water holding capacity of the soil is thereby increased.
5. The large available surface of humus has many cation exchange sites that absorb nutrients for eventual plant use.
6. Organic matter acts as a chelate and make bond to different micro-nutrient elements like Fe, Zn, Cu, and Mn. The soluble chelate helps to mobilize these micronutrients in soils and increase their availability to plants.
7. It supplies carbon to many microbes that perfom other beneficial functions in soils and growth substances produced by the micro-organisms in soil help plant growth in many ways.
8. It reduces soil erosion, shades the soil, and keeps the soil cooler in very hot weather and warmer in winter.

For maximum benefit, organic matter must be readily decomposable and continuously replenished with fresh residues. In cultivable soil, organic matter content ranges from less than 1 per cent to 15 per cent. The majority of the tropical and subtropical soils like ours have low humus content which is the consequence of its quick year round mineralization under the influence of tillage and favourable weather condition. Moreover, vegetable crops are generally cultivated intensively, and the crop residues are not sufficient to replace the organic matter lost from the soil. The soil organic matter can be replenished by adding farmyard manure, oilcakes, compost, green manure and other bulky organic residues.

Mineral values of some commonly used organic resources

Organic resources	Nutrient content (%)		
	N	P_2O_5	K_20
FYM	0.5-1.0	0.3-0.5	0.4-1.0
Compost (urban)	1.2-2.0	1.0	1.5
Compost (rural)	0.4-0.8	0.3-0.6	0.7-1.0
Slurry from gobar gas plant	16- 1.8	1.1-2.0	0.8-1.2
Dhaincha (green manure)	0.6	-	-
Cowpea (green manure)	0.7	0.1	0.6
Mustard cake	5.1	1.8	1 .1
Neem cake	5.2	1.0	1.4
Groundnut cake	7.0	1.5	1-3
Linseed cake	5.5	1.4	1.2
Mahua cake	2.5	0.8	1.8
Castor cake	5.5	1.9	1.0
Sludge	2.5	1.0	0.4
Activated sludge	4.0- 7.0	2.1-4.2	0.5-0.7
Vermi compost	1.0	0.5	1.5

Different manures and oilcakes should be applied and ploughed under the soil well in advance of seeding or planting (15 to 20 days) because of the slow availability of nutrients from the manures.

Composting is the process of reducing plant and animal refuse to a quickly utilizable form for improving and maintaining soil fertility. Raw vegetable residues are rich in cellulose and other readily decomposable carbohydrates, and have a high carbon-nitrogen ratio of 40: 1 or more. Direct application of such undecomposed, low nitrogen organic matter as manure brings about a temporary deficiency of nitrates in soil by stimulating the growth of microorganisms, which in turn compete with the crop for nitrogen, phosphorus and other nutrient elements. For this reason, the organic wastes are allowed to decompose before application and thereby the carbon nitrogen ratio is brought down to 10-12: 1 during the process of composting.

When green manuring crops are turned under the soil for decomposition, various nutrient elements in the plant body are released and returned to the soil. A fairly good quantity of organic matter is also added in this process. Leguminous green manuring crops like cowpea, dhaincha (*Sesbania* sp.), rice bean, clover, etc. are preferred to non-legume crops because legumes supply substantial quantity of nitrogen to the soil. Two important considerations with respect to green manuring are: (i) the green manuring crops should be turned under before they are matured because succulent plant materials decompose quickly, and (ii) the green manuring plants should be turned under well in advance of planting the successive crop, otherwise (a) the micro-organisms that decompose the plant materials consume soil nitrates and may depress the early crop growth, and (b) nutrients that are released from the plant materials in the course of decomposition will not be available for early crop growth. Balanced fertilization with P, K, Cu, Mo and Fe to the leguminous green manuring crops is essential for most efficient nodule activity and nitrogen fixation.

Most soils under continuous tillage for production of crops like vegetables lose organic matter at the rate of approximately 1-2 per cent per year. In this situation, returning of the crop residues judiciously is beneficial because it helps in aggregation of soil particles, improving water infiltration and aeration, in reserving plant nutrients and also acts as energy reservoir for the micro-organisms.

Macronutrient through fertilizer

In natural ecosystems, the minerals absorbed by the crops return to the soil after organic matter decomposition, and soil fertility is more or less maintained through nutrient cycling. In cultivated ecosystems like vegetable cultivation, however, all harvested biomass (product and plant residue) withdrawn from the field contains nutrients that no longer return to the soil. Hence, maintenance of soil fertility and crop yield depend on counterbalancing of fertilizer inputs. Chemical fertilizers are inorganic or synthetic organic materials of concentrated nature. They contain one or more plant nutrients in easily soluble and quickly available forms, and being concentrated, fertilizers have the advantage of smaller bulk with resultant economy and ease in storage, transport and handling.

Nitrogen: It is most often the limiting nutrient element in plant growth and the vegetable crops require more nitrogen than any other nutrients. Nitrogen is highly mobile and easily lost from the soil. In cultivable land, crops can utilize only 50-60 per cent of the available nitrogen. When the nitrogen sources are applied to the soil, they all are eventually converted to NH_4^{++} and NO_3^- forms of nitrogen. Plant roots take up these two inorganic forms of nitrogen almost exclusively. Nitrate, if left unused, is lost through downward leaching as negatively charged NO_3^- is naturally repelled by soil particles and remains free to move with soil water, but positively charged NH_4^{++} is held by soil particles and generally, are not subjected to leaching. However, both NO_3^- and NH_4^{++} forms may be consumed by micro-organisms or converted to gaseous nitrogen forms (N_2 or NH_3) and lost to atmosphere. Nitrogen use efficiency in vegetable crops can be increased by adopting the following measures:

1. Split application of nitrogen fertilizers at the time of peak requirements of the crop which decrease nitrate leaching and thereby increase nitrogen use efficiency. In general, half the dose of nitrogen fertilizers is given as basal at the time of final land preparation and the rest half is applied in different splits. However, in leguminous vegetable crops, entire nitrogen fertilizer is applied as basal dose. In other fruit vegetable crops, like tomato, brinjal, chilli, okra, etc. split applications after periodical harvests prove beneficial.
2. Nitrogen fertilizer should be applied 5-10 cm deep in the soil. When nitrogen fertilizers are applied deep in the soil, plants can take 30-35 per cent more nitrogen than when applied on the soil surface.

3. Optimum soil moisture increases nitrogen use efficiency.
4. Nitrogen fertilizers should not be applied in large quantities near the root zone because of the danger of salt damage.
5. Nitrogen use efficiency often depends on the interaction among other nutrients, particularly phosphorus and potassium. In general, when adequate P and K fertilizers are applied to soil, the most efficient use of nitrogen results. So, when P and K fertilizers are reduced, the supply of N fertilizers should also be adjusted accordingly because the resulting root system would not be able to absorb nitrogen more than its capacity.

Phosphorus: Young plants assimilate phosphorus very rapidly and thereby need water soluble forms that migrate to the initial meagre root system. Plants generally accumulate as much as 75 per cent of their phosphorus need in the early stages of growth. This emphasis's the importance of ample supply of water soluble phosphorus in the soil zone of the germinating seeds. Most soluble phosphorus ($HPO_4^{=}$, $H_2PO_4^{-}$) become fixed (precipitate and form insoluble compounds) before plant can absorb them, and hardly 10-15 per cent of the applied phosphorus is utilized by the plants. So, leaching loss of phosphorus is practically nil, and it is lost primarily by erosion of surface soil. In general, phosphorus is more available at a pH of 6.0 to 6.5 than at higher or lower pH values. In acidic soils, low availability of phosphorus is common due to formation of complex compounds of phosphorus with aluminium and iron. Phosphorus use efficiency can be increased by adopting the following measures:

1. Maintenance of soil pH at 6.5 to 7.0.
2. Phosphatic fertilizers should be applied totally as basal near the active root zone of the plant.
3. It should not be applied very deep in the soil, rather shallow band placement (2.5-5 cm below and 5-6 cm away from seeds) in rows is very effective.
4. Late application of phosphorus is not effective for plant growth.
5. Adequate moisture content in soil and combined application of N and P fertilizers increase phosphorus availability to the plants.
6. Maintenance of organic matter in soil increases P availability to the plants.
7. In cold weather, more phosphorus fertilizer should be applied as absorption of P is less in cold weather.
8. Rock phosphate or basic slag should be applied 2-4 weeks before sowing or transplanting.

Potassium: It occurs in plant body only as a mobile ion K^+ rather than as an integral part of any specific compound. In vegetable crops, luxury consumption of potassium is a very common feature. Rapid uptake of this nutrient occurs in the early stages of growth and it usually occurs faster than either N or P. So, ample potassium should be made available to the seedlings. Potassium is less mobile than nitrogen fertilizers but more mobile than phosphates, and are generally applied in full dose at the time of sowing/transplanting. Potassium fixation embraces (a) immobilization into microbial bodies and (b) becoming entrapped in the clays. It is also leached substantially through the downward movement of water. For this reason, it often falls deficient in high rainfall areas. Potassium use efficiency can be increased by adopting the following measures:

1. Split application of potassic fertilizers, particularly in loose soils of high rainfall areas.
2. Maintenance of soil pH at 6.0-7.0 by liming which reduces leaching.
3. Returning of crop residue and application of manures which add much potassium.

General fertilizer guide (Kg nutrient / hectare) for some vegetable crops for the soil with moderate fertility status

Crops	Seed source					
	Improved open pollinated variety			Hybrids		
	N	P_2O_5	K_2O	N	P_2O_5	K_2O
Potato	150	100	100			
Tomato	100	80	80	200	100	100
Brinjal	100	80	60	200	100	100
Chilli	100	60	60	150	80	80
Sweet pepper	120	80	60	200	150	100
Cauliflower	125	80	60	200	100	100
Cabbage	150	60	80	200	125	150
Onion	100	60	100			
Garlic	125	60	60			
Okra	125	75	60	200	100	100
Pumpkin	100	60	80			
Bottle gourd	60	60	80	200	100	100
Bitter gourd	100	50	50	200	100	100
Cucumber	100	60	60	150	90	90
Pointed gourd	150	60	40			
Ridge gourd	100	60	60	150	90	90
Hyacinth bean	30	60	60			
Cowpea	25	75	60			
Radish, beet, carrot, turnip	100	60	100			
Amaranth	90	50	50			
Palak	60	60	40			
Sweet potato	60	60	90			
Colocasia	100	60	80			

Sulphur: Crops require as much sulphur as they do phosphorus and therefore, sulphur is called the fourth major nutrient. Some vegetable crops like cabbage, onion, turnip, etc. have high sulphur requirement. Sulphur content in the soil is being decreased gradually because (a) modern high analysis fertilizers are free of sulphur, (b) modern plant protection chemicals are sulphur-free compounds. Organic sources of sulphur are about 70-90 per cent and rainfall also dissolve atmospheric SO_2 evolved during burning of fuels and act as a source of sulphur. Leaching losses of sulphur from the soil are very much like nitrogen and caused by percolating water through light-textured soils. Soil acidity also affects the availability of sulphur due to fixation of sulphate ion on hydrated oxide. There could be sulphide toxicity due to reduction of sulphur to hydrogen sulphide (H_2S). Sulphur can be managed in the soil through the following measures:

1. Maintenance of soil pH to 6.0-7.0 by liming.
2. Application of organic matter to the soil.
3. Application of sulphur-containing fertilizers, like ammonium sulphate, superphosphate, potassium sulphate, etc. periodically on acid and sandy soil low in humus.
4. In acute deficiency, elemental sulphur may be applied before sowing/transplanting. Application of elemental sulphur may cause soil acidity, particularly in high rainfall areas where acidity is a problem. So, at the time of sulphur application in this type of soil, application of calcium in the soil must be ensured in the form of lime and other fertilizer sources like calcium ammonium nitrate, rockphosphate, etc.

Calcium: It is needed in large quantities for cell division. It is seldom deficient except in sandy and strongly acidic soils. However, calcium is lost in a number of ways like, crop removal, leaching and soil erosion. Leaching of bases like calcium by percolating water coupled with nitrogen and sulphur fertilization contributes to natural soil acidity. Calcium can be managed in the soil by liming.

Magnesium: In light-textured soil, magnesium is leached through the percolating water. Magnesium can be managed in the soil by application of dolomite ($CaCO_3.MgCO_3$) which contains 12.0 per cent magnesium or magnesium sulphate containing 20.2 per cent magnesium.

Micronutrients through different chemicals

The main reasons of accelerated exhaustion of available micronutrients in the soil are (i) cultivation of high-yielding cultivars and hybrids which demands more macro and micronutrients, (ii) application of high dose of N, P and K fertilizers as it decreases the availability of micronutrients, (iii) inadequate application of organic matters to the soil, (iv) intensive cropping which creates micro-nutrient deficiency, (v) leaching loss of micronutrients particularly in light-textured soils of high rainfall areas, (vi) increased application of compound fertilizers because straight fertilizers may add some amount of micronutrients *e.g.,* superphosphate contains 26 ppm Cu, 50 ppm Zn, 65 ppm Mn, 9.5 ppm B and 3.3 ppm Mo. On the other hand, compound fertilizers do not contain such micronutrients, (vii) maximum application of phosphatic fertilizers decreases the availability of zinc, whereas application of ammonium sulphate or other sulphate compounds increases the availability of zinc, (viii) inadequate cultural practices decrease the availability of micronutrients because all micronutrient elements but molybdenum remain within few inches from the surface soil and better tilth increases the availability of micronutrients, and (ix) both moisture stress and excess condition influences the availability of micro nutrients. Moisture stress condition decrease the availability of Fe and B while excess moisture condition causes Zn deficiency.

Vegetable crops respond very well to the application of micronutrients under Indian condition. Generally, micronutrients are applied in four different ways:

Soil application: Micronutrient containing chemical compounds can well be applied to soil mixed with other fertilizers. Care should be taken to avoid excess application of micronutrients because optimum dose depends on soil type, crop, etc. As for example, higher rates of manganese are required for both organic soils and sandy soil that are alkaline in reaction and high rate of zinc is applied when soil is neutral to alkaline in reaction.

Seedling root dipping: This method of application is not widely practiced. Generally, 0.2-0.3 per cent solution of zinc sulphate is used for root dipping.

Seed treatment: Seeds may be sown after treating with the chemical compounds containing Cu, Fe, Mo, Zn, B, Mn, etc. Seed treatment of pumpkin and squashes with 0.5% borax or manganese sulphate increases staminate flower production and thereby yield.

Soil status and guide for soil application of chemicals for secondary and micronutrients

Micronutrients	Critical limit(mg/g)	Recommended dose of chemicals
Zinc	0.6-1.0	Zinc sulphate ($ZnSO_4$) @ 25 kg/ha with organic manure
Boron	0.5	Borax ($Na_2B_4O_7.10H_2O$)@ 10 kg/ha with basal NPK fertilizer
Molybdenum	0.2	Ammonium molybdate ($(NH_4)_6Mo_6O_{24}.4\ H_2O$ @ 1.0 kg/ha with basal NPK fertilizer
Iron	2.5-4.5	Ferrous sulphate ($FeSO_4$) @ 50 kg/ha with organic manure
Manganese	1.0	Manganese sulphate ($MnSO_4$) @ 30 kg/ha with basal NPK fertilizer
Copper	0.2	Copper sulphate ($CuSO_4$) @ 5 kg / ha with basal NPK fertilizer
Sulphur	10-12	Elemental sulphur (S) @ 25-45 kg / ha

A general guide for application of micronutrients as foliar spray

Micronutrients	Concentration of the chemicals
Iron	0.4% ferrous sulphate + 0.2% lime*
Manganese	0.4%-0.6% manganese sulphate + 0.2%-0.3% lime*
Zinc	0.2%-0.6% zinc sulphate + 0.1-0.3% lime*
Copper	0.1-0.2%,copper sulphate + 0.5% lime*
Boron	0.2% -0.3% borax
Molybdenum	0.05% sodium or ammonium molybdate

*Lime is added to neutralize the solution, otherwise leaves may get scorched. Minimum 450 litres of water must be used per hectare.

Foliar spray: Most of the micronutrients after foliar spray enter the plant body through leaves although time required for absorption is few hours to one day depending on the nutrient. Foliar application of micronutrients is widely used and at the same time suitable because of (i) convenience in application, (ii) requirement of small quantity, (iii) quick correction of deficiency, and (iv) avoidance of fixation of the micronutrients in soil. Lime should be added to neutralize the solution of ferrous sulphate, manganese sulphate, copper sulphate and zinc sulphate, otherwise foliage may get scorched. Foliar sprays may be given 2-4 times at an interval of 7 -10 days to correct the deficiency of particular micronutrient.

Integrated nutrient management for vegetable crops

Indian population is at present more than 110 crore and expected to the tune of 130 crore by 2025. Such increase in population will magnify the demand for different food production including vegetables from the rapidly shrinking cultivated land area. To produce the food, agricultural operation needs to be intensified using huge amount of fertilizer and other inputs of production. Such intensification of agriculture has its attendant problems of salinization, acidification, desertification, nutrient depletion, loss of soil biodiversity, soil erosion etc. For sustaining productivity on long-term basis because of irreparable damages it causes on natural resource base and in some cases, because of intensive cropping particularly with vegetable crops, soils have become "talcum powder" like structureless materials losing retentivity for nutrients as well as water. Besides, use of enormous fertilizer inputs substantially enhanced the production cost, deteriorated soil and water quality and increased the greenhouse gas effect making the production system unsustainable. Such intensified production system has also been implicated in over-taking the resilience of agricultural soils and those natural processes that maintain its health and quality. Sometimes the stress exceeds the resilience capacity of the soil so that restoration becomes very expensive or even impossible. Since, the long-term performance of intensive agricultural system is important for poverty alleviation and sustaining food production in the face of continuous population growth, it is needed to redo the present way of cultivation with only inorganic fertilizer and other chemicals to save the natural resource base, particularly soil, for posterity.When these environmental problems in intensive agriculture came into spotlight, organic farming was advocated as a solution for most of the problems encountered. However, the organic farming is not only a very low yielder but has many of the problems of conventional high intensive agriculture. Thus organic farming methods do not offer solution to many of the problems. To achieve sustainability in production system and maintain natural resource base in good stead, blending or integrating organic, inorganic, biological and mechanical inputs with ecological principles in mind may be an option in the concept of "integrated nutrient management" (INM) for vegetable crops.Sustainable nutrition of crops is based on the principle of a balance between availability of nutrients and requirement by the crops. In such a balanced situation, involving fine tuning towards weather and soil conditions, emphasis is given to the best time of application of nutrients, the best organic and mineral composition, the best placement of the nutrients and to the environment adapted crops/varieties. IPN enhances crop production and maintains its consistency over long periods without disrupting ecological base of farming. Once IPN is adopted, it becomes possible to cut-off the use of chemical fertilizers, other nutrient supplying chemicals and plant protection chemicals over a period of time. IPN for sustainable nutrient management may be accomplished through (a) proper crop rotation, (b) soil protection in dry weather, (c) covering the soil by a live canopy or mulch, (d) application of nutrients only as per crop requirements, (e) simultaneous application of organic matters with mineral nutrients, (f) application of quick release nutrients only as top dressing in. rows, (g) use of biofertilizers (h) integrated pest and disease management, (i) enrichment of organic matters by adding mineral additives and microbial inoculants during composting and (j) balanced application of nutrient.

Components of INM

Proper management of plant nutrients is necessary to ensure higher productivity, balanced nutrition and for checking wastage of costly fertilizer nutrients as well. The basic concept underlying the integrated nutrient management system, nevertheless, is the maintenance and possible improvement of soil fertility for sustained crop productivity on long-term basis and also to reduce fertilizer input loss and environmental degradation.

A delineation of INM components

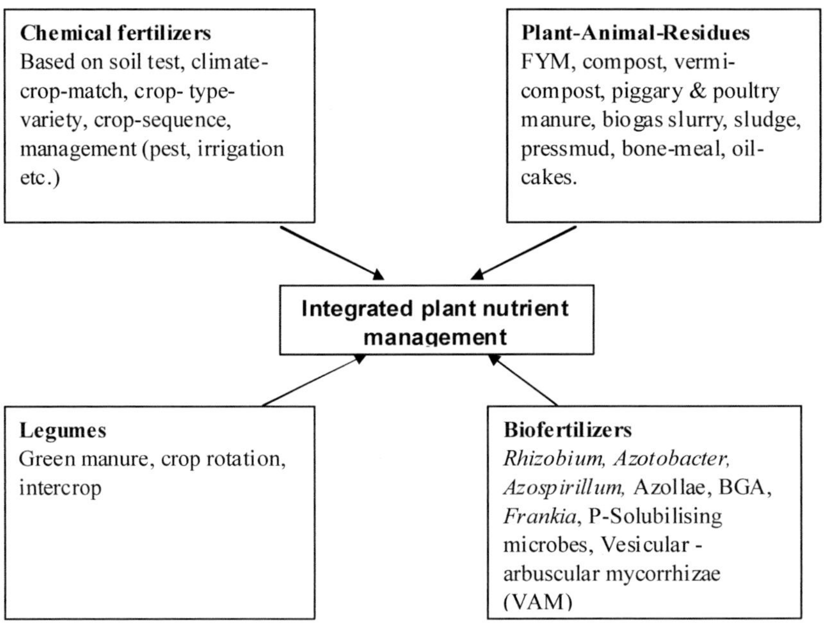

The judicious combination of organic and inorganic resources for balanced plant nutrient management is necessary since neither inorganic fertilizer nor the organic manures alone can bridge the gap between sustaining productivity and up-keeping soil health and resilience.

Biofertilizers are the cultures of microorganisms used for inoculating seed and soil or both under ideal condition to increase the availability of plant nutrients. They are cost-effective, eco-friendly and renewable sources of plant nutrients used to supplement chemical fertilizers in sustainable production system. Some common biofertilizers of agricultural importance are: *Azolla*, Blue green algae, *Rhizobium*, *Azotobactor*, *Acetobactor*, phosphate-solubilizer and sulphate-solubilizer. The free living N-fixers like *Azotobacter*, *Azospirillum* and the Phosphate solubilizers (*Bacillus polymyxa*, *Bacillus negatherium*, *Pseudomonas striata*, *Aspergillus awamonii*, *Penicillium digitatum*, etc.) are commonly used to vegetable crops through i) seed inoculation @ 1.5 kg biofertilizer dissolved in water with starch/molasses for one hectare, ii) seedling root soaking @ 1.5 kg biofertilizer dissolved in 35-40 litre of water for soaking seedlings required for one hectare for 30 minutes and iii) soil application @ 2 kg biofertilizer + 200-250 kg organic manure for one hectare during land preparation.

9. Micro-irrigation and Fertigation

In India, around 88 per cent water is being used in the agricultural sector, covering more than 80 million hectares land under irrigation. However, due to increasing demand in industrial as well as domestic sectors, pressure on the availability of water for agricultural sector has been accentuated many fold in the recent times. As per the estimate of the Central Water Commission, the utilizable water in India from surface structure is about 690 cubic km and replenishable ground water potential is 452 cubic km and this ground water should not be regarded as a long-run depletable resource. Over the years, due to frequent incidence of drought and over exploitation of ground water, the water table has been declining in many part of the country and such situations are most alarming in Punjab, Haryana and Western Uttar Pradesh. Considering its scarcity in future, planning and management of this natural resource and its optimal, economical and equitable use has become a matter of utmost urgency.

Micro-irrigation

Micro-irrigation may be an effective tool serving the purpose as this system considerably saves water besides enhancing quality and yield of the produce. Micro-irrigation is a system that provides high frequency application of water in and around the root zone of the plant with the help of emitters/drippers. It has the potential to acieve significant water savings, along with improved crop quality and yield. In this system water is applied to each plant directly to the root zone, in a limited volume of soil which maintains optimum soil moisture for crop production but avoids waste by delivering small, precise amount of water on a frequent basis only to the root zone of the plant. Dramatic improvement in irrigation efficiency has been observed with the introduction of micro-irrigation by use of spitters (micro-jets) and minisprinklers and of drip irrigation. Drip or trickle irrigation refers to the frequent but slow application of water directly to the base of the plant. With drip irrigation, it is possible to reduce soil moisture fluctuations to a minimum, enabling optimum growing condition for plant. Such low flow rate emitters may create saturation in part of the wetted volume of soil. Pressure in drip lines varies from 5 to 10 psi and length of drip lines may be as long as 600 ft but 200-250 ft is more common. Rate of water application is about ¼ to ½ gallon/minute/100 ft of row. Most of the perforated pipe designs use PVC or plastic pipe of 1-1.2 cm diameter as laterals. The perforation are punched, drilled or pierced the diameter of perforation varies from 0.07 to 0.1 cm.

This irrigation system consists of a network of pipes along with a suitable emitting device. A typical micro-irrigation system has dripper/micro-sprinkler/sprayer, distribution, lines and fittings, control head system, fertilizer tank and fittings. In India, this technology at farmer's level was introduced around 1980 although this concept of irrigation was first visualized in the early 1960 in Israel. In last 25 years or so, growth of micro-irrigation has picked up momentum although far behind than the envisaged requirement mainly because of high installation cost and lack of supporting infrastructures.

Advantages of drip irrigation

Micro-irrigation has wide applicability in horticulture, including vegetable crops. Vegetables are fast growing, vigorous and most of the feeder root systems are confined only on the upper layer of soil thus, vegetable crops are very sensitive to water stress. Different experiments indicate that drip irrigation system can save water up to 80% with increase in yield up to 88% depending on the crop, besides improving the quality of the produce at the same time saving in power, labour and inputs like fertilizers and plant protection chemicals. In total, there is a saving of 70 per cent in irrigation cost.

Comparative advantages of drip irrigation over conventional surface irrigation method.

Variables	Drip irrigation	Surface irrigation
Water saving	High to the tune of 40-100%	Less due to evaporation, run-off, percolation, etc.
Irrigation efficiency	80-90%	30-50%
Input cost	Less requirement in labour, fertilizer, pesticides and tilling	Comparatively higher requirement in all the inputs
Weed problem	Almost nil	High
Water quality	Even saline water can be used	Only normal water can be used
Disease and pest problem	Relatively less	High
Water logging, runoff	Nil	High
Water control	Better and easy	Less
Fertilizer-use efficiency	Very high and regulated supply	Heavy loss due to leaching
Range of applicability	Wide range of soil	Not suitable for sandy and undulated soil
Yield	20-100% increase	Comparatively less

Yield increase and water savings under drip irrigation over conventional surface irrigation

Vegetables	Yield increase (%)	Water saving (%)
Tomato	50	39
Brinjal	14	53
Chilli	44	62
Cabbage	2	60
Okra	16	40
Watermelon	88	46
Cucumber	45	14
Bitter gourd	39	53
Ridge gourd	17	59
Radish	2	77
Beet root	7	79

Source: Indian National Committee on Irrigation and Drainage (INCID)

Economic analysis of drip irrigation at farmer's fields indicates that cost (inclusive of subsidy) can be recovered in a period of less than 3 seasons and benefit: cost ratio is usually more than 2.5: 1 in most of the vegetable crops. The pay back period ranges from 1 to 4 years while life span of drip system is considered 8 years. The investment depends upon the kind of vegetables, spacing, quality of water required, type and discharge capacity of dripper and distance from water source. Advantages of micro-irrigation may be summarized as follows:

- Increase in irrigation and fertilizer use efficiency
- Reduce water requirement considerably
- Reduce the cost of cultivation through reducing input requirement as well as labour cost for irrigation, weeding and other inter-cultural operations
- Increase yield of crop and improve quality of produce
- Lower incidence of pest and disease and hastens maturity
- Versatility in utilization even in undulating land and in sandy soil
- Utilization of poor water quality (EC of 3 m mhos/cm) water which is unsuitable for surface irrigation in most of the crops

Limitations of micro-irrigation

In spite of technical and economic feasibility, its large scale and fast adoption is not so encouraging in India. The potentiality of drip irrigation in India is more than 27 million hectare contrary to meagre coverage of over 3 lakh hectares. Drip irrigation systems are being increasingly popular in fruit and plantation crops particularly coconut and banana and other prominent crops are mango, pomegranate, grapes, areca nut etc. The following factors have been identified as constraints in promoting micro-irrigation in India.

- High initial cost, non-availability of finance.
- Inadequate awareness and knowledge about its application, utility, method of operation and maintenance.
- Different operational problems such as
- Non-availability of spares/micro-tubes
- High cost of components
- Leakage due to damage pipes/dripper
- Blockage/clogging of emitters/microtubes
- Improper after sales service

Fertigation

In the later part of 1970, the new concept, Fertigation (fertilizers plus irrigation) has been perceived as an effective measure of applying chemicals and fertilizers to crops using modem irrigation techniques. Many crops, including most vegetable crops, extarct benefit greatly from fertilizer application through microirrigation system. Fertigation offers an opportunity for precise application of water soluble fertilizers and other nutrients to the soil at appropriate time with desired concentration. This system is capable of supplying the nutrient at desired root depth more efficiently and economically. The advantage is that the fertilizers are added in small, frequent doses, as plants need them with less chance of leaching. Fertigation has the potential to increase produce quality and yield at lower cost and using less energy. Usually nitrogen compounds particularly, calcium nitrate, ammonium nitrate, sodium nitrate, urea, diammonium phosphate, potassium sulfate, potassium nitrate and chlorate and sulphate salts of iron, zinc, copper and manganese are generally injected using drip irrigation system. Using the drip and microirrigation technique, fertilizer in liquid form is placed at the root zone of the crop along with irrigation water by achieving both fertilizer and water use efficiency. At the same time, fertilizer contamination through leaching below the plant root zone is also considerably reduced. Fertigation reduces fertilizer consumption by 30-50 per cent compared to conventional methods.

Drip irrigation used in conjunction with plastic mulches is becoming a standard practice for vegetable production in many areas of different developed country. Once a grower has installed a drip irrigation system, it makes economic and environmental sense to fertilize the crop via the system, as needed although installation cost is high. However, it does require more management and attention to details than other methods of fertilizer application. Success in using this system will depend on a sound fertility program based on soil testing and a drip irrigation system that is designed and operated properly.

Specific considerations for fertigation

- The fertilizer type used and amount required must be soluble enough to dissolve completely in the fertilizer tank water.

- The drip irrigation system needs to be completely charged or pressurized before fertigation begins. The farthest point from the pump should be at full pressure before injection of any fertilizer begins. If the system is not fully charged, air surges could occur and cause uneven application rates.
- The fertilizer should be injected ahead of the filters to ensure that any undissolved particles are filtered out before fertilizer enters the drip system.
- The period of time in which fertilizer is injected into the system must be at least as long as that required to bring the entire drip irrigation system up to full pressure which will allow each dispensing orifice in the drip line to have the same contact time with the fertilizer solution as it passes through the system.
- Any drip fertigation system should have an anti-siphon device such as a vacuum breaker or backflow preventer to protect the water supply. If backflow prevention is not in place, fertilizer that remains in the system when it is shut down will siphon back through the pump and into the water source.

Selection of a fertilizer

Requirement of fertilizer materials for fertigation depend on the actual nutrient content and solubility of the material in water. Water in which fertilizers are to be dissolved should have pH levels between 5.8 and 7.8. The solubility of some common fertilizers used in drip fertigation is given below.

Fertilizers	Nutrient composition			Salt index	Solubility (kg/100 liter of cold water)
	(%N)	(%P_2O_5)	%K_2O		
Calcium nitrate	15.5	0	0	52	102.11
Potassium nitrate	13.0	0	44.0	73	12.97
Ammonium nitrate	33.5	0	0	100	118.18
Sodium nitrate	16.0	0	0	100	73.02
Urea	46.0	0	0	75	78.19
Diammonium phosphate	16–18	46–48	0	29	42.99

To dissolve fertilizer materials having a low solubility, continuous agitation of the solution with a mechanical screw or recirculation pump may be necessary. Using warm water also increases the solubility of fertilizer. The fertilizer material must be completely dissolved, so that each liter of irrigation water contains the same amount of fertilizer. The greater the percentage of the fertilizer required for a given crop that is applied via fertigation, the more important will be proper mixing and uniformity of application. To dissolve the fertilizer, it is best to fill the nurse tank one-half full of water, and then add the fertilizer material and remaining water while constantly agitating the solution. Potassium nitrate or ammonium nitrate cause the solution to become cold as the material dissolves, so less material can be dissolved without using heat. It is recommended that that a fertilizer solution be allowed to settle prior to pumping into the system and usually settling overnight allows adequate clearing time for most fertilizer solutions. The point at which the fertilizer solution is drawn from the tank should be above the bottom, usually 12 to 18 inches which will allow enough room for impurities to settle out. Any undissolved materials usually can be removed or cleaned out of the bottom of the tank. Some fertilizers, such as some sources of potassium nitrate or calcium nitrate, have non-dissolving materials or coatings that should be skimmed off the top of the tank. Another method is to use a flocculating agent that will cause these materials to drop out of solution and settle to the bottom of the tank.

Scheduling the fertilizer application

It can be applied as much fertilizer material with fertigation as with other methods, but better production can be realized with fertigation because of the increased availability of the nutrients to the plant as a result of nutrients being placed in close proximity to the roots. At the same time, less fertilizer can be used in the drip line to get the same production.

Fertilizer requirement: Requirement of fertilizer materials depends on the nutrient content in it. This means that 3 kg of ammonium nitrate or 6.5 kg of calcium nitrate per acre will equal 1 pound of actual nitrogen per acre of nutrients being placed in close proximity to the roots. Solubility chart tells that 102.11 kg calcium nitrate or 118.18 kg ammonium nitrate can be dissolved in 100 liters of cold water.

Pre-plant application: Using a soil test as a guide, generally the recommended amounts of phosphorous and micronutrients and 20 to 40 percent of the recommended amounts of nitrogen and potassium are applied during final land preparation prior to applying the plastic mulch and drip tape. This "starter" fertilizer dose provides some nutrition to the crop during its early growth.

Injection schedule: The remaining nitrogen and potassium can be applied throughout the season as needed by the crop however, the specific schedule will vary depending on the crop. Weekly leaf tissue analysis is the best way to ensure that a crop is receiving adequate nutrients. Some general ranges of sufficient nutrients in leaves of some vegetable crops are given below.

Nutrient (%)	Muskmelon	Cucumber	Brinjal	Sweet pepper	Summer squash	Tomato	Watermelon
N	2.00–4.00	5.0–6.0	4.0–6.0	5.00–5.50	4.00–6.00	3.50–5.00	2.50–4.50
P	0.25–0.40	0.3–1.0	0.3–1.0	0.35–0.45	0.25–1.00	0.35–0.45	0.25–0.75
K	1.80–4.00	4.0–5.0	3.5–5.0	4.50–6.00	3.00–5.00	3.50–5.00	2.25–3.50
Ca	1.80–7.00	1.2–3.5	1.0–2.5	1.00–1.50	1.00–2.50	1.00–1.50	1.10–1.50
Mg	0.50–1.50	0.3–1.0	0.3–1.0	0.30–0.80	0.30–1.00	0.30–0.80	0.25–0.80
S	0.20–0.60	0.2–0.8	0.2–0.8	0.15–0.40	0.20–0.75	0.20–0.40	0.20–0.75
(ppm)							
B	20–60	25–75	25–75	25–95	25–75	30–100	30–75
Fe	40–200	50–200	50–200	40–200	40–200	45–200	40–200
Mn	20–200	25–200	25–200	20–200	25–200	20–200	25–200
Zn	20–60	20–75	20–75	20–60	20–75	20–60	20–75
Cu	4–25	5–35	5–35	5–35	5–35	5–35	4–15

Frequency of fertilizer injection can range from once a week to daily. The more frequent the application, the less potential for problems caused by application delays related to weather conditions and for injury from buildup of soluble salts. A general recommendation is to start with small amounts of N and K and increasing the rates as the season progresses and the crop demand increases. However, fertigation is yet to be adopted at a large scale in India mainly due to high installation cost, non-availability of soluble fertilizers and high cost of imported fertilizers.

10. Pollination Management

There are about 250,000 species of flowering plants on earth that require pollination. Wind, gravity, water, birds, bats, and insects are the forces that accomplish pollination. Pollination refers to the transfer of pollen from the stamen (hermaphrodite/staminate flower) to the pistil (hermaphrodite/ pistilate flower) because to produce a seed or fruit, pollen must move from the anther to a receptive stigma. Since both stamen and pistil are necessary for seed formation, they are called essential organs of a flower. Most flowers contain two other parts, the sepals and petals, which may help attract insect pollinators. A complete flower contains all four parts. If only one of the essential organs is present in a flower, it is called an incomplete flower. Vegetables may be grouped into four categories according to their manner of pollination:

Group 1: Vegetable crops with hermaphrodite sex form and predominantly self-pollinated i.e., the crops with flowers that can receive only their own pollen. *eg.* Peas, French bean, hyacinth bean, cowpea, lima beans, cluster bean, chicory, endive, tomato, etc.

Group 2: Vegetable crops with hermaphrodite sex form and predominantly self-pollinated but fair amount of cross-pollination (both air-borne and insect-borne) happens in some environmental conditions. *eg.* brinjal, sweet pepper, okra, chilli, lettuce, etc.

Group 3: Vegetable crops with hermaphrodite sex form and cross-pollinated by insect-borne pollens and mostly form seed with the pollens from genetically different plants because of self-incompatibility. *eg.* cabbage, cauliflower, broccoli, Brussels sprouts, knol khol, kale and radish,

Group 4: Vegetable crops with hermaphrodite/ monoecious/ andro-monoecious sex form and cross-pollinated by their own pollen (self-pollinated) or from pollens received from another plant (cross-pollinated) and can be divided as follows:

- Usually pollinated by air-borne pollens *eg.* garden beet, palak, corn, carrot, celery, Swiss chard, amaranth, etc.
- Usually pollinated by insect-borne pollens *eg.* onion, parsnip, muskmelon, watermelon, pumpkin, summer squash, winter squash, cucumber, gherkin, bottle gourd, bitter gourd, ridge gourd, sponge gourd, collards, mustard, rutabaga, parsley, etc.

Group 5: Vegetable crops with dioecious sex form and cross-pollinated by insect-borne pollens *eg.* asparagus, spinach, pointed gourd, giant spine gourd, ivy gourd, etc.

The pollination process

Vegetable crops with hermaphrodite sex form (peas, French bean, hyacinth bean, cowpea, lima beans, cluster bean, chicory, endive, tomato, brinjal, sweet pepper, okra, chilli, lettuce, etc) have both male and female parts in the same flower which favours self pollination. However, proper pollination both in open and green house is needed for optimal fruit set and production in these crops. Blossoms of tomato, brinjal, etc. require slight movement for sufficient pollen from the stamens to fall onto the stigma of the flower. Honey bees cause sonication due to buzz foraging over the flowers. Bumble bees cause movement by hanging upside down on the flower, fastening their jaws onto the staminal tube, and then setting the flower into vibration by activating their flight muscles which is called "buzz pollination". Such bee mediated buzz pollination and sonication cause liberation of pollens from the apical split of the anthers to effect self-pollination. Growing of tomato, sweet pepper, brinjal, etc in green house rely either on manual pollination or electrically operated vibrator for liberation of pollen and subsequently fruit set, which can be very time consuming. Using bumble bees for pollination is an effective alternative and can completely replace manual pollination. Fair amount of cross-pollination occur in okra by bees mainly due to showy petals and mucilage that attracts the insects and in chilli due to very fine dusty pollens. Some vegetable crops with hermaphrodite flower viz., onion, carrot, etc. are predominantly cross-pollinated because of dichogamy (different maturity of male and female parts in a flower), presence of nectar and strong smell in the flower. Another group of vegetable crops with hermaphrodite flower like, cabbage, cauliflower, broccoli, Brussels sprouts, radish, etc. are also chiefly cross-pollinated because of self-incompatibility (a genetically governed phenomenon when either pollens do not germinate or pollen tubes do not grow through the stylar tissue if genetic constitution of pollen and stylar tissue is same) and the flowers offer sugary liquid nectar as an added enticement for these pollinating insects.

Monoecious vegetable crops (watermelon, muskmelon, pumpkin, bottle gourd, bitter gourd, cucumber, ridge gourd, snake gourd, sponge gourd, corn, amaranth, etc.) have both staminate and pistilate flowers separately on the same plant. Andromonoecious vegetable crops (muskmelon) have staminate and hermaphrodite flowers on the same plant. Dioecious vegetable crops (asparagus, spinach, pointed gourd, giant spine gourd, ivy gourd, etc.) plants have only one sex of flower per plant. Monoecious, andromonoecious and dioecious vegetable crops generally require cross-pollination.

Many cross pollinated vegetable crops like, watermelon, muskmelon, pumpkin, bottle gourd, bitter gourd, cucumber, ridge gourd, snake gourd, sponge gourd, etc. produce pollen that is heavy and sticky and not blown easily from flower to flower which generally require insects to transfer pollens. Many insects visit flowers to collect pollen and nectar as food. Nectar is usually offered deep within the flower near the base where petals originate from around the ovary. In seeking nectar or gathering pollen, the insect accidentally brushes against anthers so that pollen grains are transferred to the stigma. So, pollination is not a deliberate behavior but an accidental one performed by the insect as they collect food. Among insect pollinators, bees are especially efficient because they eat pollen and nectar exclusively, visit many flowers of the same species during a single trip, and have hairy bodies that easily pick up pollen grains.

Cross-pollination predominantly by air-borne pollens happens in some vegetable crops which produce light pollens that are easily blown about by wind and can drift in air for long time. Flowers of these crops are either very inconspicuous (palak, garden beet, amaranth, etc.) or have special pollen receiving parts in the pistilate flower (silk in pistilate flower of corn).

When the proper compatible pollen adheres to the stigma, it germinates and a pollen tube grows through the stigma and the style to the ovary. Fertilization takes place in the ovary when the nucleus of the pollen or male germ cell unites with the nucleus of the ovule or female germ cell. Ovules become seeds and the surrounding ovary develops into the fruit.

The honeybee is the most common and effective insect pollinator of most of the vegetable crops. Honeybee activity closely coincides with the period when the flower is open. Honeybee visitation begins an hour or two after sunrise and continues until mid-afternoon. If temperatures are very warm, bee activity may decline about noon. However, honeybee visitations are generally maximum upto 10 a.m. and then decline until 2.30 to 3.00 p.m. when activity almost ceased. Bumble bees are most active in the morning and in the afternoon at temperatures between 10-30°C although they function best at temperatures between 15-25°C. Generally the number of bee visitations is more important than the length of time that each bee stayed on the flower. It has been documented that well-shaped, fully expanded fruit of watermelon occurr following eight bee visitations to a pistilate flower and fruit set is significantly reduced when only two to four bee visitations are made.

Management of bees for effective pollination

1. Vegetable farming and bee keeping should go concurrently. The honey bee is well adapted for pollination. Each colony contains thousands of individuals which are available to forage for the food (pollen and nectar) that is required to rear their replacements. Like other bees, their sense of smell, eyes, mouthparts, and numerous branched body hairs are ideally suited for finding food sources, sipping nectar, and collecting and distributing pollen. Honey bees visit a wide variety of flower types. In a single day, one bee makes 12 or more trips from the hive, visiting several thousand flowers. On each trip, it confines its visit to one plant species, collecting one kind of pollen. It has a complex communication system that enables the colony to find and collect food with maximum efficiency. These characteristics make honey bees most valuable agent for cross-pollinating crops. Bumble bee pollination has some advantages because they are active at low temperatures of 10-12°C and windy and cloudy conditions and effective in greenhouses, high tunnels and in open air.
2. At least one strong bee colony should be kept per acre of land to keep sufficiently high bee population to insure that each flower is visited by maximum number of times.

3. In order to maintain the health and activity of the bee population, any pesticide should preferably be applied to the crop when bees are not present in the vegetable field, usually at dusk or after dark.
4. Selective use of pesticides since many of the traditional classes of insecticides will have a negative impact on the hive and systemic pesticides may damage the bee population.
5. In the green house cultivation of tomato, brinjal or sweet pepper, the bee hives need supply of sugar water for the total life expectancy of the hive, since blossoms of these crops do not produce nectar.

11. Integrated Pest Management

Insect pests are one of the major factors undermining improved vegetable productivity particularly in intensive cropping system where the insect pests easily find their hosts through out the year and this situation of high biotic stresses by insect pests assumes serious proportion further with the threat of global warming. In general vegetables have low pest resistance and in general, the insect pests destroy more than one third of vegetable production with an average of about 40% yield loss which under severe infestation might cause total destruction. In the present era of chemical agriculture, different chemical pesticides are the basic armoury for insect pest control programme however, the hydrochlorin, organo phosphorous ,carbamate and other chemical compounds used as pesticides are highly toxic and pollute the environment and affect humans through their residue in food chain and in other various ways.

Majority of pesticide related health problems take place in the developing countries, because the users are not sufficiently aware of the dangers and the proper use of pesticides. Vegetables are more vulnerable to cause such health hazard because of the nature of maturity and their consumption. Many of the vegetables are harvested at very short interval and are consumed raw without much processing. So, the approach of controlling insect pests only by chemical pesticides is now posing social problems apart from being increasingly ineffective and at the same time expensive in the present complex cropping systems.

Different circumstantial evidences and backed up research findings have suggested that pesticides alone are inadequate for controlling insect pests. The use of pesticides to control pests involves significant degree of risk in polluting the environment, development of pest resistance to pesticides, pest resurgence, destruction of natural enemies and pesticide residue in the food chain. Integrated pest management (IPM) refers to the holistic approach for the management of pest population below economic threshold level (ETL) through efficient and judicious, utilization of the available resources with active participation of farmers.The output of IPM implementation would be that the farmers should become export of IPM and should be able to make their own decision in dealing with pest problems. The concept of IPM stads basically on three strategies, Prevention, Observation and the fundamental approach for a sustainable IPM towards improvement of sustainable agriculture which includes (a) biological control using one organism to control another through manipulation of pest bioagent relationship (b) use of different botanicals that are derived from plants to control insect pests (c) behavioural management through manipulation of cultivated and natural environment to preserve the natural enemy balance of the insect pests and (d) development and utilization of insect pest tolerant varieties.

Components of IPM

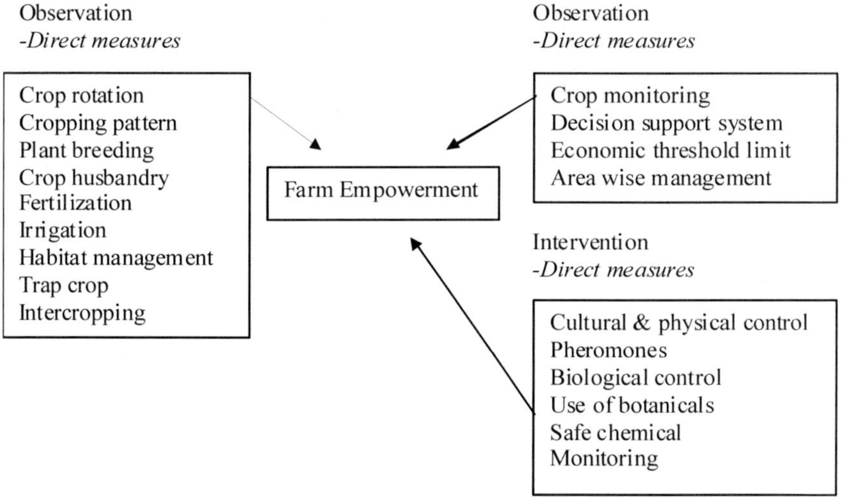

IPM is based on basic understanding and application of ecological principles integrating multi disciplinary methodologies to develop strategies for the management of agro-eco system. The objectives of IPM research in this context are (I) understanding of each component in the agro-eco system (2) to develop activity component of IPM (3) to develop basic understanding of a suitable IPM strategy against key pests of major vegetable crops and (4) to validate and improve the existing IPM technology.

During implementation of IPM approach, the farmers should be considered as a researcher in their own field becoming a part of research network supporting IPM. Therefore, the strategies to disseminate IPM research results to farm level include:

- Field training driven researches which at the same time impart training to the extension staff.
- On-farm research involving farmers' participation.
- Field demonstration through Krishi Vigyan Kendra/Govt.department /NGO for convincing the farmers regarding the usefulness of IPM in vegetable cultivation.
- Adequate Govt. interventions for making the avaiailable different biocontrol agents, botanicals, pheromone trap, etc to the farmers.

Biocontrol agents for pest management

Egg parasitoids: *Trichogramma brasiliensis* is a parasitoid specifically useful to control tomato fruit borer *(Helicoverpa armigera)* which attacks tomato, peas and beans, etc. With the onset of adult catch in pheromone trap or first flowering, release of the parasitoids as parasitized eggs @50,000/ha with repeated inoculative release at the same rate at weekly interval is very effective. Similarly, *T. chilonis* and *Trichogrammatodea bactrae* are effective to control okra shoot and fruit borer (*Earius* spp.) and diamondback moth (*Plutella xylostella*) on crucifers, respectively. Weekly release of *T. chilonis* @250,000 parasitized eggs /ha has positive effect in control of brinjal shoot and fruit borer.

Larval parasitoids: *Cotesia plutellae* and *Diadegma semiclausum* are dominant larval parasitoids of diamondback moth, among which the later one is susceptible to high temperature and is mostly active in areas having mild climate. Very high natural parasitization by these larval parasitoids indicates the importance of their role in pest suppression if conserved through proper management practices. Use of neem seed kernel extract (4%) helps control the pest with least impact to these larval parasitoids. *Campoletis chlorideae* is most common larval parasitoid of *Helicoverpa.* Usually second instar larvae are parasitized in the field. This parasitoid is mass multiplied on young larvae of *H. armigera* and *Spodoptera litura.*

Predators: Ladybird beetles *(Coccinella septempunctata* and *Chilomenus sexmaculatus)* are natural predators of aphids and other soft-bodied insects. These predators can easily be mass multiplied and used for suppression of aphids on different vegetable crops. Generally, adult beetles released @30 beetles/sq.m are sufficient to control the aphids. Green lace wings (*Chrysoperla carnea)* are commonly found to predate on soft-bodied insects like aphids, thrips, jassids, whiteflies and eggs of some lepidopteran pests. Mass release of early instars of these predators @ 50,000 eggs/ha is very effective in the management of sucking pests on vegetable crops.

Microbial biocontrol agents: Insect pathogenic bacteria, fungi and viruses are mostly used for the management of the pests of vegetable crops. Unlike natural enemies, these agents can be utilized as a prophylactic measure as done in insecticides and have an advantage of field applicability. The bacteria *Bacillus thuringiensis (Bt)* is most widely used microbial insecticide which is useful in the management of several leaf feeding lepidopteran pests (caterpillar) of vegetable crops. Several commercial formulations of *Bt* are now available in markets which are generally are used as foliar spray @1.0 kg/ha. It is particularly effective against diamondback moth on cabbage however, it is also recommended for the control of shoot and fruit borer on okra. Nuclear Polyhedrosis Viruses (NPVs) are highly target specific entomopathogens with least impact on natural enemies. The NPV's have been widely used for the management of fruit borer *(Helicoverpa armigera)* on tomato and leaf eating caterpillar *(Spodoptera litura)* on cauliflower.It is recommended to spray Helicoverpa NPV and *Spodoptera* NPV @250-300. Larval Equivalent (LE) /ha mixed with 1% jaggery and 0.1% Teepol in 250 litres of water on tomato and cauliflower, respectively in the evening hours to get the optimum control of these insect pests.

Examples of IPM way of controlling insect pests in different vegetable crops

Nursery bed management for different vegetable crops

- Soil solarization with transparent polyethylene sheet during summer is ideal to suppress the soil borne pathogens in nursery. Seed treatment with *Trichoderma* spp. @ 5-10 g/kg or soil treatment @ 10-25 g/ sq.m nursery area is effective against damping off diseases.

Brinjal

- Seed treatment with imidacloprid @ 2.5g/kg reduces the little leaf and leafhopper incidence.
- Intercropping of brinjal (2 rows) with coriander (one row) or fennel (1 row) minimizes the incidence of brinjal shoot and fruit borer.
- Integrated pest management of fruit and shoot borer includes raising the nursery under nylon net cover, removal and destruction of infested shoots along with larvae and periodical cleaning of dry leaves and old infested fruits at weekly interval along with the insect larvae starting from 25 days after transplanting and installation of sex pheromone baited plastic funnel traps at 10m distances just above canopy from 20 days after planting and need-based foliar spray of NSKE (4%) during late afternoon.

Tomato

- Seed treatment with imidacloprid @ 2.5g/kg, raising the nursery under nylon net cover, single spray of imidacloprid (Confidor) @ 0.3 ml/lit about 40 days before fruiting and subsequent spray 4% neeem seed kernel extract (NSKE) with 0.1% sticker minimizes whitefly population and reduce the incidence of tomato leaf curl virus (TLCV) and leafminor.
- Planting of two rows of marigold as trap crop with every 14 rows of tomato along with release of *Trichogramma brassiliensis* @50,000 parasitized egg/ha at flowering stage and spray of *Helicoverpa* NPV @ 350 LE/ha twice after flowering can manage the fruit borer *(Helicoverpa armigera)* satisfactorily.
- Use of synthetic pyrethroids and endosulfan alternatively with NSKE (4%) is also effective against fruit borer (*Helicoverpa armigera*).

Okra

- Seed treatment with imidacloprid @ 2.5 g/kg is effective against leafhopper in okra during early vegetative stage.
- Spraying of NSKE (3%) spray with sticker @ 0.1% minimizes the incidence of insect pests in the field.

- Spraying of endosulfan and neem formulation at their half doses at flowering stage can reduce borer (*Earius* spp.) infestation.
- Okra sown during 2nd week of June minimizes the incidence of both jassids (*Amrasca biguttula biguttula*) and shoot and fruit borer (*Earius* sp.).

Chilli

- Seed treatment with imidacloprid @ 2.5 g/kg is effective against thrips (*Scirtothrips dorsalis*) and mites (*Polyphagotarsonemus latus*, *Tarsonemus translucens*, etc.) complex during initial stage.
- Alternate spray of dicofol @ 2.75 ml/lit and wettable sulphur @ 2 g/lit is highly effective against mite.

Cucurbits

- Effective fruit fly management include deep summer ploughing to expose the dormant pupae in the field, crop rotation incorporating non- cucurbitaceous crops, food bait consisting of rotten banana or smashed pumpkin pulp (1000g) + yeast (10 g) + malathion (10 ml) + citric acid (3 g) in flat earthen pot to attract and kill the adult flies and bait spray in one plant per 10 m^2 area with 20 ml malathion (50 EC) + 20 liters water +500 g molassess keeping rest of the plants remains free from insecticides to attract adult flies and control the population.
- Spraying of carbaryl 50 WP @ 2 g/lit at cotyledon stage control red pumpkin beetles.

Cabbage

- Diamondback moth (DBM) population in cabbage can be managed by growing paired mustard rows with every 25 cabbage rows. The first rows of mustard should be sown 15 days before and second 25 days after cabbage planting. The trapped DBM larvae on mustard can be killed by sparaying 0.1% Dichlorovos or 5% NSKE. Such practices conserve the population of *Cotesia plutellae,* a dominant natural enemy of DBM.

12. Biological Management of Diseases

Many fungi, bacteria and actinomycetes have been found promising against different soil-borne pathogens causing root rot, damping off, collar rot and wilt in different vegetable crops. Important target pathogens are *Pythium* spp., *Sclerotium rolfsii, Macrophomina phaseolina, Rhizoctonia solani, Fusarium* spp., *Phytophthora* spp., *Phoma* sp. *Sclerotinia sclerotiorum, Aspergillus* spp., *Botrytis, Armillaria, Verticillium, Alternaria solani, Colletotrichum capsici and Pleospora beta.* Several species of *Trichoderma* have gained attention for their highly antagonistic property. Some of the antagonists are highly specific to a particular pathogen, but most of the fungal antagonists have broad-spectrum parasitic host range. Some antagonistic fungi also kill some plant parasitic nematodes infecting vegetables.

Potential biocontrol agents

Soil has maximum microbial biodiversity, which can be utilized for plant disease management. Several antagonists have been isolated and many of them are being utilized in management of soil-borne pathogen. The maximum biocontrol agents belong to fungal genera, viz. *Trichoderma* spp., *Aspergillus niger, A. flavus, Pythium nann, Trichothecium, Paecilomyces lilacinus, Penicillium funiculosus, Myriothecium, Corticium, Pythium cligandrum, Peniophora gigantea, Candida oleophila, Sporidesmium sclerotivorum, Coniothyrium minitans, Ampelomyces quisqualis, Chaetomium, Cladosporium, Fusarium semi tectum, Tuberculina, Phialophora, Catenaria* and *Verticillium.* Some bacterial antagonists like, *Pseudomonas* spp., *Agrobacterium radiobactor, Bacillus* spp. and Actinomycetes such as *Streptomyces griseus,* S. *rimosum* have also been utilized for management of diseases. *Trichoderma* spp. has extensively been exploited for the management of soil borne diseases.

Crops	Pathogens	Antagonists
Tomato, brinjal, chilli,bean	*Rhizoctonia solani* *T.viride*	*Trichoderma harzianum,*
Tomato, chilli, brinjal,bean, pea, cucumber	*Pythium* spp.	*T. virirde, T. harzianum, T. koningii*
Tomato, chilli, brinjal,bean, pea	*Sclerotium rolfsii*	*T. virirde, T. harzianum, T. koningii*
Pea, radish	*Pythium* spp., R. *solani*	*T. virirde, T. harzianum, T. hamatum*
Tomato, muskmelon	*F. oxysporum, P. aphanidermatum*	*T. virirde, T. harzianum*
Okra	*Macrophomina phaseolina*	*T. virirde, T. harzianum*
Cucumber	*Sclerotinia sclerotiorum*	*T. harzianum*

The best substrate is sorghum or mixture of sorghum with either rice bran or rice husk. Talcum powder is commonly used as carrier in commercial biopesticide product. The USA, Israel, Bulgaria, Australia, Sweden, Canada and India have commercialized the use of biopesticides in management of soil-borne diseases. Successful commercial use of *Trichoderma* spp. and *Pseudomonas fluorescence* was demonstrated over past ten years in most part of the country. The biopesticides are either mixture of two bioagents or alone are available with different trade names, viz. *T. viride* (Antagon TV, Trichoguard, Niprot, Bioderma Tricontron) , *T. harzianum* (Trichodex, Trichodermin, F. stop, Bio-Coat), *T. virens* (Gliogard), *T. viride* + *P. flourescens* (An tag or Combi), *T. harzianum* + *Bacillus subtilis* (Combat), *P. fluorescence* (Biomonas, Dagger G, Conquerer).

13. Mechanisms of Biocontrol

Three antagonistic mechanisms are involved in biological control of soil-borne plant pathogens, viz. competition, antibiosis and mycoparasitism. Competition occurs when there is a demand of two or more microorganisms for the same nutrient resource in limited space and time. This is most potent mechanism to control *Fusarium oxysporum* f.

sp. *melonis* by *Trichoderma harzianum*. Antibiosis occurs when production of toxic metabolites or antibiotics by the biocontrol agent has a direct effect on another organism and such toxic metabolites include gliotoxin produced by *T. virens,* viridin by *T. viride* and trichodermin by *T. harzianum*. In case of mycoparasitism or hyperparasitism, fungus is parasitized on another by chemotropic growth, recognition, attachment and degradation. Production of lytic enzymes, chitinases and â 1, 3 glucanases are most important for the lysis of pathogen cell wall. *Trichoderma* species compete aggressively, mycelium coils around pathogen hyphae and cause lysis of the cells of *Phytophthora* spp.

Flavobacterium balustinum and *Xanthomonas maltophila* are the most effective in inducing suppression of damping off causing fungus *Rhizoctonia* sp. by ISR mechanism. Many of the antagonistic bacteria also act as plant growth promoting rhizobacteria (PGPR) and systemic acquired resistance (SAR) in plants against many disease causing pathogens of vegetable crops.

Application of biocontrol agents

Soil application and seed treatment are only feasible method of using bioagents under the field condition. Seed treatment is preferred in direct sown crops, whereas soil application and seed treatment are feasible in the crops which are transplanted in the main field. Best utilization of any biocontrol agents in soil is only possible when it is supplemented with well-decomposed organic amendments, congenial moisture and protected from extreme weather conditions till it establishes in soil.

Biological seed treatment can be done by priming on seed, soaking seeds, seedling dipping, and dry power treatments depending upon the nature of biocontrol agents. Field studies clearly showed that *T. viride* is better than any other biocontrol agents in this respect. Seed treatments with *Trichoderma* @ 6-10g/kg of seed is effective provided spore concentration remain in between 106 and 109/ml. Dose for either seed or soil treatment is influenced by inoculum of pathogen present in soil and nature of seed. Soil application of *Trichoderma* @ 10-25 g along with 100 g neem cake/ sq.m is effective to manage nursery diseases of different vegetable crops.

Green manuring along with soil application of *Trichoderma* @5 kg/ha give good control of most of the soil borne diseases in *kharif* season grown vegetable crops. The *Trichoderma* must be applied soon after incorporating sunhemp preferably within three days depending upon moisture status and texture of soil.

Root dipping of vegetable seedlings by *Trichoderma* formulation @ 1% for 10-25 minutes is very effective in checking root rotting after transplanting in *kharif* season. Drenching of *Trichoderma* @ 1% near collar region should be carried out 20 days after transplanting to check root rot, collar rot and wilt incidence. Fast and vigorous seedling growth is recorded in biocontrol treated nursery beds where 20-25 days old seedlings are ready for the transplanting.

Foliar spray of *Trichoderma* on all ground portions in pointed gourd controls *Phytophthora* blight of both vine and fruit. One resident isolate of *Aspergillus niger*-V @

1% along with sticker and molasses as foliar spray against Alternaria blight and anthracnose of chilli gives satisfactory control.

Sensitivity of the bioagents with pesticides

The fungicides, which are completely insensitive to the bioagents, but sensitive to the pathogen, can be used together to increase the efficiency of disease management strategies. Pencycuran and copper hydroxide up to 400 ppm concentration is absolutely insensitive for the bioagents. The bioagents show tolerable growth after 7 days of incubation with Captan, chlorothalonil and propineb up to 200 ppm concentration.

Copper hydroxide shows synergistic effect at lower concentration, while fungistatic at higher concentration (400 ppm) for five days. Pencycuron is highly effective against *Rhizoctonia* spp. but insensitive to biocontrol agents and gives full protection when integrated both at a time. Different fungicides like, pencycuron, copper hydro oxide, captan, azoxystrobin carboxin, sulfex and antibiotic like streptocycline can safely be used in biocontrol strategies with *Trichoderma* for the management of soil-borne diseases in vegetable crops.

Imidacloprid can be integrated with *T. virens* and *T. viride* with slight adjustment in dose. Seed rate should be compensated according to soil type, insect endemic area and viral infection to get desired seedling stand of tomato. Soil application of carbofuran followed by seed treatment by antagonist can be integrated for insect pest and disease management without adverse effect on seed germination. Pesticides like cypermethrin, fenvalerate and herbicide like glyphosate are safe for integration with management by *Trichoderma.*

14. Weed Management

Agricultural production depends on various factors like, genetic make up of the crop, soil and environmental factors and different production problems that interact either to enhance output or to limit production. Of the total losses due to different production problems caused by insect pests, disease causing pathogens, nematodes and weeds one-third of the estimated crop loss is caused by weeds alone and much damage is caused during early stage of crop growth. So weed management is utmost important for successful production of vegetable crops. Intensity of weed population in different vegetable crops varies with climate, soil, management practices etc.

Weeds are no stranger to human beings and these have been recognized as a problem ever since cultivation of crops began. Jethro Tull first coined the term 'weed' in his writings "Horse Hoeing Husbandry" being the first book of agriculture in1731 in which 'weed' was termed for a plant where it is not desired. Since then various definitions of the weed has been suggested although, main feature of defining weed remains the same as the plant growing out of place and out of time and interfere with crop growth

Characteristics of Weeds

- Weeds are competitive and aggressive in nature.
- Weeds can tolerate adverse climatic condition, e.g. *Calotropis procera, Saccharum spontaneum* tolerate high temperature and low moisture condition.
- Many weeds are morphologically similar to some crops they infest, e.g. *Phalaris minor* is similar to wheat and barley; *Asphodelus tenuifolius* is similar to onion and some weed seeds are morphologically similar to crop seeds, and thus cannot be separated after seed harvest.
- Weeds have high reproductive capacity and in adverse conditions, they even can reproduce asexually, e.g. *Chenopodium album* produces 10,000 to 15,000 seeds per plant while *Phalaris minor* produces 3000-4000 seeds per plant in one season.
- Weeds can resist their control or eradication due to possession of spines, thorns, etc. that hinder their removal from a particular site. e.g. *Zizyphus mummularia, Rumex spinosus* possess spines/thorns and *Cynodon dactylon, Sorghum halepense* can regenerate easily, hence continue infestation.
- Weeds are persistent in nature due to their high longevity, prolific and perennial nature, dormancy of seeds, resemblance with crop plants, ability of reproduction through both seeds and vegetative parts. Seeds of some weeds like, Chenopodium *album, Convolvulus arvensis* can germinate even after 20-25 years of burial in soil.
- Some weeds have the tendency of setting seeds early and prior to crop plants that create serious problem in following years.
- Repeated germination of weed seeds in different phases create problem for the crop throughout its life cycle. Weeds appear in different flushes in same crop till last harvest which helps in their persistence in the field throughout crop life.
- Some perennial weeds possess extensive and deep root system of 5-7 ft in *Cyperus rotundus* and even upto 20 ft in *Convolvlus arvensis* which makes it difficult for their uprooting.

Classification of weeds

Weeds can be classified in different manner as discussed below.

Classification based on morphology

- Broad leaf: *Amaranthus spinosus*, *Eclipta alba* , *Portulaca oleracea,* etc.
- Grasses : *Echinochola crusgalli*, *Cynondon dactylon*, etc.
- Sedges: *Cyperus rotundus*, *Cyperus diformis,* etc.

Classification based on life cycle

- **Annuals:** Annual weeds which are usually small herbs with shallow roots and weak stem and propagated by seeds mature in one season
- **Rabi season annuals:** *Avena fatua*, *Phalaris minor*, *Ageratum conyzoides*, *Chenopodium album*, *Argemone mexicana*, *Cirsium arvensis*, etc.
- **Kharif season annuals:** *Dactyloctenium aegypticum*, *Echinochola colonum*, *Amaranthus spinosus*, *Amaranthus viridis*, *Solanum nigrum*, etc.
- **Biennials :** Biennials require two year completing their life cycles; in the first year they remain vegetative, and in the second year they produce flowers and set seeds. These types of weeds are mainly found in non-cropped areas. Common examples of biennials weeds are wild carrot, *Alternanthera echinata*, *Cirsium vulgare* , *Plantago* spp. , *Arctium* spp. etc.
- **Perennials :** Perennials weeds live more than two years and they reproduce both vegetatively and sexually forming seeds.
- **Simple perennials :** These perennials are propagated only by seeds e.g. *Sonchus arvensis*
- **Bulbous perennials :** These perennials possess a modified stem with scales and reproduce mainly by bulbs and seeds e.g. *Allium* sp.
- **Corm perennials :** These perennials possess a modified shoot and fleshy stem and reproduce mainly by corm and seeds e.g. Timothy
- **Creeping Perennials :** Creeping perennials are reproduced by seed, rhizome (Sorghum *halpense*), stolon (*Cynodon dactylon*), enlarged root system with numerous buds (*Convolvulus arvensis*), tubers (*Cyperus* sp.), etc.

Classfication based on habitat

- **Terrestial weeds :** These weeds are mostly found in upland or dry land condition having enough moisture but the land is not water logged e.g. *Chenopodium album*, *Solonum nigrum*, *Celosia argentea*, etc.
- **Aquatic weeds :** These weeds grow in water and complete at least a part of their life cycle in water which are of several types.
- **Free floating :** *Eichhornia crassipes*, *Pistia stratiotes*,etc.
- **Rooted floating :** *Trapa bispinosa. Ludwigia adscendens*, etc.
- **Subemersed :** *Ceratophyllum demersum*, *Hydrilla verticillata*, etc.
- **Immersed :** *Jussieua repens*, *Nelumbium speciosum*, etc.
- **Amphibious :** *Ranunculus aquatilis*, *Scirpus supinus*, etc.

Classfication based on the origin of weeds

- **Alien (Foreign origin):** *Parthenium hysterophorus*, *Argemone mexicana*, etc.
- **Apophytes (Indigenous):** *Cynodon dactylon*, *Cyperus rotundus*, *Saccharum spontaneum*, *Acalypha indica*, etc.
- **Antrophytes (Introduced by man):** *Phalaris minor*, *Corchorus acutangulua*.

Dispersal of weeds

Weeds develop various adaptive mechanisms for their continuous perpetuation and dispersal. Some seeds are smaller and lighter which are dispersed through blowing wind. Some heavier seeds are dispersed by animals and birds. Some weeds seeds possess specialized structure that extends buoyancy in the atmosphere e.g. parachute like pappus in weeds of Compositae family. Weeds of Graminae family possess fine long awns with arrow shaped pointed ends that get into clothes and cannot be removed easily and thereby dispersed to long distance. Presence of hooks and bars on surface of many fruits and seeds help in their attachment to animals and clothes of human beings. Sticky fruits are dispersed by animals and get attached to tillage equipments and thus are dispersed. Weed seeds are generally dispersed through the following modes.

- By wind
- By water current
- By birds
- By animals
- By human beings and contact with different tillage equipments.
- Along with the farm produce.
- Along with silage, dung and farm yard manure.

Harmful effects of weeds

Successful production of vegetables is adversely affected by weeds in different ways.

- Weeds are fast growing compared to the vegetable crops hence, compete with the crop for solar radiation, water, nutrients and space.
- Weeds reduce crop yield, quality and marketability of the produce.
- Weed infestation increases cost of cultivation because its control is basically a labour intensive proposition and require costly chemicals, equipments for chemical control measure.
- Some weed plants secrete harmful chemicals that may have adverse effects on crop plants, soil, human being and livestock. In India *Parthenium hysterophorus* cause

contact dermatitis and asthma. *Rhus radicans* (Poison ivy), *Ambrosia* sp. (Ragweed) cause itching, hay fever and other debilitating allergies contributing to chronic human illness and suffering.

- The perennial rhizomatous grasses like, *Imperata cylindrica, Cynodon dactylon, Sorghum halepense, Saccharum spontaneum* cause considerable damage to the land due to lack of equipment strong enough to break sods and the land some times becomes unsuitable due to heavy infestation of these weeds.
- Presence of spiny and thorny weeds like, *Amaranthus spinosus, Argemone mexicana* etc. restrict movement of farm workers in carrying out farm practices.
- Some prominent aquatic weeds are *Eichhornia crassipes, Typha angustata, Hydrilla verticillata, Ipomoea aquatica,* etc. may block drainage and impede the flow of water in irrigation channels. They also result in seepage, breakage of canal banks, flooding, silting, and inadequate irrigation water delivery to farms located at a distance from main water source.
- Certain weeds provide hiding place for insect-pests and act as host plants for certain pathogens.

Examples of some weed host for different insect pests and pathogens

Weeds	Insect pests	Crop hosts
1. Mustards	Cabbage root maggot	Cabbage, broccoli
2. Chick weed (*Stelaria media*)	Melon aphids	Most cucurbits
3. Wild morning glory (*Convolvulus arvensis*)	Aphids	Watermelon, pumpkin, some gourds (mostly ridge gourd)
4. Wild carrot (*Daucus carota)*	Carrot rust fly	Carrot, celery.
5. Night shade (*Solanum nigrum)*	Onion thrips	Onion
6. Lambsquarter (*Chenopodium album*)	Common stalk borer	Mostly tomato.
Weeds	**Pathogens**	**Crop hosts**
1. Pig weed (*Amaranthus viridis*)	Spinach mildew	Spinach, Swiss chard.
2. Night shade (*Solanum nigrum*)	Verticillium wilt, Tomato mosaic virus	Tomato
3. Night shade (*Solanum nigrum*	Potato mosaic	Potato
4. Mustards (*Brassica* sp.)	Clubroot	Cabbage, Cauliflower

Common weeds in the field of vegetable crops

Some weeds are found most frequently in the field of different vegetable crops viz., wild oat (*Avena fatua*), gullidanda/Canery grass (*Phalaris minor*), chottis wank (*Echinochola colona*), cow foot grass (*Dactyloetenium aegyptium*), swank/ maddal (*Echinochola crusgalli*), doob grass /Bermuda grass (*Cynodon dactylon*), motha /nut grass (*Cyperus rotundus*), Kans (*Saccharum spontancum*), goose grass (*Eleusine indica*), bathua (*Chenopodium album*), chaulai (*Amaranthus viridis*), tandla (*Digeria arvensis*), dodhak (*Euphorbia hirta*), fumatory/pitpapra (*Fumaria parviflora*), meadow pea/natri (*Lathyrus aphaea*), wild onion (*Asphodelus tenuifolius*), ivy leaved/bel (*Ipomoea pestigridis*), con spurry/jangli dhania (*Spergula arvensis*), dendelon/jangli gobhi (*Launea asplenifolia*), wild carrot (*Daucas carota*), milk weed (*Sonchus arvensis*), jangli palak (*Rumex dentatus*), thorny amaranthus (*Amaranthus spinosus*), night shade (*Solanum nigrum*), johnson grass (*Sorghum halpense*), orbanche (*Orbanchae aegyptica*), etc.

Methods of Weed Control

1. Preventive methods

- Care should be taken at the time of sowing of crops that weed seed free crop seed must be used.
- Movement of live stock should be checked from weed infested to clean area.
- Weed plants should be completely removed from the sides of irrigation and drainage channels.
- Use of clean farm machinery.
- Farm boundaries, roadsides, and other wastelands should be free from weeds.
- Planting of crop should only be done after cleaning of weed plants.

2. Curative or control methods

- **Mechanical methods**: Removing the weeds physically or with tools /implements is called mechanical method. Mechanical methods though traditional yet are very effective in controlling almost all the type of weeds. However, these methods are time consuming and labour intensive. Different mechanical methods adopted for controlling weed are as follows.

  a) *Tillage:* Tillage is done mainly with two objectives. to uproot the weeds growing in the field and to pulverize the soil.

  b) *Hand pulling:* Hand pulling is generally practiced to uprooting weed plants before flowering and seed setting along with their roots.

  c) *Hand weeding:* Small implements like khurpi , nirani , sickle, wheel hoe ,etc can be used for uprooting plants.

  d) *Burning:* Burning is an economical and practical means of controlling weeds particularly in organic production system.

- **Cultural methods:** In this method selection of crop is one of the most important factors to check weeds up to the great extent. The crop should be quick growing with large leaf area and good branching habit. Most of the legume and leafy vegetables grow rapidly and develop into dense canopy compared to weeds which results in effective suppression of weeds. Proper crop rotation does not allow development of several weeds particularly associated weeds with crops, and hence do not pose a serious problem. Practice of raising green manure crop is helpful in reducing weed infestation as the weeds which grow along the green manure crop are buried and subsequently decomposed into the soil. Soil solarization is also an effective method of weed control in many vegetable crops.

- **Biological methods:** Biological approaches involves various living agents such as fishes, snails, pathogens, insects, nematodes as well as botanicals i.e. bioherbicides to control weeds. Different bioherbicides like, Devine ® Biomal ®, etc are commercially available. Control of some perennial weeds by biological method has been done successfully *viz., Lantana camara* by *lace bug, Teleonemia scrupulosa*; cuscuta by *Melanogromyza cuscutae*, *Imperata cylindrica* by *Orseoliella javanica* ; *Cyperus rotundus* by a plant feeder *Bactra vermosana.* However, biological method of weed control has been very little in practical use in vegetable crops.
- **Chemical methods:** Use of herbicides for the control of weeds offers the greatest feasibility for weed control. Chemical method of weed control is less costly than other methods besides its usefulness in minimizing tillage operations.

Herbicides

Versatility in vegetation management has been extended through the use of selective agricultural chemicals. The term "Herbicide" came from the latin words *herba* meaning plant and *caldere* meaning to kill. So, herbicide is the chemical that disrupt the physiology of a plant over a long enough period to kill it or severely reduce its growth. Herbicide changes the chemical environment of plants, which can be more readily manipulated than the climatic, edaphic or biotic environment. Herbicides reduce labour and machine requirements and modify crop production technology.

The first widely used herbicide was 2,4-dichlorophenoxyacetic acid (2,4-D) which was developed by a British team during World War II with first widespread production and use in the late 1940s. It still remains one of the most commonly used herbicides in the world. However, it exhibits relatively poor selectivity and is also less effective against some broadleaf weeds, including many vinous plants, and sedges. The 1970s saw the introduction of atrazine, which has the dubious distinction of being the herbicide of greatest concern for groundwater contamination because it does not break down readily (within a few weeks) after being applied. Instead it is carried deep into the soil by rainfall causing the ground water contamination. Glyphosate was introduced in the late 1980s for non-selective weed control. It is now a major herbicide in selective weed control in growing crop plants due to the development of crop plants that are resistant to it. The combination of the herbicide with the resistant crop variety contributed to the consolidation of the seed and agro-chemical industry in the late 1990s.Many modern chemical herbicides for agriculture are specifically formulated to decompose within a short period after application which is desirable as it allows crops which may be affected by the herbicide to be grown on the land in future seasons. However, herbicides with low residual activity (i.e. which decompose quickly) often do not provide season-long weed control.

Classification of herbicides

Currently there are more than 250 herbicides in the world market. There are numerous classes of herbicides with different modes of action for killing weeds, as well as different potentials to have an adverse effect on health and the environment. Herbicides from different classes also differ in their environmental persistence and fate. Almost all herbicides can cause acute toxicity. Phenoxy herbicides are involved in acute symptomatic illnesses with relative frequency.

Different classes of herbicides

Class of Herbicide	Examples
Acetamides and analides	Alachlor, acetochlor, metolochlor, propachlor, propanil
Carbamates and thiocarbamates	Asulam, terbucarb, thiobencarb
Chlorphenoxy herbicides	2,4,-D, 2,4-DP, 2,4-DB, 2,4,5-T, MCPA, MCPB, MCPP, Dicamba
Dipyridyls	Paraquat, diquat
Heavy metals	Lead arsenate, arsenicals
Nitrophenolic and dinitrocresolic herbicides	Dinitrophenol, dinitrocresol, dinoseb, dinosulfon
Pentachlorophenol	Pentachlorophenol
Phosphonates	Glyphosate, glyfusinate, fosamine ammonium
Triazines	Atrazine, simazine, cyanazine, propazine
Urea derivatives	Diuron, flumeturon, linuron, rimsulfuron, tebuthiuron

Source: Farm Chemicals Handbook (C. Sine, ed.)

The herbicides are basically classified on the basis of a) method of application, b) mode of action and c) mechanism of action

1. Methods of application

- **Soil-applied group:** The herbicides in this group (Pre-emergent herbicide) are applied before planting and before emergences of weed or crop which are taken up by the roots of the target plant.
- **Foliar-applied group :** The herbicides of this group (Post-emergent herbicide) are sprayed on the standing plant part after the crop has emerged.

2. Mode of action

- **Contact herbicides :** These herbicides destroy only the plant tissue in contact with the chemical. Generally, these are the fastest acting herbicides but are less effective on perennial plants which are able to regenerate from roots
- **Systemic herbicides :** These herbicides are translocated through the plant, either from foliage down to the roots, or from soil application up to the leaves which can destroy a greater amount of plant tissue than contact herbicides.

3. Mechanisms of action

This classification by mechanism of action indicates the first enzyme, protein, or biochemical step affected in the plant following application. Of this herbicide

- **ACCase inhibitors :** These herbicides generally kill grasses through inhibiting the enzyme Acetyl coenzyme A carboxylase (ACCase) which is involved in the first step of lipid biosynthesis. Thus, ACCase inhibitors affect cell membrane production in the meristems of the grass plant. The ACCases of grasses are sensitive to these herbicides, whereas the ACCases of dicot plants are not sensitive to these herbicides.
- **ALS inhibitors :** These herbicides kill the plant through inhibiting the enzyme acetolactate synthase (ALS) (also known as acetohydroxyacid synthase, or AHAS)which is involved in the first step in the synthesis of the branched-chain amino acids (valine, leucine, and isoleucine). These herbicides slowly starve affected plants of these amino acids which eventually lead to inhibition of DNA synthesis. They affect grasses and dicots in the same manner. The ALS inhibitor family includes sulfonylureas (SUs), imidazolinones (IMIs), triazolopyrimidines (TPs), pyrimidinyl oxybenzoates (POBs), and sulfonylamino carbonyl triazolinones (SCTs).
- **EPSPS inhibitors :** These herbicides kill the plant through inhibiting the enzyme enolpyruvylshikimate 3-phosphate synthase (EPSPS) which is involve in the synthesis of the amino acids tryptophan, phenylalanine and tyrosine. They affect grasses and dicots in the same manner. Glyphosate is a systemic EPSPS inhibitor but inactivated by soil contact.
- **Synthetic auxin :** These organic chemicals inaugurated the era of herbicides. Synthetic auxins mimic this plant hormone. They have several points of action on the cell membrane, and are effective in the control of dicot plants. 2,4-D, Clopyralid, Dicamba , Picloram, etc. are synthetic auxin herbicides.
- **Photosystem II inhibitors:** These herbicides kill the plant through reducing electron flow from water to NADPH2+ at the photochemical step in photosynthesis. They bind to the Qb site on the D2 protein, and prevent quinone from binding to this site. Therefore, this group of compounds cause electrons to accumulate on chlorophyll molecules. As a consequence, oxidative reactions occur in excess of those normally tolerated by the cell and the plant dies. The triazine herbicides (including atrazine) are PSII inhibitors.

Organic herbicides

Organic herbicides are expensive and may not be affordable for general commercial production. They are also much less effective than synthetic herbicides but of course do not inject unnatural chemicals into the environment and can be used in a farming enterprise that has been classified as organic and organic herbicides include:

- **Spice :** Different spices are now effectively used in patented herbicides.

- **Vinegar:** The concentration of 5-20% solutions of acetic acid is effective which mainly destroys surface growth hence, respraying to treat the regrowth is needed.
- **Steam:** It is applied commercially in organic vegetable farms but now considered uneconomic and inadequate because it kills surface growth but not underground growth.
- **Flame :** It is considered more effective than steam, but environment polluting

Application of herbicides

Most herbicides are applied as water-based sprays using ground equipment. Ground equipment varies in design, but large areas can be sprayed using self-propelled sprayers equipped with a long boom, of 60 to 80 feet (20 to 25 m) with flat fan nozzles spaced about every 20 in (500 mm). Towed, handheld, and even horse-drawn sprayers are also used. Inorganic herbicides are generally be applied in big farms of the European countries and The USA aerially using helicopters or airplanes, and can be applied through irrigation systems (chemigation).

15. Management of Vegetable Crops Under Polyhouse

In the present scenario of perpetual demand of vegetables for both domestic and export purpose, it has become of utmost importance to increase the production of quality vegetables round the year and drastic curtailment of per capita land holding further necessitates such high value vegetable cultivation. Cultivation under protected structures (greenhouse/glasshouse or polyhouse) is gaining lot of importance in the recent years to meet the increasing demand of fresh vegetables throughout the year both for domestic and export market. Growing vegetables in open is season bound especially when the temperature is extreme. In protected condition, the natural environment is modified suitably for optimum plant growth, which ultimately gives higher yield and better crop quality and helps in reaping high returns especially during the off-season. Hence, cultivation of vegetable crops in protected structures during the off-season is the best alternative to get high returns.

Protected cultivation can be defined as a cropping technique wherein the micro-climate surrounding the plant is controlled partially or fully as per the requirement of the plant species during their growth period. Among different protective cultivation structures, green house/polyhouse cum rain shelter is useful in Indian context. A greenhouse is quasi-permanent structure covered with a transparent or translucent glass or various rigid or flexible plastic materials, ranging from simple homemade designs to sophisticated pre-fabricated structure which is made up of wood, bamboo, steel or even aluminium.Main purpose of developing protected structures is to create a favourable environment for the sustained growth of plant to realise its maximum potential even in adverse climatic conditions. Greenhouses are designed to optimize the penetration of incoming solar radiation. In highly specialized production practices better control of the greenhouse environment viz., temperature, relative humidity, carbon dioxide and air movement become essential. for maximum net returns.

Effective weed management in different vegetable Crops

Crops	Weed control measures	Time of application
Tomato	Metribuzin @ 0.25-0.75kg ai /ha	Post-emergence
	Butachlor or Alachlor@ 2.0 kg ai/ha	Pre-emergence, one week after transplanting
Brinjal	Pendimethalin 1.0kg ai/ha or Fluchlorin 1.5kg ai/ha or Butachlor 2.0kg ai/ha + one hand weeding 30 days after transplanting	Pre-emergence
Chilli	Pendimethalin 1.25 kg ai/ha + one hoeing 40days after transplanting	Pre-emergence
	Nitrofen @ 2.5 kg ai/ha or Oxadiazon @ 1.0 kg ai/ha	Pre-emergence
Cabbage and cauliflower	Pendimethalin @ 0.56kg ai/ha or Oxyfluorfen @ 0.25kg ai/ha or Fluchoralin@ 1.2 kg ai/ha or Nitrofen @ 2.0 kg ai/ha + one hand weeding 30 days after planting.	One day before transplanting seedlings
Knol khol	Butachlor @ 1.5kg ai/ha or Fluchloralin @ 1.0kg ai/ha or Oxyfluorfen @ 0.25kg a.i./ha	Pre-planting
Onion	Pendimethalin @ 0.75 kg ai /ha or Fluchloralin @ 1.5 kg ai/ha or Oxadiazon @ 1.0 kg ai/ha or Nitrofen @ 1.0 kg ai/ha + one hand weeding 45 days after transplanting	Pre-planting
Garlic	Pendimethalin @ 0.75 kg ai/ha or Oxadiazon @ 1.0 kg ai/ha or Fluchloralin @ 0.9 kg ai /ha + one hoeing 45 days after planting cloves	Pre-planting
Carrot and radish	Tenoran @ 2.5 kg ai/ha or Alacnlor @ 1.5 kg ai/ha	Pre-emergence, pre-planting
	Dalapon @ 1.0 kg ai/ha or Fluchloralin @ 1.5 kg ai/ha or Nitrofen 2.5 kg ai/ha	One day after transplanting
Beet and turnip	Butachlor @ 1.5 kg ai/ha or Fluchloralin @ 1.5 kg ai/ha	Pre-emergence
Pea	Alachlor @ 0.75 kg ai/ha or Tribunil @ 1.5 kg ai/ha + one hand weeding 45 days after sowing	Pre-emergence
French bean	Alachlor @ 1.0 kg ai /ha	Pre-emergence
Cowpea	Alachlor @ 2.0 kg ai/ha or Nitrofen @ 2.0 kg ai/ha	Pre-emergence
Okra	Pendimethalin @ 0.56-0.75 kg ai/ha or Alachlor @ 2.0 kg ai/ha or Fluchloralin @ 1.5 kg ai/ha + one hand weeding 45 days after sowing	Pre-emergence
Muskmelon	Alachlor @ 2.0 kg ai/ha or Fluchloralin @ 1.2 kg ai/ha	Pre-sowing
Watermelon	Butachlor @ 2.0 kg ai/ha	Pre-sowing
Bottle gourd	Alachlor @ 2.5 kg ai/ha or Fluchloralin @ 2.0 kg ai/ha or Butachlor @ 2.5 kg ai/ha	Pre-sowing
Bitter gourd	Alachlor @ 2.5 kg ai/ha or Fluchloralin @ 0.75 kg ai/ha or Butachlor @ 2.5 kg ai/ha	Pre-sowing
Pumpkin and summer squash	Butachlor @ 2.0 kg ai/ha	Pre-sowing
Ridge gourd	Alachlor or Butachlor @ 2.0 kg ai/ha	Pre-sowing
Fenugreek	Pendimethalin @ 0.75 kg ai/ha + hoeing	Pre-emergence

Historical background vis-à-vis national scenario

The greenhouse technology was in use several centuries ago by the Europeans when they used the simple structure by digging a pit and covered with transparent sheet, probably mica, to provide protection to the crops. By the early 17th century, structures made up of stone and glass was being used by orangeries. However, the commercial production of food crops, cut flowers and ornamental plants under glasshouses started from the 18th century. By the early 1900s glasshouse designs were to the level of today's traditional designs. However, the most significant breakthrough in designs came into existence with the development of plastic technology. Polyethylene polymer was first developed in the 1930s and came into agricultural use in the early 1950s in the form of plastic films and pipes for water conveyance including sprinkler and trickle irrigation. New kinds of plastic, like polyvinyl chloride, polypropylene and polyesters were developed later, which are cheaper, more durable and opaque or transparent as required for greenhouse covers and several other applications. Now- days, nearly 90% of the new green houses are being constructed by utilizing Ultra Violet (U.V) stabilized polyethylene sheets as the glazing or clading material.

Indo American Hybrid Seeds Company was the pioneer in India to utilize greenhouse technology for commercial production of seeds, ornamental plants and cut flowers in the year 1965. In the early 1960s, Field Research Laboratory of DRDO at Leh attempted solar greenhouses comprising of polyethylene sheets for vegetable production research.

The impetus of this technology in India came in the mid 1980's with the emergence of U.V. stabilized Low Density Poly Ethylene (L.D.P.E.) which was first marketed by IPCL and boosted the greenhouse research and application for raising vegetables by providing UV stabilized film and aluminium polyhouse structures. The Defence Research and Development Organization (DRDO) has provided adequate support and technological help in this project. Research on vegetable production in Leh became successful to the extent that today there are more than 1840 small and medium sized greenhouses at Leh. Underground greenhouses and soil trenches are also being used on a large scale in these remote areas and much needed vegetables are being grown throughout the year in these hostile climates.

The importance of plastics in agricultural uses got further fillip with the formation of National Committee on the Use of Plastics (NCPA) in Agriculture in 1981 under the Ministry of Chemicals and Petrochemicals. It was then transferred to Ministry of Agriculture under division of Horticulture in 1993. With a view to promote agricultural productivity with special reference to optimizing the use of available water resources, improving quality of the product and reducing post harvest losses, the Government of India had allocated funds to the tune of Rs. 250 crores with provision of subsidy during VIIIth Plan period for the horticultural crops and NCPA has been entrusted with the task of implementing the plasticulture promotional programmes like drip irrigation, plastic mulching and greenhouse technology. The NCPA had under its control 22 Plasticulture Development Centres (PDC) located at different State Agricultural Universities/Institutes which provided research support for use of plastics and introduction of new technologies viz., erection of shade nets, anti hail net, plastic packaging of seeds etc.

Application of green house technology for commercial purpose is still in its infancy in India. Very few commercial greenhouses with climate control systems under private sector in the country are at present operative for production of some high value vegetable crops like, coloured sweet pepper and hybrid seeds of sweet pepper. Although, farmers are utilizing low and medium cost greenhouses basically for raising potted plants and seedlings in the nursery. In the Gangetic plains especially in Punjab, Haryana, Uttar Pradesh and West Bengal, the farmers are using low cost poly tunnel to raise healthy vegetable seedlings so that they can be transplanted early in the fields. In Assam, efforts have been made to raise vegetable crops in the polyhouse-cum-rain shelter structures made up of locally available materials such as bamboo, timber etc. with UV stabilized film as cladding material during the long south west monsoon periods.

Principles behind protected cultivation

The green house generally reflects back 43% of the net solar radiation incident upon it allowing the transmittance of the photosynthetically active solar radiation in the range of 400-700 nm wave length which increases photosynthetic efficiency of the crops grown inside the green house/polyhouse. Moderately short to moderately long wave length (15-390 nm) ultraviolet rays damaging to the crops are absorbed by the glass or polyethylene sheet used as cladding material for green house/poly house. For this reason, excellent crop growth and high yield are generally realized in the crops grown under such protected structures. At the same time, sunlight admitted inside the green house is absorbed by the crops, floor, and other objects in the frame work of the green house which in turn emit long wave thermal radiation in the infra red region (760-2500nm) and thus temperature inside the greenhouse rise. This phenomenon of trapping the solar energy in the green house is called the "Green house effect". Such increase in the temperature of the growing environment though unavoidable, is the most important function of a greenhouse for enhancing crop growth. Enhanced temperatures inside the green house/polyhouse accelerate plant growth and sustain plant growth even when outside ambient temperature is unfavourably low. However, during summers, inside temperatures rise higher than the optimum levels which necessitate the provision of evaporative cooling / ventilation to maintain the temperature inside the structure well below 35^0C. The ventilation system can be natural or a forced one. In the forced system, fans are used which draw out hot air from inside to the tune of 7-9m^3 of air / sec / unit of power consumed with necessary charging of air inside. Various methods of cooling systems like roof shading, water film covering and evaporative cooling are generally employed. The most common methods of evaporative cooling system are "Fan and pad cooling system", "High pressure mist system" and "Low pressure mist system". However, in the cold regions, natural cross ventilation and use of agro-shade net is sufficient to maintain the desired temperature inside the poly house.

Benefits of protected cultivation

- Environment control allows cultivation of vegetable crops anywhere in the world at any time of the year even under the inclement climatic conditions when it would not be otherwise possible to grow crops in open field conditions.
- It is ideally suited for managing and producing healthy and robust seedlings even in adverse weather condition.
- Maximum realization of crop yield on the basis of per unit area and per unit input use.
- This structure is beneficial for the production of genetically engineered and micropropagated vegetable varieties and hybrids.
- Ensure production of higher quality products free from chemical residue due to effective control of the pathogens and insect pests inside the structure.
- Eficient production of high value and high quality crops for export markets which increase the income from small land holdings.
- Protected cultivation technology is efficient to generate self employment for the educated rural youth in the rural sector.

Different protected structures for raising vegetable seedlings

Diferent types of protected structures can be utilized for raising the quality seedlings

Plastic low tunnel: Low tunnels made up of UV stabilized LLDPE film of 100 micron thickness are most suitable for raising the seedlings of cole crops, solanaceous vegetables and bulb crops when the outside condition is not favourable to do so. Seedlings grown under this structure are ready for transplanting at least 10-15 days ahead than open condition.

Hotbeds: Hotbeds are in use in western counties for raising the seedlings in the early seasons. Hotbeds are similar in construction to greenhouses but are much smaller, with a low level of sophistication for controlling environmental factors.

Cold frames: A cold frame is an enclosed ground bed, sometimes sunken, with a removable sash. Heat is provided through the tapping of solar energy. Proper regulation of ventilation, shading and watering is the key to success with cold frames. Cold frame, being inexpensive, can be used in certain places of India where winter is severe. It can be used for raising early seedlings as well as for hardening or conditioning of young seedling or rooted cuttings of perennial cucurbits.

Shade house: Shade house is very useful for protecting the young seedling or rooted cuttings from the external high temperatures and light intensity and in hardening them before transplanting in the field. In fact, shade house acts as an intermediate structure between greenhouse, hotbed or cold frame and field planting.

Net house: Under tropical conditions, where artificial heating is not necessary and cooling is very difficult and expensive the roof of net house may be covered with glass or glass substitutes like plastic film and the sides are covered with wire net. This structure provides sufficient ventilation and keeps the internal temperatures sufficiently warm during winter and cold during summer months.

Important considerations for development of greenhouse

There are a number of decisions that need to be taken before starting to build a high cost greenhouse. Careful planning prior to construction is an essential first step in the development of a successful, profitable greenhouse production system. Before anything else, the site of the operation must be chosen where single and three phase power supply is assured. An east-west greenhouse orientation is generally recommended in India to get better winter light. The site should not be shaded by trees or buildings either on the same or adjacent area. The greenhouse site and adjacent areas should be relatively levelled. Levelled sites allow the construction of large greenhouses that are more economical to heat and cool on an area basis than small greenhouses. However, a levelled site will make construction easier but may provide poorer drainage. If the selected site is perfectly levelled, it may be necessary to to elevate the area where greenhouse is to be developed. After preparation, areas adjacent to the greenhouse site should have at least 2% slope away from the green house site to facilitate the runoff of rainwater. The area with heavy clayey soil should be avoided mainly because of drainage problems that clay soils can create. Sites near waste oil, chemical storage, or dump sites should be avoided. Adjacent land use is another consideration in site selection. Area adjacent to commercial vegetable growing field will increase local pathogen inoculum levels (especially airborne inoculum) and local populations of damaging insect-pests and the area nearby abandoned commercial fields, dumps, and agricultural spoil piles create the problem for greater biotic stresses in greenhouse production. Different essential services that should be available in the proposed site include-

- **Suitable water supply**: High quality water is essential for successful greenhouse production. Water obtained from a shallow or surface source may carry pathogens, nematodes, and weed seeds into the greenhouse system. Shallow wells should not pose a problem, but if a well's casing is cracked or improperly installed, it will be more prone to contamination from surface water problems than a deep well. Regardless of the water source, samples should be tested for their suitability in use. A complete water analysis should include at least pH, electrical conductivity, bicarbonates, iron, sulfur, calcium, and magnesium content.
- **Transportation and market:** Access to good roads and proximity to the markets are important considerations. Transportation requirements to the greenhouse site relate directly to the intended operation's size and marketing arrangements. Modern refrigeration and transportation reduces the need to locate the facility near markets, but shipping costs are reduced when the facility is nearer the expected markets.
- **Skilled labour:** Availability of labor is very important and labor that is conscientious and skilled, or at least trainable, is less expensive and more reliable in the long run.

Structural considerations of greenhouse

Materials used in construction: Main greenhouse construction is stand-alone or gutter connects types. The basic greenhouse glazing/covering/clading material is either glass or polyethylene sheets. Depending upon the need of environmental control, greenhouses may be installed without any control mechanism, partial control mechanism or with full control mechanism and accordingly development cost vary from low, medium and high.

Components of low cost greenhouse (size: 25m length x 4 m width)

Sl. No.	Specifications	Quantity required
1.	GI pipe-15 mm	180m
2.	GI pipe-25 mm	51m
3.	MS Flats-25 mm x 3 mm	4 kg
4.	Nuts- 6 mm & 35 mm long	4 kg
5.	Cement – concrete mixture (1:3:6)	1 cubic meter
6.	End frames (Doors)	One/Two
7.	UV stabilized LLDPE film	32 kg

Components of medium cost greenhouse (size: 25m length x 4 m width)

Sl. No.	Specifications	Quantity required
1.	All items needed for low cost greenhouse	-
	Additional items over low cost	
2.	Skirt wall – 30 cm high & 22 cm wide	-
3.	Louvers (for fans)	Two
4.	Electrical fittings (power point, distribution box, MCB etc.)	-
5.	Poly grip assembly	-
	Environmental control equipment	
6.	Inflation blower	One
7.	Fans (60 cm diameter)	Two
8.	Monoblock AC pump	One
9.	Water tank (concrete with top cover)	-
10.	Thermostat & Humistat	One each
11.	Cooling pad and fittings	-
12.	Additional UV film for double layer	40 kg

Components of high cost greenhouse (size: 25m length x 4 m width)

Sl. No.	Specifications	Quantity required
1.	UV film for double layer	40 kg
	Additional items over medium cost for environmental control equipment	
2.	CO_2 generator + distributor	One
3.	Evaporative cooling system	-
4.	Heating system	-
5.	Humidification system	-
6.	Lighting	-

Contd.

Contd. Table

Sl. No.	Specifications	Quantity required
7.	Night curtain/shading system	-
8.	Drip irrigation system	-
9.	Fertigation system	-
10.	Porous flooring	-
11.	Benches	-

Types of greenhouse

Several types of greenhouses are in use in the country which include

- Single span (24 m length x 6 or 9 m width)
- Four span (42 m length x 20 m width)
- Saw tooth type (36 m length x 24 width)
- Leh design of mud brick greenhouse (25 m length x 4 m width)
- IPCL design of wooden framed greenhouse
- Jorhat design of bamboo framed low cost greenhouse cum rain shelter
- Other localized design.

Greenhouses can be constructed in a variety of styles that range in size from small stand-alone houses with less than 3000 ft^2 of total floor area up to large gutter-connected ranges with over an acre of floor area. Small stand-alone houses typically utilize either Quonset or gable style construction and offer the advantages of low overall initial investment and flexibility of operation. However, they are generally more expensive to construct and are more expensive to heat and cool than larger houses on a square foot basis.

Greenhouse ranges are formed by connecting individual units or bays under the the eaves or gutter. Each bay's roof can be a gable or arched structure or be sloped to form an element in a saw tooth roof. On an area basis, greenhouse ranges require a lower initial cost than do smaller stand-alone houses. Greenhouse ranges are also easier to heat and cool than smaller houses because they have proportionally less wall space exposed to the outside environment.

Gable and arched bays are the most common types of construction used for greenhouse ranges. Gable types are more expensive to construct than arched types but offer the advantage of being compatible with the installation of ridge vents. Saw tooth roofs are more expensive to construct than arched or gable types but offer the advantage of allowing the use of natural ventilation techniques.

Single greenhouses are normally 20 to 35 ft wide and 90 to 120 ft long which is a good working size and easily covered by commercially available sizes of polyethylene sheets. Greenhouse ranges are constructed in sizes that are limited only by the initial cost and the type of management scheme used in the greenhouse operation. If greenhouse ranges are to be mechanically ventilated, then they should be laid out so that air does not have to travel much more than 120 feet from inlet to outlet so that temperature gradients within the greenhouse are kept to a minimum. The height of mechanically ventilated greenhouses should be as low as is feasible for comfort of the workers and to reduce the overall volume that is heated or cooled. Naturally ventilated greenhouses require the use of a ridge vent on a gable roof or a saw tooth roof. This type of roof must be sloped between 3:12 to 6:12 for proper natural ventilation. This will normally result in a higher roof than is required for mechanical ventilation.

The greenhouse structure must be designed to withstand the loads that will be imposed on it during normal use which include environmental loads such as wind, rain, and loads imposed by hanging baskets or by equipment mounted on the structure. For vegetable production, most greenhouse frames are not strong enough to perform double duty as a trellis support for vegetables so, it is best to erect a separate trellis system within the confines of the house for vegetables.

Greenhouse frames can be constructed of different materials however, aluminum frames are long lasting, corrosion resistant, light-weight and can be prefabricated which can be permanently glazed and have low maintenance requirements. Aluminum frames have a high initial cost and require the services of experienced personnel during construction. Galvanized steel frames offer high strength and long life at less expense than aluminum frames, although maintenance costs of steel frames may be higher. Metals are good conductors of heat and the heat loss or gain through an aluminum or steel frame can be significant. Wood frames have a low initial cost but have high maintenance costs and are also flammable. Painting wooden frames white will improve lighting conditions within the greenhouse Pressure treated wood with preservatives resist the decay. Single greenhouses to be used for tomatoes and cucumbers should be of such size as to provide enough space for 5 to 6 double-rows of plants. Sidewalls should be high enough to at least 5 to 6 feet to permit close placement of outside crop row to sidewall with enough room on the outside to allow access to the crop.

Tomatoes and cucumbers require trellising to maximize utilization of space and to facilitate crop management for which heavy galvanized steel pipes or "I" beams are used at the ends of the trellis to anchor the wire. Secondary supports or props will be needed down the row to prevent sagging of the wire under heavy fruit load. The same type of greenhouses can be constructed for lettuce, pepper, and herb production. Lettuce and herbs are usually produced on benches. Bench production helps facilitate crop management and provides for a cleaner product.

When designing the physical layout inside the greenhouse, flexibility should be kept in mind. Even though tomatoes, for instance may be the intended first crop, factors may change that will force one to either grow another crop or more than one crop. Row spacing requirements, trellis and irrigation design should be chalked out before the final layout is the green house.

Glazing/clading materials of greenhouse

Historically, glass was the first material used to glaze or cover the greenhouses. Glass is an excellent material if its high initial cost can be tolerated. Glass has a lifetime of 25 years or longer, has low maintenance requirements, and transmits light well. It requires skilled labor for construction and has low impact strength. High strength grades of glass should be used for glazing. Glass is heavy and requires the use of strong framing with large structural elements that can block light. In recent years, synthetic sheets and films have replaced glass as the glazing material used in most greenhouses.

The most widely used cladding material is polyethylene film which is low in cost, light in weight, available in easy-to-apply wide sections and has a high light transmittance however, it has a very short life. Ultraviolet stabilized grades, which last from eighteen months to four years, still require more frequent replacement than most other glazing materials. Other plastic materials used to glaze greenhouses are polyvinyl chloride (PVC) films and sheets, polyvinyl fluoride films, acrylic sheets, polyester films, and polycarbonate sheets. Fiberglass reinforced plastic (FRP) sheets are widely being used for glazing material. Fiberglass is composed of glass fibers embedded in a polyester resin. Fiberglass sheets are less expensive than glass but more expensive than polyethylene. They are lightweight and impact resistant comparble to that of glass and fiberglass covered greenhouses are easy to fabricate. They have a fairly high light transmittance although it is slightly less than that of glass and polyethylene.However, FRP sheets degrade when exposed to ultraviolet light and their ultraviolet resistance can be improved by the use of polyvinyl fluoride coatings and acrylic additions to the polyester resin.

Glazing materials must transmit the maximum amount of sunlight to the crop while holding minimum heat gain or loss. The use of double layers of film can add insulating value to the glazing system with only a small reduction in light transmittance. When polyethylene film is used, this system is called double-poly. Double-poly construction utilizes a small fan to inflate the area between the two layers and this dead air space greatly improves the insulating properties of the greenhouse. When compared to a single layer of polyethylene film, the double-poly system reduces heat loss or gain by 35%-40% with a 10% reduction in light transmission. A tightly inflated double-poly greenhouse is more resistant to wind damage than a single poly house and most fiberglass greenhouses however, a double-poly greenhouse requires more maintenance to keep the double poly system properly inflated and a good polylock system is required to seal the layers of polyethylene sheet together at their seams. The polylock extrusion system is the set of aluminum extrusions that will hold the poly cover on the greenhouse frame.

Ultra violet (UV) stabilized polyethylene is commonly used and replaced every three to four years. Early morning is usually the best time to install new poly sheet when wind velocity is generally minimum because even a slight breeze can cause problems. It is also best to cover a greenhouse when the temperature is cool (ideal temperature would be 10^{0}C) which will allow the plastic to expand and contract without damaging the covering. The UV stabilized linear Low Density Poly ethylene (LLDPE) is presently available in 200 micron thickness and one kg film of 200 micron thickness covers approximately 5.3 sq.m. area.

Ground cover

In the crop production systems on soils, there must be a permanent barrier between the crop and the native soil beneath the polyhouse. Native soils have populations of nematodes and pathogens that are easily introduced into greenhouse operations and very difficult to control. They can get into the greenhouse whenever the barrier between the crop and the native soil beneath the house is broken. For ground style production of vegetables, house design should minimize vertical support posts to allow effective under sterilization of the production floor area. Concrete walkways between rows in the greenhouse and between buildings not only facilitate moving materials but can improve sanitation. Walkways control the movement of traffic into and around the production area and can minimize tracking contaminated soil or plant material into the greenhouse. This is an important consideration because prevention of nematode infestation is a much useful management strategy than any curative treatment.

Sealing the greenhouse

Greenhouses may be built with an airlock entrance design. Provision of entrance porch prevents direct ingress of wind, insects, soil, and pathogen spores into the greenhouse. Such an entrance has the additional advantage of making doors easier to open and close when the fans are operating. The double entrance also prevents short-circuit air flow patterns when ventilation fans are in operation.

Growing environment

Climatic condition inside the green house is manipulated through increase in inside temperature and relative humidity and as well as reduction in the solar radiation, CO_2 level and air movement. Provision of evaporative cooling system/ ventilation during hot summer months is essential to reduce the inside temperature for proper crop growth. However, increased temperature inside the greenhouse is beneficial to those areas when ambient temperature is not conducive for crop growth. In general, the relative humidity inside the greenhouse is generally higher than the outside, mainly due to transpiration load and effective ventilation is also required to control the relative humidity levels. In most parts of the country, solar radiation is not a limiting factor for plant growth. Light control inside the greenhouse can be affected conveniently either by shading or by supplementary lighting, whenever required. Growers in northern India should be careful in monitoring light levels in winters especially during prolonged foggy conditions. In peri-urban areas, different pollutants may get deposited on the polyethylene roof thereby reducing the light transmission significantlyand this problem is very common during winter months. In such situation, it is necessary to wash the roof frequently to maintain adequate light levels inside the greenhouse. Crops assimilate CO_2 from the atmosphere for photosynthesis and CO_2 concentration inside the greenhouse is always higher than the ambient early in the morning for which plant growth is generally vigorous inside the green house. With the onset of sun, this level quickly depletes and goes down the normal level during the day if adequate air exchanges are not maintained. CO_2 enrichment inside the green house, if required, is generally accomplished by burning fuel like propane or installation of a CO_2 generator.

Growing media

Several growing media can be used to produce vegetables in a greenhouse which include conventional ground culture in the soil to sophisticated hydroponic production which include nutrient film technique (NFT) and various versions of traditional NFT, perlite, rock wool, peat and bark mixes, lay-flat bags, upright bags, trough culture, pot culture.. The choice of growing media entirely depends on grower's preference and on marketing constraints since some markets particularly in developed countries may demand true NFT and not soil-produced vegetables. Each cultural system has its own set of requirements for management and successful crops are routinely possible with any system as long as production details are understood. For example, NFT requires precise management of fertilizer and irrigation programs. NFT system components are reusable, therefore, the system components are relatively inexpensive when amortized over many crops. A major disadvantage of NFT is that it is a recirculating system and the fertilizer solution circulates from the plants to the reservoir in a closed-system fashion. If a disease causing pathogen is introduced into the system, it can easily infect all of the plants in the greenhouse. Therefore, frequent flushes and extreme vigilance are required to help provide sanitation in the NFT system. Perlite and rock wool systems also rely on precise water and fertilizer management but are not in closed systems. Crops are grown in individually wrapped slabs of rock wool or bags/pots of perlite and receive the water and fertilizer from a micro system. Excess water and fertilizer solution leaches from the media and is removed from the greenhouse. The rock wool or perlite, however, must be replaced after every season or two. This would be less costly on a per year basis if several successive crops are grown in the same media. Ground culture is the most common production method mainly because of the involvement of comparatively less cost. However, any production system requires diligent sanitation between crops which involves removing old plants and bleaching, fumigating, or steaming production system components and growing surfaces. Different management operations include formulating fertilizer solutions, monitoring irrigation systems, checking greenhouse operations such as heating and cooling, pruning and tying the crop, pollinating different crops like tomato, sweet pepper, brinjal, etc, spraying for pest control, cluster pruning in tomatoes, harvesting, grading, packing, transporting, accounting book work, etc. and all of these operations must be done precisely on time. Therefore, the grower must be committed to doing a good job of management or must have competent labor force for these operations.

Greenhouse production of some high value vegetable crops

It has already been discussed that several production methods can be followed to produce vegetables in a greenhouse which include planting directly in the soil within the greenhouse, using containers with soil less mixes, nutrient film technique (NFT), rock wool production , production in perlite, etc. The last three techniques are called hydroponic production because all of the nutrients required by the plant are supplied through irrigation water.

Selection of crop

By and large, all vegetables can be grown in greenhouse condition. However, growing high value crops like cucumber, sweet pepper, tomato, okra, lettuce, broccoli, savoy cabbage, red cabbage, Chinese cabbage etc. to get more return per unit area. Each vegetable has specific cultural requirements for successful production.

Selection of suitable variety

The varieties selected for specific crop must be adaptable to particular growing zone and must meet the requirement of local as well as export market demand. Before going to large scale production of a particular crop, screening of suitable variety is a prerequisite. In general, cucumber varieties recommended for growing in greenhouses are of parthenocarpic type (produce fruit without pollination). In fact, care should be taken to avoid pollination by non-parthenocarpic cucumbers because this would result in bitter, misshapen fruits. Tomatoes selected for greenhouse production are indeterminate types. Indeterminate tomatoes continue to grow and set fruit throughout their life cycle and varieties selected for greenhouse production fall into two broad categories: large and cluster types. The larger types are harvested singly while the cluster types are harvested in clusters of four to seven fruit and are sold with the cluster stem still attached. Important characteristics of tomato suitable for green house production include high yields, freedom from fruit cracking, disease resistance, high percentage of number one fruits and freedom from green shoulder. Desirable varietal characteristics of sweet pepper are uniformity in shape and fruit colour (red, yellow, violet, etc.); tolerant to viral disease, high yielding ability and good fruit quality. Many different types and varieties of lettuce can be grown under greenhouse conditions which are loose-leaf, butter head and romaine or cos types which are preferred than the crisp head types.

Cultural requirements

1. Tomato

Tomato is well suited to conventional ground culture in soil. Hwever, hi-tech production system, seeds can be sown into various soil-less mixes as well as into rock wool or foam cubes and sanitized or new trays with drainage holes are used. Seedlings after emergence will need to be placed under optimum light (1000 to 1500 foot candles) and temperature regimes (24-26°C). All greenhouse environmental control systems need to be checked before transplanting which includes thermostats or controllers, alarm systems, fan motors and belts, furnaces, blowers, heat distribution tubing, and vents. Transplanting of seedlings can be done in about three weeks after sowing in medium which is thoroughly moistened with water or dilute nutrient solution. If bags of perlite are used for production, they should be moistened and the drippers should be placed in each bag before slits are cut to allow excess solution to drain away. For perlite bag or rock wool culture, seedlings are transplanted 45- 60 cm apart in the row with between-row spacing of 120 cm. Seedlings are transplanted at 90 cm apart between rows and 45 cm between plants in the soil system. However, spacing can vary widely depending on individual preferences and training system used. Tomatoes are usually transplanted in the greenhouse in the month of August-September for autumn-winter crop production. The irrigation system is then operated to apply nutrients to the blocks. Operation of the irrigation system might need to be manual or by time clock early in the season (a week or two). After this, the system can be operated by the starter tray. This procedure ensures that the seedlings receive enough nutrient solution while the roots are being established in the slabs through the irrigation emitter placed on top of the block with the flow of water directed at the block. High temperatures in some of the regions can harm the crop in some region. The grower should try to keep temperatures at 25-26°C as much as possible. Using fans, cool pads and shading the house are all methods that can help to reduce the temperature. Cooling pad thermostat should be set at about 26^{0}C. Night temperatures should be maintained at or above 16-17°C and both higher and lower temperatures than this range can result in fruit quality problems. High temperatures can result in flower abortion. Pulsing outside air into the greenhouse by a timer on the jet-fan vent will help to reduce humidity during the night. Plants need to be inspected for insects or disease and management programmes should be initiated.

The indeterminate tomato varieties grown in greenhouse must be trellised and prunned on regular basis. Some varieties may grow to very long height over 7-8 months production period particularly in temperate countries. Automatically, the greenhouse will not be able to accommodate the plants of such varieties vertically. Therefore, the plants are trellised up a string to a cable suspended about 8 feet above, at which time the plants are lowered and leaned. Sufficient string is used with each plant so that the string can be unwound and extended as the plants are lowered and leaned. As the plants continue to grow, they are periodically lowered from the overhead cable and leaned.

As soon as flowers appear, pollination should begin. The flowers must be vibrated or shaken each day in order to ensure pollination. This can be done with an electric tomato pollinator, an old electric toothbrush, a modified door bell motor or a small lawn blower. The best time to do this necessary operation is between 10 a.m. and 3 p.m. During overcast or rainy days, pollen will not be dry enough to adequately pollinate the crop; therefore, plants must be shaken during sunny dry days. Bumblebees which are more efficient for pollination but but less aggressive, than honeybees can also be utilized for pollination purpose and one bee hive generally lasts for eight to 12 weeks.

Prunning of crop and thinning of fruits must be done throughout the crop cycle. Lateral branches should be removed as soon as they reach to 2 - 3 inches in length from leaf axils while the main stem is allowed to grow continuously. Fruit must be thinned as they form in the flower clusters. At first or second cluster, the plants should be leaned slightly by moving or twine down the trellis cable a foot or so which will prepare the plant stem for the leaning and lowering process so that stem breakage is reduced. The variety will determine how many fruit should be left at each flower cluster and the fruits need to be harvested at least three times weekly for most of the season.

2. Cucumber

Cucumber is a warm season crop with required growing conditions of 26 to 28°C and plenty of sunlight. The plants are indeterminate in growth habit, continually producing fruit on new growth. Greenhouse cucumbers are more sensitive to low temperatures than tomatoes and minimum temperatures should be no lower than 18°C for sustained production. Prolonged temperatures above 35°C should also be avoided as fruit production and quality are reduced at high temperature condition.

Cucumber can also be grown successfully grown in soil however, different types of production systems and substrates (trough, NFT, bag culture, and rock wool) can be used in high-tech greenhouse production. Cucumbers are ordinarily established in the greenhouse as transplants using rock wool or foam blocks. Cucumber seed germinates rapidly (2-3 days) at an optimum germinating temperature of 28 - 29°C temperature. Once the seed has germinated in the germination room, the temperature should be lowered to 25°C for proper growth of the seedlings. Seedlings are usually ready for transplanting in two to three weeks under optimal conditions.

Cucumbers are vining crops, so they are grown with a trellis system. Several trellis systems have been developed to produce greenhouse cucumbers. These systems, using either one or two wires strung approximately 8 feet above the crop, are referred to as single and V-cordon systems. Both single and double rows of plants can be used with the single cordon system. In either case, the plants are trained up a string to the suspended wire where the plant is draped over the wire. At this time, the plant may be pinched and two laterals are allowed growing down to the ground, or the primary stem may be allowed to continue growing to the ground. In either case, all other laterals are removed as they appear and at least one fruit is allowed to develop at each node. In the V-cordon system, two wires are suspended so that the plants can be alternately strung to one wire or the other. This results in a V-shaped system of plants growing to the respective wires. In other ways, they are treated the same as the single wire system in terms of pruning laterals and draping over the wire. In addition, intra row spacing will differ between the two systems. In the single cordon system, plants are spaced 30-45 cm apart. If a single wire with a double row is used, the plants are spaced 45-60 cm apart with 60 cm between the double rows. With the V-cordon system, the plants are spaced 30 cm apart with plants alternatively strung to one wire or the other. Row to Row spacing for the double-row single-wire system is 180 cm; for the single-row single-wire system is 120cm and for the V-cordon system is 150 cm.

The most common pruning system for either vertical cordon or V-cordon trained plants is known as the umbrella system. In this system, all lateral branches are pruned out as they appear until the main stem reaches the overhead wire. The growing point of the main stem is removed when one or two leaves have developed above the wire. Two lateral branches near the top of the plant are allowed to grow and are trained over the overhead wire resulting in these two branches growing downward and the growing point of each lateral is removed when the branches come near to the ground. Fruits are developed at the node of each leaf. The fruits on the base 30 inches of the main stem should be pruned off as soon as they appear allowing the plant to vigorously produce early vegetative growth which is essential for maximum fruit production. Fruits above the basal 30 inches of the main stem are then allowed to develop. The productivity of the laterals is generally less than the main stem.

Fruit matures continuously as the plants grow and from anthesis to marketable size of fruit, it will take 10 to 14 days depending upon the cultivars. Therefore, continuous harvesting usually is done three times in a week. Harvested fruit must be protected from desiccation and bruising because the skin is rather tender and prone to water loss. Greenhouse cucumbers are individually shrink-wrapped to restrict desiccation and extend shelf life.

3. Sweet pepper or Capsicum

Capsicum plants are slow-growing particularly in low temperature condition, and need a high temperature to develop at a reasonable rate. They can be grown perfectly in soil, growing media or NFT, but they are sensitive to salt and require good water. Capsicum is prone to disorders like blossom-end rot and cracking, as well as to some pests and virus diseases.

For soil-grown crops sometimes broadcast sowing is applied with 1-2 gram per m^2. However, normally sowing is done in trays with 10 gram per m^2 for scattered sowing, or 40 gram per m^2 for space-sowing and optimal temperature for pepper seed germination is between 24 - 25°C. The young seedlings require good light conditions hence, after germination the covering material is removed and the temperature lowered to 23°C, although the temperature range of 18 - 25°C is good for the seedlings. The seedlings can be pricked out when the cotyledons are fully expanded after 12 days at high temperature, or after 18 days at low temperature condition.

Capsicum growth is very much dependent on temperature particularly the root temperature which must be sufficiently high (20-22°C). Lower root temperature increases the risk of *Pythium*, while too high root temperature will cause too much plant growth and insufficient flower production.So for growing under cold conditions, the use of benches or bottom heat is beneficial. Capsicum transplants require a minimum day temperature of 23 - 24°C, and a night temperature of 18 - 19°C. A relative humidity of 75 percent is ideal for plant growth but overhead spraying is not generally advised in temperate conditions as it increases the risk of fruit rot caused by *Botrytis* sp. however, over head misting is generally done in tropical conditions to reduce the inside temperature. CO_2 enrichment to 700 ppm will promote growth of the seedlings from the stage where true leaves are expanded. Good light conditions are essential and artificial lighting supports plant growth, but its economis feasibility depends on natural light levels, prices etc.

Plants are ready for transplanting five weeks after sowing in summer, or up to eight weeks under winter. The plant density in the greenhouse depends on the training system to be applied later (single-row, V-system, etc.) however, it is stem density rather than plant density that matters. Optimal stem density is around seven or eight stems per m^2 in summer (may be higher in a new glasshouse), and around six stem per m^2 in unfavourable light conditions in winter. It is most common to have two main stems per plant, for which 3 - 4 plants/m^2 should be planted. Good soil structure and moisture content are important, and high salt or high nutrient levels are unfavourable. The young plants require good calcium and nitrate supply. The roots will establish within a week, and preparations for plant training can start.

In growing media, nutrients should be available as soon as the seedlings start growing. In peat, liquid-feeding should start four to six weeks after pricking out. The plants are often supported by two small sticks without an elastic band.

The plants initially develop one single stem and after 9-13 leaves, a terminal flower develops and the plant branches into two. Sometimes three or four shoots develop naturally rather than two and also two flowers may develop rather than one. The flower in the first branching is the crown bud. Each shoot will branch further after one or two leaves, while producing flower buds at the point of division and the plant grows further in this way. If no pruning is applied, the later side-branches will not develop, thus naturally avoiding a too bushy plant shape. Plants should be pruned to the two strongest stems, and these should be supported by tying to a wire suspended 8 feet above the plants. All lateral stems are removed for the first five to seven nodes above the ground. Above this, plants are allowed to branch from the two main stems. One to two fruit will set per node. Plants are often vibrated or trellis wires tapped to ensure proper pollination and fruit set. In older plants, the pruned side shoots will carry fruit. Training and pruning should be done every three weeks, or even every two weeks in periods of fast growth. In summer it is better not to prune leaves in the upper 15 cm of the plant in order to provide shading to the fruit. At the end when the last fruit are set the plants can be topped to avoid stem breaking and to improve fruit size.

Pollination is not markedly improved by using an 'electric bee' or by spraying plant hormones, but pollination is clearly better when honey bees or bumble bees fly in the greenhouse. Bees increase the number of seeds in capsicum fruit, and such fruit attract more assimilates, and therefore grow faster or bigger with less deformity.

Day temperature and light are not so important for pollination itself, but high light and low day temperature are beneficial for setting and first growth of the new fruit. In aged crop, fruit set can be difficult especially when the light levels go down. Low fruit set particularly due to low light condition encountered in temperate climate can be overcome by several ways. The most practical way is by dropping the night temperature to 18°C keeping the day temperature at 20-22°C however, if fruit set need further help the day temperature can also be dropped to as low as 18 °C. It is good if the root temperature can be dropped to 18°C as it will give more flowers. Very low night temperature of 10-12°C generally ensures a high percentage of fruit set, but it induces fruit with a few seeds only, which will abort or remain small. About 18°C night temperature (and root temperature) is optimal for fruit set. Other way to help increasing fruit set is increasing the CO_2 concentration to 700-1000 ppm. Fruit set takes about a week, and is normally completed in five to eight weeks after planting. In an old plant stand, it is important to allow for a long fruit growth period (6-9 weeks for green fruit, 9-12 weeks for red) and to get the last flush set as soon as possible. This can best be done by harvesting the plants empty, even harvesting green, and then setting night temperature very low (even as low as 14°C) until fruit set is completed and raising the CO_2 level will help too. Fruit ripening can be speed up at higher temperature (24°C average) condition after the harvestable-green stage.

Too much fruit load on a plant may cause decaying of roots, and as a consequence *Pythium* may attack causing root rot. The number of fruit a plant can raise depends on light, temperature and plant size. Young plants can support only 12 fruit per m^2 under poor conditions, and over 24 fruit per m^2 under good light conditions (i.e. four to eight fruit per plant, or two to four per stem). An older crop can have over 40 fruits per m^2 at a time. Mature plants control fruit load themselves by fruit set and fruit thinning. In young plants, however, fruit load must be controlled by the grower, in order to achieve a reasonable fruit weight (e.g. minimal 150 gram per fruit). The best method to correct the number of fruit is by removing excess fruits which improves fruit quality (as inferior fruit are taken out) and fruit size (as remaining fruit experience less competition). The earlier the fruit thinning is done, the more the other plant parts (fruit, shoots and roots) will benefit however, if too few fruit are left on the plant, these are more at risk for fruit cracking. Secondly, fruit load can be corrected by raising the temperature just after fruit set which reduces the amount of assimilates available causing abortion of some new fruit. However, this method does not have the positive effects on fruit quality.

About 5-9 weeks after fruit set depending on the variety, fruit will reach their full size to get firm and green fruits (the 'harvestable-green' stage). Further ripening on the plant until fully-coloured condition generally takes another 2-4 weeks. The total period from blossom to full-colour ripe fruit is 7-12 weeks. Green fruit are incapable of ripening after removal from the plant however, when harvested a week after the harvestable-green stage the fruit will turn gradually into colour, which is not desired. The right stage for red fruit is when they have reached full colour and are still firm. Overripe fruits are soft, and have a low keeping quality. Harvesting green fruit will result in an earlier unloading, and thus earlier setting of new fruit. Green harvesting can be recommended for the first flush when the plants are still small. If only some fruit are harvested green, the other fruit are more at risk of cracking, so for that reason it is better to harvest all first fruit green at once. On the other hand, when all fruits from one stem are harvested at the same time, the sudden change in fruit load may cause a too vigorous growth of that stem producing too coarse flowers which will develop in low quality fruit. So the last ripe fruit should be harvested after some new fruit have set on that stem. The effect of harvesting on shoot growth can be used to achieve equal length of the stems on one plant. Capsicum fruit must be harvested with a very sharp, small knife, to get a smooth stem-end appearance, and so to minimize damage to other fruits.

Nutrient application through drip system

Generally in high-tech greenhouse production system, fertilizers are applied to the crop through drip irrigation system (fertigation). While the water moves through the laterals, it drops through drippers drop by drop to the individual plant in a rock wool, perlite, or bag system or to the channels in an NFT system. It is extremely important to choose an emitter that will not easily clog and one that will deliver adequate solution to the crop in a short period of time. Therefore, it is recommended that the drippers and tubing have at least a 0.05 inch inner diameter, especially for rock wool, bag, or trough culture. Applicators that position the dripper or tube end on a stake above the media surface are desired because they can be easily inspected for flow rate or clogging problems. Emitter flow rate is a critical aspect of the design of the irrigation system. Growers must be sure that the pump capacity, pipe sizing, pressure regulators and proportioner pump flow rates are all taken into consideration, so that an adequate and uniform flow rate is realized over the entire house. Systems might need to be reduced into smaller zones to achieve these goals. It is extremely important to design the irrigation system correctly before placing the first plant into the house. In any fertigation system, the basic components are: stock tanks for fertilizers, a water source, a method for mixing fertilizer and water in correct proportion, and a pump to move the water and fertilizer mix to the plants. In all systems, it is important to be sure the fertilizer is formulated properly. Through proper mixing of correct amount of fertilizer so that the plants will receive the right concentration (ppm) of the various nutrients. Most concentrates can be made up to be diluted 1:100 for ease of operation. Certain nutrients cannot be mixed in concentrated forms with other specific nutrients because insoluble precipitates may form. For example, potassium phosphate will form a precipitate (calcium phosphate) if mixed with calcium nitrate in the same stock tank. Therefore, all fertigation systems that rely on stock solutions will need to have at least two stock tanks. The pH of the solution is critically important when plants are grown hydroponically. The nutrient solution has no buffering capacity like that in solid media. This means that pH can readily change during the production cycle and must be monitored and adjusted to 5.5-6.5. Any fertilizer delivery system should be designed to facilitate cleaning after each crop. The delivery system should be easily accessed during and after cropping for repairs and for cleaning. The ends of the sub-mains in the greenhouse should be fitted with caps or valves for flushing. The most helpful manner to determine the effectiveness of the fertigation program is to analyze the plant tissue for nutrient contents which can help detect deficiencies and toxicities (mentioned in the chapter Fertigation).

Nutrient requirements of tomato in greenhouse system

Fertilizers	A Seeding to first fruit (g per 100 litre)	B First fruit to termination (g per 100 litre)
Magnesium sulfate	50	50
Monopotassium phosphate	27	27
Potassium nitrate	20	20
Potassium sulfate[1]	10	10
Calcium nitrate	50	68
Iron (Fe 330)	2.5	2.5
Micronutrient stock	15ml	15ml

Note: Potassium sulfate is optional; 15 ml from the stock solution of 450 ml is used for each 100 L final nutrient solution

Nutrient requirements of cucumber in greenhouse system

Fertilizers	A Seeding to first fruit (g per 100 l)	B First fruit to termination (g per 100 l)
Magnesium sulfate	50	50
Monopotassium phosphate	27	27
Potassium nitrate	20	20
Calcium nitrate	68	136
Chelated iron (Fe 330)	2.5	2.5
Micronutrient stock	15 ml	15 ml

Note: 15 ml from the stock solution of 450 ml is used for each 100 l final nutrient solution

Micronutrient requirements of tomato in greenhouse system

Chemical compounds	grams per packet
Boric acid (H_3BO_3)	7.50
Manganous chloride ($MnCl_2$ x $4H_2O$)	6.75
Cupric chloride ($CuCl_2$ x $2H_2O$	0.37
Molybdenum trioxide (MoO_3)	0.15
Zinc sulfate ($ZnSO_4$ x $7H_2O$)	1.18

Note: One packet (15.95 g) plus water to make 450 ml stock solution (heat to dissolve). Use 15 ml for each 100 litre final nutrient solution

Maximum desirable concentrations (ppm) for specific salt ions in water for greenhouse capsicum production.

Elements	Maximum desirable concentrations (ppm)
Nitrogen (NO_3 - N)	5
Phosphorus (H_2PO_4 - P)	5
Potassium (K^+)	5
Calcium (Ca^{++})	120
Magnesium (Mg^{++})	25

Contd.

Contd

Elements	Maximum desirable concentrations (ppm)
Chloride (Cl^-)	100
Sulphate (SO_4^{--})	200
Bicarbonate (HCO_3^-)	60
Sodium (Na^{++})	30
Iron (Fe^{+++})	5
Boron (B)	0.5
Zinc (Zn^{++})	0.5
Manganese (Mn^{++})	1.0
Copper (Cu^{++})	0.2
Molybdenum (Mo)	0.02
Fluoride (F^-)	1

Note: 1 ppm is equal to 1 milligram/litre, so to convert ppm to millimoles, the ppm concentration is to be divided by the molecular weight of the element that is to be working with. For example:

- 1 ppm of NO_3 = 1 mg/litre

  1 mg/ litre of NO_3 / 62 mg/mole = 0.016 millimoles of NO_3 in one litre
- 1 ppm of Fe = 1 mg/litre

  1 mg/litre of Fe / 56 mg/millimole = 0.018 millimoles of Fe in one litre.
- 1 ppm of magnesium (Mg) = 1 mg/litre

  1 mg/litre of Mg / 24 mg/millimole = 0.041 millimoles of Mg in one litre.

As these examples show, a solution containing 1 ppm of various elements or molecules will contain different mole or millimole amounts of these same elements.

- To convert millimoles to ppm:

  ppm = millimoles/litre x molecular weight (mg/millimole)

  Example:

  ppm NO_3 = 0.016 millimoles of NO_3 in one litre x 62 mg/millimole

  = 1 ppm NO_3

Calculating the required amounts of the various fertilizers is dependent on the volume of water to be used. This is determined by the volume of the stock tank (e.g. 200 litres) multiplied by the injection ratio (e.g. 100:1 or 200:1). For example, using a 200 litres fertilizer concentrate stock tank, and a 200:1 injection ratio, the volume of water that will be used to calculate the amount of fertilizer to add will be: 200 litres (stock tank volume) x 200 (injector ratio) = 40,000 litres

Management of insect-pests and diseases

Modern greenhouse design has tried to minimize or manage insect-pests and disease problems below economic threshold level so that plants can be grown pesticide free under many conditions. The following sanitation and other cultural practices are strongly recommended to manage insect-pests, mites as well as diseases.

1. A greenhouse should be thoroughly cleaned before planting a new or first-time crop by means of burning, burying, or hauling away all leftover roots and other plant parts so that there is no chance that insects in the egg, larval, nymphal, pupal, or adult stages could remain inside the greenhouse. Crop residues must be removed immediately after the final harvest.

2. In addition to removing all plant debris, growers should thoroughly clean or sterilize the greenhouse to make it as insect-free as possible through washing walls, steam cracks and crevices and fumigating with appropriate insect bombs, or using space sprayers or foggers. Broad-spectrum insecticides can also be used to spray cracks, crevices, walls, and other potential hiding places. Attempts should be made to spray around doors, windows, and other possible points of entry. In hot weather, closing up the empty greenhouse for a week will help eradicate insect pests

3. Some of the innovative managements include use of ultra-fine screening (50 mesh nylon net) over all openings to prevent harmful insects from entering the house. Plants should, however, be inspected daily for insect infestations. If pesticides are to be used, care must be taken in selecting materials and time frame of wetting period to harvest the crop. Products that are registered for field use may not be registered for greenhouse use. Moving fan jets to ground level and automating the watering and fertilizing process eliminate free water on plant surfaces, which reduces disease incidence.

4. Sanitation must be practiced not only during preplant times but also throughout the growing period. All plant parts generated by pruning should immediately be disposed. Culls (undesirable) or overripe fruit should be removed from the greenhouse and surrounding areas because insects are often attracted to and can live for long periods on these plant materials.

5. Planting outdoor crops near greenhouses should be avoided which may serve as host plants for insect pests and may be a source of diseases as well. Weeds should not be allowed to grow within the greenhouse or to become established around the greenhouse. A 10 - 30-foot vegetation-free zone around the greenhouse can be created with a heavy-duty geotextile weed barrier covered with gravel.

6. A clean stock program will aid in keeping insect-pests out of the greenhouse. Plants coming from other greenhouses should be carefully inspected for insects and diseases and temporarily quarantined until it is clear that the plants are free of pests. Greenhouse workers should avoid wearing yellow clothing because it is highly attractive to insect like whitefly which is the main carrier of viral diseases.

7. Addition of an 'air lock' type entrance to each production house so that wind-carried insects, soil etc., do not enter the production area directly from the outside

8. Prevention of un-sterilized soil from being carried into the production space.

9. Restricted access by visitors to production and transplant houses.

10. Raising transplants at a height of at least 1 foot above the ground to minimize dust or splashing soil contamination of plants.
11. Prohibition of cigar, cigarette, snuff, and chewing tobacco use by workers involved in production areas, so as to minimize contamination by viruses present in tobacco.
12. Rigid hand-scrubbing rules for personnel involved in pruning, pollinating, tying, or harvesting activities.
13. Filters on all air intakes to restrict air-blown soil and vector-insect entry.
14. Periodic tool, walkway, and bench surface treatment with disinfectants.
15. The use of disease resistant varieties when available helps to reduce disease problems.
16. Administration of biological control measures in the greenhouse environment as far as possible and if it is not feasible to use biological controls and if biorational pesticides do not offer sufficient control, then only chemical pesticides may be used with utmost care and precision.

Harvesting operations

The primary goal of harvest and post harvest handling operations is to maintain vegetable quality as close as possible to harvest condition through subsequent handling operations. Maintaining high quality is the thread that links each of the components in the harvest/handling system, and carelessness at any stage can quickly change the grade.

Crops must be harvested at the optimal stage of maturity for the intended use. Proper harvesting techniques are also essential for minimizing mechanical injury. Frequent sanitizing of hands/gloves and clippers can assist in reducing damage and disease problem. A reusable picking container must be selected that has smooth inner surfaces, is not too deep and is made of an easily cleanable material, such as plastic. Harvest operations should be done during the cooler part of the day and after harvest, the product should be protected from the sun to prevent temperature increase. Harvested material should be pre-cooled as soon as possible. Provision should be made for gently transporting the harvested product to the packing/storage area. Prior to reuse, the containers should be cleansed and sanitized to minimize the spread of decay pathogens.

Harvesting of tomato, pepper and cucumber should be a continual process throughout the growing cycle. Tomatoes and peppers are harvested with stems attached and are packed that way to avoid bruising and damage. Care should be taken so that tomatoes are not stored or transported with other vegetables such as peppers, lettuce or cucumbers because ethylene that is evolved during ripening tomatoes can damage other vegetables.

16. Post Harvest Handling and Storage of Vegetables

Horticultural development in in India has not yet taken proper shape as adequate attention was not paid so far to post-harvest handling, storage, marketing and processing ofdifferent horticultural crops including vegetables. Development has so far been confined mainly to production enhancement of different horticultural crops. Impact of increased production of vegetables is greatly lowered because of huge spoilage that takes place during the period between harvesting and consumption. It has been estimated that about 30 per cent of vegetables is rendered unfit for consumption every year on account of spoilage resulting in the draining of crores of rupees. A complex series of metabolic adjustments occur in the harvested vegetables which are influenced by dislocation of supply of nutrients, water and plant hormones from the parent plant to the harvested product. Basic respiratory process is necessary to supply the energy required to maintain the essential repair reactions upon which the integrity and independent existence of the organ depend. Because of this maintenance metabolism, the harvested vegetables continue their normal development which leads to the inevitable physiological consequences like, ripening, sprouting, toughening, senescence, etc. In addition, the harvested vegetables have the ability to respond metabolically to environment in which they are stored, to wounding and to the microbial attack. The whole process leads to post-harvest deterioration of the produce. In the present situation, where per capita consumption of vegetables is still very low, reduction of post-harvest losses of vegetables is of vital importance. The following advantages can be realized due to minimization of post-harvest losses and subsequent storage of the produce.

1. It helps in extending the period of marketing, preventing stagnation of the produce and onward crash in prices during peak harvest time.
2. It helps in regulated supply of high quality produce to the consumers, which ensures higher remuneration to the producer and longer availability of nutritious and healthy produce to the consumer at reasonable price.
3. The growers will naturally be inclined to produce more because they can reap profit for the produce which would have gone to waste otherwise.
4. Wastage of energy for production and marketing of the produce will be reduced.
5. It increases the food supply and provides better nutrition to the people with the same amount of non-renewable resources such as, land, water, energy, capital and labour.
6. It will reduce colossal waste, the problem of garbage disposal and consequent pollution.

Nature and causes of loss of harvested vegetables

Fresh vegetables are perishable in nature because they are composed of living tissues. A number of physiological and biochemical processes take place in the freshly harvested vegetables which contribute to different changes within the product resulting in the deterioration of the quality, condition and appearance of the vegetables. Again, all vegetables at the time of their harvest are covered with countless bacterial and fungal spores, and some of which can cause decay of the fresh produce. So, nature and causes of losses of the fresh vegetables may be categorized as: (i) losses due to physiological processes, (ii) losses due to biochemical reactions, and (iii) losses due to microbial invasion.

Losses due to physiological processes

Different physiological processes like transpiration, respiration, senescence, sprouting and seed germination which are continued in fresh vegetables are the main causes of this type of loss.

Transpiration: Majority of the vegetables contain more than 80 per cent water and some of them like tomato, cabbage, spinach, palak, asparagus, French bean, lettuce, cucumber, etc. contain as high as 90-95 per cent water. Vegetables lose water mainly by transpiration through stomata *(e.g.,* leafy greens). Other paths of water loss are stem scars *(e.g.,* tomato), lenticels *(e.g.,* potato) and cracks that are resulted from mechanical injury. Minimum water loss causes many vegetables wilted or shrivelled which make the tissue tough or mushy, less crispy and palatable and eventually render the produce unmarketable. This happens to several leafy greens and root crops within a few hours after harvest. Tissue structure of the vegetables and surface area to volume influence the rate of water loss. For this reason, water loss is the least in tomato, moderate in cabbage and sweet pepper, high in summer squash and very high in French bean, different leafy greens, etc. Water loss is also associated with degradative effects on pigments and vitamins. Hydrolysis of protein also occurs under wilted condition.

Respiration: In general, there occurs a sudden spurt in the rate of respiration in the vegetables immediately after harvest followed by a gradual decline with the advent of senescence. Enzymatic oxidation of sugars and degradation of other compounds like, protein, lipids and organic acids are also channelised through respiration. Respiratory rate of vegetables is closely related to their storage life. Higher the rate of respiration, shorter will be the shelf-life of the produce and *vice versa.* Respiration in the harvested vegetables depends on temperature of the storage atmosphere.

Respiration results in reduction of both reducing and non-reducing sugars, and even protein content. Vegetable crops may be classified according to the rate of respiration by the freshly harvested edible plant parts, and respiratory activity is evaluated by the evolution of CO_2 per kg of the produce per hour. Carbon dioxide output may be less than 40 mg/kg/hour in vegetables with low respiratory activity, while it may be more than 120 mg/kg/hour in vegetables with very high respiratory activity. Vegetables which respire less after harvest deteriorate less rapidly and can be stored for relatively long period if other factors remain in favour.

Respiratory activity of the produces	Crops
i) Very low	Potato, onion
ii) Low	Cabbage, sweet potato, turnip, cucumber
iii) Moderate	Chilli, carrot, tomato, beet, sweet pepper
iv) High	Hyacinth bean, French bean, peas, lettuce, lima bean, radish
	Green onion, muskmelon, watermelon, cauliflower, okra, leafy greens, broccoli, asparagus, parsley

Source: Desai, U.T., Kadam, S.S. and Salunkhe, O.K. (1993) Post-harvest handling, storage and processing of vegetables, (In) : *Vegetable Crops* (eds. T.K. Bose, M.G. Som and J. Kabir), Naya Prokash, Calcutta.

Ripening: It is the sequential changes in sensory factors, colours, texture and taste of the fruit marking the completion of fruit developmental phase and subsequent commencement of senescence phase. The significant physical change that occurs during ripening is the softening of tissues and during softening, pectin which is derived from pectose remains as a colloid in the fruit juice in course of ripening process. Ripening makes those vegetables unfit for marketing which are generally consumed at their immature stage. The vegetables which are consumed at their ripe stage are spoiled due to over-ripening. During over-ripening, pectin is decomposed to pectic acid and alcohol.

Senescence: It is the last stage of development of the tissue when anabolic biochemical processes give way to catabolic processes leading to death of the tissue. Increased membrane permeability, respiration, phosphorylation and shift in protein synthesis lead to deteriorative processes and eventually senescence. During senescence, the most obvious change is the degradation of chlorophyll and development of carotenoid pigments. Excessive senescence of the vegetable tissues causes serious loss of the produce. Senescence process invariably evolves ethylene which autocatalyzes several physio-biochemical reactions including increased respiration and ripening. Ethylene induces changes in permeability of the mitochondrial membrane, thus facilitating the increased ATP movement and thereby initiating several reactions including respiration.

Sprouting: It is the active resumption of accelerated growth from the storage tissues (tuber, bulb, corm, etc.) which cause deterioration of potato, onion and garlic in storage. Sprouting is increased with the increase in temperature and relative humidity of the storage.

Seed germination: Seed germination in some vegetables like chow-chow *(Sechium edule)*, cowpea, French bean, etc. sometimes occur in the harvested fruits itself which considerably lowers the product quality.

Losses due to biochemical reactions

Different biochemical reactions in compounds in the harvested vegetables lead to significant losses of the produce in storage and some of this biochemical reactions are-

Enzymatic carbohydrate conversion: Conversion of sugar to starch is very much detrimental to most of the vegetable crops, and conversion of starch to sugar is harmful for potato and sweet potato.

Enzymatic change in flavour: It usually lowers the product quality.

Enzymatic change in vitamin content: In general, vitamin C content decreases during storage probably due to activity of the enzyme, ascorbic acid oxidase.

Synthesis or destruction of pigment: During senescence and ripening, chlorophyll degrades with subsequent development of carotenoid pigments. These changes are detrimental to vegetables which are consumed at green, tender stage and are beneficial to the vegetables which are consumed at ripe stage.

Softening of tissues: It is the physical change in the produce, mediated by different enzymes and is associated with ripening and senescence processes. It is usually detrimental to most of the vegetables.

Toughening and spongyness: It generally occurs in green beans, lettuce, celery, etc. due to development of fibre which greatly reduces the market value of the produce.

Fat auto-oxidation: Fat content in vegetables though very low, yet is important from quality and storage point of view. Rancidity due to fat auto-oxidation imparts undesirable flavour to the stored product.

Millard reactions: These non-enzymatic reactions take place between nitrogenous compounds and sugar, nitrogenous compounds and organic acids, sugar and organic acids or organic acids among themselves which may independently cause browning of the product.

Losses due to microbial invasion

Decay caused by different moulds (fungi) and soft rot by bacteria are important causes of spoilage of the harvested vegetables. Tender and succulent nature of the vegetables makes them easily invadable by the micro-organisms. Again, majority of vegetables are grown near the soil line which makes easy reach for the microbes to the vegetables. Besides, rough and careless handling, cuts and bruises make easy entry of the decay causing organisms. The most common fungi causing rots in vegetables are *Alternaria, Diplodia, Phomopsis, Fusarium, Rhizopus, Botrytis. Aspergillus, Penicillium,* etc. and the most common bacteria causing soft rots are *Erwinia, Pseudomonas* and *Ceratocystis.* High temperature and relative humidity favour the development of these post harvest decay causing micro-organisms. Some of the important storage diseases are: neck rot of onion caused by *Botrytis allii,* black mould of onion by *Aspergillus niger, dry* rot of garlic by *Fusarium oxysporium* and *F. solani,* soft rot of cabbage by *Erwinia carotovora,* storage rot of brinjal by *Botrytis cineria.* etc.

Factors responsible for post-harvest losses

- Inadequate knowledge of maturity standards and harvesting of vegetables.
- Improper handling of the produce during harvest.
- Lack of pre-cooling facilities immediately after harvest to remove field heat.
- Improper curing which leads to huge loss of potato and onion.
- Non-availability of adequate multipurpose cold storage facilities.
- Lack of post-harvest treatment with fungicides.
- Inadequate use of packaging materials.
- Lack of quick, efficient and cost-effective transport systems.

- Non-utilization of packing materials like SO_2 releasing pads, and ethylene absorbants.
- Seasonal stagnation of the produce.
- Limited utilization (less than 1%) of the produce for processing.

Measures to control post-harvest losses

Pre-harvest operations: Proper pre-harvest cultural operations help in prolonging the shelf-life of vegetables and some of the pre-harvest operations are mentioned.

- Avoidance of heavy application of nitrogenous fertilizers with essential supply of potassic fertilizers to improve keeping quality of vegetables because potassium activates or stimulates enzyme activity that are involved in carbohydrate and protein metabolism and is essential for cell organization and structure of cell walls.
- Irrigation should be discontinued 15 days before harvesting potato, onion, garlic, sweet potato, elephant foot yam, yam, etc. for obtaining tubers/bulbs with maximum storage life as it checks sprouting.
- Pre-harvest spray of maleic hydrazide (2000-3000 ppm), a growth inhibitor in combination of Difolatan (0.25%), a fungicide 15 days before harvest reduces storage loss and inhibits sprouting in onion.
- Pre-harvest spray of maleic hydrazide (500-2500 ppm) inhibits sprouting and thereby reduces losses of potato in storage. It seems that effect of growth retardant like, maleic hydrazide may be a consequence of the depression of gibberellin synthesis.
- Pre-harvest spray of maleic hydrazide to carrot inhibits both sprouting and rooting during storage at 0°C.
- Pre-harvest spray of maleic hydrazide (3000 ppm) 3 weeks before harvest is effective to inhibit sprouting and prevents weight loss and discolouration of cloves in garlic.
- Three pre-harvest sprays of Difolatan (0.2%) at an interval of 10 days are effective in controlling post-harvest diseases of tomato. In tomato, foliar sprays of chelated iron before flowering lower the decay percentage and weight loss in storage.
- Blanching in cauliflower (tying of leaves to protect the curd from scorching sunlight) ensures good quality and comparatively long storability of the curd.
- Mulching also somewhat prolongs the shelf-life of vegetables because mulching materials obstruct reflection of soil heat to the produce and thereby reduce field heat in them which checks deterioration.

Harvesting techniques

Development of vegetables starts with the formation of the edible plant parts like, setting of fruits, emergence of seedling, the swelling of a root, tuber or bulb, the elongation of a stalk or petiole, etc. Horticultural maturity in vegetables reaches when the edible portion has attained some desirable conditions. Harvest is a single deliberate operation to separate the edible plant part from its growth medium. Time of harvest is a complicated consideration as great differences occur in the rate of development and maturation of different vegetables. Harvesting should always be done at proper stage of maturity because it not only determines the quality of the product but also prolongs its shelf-life. Tomato harvested at breaker stage ensures good quality, better shelf-life and proper development of colour. Harvesting should be done with care to avoid damage, bruising and cracking of the product which checks easy entry of the decay causing microorganisms. Harvesting during hot period of the day should be avoided as immense field heat causes wilting and shrivelling of the produce. Evening harvested tomato has the longest shelf-life, uniform ripening and better quality. Harvesting during or immediately after rain should not be done because most favourable conditions for multiplication of the decay causing micro-organisms are provided at that time.

Handling of the produces

Washing, trimming and grading: After harvesting, the produce should be thoroughly washed with clean water, preferably running water or by sprayers to reduce the load of micro-organisms and to remove field heat from the produce. Rotten, diseased, insect damaged and discoloured leaves of the vegetables should be trimmed off and at the same time damaged vegetables should be discarded. The harvested vegetables should be graded in accordance with shape, size, colour, maturity, freedom from disease and pest attack, etc. which help the storage process convenient and prevent the healthy produce from being contaminated by decay causing micro-organisms.

Precooling: It is the process of rapid removal of field heat from freshly harvested produce before their packaging or transportation. It reduces weight loss and shrinkage, causes delay in ripening and retards ethylene production which restricts enzymatic and respiratory activity and also reduce refrigeration load while storing at low temperature. Prompt cooling to required temperature also inhibits growth of decay causing micro-organisms. Pre-cooling is particularly important when the vegetables are transported to long distances. Pre-cooling of vegetables with cold water mixed with some fungicides to 10-12°C increases shelf-life by reducing respiration and ripening rate with ensuring good surface colour. Various methods of pre-cooling are hydrocooling, air-cooling, room cooling, forced air-cooling, top icing and vacuum cooling. In Indian condition, pre-cooling of vegetables is generally done by immersing the produce in water. Water is an excellent medium to which heat from the vegetables is transferred. This hydrocooling will be more effective if ice-cold water is used. Most modem method of pre-cooling is vacuum cooling. In the developed countries, pre-cooling is mostly done by keeping the produce in refrigerated trucks and by placing ice in the packages.

Curing: It is a post-harvest treatment of tuber and bulb crops particularly potato and onion. In this treatment, the produces are exposed to relatively high temperature and relative humidity conditions at farm level which cause suberization of the outer tissue followed by development of periderm in potato, sweet potato and other tuber crops and drying of scale leaves and compaction of neck in onion and garlic. The protective layer due to formation of special cell wall thickening may be of two types: primary and secondary, *i.e.,* periderm. The primary protective layer is formed as a result of lignification and suberization of parenchyma cells in this region or of cells arising from them by irregular cell division. Different cell wall materials like, cellulose micro fibrils, wound gums, etc. are deposited in the cell wall. These tough coverings act as effective barrier against microbial invasion and water loss by transpiration. Proper curing can considerably reduce spoilage of potato, onion and other bulb and tuber vegetables like garlic, sweet potato, cassava, elephant-foot yam, yam, etc. in storage. Curing of onion and potato in perforated plastic crates for 5 days at room temperature is the most effective for reduction of losses due to rotting in storage.

Waxing: Application of fungicides and sometimes growth substances and other chemicals along with edible wax is quite helpful in prolonging the shelf-life of different fruit vegetables. Waxing is particularly helpful for tender fruit vegetables like okra, ridge gourd, pointed gourd, bitter gourd, snake gourd, brinjal, etc. Fruits of bitter gourd and snake gourd dipped in solution of wax emulsion (6-12%) + sodium phenyl phenoate for 30-60 seconds prolong the shelf-life up to 3 and 14 days, respectively at room temperature condition. In okra, fruit dipping in wax emulsion (12%) + cycocel (100 ppm) is effective in minimizing the physiological losses in weight and in retention of chlorophyll up to 6 and 12 days at 32°C and 10^0C conditions, respectively. Coating of muskmelon with wax emulsion and thiabendazole, a fungicide increases storage life up to 7 days.

Post-harvest treatment of chemicals

Plant protection chemicals: Control of decay causing micro-organisms is of prime importance in storage. Pre-harvest sprayings along with post-harvest application of fungicides and antibiotics significantly prevent the decay causing fungi and bacteria. Fungicides like Benlate, Thiabendazole, Benomyl, Dithane M-45, Captan, etc. are very effective against many decay causing pathogens. Application of lime and formalin is sometimes effective. Dusting of lime @ 12 kg/ha at cut surface of leaf blade during harvesting effectively reduces black mould of onion. Application of formalin (0.3%) dip is effective against dry rot of garlic. Benomyl is effective in preventing growth of most damaging fungi in tomato. Bacterial soft rot of packed spinach and lettuce can be checked by dipping in Streptomycin (1000 ppm). Application of vitamin K_5 (50-500 ppm concentration) markedly inhibits the activity of a number of micro-organisms and prolongs shelf-life of vegetables.

Antitranspirants: Both transpiration and photosynthesis show close dependence upon leaf resistance. Since CO_2 entering the substomatal cavity has to traverse the same pathway as the escaping H_2O molecules during transpiration, there arise corresponding resistance for CO_2 and H_2O diffusion. Photosynthesis is more likely limited by the diffusive resistance, the internal process than by stomatal capacity. The use of antitranspirants as an aid to increased water conservation and consequently enhanced water use efficiency is based on the anticipation that chemically induced increased diffusive resistance cause transpiration to decline more than photosynthesis. Application of antitranspirants (Mobileaf, Kaoline, etc.) is particularly beneficial for green leafy vegetables.

Growth substances: Deteriorative physio-biochemical processes that go on inside the harvested vegetables are mediated by indigeneous growth hormones, principally auxin, gibberellins and cytokinins. Exogenous application of growth substances can alter different metabolic process, and some of these alterations are beneficial to extending the shelf-life of vegetables. Yellowing due to degradation of chlorophyll is one of the main factors that limit the shelf-life of cauliflower, cabbage, broccoli and Brussels sprout. Senescence which causes yellowing can effectively be retarded by pre or post-harvest application of BA (N6-benzyladenine), a synthetic cytokine compound at lower concentrations (5-15 ppm). Cytokinin controls development of chloroplast through stimulating the conversion of proplastids into chloroplasts with grana. Auxinic growth substances like 2,4-D, 2,4,5-T have profound effect on delaying leaf abscission of cauliflower when applied before or after harvest. Ripening of tomato can be delayed and degradation of chlorophyll in leafy greens can be checked by post-harvest application of gibberellins. Growth inhibitors like, MH (maleic hydrazide), CIPC (chloro-iso-propyl carbamate) and MENA (methyl ester of naphthalene acetic acid) are effective in controlling sprouting of tuber and bulb crops. Potatoes treated with 1% aqueous emulsion of CIPC and kept in cold storage do not have any sprouting for 5 months. Synthesis of solanine in potato is inhibited by Alar (N, N-dimethyl aminosuccinamic acid). Cycocel (2-chloroethyl-trimethyl ammonium chloride), another growth retardant can prolong the shelf-life of lettuce, asparagus, broccoli, etc. However, effective concentration and right stage of application is very important for obtaining maximum benefit.

Maleic hydrazide at (1000 ppm) when applied 3 times at weekly interval reduces bulb sprouting of onion in storage. In broccoli, post-harvest dip in 20 ppm BA for 15 seconds and held at 25°C under polyethylene cover prolongs the storage life by retaining green colour. In Brussels sprouts, spraying with GA (100 ppm) extends the shelf-life to 9-10 days at 20°C. In head lettuce, application of 8 ppm kinetin for 30 seconds and packed in perforated polyethylene bags and stored at 6°C and 90% relative humidity retard senescence. In tomato, application of cycocel 100 ppm decreases fruit decay and weight loss. Post-harvest treatment with BA at (5-20 ppm) increases the shelf life of radish. In okra, fruit dipping in the solution of 100 ppm GA for 10 minutes and packed in polyethylene bags extend the shelf-life up to 9 days. In pointed gourd, fruit dipping in 100 ppm GA or MH solution extends the shelf-life up to 7 days in ambient condition. Application of BA at (50 ppm) concentration is effective in reducing decay loss of ridge gourd stored at 0.5°C temperature.

Other Chemicals: In cucumber, application of borax (4%) reduces fruit decay when stored at room temperature. In tomato, potassium permanganate ($KMnO_4$) increases the length of storage life by absorbing ethylene from around the stored fruits. Broccoli, dipped in sodium benzoate solution (0.01 M) for 3 minutes and then storing at 20°C effectively reduce the rate of ethylene production, respiration and subsequent yellowing. Okra, dipped in the solution of ascorbic acid (250 ppm) and then packed in polyethylene packages prolongs the shelf-life up to 12 days at 32°C. Post-harvest application of calcium nitrate or chloride in the form of dip or pressure infiltration extends the shelf-life of tomato and other vegetables mainly by reducing weight loss and checking CO_2 and C_2H_4 production. In pointed gourd, fruit dipping in sodium benzoate solution (250 ppm) extends the shelf-life up to 7 days in ambient condition.

Ethylene absorbants: Ethylene, a volatile plant hormone, is usually synthesized and eventually emitted from the tissue during senescence or ripening. This emitted ethylene also acts as an autocatalyst to these physiological processes which lead to further deterioration of the produce. So, fast removal of the emitted ethylene from around the produce checks these deteriorative physiological processes and thereby prolongs the shelf-life of the vegetables. Several chemical formulations like brominated activated carbon, celite with $KMnO_{4,}$ $KMnO_4$ on vernl iculite have been proved to be effective as ethylene absorbants. Application of $KMnO_4$ (1000 ppm) in paper lining can effectively be used as cushioning material for packaging tomato which acts as ethylene absorbent.

Packaging

It is an important consideration in prolonging the shelf-life of vegetables because of the protection provided by the packages from external injuries and from being shrivelled. Adequate packaging coupled with effective marketing can reduce the post-harvest losses considerably. So, the two main functions of packaging are: to assemble the produce in convenient units for handling and to protect the produce during transportation and marketing operations. The common packages for vegetables in India are wooden boxes, wire-bound boxes, bamboo baskets, arhar *(Cajanus cajan)* stick baskets, corrugated fibreboard boxes, plastic crates and jute bags. Cheap materials like, polyethylene films, paperboard boxes lined with polyethylene can effectively be used in packaging. Wrapping of vegetables with newspaper and tissue paper may be useful because it reduces physiological loss in weight. Plastic crates are better in case of bulk transportation from the point of view of less bruising loss. Polyethylene packaging reduces decay and softening of the produce but the thickness and permeability of the film to CO_2, O_2 and water vapour need to be standardized for each vegetable. Handling of the produce during packaging should be prompt to minimize deterioration.

Transportation

Most of the post-harvest losses of vegetables occur during transportation and distribution. It is estimated that spoilage of the vegetables takes place to the tune of one-third between farm and roadhead and another one-third between road head and consumer. In India, transportation network for perishable commodities like vegetables is very weakly developed. The concept of refrigerated transport for vegetables is yet to be developed on commercial basis. Spoilage of the produce during transportation occurs mainly because of decay owing to uncontrolled temperature and relative humidity condition in the carriages. This deterioration in transit is enhanced due to inadequate and loose packaging.

Marketing

Interests of the producers as well as consumers are poorly served due to present faulty marketing systems for vegetables in India. Moreover, huge quantities of wastes are also dumped due to poor marketing.

Storage

Storage is one of the most important aspects of the post-harvest handling of vegetables. Storage extends the period of availability of fresh vegetables by arresting the metabolic breakdown and decay by controlling the temperature, relative humidity, concentration of certain gases in the storage atmosphere, chemical treatment and by irradiation. Forms of storage range from primitive on-farm storage to highly sophisticated controlled environment storage.

On-farm storage: The time lag between harvest and marketing cannot be kept to minimum in Indian condition owing to poor transport network. In this situation, on-farm storage of some hardy vegetables like cabbage, beet, carrot, turnip, etc. for very short period sometimes checks post-harvest losses to some extent. The crops are generally placed in piles on the ground and covered with straw at first and then with soil. Sometimes, vegetables are stored in the farm in trench or pit.

Home storage: Limited quantity of some vegetables like ripe pumpkin, ripe winter squash, sweet potato, yam, well cured potato, onion and garlic, etc. which keep well in relatively high temperature and low relative humidity condition in storage can be kept in the shady and cool comer of the house for quite a long time.

Low cost storage without refrigeration

Ventilated storage structure for onion and garlic: Under Indian condition, onion and garlic are stored in hut-like ventilated storage structure kept in open where considerable storage loss due to drying, rotting and sprouting occurs. A considerable reduction of spoilage of onion and garlic can be achieved by modifying the existing storage structure with the introduction of top and pile ventilation which is accomplished by perforated bamboo pipes and by keeping the storage structure under shed. Storage of onion in perforated plastic crates stacked in a ventilated room or shed reduces the storage loss to a considerable extent.

Zero energy cool chamber: Based on the principle of evaporative cooling, this cool chamber achieves its objective by maintaining low temperature and high humidity as compared to the ambient conditions outside. The cool chamber is provided with a wet porous bed and the passage of air which causes cooling and humidification by evaporation of water. Vegetables are kept in plastic crates inside the cool chamber to avoid crushing and to prevent direct contact of the vegetables with wet bricks. Due to low temperature and high humidity inside the chamber, the shelf-life of vegetables is increased considerably.

Cold storage: Cold storage or refrigerated hold has nowadays become a part of the modem living. Marketability of perishable produce like vegetables is closely linked with the development of cold storage. Low temperature storage is the best method of slowing down metabolic processes and decay of vegetables and thereby most effectively extends the shelf-life of the produce. It is always desirable to retard the physio-biochemical processes that go on inside the harvested vegetables with a view to preserve the quality and nutritive value of the produce. Low temperature has profound effect on the rate of respiration of the harvested vegetable and on the rate of multiplication of the decay-causing micro-organisms. Higher the temperature more is the rate of respiration and at the same time multiplication of the micro-organisms. Low temperature also reduces the moisture loss from the produce because moisture holding capacity of air at low temperature is much less than it is at warm temperatures. Besides low temperature, relative humidity of the storage chamber is another important factor. Low relative humidity of the storage atmosphere leads to rapid moisture loss from the produce while higher relative humidity favours microbial growth. In the cold storage, both temperature and relative humidity are considered simultaneously and in fact, the requirements are specific for each vegetable.

Recommended storage temperature and relative humidity for different vegetables

Vegetables	Temperature (0°C)	Relative humidity (%)	Storage life (Weeks)
Brinjal	10.0-11.1	92	2-3
Tomato (unripe)	8.9-10.0	85-90	4-5
Tomato (ripe)	7.2	90	1.0
Chilli (green)	7.2	85-90	3-5
Chilli (ripe)	5.6-7.2	90-95	2.0
Potato	3.0-4.4	85	34.0
Cabbage (early)	0.0-1.7	92-95	4-6
Cabbage (late)	0.0-1.7	92-95	12.0
Cauliflower (late)	0.0-1.7	85-95	7.0
Bitter gourd	0.6-1.7	85-90	4.0
Cucumber	10.0-11.7	92	2.0
Muskmelon	1.7-3.3	85-90	1.5
Pumpkin	10.0-12.0	70-75	24-36
Wintersquash	12.8-15.6	70-75	24-36
Watermelon	7.2-15.6	80-90	2.0
Beet (topped)	0.0-1.7	90-95	8-14
Beet (bunched)	0.0	90	1.5
Carrot (topped)	0.0	95	20-24
Radish (topped)	0.0	88-92	3-5
Turnip	0.0	90-95	8-16
Hyacinth bean	0.6-1.7	90	3.0
Lima bean	4.4-7.2	90-95	1.5-2.0
Pea	0.0	88-92	2-3
Garlic	0.0	65	28-36
Onion (bulb)	0.0	70-75	20-24
Onion (leaves)	0.0	90-95	2.0
Taro	11.1-12.8	85-90	21
Cassava	0.0-1.7	85	23
Sweet potato	10.0-12.8	80-90	13-20
Yam	26.7	66-70	3-5

Contd.

Contd.

Vegetables	Temperature (0°C)	Relative humidity (%)	Storage life (Weeks)
Ginger	7.2-10.0	75	16-24
Okra	8.9	90	2.0
Sweet corn	0.6-1.7	90-95	1.0
Celery	0.0-0.6	92-95	8.0
Lettuce (head type)	0.0	90-95	3.0
Lettuce (leaf type)	0.0	95	1.0
Coriander leaves	0.0-1.7	90	5.0
Asparagus	0.0	95	3-4

Source: Pantastico, Er. B., T.K. Chattopadhyay and H. Subramanyam (1975) Post-harvest Physiology, handling and Utilization of Tropical and Subtropical Fruits and Vegetables (ed. Er. B. Pantastico), AVI Publishing Co., Westport, Connecticut, USA.

Controlled atmosphere (C.A.) storage: It is storage of products under modified or controlled atmosphere where atmospheric composition surrounding the commodity inside the storage chamber is different from that of normal air. Usually, reduction of oxygen and/or elevation of carbon dioxide concentration inside the storage chamber are maintained. This storage markedly retards ripening, respiration and sensecence and thereby prolongs shelf-life of vegetables. This storage proves beneficial for vegetables that deteriorate rapidly or those that complete ripening after harvest and when low temperature is unfavourable for a particular vegetable.

Hypobaric storage: It refers to storage of vegetables under low atmospheric pressure where the products are kept in vacuum, tight and refrigerated chamber. Low-pressure inside the storage chamber reduces oxygen tension which decreases respiration and facilitates the removal of ethylene and volatile compounds produced by the vegetables resulting in retardation of ripening and senescence.

Irradiation: Preservation and extension of shelf-life of the fresh produce by irradiation is of considerable interest at present. Irradiation is found useful to extend the shelf-life of the produce by restricting the growth of decay-causing micro-organisms and by slowing down the rate of metabolism in the produce. In addition to this, irradiation has been found to be of great practical significance in controlling sprouting of potato and onion. Irradiated produce also maintains its quality for prolonged period.

Processing: Different vegetables processed into various durable products, particularly during surplus production and seasonal super-abundance can effectively reduce colossal post-harvest losses. Vegetables are, in general, processed to manufacture different products like canned products (potato, pea, beans, okra, cauliflower, carrot, sweet potato, drumstick), dehydrated products (potato, pea, onion), juices (tomato, watermelon), jams (tomato, watermelon, muskmelon), jelly (watermelon), pickles (chilli, carrot, cucumber, cabbage, pumpkin, cauliflower, turnip, beet), sauerkraut (cabbage), preserve and candy (wax gourd, carrot, beet), frozen product (French bean, sweet pepper, potato) and different tomato products (puree, paste, sauce, ketchup).

17. Marketing and Export of Vegetables

Marketing includes all the steps from the time the produce is ready for harvest until it is reached to the consumer. Momentum in vegetable cultivation depends as much on improved technologies and adequate supply of inputs as on proper marketing of the produce. However, disorganized feature of vegetable marketing is disincentive to vegetable production in our country.

Important steps in handling the produce during marketing

Preservation of quality of vegetables is the most important criterion for achieving success in vegetable marketing. Quality of the product can be improved considerably by improved agro techniques, timely harvesting, grading, packaging and other handling methods. Physical handling of the product which is very important from their marketing point of view is discussed below.

Harvesting: Harvesting should all the time be done at proper stage of horticultural maturity because it not only determines the quality of the product but also prolongs its shelf-life. Generally, no hard and fast rules are followed with regard to the time of harvesting. However, optimum maturity of pea, beans, lettuce etc. is often predicted by heat unit in some developed countries.

Pre-cooling: It is done for rapid removal of field heat from the freshly harvested vegetables before packaging or transportation to delay ripening of fruits and shrinkage of the produce. In our country, it is done mostly by immersing the harvested vegetables in water to remove the field heat from the produce which retard ripening and senescence process. Other improved methods of pre-cooling practised in the developed countries are: placing the produce in refrigerated van, cooling with ice water, placing ice in packages and vacuum cooling.

Waxing: Some vegetables like, tomato, brinjal, sweet pepper, cucumber, muskmelon, carrot, etc. are often waxed with a water emulsion by dipping or spraying to retard the moisture loss from the product and at the same time, to improve their lustre. Sometimes, fungicides and antibiotics are also applied with the wax emulsion. However, this practice of keeping the product sound and lusturous is generally not followed in our country.

Grading: Grading refers to the marketing function which makes orderly marketing and ensures higher price for the well-graded products. Moreover, selling of well-graded products is considered to be an honest dealing and it also simplifies the whole marketing process. It basically means sorting of the unlike lots of the produce into different lots according to quality specifications like, size, shape, colour, stage of maturity, freedom from disease and pests, etc. Each lot has substantially the same characteristics so far as quality is concerned. Standardization means making the quality specifications of the grades uniform among buyers and sellers over space and time. The characteristics *i.e.* quality specifications are the bases of which products are standardized and are called "Grade Standards".

International Organization for Standardization (ISO) which came into existence in 1947 and Codex Alimentarious Commission (CAC) established in 1963 are the two prominent international organizations engaged in standardization of agricultural products. Grading system in India started as early as 1937 with the introduction of Agricultural Produce (Grading and Marketing) Act. At present the Directorate of Marketing and Inspection under the Ministry of Agriculture exercise control on quality of agricultural commodities that are graded under "Agmark" (Agricultural Product Grading and Marketing) before they are exported. At present only three vegetables, potato, onion and garlic comes under such grading only for export. However, proper follow up programmes are not there to introduce these grades in the domestic markets.

Preparation for the market: This is an important step in marketing as it improves the external quality of the produce. In other words, the produce can be presented to the consumer in attractive form. Many vegetables require some special preparation like, cleaning, trimming, tying in bunches, etc before they are offered for sale in attractive manner.

Packaging: It is an important consideration in the process of marketing. Proper packaging makes transportation and retail sale of the vegetables more convenient. It also increases the shelf-life of the products because of the protection provided by the packages from external injuries and from being shrivelled. Adequate packaging coupled with effective marketing can reduce the post harvest losses of the vegetables. Packages also provide a measure of the contents which is also very important from the marketing point of view. Nowadays, alternative packaging materials which are recyclable are receiving attention. Corrugated fibreboard (CFB) boxes, paperboard boxes lined with polyethylene, polyethylene film, plastic crates and moulded paper trays are suitable alternatives. Processing industries and some co-operative societies are regularly using plastic crates for transportation of vegetables from the field as well as for storage. Wider adoption of modern packaging is limited in India due to (a) high cost of the packaging materials, (b) involvement of additional labour charges for packaging and pre-packaging operations, (c) shortcoming in the system of collection of the used packaging materials and (d) lack of information about the specification of different packaging materials for different vegetables.

Transportation: Timely and speedy delivery of vegetables with minimum spoilage and at reasonable cost is the essential feature of effective vegetable marketing. Vegetable production areas can effectively be extended if such potential areas can be brought under good transportation network.

18. Export Trade of Vegetables in India

Vast export potential exists in India for fresh as well as different processed products of vegetables. India has been mainly exporting fresh onion, different other fresh vegetables, canned vegetables, dehydrated vegetables and various other products like paste and ketchup (tomato), frozen vegetables, pickles, juices and powdered vegetables (onion and garlic) mainly to Middle East countries like Baharin, Qatar, Muscat, Tehran, Abudhabi, Sharjah and Doha for many years. In vegetable export, main focus needs to be concentrated on neighbouring countries like, Bangladesh, Sri Lanka and Mauritius, which have been traditional outlets for onion and potato export. Other countries to which vegetables and its different products are exported are Malaysia, Singapore, Sri Lanka, Pakistan, UK, Denmark, New Zealand and Germany. Middle East markets are receiving hardly 5% of fresh vegetables from India though tremendous potential exists in increasing vegetable export to Middle East.

Apart from fresh onion, other prime vegetables which are exported in fresh condition are okra, bitter gourd, bottle gourd, chilli, potato and garlic. Different other fresh vegetables, like tinda, taro, radish, carrot, peas, cauliflower, cabbage, ginger, tomato, French bean, pumpkin, asparagus, celery, brinjal, pointed gourd, drumstick, elephant-foot yam, broccoli, sponge gourd, sweet pepper, sweet com and lima bean are also exported in good quantities. Fresh onion is mainly exported to Bangladesh, United Arab Emirates, Singapore, Malaysia, Saudi Arabia, Sri Lanka, and Mauritius. Non-traditional vegetables like asparagus, celery, sweet pepper, sweet corn, lima bean and organically grown vegetables have good export potential in both fresh and processed forms to Europe and South Asia.

There is continued demand for frozen vegetables and tomato products in the foreign countries. Tomato paste and ketchup are the potential items of export particularly to Middle East, UK, Africa, Australia and Italy. Canned vegetables are mainly exported to Germany, Denmark, UK and New Zealand and dehydrated vegetables to Germany, UK and New Zealand. There is good demand for canned peas, beans and gherkin in Europe. Among dehydrated vegetables, peas, onion and garlic are important and among frozen vegetables, prospects for peas, cauliflower, French bean, carrot and okra are bright. Chilli oleoresin is another important export item to USA, Europe and Japan. The Agricultural and Processed Food Products Export Development Authority (APEDA) has identified onion, potato, okra, bittergourd, bottle gourd, chilli, asparagus, celery, sweet pepper, sweet com, green peas, French bean, gherkin and cherry tomato as potential vegetables for export from India.

Main constraints of export

- Non-availability of air cargo space and high freight rates from India to Gulf countries and UK compared to Kenya, Jordan, Lebanon, etc. are the two major constraints for increasing export of vegetables from India.
- High domestic demand of vegetables and prices hinder regular export.
- Domestic production base has not been geared in the form of developing export oriented production zone to meet export demand which reveal lack of linkage between production base and export.
- Many high value vegetables which are in demand in Europe are not generally cultivated in India and even the seeds of these crops are often not available in domestic market.
- Meagre concept of export-oriented quality production.
- Lack of capital inflow for development of infrastructure for high-tech vegetable farming and post-harvest management.

- Inadequate storage facilities at the main airports (Mumbai, Delhi and Thiruvanantapuram) for export
- Lack of refrigerated shipment facilities.
- Inadequacy in packaging and transportation.
- India is located far from European and American markets.
- Inadequate export-oriented research programmes in the Agricultural Universities and ICAR institutes.

High Value Vegetables for Export

Great demand exists in the International market for some of the high value vegetables. Due to low labour cost and varied agro-climatic condition, India can go for export orinted production of many such vegetables mentioned below at a cost much lower than the developed countries.

- French bean and lima bean: Fresh for domestic market and Gulf market and processed and frozen for European market.
- Sweet corn: Fresh for domestic and Gulf market and processed for world market.
- Asparagus: Fresh for domestic and Gulf market and processed for world market.
- Celery: Fresh for domestic and Gulf market.
- Sweet pepper: Fresh for domestic and Gulf market.
- Brussels sprout: Fresh for European market.
- Broccoli: Fresh for European market.

Steps to be taken to boost up export

- Export policy should be long-term and consistent and frequent changes in export policy should be avoided in view of likely adverse effect on foreign markets and export orinted growers.
- If the domestic prices are high even then a minimum quantity of vegetables should be exported to retain the export market. Export market must be preserved even in periods of low international prices by giving cash incentives.
- Adequate quality control measures should be taken while exporting vegetables to international market through proper functioning of the Quality control laboratories.
- There should be close linkage between exporter and producers of exportable surplus so that producer gets reasonable prices and the quality of products to the international market is assured.

- Export policy should be oriented towards generating more employment, especially amongst the small and marginal farmers, tribals and agriculture labourer. Export oriented vegetable production on contractual basis need to be operated through cooperatives or self-help groups among the small and marginal farmers.
- It is necessary to develop market intelligence set up for export trade through research and development activities.
- Ventilated and temperature controlled transportation should be introduced and highest priority should be given to these perishable commodities.
- Modern storage facilities should be provided at the collection centre and ports having tier system of storage with proper air circulation.
- Air freight rates should be lowered and more cargo space should be provided for vegetable export.
- Off-season vegetable cultivation particularly in protected condition should be encouraged to avoid competition and to catch season of demand.

19. Seed Technology

Seed is the sexual propagating unit that arises through various events in the reproductive process of plants. Botanically seed is the mature ovule containing three basic components namely, (a) embryo, the new plant in miniature, (b) food storage tissue (endosperm, perisperm, and cotyledon) and (c) seed coats, the protective coverings. So, seed, strictly speaking, is the embryo embedded in the supporting food storage tissues which are enveloped by seed coats. It is the marvel of nature because of (a) suspended animation of the embryo after great upsurge of metabolic activity following fertilization, (b) capability of awakening *i.e.,* resurgence of metabolic process and developmental activities upon getting favourable condition which lead to germination and subsequent establishment of the plant and (c) storage of all information regarding the characters of the future plant in the chromosomes of the embryo in the form of genes.

Factors affecting viability and longevity of seeds

Seed viability is the capability of the seed to germinate and produce normal seedlings hence, synonymous to germinability. A seed having low degree of metabolic activity (enzyme activity, respiration, cell division, etc.) may not germinate even if it is still viable. Viability synonymous to germinability is the most important from commercial point of view and it remains at its highest at the time of physiological maturity of the seeds and decline gradually with ageing. Seed longevity is the time span upto which seed viability can be retained after harvest and it is synonymous to storability of seeds. These two very closely related intrinsic seed technology phenomena are affected by almost same factors discussed below.

Genetic factors

The physical condition and physiological state of seeds greatly influence their life span, and these two preconditions are governed by the genetic factors. For this reason, certain seeds are short-lived while some remain alive for many years. On the basis of the relative stability of the seeds to moisture content, the seeds are grouped into two categories.

Orthodox: These seeds can be dried to a low moisture level and would show loss of viability with the rise in moisture content. This category of seeds is usually long-lived. Most of the vegetable seeds fall under orthodox category. Longevity among the orthodox seeds varies with the kind/variety of seeds. Seeds with high starch and protein content are more hygroscopic than those with high fat (lipid) content. Oily seeds are least hygroscopic. Due to reduced hygroscopicity, moisture is hold tightly in the seeds of high lipid content resulting in additional water to become excessive and contributes to their fast deterioration. Moisture is released quickly in the atmosphere from the starchy and proteinaceous seeds due to their more hygroscopicity resulting comparatively slow deterioration. Seeds with high starch and protein content are comparatively long-lived than seeds of high lipid content. Seeds of onion, cabbage, cauliflower, knol khol, okra, French bean, Indian spinach having high oil content are comparatively short-lived. Cotyledon of pumpkin, summer squash also contains considerable oil but longevity of those seeds is comparatively more probably due to their leathery seed coat which hinder moisture uptake by the seeds. Similarly, tomato seeds with high lipid content are also long-lived. The physiological basis for differences in longevity between species is not well understood.

Grouping of vegetable seeds according to their longevity in normal storage condition

Longevity	Vegetable seeds
Up to 1 year	Onion, French bean, etc
Up to 2 year	Okra, carrot, chilli, sweet pepper, cauliflower, cabbage, broccoli,etc
Up to 3 year	Radish, beet, peas, ridge gourd, turnip, palak,etc
Up to 4 year	Tomato, brinjal, pumpkin, summer squash, bottle gourd, bitter gourd, watermelon, muskmelon, cucumber, spinach, etc

This grouping clearly demonstrated that some seeds deteriorate much faster than others in storage due to ageing, the irreversible process. Hence, longevity of some seeds can be extended comparatively more by manipulating storage environmental condition.

Recalcitrant: These seeds require relatively high seed moisture content for longevity and when dried below certain critical moisture level, they rapidly lose viability. These types of seeds lose their viability rapidly in open-air storage. These seeds germinate normally when they are sown quickly after extraction from fruits. Very few vegetable seeds are recalcitrant in nature *e.g.,* seeds of pointed gourd, chowchow, etc. Very fleshy seeds of chowchow *(Sechium edule)* can be stored along with the intact ripe fruits for some period of time but excised seeds lose viability very soon. For this reason, the whole ripe fruit along with the seed in tact is normally planted directly for propagation.

Initial seed quality

Seed lots having plumpy and vigorous seeds store longer than that of deteriorated lots. Even at the best storage conditions, the initial quality of the seed cannot be improved (except for dormant seed) but can only be maintained.

Seed moisture content

The most important factor influencing seed viability and longevity is the moisture content of the seed. The concentration of water in the seed tissue directly affects the rate of deterioration at a given temperature. Again, moisture content is determined by the relative humidity of storage and by seed characteristics, largely lipid content. The rate of deterioration increases as the seed moisture content increases. Higher moisture content enhances the biological activity, particularly respiration, which in turn causes awakening of the seeds. With the initiation of awakening *i.e.,* start in metabolic activities, seed viability gradually deteriorates.

Higher moisture content of the seeds causes excessive heating due to enhanced respiration and other biological activities. Besides, high moisture content promotes storage fungi and insect activities causing further deterioration. The seed moisture content eventually comes into equilibrium with the relative humidity of the storage atmosphere. This relationship is determined by placing seeds in a range of known relative humidities produced in closed containers with the use of saturated salt solutions. Difference between the species/kind of seeds in this aspect is largely attributed to seed composition, particularly lipid content, as lipids have little affinity for water.

Effect of seed moisture content on seeds in storage

Seed moisture content	Effect on seed
33-60%	Germination occurs on imbibation, quick loss of viability
18-20%	High respiratory rate, high heating with inadequate ventilation, high activity of moulds and storage pests, seeds susceptible to mechanical damage in handling
13-17%	Still high respiratory rate, moulds and insects can be damaging fumigation can affect seed viability, resistant to mechanical damage.
10-12%	Seeds can be stored reasonably well for 6-18 months depending on the kind of seed in open storage in temperate climate, insects can be a problem, and fumigation can safely be done.
8-10%	Seeds can safely be stored for 1-3 years in open storage, very little insect activity, seeds susceptible to mechanical damage.
4-8%	Safe moisture content for sealed storage.
0-4%	Excessive desiccation is damaging to seeds, seeds may lose viability.

Source: Ramamoorthy *et al*. (1989)

The overwhelming importance of seed moisture on seed longevity can be conceived from the thumb rule of Harrington (1960). For each 1% increase in seed moisture between 5-14%, the longevity of the seed is halved. This range of 5 to 14% is fixed in the thumb rule because speed of ageing may increase below 5% moisture and fungi and stored pests may be active above 14% moisture. Seed drying up to 10% moisture content appears quite satisfactory in open storage condition. For storage in sealed containers, it is necessary to dry the seeds up to 4-8% moisture content depending on particular kind of seed.

Environmental factors

Temperature and relative humidity are the principal external factors that influence seed longevity. The effects of relative humidity and temperature are highly interdependent. Most of the crop seeds lose viability rapidly at relative humidity approaching 80% and temperature between 25° to 30°C. Most of the seeds can be kept for more years at relative humidity of 50% or less with the storage temperature of 5° to 10° C.

Relative humidity: Seed moisture content and relative humidity of the storage environment are closely correlated because of the hygroscopic nature of the seed. The hygroscopic nature of the seeds allows them to adjust their moisture content in equilibrium with the relative humidity of storage environment. The equilibrium is attained when the seed has no further tendency to absorb or lose moisture. Through this correlation, advantage of dry seed is nullified when they are stored in the environment with high relative humidity

Different seeds do not have the same equilibrium moisture content at the same relative humidity and temperature due to difference in seedcoat structure and chemical composition of seed. It is clear that ultimate moisture content of the seed in storage is determined by the moisture equilibrium due to interaction between seed moisture content and relative humidity of the storage atmosphere.

Temperature: In general, rise in temperature by 10^{0}C stimulates seed respiration by 2.6- to 3-fold up to a certain point. Since seeds are poor conductors of heat this in turn brings about further rise in seed temperature. It has been established that high respiration associated with enhanced biological activities deteriorates seed viability. According to a thumb rule (Harrington, 1960), for each 5^{0}C increase in seed temperature between 0^{0}C and 50^{0}C the life of the seed is halved. From the other way of the thumb rule, a 2^{0}C decrease in storage temperature, storage potential of the seeds are doubled. Hence, subfreezing temperature condition increases longevity of most of the seeds. Temperature affects seed longevity more adversely at high seed moisture level (high storage humidity).

Moisture-Temperature relationship: Seeds attain rather specific and characteristic moisture content when subjected to a given level of atmospheric humidity at a particular temperature. In general, the higher the temperature in the storage atmosphere the lower should be the humidity and seed moisture content for safe storage. Conversely, lower the temperature the higher will be the tolerable moisture content in the seeds for safe storage. For this reason, 4-8% seed moisture content is ideal for sealed storage in any temperature condition (definitely not beyond 50°C). But for somewhat high moisture level, reduced storage temperature is essential. For this reason, longevity of the short-lived seeds like onion, French bean, etc. is increased in cold storage condition.

Gaseous composition of storage atmosphere: An atmosphere of higher O_2 concentration than that of air hastens deterioration of seeds and reduced O_2 pressure in storage environment tends to increase the seed viability. An atmosphere of N_2 and CO_2, tends to delay the seed deterioration process. So, changing the composition of storage atmosphere by creating vacuum and increasing CO_2 or N_2 level extends longevity of the seeds.

Activity of organisms associated with seeds in storage

The bacteria, fungi and storage insect-pests cause shortening of seed longevity. Most seed-borne pathogens are detrimental to seed quality, especially with respect to germination potential. Seed-borne fungi may kill the seed before germination but more frequently it reduces the germinability by pre-emergence killing and seedling blight. Increase in storage temperature and relative humidity enhances both activity and multiplication of seed borne pathogens and storage insect pests.

Maintenance of genetic purity

Advantages of breeding superior varieties or hybrids cannot be realized if seeds of the open pollinated variety and parents of the hybrids are not multiplied keeping their genetic purity. If genetic purity is not maintained identity of the concerned variety and the parental lines of the hybrid is lost. Any open-pollinated variety, may it be of self - or cross-pollinated crop, will show genetic deterioration if grown for a number of years by the farmers due to many reasons and such deterioration is more rapid in cross-pollinated crop due to their out crossing behaviour. If genetic purity of the variety is not maintained, production of pure seeds of that variety will not at all be possible. Hybrid seeds of both self- and cross-pollinated crop must be supplied every year to the farmer. In this case, genetic purity of the parents must be maintained strictly and thereafter controlled hybridization is rendered to for the production of hybrid seeds.

Factors responsible for deterioration of a variety or parents of a hybrid

Genetic shift: Quantitative characters are governed by many genes and considerably affected by environment. Specific difference in environmental conditions may have a greater effect on some genotypes than on the others or there may be a change in the order of merit of a series of genotypes when measured under different environments. Genetic shift is the change in genetic makeup of a particular genotype if grown over a long period in areas outside their adaptable environmental conditions (temperature regime, photoperiod, elevation, soil fertility, etc.). Such genetic shift may give rise differential growth response which may not be in conformity to the varietal/parental characters. Since such developmental variation in the plants of a particular variety/parental line involves their genetic makeup, it may lead to deteriorate a particular variety or parental line of a hybrid. The tendency of genetic shift is likely to be greater in the genotypes bred for tolerance to abiotic stresses like, cold, drought, soil salinity, etc.

Seed production of a variety or parental; line should be carried out in the region of its adaptation to avoid the induction of genetic shift in the varietal population.

Natural crossing: Variety denotes an assemblage of cultivated plants which on reproduction (sexually or asexually) retain their specific distinguishing features. Such retention of the specific characters of a variety or parental line in the sexual progenies is correlated with unalteration of their genetic makeup which in turn depends on controlled pollination during reproduction (i.e., at the time of seed production). Natural crossing refers to transfer of pollens from one flower to the stigma of the others and if it results in cross fertilization (pollens from the plants of different genetic makeup which are not in conformity to those of the particular variety) leads to the change in genetic constitution of the seeds produced. Obviously, natural crossing is the major source of genetic contamination and variety deterioration in cross-pollinated (cabbage, cauliflower, knolknol, broccoli, cucurbits, radish, carrot, etc.) and often cross-pollinated (okra, chilli, brinjal, etc.) crops. In predominantly self-pollinated crops (tomato, pea, beans, etc.) showing limited natural out crossing propensity, it is not a serious source of genetic contamination and variety deterioration. Genetic deterioration of a variety or parental line due to natural out crossing happens due to pollen contamination from the plants like, (a) other variety of the same crop, (b) other cross compatible crop (c) off-types (off-types differ in the characteristics and genetic makeup from the plants of the variety or line), (d) diseased plant (natural crossing with the pollens of diseased plant (susceptible) may induce disease susceptibility character to the plants of the concerned variety) and (e) mutants.

To control natural out crossing specific land requirements are fixed for the seed production of particular crop to avoid contamination of volunteer plants grown from the seeds that remains in the field from the previous crop. However, strict adherence to the prescribed isolation distance for the particular crop mainly prevent natural out crossing. Isolation distance for a particular crop is generally fixed based on different criteria like, natural out crossing percentage, floral adaptation for self and cross pollination, presence of nectar in the flower, floral adaptation for the visit of pollinizer insects, nature of pollen, etc.Naturally, isolation distance is comparatively high for cross-pollinated crops though it varies widely with the crops (1000-1600 m for cabbage, cauliflower, knol khol ,Chinese cabbage, radish, turnip, beet and palak; 800-100 m for onion, carrot; 400-500 m for different cucurbits; 300-400 m for amaranth), moderate for often-cross-pollinated crops (150-200 m for brinjal; 200-300 m for chilli, sweet pepper, okra) and low for self-pollinated crops (25-30 m for tomato, peas and beans, fenugreek, lettuce, etc.). Isolation also prevents deposition of pathogens to the seed crop by wind and insects from neighbouring crop field, and help avoiding mechanical mixtures during sowing/planting, harvesting, processing and handling of seeds

Spontaneous mutation: Sudden heritable change in the characteristics of some of the plants of the varietal or parental line population may occur naturally. Such plants having changed genetic makeup and showing deviated characters are called mutants. Spontaneous mutation in most of cases is minor and very difficult to detect. Spontaneous mutation generally is not a serious factor of genetic deterioration of a variety or parental line of self pollinated crop showing limited natural cross pollination. Spontaneous mutants cause genetic deterioration of a variety by (a) contamination of the varietal plants with the pollens from the mutants, and (b) their presence in the pure seed lot as mechanical mixture.

Spontaneous mutation may cause genetic deterioration more seriously in random mating population of a variety or parental line of cross-pollinated crops. Such mutation may produce new allele not present in the population and due to random mating (each individual plant of the population has equal opportunity of mating with any other individual of that population), frequency of existing alleles may be changed which lead to the development of genetically altered plants in the population.

Presence of genetic variation in the variety: Both gene and genotype frequencies of the random mating population of a variety of cross-pollinated crop generally remain constant under controlled reproduction in spite of considerable heterozygous balance in such population. If gene frequency of the population is changed with the introduction of new allele(s) by out crossing, deterioration of that variety of the cross-pollinated crop will be initiated.

In the often-cross-pollinated crops (okra, chilli, brinjal, etc.), the population of a variety is neither random mating nor highly heterozygous, but quite appreciable out crossings in these crops cause minor genetic variation in some plants in the population of the variety or parental line. This heterozygosity is not balanced as in the case of random mating population of cross-pollinated crop. Such minor variation in the plants of the varietal population may act as a source of genetic deterioration of that variety through (i) contamination of the varietal plants with the pollens from the genetically varied plants in the seed plots and (ii) their presence in the pure seed lot as mechanical mixture.

In some cases, the variety or parental line has not been stable before its release or use for producing hybrid seeds due to faulty breeding procedures. Such unstable variety or parental line show segregation during reproduction (seed production) and within few cycles of seed production, the seeds that are produced in the name of the particular variety or parental line will give rise to the plants having different characters.

Mechanical mixtures: It is not the cause of direct genetic deterioration of a variety but is very important source of varietal deterioration because (a) plants produced from the mixed seeds in the seed lot of the variety or parental line may cause pollen contamination in the next cycles of seed production, and (b) mixed seeds reduce physical purity of the seed lots. Mechanical mixtures may take place with the seeds of the plant types of the same crop namely, (i) other varieties, (ii) volunteer plants, (iii) off-types, (iv) mutants and (v) diseased plants.

Susceptibility to diseases: A newly bred variety may get deteriorated due to its susceptibility to particular disease. Very often, such varieties are not put in the seed production programmes. Similarly, such parents are not employed in the production of hybrid seeds.

Grow out test and roguing

Varieties or parental line under seed production should periodically be tested by grow out tests to make sure that they are being maintained in their true form. In fact, grow out test in combination with proper rouging is the most important step in the maintenance of purity of the parental lines of a hybrid.

Roguing generally refers to removal of off types and undesirable types from the seed plot or grow out test plot as these are potent sources of genetic contamination of a variety or parental line. "Off-types" are the plants of the particular crop differing in their characteristics from those of the concerned variety or parental line. Off-types may arise in the varietal population due to (a) segregation during seed production cycles owing to varietal unstability, and (b) spontaneous mutation.

In tomato, brinjal, chilli, sweet pepper, etc deviated characters of the off-types are apparent in the foliage, fruit shape, fruit colour, etc. In okra, the off-types are distinguishable on the basis of plant height, leaf shape, pigmentation, flower size and shape, fruit shape, etc. In cucumber, they are mainly detected by fruit colour, shape of spines in fruits and their colour. In muskmelon, such detection is generally done on the basis of fruit colour and fruit netting. In pumpkin, off types are detected on the basis of ovary shape, fruit shape, ribbing in fruit and fruit colour. In watermelon, it is done on the basis of fruit shape and colour. Similarly, off-types in cabbage and cauliflower are detected according to the time of heading/curding and shape of head/curd.

"Undesirable plants" are the plants of the particular crop which conform to the concerned varietal characteristics but having some undesirable features hence, are not acceptable in the seed plot.

While roguing out off-types and undesirable types, volunteer plants, undesirable crop plants, other crop plants, and obnoxious weeds must be removed from the seed plot or grow out test plot. In most of the vegetable crops, 2-4 roguings are required depending on the crop. In tomato, brinjal, chilli and okra generally, two roguings are done: first before flowering on the basis of foliage characters, pigmentation, plant height etc. and second at fruiting stage on the basis of fruit shape and colour. Similarly, leguminous and leafy vegetables are rogued out twice. Root and bulb crops are rouged out thrice, first before flowering on the basis of the characters of the top, second before replanting of roots or bulbs according to their shape, size and colour, and third at flowering as per bolting behaviour. The cucurbits also require three roguings, first before flowering on the basis of vine and leaf characters, second at flowering according to flowering pattern and shape of ovary and third at fruiting stage according to fruit shape, colour, spine colour (for cucumber), netting pattern (for muskmelon), peduncle character (for pumpkin). The cole crops are rogued out four times, first on the basis of time of heading/curding/stem bulging, second on the basis of shape and size of mature head/curd/stem knob (in knol khol), third just before initiation of flower stalk and fourth at flowering according to flowering behaviour. Roguing in both seed and grow out test plot must be carried out by experienced person preferably with vegetable breeding back ground and familiar with different characteristics of the crop.

References

Botany (1992) (J.Ben Hill, Henry. W. Popp and Alvin, R.Grove Jr.) Tata McGraw-Hill, New Delhi.

Crop Physiology (1978) (ed. U.S. Gupta), Oxford & IBH, New Delhi.

Cropping Systems- Theory and Practice (1989) (B.N.Chatterjee, S. Maiti and B.K. Mandal), Oxford & IBH.

Farm Chemicals Handbook (1998) (Ed. C. Sine), Willoughby, OH: Meister.

Fertilizer Use (1960) (A. Jacob and H.V. Unexhull), Varlags Gesellschaft fur Ackerban Mbh, Hannover.

Fundamentals of Horticulture (1964) (J.B, Edmond, A.M. Musser and F.S. Andrews) The Blakiston Co., INC., Toronto.

Glossary of Horticulture (2004) (P.Hazra, S.K.Ghosh, T.K.Maity, M.K.Pandit and M.G.Som) Kalyani Publishers, Ludhiana (2nd Edn.).

Hawk's Physiological Chemistry (1979) (Ed. Oser. L. Bernard), Tata McGraw Hill.

Irrigation Principles and Practices (1979) (E.V. Hasen, O.W.Israelsen and G.e.Stringham), John Wiley & Sons, New York.

Knott's Handbook for Vegetable Growers (1980) (Donald N. Maynard and George J. Hochmuth), John Wiley & Sons, New York.

Modern Technology for Vegetable production (2011) (P. Hazra, A. Chattopadhyay, K. Karmakar and S. Dutta), New India Publishing Agency, New Delhi.

Plant Growth and Development (1981) (A.Carl. Leopold and Paul, E. Kriedemann) Tata McGraw-Hill, New Delhi.

Plant Propagation (1989) (M.K. Sadhu) Wiley Eastern Ltd., Calcutta.

Post-harvest Physiology, Handling and Utilization of Tropical and Subtropical Fruits and Vegetables (1975) (Ed. E.B.Pantastico), AVI Publishing Co., USA.

Potassium- A Product of Nature (1985) (G. Kemmler and H.Hobt), Kali-und Salz Ag, FR Germany.

Production Technology of Vegetable Crops (1993) (K.G. Shanmugavelu), Oxford & IBH.

Seed Technology (1980) (R.L.Agrawal), Oxford & IBH Publishing Co., New Delhi.

Soil Fertility and Fertilizer Use (1989) (S. Tandon), IFFCO.

Soils: An Introduction to Soils and Plant Growth (1990) (R.L.Donahue, R.W. Miller and J.C.Shickluna), Prentice-Hall of India.

Techniques in Seed Science and Technology (1992) (Eds. P.K.Agrawal and M.Dadlani), South Asian Publishing Co., New Delhi.

Technology for Vegetable Production and Improvement (1999) (P. Hazra and M.G.Som) Naya Prokash, Kolkata, West Bengal.

Textbook of Vegetables, Tubercrops and Spices (2001) (eds. S.Thamburaj and Narendra Singh), I.C.A.R., New Delhi.

The Nature and Properties of Soil (1980) (N.C.Brady), MacMillan Publishing House, INC., USA.

The Physiology of Vegetable Crops (1997) (ed. H.C. Wien) CAB International, UK, USA.

Tuber Crops (1988) (S.P.Ghosh, T.Ramanujam, J.S.Jos, S.N.Moorthy and R.G.Nair), Oxford & IBH.

Unexploited Tropical Vegetables (1988) (P. Indira and K.V. Peter), Publication Unit, Directorate of Extension, Kerala Agricultural University.

Varieties of Vegetable Crops in India (2005) (P.Hazra, M.K.Banerjee and A. Chattopadhyay), Second Edition, Kalyani Publishers, Ludhiana.

Vegetable Crops (1979) (H.C.Thompson and W.C.Kelley), Tata McGraw Hill.

Vegetables Crops (2007) Gopalakrishnan, T.R., New India Publishing Agency, New Delhi.

Vegetable Crops in India (1986) (Eds. T.K.Bose and M.G.Som), Naya Prokash.

Vegetable Growing in Homegardens of Tropical and Subtropical Areas (1988) (P.F.Konokov and V.Kiran), Mir Pub., Moscow.

Vegetable Science (1949) (H.D. Brown and C.S.Hutchison), J.B. Lippincott Co., New York.

Vegetable Science (2006) (P. Hazra and M. G. Som) Kalyani Publishers, Ludhiana.

Vegetable Science and Technology in India (2006) (V. Swarup). Kalyani Publishers, Ludhiana.
Vegetable Seed Production and Hybrid Technology (2009) (P. Hazra and M. G. Som), Kalyani Publishers, Ludhiana.
Vegetables for the Tropical Region (Prem Nath), ICAR.
Vegetative Propagation of Plants (1985) (K.K. Nanda and V.K. Kochhar) Kalyani Publishers, New Delhi.

Chapter – 4

Tropical and Subtropical Fruits

T. Radha and Lila Mathew

Fruits play a very significant role in the economic development, nutritional security, employment generation and overall economic growth of a country. Tropical and subtropical fruits can be broadly defined as those originated and commercially grown in the tropics or subtropics, which are evergreen and perennial in most cases. They possess a limited degree of frost resistance and practically non-existent below 10°C. Distinction between tropical and subtropical fruits is based mainly on sensitivity to temperature. Tropical species are sensitive to temperature below 20°C and require a climate with average mean temperature higher than 10°C during the coldest month and additionally many of them require humid environmental conditions. Typical examples of tropical fruits are traditional ones of South-East Asia like mangosteen, durian and rambutan. Most of the fruits can be cultivated equally well in either the tropics or the subtropics and the classic examples include banana and mango. Hundreds of tropical and subtropical fruits exist, but only about 50 are well known through out the world. With respect to production and importance, these fruits can be divided into three categories with a little overlapping.

Major fruits

These are cultivated commercially in many countries and have demand in both local and export-import markets. Main examples include banana and plantain, citrus, mango and pineapple. The world production and the major producing countries of the fruits of this group are given also where.

Minor fruits

These are not cultivated extensively but many of them have importance and demand in a number of countries. Examples are Avocado, Guava, Jack fruit, Litchi, Longan, Mangosteen, Papaya, Passion fruit, Rambutan, Sapota, Annonas, Bread fruit, Carambola, Durian, Langsat, Longan, Macadamia and White sapota.

Production of major tropical and subtropical fruits

S.No.	Fruits	World production (1000 t)	Major producing countries
1.	Banana	58,687	Nigeria, Mexico, Costa Rica, Columbia, Brazil, India, Indonesia, Philippines, Spain, Papua New Guinea
2.	Mango	24,975	India, Indonesia, Philippines, Thailand, Mexico, Brazil, Australia
3.	Citrus	66,055	Brazil, US, India, Mexico, Spain, China, Italy, Egypt, Pakistan, Greece, S. Africa
4.	Grapes	67,206	Italy, France, USA, Spain, China, Turkey, Iran, Argentina, Australia, Germany and Chile
5.	Pineapple	13,455	Philippines, India, Indonesia, China, Brazil, US, Mexico, Nigeria

Wild fruits

They belong to diverse botanical families. These are not cultivated commercially in any country. Characterization, conservation, selection, breeding and popularization of these fruits/ species are need of the day.

Indian scenario

India is blessed with a wide range of agro climatic zones and soils which offer great potential for production of a variety of fruits throughout the year. The fruit industry in India exhibited considerable progress during the recent decades and it emerged as the largest producer of fruits in the world, contributing nearly 10% of the total world fruit production. The per capita availability of fruits in India is estimated 62 g, while as per FAO norms, 85 g/day is the requirement per capita fruit consumption. The targeted fruit production of India by 2020 is 200 Mt from the 47.7 Mt (2002-03) and there exists great scope for development of fruit industry in the country. Major thrust should be given for increasing production and productivity of fruits. To achieve this goal, the identified research priorities include collection and conservation of germplasm, measures to overcome chronic production problems, developing suitable horticulture based farming systems, developing high yielding varieties, faster multiplication of quality planting materials, introducing and popularizing suitable high quality fruit species from elsewhere and reducing post-harvest losses.

1. Botanical Background/Classification

Tropical and subtropical fruits belong to different families. The list shows 137 families of which the most important ones are given below. Botanically a fruit is a developed ripened ovary with or without other parts of flowers like pedicel, sepal and receptacle . Different types of fruits exist among this group like berries (grapes, papaya, banana), drupes (mango), pomes (loquat), capsules (durian), nutlets (litchi), hesperidium (citrus fruits), syncarpium (pineapple), sorosis (jack fruit) and nuts (pistachio, peacan, hazelnut, walnut).

Important families of fruits

S.No.	Families	Fruits
1	Anacardiaceae	Mango
2	Annonaceae	Custard apple
3	Bombacaeae	Durian
4	Bromeliaceae	Pineapple
5	Caricaceae	Papaya
6	Clusiaceae (Guttiferae)	Mangosteen
7	Lauraceae	Avocado
8	Meliaceae	Langsat
9	Moraceae	Jack fruit, Bread fruit
10	Musaceae	Banana and plantain
11	Myrtaceae	Guava
12	Oxalidaceae	Carambola, Bilimbi
13	Palmaceae	Date
14	Passifloraceae	Passion fruit
15	Profeaceae	Macadamia
16	Rosaceae	Loquat
17	Rutaceae	Citrus fruits
18	Sapindaceae	Litchi, Longan, Rambutan
19	Vitaceae	Grapes
20	Sapotaceae	Sapota, Egg fruit

Classification systems based on different criteria as shown below are also used to group fruit crops.

- Climatic requirement
- Photoperiodic responses
- Tolerance to shade
- Photothermal requirement
- Drought tolerance
- Salt tolerance
- Soil reaction
- Specificity to soil texture
- Soil topography
- Horticultural classification based on bearing habit, longevity of plants, rate of respiration during ripening, size of fruits and nutritional aspects

(*Refer:* Radha and Lila Mathew, 2007. *Fruit Crops*. Horticulture Science Series. No. *3*)

2. Nutritive and Medicinal Value of Fruits

Fruits along with vegetables are called 'Protective foods' due to their richness in vitamins and minerals which help to improve human health. Fruits act as reserves of Vit. A, B complex and C, minerals like Ca, Fe and P and organic acids like citric, tartaric and malic. Fruits are easily digestible and contain enzymes as bromelin in pineapple and papain in papaya. These are digestive stimulants and good sources of roughage which help in easy bowel movement preventing constipation and related problems. Fruits are energy giving materials with high calorific value. These are also rich in pectin, fibre and cellulose which promote intestinal motility. Fruits are good sources of antioxidants also. Nutritive significances of some of the fruits are given below and composition of the most important crops are also furnished.

Banana	The highest calorie content. Good source of Vit. A, B and C. Low in protein and fat and rich in potassium. Easy to digest, recommended for baby food.
Mango	Rich in Vit. A and C. Contains minerals like Fe, Ca and P. Organic acids present are citric, malic and oxalic. Sugars include sucrose, glucose, fructose and maltose.
Pineapple	Juice has a cooling and refreshing effect. Rich in Vit. C, A and carbohydrates. Good source of Ca, P, Fe, K and Mg. Contains enzyme bromelin.
Avocado	Rich in fat and has high energy value. Fruits for diabetic patients. Rich in Vit. A, B and C and K, P, Mg and Fe.
Citrus	Refreshing fruits. High content of Vit. C. Rich in pectin, essential oils and glucosides.
Guava	Poor man's apple. Very rich in Vit. C. Contains iron, P and calcium. Rich in pectin content.
Mangosteen	Very sweet pulp with glucose as major sugar component, imparts instant energy. The richest source of natural xanthone compounds, fruit rind contains tannins and anthocayanin. Fruits are cooling and refreshing .
Papaya	Good source of carotene and Vit. C, potassium and calcium. Contains digestive enzyme papain

Tropical and subtropical fruits and their plant parts like bark, roots and leaves are known for their medicinal values also. Almost all the fruits help in correcting problems due to nutritive deficiency. Fruit juices are recommended to treat ailments like scurvy, night blindness, asthma, fever, anaemia and ulcers. Fruits contain natural sugars which help in brain functioning and stimulation. Following are a few of the known medicinal properties of important tropical and subtropical fruits.

Pineapple

Reported to have capacity to arrest multiplication of cancerous cells and to control enhancement of platelet count in blood. Juice is diuretic and also used for gargling in sore throat and as an antidote for sea sickness. Immature fruits have abortification property.

Amla

Widely used in Ayurvedic medicines for treatment of different diseases. Fruits are antiscorbutic, diuretic, laxative and cooling. They have high antioxidant property. Dried fruits are used for treatment of chronic dysentery, bronchitis, diabetes, jaundice, dyspepsia, anaemia and haemorrhages. Amla fruits are major ingredients of hair tonics also.

Mangosteen

Medicinal value is enormous. Used against dysentery, women's complaints, urinary disorders, cancer, arthritis, liver complaints and skin problems. Fruit juice is a health tonic with plenty of antioxidants. Reported to have antibacterial, antifungal and astringent properties.

Jamun

Fruits are used for treatment of diabetes, dysentery, heart and liver complaints. It lowers blood pressure. Seed powder is used in treatment of diabetes which helps to reduce urine sugar very quickly and permanently. Seeds contain an alkaloid Jambosin and a glycoside Jambolin or Antimellin which reduces or stops the diastatic conversion of starch into sugars.

Jack fruit

Used for curing inflammations, constipation, wound healing and skin diseases. 'Lectine', a natural protein from fruit is used in cancer treatment. Extract from jack fruit, 'Jacaline' inhibits growth of HIV infection *in vitro.*

Papaya

Green fruits are diuretic and mild laxative. Used in curing piles, dyspepsia of spleen and liver, digestive disorders, diphtheria and skin blemishes.

Grapes

'Grape therapy' to treat cancer is historically known due to presence of 'resveratrol' with anti-carcinogenic activity. Red wine consumption is associated with lower bad cholesterol and higher good cholesterol contents.

Olive

Table olive and olive oil have high medicinal values. Olive oil consumption helps to reduce bad cholesterol level and to raise good cholesterol level content. Olive oil plays important role in diet, beauty therapy and medicines.

Passion fruit

Good source of Vit. C and K and dried flowers contain an alkaloid passiflorin, used for relieving pain and inducing sleepness.

Pomegranate

Juice is used for leprosy patients and for treatment of dysentery and diarrhea. Fruits are mildly laxative and good for curing inflammation of stomach.

Date

Rich in iron and recommended against anaemia. Has high tannin content and is used as an astringent in intestinal complaints and good for sore throats, colds and bronchial cough.

Bread fruit

Helps to reduce high blood pressure.

Carambola

Valued as a diuretic agent.

Mango

Used against night blindness and eye sight problems.

Citrus Fruits and others rich in Vit. C

Prevent colds and similar infections. Juice is given in fevers and jaundice and also for curing anaemia, scurvy and beriberi. Bitter glucoside 'Naringin' provides prevention against malaria.

The main method of consumption of most tropical and subtropical fruits is in fresh form. One of the classic exceptions to this is bread fruit, eaten only after cooking. Processed products from fruits are jams, jellies, squashes, juices, marmalades and candy. Pickles, chutneys, salads and dehydrated products are also prepared out of different fruits. Wine making (grape) and oil industry (olive) also have significant role in selected fruit culture. Apart from these consumptive uses, several fruits are used for various purposes including religious, textile and paper industries, handicrafts, construction, furniture making and also as firewood source.

3. Genetic Resources

Most of the important fruit crops originated from tropical and subtropical regions of America (papaya, avocado, pineapple, and guava) and Asia (citrus fruits, mango, banana, and litchi). The macadamia nut originated from Australia and date is the only one among the tropical and subtropical fruits originated in African continent. No tropical or subtropical fruits have their place of origin in Europe.

Plant Genetic Resources (PGR) form a major component of the biological diversity and these are one of the most valuable assets available to mankind. The genetic diversity of crops represented by traditional local cultivars and wild relatives are disappearing rapidly during recent decades. Effective conservation and sustainable use of PGR are being given top priority in many countries. Environmental destruction, over-exploitation, replacement of traditional cultivars and modernization of agriculture are a few factors responsible for rapid loss of genetic diversity. Many countries took steps to face these problems and are involved in activities to conserve the genetic diversity of plants including that of tropical and subtropical fruits, since they are an important asset to the countries in which they are produced, both nutritionally and economically. The Biodiversity International (IBPGR) provides support for collection, conservation, ecogeographic and systematic surveys of fruits, regionally and locally important. Though more than 700 edible fruits are native to humid tropical areas of the world, only about 30 are cultivated and marketed on a commercial scale around the world.

Asia, the Pacific and Oceania (APO) region contains about 45 countries and is the most important area with respect to genetic diversity of tropical and sub tropical fruits. This region holds rich diversity with respect to fruits which play significant role in providing food for local consumption. It is reported that about 500 species out of a total 700 of fruits of this group occur in this region. Important native fruit diversity in this region is represented by fruits like citrus, banana, mango, rambutan, durian, litchi, jackfruit, mangosteen, longan, langsat, jujube and carambola. Minor fruits of local importance include aonla (*Emblica officinalis*), bael (*Aegle marmelos*), wood apple (*Feronia lumonia*), jamun (*Syzygium cuminii*), phalsa (*Grevia subinequalis*) and tamarind (*Tamarindus indica*). India has a very rich biodiversity as far as tropical and sub tropical fruits are concerned. The most important among them is mango. India is the world's richest mango germplasm centre where more than 1000 vegetatively propagated varieties/types exist. They exhibit wide variation with respect to fruit shape, size, skin colour, time of maturity, stone size, pulp quality, yield and bearing regularity. *Mangifera* genus consists of 41 species and wild *M. indica* occurs in North Eastern India.

Underutilized fruits like ber, aonla, jackfruit, bael, custard apple, jamun, karonda and tamarind are also spread over the country and a good number of germplasm accessions of majority of these are maintained in different centres for future use.

Important tropical and sub tropical fruits of the world

S. No.	Scientific name	Common Name	Family
1	*Aegle marmelos*	Bael	Rutaceae
2	*Ananas comosus*	Pineapple	Bromeliaceae
3	*Annona atemoya*	Atemoya	Annonaceae
4	*Annona cherimola*	Cherimoya	Annonaceae
5	*Annona muricata*	Sour sop	Annonaceae
6	*Annona reticulata*	Bullock's heart	Annonaceae
7	*Annona squamosa*	Custard apple	Annonaceae
8	*Artocarpus altilis*	Bread fruit	Moraceae
9	*Artocarpus heterophyllus*	Jack fruit	Moraceae
10	*Averrhoa bilimbi*	Bilimbi	Oxalidaceae
11	*Averrhoa carambola*	Carambola	Oxalidaceae
12	*Bassia latifolia*	Mahua	Sapotaceae
13	*Buchanania lanzan*	Chironji	Anacardiaceae
14	*Caratonia siliqua*	Carob	Fagaceae

Contd.

Contd. Table

S. No.	Scientific name	Common Name	Family
15	*Carica papaya*	Papaya	Caricaceae
16	*Carissa carandas*	Karonda	Apocynaceae
17	*Casimiroa edulis*	White sapote	Rutaceae
18	*Chrysophyllum cainito*	Star apple	Sapotaceae
19	*Citrus* sp.	Citrus	Rutaceae
20	*Cyphomandra betacea*	Tree tomato	Solanaceae
21	*Dimocarpus longan*	Longan	Sapindaceae
22	*Diospyros discolor*	Velvet apple	Ebenaceae
23	*Diospyros kaki*	Persimmon	Ebenaceae
24	*Durio zibethinus*	Durian	Bombacaceae
25	*Emblica officinalis*	Aonla	Euphorbiaceae
26	*Eriobotrya japonica*	Loquat	Rosaceae
27	*Feronia limonia*	Wood apple	Rutaceae
28	*Ficus carica*	Fig	Moraceae
29	*Flacourtia* spp.	Lovi lovi	Flacourtiaceae
30	*Garcinia gummigutta*	Malabar Tamarind	Clusiaceae
31	*Garcinia mangostana*	Mangsteen	Clusiaceae
32	*Grewia subinaelqualis*	Phalsa	Tibiaceae
33	*Hylocereus undatus*	Dragon fruit	Cactaceae
34	*Lansium domesticum*	Langsat	Meliaceae
35	*Litchi chinensis*	Litchi	Sapindaceae
36	*Macadamia* spp.	Macadamia	Proteaceae
37	*Malpighia punicifolia*	West Indian Cherry	Malpighiaceae
38	*Mangifera indica*	Mango	Anacardiaceae
39	*Manilkara achras*	Sapota	Sapotaceae
40	*Morinda citrifolia*	Noni	Rubiaceae
41	*Morus alba*	Mulberry	Moraceae
42	*Musa* spp.	Banana	Musaceae
43	*Nephelium lappaceum*	Rambutan	Sapindaceae
44	*Olea europaea*	Olive	Oleaceae
45	*Opuntia ficus-indica*	Cactus pear	Cactaceae
46	*Passiflora* spp.	Passion fruit	Passifloraceae
47	*Persea americana*	Avocado	Lauraceae
48	*Phoenix dactylifera*	Date	Palmae
49	*Phyllanthus acidus*	Star gooseberry	Euphorbiaceae
50	*Physalis peruvia*	Cape gooseberry	Solanaceae
51	*Pouteria campechiana*	Egg fruit	Sapotaceae
52	*Psidium guajava*	Guava	Myrtaceae
53	*Punica granatum*	Pomegranate	Punicaceae
54	*Salvadora oleoides*	Pilu	Salvadoraceae
55	*Spondias* spp.	Indian hog plum	Anacardiaceae
56	*Syzygium cuminii*	Jamun	Myrtaceae
57	*Syzygium jambos*	Rose apple	Myrtaceae
58	*Syzygium uniflora*	Surinam cherry	Myrtaceae
59	*Tamarindus indica*	Tamarind	Leguminosae
60	*Vitis vinifera*	Grape	Vitaceae
61	*Zyziphus mauritiana*	Ber	Rhamnaceae

4. Plant Propagation

Propagation of plants can be defined as the controlled reproduction or multiplication or perpetuation of selected plants by man. There are two major methods of plant propagation viz., sexual (seed) and asexual (vegetative).

Sexual / Seed propagation

Sexual propagation refers to recombination of genetic materials resulting in progeny that differs from each other and from their parents. A seed is a young plant, ready to grow under favourable conditions. Capacity of a seed to germinate is termed as viability. Viable embryo, favourable internal and external conditions are prerequisites for obtaining a seedling. A seed in general has three basic parts like embryo, storage tissues and seed coat or coverings. Embryo is the new plant resulting from union of male and female gamates as an aftermath of fertilization. Seedlings are not 'true to type' and variation occurs in growth, yielding capacity and quality of fruits. Seedlings exhibit long pre bearing age. At the same time, seedlings are more hardier and long lived when compared to vegetatively propagated plants. Seeds are inevitable for hybridization programmes and for root stocks to do budding and grafting. The external conditions to be favourable for seed germination include moisture, temperature, oxygen and the medium. A few seeds exhibit dormancy which may be due to embryo condition, seed coat condition or due to physiological reasons. A dormant seed will not germinate unless the condition is eliminated by some means. Mechanical and chemical methods to soften the seed coat help to break dormancy due to seed coat barriers. Mere washing and soaking in water also help to improve germination in seeds with hard seed coats. Hormonal treatment especially with GA is beneficial for improving germination. Time taken for seed germination varies from species to species. For example, mango stones require about one month to germinate whereas guava seeds require only 2-3 weeks. In jackfruit, the phenomenon of vivipary, in which seeds germinate even when they are inside the fruit/attached to the mother plant is very common. Seeds which can be dried and stored are called Orthodox type as in cereals whereas seeds which cannot be dried and stored retaining viability, are called recalcitrant as in mango.

Apomixis is the phenomenon of development of seeds without the complete sexual process resulting in 'true-to-type' condition in the progeny. Such seeds are called as apomictic seeds. Apomixis may be of different types.

Recurrent: Here the female gametophyte develops into embryo without meiosis.

Adventitious embryony: Embryos arise from a cell or group of cells in the nucellus or integuments. The sexual embryo develops through the normal way and the seed contains both sexual and apomictic embryos and condition is termed as polyembryony. This phenomenon is seen in mango varieties like Chandrakaran, Bappakkai, Kurukkan, Goa, Muvandan and Vellaikolamban and in certain Citrus species. These nucellar seedlings have horticultural uses as rootstocks for grafting due to their uniformity.

Non-recurrent: Embryo arises directly from egg nucleus without fertilization and the embryo remains haploid and this condition is not consistent.

Vegetative apomixis: In certain plants, instead of flowers, vegetative buds or bulbils develop in the inflorescence. Eg. *Alpinia sp.*

Parthenogenesis: is a specific condition in which viable seeds develop without pollination and fertilization. All parthenogenetic fruits will be parthenocarpic also, which indicates development of fruits without fertilization. Mangosteen is the classic example for parthenogenetic species.

Vegetative propagation

Also called as clonal propagation and is by means of vegetative parts of plants like roots, shoots or leaves. The most important advantage of vegetative propagation is production of 'true-to-type' progenies with short prebearing age. Exploitation of root stocks for desirable characters, short stature in general, rapid multiplication and conversion of inferior quality plants into superior quality through top working are some of other advantages.

Methods of vegetative propagation

There are two major types of asexual propagation viz., based on plants on their own roots and based on roots of other plants.

Methods based on own roots

Specialized structures seen naturally as suckers, runners, bulbs, corms, rhizomes, tubers, crown, offsets and slips. These are either separated from the mother plants and used for planting (by separation) or are divided into pieces and each piece gives rise to a new plant (by division), mostly in stem modifications like corm, rhizome and tubers.

Cuttings: A cutting is a piece of vegetative tissue, stem, root or leaf, that when placed under suitable environmental conditions regenerates the missing parts and produces a self-sustaining plant. This method is less expensive, quick, simple and can be used for making large number of plants from a few plants. Root cuttings can be used only in certain plant species which have regeneration capacity from roots like curry leaf and breadfruit. Leaf cuttings are prepared from leaves with thick, fleshy lamina with regeneration capacity. Stem cuttings are the most commonly used in a wide variety of plant species and based on maturity of wood used, they may be hardwood, semi-hardwood, softwood or herbaceous. Grape vines can be readily rooted using hardwood cuttings. Fig and pomegranate can also be multiplied through such cuttings. Rooting of cuttings can be induced and improved by treating with growth regulators and Indole Butyric Acid (IBA) is the most commonly used one for this. The concentration and method of treatment are to be standardized for each plant species.

Layering: In this method, stems are induced to produce adventitious roots while they remain attached to the parent plants. Different methods of layering are practiced such as tip layering, simple layering, trench layering, mound layering and air layering. In tip and simple layering methods, roots are induced at suitable points on selected shoots. In mound and trench layering, the mother plants are totally converted into a number of rooted layers. Air layering is the most common method followed commercially in a number of fruit crops like guava, West Indian cherry and litchi. In this case, roots are induced at girdled points on selected shoots by covering with suitable moist medium like sphagnum moss.

Methods based on roots of other plants

Propagation through joining two plants/parts so that they form a graft union and grow as a single intact plant comes under this group, which includes budding and grafting. The plant part below the graft/bud union contributing the root and basal portion of the resultant plant is called as rootstock and the upper part, collected from the choice variety and joined with rootstock is called scion. In budding, a detached bud and in grafting a shoot portion with multiple buds collected from the choice variety are used as the scions. Rootstocks and scions should be compatible to get successful graft/bud union. Botanical relationship between the stock and scion determines compatibility to a large extent. There are several methods of grafting and budding mainly based on preparation of stock and scion. In all the methods, the objective is to bring cambium layers of the stock and scion together and to hold them tightly while the union forms. The common grafting methods are inarching (approach grafting), soft-wood grafting, veneer grafting, side grafting, whip and tongue grafting and stone or epicotyl grafting. Common budding methods practiced are shield-budding (T-budding), patch budding, forkert budding, ring budding, flute budding and clip budding.

Among the budding methods, the most commonly followed are 'T' budding and patch budding. In the former, the rootstock is prepared in the shape of alphabet T and in the latter, rootstock is prepared by removing a patch of bark. In all budding methods except in chip budding, scion is prepared only by taking the bud with bark, whereas in chip budding a piece of wood portion is also attached with the prepared scion. This method is followed in grapes commonly.

Different methods of grafting are followed commercially for multiplication of fruit crops. Among the methods, inarching or approach grafting is the good old method for many types of fruit crops. In this method, the scion is detached from the mother plant only after establishment of the graft union. Rootstocks for this purpose are usually one year old seedlings. Though there exists practical difficulties in doing this method like availability of low lying branches on mother tree and watering of the rootstocks while attached to the scion tree, even now, this is preferred by nursery men in mango, sapota and jack . Softwood grafting and epicotyl (stone) grafting are now followed in most of these fruit crops since, scion shoots can be taken from the choice trees and grafting can be done in nurseries. Availability of young seedlings for rootstock purpose, (10-30 days old in stone grafting, 1-3 months old ones in soft wood grafting) is a limiting factor in these methods. Total duration for obtaining ready-to plant grafts in approach method is also more (18-20 months) when compared to other methods.

Micropropagation

Also called as tissue culture or *in vitro* multiplication which refers to the multiplication of plants, in aseptic condition and in artificial growing medium, starting from very small plant parts. Rapid multiplication of true-to-type plants throughout the year is the most important advantage of micro propagation. Production of a large number of plants utilizing small area, availability of virus-free plantlets, facility for long distance transport and long term storage of clonal materials are some of the other advantages of this method of multiplication. The main disadvantages are the high initial cost involved in the establishment of laboratories and requirement of skilled manpower for carrying out the works.

Meristem culture, shoot tip culture, callus culture, cell culture, embryo culture, anther culture and protoplast culture are the major methods followed in different crop plants. The starting material for establishing a culture, called as 'explant' should be standardized for each crop. The set of operations starting from selection of explant upto planting out from the laboratory is together referred as 'protocol' and standardization of protocol for each crop species is cumbersome and time consuming. In tropical and sub tropical fruit crops, protocol for tissue culture is standardized and utilized commercially only in a few species like banana, pineapple, papaya and some citrus fruits.

Top working

Top working refers to the method of rejuvenation of an unproductive tree or low quality fruit plant into a choice variety with high yielding capacity of good quality fruits. The unproductive trees are headed back and grafting/budding with scions from choice variety is done on new shoots arising or on to the main trunk by bark/cleft grafting methods. Top working is being recommended for rejuvenation of old unproductive trees of mango, aonla and custard apple. In dioecious species like Malabar tamarind (Garcenia), sex of the plant can be changed from unproductive male to productive female through top working methods.

Rootstocks

Rootstocks are used for budding and grafting operations, which contribute the root and basal parts of the resultant plants. In majority of the cases, seedlings are used as rootstocks (mango and jack) and in certain crops vegetatively produced rootstocks – clonal rootstocks – are used (grapes). Influence of rootstocks on the behaviour of the resultant plants is well established in some fruit species. This influence is termed as the stock-scion relationship. One can exploit the influence of rootstocks in terms of desirable characters and go for selective use of them for specific purposes. Rootstock research is well developed in citrus and grapes among the tropical and subtropical fruits. Influence of rootstocks on plant growth is mainly with respect to size and vigour of plants, fruitset and yield, fruit size and quality, pest and disease resistance, frost resistance and resistance to adverse soil conditions.

Based on influence of rootstocks on size and vigour of plants – which is the most elucidated one – they can be grouped as dwarf and vigorous rootstocks. This is well established in temperate fruits like apple when compared to tropical and subtropical crops.

Rootstocks for specific purposes

S.No.	Crops	Rootstock	Purpose
1.	Mango	Vellaikolomban Mylepelian	Dwarfing
		Muvandan	Vigorous
		Kurukkan, Nileshwar Dwarf	Salt stress
2.	Citrus	Sour orange	Semi-dwarf , Cold hardy Improves fruit quality
		Trifoliate orange	Dwarfing, Cold hardy Tolerant to nematodes and tristeza virus
		Rangpur lime	Vigorous, hardy, Tolerant to drought, Tristeza virus
		Rough lemon	Vigorous, Drought tolerant Early and high yield Tolerant to tristeza
		Troyer and Carrizo citrange	Semi-dwarf, cold hardy. Resistant to soil borne diseases and nematodes Tolerant to tristeza
		Cleopatra mandarin	Semi vigorous, Cold hardy Resistant to soil borne pathogens, Tolerant to Tristeza
3.	Grape	Dogridge	Nematode resistant salinity tolerant
		Salt creek	Salinity tolerant
		Freedom Harmony	Nematode resistant
		Riparia Glorie St. George Teleki 5-A	Phylloxera resistant
		Ruggeri 110 Richter	Drought tolerant
4.	Guava	Chinese guava	Wilt resistant
		Psidium puncilum	Dwarfing
		Aneuploid No.82	Dwarfing

Contd.

S.No.	Crops	Rootstock	Purpose
5.	Sapota	Khirni (Rayan)	Most compatible
		Sapota seedlings *Manilkara kanki* Star apple *Bassia longifolia* *Madhuca latifolia*	
6.	Avocado	Mexican rootstocks	Cold resistant, Dwarfing
		Duke series (Duke-7)	Root rot resistant
7.	Ber	Ber seedings *Zizyphus rotundifolia* *Z. nummuleria*	Dwarfing

Specific rootstocks identified for manipulating the characters of grafted/budded plants are given below.

Though in each species, a number of vegetative propagation methods may be possible, the commercially viable methods are based on overall success (%), economics and easiness.

The commonly and commercially followed propagation methods in different tropical and subtropical fruit crops are given below.

S.No.	Fruit crops	Methods of propagation	Details
1.	Mango	Approach grafting, Stone grafting, Softwood grafting	Local varieties and polyembryonic types are used as rootstocks. Vellaikolamban is a known dwarfing rootstock.
2.	Banana	Suckers, TC plants	Sword suckers with narrow leaves and well developed base
3.	Pineapple	Suckers, Slips, TC plants	Medium to large sized suckers (500-1000 g/one)
4.	Mangosteen	Seeds	Parthenogenesis exists and seedlings are true-to-type
5.	Papaya	Seeds	Commercially produced under controlled conditions
6.	Sapota	Approach grafting, Softwood grafting	Rootstocks used are Khirni seedlings (*M. hexandra*), sapota seedlings, star apple seedlings

Contd.

S.No.	Fruit crops	Methods of propagation	Details
7.	Avocado	Seeds, Air layering, Veneer grafting	Seedlings are not true-to-type. Grafting on its own rootstocks.
8.	Citrus fruits		
	Acid lime	Seeds,	Polyembryonic seeds
	Pummelo	Air layering	
	Sweet Orange	Seeds, Shield/patch budding	Rough lemon, Rangpur lime and *C. jambhiri* are used as roostocks
	Mandarin orange	Seeds, Shield budding	Rough lemon, Rangpur lime, Karna Khatta and Troyer Citrange etc. are used as rootstocks
	Sweet lime	Hardwood cuttings, Air layering	
	Rangpur lime	Seeds	
	Lemons	Shield budding Stem cuttings	Trifoliate, orange is the common rootstock
	Grape fruit	Shield budding	Jatti Khatti, Karna Khatta and Citranges are used as rootstocks
9.	Grape	Hardwood cuttings Clip budding	Specific rootstocks are used for different purposes Eg: Dogridge – Nematode and salt resistance Salt creek – salinity resistance
10.	Litchi	Air layering	One year old shoots are used
11.	Olive	Stem cuttings	Shoots of 1-2 year old are used for preparation of cuttings
12.	Passion fruit	Stem cuttings Micrografting	*in vitro* shoot tip grafting is followed for mass multiplication of hybrids
13.	Persimmon	Seeds Soft wood and semi hardwood cutting	Mist chamber is necessary for rooting of cuttings
14.	Pomegranate	Stem cuttings Air layering	
15.	Aonla	Seeds Inarching Patch budding	Very long juvenile period Own rootstocks
16.	Ber	Seeds Patch/shield budding	Different species of *Ziziphus* are used as rootstocks

Contd.

Contd. Table

S.No.	Fruit crops	Methods of propagation	Details
17.	Custard apple	Seeds	Viable for 3-4 years Natural variability is very high
18.	Date palm	Seeds Offshoots (suckers)	Rooted suckers of 2-5 years old are used
19.	Fig	Hardwood cuttings Air layering	Shoots of 1-2 year old are used
20.	Jack fruit	Seeds Approach grafting Epicotyl grafting	Long juvenile period
21.	Jamun	Seeds Patch/shield budding	20-50% polyembryony and apomictic seedlings can be used. 1-2 year old jamun seedlings are used as rootstocks.
22.	Guava	Air layering	Shoots of previous years growth are used

5. Design and Orchard Management

Establishment of orchards should be done with great care since it is a long term investment and lack of planning and wrong practices lead to negative consequences. In general, the major factors to be considered in the context include site selection, lay out and systems of planting, planting distances, selection of planting materials, planting, soil management practices, nutritional management, irrigation training, pruning and weeding. (All these factors are discussed in the Volume on *Fruit Crops* of Horticulture Science Series Published by NIPA in a detailed manner).

Canopy management

Canopy Management of fruit trees also should be given top priority for successful orcharding. By canopy management of trees, increased productivity per unit area of the land with easiness in cultural operations is envisaged. Fairly good knowledge about fruiting habit of a tree and influence of environmental factors on productivity are pre-requisites for effective canopy management. Canopy of a tree refers to the physical composition comprising of stem, branches, shoots and leaves. Number and size of leaves, architecture of stem, branches and shoots determine the canopy density of a tree. Canopy management deals with development and maintenance of tree structure in relation to size and shape for maximum productivity and quality. Size, shape and volume of canopy are influenced by climate, planting density, root stock, method of propagation, training, pruning, regularity of bearing, soil characters, nutrition, irrigation, growth regulators used, diseases and pests and other environmental factors.

Basic principles in canopy management

- Maximum light utilization
- Avoiding conditions which build up micro-climate congenial for pests and diseases
- Convenience in carrying out cultural operations in the orchards, maximizing productivity and quality.
- Achieving the required canopy architecture in an economic way.

Light plays a very important role in flower induction and fruit development, mainly through carbohydrate synthesis. Light interception influences fruit yield whereas light distribution determines fruit quality. Light exposure influences flower bud differentiation, fruit set, fruit colour and quality. Any canopy management practice should aim at reduction of excessive canopy shading and increasing air circulation in fruiting region.

Commonly used practices to achieve these goals include control of tree vigour, reduction of canopy shading, training and pruning systems to increase light interception and distribution. Opening of tree canopies and diagonal orientation of growth by pruning and training, reducing canopy vigour using dwarfing rootstocks, growth retardants, suitable nutrient management and water management practices are commonly adopted to obtain a desired productive canopy for trees in orchards. Effective canopy management is being practiced by one or the other above methods in mango, guava and grapes.

Precision farming

Precision farming identifies critical factors where yield is limited by controllable factors. It is a more precise farm management made possible by modern technology(s). Variations occurring in crop or soil properties are recorded, mapped and then management actions taken to correct them. Adoption of site-specific management systems using remote sensing, GPS and geographical information system (GIS) is possible following development of geomatics technology in recent past. This is a refinement of a good whole field management where management decisions are adjusted to suit variations in resource conditions. Precision farming requires special tools and resources for recognizing inherent spatial variability associated with soil characteristics, crop growth and for prescribing the most appropriate management strategy on a site specific basis. Conventional agriculture is based on uniform application of fertilizer, herbicide, insecticide, fungicide and irrigation, without considering spatial variability. The new concept of precision farming has emerged to alleviate the ill effects of over and under usage of inputs. In this modern farming system, three aspects are covered viz; data collection, analysis or processing of recorded information and recommendations based on available information. Development and implementation of the modern farming system require detailed discussion and experimentation though the potential of this technology is already demonstrated. Under Indian conditions, precision farming can be adopted after proper experimentation to find out applicability in Indias agricultural scenario.

Bioregulators in fruit production

Plant growth substances or plant regulators are organic chemical compounds other than nutrients which in low concentration promote/inhibit or modify any physiological functions of the plant. They are readily absorbed at the site of application and moved rapidly through the tissues to the site of action.

Plant hormones or phytohormones are also organic substances produced by the plants in minute quantities, regulate the physiological processes at a site remote from its place of production. Plant growth regulators are synthetic substances whereas phytohormones are natural ones produced by the plants. There are five groups of bioregulators: auxins, gibberellins, cytokinins, ethylene and abscisic acid.

Auxins: This group of organic substances at low concentrations promotes growth along the longitudinal axis. Auxins cause cell elongation by loosening of the cell wall. Other physiological effects of auxins are secondary growth – i.e. formation of secondary vascular cambium and promotion in cambium activity, rooting of cuttings, apical dominance, reduction in fruit drop, increase in fruit size, induction of parthenocarpy, organogenesis and weed control.

Gibberellins: Gibberellins stimulate predominantly cell elongation. Gibberellins are widely used for breaking seed dormancy, acceleration of seed germination, stimulation of bud break, increasing lateral branching, vegetative growth and yield, stimulation of fruit set, increasing fruit size and quality, delaying senescence and fruit ripening.

Cytokinins: Cytokinins known as juvenile hormones induce cell division in plants. They also stimulate cell enlargement in cotyledons, initiate bud break, radicle growth, accelerate chloroplast differentiation and chlorophyll synthesis.

Ethylene: Known as ripening hormone, ethylene is gaseous in nature. Ethylene promotes germination, proliferation, epinasty, root formation, abscission, induces flowering, male sterility, stimulates early production of female flowers and inhibits growth.

Abscisic acid: These are regarded as inhibitors which control abscission, senescence and water relations in plants.

Apart from the above major groups, the steroid hormones jasmonic acid, fusicoccin, brassinosteroids and oligosaccharides were also discovered from plants and proved to have hormone like properties. Unlike animal hormones, which have specific regulatory functions, plant hormones have widely overlapping functions. These hormones are distinctive both in chemical characteristics and in their function but the behaviour and response of each crop are highly variable according to the composition and concentration of compounds.

Besides these naturally occurring plant bioregulators, an array of synthetic chemicals, capable of altering many aspects of growth and development in plants are also available. Among them, important ones include synthetic auxins: indole butyric acid (IBA), phenoxy acetic acid, naphthalene acetic acid (NAA), Beta-naphthoxyacetic acid (BNOA), 4-chlorophenoxy acetic acid (4-CPA), 2,4,dichlorophenoxy acetic acid (2,4-D), 2,4,5-

trichlorophenoxy acetic acid, benzothiazole-2-oxyacetic acid (BOA) and 4-{(4-chloro-o-tolyl) oxy} butyric acid (MCPB); synthetic cytokinins – kinetin and benzyl adenine.

Plant growth retardants, which are synthetic organic chemicals which cause retardation of cell division steps in pathways of hormone biosynthesis, have also got significant applications in fruit culture. Important ones are GA biosynthesis inhibitors, eg. triazoles like paclobutrazol (cultar pp 333), uniconazol (S-3307, XE-1019), chloromequat chloride (cycocel or CCC), mepiquat chloride, ancymidol, flurprimidol (EL 500), norbornanodiazetines viz., tetcyclasis (BAS 106), AMO-1618, Phosphon-D, Maleic hydrazide and B-Nine.

Applications / uses of plant bioregulators in various stages of fruit culture are discussed here.

Propagation: In the field of propagation, plant regulators are used for breaking seed dormancy, inducing rooting of cuttings and layers, stimulation of growth of root stocks and cambial activity of graft union and micropropagation of fruit plants.

Seeds of many fruit plants require an extended period of long days and low temperature before germination. Breaking of this rest period/reducing the period of stratification can be achieved by application of gibberellic acid in temperate fruit crops like apple and peach. Germination of apricot seeds is enhanced by soaking them in kinetin (5 ppm) or GA (500 ppm) for 24 hours. Germination is accelerated and growth and vigour are hastened in citrus by treating the seeds with gibberellic acid (40 ppm). Pre-sowing soaking of seeds in GA solution stimulated germination in citrus fruits viz., sweet oranges, acid lime, Cleopatra mandarin and trifoliate orange, litchi and grapes (1500 ppm).

Irregular or uneven germination of cherimoya (*Annona cherimoya*) seeds can be overcome by soaking seeds in GA (500 ppm) for 24 hours. Soaking the seeds in GA_3 solution promotes germination (%), early germination and seedling growth in sapota (400 ppm), aonla (250-500 ppm), jackfruit (500 ppm) and ber (200 ppm).

Cuttings are used as a propagating tool in some fruit crops. Widely used plant growth regulator in rooting of cuttings is auxins, especially IBA which has low degradability by auxin destroying enzymes. Others are NAA, IAA, 2,4-D and 2,4,5-T. GA has antagonistic effect on rooting of cuttings. In temperate fruit crops like apple, peach, plum and pear, rootstocks are multiplied through cuttage. The difficult to root stocks like M9 in apple is treated with IBA (1000-2000 ppm) for quick root initiation.

In citrus, sour orange rootstocks were multiplied by stem cuttings and the highest rooting (%) was obtained with IBA (2000 ppm). Dormant one year old wood cuttings are treated with IBA or NAA to promote rooting in grapes.

Etiolation and use of root promoting substances like IBA promote rooting in avocado cuttings. Use of bottom heat (26°C) in cutting beds and spraying NAA and cytokinin once in a week improve rooting of cuttings in avocado. Rooting can be enhanced by girdling and application of IBA (5000-10000 mg/kg talc) in litchi.

Preconditioning of pomegranate shoots in June-July by girdling/etiolation and treating the cuttings with IBA solution, either quick dip (5000 ppm for 30 seconds) or slow dip (50 ppm for 18 hr) methods and planting under mist enhance rooting(%).

Citrus fruits viz. Washington Navel orange, Cleopatra mandarin, sweet lime, lemons and citron, cuttings root profusely with IBA (1000 ppm) treatment and also recorded the highest (%) of survival. In pummelo, the highest rooting (%) and more number of roots were obtained by treating semi hardwood cuttings in NAA (1000 ppm) solution.

In sapota, jackfruit and guava, rooting of cuttings is possible. Shoots are ringed, etiolated (20-25 days) and treated with ethephon before taking cuttings. Cuttings are treated with IBA (5000 ppm) and kept under mist for better rooting.

Growth regulators yielded excellent results in layering especially in stooling (mount layering) and air layering techniques. Auxin application improved rooting and survival of layers produced by mound layering in apple. In litchi, pretreatment of stock plants with cycocel (100 ppm) or ethrel (250 ppm) and treatment of air layers with IBA (3000-5000 ppm) gave the best rooting and survival (%). Treatment with IBA (10,000 ppm) in lanolin paste produced more number of roots in air layers of pomegranate.

in vitro culture

Aseptic culture of plant cells gained considerable importance in research and application in commercial scale. The major areas include micropropagation, embryo culture, anther culture, protoplast fusion and development of transgenic plants.

Rapid multiplication of planting material is possible through micropropagation. It is standardized in many fruit plants and commercially adopted in banana and pineapple. Virus free plants could be evolved through shoot tip culture especially in citrus. Shoot tip grafting (STG) is also done in citrus with the above intension. Micrografting is successful in apricot, almond, pistachio nut, passion fruit and citrus.

Embryo culture is standardized in longan, avocado, olive, papaya, grapes and banana. Protoplast culture is reported in apple, plum, grapes, citrus and banana.

Genetic transformation of apple rootstock M26 was done successfully using suitable vectors. Transgenic plants were also developed in pear, peach, apricot, citrus, papaya and banana.

Growth regulators like BA, NAA, 2,4-D, kinetin, IBA, zeatin, BAP were used in different media and its dosage was standardized for the success of above programmes.

Plant growth and development

Endogenous hormones control different growth and development processes in plants. Any alteration for the benefit of man is possible by manipulating this hormone level in the plant system through exogenous application of growth regulators.

Dormancy is observed in various plant parts like seeds, buds and bulbs. It is a natural phenomenon for regulation of plant growth and development and to overcome adverse climates. Actually the endogenous inhibitor-promoter hormone balance determines the onset and termination of rest. Exogenous application of GA and cytokinins increases the level of endogenus growth promoting hormones and decreases the level of ABA. Breaking of rest is of great significance in seed germination, breeding programmes and crop regulation. Termination of bud rest by GA application is a practice in apple and peach orchards. Under normal condition, these crops require 1-1½ months chilling for termination of rest. GA sprays successfully reduce the chilling period requirement.

Prolonging rest period in seeds and buds are advisable in crops like citrus and grapes. In certain localities, these plants are grown in such a climate where the night temperature falls below minimum level for normal growth. Seedling growth/shoot growth is often undesirable at this time as it may suffer injury. Growth retardants ABA, CCC and SADH applied to foliage delayed the bud break.

Controlling vegetative growth is another area of interest. Bioregulators can be used for enhancing or scaling up plant size. GA spraying makes the plants taller and vigorous. Plant growth retardants are recommended for controlling height and vigour in fruit crops. It is much appreciated in high density planting and will facilitate better management practices. In apricot, litchi and mango, application of paclobutrazol and SADH reduced growth and vigour of plants. Ethrel treatments were also effective in controlling growth in grapes, avocado and mango.

Flowering

Regulation of flowering is achieved in many fruit crops by manipulating the endogenous hormone level which is triggered by environmental factors like temperature and photoperiod. GA is capable of substituting certain specific environmental conditions that regulate flowering. Flowering in long day plants is controlled by the endogenous GA levels while application of CCC and SADH promoted flowering in short day plants.

In sweet and sour cherries, GA_3 is applied to prevent excessive flowering to minimize competitive effect of early fruiting on vegetative growth. Gibberellic acid (GA_3) also inhibits flowering in several stone fruits.

For many years, it is reported that ethylene stimulates flowering of Bromeliads and now ethylene is used commercially to stimulate flowering in pineapple. In litchi, ethylene or NAA induces/regulates flowering, which replace the requirement of girdling. Flowering is enhanced in grapes and lemon by CCC treatment while paclobutrazol induced regular flowering in mango. Soil application of cultar (PP 333-paclobutrazol) at 10 g a.i. in five litres of water at 60 cm away around the tree trunk in the ring of 15 cm depth during August resulted regularity of bearing in mango varieties Alphonso and Kesar. Benzothiazole-2-oxyacetic acid (BOA) hastened flower initiation and promotion of flowering in lower nodes of papaya plant.

Endogenous growth regulators, photoperiod and temperature are the major determinants of sex expression in flowers. Sex expression is primarily controlled by the endogenous ethylene – GA balance and accelerated by environmental factors. External application of bioregulators can alter sex of flowers but results are not consistent. Spraying 100-125 days old papaya plants with GA_3 (25 ppm), ethrel (100 ppm), TIBA (50 ppm) or SADH (500 ppm) enhanced female flower production. It is reported that ethylene applied at a certain time of pollen development causes male sterility in plants. In persimmon, pollination can be replaced by GA (25 ppm) spraying at blooming stage. This treatment increased production of seedless fruits and also reduced fruit drop.

Fruit set and development

Fruit set and further physiological events like fruit size, quality, maturation and seedlessness are regulated by hormones. Under unfavourable environmental conditions, low synthesis of growth promoters and/or excess production of inhibitors reduce or prevent fruit set. It can be increased by the exogenous application of bioregulators.

GA (10 ppm) treatment is effective in improving fruit set in cranberry, blue berry and phalsa. 4-CPA treatment served as an alternative to the costly techniques of caprification and cross pollination for improving fruit set in fig. Treatments of GA, 4-CPA, 2,4,5-T or B-NOA improved fruit set in pears, strawberry and cherry. In grapes, 4-CPA application replaced the cultural practice of girdling for improving fruit set and berry size. Increase in fruit set can be achieved by spraying IAA (40 ppm) or NAA (20 ppm) in certain litchi varieties. NAA (15-30 ppm) sprays increased fruit set in mango, ber and sapota. Promalin (GA_{4+7} + BA) a combination chemical significantly influenced fruit set in certain fruit crops. It also increased yield and improved shape of fruits.

Fruit drop, at immature or mature stages is a serious problem in fruit crops. Application of NAA, 2,4,5-TP, SADH in apple and NAA and SADH in pear at full bloom stage were highly effective in controlling fruit drop. Moreover SADH treatment prevented development of water cores and cold storage scales in apple fruits. Spraying 2,4-D (10 ppm) with aureofungin (20 ppm), 2,4-Trichloro phenoxy propionic acid (60 ppm), or GA reduced fruit drop in citrus fruits. Bioregulators NAA (25 ppm) or GA (20 ppm) applied twice, at fruit set and one month later effectively controlled fruit drop in ber. In mango, application of 2,4-D (10-15 ppm) or CCC (200 ppm) immediately after set reduced fruit drop.

Fruit thinning is practiced in grape varieties viz., Thompson Seedless and Perlette where more number of berries in a cluster reduces berry size, increases compactness leading to berry cracking and rotting at harvest. Here prebloom spray of GA is given extensively for reducing fruit set and obtaining loose and attractive clusters. Application of sodium 4,6-dinitro-0-cresol (DNOC, Elgetrol) at full bloom or NAA at post bloom stages caused satisfactory thinning in apple, plum, prune, apricot, pear and sweet cherry. Growth regulators are employed for producing larger and attractive fruits which gain the consumer appeal. They are also used to hasten or delay fruit maturation to meet peak demand of fruits and for altering harvesting period to avoid unfavourable environmental factors. Exogenous application of auxins increased fruit size and altered fruit maturity in plums, strawberry and oranges. NAA (200-300 ppm) increased fruit size in pineapple. Application of 2,4,5-T at beginning of second stage of growth (double sigmoid) increased fruit size and hastened maturity in fig.

Gibberellic acid is used commercially in the seedless grape industry since 1960 to alter the physiological events, rachis cell elongation, flower thinning and berry enlargement. A prebloom spray is given to promote elongation of rachis which provides larger and loosened frame work for the grape bunch. GA application at anthesis stage reduced berry set (prevents bunch rot) and increased berry size. A combination of GA and brassinosteroids or CPPU (2-chloropyridyl phenyl urea), a cytokinin derivative applied as post bloom spray induced berry enlargement in grapes and increased fruit size and weight in apple. Application of GA (10 ppm) increased fruit set in phalsa while GA (20 ppm) or 2,4,5-T (5 ppm) improved fruit retention, fruit size, weight, yield and quality.

Alteration in the period of fruit maturity and uniformity in ripening is much important in fruit industry for meeting early/late market demand, staggering fruit harvest and to avoid effects of unfavourable weather. Ethephon sprays during fruit development in grapes, strawberry and cranberry resulted early fruit maturity, improved colour, quality and uniform ripening. This facilitates single harvest of the fruits and also reduces labour cost.

A preharvest spray of ethrel (250 ppm) at colour turning stage induced early maturity in ber. Uneven ripening can be overcome by ethylene treatment in fruits like citrus and it can be hastened in mango and banana. In grapes, proportion of green berries at harvest can be reduced by ethrel (250 ppm) spray at colour break stage of berries. Ethephon applied to citrus fruits prior to storage ensures post harvest degreening. Dipping lemon fruits in ethrel (1000 ppm) solution resulted in uniform yellowing.

Pre-harvest sprays of benzyl adenine (BA) retained the freshness and storability of fruits. Retention of pedicels in cherry and calyxes in strawberry are reported with BA application which fetches high market value.

Growth regulators are also used to prevent fruit russet (a superficial disorder in which the exocarp is interrupted), internal browning and frost injury. Preharvest spray with GA is effective in controlling fruit russets in apple and grapes, while a combination of GA and ethephon reduced internal browning in fruits like apple and peach. Treatment of GA in pear and 2,4,5-T in apricot is useful for increasing tolerance of fruits to low temperature. Mango malformation can be controlled by spraying NAA (200 ppm) + methyl-2-benzimidazol carbamate (MBC-0.1%) at flower bud differentiation and panicle emergence stages. Foliar spraying of paclobutrazol (10 or 25 ppm) at full bloom and one month after fruit set reduced incidence of spongy tissue in Alphonso mango.

Fruit harvesting

At the time of maturity, fruits of some species are seen attached to the stem tenaciously which prevent mechanical harvesting. Shaking of vines or branches of such trees where fruits are held firmly may cause crushing or bruising of fruits. In this context, use of chemicals inducing abscission facilitate easy harvesting. Pre harvest spray of ethrel resulted uniform maturation of fruits with accelerated ripening. Promising results were obtained in fruit crops like apple, pear, plum, cherry and grapes. Here, concentration of the chemical is to be standardized to avoid complete abscission. Application of cyclohexamide at pre harvest stage resulted easy harvest in citrus fruits. A mixture of one % glycerin + ethephon is one of the best formulation for abscission. Ethylene releasing compounds promote abscission in olive fruits and black berries.

Post harvest management

Even though this is a wide area to be exploited, the only commercial application of hormones is in the field of ripening. Ripening processes in fruit is associated with changes in colour, membrane permeability, hormone level, nucleic acid metabolism, production of volatiles, respiratory and hydrolytic enzymes and cell wall softening. Ethylene, the ripening hormone plays the vital role in fruit ripening. Auxin induced ethylene production or external application of ethephon can trigger ripening process. But gibberellins delay ripening in fruits by retarding the pigment degradation and activity of ripening enzymes. GA_3 (100-200 ppm) or calcium nitrate (1%) delay ripening and increase shelf life in citrus fruits. In grapes, preharvest spray of CCC (1000 ppm) with captan (0.2%) or benomyl (0.2%) improved quality and storage life. Dipping mango fruits in GA (200 ppm) or cycocel (400 ppm) delayed ripening. Gibberellic acid and kinetin delay ripening process in mango fruits by inhibiting chlorophyll degradation, carotenoid degradation and retardation of activities of hydrolytic enzymes.

Defoliation

Chemical defoliation has limited use in fruit culture. In grapes, manual defoliation is done prior to pruning/which usually hinders the efficiency of pruners. In such cases, particularly in tropical region, chemical defoliation is advised prior to pruning by spraying ethephon (1000 ppm). This will also induce early and uniform bud break.

Parthenocarpy

Growth regulators are used for induction of parthenocarpy in fruits. In seeded fruits, pollen grains provide stimulus for ovary growth and prevent ovary abortion. Synthetic auxins induce unpollinated ovaries to develop into full sized fruits as in fig, avocado and strawberry. Pollinizers are essential for adequate pollination and fruitset in many temperate fruit crops viz., apple, pear, peach, plum, almond and apricot. Pollination can be replaced by application of GA at blooming stage. GA induced parthenocarpic development of fruits is obtained in the above fruits and also in grapes, mandarin, grapefruit and guava. In grapes and fig, parthenocarpy is also reported by cytokinin sprays. Combination sprays of bioregulators are also effective. Auxin and GA induced parthenocarpy in blue berry and cherry while auxin, GA and cytokinin combination produced such fruits in apple cv. Cox's orange pippin.

As weedicide

In India, weeds cause about 35 % loss in yield. Plant growth regulators are often regarded as herbicides in many fruit orchards. Auxins are widely used as weedicides. Dicotyledonous weeds are very sensitive to high auxin level. Auxin derived herbicides are phenoxy acetic acid analogous (2,4-D, 2,4,5-T, CPA and MCPA) and benzoic acid analogs (chloramben, picloram, dicamba and substituted picolinic acid). Among these, 2,4-D is the most commercially used herbicide. It is economic, non-corrosive, effective at low concentrations, highly selective, easy to formulate and safe to handle. Apart from these, herbicides like TCA, fernoxone and dowpan for monocots and paraquat, diquat, bromacil,

Nutritive Value of tropical and sub tropical fruits

Fruits	Protein	Fat	CHO	Ca	P	Fe	Vit.A	VitB1	B2	B3	Vit.C
	g.	g.	g.	mg	mg	mg	Univ. Unit.	mg	mg	mg	mg
Aonla	0.5	0.1	14.1	0.05	0.02	1.2	-	30	0.2	-	600
Avocado	1.7	22.8	0.8	0.01	0.08	0.7	-	-	-	-	13
Banana	1.3	0.2	36.4	0.01	0.05	0.4	traces	150	0.3	30	-
Bael	1.3	0.2	30.6	0.09	0.05	0.3	186	12	0.9	11 91	15
Ber	0.8	0.1	12.8	0.03	0.03	0.8	55	70	-	-	-
Breadfruit	1.5	0.2	17.9	0.04	0.03	0.5	15	-	-	-	-
Custard apple	1.6	0.3	23.9	0.02	0.04	1.0	traces	-	-	-	-
Dates (Persian)	3.0	0.2	67.3	0.07	0.08	10.6	600	90	0.8	30	traces
Durian	2.8	3.9	34.1	0.01	0.05	1.0	20	-	-	-	-
Fig	1.3	0.2	17.1	0.06	0.03	1.2	270	-	0.6	50	2
Grape	0.8	0.1	10.2	0.03	0.02	0.4	15	40	0.3	10	3
Grapefruit	1.0	0.1	10	0.03	0.03	0.2	-	120	0.3	20	31
Guava	1.5	0.2	14,5	0.01	0.04	1.0	Traces	30	0.2	30	299
Jackfruit	0.1	0.8	18.9	0.02	0.03	0.5	540	30	0.4	-	-
Jamun	0.7	0.1	19.7	0.02	0.01	1.0	-	-	-	-	-
Lime	1.5	1.0	10.9	0.09	0.02	0.3	26	20	0.1	-	63
Litchi	0.7	0.3	9.4	0.21	0.31	0.03	42	87.3	-	12 2.5	traces
Loquat	0.7	0.3	10.2	0.03	0.02	0.7	-	-	-	-	-
Mango	0.6	0.1	11.8	0.01	0.02	0.3	4800	40	0.3	50	13
Mangosteen	0.5	0.1	14.3	0.01	0.02	0.2	-	-	-	-	-
Orange	0.9	0.3	10.6	0.05	0.02	0.1	350	120	0.3	60	68

Contd.

Contd. Table

Fruits	Protein	Fat	CHO	Ca	P	Fe	Vit.A	VitB1	B2	B3	Vit.C
	g.	g.	g.	mg	mg	mg	Univ. Unit	mg	mg	mg	mg
Papaya	0.5	0.1	9.5	0.01	0.01	0.4	2020	40	0.2	250	46
Persimmon	0.8	0.2	19.0	0.01	0.01	0.3	1710	-	-	-	-
Pineapple	0.6	0.1	12.0	0.02	0.01	0.9	60	-	-	120	63
Pomegranate	1.6	0.1	14.6	0.01	0.07	0.3	-	-	-	100	16
Roseapple	0.7	0.2	9.7	0.01	0.03	0.5	-	-	-	50	-
Woodapple	7.3	0.6	15.5	0.13	0.11	0.6	-	-	-	170	-

dinoseb, simazin and atrazin for various other type weeds are also used in orchards for controlling weeds. Since many of these herbicides are toxic chemicals, the grower should take utmost care while handling and spraying them in the orchards.

6. Physiology of Fruit Trees

Growth and development of many fruit plants consist of two distinct phases, vegetative and reproductive and an overlapping one- the transition phase. These are primarily controlled by genes and to a certain extent, variation in physiological and morphological traits occurs due to environmental factors. The genetic make up or heredity is responsible for the differences in the growth pattern between species and cultivars and plays a vital role in adaptability to a particular environment.

On set of vegetative/juvenile phase is with germination of a seed. It is characterized by three important processes viz. cell division, cell enlargement and initial stages of cell differentiation. This leads to formation of plant parts; the terminal leader, branches, leaves and roots. Physiological and chemical functions of these plant parts are not similar. Terminal leader will be always active and in juvenile state. It has maximum functional leaves and by utilizing nutrients and sunlight synthesize food for growth and development of plant. That is why lower branches of a tree become less vigorous and less active with many dead tissues.

Internal physiology of fruit trees with regard to growth, development, dormancy and shedding of leaves are almost similar even though certain external factors play a role. In short lived fruit plants like banana and pineapple, they are regarded as monocarpic in nature as they die after flowering and fruiting. Perennial trees exhibit a different pattern as temperate crops like apple exhibit a biennial growth cycle. During winter, leaves shed and tree becomes physiologically inactive or dormant. In spring season after the onset of rain or irrigation, shoots, leaves and roots grow fast, tree manufactures enough carbohydrates, flower buds appear and fruiting occurs. Such trees respond well to pruning as in grapes. In mango, usually an 'on' year is followed by an 'off' year, indicating alternate or biennial bearing tendency. In off year, there will be more vegetative growth and carbohydrate level increases and this reserve food is utilized for flowering in the next year. In some cultivars, more than one 'off' year is required to make up carbohydrate level.

Fruit plants exhibit a juvenile period, but length of this phase is determined by genes and modified by environmental factors. Juvenile morphology is distinct in some fruit plants. Juvenile types of stem growth, thorniness and leaf form persist for 4 to 10 years. Citrus plants exhibit thorniness during juvenile phase. Some plants exhibit a typical branching pattern. Unique leaf forms are seen in jackfruit and citrus.

Juvenile wood exhibits a strong growth rate. This may be a reason for the fast rooting of juvenile cuttings. Juvenile cuttings contain high amount of endogenous auxins as the enzymatic auxin oxidation is less in this part compared to aged wood.

In many plants, transition from juvenile to mature condition takes place gradually. Then flowering and fruiting occur and in this stage, growth rate, geotropism and apical dominance are depressed. Ultimately the entire branch becomes senescent. In monocarpic fruit plant, death of whole plant occurs after fruiting. In tree fruits, the branch again switches over to vegetative phase, controlled by heredity, endogenous hormones and external factors.

Structure and mechanism of growth

A fruit plant basically comprises of three parts: roots, stem and leaves. Fruit trees have tap root system while annual or biennial crops like banana and pineapple have fibrous roots. Root hairs present in the root absorb water and minerals from soil through endodermis. From there, it enters the xylem vessels and then through capillary movement to the tracheids and vessels of other cells of the tree.

The trunk or stem consists of three parts – pith in the centre, the wood surrounding pith and the bark covering the stem. Pith is composed of several minute cells connected with bark and medullary rays which contact with external air. The woody stem has xylem, phloem and cambium tissues. Xylem is the wood portion, inner part of the bark is phloem and cambium is a thin meristematic layer between wood and bark. Outside the phloem, corky tissue of bark is seen which protects inner tissues of the stem. Water and dissolved nutrients are carried from root to upper parts through xylem vessels while phloem transports the manufactured food in both upward and downward directions.

A branch has a similar structure and it is the prolongation of stem. Twigs arise from branch and leaf is an expansion of twig. Stomata present in leaves are responsible for gaseous exchange which helps in photosynthesis. Transpiration in plants causes capillary movement of water in plants.

A bud may be terminal or lateral in position on the twig. It is the undeveloped shoot or flower or both. Buds may be latent, dormant or resting, situated at nodes in shoots or twigs. In many fruit plants, water sprouts are seen which arise from latent buds.

Reproductive organ in a tree is the flower bud. Flowers are produced on terminal or lateral branches in the leaf axils as in mango or sapota or on main stem as in jackfruit. After pollination and fertilization, fruit set is noticed. Each and every fruit tree has its own flowering and fruiting habits. Parthenocarpic fruit development is seen in banana, pineapple and in certain types of citrus plants. Presence of seeds helps to increase fruit size. Heavy fruiting in a year causes fruit drop due to insufficient amount of carbohydrates and water deficiency. Thinning is done in fruits which will increase the size of fruits.

Important processes taking place during growth and development of a plant are photosynthesis (production of food), respiration (release of energy), transpiration (upward movement of water) and absorption (taking off raw material and translocation).

Movements of organic and in-organic solutions and photosynthetic products take place through translocation. Water and minerals are transported through xylem vessels whereas phloem translocates carbohydrates in all directions. Some physiologists opine that both tissues can transport hormones and nutrients. Movement of solution through xylem vessels is controlled by a water potential gradient in response to rate of evaporation from stomata / leaves. Mechanism of translocation in the phloem vessels is not only due to action of gradients but still other theories are put forwarded by physiologists.

Apart from roots, water is absorbed through epidermal cells situated in leaves and other parts. Foliar sprays of micronutrient and hormonal solutions are absorbed in this manner.

Translocation patterns of different group of growth hormones are not similar. Polar nature of translocation is noticed in endogenous and synthetic auxins. They are transported along the longitudinal axis of plant. It moves from apical to basal regions through plasma membranes and cell walls. Non polar movement is observed in case of gibberellins and abscisic acid. If they are applied in any one part, the effect is seen in all regions of shoot and root indicating their movement in all directions through xylem and phloem sap. Mechanism of translocation is not clear in case of cytokinins.

Mechanism of flowering and fruiting

Flowering is a crucial and complex biochemical process. Large number of fruit plants flower in a particular season, but fruit crops like pineapple and banana do not show such a pattern. Flowering is controlled by internal factors like C/N ratio, flowering hormones and external systems like photoperiod, temperature, nutrition and moisture . Some subtropical fruit plants require a cold treatment- vernalization - for flowering. Many perennial fruit species need chilling temperature to flower in spring. It is established that when the plant attains physiological maturity and under favourable conditions, a stimulant, designated as 'florigen' is formed in the leaves and translocated to meristem and induces flowering. Rise in DNA and RNA synthesis is recorded on attaining floral stage in some fruit species. Apart from nucleic acids, phyto hormones viz. auxin, gibberellins, ethylene and abscisic acid also have influence on flowering. External application of these hormones induces flowering in fruit crops like pineapple, citrus and mango. Such a response has not been seen to occur in many short day and long day plants which need further research and clarification.

Mechanism of fruiting involves development of ovary, pericarp or the enclosing tissue. Pollination and fertilization are necessary for fruit development. In crops like banana and certain types of fig, pollination is not a must as the berry development is through vegetative parthenocarpy. Stimulative parthenocarpy or stenospermocarpy is present in grapes where stimulus of pollen is required for fruit growth. In normal case, ovule develops into seed, ovary into fruit while peduncle develops into stem of fruit. Sometimes, receptacle develops into edible part as in strawberry and fig, aril in case of litchi, rambutan and mangosteen and perianth in jackfruit .

Unfruitfulness is seen in some fruit plants due to monoecious or dioecious nature, self or cross incompatibility, defective pollen and pistils. Environmental factors like high or low temperatures, excessive rainfall, heavy winds, tree growth and nutritional aspects viz., sick trees, poor nutrition, improper management practices are also responsible for poor fruit set. Flower and fruit drop due to hormonal imbalance, lack of proper carbohydrate/ nutrients, adverse climatic condition, deficiency in moisture and incidence of pests and diseases are prevalent with many fruit plants.

Factors associated with growth and flowering of fruit plants

Growth, development, flowering and fruiting of plants are influenced by both **external** and **internal** factors. Atmosphere covering the orchard comprises of macro and micro environmental elements. Major factors affecting growth of fruit plants are light, temperature, altitude, rainfall, soil and minor ones like wind velocity, relative humidity and topography. All these factors are interdependent in their action / influence on development of fruit crops.

Light, an important factor is the solar energy spreading as electromagnetic radiation. This is transmitted as waves of spectrum of light particles known as photons. Wave length is measured in terms of micron (1 micron = 10^{-6} millimeter). Light energy comprises three components – invisible, visible and infra red fractions. Invisible fraction of light includes gamma rays, x-ray, cosmic and ultra violet rays. Infra red is mainly responsible for transmitting heat energy. The visible wave length ranges 370-760 n/μ. In this range, the red and far red lights transmit heat and light and are beneficial for dark conversion of phytochromes. This fraction of visible light emits seven colours known as VIBGYOR. In this spectrum red, light has longer wave length (760 n/μ) while violet has the shortest (370 n/μ). Red light has a major role in the synthesis of carbohydrates in fruit crops. Intensity of light can be measured in meter candle (Lux).

Most of the tropical evergreen fruit plants absorb higher quantity of solar energy and utilize it, if other environmental factors like temperature and moisture are favourable. Under tropical desserts and arid zones, since the supply of moisture is limited, leaf shape is modified to compact leaves with thick cuticle as in date palm or to pines. The quantum of absorption of light energy is reduced.

In subtropics, there exists more ultraviolet and blue violet rays. Radiation is more in hilly regions than in plains. Since the sunlight is vertical here, plants have maximum dry matter production. Long day plants are grown in sunshine period while short day plants during winter.

Photo morphogenic influence generates both visible and invisible changes in plants. Mechanism of induction of morphogenesis by light is mediated through enzyme activation, synthesis of hormones like gibberellic acid, kinetins and abscisic acid and alteration in membrane permeability. Certain wave lengths of light (eg. Red light) reduce phytochromes and this brings a rapid change in the membrane permeability which brings about the following developmental activities in fruit plants.

Influence of light on germination of seeds of fruit plants is obvious. There are three types of responses, i) complementary, ii) obligatory requirement and iii) no response to light. In plants, root development is influenced by light. Blue light improves growth of roots and root hairs while red light retards the growth and affects absorption efficiency of roots.

Favourable light intensity and duration promote stem elongation and growth. Even though the rate of photosynthesis is high in yellow or red band of spectrum, high rate of respiration and biodegradation of auxins cause some photothermal adverse effects on growth. But wave lengths up to yellow are congenial for stem growth in fruit plants.

Comparatively longer duration of light and wave lengths between 560 to 660 n/μ favour leaf growth in fruit plants. Higher light intensity in tropical rain forest resulted in the production of dense and thick leaves while lower light intensity under temperate regions induces thinner and transparent leaves.

Various pigments like chloroplast, anthocyanin, carotenoids and lycopene are synthesized in fruit plants. Among them chloroplast is the most important one and the highest rate of synthesis is found under wave length of 440-660 n/μ.

Light interception and duration affect flowering in fruit trees viz. flower bud differentiation and fruit set. Failure of flowering in mango trees with dense canopies is reported. Response of plants to various duration of light period is known as photoperiodism. This effect is more pronounced in places where length of day is much longer than night. Tropical fruit plants which are grown near to equatorial regions are not influenced by photoperiodism. But in subtropical fruit plants like grapes, light intensity has got influence on flower bud formation. Based on this, fruit crops are classified into long day, short day and day neutral plants.

Intensity and quality of light influence fruit development and maturity. Fruits produced on the outer whorl of canopies in mango, citrus, guava and sapota receive full sunlight and exhibit early maturity. Light also influences the carbohydrate synthesis and biomass accumulation. Blue and red light spectrum is the most effective for the synthesis of biomass. But under high intensity of light, higher rate of respiration and heat generation cause degradation of biomass. In mango, fruit yield is related to light interception whereas fruit quality is a function of light distribution.

Light has got a pronounced influence on the synthesis of phytohormones. Flower inducing hormones like florigen are produced during dark period.

Tropic movement of plants towards or away from plants is known as phototropism. Blue light of visible spectrum is more effective in inducing tropic movements in plants.

The light utilized by plants for the photosynthesis corresponds to 400-700 n/μ of visible spectrum. Photosynthesis in grapes rapidly increases upto the light intensity of 5000 ft (200 watt/m^2). Trees with dense canopies receive less sunlight in the inner whorls resulting in a reduced rate of photosynthesis.

Temperature

It is one of the crucial factors for plant growth, development and flowering. According to the temperature requirement and locality, fruit plants are classified into tropical, subtropical, temperate and arid crops. Most of the fruit plants can be grown in a temperature range of 5 to 50°C, even though each one has its own requirement and the optimum range lies in between 15-30°C.

Heat unit requirements from flowering to fruit maturity are ascertained in many fruit crops. Heat unit requirement is very specific in grape and temperate fruit varieties. Temperature has got definite effects on morphogenesis. A rise in temperature will increase the rate of all biological and biochemical processes at the expense of accumulated photosynthates. Flower induction, fruit set, fruit drop and leaf fall during winter months. are influenced by temperature.

Responses of fruit crops to temperature are known as thermoperiodicity. In these reactions, temperature of dark period is more important than light period for fruit bud differentiation. Induction of reproductive phase by temperature in fruit plants is termed as thermal induction.

Exposure to adverse temperature causes various injuries viz., chilling injury, freezing injury, desiccation, direct thermal injury and metabolic disturbances.

Altitude is another factor that affects growth of fruit plants. It determines what kind of fruit crops can be grown in a region. Based on the altitude and rainfall distribution, there are four fruit zones in South India; hill fruit zone (> 1500 m altitude), warm humid zone of lower hill slopes (600-1500 m), high rainfall and humid zone (up to 600 m amsl) and dry, arid zone with low rainfall.

Rainfall or moisture is one of the essential components for growth and development of fruit plants. Water is the major constituent of plants and deficiency in moisture results in declined growth rate.

Soils around the plant play an important role in fruit production. Fruit plants can be grown in different kinds of soil like alluvial, laterite, loamy, sandy loam, clayey, red or black soils. Whatever be the type of soil, the nature, depth, organic as well as inorganic contents, drainage are some factors which influence growth of fruit crops. A soil pH range of 5.5 to 6.5 is the best for fruit culture.

Soil is a complex mixture of primary and secondary elements and organic matter. Nutrient needs of fruit trees can be known from soil analysis and leaf analysis and these techniques were standardized in most of the fruit plants. Soil fertility, nutrient supply and orchard management practices should be done for getting good vegetative growth and yield in fruit crops.

Among the internal factors affecting growth and development, relative proportion of carbohydrates and nitrogen is important. A balance of carbohydrate and nitrogen (C/N ratio) should be maintained for better production. Other elements in this group are bioregulators, enzymes and cofactors.

References

Bal, J.S. (1997) Fruit Growing. Kalyani Publishers.

Bhargava, B.S. and Chadha, K.L. (1993) Leaf nutrient guide for fruit crops. Advances in Horticulture. Vol.2. (Eds.) K.L.Chadha and O.P. Pareek. Malhotra Publishing House, New Delhi.

Bose, T.K., Mitra, S.K. and Sanyal, D. (2002) (Eds.) Fruits – Tropical and Subtropical. Naya Udyog, Calcutta.

Chadha, K.L. (1998) Improvement in tree fruit and plantation crops. *Indian J.Hort.* 55 (4): 265-296.

Chadha, K.L. (2001) Handbook of Horticulture. Indian Council of Agricultural Research, New Delhi.

Chattopadhy, T.K. (1994) A Textbook on Pomology. Kalyani Publishers, New Delhi.

Chundawat, B.S. (1997) Nutrient Management in Fruit crops. Agrotech Publishing Academy, Udaipur.

Chundawat, B.S. (2001) Integrated nutrient management in tropical and sub tropical fruits. *Indian J. Hort.* 58 (1&2): 59–69.

Devi, S.L. and Mruthunjaya, S. (2000) Nutritional Disorders and their Management in Fruit crops. Krishi Vikas Publications, Bangalore.

Gangwar, A.C. and Rai, K.N. (1995) Emerging Agricultural Export Opportunities in India. *Bihar J. Ag. Marketing*. 3: 370-377.

Pathak, R.S. and Pathak, S. (2001) Fruit Production in problematic soils. *Indian J. Hort.* 58 (1&2): 16 - 22.

Radha, T. and Mathew, L. (2007) Fruit Crops. (Ed). K.V.Peter. New India Publishing Agency, New Delhi.

Rajan, S. and Markose, B.L. (2007) Propagation of Horticultural Crops. (Ed). K.V.Peter, New India Publishing Agency, New Delhi.

Shikhamany, S.D. (2001) Canopy management of the tropical and subtropical fruit crops. *Indian J. Hort.* 58 (1&2): 28-32.

Singh, S.P. (1995) Commercial Fruits. Kalyani Publishers, Calcutta.

Chapter – 5

Spices Biotechnology

K. Mary Mathew, K. Pradeep Kumar, Swapna Sasidharan, C.S. Anisha Sherin Jose, Ramya Lakshmi and Y.S. Rao

India known as the legendary land of spices, is the foremost in their global production, consumption and export. The history of Indian spices dates back to the beginning of the human civilization. There are references about Indian spices and their uses in the Vedas (6000 BC), by 'Manu', the lawgiver in 4000 BC, by the Babylonians and Assyrians (around 3000 BC), and in the Old Testament (1000 BC) of the Bible (Sivaraman and Peter, 1999). From time immemorial, spices are sought like gold and are one of the greater factors in human progress having done much to change the course of history and geography besides promoting international relations. Quest for spices created a furore comparable only to the Crusades, and was one of the dominant factors in European history during the middle ages and as late as the sixteenth century. Spices played a prominent part in all the civilizations of antiquity, in ancient China, India, Babylon, Egypt, Greece and Rome.

According to Hirasa and Takemasu (1998) the term 'spice' can be defined as the dry part of a plant, such as roots, leaves and seeds, which impart to food a certain flavour and pungent stimuli. By clubbing spices and condiments into one group, the International Organization for Standardization (ISO) illustrated that the term spice or condiment applies to "such natural plant or vegetable products or mixtures or thereof, in whole or ground form, as are used for imparting flavour, aroma and piquancy to and for seasoning food" (Manay and Shadaksharaswami, 1997). International Spice Group suggested more or less a similar observation. That is 'spices are any of the flavoured or aromatic substances of vegetable origin obtained from tropical or other plants commonly used as condiments or employed for other purposes on account of their fragrance, preservative or medicinal qualities' (Sivaraman and Peter, 1999).

Spices constitute an important group of agricultural commodities, indispensable to the culinary art of any part of the world. Spices production is geographically concentrated in developing countries and contributes significantly to their foreign exchange. No country in the world grows as many kinds of spices as India, according to Rosengarten (1969). Almost all the states and union territories of India grow one or the other spices. Because of the varying climates - from tropical to sub-tropical to temperate-almost all spices grow splendidly in India. Under the act of Parliament, a total of 52 spices are brought under the purview of Spices Board though 109 spices are notified in the ISO list and as per the report of Bureau of Indian Standards, 63 spices are produced in the country. Export of spices has been one of the major sources of revenue for Indian economy. India cultivates spices in 3.2 million ha with an annual production of about 5.9 million tones, contributing to nearly 60% of its production globally, 41% of the trade in terms of quantity, but only 40% in terms of value. Domestic consumption accounts for 90-95% of the total production and the rest is exported (Anon 2006). List of spices under the purview of Spices Board, India is given in Table 1.

1. List of Spices

Table 1: List of Spices (Courtesy: www.indianspices.com)

S. No	Common Name	Scientific Name
1.	All Spices	*Pimenta dioica* (L) Merr.
2.	Aniseed	*Pimpinella anisum*
3.	Asafoetida	*Ferula asafoetida* L.
4.	Basil	*Ocimum basilicum* L.
5.	Bay Leaf	*Laurus nobilis* L
6.	Bishop's weed	*Trachyspermum ammi* L.
7.	Black pepper	*Piper nigrum* L
8.	Cambodge	*Garcinia cambogia* (Gaertn) Desr.
9.	Caper	*Capparis spinosa* L.
10.	Caraway	*Carum carvi* L.
11.	Cassia	*Cinnamomum cassia* Blume.
12.	Celery	*Apium graveolens* L.
13.	Chillies	*Capsicum annuum* L.
14.	Cinnamon	*Cinnamomum zeylanicum* Breyn
15.	Clove	*Syzygium aromaticum* (L)
16.	Coriander	*Coriandrum sativum* L.
17.	Cumin	*Cuminum cyminum* L
18.	Curry leaves	*Murraya koenigii* (L) Sprengel
19.	Dill	*Anethum graveolens* L.
20.	Fennel	*Foeniculum vulgare* Mill
21.	Fenugreek	*Trigonella foenum-graecum* L.
22.	Garlic	*Allium sativum* L.
23.	Ginger	*Zingiber officinale* Rosc.
24.	Greater galangal	*Alpinia galanga* Willd.
25.	Horse radish	*Armoracia rusticana* Gaertn
26.	Hyssop	*Hyssopus officinalis* L.
27.	Juniper berry	*Juniperus communis* L.
28.	Kokam	*Garcinia indica* Choisy
29.	Large cardamom	*Amomum subulatum* Roxb
30.	Long pepper	*Piper longum* L.
31.	Lovage	*Levisticum officinale* Koth.
32.	Marjoram	*Marjorana hortensis* Moench
33.	Mint	*Mentha piperita*

Contd.

Contd. Table 1

S. No	Common Name	Scientific Name
34.	Mustard	*Brassica juncea* L. Czem
35.	Nutmeg & Mace	*Myristica fragrans* Houtt.
36.	Oregano	*Origanum vulgare* L.
37.	Parsley	*Petroselinum crispum* Milli.
38.	Pomegranate Seeds	*Punica granatum* L.
39.	Poppy seed	*Papaver somniferum* L.
40.	Rosemary	*Rosemarinus officinalis* L.
41.	Saffron	*Crocus sativus* L.
42.	Sage	*Salvia officinalis* L.
43.	Savory	*Satureja hortensis* L.
44.	Small cardamom	*Elettaria cardamomum* Maton
45.	Star Anise	*Illicium verum* Hook
46.	Sweet flag	*Acorus calamus* L.
47.	Tamarind	*Tamarindus indica* L
48.	Tarragon	*Artemisia dracunculus* L.
49.	Tejpat	*Cinnamomum tamala (Buch Ham)*
50.	Thyme	*Thymus vulgaris* L.
51.	Turmeric	*Curcuma longa* L.
52.	Vanilla	*Vanilla planifolia* Andr.

2. Biotechnological Interventions in Spices

According to the article 2 of the United Nations Convention on Biological Diversity, Biotechnology is "any technological application that uses biological systems, living organisms or derivatives thereof, to make or modify products or processes for specific use". Practical application of the theories of Biotechnology in various aspects of day to day life has been the subject of research for many decades now. Biotechnological research on spices encompasses the following applications.

- Micro-propagation and large scale production of Tissue Culture (TC) plants.
- Other Tissue Culture techniques for Crop Improvement and metabolite production
- Molecular marker studies for
    - i) Genetic Diversity analysis
    - ii) Marker-Assisted Breeding
    - iii) DNA fingerprinting and Barcoding for Quality assurance
    - iv) Disease Diagnostics
    - v) Genetic Engineering
    - vi) Detection of Adulterants

3. Micropropagation

Micropropagation, one of the largest established uses of tissue culture and in turn plant biotechnology, is a relatively efficient, safe and often economical method of *in vitro* plant propagation, if judiciously applied. Micropropagation is a recognized means of producing millions of genetically identical plants under aseptic conditions. Large scale production, economy of time and space, clonal consistency and uniformity, freedom from seasonal constraints, possibility for short and long term storage of valuable germplasm, and maintenance and production of disease-free plants are some of the major advantages that micropropagation offers over conventional methods of propagation. Micropropagation techniques employed in spice crops like cardamoms and vanilla (Nadgauda *et al.,* 1983, Rao *et al.,* 1993, Mary *et al.,* 1999, Rao *et al.,* 2006, Rao *et al.,* 2012a, Rao *et al.,* 2012b) were aimed at a breakthrough in production required to meet the growing demands of spices for domestic consumption and export. This *in vitro* technique for rapid multiplication of planting material is beneficial in other spices as well where seed propagation techniques are time-consuming and inconsistent in outcomes (Table 2). For spice crops producing recalcitrant seeds as in clove, micropropagation offers effective solutions for rapid multiplication as well as for maintaining planting material resources. Tree spices present a unique problem for plant breeding because of their long juvenile phase and cross pollination induced heterogeneity which pose serious handicaps for rapid improvement and assessment of the available varieties (Nair *et al*., 1977, Mary *et al.,* 1998, Rao *et al.,* 1999, Rao *et al.,* 2002). Breakthrough in production can be achieved by planting *en masse,* high yielding elite variants multiplied through micropropagation techniques.

Since its adoption in 1960s, there is considerable progress in micropropagation technology for propagation of agricultural crops and spices as well. Micropropagation through enhanced axillary branching is the most widely and increasingly applied technique as a quicker alternative to conventional methods of vegetative propagation. With the exception of a few spice crops, *viz*., cardamom (small and large), vanilla, ginger, turmeric and certain herbal spices, these techniques have not been extensively used for propagating spice plants on a commercial level. Unlike many herbal and perennial spice crops for which efficient micropropagation protocols are developed with relative ease, tree spices have always shown slow response in tissue cultures (Nair and Ravindran, 1988). Protocols for micropropagating cinnamon, clove, garcinia, and tamarind and curry leaf have been standardized (Mary *et al.,* 1998, 1999, Rao *et al.,* 1999, 2008). Details of published reports on micropropagation and tissue culture for several different spice crops are shown in Table 2.

Table. 2: Details of *in vitro* techniques developed for different spice crops

Spices	Type of culture	References
Allium sativum	Callus regeneration	Kehr and Schaeffer (1976)
		Bhojwani (1980)
	Meristem culture	Wang and Huang (1974)
Amomum subulatum	Direct regeneration	Pradip Kumar (2007)
		Rao *et al*. (2006)
Ammi majus	Somatic embryogenesis	Grewal *et al*. (1976)
Anethum graveolans	Adventitious shoots	Sehgal (1968)
Apium graveolans	Direct embryogenesis	Reinert *et al.* (1966)
Embryonic cultures		Al-abata and Collin (1978)
Armoracia rusticana	Morphogenesis	Meyer and Milbrath (1977)
Artemesia dracunculus	Direct regeneration	Garland and Stoltz (1980)
Brassica juncea	Direct regeneration	Elmbeuser *et al.* (1978)
Capsicum annuum	Anther culture	George and Narayanaswamy (1973)
Carum carvi	Somatic embryogenesis	Ammirato (1978)
Cinnamomum veerum	Shoot tip culture	Rai and Chandra (1987)
Coriandrum sativum	Somatic embryogenesis	Steward *et al.* (1970)
Crocus sativus	Callus regeneration	Ding *et al*. (1981)
Cuminum cyminum	Adventitious shoots	Jha *et al.* (1982)
Curcuma longa	Shoot bud culture	Nadgauda *et al.* (1978)
Elettaria cardamomum	Micropropagation	Nadgauda *et al*. (1983), Reghunath and Gopalakrishnan (1991)
Foeniculum vulgare	Shoot tip culture	Garcia Rodriguez *et al.* 1978
Mentha piperita	Callus culture	Lin and Staba (1961)
Mentha spicata	Micropropagation	Chambon *et al.* (1985)
Mentha arvensis	Callus culture	Bhaumik and Datta (1988)
Mentha species	Meristem culture	Geslot *et al.* (1989)
Ocimum basilicum	Micropropagation	Ahuja *et al.* (1982)
Oreganum vulgare	Callus culture	Neenakumari and Saradhi (1992)
Papaver somniferum	Somatic embryogenesis	Nessler (1982)
Piper longum	Callus regeneration	Bhat *et al.* (1992)
Piper nigrum	Shoot tip culture	Mariska *et al*. (1989)
	Direct regeneration	Nazeem *et al.* (1990)
		Philip *et al.* (1992)
Rosemarinus officinalis	Shoot tip and axillary bud culture	Misra and Chaturvedi (1984)
Syzygium aromaticum	Callus cultureNodal cultures	Mariska *et al*. (1989)
Tamarindus indica	Single node culture	Mary Mathew and Hariharan (1990)
Thymus vulgaris	Axillary bud proliferation	Furmanova and Olzowski (1980)
Trigonella foenum- graecum	Callus culture	Subramanian *et al.* (1968)
Vanilla planifolia	Axillary bud proliferation	Jarret and Fernandez (1984)
	Direct shoot regeneration	Zuraida *et al.* (2013)
	Bud Regeneration	Philip *et al.* (1986)
		Rao *et al*. (1993), Mary *et al.* (1999)

Contd.

Contd. Table 2

S. No	Species	Type of culture	References
	Zingiber officinale	Adventitious shoots	Hosoki and Sagawa (1977)
		Shoot tip culture	Wang and Hu (1980)
		Organogenesis	Kuruvinashetty *et al.* (1982)
			Inden *et al.* (1988)
			Rao *et al.* (1996)
		Micropropagation	Kavyasree (2011)

4. Protocols for *in vitro* Multiplication of some Spices

a) Cardamom

Surface sterilization and preparation of explants

Cut sprouting rhizome buds, at least 8 cm long and dip the ends in fungicide immediately. Clean the explants, remove the scales and suspend the explants in an antibiotic solution preferably chloramphenicol (0.2 %) for 1 hour. Rinse thoroughly, blot dry and surface sterilize in 0.8% mercuric chloride for 10 -12 minutes. Suspend in 0.2 % antibiotic solution prepared in sterile distilled water. Blot dry, trim base and inoculate.

Culture medium

Murashige and Skoog (1962) supplemented with 2mg/l BAP, 0.1mg/l IAA, 3% sucrose and 8 gm/l Agar.

Primary culture establishment

Inoculate the prepared explants on to the above medium. Incubate the cultures under 16 hour photoperiod (1000 lux) at 25±2°C. Axillary buds start sprouting 3rd to 4th week of culture initiation.

Sub culture cycles: Separate the axillary buds from mother tissue when they attain a length of 2-3cm length sub culture onto fresh medium of same composition. Multiple shoots which start to appear within 6-8 weeks of culture are then separated from clusters and transferred to fresh medium when they attain an average length of 2 cm.

Multiplication ratio : 1: 1.25

Shoot elongation and rooting

Shoots could be elongated and rooted on the same medium by prolonging the culture period without any subcultures.

Hardening

Plantlets 6-8 cms long with 2-3 leaves and good roots can be selected for primary hardening. For this plantlets are first cleaned free of agar carefully under tap water and suspended thereafter in fungicide solution preferably bavistine (0.1%) for 15 minutes. Transfer to micropots containing soilrite and place under mist chamber in green house.

Secondary hardening is done when plantlets are well established in soilrite showing 2 or 3 newly sprouted leaves within 6-8 weeks. Transfer primary hardened plants to polybags for secondary hardening as practised in seedling nurseries. Transfer to field conditions, when the plantlets attain a length of 25-30 cm and are robust. Perform field preparation, planting and management as per the recommended package of practices for cardamom, for successful establishment and growth of plantlets.

b) Vanilla

Surface sterilization and preparation of explants: Nodal sections from greenhouse maintained plants were used as explants. Surface sterilization was carried out with 0.08% $HgC1_2$ for 12 minutes preferably under vacuum. After thorough rinsing the explants were suspended in pre-sterilized Streptomycin solution (0.1%) for 15 minutes, trimmed and blot dried the explants prior to inoculation.

Primary culture initiation and establishment

Medium: Murashige and Skoog (1962) / Knudson- C (1946) + 1mg/l BAP + 0.1 mg/ l IAA + 0.1 mg/l D- Biotin + 0.1 mg/l Ca pantothenate + 3% sucrose 7.5 gm/l Agar. pH – 5.8 Multiplication ratio: 1: 2.5. Further subculture: At intervals of 40- 45 days on the same medium.

Shoot elongation and rooting

Elongation and rooting of shoots could be facilitated on MS + 0.1 mg/l KN + 0.1 mg/l NAA + 2% Sucrose + 0.65% Agar Duration of Culture: 50-60 days

Hardening and acclimatization

Suspend ex agar plants in bavistine solution (0.1%) for 15 minutes. Transfer to micropots containing soilrite and place under mist chamber in green house. Plantlets should be sprayed with a bio-agent preferably Pseudomonas (10 ml / litre water) once in 15 days to prevent any disease incidence. A prophylactic spray of Bavistine (2 g/ litre water) may also be given at well-spaced intervals of about 15 days before *Pseudomonas* sprays.

Secondary hardening

After primary hardening, the plantlets should be transferred to potting mix and allowed to get acclimatized in the secondary nursery for 2 months (or) uptil the plantlets attain a length of at least 30cm. Plantlets should be sprayed with Pseudomonas and one or two rounds of NPK at a concentration of 2 g./litre water. Transfer to field conditions, when the plantlets attain a length of 25-30 cm and are robust. Field preparation, planting and management should be performed as per the recommended package of practices for vanilla, for successful establishment and growth of plantlets.

c) Large Cardamom

Surface sterilization and preparation of explants

Cut sprouting rhizome buds, at least 8 cm long and dip the ends in fungicide while collection. Remove the scales and suspend the explants in chloramphenicol (0.2 %) for an hour. Rinse and surface sterilize in 0.8% mercuric chloride for 10 minutes. Suspend in 0.2 % antibiotic solution prepared in sterile distilled water. Blot dry and inoculate.

Culture medium

Murashige and Skoog (1962) supplemented with 2mg/l BAP, 3% sucrose and 8 gm/l Agar.

Primary culture establishment

Inoculate the prepared explants on to the above medium. Incubate the cultures under 16 hour photoperiod (1000 lux) at 25±2°C. Axillary buds start sprouting 3rd to 4th week of culture initiation.

Sub culture cycles: The axillary buds-after attaining 2-3cm length were separated from explants and sub cultured onto fresh medium of same composition. Multiple shoots started to appear within 6-8 weeks of culture. Shoots were transferred to fresh medium when they attained an average length of 2 cm. Multiplication ratio: 1: 1.75.

Shoot elongation and rooting

Shoots could be elongated and rooted on MS medium containing 0.1mg/l NAA or IAA or by prolonging the culture period without any subcultures.

Hardening

Procedure for cardamom can be adopted for primary hardening. Transfer primary hardened plants to polybags for secondary hardening. Transfer to field conditions, when the plantlets attained a length of 25-30 cm and are robust. Perform field preparation, planting and management as per the recommended package of practices for large cardamom, for successful establishment and growth of plantlets.

d) Herbal Spices

i. Thyme

Explant: Nodal micro-cuttings from field grown plants

Pre-treatment: Systemic fungicide (0.1% bavistine)

Surface sterilization: 0.05% Mercuric Chloride

Medium: MS + 0.5 mg/l BAP + 0.1mg/l IAA

Average shoot multiplication rate: 1: 3.2

Rooting: MS medium + 0.5 mg/l IBA + 0.5 mg/l NAA

Hardening: Plant in net pots containing soilrite

Hardening period: 6-8 weeks

ii. Oregano

Explant: Nodal micro-cuttings from field grown plants

Pre-treatment: Systemic fungicide (0.1% bavistine)

Surface sterilization: 0.04% Mercuric Chloride

Medium: MS + 0.5 mg/l BAP + 0.1mg/l NAA

Average shoot multiplication rate: 1: 3.5

Rooting: Rooting observed in medium same as the culture medium

Hardening: Plant in net pots containing soilrite

Hardening period: 6-8 weeks

iii. Salvia

Explant: Nodal micro-cuttings from field grown plants

Pre-treatment: Systemic fungicide (0.1% bavistine)

Surface sterilization: 0.04% Mercuric Chloride

Medium: MS + 1 mg/l BAP + 0.1mg/l NAA

Average shoot multiplication rate: 1: 1.5

Rooting: MS + 1 mg/l IBA

Hardening: Plant in net pots containing soilrite

Hardening period: 6-8 weeks

iv) Mint

Explant: Nodal micro-cuttings from field grown plants

Pre-treatment: Systemic fungicide (0.1% bavistine)

Surface sterilization: 0.05% Mercuric Chloride

Medium: MS + 0.5 mg/l BAP + 0.1mg/l IAA

Average shoot multiplication rate: 1: 3.8

Rooting: MS medium + 0.5 mg/l IBA + 0.5 mg/l NAA

Hardening: Plant in net pots containing soilrite

Hardening period: 6-8 weeks

5. Factors Affecting Large Scale Propagation Systems

Profitable tissue culture ventures could be started with minimum input and maximum outputs. For this, pragmatic methods should be adopted

- For effective utilization of manpower to minimize expenditure on wages and, salaries
- To minimize overhead expenses
- To reduce medium cost by using less expensive, quality raw materials.

With this perspective, the efforts made by Ganapathy *et al.* (1995) in banana followed by Rao *et al.* (2006, 2012) in vanilla and large cardamom micropropagation systems are worth mentioning. The former used ordinary chemical grade sugar in place of analytical grade and ordinary water in place of double distilled water. The latter, in vanilla and large cardamom micropropagation systems used ordinary sugar cubes and micro filtered water in tissue culture medium, which proved equally good. It is advisable to incorporate such changes only after sufficient trials are made. Similarly viability of the protocol is to be ascertained for commercial multiplication programmes. Primary hardening is another crucial step in the successful establishment of tissue culture. Infrastructure required for primary hardening like climate controlled green house should be more sophisticated and day to day inspections of the plantlets are required.

Increasing the multiplication rate of cultures results in reduction of the time spent in culture to obtain the required number of plants. The entire *in vitro* rooting step is often unnecessary, the shoots being treated as micro-cuttings and transformed to the soil medium directly thereby saving on the extra step.

In cross pollinated crops a lot of variability exists in the natural population. In contrast, through micropropagation uniformity in the plant population and thereby uniform yield in the plantation can be achieved. If the mother plant is a good yielder associated with quality characters, same yield potential and quality characters will be maintained, in the offspring. Tissue culture plants show exactly the characteristics of the mother plant. The farmers should themselves select elite mother plants, adaptable to their climate and make arrangements with the commercial tissue culture labs to get these elite plants multiplied. If the history of the mother plants is not known, then the farmers may encounter problems like non-adaptability to the particular locality, climate and lesser yield. In other words, we can say that the tissue culture technology is only a tool to multiply the plants in sufficient number and has no role in increasing or decreasing the yield or other parameters by itself. The higher yields completely' depend on the potentiality and adaptability of mother plants.

Another factor influencing quality of tissue culture plantlets is the occurrence of variability, which may be either due to formation of callus during micro propagation process or prolonged culture on the media. Callusing at any stage should be avoided during micro propagation. The culture duration of cycles also should be restricted either to ten subcultures for a period of one year from the primary culture initiation to rooting unlimited number of subcultures could lead to serious disorders like small sized flowers and fruits or weak rooting systems during field trials. Similarly, excess dosages of cytokinin lead to production of undesirable off types.

Since the possibility of occurrence of somaclonal variations during the *in vitro* process cannot be ruled out, multiplication of shoots should be carried out using the axillary branching method, which is least prone to variations. Notwithstanding, the Department of Biotechnology (DBT), New Delhi has recognized the need for a certification programme for tissue culture plants since inadvertent multiplication of undesirable variant plants as well as virus infected plants will not only result in poor performance, but also in undesirable spread of viruses wherever such plants are grown. Therefore it has become mandatory that commercial units should get accreditation or assure clonal uniformity and virus free nature of tissue culture raised plants through various referral/ accredited laboratories or certified facilitation centres.

6. Field Performance of Tissue Cultured Plants

Indian Cardamom Research Institute, Myladumpara, Spices Board of India laid out field trials in farmer's field utilizing TC plants for small and large cardamoms and vanilla. On similar lines large scale farm trials were taken up in farmers' field for TC plants of black pepper as a network project by Kerala Agricultural University, Indian Institute of Spices Research, Calicut and Spices Board, Cochin. The Department of Biotechnology, New Delhi, extended financial support for all these programmes.

Small Cardamom: Use of tissue culture gained much importance in recent years. The Indian Cardamom Research Institute (lCRI), Spices Board, Myladumpara in collaboration with Department of Biotechnology, New Delhi took up large scale field evaluation trials spread over the cardamom growing tracts under the project entitled "Tissue Culture Cardamom - Product Plan" during 1989 - 1995. In all, 100 units having an area of one ha under each unit, with 50 units in Kerala, 35 in Karnataka and 15 in Tamil Nadu were planted with TC plants.

In each unit, 80% of the area was filled with TC plants, while remaining 20% with open pollinated (OP) seedlings (and a total of 1.6 lakh Tissue Culture plantlets of small cardamom) were planted totalling 102.4 ha. Observations were made on the tiller and panicle production and' also yield. The primary observation indicated that there was a setback in the establishment of tissue culture plants in the initial stage due to delay of tillering at initial stage. After the establishment, the TC plants produced more number of new tillers and performed well. The increase in mean number of tillers in TC plants 'when compared to OP seedlings ranged 0-30; with an average of 14. There was a noteworthy increase (about 20 %) in panicle production rate in TC plantlets compared to OP seedlings. As a whole about 25-37% increase in yield was recorded in tissue culture plantlets as against their seedling counterparts. It was the first ever large-scale demonstration of Tissue Culture planting of a commercial crop in farmers' fields in India. Subsequently the Spices Board supported a programme for covering 400 ha with tissue culture cardamom in a phased manner.

Vanilla and large cardamom: Success in the aforesaid project led to initiation of two more such evaluation cum demonstration projects by Spices Board in collaboration with DBT, New Delhi for vanilla and large cardamom in 1996. Under these projects, planting of 2 lakh tissue culture plantlets of vanilla in 150 ha in Kerala, Karnataka and Tamil Nadu and 2.8 lakh tissue culture plants of large cardamom in 70 ha in Sikkim and Darjeeling district of West Bengal were accomplished.

In 2002-2003, a net Work project entitled "On farm Evaluation of Tissue Culture Derived Black Pepper" was/initiated involving Kerala Agricultural University (KAU), Vellanikkara, (ICAR) Indian Institute of Spices Research (IISR), Calicut and Spices Board with the financial assistance from DBT, New Delhi, to plant 100 ha area in Kerala, Kamataka, Tamilnadu, Andaman and Nicobar Islands, and North Eastern states.

7. Other Tissue Culture Techniques

Production of haploids (Anther and Ovule culture)

Haploid plants are defined as those having only one complete set of chromosomes or half the chromosome number of the normal plant. When haploids are subjected to doubling of chromosomes homozygous fertile plants or pure lines are obtained which can be of great use in practical breeding applications. One of the methods for production of haploids is by *in vitro* culture of microspores / anthers or ovules / ovaries.

There have been only limited research on induction of haploids in spice plants compared to other agronomically important crops. Reports have been published on the induction of haploids in chilli (George and Narayananswamy, 1973) and triploid callus production and recovery of diploid embryos from callus in parsely (Chand and Roy 1981) which suggest that these can be followed as model systems for utilization in other spice crops as well. Ferrie and Waterer (2009) have used double haploid (DH) technology to create and to evaluate DH lines in replicated field trials on a range of spice crops, cumin, dill, fennel, anise and caraway. Ferrie *et al.* (2011) also claimed to have identified a DH line of annual caraway that produced substantially higher seed yields than the parental lines. Anther/ ovary culture techniques have to be exploited and utilized effectively for genetic improvement programmes of spices in order to generate isogenic lines in a shorter time in comparison with the extended classic plant breeding methods.

Meristem culture

One of the most widely used and commercially successful application of Tissue Culture is the multiplication of virus free plants by meristem culture. The technique involves the excision of the organized microscopic apical meristem of the shoot and its subsequent *in vitro* culture. The explants comprise the apical dome and a few leaf primordia and exclude any differentiated provascular or vascular tissues. Since the terminal region of the shoot meristem, above the zone of vascular differentiation, is unlikely to contain pathogens, the major advantage is that it has the potential for excluding pathogens that may have been present in the donor plants. A second advantage is the genetic stability inherent in the technique, since plantlet production is from an already differentiated apical meristem without the intervention of a callus phase thereby minimizing somaclonal variation. Despite its abundant prospects in commercial micropropagation industry for obtaining disease free plantlets in spice crops such as cardamom, large cardamom, vanilla, ginger, turmeric etc where virus, fungal and bacterial diseases spread easily through conventional vegetative propagation methods, reports on full-fledged protocols through meristem cultures are limited. Nevertheless, a few reports on ginger, through meristem tip culture to obtain pathogen free plant materials for commercial utilization and also removal of nematodes have been documented (Bhagyalakshmi and Singh 1988, Pillai and Kumar, 1982; Smith and Hamill, 1996; Rout *et al.*, 2001). Simultaneous elimination of Cucumber mosaic virus and Cymbidium mosaic virus infecting *Vanilla planifolia* through meristem culture to a frequency of 75% was reported by Retheesh and Bhat (2010). Meristem culture can also be utilized for long term storage of disease free clonal germplasm of spice crops through cryopreservation techniques as detailed by Nirmal Babu *et al.* (2012).

Somatic embryogenesis

Somatic embryogenesis is a process where a bipolar embryo like structure having pro-embryogenic potential, is generated from a somatic cell or a group of somatic or vegetative cells that are not normally involved in the development of embryos. Unlike zygotic embryos no endosperm or seed coat is formed around a somatic embryo. Somatic embryogenesis occur directly when embryos develop directly from explant tissue creating an identical clone while indirectly it occurs from unorganized callus tissue. This technique is generally preferred to direct or indirect organogenesis as it produces bipolar embryos which are quantitatively advantageous and easier to establish. With its potential for large scale propagation and prospects for automation, this technique could be useful for crop improvement programmes, in genetic transformation studies and production of somaclonal variants. Some of the earliest reports on somatic embryogenesis were in mid 1960s with the pioneering works of Maheswari and Gupta (1965) in *Foeniculum vulgare,* a seed spice. Somatic embryogenesis has been reported in several spice crops like Parsley (Vasil and Hildebrandt, 1966), Coriander (Steward *et al.,* 1970), Garlic (Abo El-Nil 1977), Caraway (Ammirato, 1978), poppy (Nessler, 1982), black pepper (Nair & Gupta 2006), cardamom (Manohari *et al.,* 2008), ginger (Kakkar *et al.,* 1993, Lincy *et al.,* 2009), lovage (Park *et al.,* 2010) to name a few.

Somatic embryogenesis which may also involve secondary cycle of embryogenesis, offers a number of potential applications like,

a. Multiplication of selfed embryos to minimize procedures like bud pollination for maintenance of inbred lines of self incompatible species.

b. Rapid cloning of valuable breeding stocks at the earliest possible stage of life cycle after crossing.

c. *In vitro* selection and screening of genotypes for whole plant characters at the earliest possible stage of life cycle.

Synthetic seeds

Synthetic seeds are artificially encapsulated somatic embryos, shoot buds or any other tissue that possess the ability to regenerate into a plant under *in vitro* or *ex vitro* conditions and that retain this potential after storage. Synthetic seeds are enclosed within an artificial medium that supplies nutrients encased in an artificial seed covering. Though a very large number of somatic embryos can be produced in bioreactors germination of, the embryos is not efficiently done in liquid medium. Transfer to a solid medium to induce rooting and further transfer to soil have to be carried out thereby increasing the costs of production. To overcome this hurdle in efficient plant production systems, the concept of artificial-seeds or synthetic seeds was developed so as to transfer embryos and germinate them directly in the soil. The artificial seed in effect acts as the natural seed in its structure and several of its functions.

Synthetic seed production was first reported in the spice, celery (Redenbaugh *et al.*, 1984). Subsequently, encapsulation of somatic embryos or vegetative parts like shoot tips and axillary buds in several spices received considerable attention in recent past. Encapsulation of embryoids or shoot tips in sodium alginate was successfully carried out in coriander (Chen *et al.*, 1991, Stephen and Jayabalan 2000). Likewise production of synthetic seeds was reported in ginger, turmeric cardamom, large cardamom and vanilla (Ganapathy *et al.*, 1994, Ravindran *et al.,* 1996, Sharma *et al.,* 1994, Charles 2007, Pradeep Kumar, 2007). High frequency plant regeneration from *Allium sativum* encapsulated calli and bulblets were reported (Kim and Park 2002; Bekheet, 2006). *In vitro* plant regeneration from encapsulated somatic embryos of *Piper nigrum* was reported by Nair and Gupta (2007). Production of disease free encapsulated shoot buds of *Zingiber officinale* and their conversion into plantlets were detailed by Sundararaj *et al.* (2010). Synthetic seeds have the following advantages to score over present techniques of planting material multiplication in terms of space, time and overall costs

- Allows handling and storage and transportation with relative ease
- It has potential for long term storage without losing viability
- Clonal fidelity can be maintained
- Can be utilized as a channel for improved varieties or plant lines produced through biotechnologies to be delivered directly to green house or field
- Can be utilized for mass propagation of elite plant varieties

Protoplast culture

One of the most significant developments in the field of plant tissue culture is the isolation, culture and fusion of plant protoplasts (Cocking 1972). The technique is important because of its far reaching implications in studies of plant improvement by cell modifications and somatic hybridizations (Bajaj, 1974). One of the main applications of protoplast technology is somatic hybridization or fusion to produce plants carrying genetic information from sexually incompatible species. Protoplast fusion results in a nuclear hybrid or a cytoplasmic hybrid. When cells of distant species are fused, chromosome elimination occurs or a whole chromosome set of one parent is excluded. Many variant plants can thus be obtained. In rare cases, hybrid plants may express normal phenotypes and contain genetic information from both parent species (Dudits *et al.,* 1980). Protocols for protoplast isolation and culture were standardized in a few spice crops like *Capsicum annuum* (Saxena *et al.,* 1981) and *Foeniculum vulgare* (Miura and Tabata 1986), *Brassica juncea* (Lei *et al.,* 1991), *Piper nigrum* and *Piper colubrinum* (Shaji *et al.,*1998), *Elettaria cardamomum* and *Zingiber officinale* (Geetha *et al.,* 2000), *Vanilla planifolia* (Mary *et al.,* 2012). The success of protoplast isolation and regeneration opens up new avenues for future programmes for gene transfer such as improved agronomic traits and stress & disease tolerance and through somatic hybridization in economically important spice crops.

8. Secondary Metabolites Production

Secondary metabolites are compounds produced by plants at various sites within the cell and stored primarily within the vacuole. Different classes of these compounds constitute the bioactive compounds in several medicinal, aromatic, colorant, and spice plants and/or functional foods. Plant cell culture technology is potentially valuable for studying biosynthesis of secondary metabolites and an attractive alternative source to whole plant for the production of commercially important plant products. Biotechnological approaches can be used as an alternative to overcome the limited availability of biologically active, commercially valuable, and medicinally important secondary metabolite compounds from spices.

Since many years, plant products are being used in pharmaceuticals, cosmetics and in food industry. Through tissue culture, it has been possible to produce a range of components including fragrances and flavours from spices.

Secondary metabolite production was reported in *Allium sativum* (Madhavi *et al.*, 1991), *Capsicum annuum* (Lindsey and Yeoman, 1985; Johnson *et al.*, 1991), *Crocus sativus* (Sharma *et al.*, 1991), *Cuminum cyminum* (Agarwal and Patwardhan, 1990), *Rosemarinus officinalis* (Hippolyte *et al.*, 1991) and *Vanilla planifolia* (Romagnoli and Knorr, 1988). Karalija and Paric (2011) reported secondary metabolite production by shoot culture of herbal spice *Thymus vulgaris*. Vanilla flavour, could become a good example of a high value, low volume product, if derived from plant cell culture. High cost and relatively uneven supply of quality natural vanilla can be addressed by developing an *in vitro* bio-production system.

Capsaicin from Capsicum (Johnson *et al.*, 1991), crocin, picrocrocin and saffranol from Saffron (Venkataraman, 1996) are examples of active principles, extracted from spice crops through biotechnological means. Capsanthin from paprika and curcumin from turmeric have good demand in food industry being natural colourants. Protocols are being worked out for their production in suitable medium.

Nutmeg and mace are the two important spices obtained from nutmeg tree. Nutmeg is the kernel while mace is the aril that surrounds the seed. *In vitro* proliferation of mace tissue was reported (Nirmal Babu, *et al.*, 1992). Since mace is the source of anti-carcinogenic compound myristicin, this technique can be explored for *in vitro* production potential of myristicin.

Most of the reports on saffron deal with *in vitro* proliferation of stigma (Ding *et al.*, 1981; Homes *et al.*, 1987; Ilahi *et al.*, 1987; Isa and Ogasawara 1988; Ahuja *et al.*, 1994), *in vitro* synthesis of colour components and metabolites (Sano and Himeno, 1987; Himeno *et al.*, 1988; Koyama *et al.*, 1987, Sharma *et al.*, 1991; Bhaagyalakshmi *et al.*, 1995) and *in vitro* metabolite production (Venkataraman *et al.*, 1996).

Though the feasibility of *in vitro* production of spice principles is demonstrated, methodology for-scaling up and reproducibility need to be developed before it reaches commercial levels. Once standardized this technology will have tremendous potential in industrial production of important compounds such as capsaicin, vanilla, crocin, picrocrocin, safranol, myristicin, anethole, menthol, curcumin etc.

9. *in vitro* Germplasm Conservation

Since India is renowned for its variety and diversity of spices, genetic diversity in spices is our national legacy and needs to be conserved. Germplasm conservation is the conservation of genetic resources for an organism. For plants, the conventional method of germplasm conservation is through seed storage, collection in a field repository or nursery *in situ*. Maintenance of germplasm in an *in vitro* gene bank (*ex situ*) is an alternate and safe method of conservation when compared to conventional methods. Exchange of germplasm in the form of plantlets in tissue culture medium is desirable, since material will often be free from pathogens, thus, reducing risk of contamination during transport and handling. Cultures can be induced into a state of slow' growth by culture at reduced temperature or in the presence of growth regulators or osmotically active additives (Staritsky *et al.*, 1985; Withers, 1987).

In vitro storage of multiple shoot cultures of ginger at ambient temperatures was reported (Dekkers *et al.,* 1991). *In vitro* conservation protocols were standardized for medium term conservation in spices like cardamom, ginger, turmeric, vanilla and herbal spices (Nirmal Babu *et al.,* 1996). The National Bureau of Plant Genetic Resources (NBPGR), New Delhi has over 702 accesions of various spices, medicinal and aromatic crops in its cryo gene bank. A detailed account of cryopreservation of genetic resources of spices has been reported by Nirmal Babu *et al.*, 2012). They have also mentioned about the advantages of DNA banks. However, though it allows gene isolation, cloning and transfer, it does not allow regeneration to whole plants. *In vitro* conservation however requires a high degree of expertise in the maintenance and management of collections of biological materials at ultra-low temperatures, or as freeze dried material, to secure their long-term integrity and relevance for future research, development, and conservation (Nirmal Babu *et al.,* 2012).

10. Somaclonal Variation

Somaclonal variation is the variation observed in plants that have been produced by plant tissue culture methods. The benefit of somaclonal variation is crop improvement as variation leads to the creation of additional genetic variability. However, somaclonal variation becomes undesirable in horticulture and forestry industries where tissue culture is employed for rapid propagation of elite genotypes which require clonal uniformity.

Attempts on induction of variability through somaclones in spices such as black pepper (Shylaja *et al.,* 1994, Nazeem *et al.,* 1996) resulted in identification of tolerant lines to *Phytophthora* foot rot resistance through *in vitro* selection of calli as well as generation of somaclones using crude culture filtrate and toxic metabolite isolated from *Phytophthora capsici.* Organogenesis and plantlet formation were achieved from callus cultures of turmeric (Shetty *et al.* 1982, Nirmal Babu *et al.,* 1996), and molecular and morphological variations were studied (Nirmal Babu *et al.,* 2003). Nadgauda *et al.* (1982) had also reported plant regeneration from leaf base callus of turmeric and random amplified DNA analysis of regenerated plants showed variation at DNA level. Variants with high curcumin content were isolated from tissue cultured plantlets. Root rot disease tolerant clones of turmeric cv. Suguna were isolated using continuous *in vitro* selection technique against pure culture filtrate of *Pythium graminicolum* (Gayatri *et al.,* 2005). The reactions of tissue cultured plant cells to fungal pathogen were reportedly very similar to responses of the whole plants in field.

11. Molecular Marker Studies

Genetic diversity analysis, DNA fingerprinting, marker-assisted selection and breeding

Of late, there has been an increasing emphasis on usage of molecular markers for characterization of genotypes, DNA finger printing, phylo-genetic analysis, varietal typing, establishing trueness to type of tissue cultured plants, identification and cloning of important genes, marker assisted selection and understanding of inter relationships at molecular level. Molecular markers have proved valuable in crop breeding, especially in studies on genetic diversity and gene mapping. The commonly used polymerase chain reaction (PCR) based DNA marker systems for diversity analysis are Random Amplified Polymorphic DNA (RAPD), Amplified Fragment Length Polymorphism (AFLP), Restriction Fragment Length Polymorphism (RFLP), Inter Simple Sequence Repeats (ISSR), and simple sequence repeats (SSRs) or microsatellites. Another group of markers *viz*., Expressed sequence tags (EST), have helped to generate a vast amount of sequence data from various plant species; SSRs mined from these data are useful as molecular markers and since they are derived from transcripts, they would be useful for assaying the functional diversity in natural populations or germplasm collections and could prove useful for marker-assisted selection.

Identification and characterization of germplasm has been an important link between the conservation and utilization of plant genetic resources. RAPD, RFLP and ISSR polymorphisms have been used to characterize germplasm collections comprising important cultivars, varieties and related genera of cardamom to develop fingerprints and to study their inter-relationships (ICAR-IISR '2002, 2003). The studies on cardamom indicated that Kerala and Kamataka cardamom populations are divergent. Likewise, observations revealed that *Amomum* sps is the closest to cultivated cardamom among several Zingiberacean genera studied (Jayakumar *et al.*, 2005). Identification and genetic variation within eight high yielding varieties of ginger through RAPD markers was reported by Harisaranraj *et al.* (2009). They observed that analysis of genetic relation through RAPD markers provided a reliable method for identification of varieties than morphological characters.

Genetic diversity analysis in small and large cardamoms have been undertaken by Indian cardamom Research Institute (ICRI), Spices Board in collaboration with ICAR-Indian Institute of Spices Research (ICAR-IISR) and is presently being continued by ICRI. Small cardamom germplasm is being maintained at the ICRI headquarters, Myladumpara, Kerala, where around 800 accessions of cardamom are conserved. Large cardamom germplasm is being maintained at ICRI regional research station at Sikkim where about 300 accessions are conserved. Microsatellites and Inter Simple Sequence Repeats (ISSR) markers are being used for studying the diversity within the germplasm to create a core collection by eliminating duplications of accessions. Till date, 150 accessions of small cardamom and 50 accessions of large cardamom have been analysed (Remya Lekshmi *et al.*, 2013).

Several ESTs have been developed in Zingiberacean spices such as ginger, turmeric etc. Recent advances have generated a large number of expressed sequence tags (ESTs) that has been made available in public database (NCBI) thereby offering an opportunity to develop EST derived SSR markers by data mining in these crops (Joshi *et al.*, 2010, Koo *et al.*, 2013).

In perennial orchid spice vanilla, which lacks classical phenotypic markers, molecular marker techniques like RAPD and AFLP, were used to estimate genetic similarities and interrelationships in cultivars, seed progenies, somaclones and inter-specific hybrids. To study existing variability in cultivated vanilla, species inter-relationships, identification of inter-specific hybrids and fingerprinting of important genotypes Minoo *et al.* (2002) studied isoenzyme, RAPD, AFLP polymorphism supplemented by morphological characters. Genetic diversity studies of limited samplings of cultivated *V. planifolia* in the Indian Ocean islands and elsewhere (Besse *et al.*, 2004 and Divakaran *et al.*, 2007), have already been documented. Microsatellite markers were developed in vanilla and primer pairs successfully isolated were used to assess the intra- and inter-specific polymorphism within a sample of cultivated vanilla including *V. planifolia* and *V. tahitensis* accessions (Bori *et al.*, 2008).

DNA fingerprinting is the process of cross matching two strands of DNA. In plants, it has become one of the powerful and widely known applications of biotechnology today for identifying cultivars and for germplasm diversity studies. Molecular markers have been used by several workers to distinguish different cultivars. In seed spice celery, Yang and Quiros (1993) used RAPD to identify cultivars; Aquadro *et al.*, (2006) and Wang *et al.* (2011) utilized SSR and SRAP markers to study celery diversity. Sasaki *et al.* (2004) applied single-nucleotide polymorphism (SNP) analysis of the trnK gene to identify curcuma plants. Over 96 cultivars and related species of turmeric as well as 96 accessions of ginger were analysed using RAPD profiling and interrelationships studied between turmeric cultivars and related species (ICAR-IISR 2004). Generation of inter specific hybrids between *Vanilla planifolia* Andrews and other vanilla species and their validation using molecular markers was reported by Divakaran *et al.* (2006) and Rao *et al.* (2012).

DNA markers can be used to detect genetic dissimilarities in micropropagation systems that might not be so easily evident from other commonly used techniques. Morphological characters coupled with RAPD profiles indicated genetic fidelity among randomly selected micropropagated plants of Black pepper varieties Subhakara and Aimpiriyan, indicating that micropropagation techniques can undoubtedly be used for commercial multiplication (Nirmal Babu *et al.*, 2003). Genetic uniformity of micropropagated *Piper longum* using RAPD profiling was reported earlier by Ajith (1997). Earlier studies by Rout *et al.* (1998) indicated that RAPD profiles did not show any polymorphism among the micropropagated plants. Ajith *et al.* (1997) and Paraani *et al.* (1997) standardized RAPD finger printing of select micropropagated plants of *Piper longum* for conservation.

Marker assisted breeding, as another tool or an auxiliary means for conventional breeding allows selection of desirable traits and selection against undesirable traits that are incorporated especially during backcrossing. Large sequence dataset for gene discovery, molecular marker development and marker-assisted selection and breeding can be efficiently generated by high throughput transcriptome sequencing. Much work on these aspects have been done in spice crops *viz.* celery (Fu *et al.*, 2013), sesame (Wei *et al.*, 2011) and red pepper (Lu *et al.*, 2012) to name a few. Likewise, marker assisted selection (MAS) of resistant varieties is a reliable and faster method of selecting the right varieties for cultivation. Swetha and Subramanian (2008) worked on isolation and molecular analysis of R-gene in resistant *Zingiber officinale* (ginger) varieties against *Fusarium oxysporum* f. sp. *zingiberi*.

12. Disease Diagnostics

With the advent of the PCR technique and DNA based molecular markers, tremendous progress is made in identification of diseases and characterization of plant pathogens using various molecular techniques like Western blotting, nucleic acid hybridization and PCR in different spices. PCR based molecular techniques are highly sensitive and are therefore, considered as rapid methods for the detection of viruses and are being employed for the detection and confirmation of virus diseases before and after appearance of symptoms. Diagnosis of viral diseases or virus indexing through molecular methods is being extensively studied to prevent spread of infection in plantations.

Johnson *et al.* (2005) reported a method for isolation and reverse transcription of high quality RNA from *Piper* species. Bhat and Siju (2007) developed single-tube multiplex RT–PCR (reverse transcription-polymerase chain reaction) for the simultaneous detection of *Cucumber mosaic virus* (CMV) and *Piper yellow mottle virus* (PYMoV) in black pepper the two most important viral diseases of pepper. Another procedure for total RNA isolation from cardamom and detection of CdMV through RT-PCR using primers targeting the conserved region of coat protein was standardized and subsequently validated by testing more than 50 field cardamom samples originating from Karnataka and Kerala states (Biju *et al.*, 2010).

Biotechnologies are employed to characterize several virus strains affecting spices. Several reports on analysis and characterization of fungal pathogens affecting spices were documented. Jacob *et al.* (2003) observed high genetic diversity in the coat protein and 30 untranslated regions among the geographical isolates of Cardamom mosaic virus from South India. Bhat *et al.* (2005) reported isolation and sequencing of CMV coat protein gene infecting black pepper. Bean yellow mosaic virus (BYMV) associated with *Vanilla planifolia* was characterized by Bhadramurthy *et al.* (2011). Variability in Cardamom mosaic virus isolates occurring in India using coat protein gene sequence was investigated by Siljo *et al.* (2013).

13. Genetic Engineering

Genetic engineering is the process of transferring individual genes between organisms or modifying the genes in an organism to remove or add a desired trait or characteristic using recombinant DNA technology and such organisms that have undergone genetic modification through this process are referred to as genetically modified organisms (GMOs). Application of genetically modified plants would become one of the major inputs to accomplish increased agricultural production at least within a few decades to come. So far, 21 transgenic crop varieties have been either released or recommended for general cultivation. Many countries approved the large scale use of transgenic crops. India also started field experiments in open environments using transgenics of rape seed, tomato and cotton since 1995. Research on transgenic spices has also been initiated, though it is restricted to greenhouse level for now. Production of transgenic varieties may be the probable answer to solve *Phytophthora* foot rot disease of black pepper or rhizome rot of ginger. *Agrobacterium* mediated genetic transformation of pepper and transient expression of gus (glucuronidase) in ginger callus were attempted at ICAR-IISR, Calicut. The stable nature of the transgenic varieties is still a matter of confirmation. The concerns over the, bio-safety of growing genetically modified organisms (GMOs) and consumer acceptance of genetically modified food product are issues which are still not completely settled. Biosafety concerns have led to the development of regulatory regimes in various countries for research, testing, safe use and handling of GMOs and products thereof. India is one of the earliest countries to establish a biosafety system for regulation of GMOs. The Government has set out rules and regulations for handling Genetically Modified Organisms (GMOs) including transgenic plants in 1989 itself, through its Environment Protection Act, 1986 (EPA).

Agrobacterium mediated gene transfer systems in spices *viz. Piper nigrum* was reported as early as in 1994 by Sasikumar and Veluthambi. They reported that kanamycin at 50μg completely inhibited callus formation and growth and hence is the minimum concentration needed to select transformed tissues of black pepper. They have also obtained primary transformants for kanamycin resistance in the cotyledons using *A. tumefaciens* binary vector strains LBA 4404 and ERA 105. Sim *et al.* (1995) reported *Agrobacterium* mediated transfer of GUS to black pepper. Work was also undertaken to confer resistance to Black pepper against foot rot fungi *Phytophthora capsici.* The relevant genes were studied in the wild variety *Piper colubrinum*. Asha and Rajendran (2009) reported transformation of black pepper by transferring the complete exogenous complement of DNA from *P. colubrinum* into *P. capsici* via pollen tube pathway. The plantlets generated showed molecular characteristics of both parents.

Work is also under progress to confer resistance to Pepper Yellow Mottle virus (PyMoV) using antisense gene technology. Bhat *et al.* (2005) initiated the work in ICAR-Indian Institute of Spices Research and the putative transgenics developed – IISR Subhakara and Panniyur I are under evaluation. (IISR 2011-12). Work on similar lines was being carried out in Chillies (Lee *et al.*, 1993, 2004; Dong *et al.*, 1995, Zhu *et al.*,1996, Cai *et al.*, 2003); Small cardamom (Backiyarani *et al.*, 2005) against cardamom mosaic virus.

Suma *et al.* (2008) reported successful transformation of ginger for GUS activity using agro bacterium mediated gene transfer in somatic embryos. *Agrobacterium* based DNA transfer system has advantages in plant transformation due to its simplicity, precision, integration of DNA sequence with defined ends, linked transfer of genes of interest along with the transformation marker, high frequency of stable transformation with many single copy insertions, and the ability to transfer long stretches of T-DNA (Veluthambi *et al.*, 2003). Further works are being done to study the application of agrobacterium mediated transformation to control soft rot disease in ginger. Wild ginger varieties have been shown to exhibit resistance to fungi *Pythium aphanidermatum* which causes soft rot. This resistance mechanism is being studied to isolate the responsible genes so that it can be transferred to the cultivars to confer resistance. Ashwati and George (2007) evaluated and characterised resistance gene candidates in ginger. Nirmal Babu *et al.* (2005) reported successful transformation of Black pepper osmotin gene, conferring resistance to Phytophthora.

A few putative genomic and cDNA fragments associated with resistance related genes have been isolated and programmes on isolating, cloning and validation of full length genes are in progress (ICAR-IISR 2004, 2005). Molecular cloning of cDNA fragments encoding the defence related protein B-1,3-glucanase in black pepper (*P. nigrum* L.) and methyl glutary CoA reductase in *Piper colubrinum* was attempted to isolate resistance genes in black pepper (Sharma *et al.,* 1996; Girija *et al.,* 2005). Recently, Nair and Thomas (2013) have sequenced the resistance gene ZzR1 from *Zingiber zerumbet,* responsible for resistance to soft rot in ginger. The transgenic works on spices till date have been reviewed by Babu *et al.* (2013). Table 3 lists some of the genes from different spices which have been characterized.

Table 5: Genes from spices characterised for further studies

Spice crops	Gene characterised	References
Piper colubrinum	Osmotin	Mani and Manjula, 2010
	Partial resistance gene analogs	Tiing *et al.*, 2012
Piper nigrum	Partial resistance gene analogs	Tiing *et al.*, 2012
Zingiber species	Resistance gene candidates	Nair and Thomas, 2007
Zingiber species	R gene against *Fusarium oxysporum*	Priya and Subramanian, 2008
Zingiber zerumbet	ZzR1	Nair and Thomas, 2013
Zingiber officinale	ZoPR5	Prasath *et al.*, 2011
Curcuma amada	CaPR5	Prasath *et al.*, 2011
Curcuma longa	NBS – LRR - Resistance gene candidates	Joshi *et al.*, 2010
Curcuma zedoaria	R gene against *Pythium aphanidermatum*	Basudeba *et al.*, 2013

14. Identification of Adulterants

The upsurge in both value and consumer demand of spices has increased the demand for better quality spices from suppliers and retailers. This has encouraged illegal or fraudulent practises (e.g. adulteration), which makes continuous monitoring a necessity to identify such malpractices. Though well-established, sensitive and accurate analytical methods such as thin layer chromatography (TLC), high performance liquid chromatography (HPLC) and spectrophotometric methods (e.g. UV/Vis spectrophotometry) etc. are prevalent in identifying spice adulterants, all these methods are time-consuming, costly and may require specialised personnel for system operations. DNA-based methods have the potential to complement these approaches (Lum and Hirsch 2006). They are relatively simple, rapid, and highly sensitive and have the potential to be developed for quantitative analysis. Moreover large-scale detection of plant-based adulterants can be carried out using commercial PCR kits.

Sasikumar *et al.* (2005) reported an efficient, relatively original method for the detection of extraneous *Curcuma* sp. contamination in powdered market samples of turmeric using molecular markers (RAPD), which are not easily discriminated by other analytical techniques routinely used. The development and application of sequence characterised amplified region (SCAR) markers for the detection of plant based adulterants (dried red beet pulp and powdered *Ziziphus nummularia* fruits) in traded ground chilli was reported by Dhanya *et al.* (2011). Dhanya and Sasikumar (2010) in an exhaustive review detailed various DNA based methods and their application in biological adulterant detection and authentication of a wide range of spices such as black pepper, chilli, turmeric, seed spices and herbal spices.

DNA barcoding is assuming significance as a new tool for quality assurance and food traceability in many agri-food commodities including spices. It is based on the analysis of short, standardized and universal DNA region/s (named 'barcode sequence/s'), and has been proposed as a universal tool for species identification (Hebert *et al.,* 2003; www.barcoding.si.edu). DNA barcoding uses state-of-the art biotechnology to help identify plant material in a rapid, accurate and cost-effective manner (Newmaster *et al.,* 2013). Reports on spices are mostly on adulteration of traded products and food traceability. Parvathy *et al.* (2014) reported DNA Barcoding techniques used to detect chilli adulteration of traded black pepper powder. De Mattia *et al.* (2011), evaluated the efficacy of a DNA barcoding approach as a tool for identifying herbal spices most commonly used as kitchen spices such as Mint, Basil, Oregano, Sage, Thyme and Rosmary. They also stated that definition of a universal identification system based on barcoding would allow not only the characterization of spices starting from a small plant fragment, but also the traceability of commercial spice products sold in powder form.

15. Future Prospects

Application of biotechnology with its extensive and convincing benefits can make tremendous impact in spice industry especially in the fields of crop improvement, food safety and quality with greater speed, efficiency and flexibility. Biotechnology holds immense prospects for crop improvement in terms of production and productivity both quantity and quality-wise. Modern biotechnology allows transfer of one or a few desirable genes, thereby permitting scientists to develop crops with specific beneficial traits and reduce undesirable traits. Wild relatives of crops have untapped potential in that it exhibits resistance to many problems being faced in fields and plantations. Resistance to biotic factors like pathogens, abiotic factors like drought, salinity, nutrient constraints and effect of weather inconsistencies can be rendered ineffective by judicious use of genetic engineering techniques. Identification, isolation and characterization of genes responsible for such adaptations and their insertion into cultivars could transfer desirable agronomical traits to them. Similarly, development of efficient micro propagation protocols provide faster and reliable method of increasing the production of spices. A multidisciplinary approach involving various biotechnologies, molecular breeding, genetic engineering and their integration with conventional breeding can undoubtedly develop improved varieties of spice crops. Modern biotechnology also offers effective techniques to address food safety concerns. Biotechnologies can be used to decrease the time necessary to detect food borne pathogens, adulterants, toxins and contaminants. Biotechnology holds significant promise to the future but with certain amount of risk associated with any new technology and therefore it must be carefully regulated so that the maximum benefits are reaped with the least risk.

References

Abo El-Nil MM, (1977) Organogenesis and embryogenesis in callus cultures of garlic (*Allium sativum* L.). *Plant Sci. Lett.* 9: 259-264.

Acquadro A, Magurno F, Portis E and Lanteri S (2006) dbEST-derived microsatellite markers in celery (*Apium graveolens* L. var. dulce). *Molecular Ecology Notes* 6: 1080–1082.

Agarwal S and Patwardhan.M H. (1990) Formation of cumin aldehyde from *in vitro* cultures of cumin *(Cuminum cyminum* L.var. *Marthiana*). *Current Science* 52: 645-646

Ahuja A Koul S and Koul B L. (1994) Saffron (*Crocus sativus* L) tissue culture approaches for regeneration and improvement. International conference on current progress on Medicinal and Aromatic Plant Research 1995. Calcutta. India .p.28-29.

Ahuja A, Verma M and Grewal S. (1982) Clonal propagation of ocimum species through tissue culture. *Indian J. Exp. Biology* 20: 455-458.

Al -Abata S and Collin A. (1978) Control Of embryoid- development in tissue culture of celery. *Annals of Botany*, 42: 773-782.

Ajith, A., Paranima Rao, C.S., Lalitha R and Balakrishna, P. (1997) Micropropagation and genetic fidelity studies in *Piper longum*. (In) Biotechnology of Spices, Medicinal and Aromatic Plants, Indian Society for Spice, Calicut, India pp. 94-97.

Ammirato, P.V. (1978) Effect of 8- azaguanine in the development of somatic embryos from cultured caraway cells. *Plant Physiology*. 61 : p.46.

Anon, (2006) Annual Report (2006) Spice Board, Cochin, India

Asha, S. and Rajendran, P.C. (2010) Putative transgenic plants through plant transformation against Phytophthora foot rot in black pepper (*Piper nigrum* L.) *Asian. J. Biosci.,* 4 (2): 135-141

Ashwati Nair, R. and George Thomas (2007) Evaluation of resistance gene (R gene) specific primer sets and characterization of resistance gene candidates in ginger (*Zingiber officinale* Rosc.) *Current Science* 93 (1): 61-66

Backiyarani, S., Manohari, C. Jebsingh, T., Jacob, T. and Usha, R. (2005) Cloning of coat protein gene of Kursuppara isolate of Cardamom mosaic virus for developing transgenic virus-resistant cardamom. (In) Recent trends in horticultural biotechnology, Vol. II. ICAR national symposium on biotechnological interventions for improvement of horticultural crops: issues and strategies, Vellanikkara, Kerala, India, 10-12 January, 2005; New Delhi: New India Publishing Agency, 693-698p.

Bajaj Y P S. (1974) Potentials of protoplast culture work in agriculture. *Euphytica*. 23: 633-649.

Basudeba, K., Nanda, S., Nayak, P.K., Nayak, S. and Joshi, R.K. (2013) Molecular characterization and functional analysis of CzR1, a coiled-coil-nucleotidebinding– site-leucine-rich repeat R-gene from *Curcuma zedoria* Loeb. That confers resistance to *Pythium aphanidermatum*. *Physiol. Mol. Plant Pathol*. 83:59-68.

Bekheet SA. (2006) A synthetic seed method through encapsulation of *in vitro* Proliferated bulblets of garlic (*Allium sativum* L.). *Arab J Biotechnol,* 9:415 - 26.

Besse , P., D. Da Silva , S. Bory, M. Grisoni, F. Le Bellec and Duval, M.F. (2004) RAPD genetic diversity in cultivated vanilla: *Vanilla planifolia* , and relationships with *V. tahitensis* and *V. pompona*. *Plant Science* 167: 379-385.

Bhadramurthy, V, Bhat, A.I. and Biju, C.N. (2011) Occurrence and coat protein sequence based characterization of Bean yellow mosaic virus (BYMV) associated with vanilla (*Vanilla planifolia* Andrews) in India. *Archives Phytopathol and Plant Protection* 44:1796-1801

Bhagyalakshmi B and Singh NS (1988) Meristem culture and micropropagation of a variety of ginger (*Zingiber officinale* Rosc.) with a high yield of oleoresin, *J Hortic Sci Biotech* 63:321-327.

Bhagyalakshmi N, Ravisankar G A and Venkataraman LV. (1995) *In vitro* stigma regeneration, production of stigma compounds in suspension cultures and enzymatic markers during somatic embryogenesis of saffron. (In) *All India Symposium on Recent Advances in biotechnlogical Applications on Plant Tissue and Cell culture, CFTRI,* Mysore, p.10

Bhat A I, Haresh P S and Madhubala R. (2005) Sequencing of coat protein gene of an isolate of Cucumber Mosaic Virus infecting black pepper. *Indian Journal of Plant Biochemistry and Biotechnology* 14:37-40.

Bhat S R, Kacker A and Chandel K P S. (1992) Plant regeneration from callus cultures of *Piper longum* L, by organogenesis. *Plant Cell Reports*. 11:525-528.

Bhat, A. I. and Siju, S. (2007) Development of a single-tube multiplex RT–PCR for the simultaneous detection of Cucumber mosaic virus and Piper yellow mottle virus associated with stunt disease of black pepper, *Current Science* 93(1): 973.

Bhawmik C and Dutta P.C. (1988) Development of Japanese mint tissue culture method. *Indian Perf.* 32:125-129.

Bhojwani S.S. (1980) *In vitro* propagation of garlic by shoot proliferation. *Sci. Hort.* 13:47-52.

Biju, C.N., Siljo, A and Bhat, A.I. (2010) Survey and RT-PCR Based Detection of Cardamom mosaic virus Affecting Small Cardamom in India. *Indian J. Virol*. 21:148-150.

Bory S, D. Da Silva, A.M. Risterucci, M. Grisoni, P. Besse and Duval, M.F. (2008) Development of microsatellite markers in cultivated vanilla: Polymorphism and transferability to other vanilla species, *Sci. Hort* 115 420–425.

Cai, W.Q., Fang, R.X., Shang, H. S., Wang, X., Zhang, F.L., Li, Y. R. Zhang, J.C., Cheng, X.Y., Wang, G. L. and Mang, K.Q. (2003) Development of CMV and TMV Resistant Chili Pepper: Field Performance and Biosafety Assessment. *Mol. Breed.,* 11(1):25-35.

Chambon C, Poupet A and Maia N (1985) Problems poses par la multiplication vegetative in vitro actes du colloque lesmenthes en France. Aspects scientific economique et Industrial. lyen, 20.21 Mars., pp 51-60.

Chand S and Roy S C. (1981) Induction of organogenesis in callus cultures of *Nigella sativa* L. *Annals of Botany* 48: 1-4.

Charles K and Rao Y S. (2007) A contribution to the *in vitro* studies on ginger. Ph.D thesis, Andhra University, Waltair.

Chen C H (1976) Vegetative propagation of celery plant by tissue culture. Proc. S. Dakote Acad. Sci. 55:44-48.

Chen RR, Zhang JT, Li BP, Guo SS, Hao JP and Zhou XM (1991) Studies on the production of artificial seeds of Coriander. *Chin. J. Biotechnol*. 7: 127-134.

Cocking. E.C. (1972) Plant protoplasts- isolation and developments. *Ann. Rev, Plant Physiol.* 23: 29-50.

De Mattia Fabrizio, Ilaria Bruni, Andrea Galimberti, Francesca Cattaneo, Maurizio Casiraghi and Massimo Labra, (2011) A comparative study of different DNA barcoding markers for the identification of some members of Lamiacaea, *Food Research International* 44: 693–702

Dekkers A J, Rao A N and Goh C J. (1991) *In vitro* storage of multiple shoot cultures of ginger at ambient temperatures of 24-29 C. *Sci. Hortic. (Amst.)* 47: 156-167.

Dhanya, K. and Sasikumar, B. (2010). "Molecular Marker Based Adulteration Detection in Traded Food and Agricultural Commodities of Plant Origin with Special Reference to Spices", *Current Trends in Biotechnology and Pharmacy*, 4(1): 454-489.

Dhanya, K., S. Syamkumar, S. Siju and Sasikumar, B. (2011) SCAR markers for adulterant detection in ground chilli", *British Food Journal,* 113(5):656 – 668

Ding HJ, Bai S H, Wu Y and Fan X P (1981) Induction of callus and regeneration of plant lets from corm of *Crocus sativus, Acta Bot. Sin.*23: 419-420.

Divakaran, M., V. N. Jayakumar, S.S. Veena , J. Vimala , A. Basha , V. Saji , K. Nirmal Babu and K. V. Peter (2007) Genetic variations and interrelationships in *Vanilla planifolia* and few related species as expressed by RAPD polymorphism. Genetic *Resources and Crop Evolution* 55:459-470.

Divakaran, M., K. Nirmal Babu, P.N. Ravindran, and K.V. Peter (2006) Interspecific hybridization in vanilla and molecular characterization of hybrids and selfed progenies using RAPD and AFLP markers. *Scientia Horticulturae* 108: 414-422.

Dong, C., Jiang, C., Feng, L., Li, S., Gao, Z. and Gou, J. (1995) Transgenic tomato and pepper plants containing CMV sat-RNA cDNA. *Acta Hortic.* 402:78-86.

Dudits D, Feger O, Hadlaezky G, Konoz O, Lazar G B and Howrath G.(1980) Intergeneric gene transfer mediated plant protoplast fusion, *Mol Gen. Genetics.* I 79: 283-288.

Elmbeuser H A, Lein C and Neumann K H. (1978) An investigation of the relationship between phytohormone content and growth of brassica species in tissue culture. *Z. Pfenzen Physiol* 88:25-32.

Ferrie A M R and Waterer Doug (2009) Development of Improved Spice Crops using Double Haploid Technology, Final Report, ADF Project No. 20050711 (PBI/NRC, Saskatoon), Dept Plant Sciences, University of Saskatchewan, Saskatoon

Ferrie A.M.R, T.D. Bethune, G.C. Arganosa and D. Waterer (2011) Field evaluation of doubled haploid plants in the Apiaceae: dill (*Anethum graveolens* L.), caraway (*Carum carvi* L.), and fennel (*Foeniculum vulgare* Mill.) *Plant Cell Tissue and Organ Culture*, 104(3):407-413.

Fu N, Wang Q and Shen H-L (2013) De Novo Assembly, Gene Annotation and Marker Development Using Illumina Paired-End Transcriptome Sequences in Celery (*Apium graveolens* L.). PLoS ONE 8(2): e57686. doi:10.1371/journal.pone. 0057686. Editor: Michael N. Nitabach, Yale School of Medicine, USA.

Furmanowa M and 0lszowska.(1980) *Thymus vulgaris* L. Propagation through tissue culture. *Acta Pol. Pharni.* 37: 243-247. .

Ganapathy T.R., Bapat V. and and Rao P.S. (1994) *In vitro* development of encapsulated shoot tips of cardamom. *Biotech. Techniq.* 8(4):239-244.

Ganapathy T.R., Mohan J.S.S., Suprasanna P, Bapat V.A. and Rao P.S., (1995) A low cost strategy for *in vitro* propagation of banana. *Current Science* 68: 646-650.

Garcia-Rodriguez. M.J., Paupadrin C. and Saussay R. (1978) Sur la formation d'tissue secreteur et la synthese d" anethole par des tissues de fenouil *(Foeniculam vulgare* mill var. dulce) cultives *in vitro. CR Acad. Sci. Paris.* 287: 693-696.

Garland P. and StoltzL P. (1980) *In vitro* propagation of tarragon *(Artemesia dracunculus var.sativa) Hort Sci.* 15: 739.

Gayatri M.C., Roopa D.V. and Kavyashree R. (2005) Selection of turmeric callus for tolerant to culture filtrate of *pythium graminicolium* and regeneration of plants. *Plant Cell Tissue and Organ Culture* 83:3310.

Geetha S.P., K. Nirmal Babu, J Rema, P N Ravindran, and K.V. Peter, (2000) Isolation of protoplasts from cardamom (*Elettaria cardamomum* Maton) and ginger (Zingiber officinale Rosc), *J Spices & Aromatic Crops* 9(1): 23-30.

George L. and Narayanaswamy S. (1973) Haploid capsicum through experimental androgenesis, *Protopl.* 78:467-470.

Geslot A., Christian C., Gilles M., Mohamed E.L. Maataou and Pierre Neville. (1989) Multiplication *In vitro* de menthes associant la culture d apex et le microbouturage. *Bull. Soc. Bot.* FR 136: 31-38.

Girija D., Beena P.S. and Nazeem P A (2005) Molecular cloning of a cDNA fragment encoding the defense related protein β-1, 3-glucanase in black pepper (*Piper nigrum* L.). *Proceedings of the Kerala Science Congress*, January 2005, KFRI, Peechi, Kerala. 81-82p.

Grewal S., Sachdeva V. and Atal C.K. (1976) Regeneration of plants by embryogenesis from hypocotyls cultures *of Ammi majus* L. *Indian J. Exp.Biol.* 14:716-717.

Harisaranraj, R., K. Suresh and Saravanababu, S. (2009) DNA Finger Printing Analysis among Eight Varieties of *Zingiber officinale* Rosc. By Using RAPD Markers, *Global Journal of Molecular Sciences* 4 (2): 103-107.

Hebert, P.D.N., Cywinska, A., Ball, S.L. and DeWaard, J.R. (2003) Biological identification through DNA barcodes. *Proceedings of the Royal Society of London*. Series B, 270:313-321.

Himeno H, Matsushima H and Sano K.(1988) Scanning elecron microscopic study on the *in vitro* organogenesis of saffron stigma and style like structures. *Plant Science*. 58:93-101.

Hippolyte L, Merin B, Baccon J C and Jonard R. (1991) Influence of maintenance medium and sucrose concentration on the rosemarinic acid production by cell suspension culture of sage (*Salvia officinalis* L.) *CR. Acad. Sci. Ser.,* 313:365-371.

Hirasa, K. and Takemasu, M. (1998) Spice Science and Technology. Marcel Decker Inc. New York.

Homes J, Legros M and Lazul. (1987) *In vitro* multiplication of *Crocus sativus* L. *Acta Hortic.* 212:675-676.

Hosoki T. and Sagawa Y. (1977) Clonal propagation of *ginger. Hort.Sci.* 12: 451-452.

IISR. 2002. Annual Report (2001-2002) Indian Institute of Spices Research, Calicut

IISR. 2003. Annual Report (2002-2003) Indian Institute of Spices Research, Calicut

IISR. 2004. Annual Report (2002-2003) Indian Institute of Spices Research, Calicut

IISR. 2005. Annual Report (2002-2003) Indian Institute of Spices, Research, Calicut

IISR. 2012. Annual Report (2011-2012) Indian Institute of Spices Research, Calicut

Inden H, Asahira T and Hirano A. (1988) Micropropagation of *ginger. Acta hort.* 230: 177-184.

Isa T. and Ogasawara T. (1988) Efficient regeneration from callus of saffron *(Crocus sativus) Japanese Joumal of breeding* 38(3): 371-374.

Jacob T., Jebasingh T., Venugopal M.N. and Usha R. (2003) High genetic diversity in the coat protein and 30 untranslated regions among the geographical isolates of Cardamom mosaic virus from South India. *J Biosci.* 28(5):589–95.

Jarret R.L. and Fernandez, R. (1984) Shoot tip vanilla culture for storage and exchange. Plant Genetic Resource News Letter. 57: 25-27.

Jayakumar VN, Nirmal Babu K and Venugopal M N. (2005) Molecular characterization of small cardamom *(Eletteria cardamomum* Maton) collections and related genera using RAPD, PCR-RFLP and IISR markers. National symposium on Biotechnological Interventions for improvement of Horticultural crops: issues and strategies, 10-12, January 2005, Kerala Agricultural University, Thrissur, Kerala (abstracts: p. 122).

Jha T.B., Roy S.C. and Mitra G.C. (1982) *In vitro* culture of *Cuminum cyminum* regeneration of flowering shoots from calli of hypocotyl and leaf explants. *Plant Cell Tissue Organ Culture,* 2:11-14.

Jiby, M.V. and Bhat, A.I. (2011) An Efficient Agrobacterium-Mediated Transformation Protocol for Black Pepper (*Piper nigrum* L.) Using Embryogenic Mass as Explant. *J. Crop Sci. Biotech.* 14 (4): 247-254.

Johnson G K, Ganga G, Sandeep Varma R, Sasikumar B and Saji K V. (2005) Identification of hybrids in black pepper (*Piper nigrum* L.) using male parent-specific RAPQ markers. *Current Science* 88:1-2.

Johnson T S, Ravisankar G A and Venkataraman LV. (1991) Elicitation of capsaicin production in freely suspended cell~ and immobilized cell cultures of *Capsicum frutescenc* Mill. *Food Biotech,* 5:197-200.

Joshi R. K, Ananya Kuanar, Sujata Mohanty, Enketeswara Subudhi, and Sanghamitra Nayak (2010) Mining and characterization of EST derived microsatellites in *Curcuma longa* L. *Bioinformation* 5(3): 128–131.

Joshi, R.K, Mohanty, S., Subudhi, E. and Nayak, S. (2010) Isolation and characterization of resistance gene candidates in turmeric (*Curcuma longa* cv. Surama). *Genet Mol. Res.*, 9(3):1796-1806.

Kackar, A., Bhat, S.R., Chandel, K.P.S. and Malik, S. K., (1993) Plant regeneration via somatic embryogenesis in ginger. *Plant Cell Tissue and Organ Culture* 32:289–292.

Karalija, E. and Pariæ, A, (2011) The effect of BA and IBA on the secondary metabolite production by shoot culture of *Thymus vulgaris* L. *Biologica Nyssana*, 2 (1): 29-35.

Kavyashree R. (2011) *In vitro* propagation of ginger variety Suruchi. *Advances Biotech* 1(1): 17-19.

Kehr A E and Schaffer G.W. (1976) Tissue culture and differentiation of garlic. *Hort. Sci.* 11: 422-423.'

Kim MA and Park J.K. (2002) High frequency plant regeneration of garlic (*Allium sativum* L.) calli immobilized in calcium alginate gel. *Biotechnol Bioproc Eng,* 7:206-211.

Knudson L, (1946) A new nutrient solution for the germination of orchid seed. *American Orchid Society Bulletin* 14: 214–217.

Koo Hyun Jo, Eric T McDowell Xiaoqiang Ma, Kevin A Greer, Jeremy Kapteyn, Zhengzhi Xie, Anne Descour, HyeRan Kim, Yeisoo Yu, David Kudrna, Rod A Wing, Carol A Soderlund and David R, (2013) Ginger and turmeric expressed sequence tags identify signature genes for rhizome identity and development and the biosynthesis of curcuminoids, gingerols and terpenoids. BMC Plant Biology, 13:27 doi: 10.1186/1471-2229-13-27

Koyama A. Ohmori Y, Fujioka H, Miyagawa H, Yamaski K and Konda H (1987) Formation of stigma like structures and pigments in culture tissues of *Crocus sativus. Jap. J. Pharmacog.* 41:226-229.

Kuruvinashetty M S, Haridasan P and Iyer R D. (1982) Tissue culture studies in *turmeric.* (In) Ginger and turmeric, (eds) M K Nair, T Premkumar, P N Ravindran and Y R Sharma. CPCRI, Kasargod, pp. 29-41.

Lee S, J., Kim, B.D. and Paek, K.H. (1993) *In vitro* plant regeneration and Agrobacterium-mediated transformation from cotyledon explants of hot pepper (*Capsicum annuum* cv. Golden Tower). *Korean J. Plant Tissue Cult*, 20:289–94.

Lei J J, Chen S R, Guo Y L and Huang J H. (1991) Development of plantlets from leaf protoplast culture in *Brassica juncea. Acta Bot. Sin.* 33: 91-97

Lin M and Staba J (1961) Peppermint and spearmint tissue cultures: 1. Callus formation and submerged culture. *Llyodia* 24: 139-145.

Lincy A.K., Remashree, A.B. and Sasi Kumar, B. (2009) Indirect and direct somatic embryogenesis from aerial stem explants of ginger (*Zingiber officinale* Rosc.) *Acta Bot. Croat.* 68(1):93-103.

Lindsey K. and Yeoman M.M. (1985) The synthetic potential of immobilized cells of *Capsicum frutescence. Planta*.162:495-501.

llahi I., Jabeen M.D. and Fiidou N.C. (1987) Morphogenesis *with Saffron* tissue culture. *Journal of Plant Physiology* 128:221-223.

Lu FH, Cho M.C. and Park Y.J. (2012) Transcriptome profiling and molecular marker discovery in red pepper, *Capsicum annuum* L. TF68. *Molecular Biology Reports* 39:3327–3335.

Lum, M.R. and Hirsch, A.M. (2006) Molecular methods for the authentication of botanicals and detection of potential contaminants and adulterants. *Acta Horticultura*e, 720:59-72.

Madhavi D.L., Prabha T.N., Singh N.S. and Patwardhan M.V. (1991) Biochemical studies with garlic *(Allium sativum)* cell cultures showing different flavour levels. *J. Sci. Food Agric.* 56:15-24.

Maheswari S.C. and Gupta RP. (1965) Production of adventitious embryoids *in vitro* from stem callus of *Foeniculam vulgare, Planta* 67:384-386.

Manay, S. and Shadaksharaswami, M. (1997) *Foods-facts and Principles.* New Age International Publ. New Delhi. pp. 321-330.

Mani, T. and Manjula, S. (2010) Cloning and characterization of two osmotin isoforms from *Piper colubrinum. Biol. Plantarum.*, 54 (2): 377-380.

Manohari, C., Backiyarani, S., Jebasingh, T., Somanath, A. and Usha, R. (2008) Efficient plant regeneration in small cardamom (*Elettaria cardamomum* Maton) through somatic embryogenesis, *Indian Journal of Biotechnology*, 7: 407-409

Mariska I,Gatidan E and Sukmadjaja D. (1989) Clove. *Edisikhusus Lett.* (Indonesia) 5: 4549.

Mary Mathew K and Hariharan M. (1990) *In vitro* multiple shoot regeneration in *Syzygium aromaticum* (L) Merry & Perry. *Annals of Botany* .65: 277-279.

Mary Mathew K., Y.S. Rao, K. Pradip Kumar, Saly Kutty Joseph. R. Lakshman and K.J. Madhusoodanan (1999) Micropropagation of curry leaf (*Murraya koenigii* L.). *J. of Spices and Aromatic Crops* 8(1):77-79.

Mary Mathew, K., Y.S. Rao, K. Pradipkumar, K.J. Madhusoodanan and S.N. Potty (1999) *In vitro culture systems in vanilla*. Proc. of National Conference on "Plant Tissue Culture and biotechnology – emerging trends" University Press (India) Limited, Hyderabad, pp.171-180.

Mary Mathew, K., Y.S. Rao, K. Pradipkumar, Salykutty Joseph, R. Laxmanan and K.J. Madhusoodanan (1998) *In vitro* culture system for clove (*Syzygium aromaticum* L). Proceedings of XIII PLACROSYM, Allied Publishers Limited, New Delhi.

Mary Mathew. K, R Lakshmanan, Rao Y S, Joji Mathew and Pradip Kumar, K, (2012) Isolation and culture of protoplasts of *Vanilla planifolia* Andrews and *Vanilla wightiana* Lindle- a wild relative. *Crop Improvement* 39 (1): 31-35.

Masceranhas A F, Nair S, Kulkarni V M, Agarwal D C, Khuspe S S and Mehta U J. (1987) Tamarind. In: Cell and Tissue Culture in F, Vol. 3(eds). J.M. Bonga & D.J. Durzan. Martinus Njhoff, Dordrecht, pp.316-330.

Meyer M M and Milbrath GM. (1977) *In vitro* propagation of horse radish with leaf pieces. *Hort Sci.* 12:544-546.

Minoo D (2002) Seedling and somaclonal variation and their characterization in vanilla. Ph.D thesis. Calicut University, Kerala, India.

Misra P and Chaturvedi HC.(1984) Micropropagation of *Rosemarinus officinalis* L. *Plant Cell Tissue Organ Culture.* 3: 163-168.

Miura Y and Tabata M. (1986) Direct somatic embryogenesis from protoplasts of *Foeniculum vulgare. Plant Cell Reports,* 5: 310-313.

Murashige, T. and Skoog, F. (1962) A revised medium for rapid growth and bioassays with tobacco tissue cultures. *Physiol Plant* (15)473-479.

Nadgauda R S, Hendre R R and Jagannathan V. (1978) Rapid multiplication of turmeric *(Curcuma longa* L) plants by tissue culture. *Indian Journal of Exp. Biol.* 16: 120-122.

Nadgauda R S, Khuspe S S and Mascarenhas A F. (1982) Isolation of high curcumin varieties of turmeric from tissue culture. In: Proceedings of V annual symposium on plantation Crops (ed) R.D. Iyer, pp: 143:144, CPCRI, Kasargod.

Nadgauda RS, Masceranhas AF and Madhusoodanan K J. (1983) Clonal multiplication of cardamom *(Elettaria cardamomum* Maton) by tissue culture. *J. Plant. Crops.* 11: 60-64.

Nair MK and Ravindran PN. (1988) Tissue culture of pepper and tree spices: Potential priorities and applications. *Proc. Nat. Sym. on Plant Tissue Culture.* pp. 163-169.

Nair MK, Premkumar T, Shanna Y R and Ratnambal MJ. (1977) Prospects and problems of tree spices cultivation in India. *Indian Spices* i4: 2-9.

Nair RR and S. Dutta Gupta, (2006) High frequency plant regeneration through cyclic somatic embryogenesis in black pepper (*Piper nigrum* L), *Plant Cell Reports* 24: 699-707.

Nair RR and DS Gupta (2007). *In vitro* plant regeneration from encapsulated somatic embryos of black pepper (*Piper nigrum* L.). *J. Plant Sci.* 2: 283-292.

Nair, R.A. and Thomas, G. (2013) Molecular characterization of ZzR1 resistance gene from *Zingiber zerumbet* with potential for imparting *Pythium aphanidermatum* resistance in ginger. *Gene*, 516: 58-65.

Nazeem P A, Joseph I, Rani T G, Valsala P A, Philip S and Nair G S.(1996) Tissue culture system for *in vitro* pollination and regeneration of plantlets from *in vitro* raised seeds of *ginger-Zingiber officinale* roes. International symposium on medicinal and aromatic plants ISHS; *Acta Horticulturae* 426.

Nazeem P A, Joseph Land Nair G S. (1990). *In vitro* plantlet regeneration in black pepper *(piper nigrum* L) *Proc. of National Sym. on Current Trends in Biotechnology*, pp 29.

Neenakumari and Saradhi. PP. (1992) Regeneration of plants from callus cultures of *Oreganum vulgare* L. *Plant Cell Reports.* 11: 476-479.

Nessler CL. (1982) Somatic embryogenesis in the opium poppy *(Papaver sornniferum) Physiol. Plant.* 55: 453-458.

Newmaster Steven G., Meghan Grguric, Dhivya Shanmughanandhan, Sathishkumar Ramalingam and Subramanyam Ragupathy, (2013) DNA barcoding detects contamination and substitution in North American herbal products, BMC Medicine 11:222, http://www.biomedcentral.com/1741-7015/11/222.

Nirmal Babu K. (2003) Molecular characterization and preparation of molecular maps in black pepper. Final Report, NATP (CGP), ICAR, New Delhi, pp 50.

Nirmal Babu K., John Zachariah T., Minoo D., Shamsudeen K and Ravindran P.N. (1992) *In vitro* proliferation of nutmeg aril (mace) by tissue culture. *J. Spices and Aromatic Crops* 1:88-93.

Nirmal Babu K., Ravindran P.N. and Sasikumar B. (2003) Field evaluation of tissue cultured plants of spices and assessment of their genetic stability using molecular markers. Final report submitted to DBT, New Delhi, pp 94.

Nirmal Babu K., Samsudeen K. and Ravindran P.N. (1996) Biotechnological approaches for crop improvement in Ginger, *Zingiber officinale* Rose. In: *Recent Advances in Biotechnological Applications on Plant Tissue and Cell Culture.*

Nirmal Babu K., Suraby E.J., Cissin J., Minoo D., Pradeep Kumar T., Parthasarathy V.A. and Peter K.V. (2013) Status of transgenics in Indian spices. *Journal of Tropical Agriculture*51(1-2):1-14.

Nirmal Babu, K., G. Yamuna, K. Praveen, D. Minoo, P.N. Ravindran and K.V. Peter (2012) Cryopreservation of Spices Genetic Resources, Current Frontiers in Cryobiology, Prof. Igor Katkov (Ed.), ISBN: 978-953-51-0191- 8, InTech, http://www.intechopen.com/books/current-frontiers-in-cryobiology/ cryopreservation of- spices-genetic-resources

Parani M., Anand A. and Parida A. (1997) Application of RAPD finger printing in selection of micropropagated plants of *Piper longum* for conservation. *Current Science* 73(1): 81-83.

Park Woo Tae , Yong Kyoung Kim, Md. Romij Uddin, Nam Il Park, Su Gwan Kim, Sook Young Lee and Sang Un Park(2010) Somatic embryogenesis and plant regeneration of lovage (*Levisticum officinale* Koch) *Plant Omics J* 3(5):159-161.

Philip V.J., Joseph D. and Nainar SAZL 1986) Clonal propagation of *Vanilla planifolia* using tissue culture. *J. Pl. Physiol.*122: 211-215.

Philip V.J., Joseph D., Triggs G.S. and Dickinson N.M. (1992) Micropropagation of black pepper *(Piper longum* L.) through shoot tip cultures. *Plant Cell Reports.* 12: 41-44.

Pillai S.K. and K.B. Kumar (1982) Note on the clonal multiplication of ginger *in vitro. Indian Journal of Agric. Sc*, 52: 397-399.

Pradipkumar K. (2007) A contribution to the *in vitro* studies on large cardamom *(Amomum subulatum* L Roxburgh). Ph. D. Thesis. Andhra University, Waltair.

Prasath, D., El-Sharkawy, I., Tiwary, K.S., Jayasankar, S. and Sherif, S. (2011) Cloning and characterization of PR5 gene from *Curcuma amada* and *Zingiber officinale* in response to *Ralstonia solanacearum* infection. *Plant Cell Rep.*, 30(10):1799-1809.

Priya, R.S. and Subramanian, R.B. (2008) Isolation and molecular analysis of R-gene in resistant *Zingiber officinale* (ginger) varieties against *Fusarium oxysporum f.sp. zingiberi*, *Bioresource Technol.*, 99:4540-4543.

Rai, V.R.S. and Jagadishchandra, K.S. (1987) Clonal propagation of *Cinnamomum zeylanicum* Breyn by tissue culture. *Plant Cell Tissue and Organ Culture*,9:81-88.

Rao, Y.S., Biswas, A.K., Gupta, U., Varadarasan, S., Gopakumar. B, Potty, S.N. and Thomas, J. (2005) Tissue Culture - Large Cardamom: Field Performance. *Spice India* 16 (11):15-22.

Rao, Y.S., Mary Mathew, K., Pradipkumar K., Madhusoodanan, K.J. and Potty, S.N. (1996) Economics of *in vitro* propagation of vanilla and ginger. (In) Proc. of PLACROSYM XII, 209-213.

Rao, Y.S., Mary Mathew. K., Pradipkumar, K., Lakshmanan R., Joji Mathew and J. Thomas. (2012) Cost Projections of Commercial production of Tissue Culture Plants of Vanilla (*Vanilla planifolia* Andrews). Advances in Plant Sciences 2012 b, 25 (II): 367-371.

Rao, Y.S., Mary Mathew K., Pradipkumar. K., Madhusoodanan, K.J., Potty, S.N. and George, C.K. (1999) Practical Manual for Micropropagation of Some Selected Spices, Indian Cardamom Research Institute, Spices Board, Cochin-682 025 pp 1-60. Technical Publication.

Rao, Y.S., Mary Mathew K., Pradipkumar K., Salykutty J., Laxmanan R., Madhusoodabnan K.J. and Potty S.N. (1997) Tissue culture studies on tree spices. (In) (Edison S, Ramana KV, Sasikumar B, Nirmal Babu K and Santhosh JE (eds.) *Biotechnology of Spices, Medicinal and Aromatic Crops.* Indian Society for Spices, pp 39-44.

Rao, Y.S., Mary Mathew. K, Pradipkumar, K., Lakshmanan, R., Joji Mathew and Thomas, J. (2012a), Cost Projections of Commercial production of Tissue Culture Plants of Vanilla (*Vanilla planifolia* Andrews) *Advances in Plant Sciences*, 25 (II): 367-371.

Rao, Y. S., Mary Mathew K., Madhusoodanan, K.J. and Naidu, R (1993) Multiple shoot regeneration in Vanilla (*Vanilla planifolia* Andr). *J. of Plantation Crops,* 21: 351-354.

Rao, Y.S., Mary Mathew K., K. Pradipkumar, Lakshmanan, R. and Thomas, J. (2006) Large scale production, costs involved and field performance of TC plants of large cardamom (*Amomum subulatum* Roxb), *J. of Plantation Crops*, 34:(3):591-596.

Rao, Y.S., Mary Mathew K., Lakshmanan, R, Madhusoodanan K.J. and Pradipkumar, K. (2008) Development of *in vitro* protocol for plantlet production in Garcinia, a tree spice, Forest Biotechnology in India (Eds. Ansari S.A., Narayanan C. and Mandal A.K.), Satish Serial Publishing House, New Delhi. pp. 183-188.

Rao, Y.S., Mary Mathew. K., Kuruvilla, K.M., Lakshmanan, R., Gilu Lizy George and Potty, S.N. (2002) Practice and promise of tree tissue culture, Plantation Crops Research and Development in the new Millennium, *Proc. of PLACROSYM-XIV*, Hyderabad. pp 262-266.

Ravindran, P.N., Peter K.V., Nirmal Babu, K., Rema J., Shamsudeen K, Minoo D, Geetha S P, Sajina A, Mini P M, Manjula C and John C Z. (1996) Biotechnological approaches in spice crops - present scenario and future prospects. (In) Biotechnology for Development (eds. MR. Das and Sathish Mundayoor) STEC, Kerala, pp 175-197.

Redenbaugh, K., Brian, D.P., James, W., Mary, E., Peter R. and Keith A.W. (1986) Somatic seeds: Encapsulation of asexual plant embryos, *Biotechnology* 4: 9-83.

Redenbaugh, K., Nichol. J., Kossler, M.E. and Pasch, B.D. (1984) Encapsulation of somatic embryos for artificial seed production. *In vitro.* 20:256.

Reghunath, B.R. and Gopalakrishnan P.K. (1991) Successful exploitation of *in vitro* culture techniques for rapid clonal multiplication and crop improvement in cardamom. Proc. of the Kerala Sci. Congress., 1991, pp 70-71.

Reinert J, Backs-Huseman D and Krosing M. (1966) Faktoreader embryogense ingwebekulturen aus kultreformer von umbelliferen. *Planta* 68: 375-378.

Remya Lekshmi, Sherin Jose, Anisha C.S., Swapna Sasidharan, Mary Mathew, K. and Rao, Y.S. (2013) Genetic Diversity in large cardamom (*Amomum subulatum* Roxb) using ISSR markers. *Ad Plant Sci* 26 (II): 347-351.

Retheesh, S.T. and Bhat, A.I. (2010) Simultaneous elimination of Cucumber mosaic virus and Cymbidium mosaic virus infecting *Vanilla planifolia* through meristem culture. *Crop Protection*, 29:1214-1217.

Retheesh, S.T. and Bhat, A.I. (2011) Genetic transformation and regeneration of transgenic plants from protocorm-like bodies of vanilla (*Vanilla planifolia* Andrews) using *Agrobacterium tumefaciens*. *J. Plant Biochem. Biotechnol*. 20: 262-269.

Romagnoli, L.G. and Knorr, D. (1988) Effects of ferulic acid treatments on growth and flavour development of cultured *Vanilla planifolia* cells. *Food Biotech.*, 2: 93-104.

Rosengarten (1969) The Book of Spices. Livingston Publication Company. Philadelphia

Rout, G.R., Das, P., Gael S. and Raina S.N. (1998) Determination of genetic stability of micropropagated plants of ginger using random amplified polymorphic DNA (RAPD) markers. *Bot. Bulletin of Academica Sinica* 39(1): 23-29.

Rout, G.R., S.K., Palai, S. Samantaray and Das P. (2001) Effect of growth regulator and culture conditions on shoot multiplication and rhizome formation in ginger (*Zingiber officinale* Rosc) *in vitro. In vitro* Cell. *Dev. Biol- Plant,* 37: 814-819.

Sano, K. and Himeno, H. (1987) *In.vitro* proliferation of saffron *(Crocus sativus* L) stigma. *Plant Cell Tissue and Organ Culture* 11: 159-166.

Sasaki Y., Fushimi, H. and Komatsu, K. (2004) Application of single nucleotide polymorphism analysis of the tmK gene to the identification of Curcuma plants. *Biological and Pharmaceutical Bulletin* 279(1): 144-146.

Sasikumar, B. and Veluthambi, K. (1994) Strain and vector specificity in agrobacterium black pepper interaction. Abst. International symposium on Plantation Crops.p-74.

Sasikumar., S. Syamkumar, Remya, R. and John Zachariah, T. (2005) PCR Based Detection of Adulteration in the Market Samples of Turmeric Powder, Food Biotechnology Volume 18(3):299-306.

Saxena, P., Gill R., Rashid, A. and Maheswari, C.S. (1981) Isolation and culture of *Capsicum annum* L. and their regeneration into plants flowering *in vitro. Protoplast.* 108: 357-360.

Seghal, (1968) *In vitro* development of neomorphs in *Anethum graveolens. Phytomorphology.* 18:509-514.

Shaji, P., Anandraj, M. and Sharma, Y. R. (1998) Comparative study of protoplast isolation and development in *Piper nigrum* (black pepper) and *P. colubrinum.* (In) N M Mathew and C.K. Jacob (eds). Developments in Plantation Crops Research, Allied Publishers, New Delhi. pp 51-53. Allied Publishers, New Delhi. pp 51-53.

Sharma, K.S., Sharada, K, Maesato, K., Hara T. and Sonada, Y. (1991) Chemical and sensory analysis of saffron produced through tissue cultures of *Crocus sativus. Plant Cell Tissue Organ Culture,* 26: 11-16.

Sharma, T.R., Singh B.M. and Chauhan, R.S. (1994) Production of encapsulated buds of *Zingiber officinale* Rose. *Plant Cell Rep011f* 13:300-302.

Sharma, Y.R., Anandaraj, M., Philip, S. and Bindu, M.R. (1996) Development of resistance to *Phytophthora capsici* in *blackpepper* (*Piper nigrum* L.) through biotechnological approaches: present status and future strategies. In: Proceedings of the National Seminar on Biotechnology of Spices and Aromatic Plants, Calicut. 24-25, April, 1996, pp. 19.

Sherin, Jose, Swapna Sasidharan, Remya Lekshmi, Ragina, V.C., Anisha, C.S., Mary Mathew, K. and Rao, Y.S. (2013) Virus indexing of black pepper (*Piper nigrum* L) against CMV and PYMoV. *Ad. Plant Sci.* 26 (II): 317-320.

Shetty, M.S.K., Hariharan, P. and Iyer R.D. (1882) Tissue Culture studies in turmeric. In: *Proc. Nat Seminar on Ginger and Turmeric*, Calicut (M.K. Nair, T.Premkumar, P.N. Ravindran and Y.R. Sharma, Eels) pp 39-41, CPCRI, Kasaragod.

Shylaja, M.R., Nair, S.G., Nazeem, P.A., Mallika, V.K. and Mathew, M.K. (1994) *In Vitro* screening of black pepper for tolerance to *Phytophthora capsici*. PLACROSYM XI, Calicut: pp24-30.

Siljo, A., Bhat, A.I., Biju, C.N. (2013) Analysis of variability in Cardamom mosaic virus isolates occurring in India using symptomatology and coat protein gene sequence. *Journal of Spices and Aromatic Crops*, 22: 1.

Siljo, A., Bhat, A.I., Biju, C.N. and Venugopal, M.N. (2012) Occurrence of Banana bract mosaic virus on cardamom. *Phytoparasitica*. 40:77-85.

Sim, S.L., Jafar R., Grierson, D., Power, J.B. and Davey, M.R. (1995) Applications of molecular biology and genetic manipulations to pepper *(Piper nigrum)* breeding: present state and prospects. In: Abstracts FAO/IAEA International symposium on the use of induced mutations and molecular techniques for crop improvement, Vienna, Austria.

Sivaraman, K. and Peter, K.V. (1999) An overview of spices development in India. *Indian Jour. Arecanut Spices and Medicinal Plants* l (1):1-11.

Smith, M.K. and Hamill, S.D. (1996) Field evaluation of micropropagated and conventionally propagated ginger in subtropical Queensland. *Aust. J. Exp. Agric.,* 36: 347-354.

Stephen, R. and Jayabalan, N. (2000) Artificial seed production in coriander (*Coriandrum sativum* L.). *Plant Tissue Cult.* 10:45-49.

Steward, F.C. Ammirato, P.V. and Mapes, M.O.(1970) Growth and development of totipotent cells: some problems, procedures and perspectives. *Annals of Botany*, 34: 761-787.

Stratisky, G., Dekkers, A.J., Louwars, N.P. and Zandwoort, E.A. (1985) *In vitro* conservation of aroid germplasm at reduced temperatures and under osmotic stress. (In) *Plant tissue culture and its Applications* (Eds.) L.A.Withers & P G. Alderson (Butterworths) pp.277-284.

Subramaniam, M.K., Royan Subramaniam, S., Gopinath, P.M., Gupta, K.C. and Vasantha. (1968) Histogenesis, organogenesis arid morphogenesis in callus cultures of *Trigonella foenum - graecum* L. and *Vigna unguiculata* (L) Walp. *Curr. Sci.* 37: 398-399.

Suma, B., Keshavachandran, R. and Nybe, E.V. (2008) *Agrobacterium tumefaciens* mediated transformation and regeneration of ginger (*Zingiber officinale* Rosc). *Journal of tropical Agriculture* 46 (1-2): 38-44

Sundararaj, S.G., Agrawal, A. and Tyagi, R.K. (2010) Encapsulation for *in vitro* short-term storage and exchange of ginger (*Zingiber officinale* Rosc.) germplasm. *Sci. Hortic.* 125:761-766.

Swetha Priya, R. and Subramaniam, R.B. (2008) Isolation and molecular analysis of R-gene in resistant *Zingiber officinale* (Ginger) varieties against *Fusarium oxysporum* f.sp. *zingiberi*. *Bioresour Technol.* 99(11):4540-4543.

Tiing, L.E., San, H.S., Eng, L. and Det, P.A. (2012) Cloning and Characterization of Resistance Gene Analogues (RGAs) from *Piper nigrum* L. cv. Semongok Aman and *Piper colubrinum* Link. *Int. J. Biosci. Biochem. Bioinfo*, 2:342-348.

Vasil, I.K. and Hildebrandt, A.C.(1966) Variations in morphogenic behaviour in plant tissue cultures. 11. *Petroselinum hortense. Amer. J. Bot.* 53: 869-874.

Veluthambi, K., Aditya, K.G., and Sharma, A. (2003) The current status of plant transformation technologies. *Curr. Sci.*, 84(3):368-337.

Venkataraman, L.V. (1996) Biotechnological approaches for.production of saffron and capsaicin- A perspective. In: *Proceedings of National Seminar on Biotechnology of Spices and Aromatic Plants*. Calicut, April, 24-25.1996. pp.26.

Wang, P.J. and Huang, L.C. (1974) Studies on the shoot meristem cultures of *Allium sativum. Chinese J. Hort.,* 20:79-87.

Wang, P.J. and Hu, C.Y. (1980) Regeneration of virus free plants through *in vitro* cultures. In: *Advances in Biochemical Engineering - Plant Cell Cultures* 11. (Ed) A. Fisher, Springer-Verlag, Berlin, pp 61-69.

Wang S, Yang W and Shen H (2011) Genetic diversity in *Apium graveolens* and related species revealed by SRAP and SSR markers. *Scientia Horticulturae* 129:1-8.

Wei W, Qi X, Wang L, Zhang Y, Hua W, *et al.* (2011) Characterization of the sesame (*Sesamum indicum* L.) global transcriptome using Illumina paired-end sequencing and development of EST-SSR markers. *BMC Genomics* 12: 451.

Withers, LA. (1987) Long term preservation of plant cells tissue and organs: *Oxford Survey of Plant Molecular and Cell Biology* 4:221-272.

Yang, X. and Quiros C (1993) Identification and classification of celery cultivars with RAPD markers. *TAG, Theoretical and Applied Genetics* 86: 205–212.

Zhu, Y.X., Ou-Yang, W.J., Zhang, Y.F. and Chen, Z.L. (1996) Transgenic sweet pepper plants from *Agrobacterium*-mediated transformation. *Plant Cell Rep.*, 16:71–75.

Zuraida, A. B. Rahman, Kamarulzaman Hassan Fatin Liyana Izzati, Othman Ayu Nazreena, Wan Sembok Wan Zaliha, Che Mohd Zain Che Radziah, Zainal Zamri and Subramaniam Sreeramanan. (2013) A simple and efficient protocol for mass propagation of *Vanilla planifolia*. *American Journal of Plant Sciences* 4: 1685-1692.

Chapter – 6

Plantation Crops – Mechanization

Shinoj Subramanian

Mechanization is widely recognized as a means to reduce drudgery, reduce cost and improve efficiency of farming operations. Mechanization is also significant since labour is getting short due to urbanization and increased educational level of rural mass. It stabilizes farm system by work timeliness, increases yield by working precisely, makes multi-crop system feasible for farm management and causes improvement of health by release from heavy work and improvement of living conditions. It also releases from gender gap by saving time spent to farm work of woman and level up rural development by spreading of engineering knowledge etc.

Mechanization of plantation crops should be planned in the early planting stage itself as many mechanized farm operations require well connected pathways, roads etc. for access of machineries. Machinery made available for the plantation industry should be hardy in design. Having complicated hydraulics, and hydrostatic devices, transmissions with complex on board electronics attached will pose problems to the estates in maintaining this equipment at optimum levels of efficiency. Machinery(s) required for almost all plantation crops are similar except specialized equipments for crop specific jobs. The operations, which can be mechanized in the production and processing of plantation crops are classified into following broad categories:

1. Land Preparation
2. Tillage
3. Planting
4. Field Maintenance
5. Vegetation Control
6. Animal/Bird Control
7. Insects and Diseases Management
8. Irrigation
9. Manuring
10. Harvesting
11. Collection and Transporting
12. Integrated Mechanization

1. Land Preparation

Land preparation involves clearing of the land, leveling and grading and tillage operations.

Clearing of the land

Land clearing involves removal of existing native cover, including trees, bushes and boulders from land surface. The slash-and-burn practice of land preparation that farmers use traditionally causes land degradation and human health hazards. As an alternative to slash-and-burn, a mechanized, fire-free method of land preparation was evaluated on smallholdings (Denich *et al.*, 2004). Use of machinery like tractor propelled choppers for cutting fallow vegetation and chopping it for mulch eliminate the need for hard labour and fire for land clearing and it increases labour productivity. To eliminate or minimize competition for nutrients, moisture and light between the existing vegetation and the new plantation, felling is required. This operation uses crawler bulldozers equipped with front-end sharp angle or straight blades to cut and fell single trees at or near ground level (FAO, 1986). On level sites, trees should be felled in the direction, the trunk is leaning or in the direction in which most of the branches are pointing. On slopes, trees should be felled along the contour line or uphill. In large trees with a broad root system, roots on the side of the felling direction should be cut first. Roots are then cut on the opposite side from which the tree falls, and the hole is refilled. The tractor should approach the tree slowly with the blade fully raised until contact is made with the trunk and then the tree should be pushed with the engine set at maximum speed and full power. This operation usually requires some sort of excavation for exposing and cutting the broad root systems of large trees which result in extensive soil disturbance. Caution should be exercised when using this method on vulnerable sites like sloping lands with erodable soils. The heavy machinery and blades required for this operation usually cause soil compaction and should be avoided on shallow and stony soils. Vegetation clearing by chain technique employs two crawler tractors with front-mounted blades or rakes pulling a heavy anchor chain. In areas with large trees, the operation requires additional lead bulldozers or follow-up tractors equipped with a tree stringer to push over big trees with large root systems. The two tractors should move forward at the same steady speed, dragging the chain in a U-shape on the ground felling the trees inside the chain. Width of the operation area and the distance between the two parallel tractors should vary from 15 to 25 m depending upon height of the trees. In this way, the trees enclosed inside chain start falling from the back towards the centre of the area.

Clearing of vegetation must be followed by breaking the land to get a workable seedbed into which a crop can be seeded. Land breaking includes removal of roots, stumps and rocks. Natural vegetation should be retained in drainage paths (creeks and gullies) to reduce the risk of soil erosion. Roots are loosened by ploughing or disking the land. After breaking the soil, loosened roots are raked into long rows (windrows) using a flex harrow, side-delivery root rake or rotary-drum root rake. Depending on the field, rocks may have to be removed so the soil can be cultivated. Brush and stumps are piled into windrows using a brush-rake (bulldozer blade with spikes along the bottom edge) and converted into organic manure using composting techniques. Burning is also an alternative, however it is not recommended due to environmental pollution.

Leveling and grading

The practice of land grading or land leveling consists of shaping the soil surface within a field to improve surface drainage and eliminate areas where surface water may pond. It is also necessary to make suitable field surface to control flow of water to check soil erosion. Usually the field is not graded to a true level surface, but a gentle uniform slope is maintained to meet the requirements of irrigation and drainage.

Land grading is beneficial in unirrigated areas to conserve moisture, reduce soil erosion and to provide for surface drainage. In low rainfall areas, land grading helps in more infiltration of water and assures more even moisture distribution. Grading eliminates small cuts and furrows, which have potential to get converted to channels or gullies and cause soil erosion in long run. Criteria for land grading are influenced by the characteristics of soil profile, prevailing land slope, rainfall characteristics, cropping patters, methods of irrigation and other special features of the land including preference of farmer. Any land development programme should consider location of boundaries, irrigation and water supply system, drains and farm roads. Land grading and leveling are usually done by cutting high spots and filling low spots. Fertile top soil need to be conserved by scrapping and replacing. The activity requires use of cultivation and land leveling equipment like scrapers and tractor drawn equipments. Rough grading is done by bulldozers, tractor drawn carrier type scrappers and small scrappers drawn by tractors. Land smoothing is done by tractor drawn land planes.

Both crawler and rubber-tyred wheel type tractors are used for land grading. They vary in size from medium size farm tractors to heavy crawler tractors. The crawler tractor is employed when heavy earth movement is required. The crawler track gives superior traction on varying soil and ground conditions. It is more capable than wheel tractor in loading an attached scrapper and cutting and pushing of earth when equipped with dozer blade. Since its speed is limited, it is not very suited when more haul distances are involved. Rubber-tyred tractors are suitable for earth moving when the emphasis is on speed rather than drawbar pull. They can be used with advantage on land grading jobs requiring long distance hauling.

Bulldozers consisting of crawler tractors equipped with dozer blades, are frequently employed in cutting and pushing earth to short distances. They are suitable for rough grading when the haul distance does not exceed 25 m. Tractor drawn scrappers are a variety of scrappers in land grading. They range in size from the terracer blade to heavy carrier type scrappers. The carrier type scrapper is widely used for large scale land grading operations.

Tractor drawn disc ridger is an effective implement for making border ridges. It consists of four discs in two gangs so assembled as to gather soil into ridge as the implement is pulled forward. The disc gangs can be moved in or out on the bar to make larger or smaller ridges. The V-ditcher is a useful implement for the construction or cleaning of field channels for irrigation and drainage. Operation of the V-ditcher in constructing a channel is much like that of an over-size mould board plough is used to make a furrow where the line is staked. The plough is run again on the same furrow, but in the opposite direction, throwing the soil out on the other side. The V-ditcher is then used in the furrow to enlarge it into a channel.

Drainage excavators equipped with V-shaped cutter blades and tractor mounted digger loaders are frequently employed for constructing large size irrigation and drainage canals. Dragline excavators with bucket type cutter heads are used for digging canals and for cleaning them. Self-propelled trenching machines are used for making trenches for laying tile lines and underground pipeline irrigation systems.

The choice of equipment for land grading and field layout depends on the quantity of earthwork involved, type of power available, economic feasibility of the machine and size of farm (Michael, 1999).

2. Tillage

Tillage helps in incorporation of plant residues, fertilizers etc., in seedbed preparation, weed control and soil and water conservation. Tillage is classified by the quantity of soil disturbance that occurs during the event. Primary tillage often inverts the surface soil and buries most of the plant residues. These are often the first operations done for land preparation. In primary tillage, the soil is cut and shattered with relatively deep penetrating tools (15 to 75 cm) and leave a rough surface texture. Ploughs, listers, bedders and rotary tillers are the main implements used for primary tillage. Secondary tillage implements pulverize, level and firm the top 5 to 15 cm of soil, which disturb less soil and bury less residue. Disk harrows, cultivators and rotary tillers are the main implements used for secondary tillage. With the introduction of powerful tractors and modern soil tilling implements, farmers' ability to stir the soil has increased dramatically. This has spawned concerns about soil degradation, most notably organic matter depletion, compaction and erosion.

Primary tillage implements

There are different types and designs of ploughs like mould board plough, disc plough, rotary plough, listers, chisel or subsurface plough, sub-soiler and sweeps ridger used for primary tillage. Mould board ploughs typically bury 90-95% of the residue, where as disc ploughs bury 80-90% of residues. Chisel ploughs are residue friendly, leaving 70% or more of the residues on the soil surface. Spacing of the chisels and speed of the tractor determine the actual amount. Subsoilers are used to break through compacted tillage zones to improve water infiltration and root penetration. These plows maintain 90% or more surface residues.

Mould board plough

The mould board plough cuts loose furrows slice, inverts the furrow slice more or less in pulverized form. It is used for covering green manure crops. It is also suited for covering or inverting grass into the soil immediately after the rains. Tractor operated mould board ploughs are either trailed type, mounted type or semi-mounted type. Trailed type is also known as pull type of mould board plough and is a complete unit in itself supported on two or three wheels. In such units, depth control levers are provided to regulate depth of plough. Mounted type ploughs are attached to a tractor. Use tractor lift for lowering or lifting. Semi-mounted type is partially supported by the tractor three-point linkage and partially by wheels.

Different soil conditions require mould boards of varying shapes and sizes to do a good job of ploughing. Texture of soil, amount of soil moisture and extent of vegetative cover on the surface determine the amount of soil pulverization. Extent of pulverization depends on shape of mould board. A long gradual curving mould board turns the furrow gently and does not break the soil much. Short, abruptly curved mould boards twist and shear the soil and pulverize it. There are different types of mould board ploughs classified function wise. General purpose ones are the best for all round general farm use to give thorough pulverization. Stubble type is adapted for ploughing an old ground where good pulverization is desired. It has relatively short and broad mould board that is curved rather abruptly near the top. Sod or breaker type is used in tough sod (grass land) where it is desired to turn the furrow slice completely so that the grass may not continue to grow. Slat type is preferred for more sticky soils where it is difficult to get the mould board to scour. High speed mould board ploughs are used on tractor ploughs for general farm use.

Disk ploughs

Disk ploughs are favoured in areas where the climate is dry and where the soil is rough and stony as friction is reduced due to its rolling bottom instead of the sliding bottom of the mould board plough. Such soil conditions do not permit the operation of mould board ploughs to good advantage. Disk ploughs also work well in heavy clay, hard pan and loose sandy soils. It leaves the trash on top of the ground to conserve moisture. It is also preferred for land infested with heavy growth of vegetation and for land requiring deep ploughing for reclamation purposes. Contrary to the mould board plough, penetration of disk plough depends mainly on weight of plough as a whole. Tractor drawn disk plough weighs 180 to 540kg / disk (Ojah and Michael, 1998). Tractor drawn disk ploughs may have one to seven disk blades attached to a common axis. When the plough is pulled, the individual disk rotates on its own axis. The furrow slice rides along the curvature and is pulverized to some extent before being thrown in getting more pulverization and better scouring of furrow slice. In order to cut a deeper furrow slice, the tilt angle of the disk is reduced, that is, the disk is set in a more vertical position. The other method of increasing the penetration is by adding weights to plough frame. If the soil condition is favorable, the tilt angle should be increased to achieve better turning of furrow slice. If the soil condition is not favorable, increasing the disk angle improves penetration but the width of cut is reduced.

Other primary tillage tools

Listers are preferred in dry land farming areas and where the soil is crumbly. A lister works much like a mould board plough, except that it has double mould boards and throws the soil in two directions. This leaves a furrow in which the crop is planted. The sub-soiler is a tool that breaks the ground at greater depth, breaks the hard pan, and does not pulverize the surface soil as much as other tools. This is provided with a heavy beam to which a sweep or chisel is attached. When it is pulled through the soil, it breaks open the soilup to considerable depth. Trash and mulch are left on top. Sub-soiling improves the moisture and air permeability of soil. Sweeps are sub-surface tillage tools that work at a shallow depth and do not turn the soil. For the best results, the sweeps are staggered to permit trash to flow through without plugging. The sweeps should be set to operate almost flat and their sharpness should be maintained. A ridger is required for row planted crops. It is used for making field furrows or channels, earthing up and similar other operations. It generally has a V shaped share either rigidly fixed or hinged to mould boards. Depending upon size of furrows, it is also known as a double mould board or middle buster plough. The short beamed ridger is provided with a guage wheel attached to the front end of the beam. This guage wheel facilitates movement and control of ridger.

Secondary tillage implements

Harrows, cultivators, hoes, clod brakers and levelers. come under secondary tillage implements. The function of disk harrow is to pulverize the soil, leaving the surface mulch and a compact sub-surface. The harrow has concave blades which can be made to cut either shallowly or deeply by mechanically or hydraulically controlled mechanisms. Disk harrows are made with either double action or single action gangs. Double action gangs may have either tandem disk arrangements or offset disk arrangements. Single action disk harrow tends to ridge the soil whereas the other type counteracts the ridging effect. The tandem harrows have four gangs. The two front gangs throw the soil outward and the rear gangs shift the soil inward. The offset disk harrow has only two gangs placed one behind the other. Disks of both gangs face in opposite directions, thus throwing the soil in both directions. No soil remains uncut by the offset disk harrow.

Another type of harrow is drag harrow, which can be classified into two, viz. spike tooth and spring type. Spike tooth harrows are either rigid or flexible. The flexible type tractor drawn can be rolled up for transporting. The basic frame of a harrow may be triangle. It has pointed steel pegs about 23 cm long with their pointed ends towards ground. Each peg is rigidly clamped with the help of a U-bolt to the cross bar of the frame. Spring tyne tractor drawn harrows have looping, elliptical or spring like tynes. They penetrate more comparing to spike tooth harrows and is suitable for soils where obstructions like stones, roots and weeds. are hidden below the soil surface.

Cultivators are used for manipulating the soil after the planting is done to control weeds, to prevent surface evaporation losses, to maintain soil in good tilth, to achieve rapid infiltration of rainfall and to ensure adequate soil conservation. Tractor drawn cultivators are mostly mounted, generally front mounted types.

Care and maintenance of tillage implements

Sharpening of working edges of all types of ploughs are important to keep them efficient. Wheel of tractor drawn ploughs should be properly lubricated during working season. Nuts and bolts should be checked for proper tightening. While transporting the plough, precaution should be taken not to cross the ditches at high speeds, otherwise plough wheels get damaged. Plough levers should be occasionally greased to prevent rusting. As far as possible, the plough should be covered during rainy season or at least the metallic surfaces should be coated with a thin layer of grease. A mixture of used crank case oil and grease provides a good coating material. Before the start of the season, ploughs should be checked for adjustment, sharpness of cutting edges, repairs of the broken parts, replacement or repairs of the worn out parts and replacement of missing parts like nuts, bolts, pins and grease nipples. It is always helpful to store a few important items during the idle season so that no difficulty is encountered particularly during the working season, if any part is damaged due to unforeseen circumstances.

3. Planting

If the terrain is not too steep and other soil conditions are congenial, mechanized planting would be economical. Hand planting is more satisfactory in small areas, on steep terrain, and on extremely rough rock- or debris-covered land. Hand planting also is necessary where size of seedlings prevents proper planting by machine. Mechanical planting usually is not warranted for less than 1,000 seedlings because of the time involved in acquiring, transporting and using the equipment.

The planting machines are of three general types (Slusher, 1993). The floating type is attached to a tractor by a three-point hitch so the entire machine can be raised from the ground by hydraulic lift on the tractor. The semi-floating type has its front end carried by the tractor and its back end carried on wheels; it cannot be lifted by the tractor. The trailer type has all or nearly all its weight carried on its own wheels.

Most planting machines consist of a rolling coulter, a trencher, an operator's seat and packing wheels. The rolling coulter is a disc-like cutting wheel that cuts through the ground surface and severs old roots, trash and other debris. It also serves to automatically raise the trencher over buried rocks and logs. It should cut at a depth slightly below the trencher point. The trencher on many planters is essentially a moldboard plow with the moldboard cut away and replaced by metal sides moving parallel with the landside. The trencher's purpose is to make an opening in the soil sufficiently wide and deep to receive roots of the seedling tree in a way that they are not twisted, L-shaped, or crowded into a single plane. The front of the trencher is shaped so that it lifts the soil from the trench upward and to the side, where it remains until the trencher has passed. It is deposited back into the trench from above, falling downward among the roots of tree being planted, carrying the roots downward while maintaining good distribution of the branch rootlets. Packing wheels usually are two wheels following the trencher that are set on an angle to force the furrow to close around the seedling and to pack the soil around the tree. Weight of the operator normally provides the weight necessary for proper soil compaction.

Tractor speed affects quality of planting. It is important to have a tractor with enough power to pull steadily at slow speeds. Some factors that determine planting rates are soil quality and condition; presence of stumps and stones size and condition of planting stock; and skills of the planter operator.

Different designs of planters are available suited to specific requirements. A prototype trailed type transplanter for oil palm seedlings with a configuration consisting of main chassis, seedling bin, seedling planting assembly, operator compartment and associated hydraulic system is designed and developed at the Department of Biological and Agricultural Engineering, University Putra Malaysia (UPM), Malaysia. The transplanter is trailed from a 4-wheel tractor having at least 85 hp (63.4 kW) engine size and equipped with a seedling bin that could accommodate 20 oil palm seedlings / operating trip. Two operators are required in the involved transplanting operation; a driver for the tractor and an operator for transplanter. The driver drives the tractor-transplanter combination in the field, while the operator operates the hydraulic control system of transplanter to integrate preparations of planting hole, placement of seedling in the prepared hole, covering of seedling in the prepared hole, and compacting of soil around the planted seedling (Pebrian and Yahya, 1993).

The post hole digger is a power operated auger for digging of holes and pits. It consists of a shaft with pointed bits, over which a metal strip (blade) is welded in a spiral form to form a large size screw. It is either operated by an auxiliary engine or with PTO shaft of a tractor through gear box. Power tiller operated post hole diggers are also available. The augers are available with diameter ranging from 150 to750 mm.

4. Field Maintenance

Field maintenance operations include overall Monitoring of the field, Vegetation Control, Animal/bird Control, Insect and Disease Control, fertilizing, disaster management/reclamation and Irrigation.

Field monitoring

A regular program of monitoring at specific time periods ensures the success of a plantation. Plantations should be evaluated more frequently during the first growing season. All monitoring should be based on a check list and suitable corrective actions should be taken as and when required. During evaluation process, crops ready for harvesting, harvested area, infestations by insect, pest, disease, animal, competing vegetations and climatic parameters. need to be noted down. In advanced systems, satellite remote sensed data are used for such applications. The advantages of using satellite remote sensed data are that it can be frequently collected for large areas and that the data are unbiased. It, therefore, lends itself well to agricultural applications (Krishna Rao *et al.*, 1999; Narciso and Schmidt, 1999; Noonan, 1999; Sirvastva *et al.*, 1999). The use of satellite imagery to monitor area harvested by small scale sugarcane farmers was discussed by Gers and Schmdit (2001), which can be extended to many other crops. In most of plantation crops being cultivated in remote locations where large area is covered, use of satellite imagery is more promising.

5. Vegetation Control

Success of a planting will often be determined by control of competing vegetation before and after the crop planting. Good site preparation will get ride off weeds, but weed competition may need to be controlled for at least until the crop is well established.

Herbicide application: Herbicides are often the most effective method for follow-up of weed control. Robotic systems are employed for herbicide spraying without harming the crop. Lee (2003) tested robotic weed control systems for spraying in-row weeds and found that it correctly identified and did not spray 75.8% of the tomato plants and correctly sprayed 47.6% of weeds. A real-time robotic weed control system was demonstrated (Slaughter *et al.*, 2000) in commercial cotton fields correctly spraying 89% of nutshedge weeds between cotton plants in the seedline while correctly identifying and not spraying 79% of the cotton plants. The weed removal efficiency of this robotic system was comparable to or better than hand-hoeing crews which also make misclassifications and leave weeds behind, eliminating only about 65 to 85% of the weeds on an average. In addition, a significant reduction in off-target chemical application was observed, requiring only 25% of the chemical normally applied by a precision band-spray system. This technology provides significant environmental and economic benefits, while optimizing agricultural production.

Thermal weeding: Flame weeding is a type of thermal weed control in which propane gas burners to produce a carefully controlled and directed flame that briefly passes over the weeds. The intense heat sears the leaf, causing the cell sap to expand and disrupt cell walls. Flame weeders come in a range of human and tractor-powered models. Market-farming equipment options include hand-held single-torch flamers, as well as push-wheeled multiple-torch flamers mounted under a flame hood. Tractor-powered kits are available in 2, 4, 6, and 8-row models, with or without a flame hood; other options include a complete toolbar setup with accompanying cultivator attachments for between-row mechanical cultivation. In the 1980s and 1990s, flame weeding made a rapid comeback as a non-chemical weed control technique, especially among organic farmers (Steve Diver, 2002)

Infra-red weeders are improved version of flame weeders, heated by a propane torch, but the flame is directed toward a ceramic element or steel plate that radiates at temperatures of 1800 to 2000°F. The danger associated with an open flame is thereby minimized. The mechanism of weed control is the same as in flame weeding; cell contents plasma and proteins are disrupted and the plant wilts down and dies. Infra-red heaters are available in hand-held, push-wheeled and tractor-mounted models. Some of the tractor models feature the injection of forced air to increase the effect. While infrared weeders appear to be a promising new thermal weed control tool, the equipment expense remains prohibitive for many smaller-scale market farmers.

Kerpauskas *et al.* (2006) reported the possibilities of weed control by water steam application using a mechanical device as an improvement over flame methods. It was found that thermal weed control by water steam offers an efficient, reliable, ecological and economical means of weed extermination. The method was found suitable for destroying short life weeds and not for perennials as they can re-sprout.

Plasticulture: Light exclusion techniques are widely used in small scale horticultural organic crop production. Materials used include black plastic, carpet, straw, cardboard, tree bark and wood chippings. The effects of such mulches on weeds, pests and crop yields, have recently been studied by several workers. In their study, Lennartsen et al. (1990) compared different surface mulches for their effect on clearing an established grass pasture and subsequent crop yields for organic horticultural produces.

Mechanical weeding: Mechanical devices are available for physical removal of weeds. Control may be suitable for some post-planting situations. Inter-row cultivator is one among mechanical devices. Different designs of inter-row cultivators are available in the market. The basic requirements for successful inter-row cultivations for weed control are: i) cutting or uprooting weeds, and then either completely bury them or leave them on the soil surface for desiccation; ii) Protecting the crop plants; iii) Provision for controlling implement direction and implement depth; iv) Maintenance or improvement of soil conditions.

Cultivators require careful adjustment to ensure optimal performance. To kill a maximum number of weeds, cultivators should be operated as close to crop rows as possible without injuring the crop. The effective use of cultivators requires a fair amount of experience and careful observation. The effectiveness of cultivation is directly influenced by cultivation depth and degree of soil moisture. Cultivation that is too shallow may spare weeds and cultivation that is too deep increases the risk of crop damage. Working depth can be adjusted by means of wheels attached to the frame or the three-point hitch. Use of weights and reduction of tractor speed can also increase the operating depth. Cultivating when the soil is too wet leads to clod formation and may not destroy weeds. Optimal level of soil moisture depends on cultivator type. Rotary hoes and harrows are the best suited to moist soils.

In general, two types of cultivators are required for effective weed control: one at pre-emergence or early post-emergence and a second later in the season. The vulnerability of crops to damage depends on their growth stage. For example, legumes are the most vulnerable at the hook stage, when cultivation reduces yield. Cultivation is not only effective in controlling weeds; it also benefits the crop by breaking up the surface crust, aerating soil, stimulating activity of soil microflora, reducing the evaporation of soil moisture and facilitating infiltration of rainwater.

Shallow disking or rototilling between rows is effective if care is used to avoid damaging the trees and their root systems. Mowing reduces weed maturation and its seed production and minimizes rodent habitat, but it may also stimulate grass root growth and intensify competition for soil nutrients and water. Hand or mechanical cutting of woody vegetation may effectively release young weed seedlings, but repeated treatments may be needed due to stump sprouting.

6. Animal/Bird Control

Most plantings experience some type of animal damage (e.g., browsing, rubbing and feeding), however the severity of that damage will vary across planting sites and between crops. Many different techniques can be employed to discourage severe wildlife damage, in which mechanical devices and methods are proven for their performance. Fences-both electric and ordinary and tree shelters. create a physical barrier between the seedling and animal, where as ultrasonic, sonic, visual scare devices and roost inhibitors repel animals. Repellents rely on fear, conditioned avoidance, or taste to discourage animal browse. Repellents work for short periods, but their effectiveness is reduced with time. These devices remain intact to be effective and maintenance is often required.

Fencing: An efficient fence system should be designed by considering factors like purpose for which it is installed, species of animals to be protected, location in the farm for the best performance (Michael *et al.*, 1990). A permanent fencing is required with good quality building materials all around the plantation where as a low cost temporary fencing is enough for nurseries. There are different types of fences suiting to different requirements.

Woven wire fences consist of a number of horizontal lines of smooth wire held apart by vertical wires called stays. The distance or spacing between horizontal line wires vary from as close as 38 mm at the bottom for small animals, to as wide as 230 mm at the top for large animals. In general, the spacing between wires gets wider as the fence gets taller. Woven wire is available in many combinations of wire sizes and spacings, as well as a number of horizontal line wires and fence heights. The height of most woven wire fencing materials ranges from 660 to 1220 mm. The fence height should be selected based upon the animals size and their jumping ability. Stay wires should be spaced 152 mm apart for small animals and 305 mm for large animals.

Barbed wire fences are made of two or more strands of smooth, galvanized-coated, steel wire twisted together with two or four barbs spaced every 10 to 13 mm. They are generally classified as either a standard or suspension barbed wire fence. Standard barbed wire fences usually have three to five strands of barbed wire stretched between posts that are spaced between 4.5 to 7.5 m apart. The suspension fence has 4 to 6 strands of wire stretched tight so there is no more than 76 mm of sag between posts. Depending upon the topography, line posts are generally spaced between 24 to 36 m apart. The wires are held apart by twisted wire stays spaced 4.8 m apart. Wind or animals hitting the fence cause it to sway back and forth. This swaying motion keeps animals away from the fence.

High-tensile fences are constructed mostly with 12 1/2 or 14 gauge Class 3 wires which have tensile strengths and flexibility to bend, wrap, tie in knots or clamp with crimping sleeves. Wires are held in tension along wood, fiberglass, insulated metal posts or a combination of posts and battens or droppers. Tension in the wire is maintained by permanent in-line strainers. A tension indicator spring is used to indicate wire tension. High-tensile wire fences offer several advantages over conventional fencing like ease to construct, durability, low cost and less maintenance.

Electric fencing: Electric fences can be an effective, safe and inexpensive means of providing both temporary and permanent fencing if they are constructed properly and energized with a properly sized controller. Electric fencing does not need to be strong because it seldom comes under pressure, but it must be well designed and constructed to absorb the impact of animals. Several advantages of electric fencing are low cost, inexpensive to operate and can be used to extend the life of old permanent fences. Electric fencing is based on a simple principle that the energisers have two terminals viz., "Fence" and "Earth/ Ground". A thin, insulated wire is charged with electric energy from the "Fence" terminal. A second, thin wire is connected to an earth. Whoever/whatever closes the circuit between the two terminals receive a substantial, safe (low amperage/high voltage) electric shock. It is very important to keep weeds and grass cut away from the fence, especially when using low impedance controllers. If grass and weeds are allowed to touch most polywires, the charge produced from low impedance controllers cause the small-diameter wires to burn in two. Polywires with stainless steel wires are more durable, but electric conductivity is lower. Aluminum conducts electricity better, but it breaks more easily. Aluminum, stainless steel and high-tensile wire can also be used. To effectively train animals to stay away from an electric fence, the animals need to see the wire as they feel the shock. Tying pieces of white cloth or brightly colored plastic ribbon will help make these wires more visible.

Safety precautions are to be keenly observed while operating an electric fence. Some of them are (McKillop *et al.,* 1995):

- Always seek planning approval from appropriate authorities before installing a fence alongside any road, railway or telecommunication line.
- Always display warning signs at less than 50 m intervals along a fence where there is public access.
- Always ensure the fence is well insulated, to prevent fence posts or any other structure becoming electrified.
- Always consider the risk of an animal or person becoming entangled in live fence wires.
- Always charge the battery in a well ventilated area and in accordance with manufacturer's instructions.
- Always disconnect the battery from the fence energiser when charging from the mains.
- Always disconnect the power supply before working on the fence.
- Never site an energizer earth electrode within 10 m of any mains earthing system.
- Fence wires must not be attached to power line poles.
- Never connect more than one energizer to the same fence.
- Never electrify barbed wire.
- Never install a fence under and parallel to an overhead power line as the induced voltage generated on the electric fence conductor wire may be hazardous if crossing a power line, do so at right angles.
- Never install an electric fence above the recommended height (currently 2 m) when crossing a power line.

- Never attach electric fence wires to mains power line poles.
- Never overcharge the battery.
- Electric fence wires when crossing a public highway must be at least 5.45 m above the ground.

Repellents: Ultrasonic pest control works the best when the targeted animals are bats, rats, mice, rodents, squirrels, raccoons or other small to medium sized animals. The irritating sound waves produced by ultrasonic sources will not harm animals, but will attack them aggressively in any enclosed or semi-enclosed areas where they roost. The adjustable variations in pitch and sequencing make it difficult for animals to acclimate to any constant sound pattern.

Some of the sonic bird control devices feature birds' distress cries on a microchip. The birds perceive danger when they hear these sounds. They become agitated and disoriented: they think their flock mates are in trouble so they flee the area. Flock migration, feeding, roosting and loafing habits help determine the best location and settings for birds that gather in large numbers. Such devices are species-specific and programmable. The birds have to be targeted with a species-specific approach - not scattershot. There will be different bird sounds recorded on the unit. In addition to programming the selected birds, the volume, time off periods (between sound blasts), and whether the bird sounds play in random or sequential order can be programmed. These programming options make it virtually impossible for birds to get used to any one pattern.

7. Insects and Diseases Management

All plantations experience some degree of insect and disease damage. Well tested and proven insecticides and pesticides are available to control these. There are a number of sprayers available for effective application of these, which distribute the chemical uniformly on plant surface with provisions to vary the quantity applied. The liquid should be broken into fine droplets with provisions to control droplet size since if the droplets are fine, it may be carried away with wind, and if too big, the spray may not cover entire plant. The sprayer unit essentially consists of a pump, tank for storing the liquid, agitator, air chamber, pressure regulator and pipes and nozzles. Agitator is to mix the chemical occasionally to prevent settling and air chamber is to control pulsation of the fluid when piston pump is used. The pressure regulator helps in adjusting the spraying pressure.

In areas where there is shortage of water and those farms, which solely depends on rainfed agriculture, the chemicals are applied in powder form, which does not require any dilution using equipments known as dusters. This operation is called dusting. Since the adhering properties of powder is inferior to that of diluted chemicals, this method is the best suited for humid climates.

There are mainly three types of sprayers viz., manual sprayers, power sprayers and aerial sprayers. All such types of sprayers can be further classified into low pressure sprayers and high pressure sprayers. Hand operated sprayers are suitable for small holdings and power operated sprayers are suitable for large areas and aerial spraying for massive spraying.

Manual sprayers: Bucket type sprayer, knapsack sprayer, compression sprayer, rocker sprayer and foot sprayer. are the most common manual sprayers. Bucket type sprayer consists of a single acting or double acting pump which sucks chemical from a bucket, convey through a flexible piping and spray through a nozzle fixed to the end of a metallic handle. Knapsack sprayer can be mounted in the back of the operator so that he/she can spray with right hand while operating handle of pump with left hand. An air chamber is provided to maintain uniform pressure. Compression sprayer consists of a tank partly filled with chemical and remaining portion filled with compressed air by operating the handle provided at top. Pressurized chemical is sprayed through a nozzle. The air filling needs to be done at frequent intervals. Rocker sprayer is a long lever high pressure sprayer consisting of an air chamber, piston pump and a lever fixed to a wooden platform held tightly to the ground by foot pressure of operator. The piping and filter are placed in a bucket-containing chemical and the chemical is sucked when the piston pump is operated by moving the long lever to and fro. The discharge line is held by another operator for applying chemical to the target. There is also provision for two discharge lines to increase field capacity. The spray chemical can be delivered up to a height of 10 m. and this sprayer is suitable for spraying tall trees like coconut, arecanut and rubber plantations. Foot sprayer is also similar to rocker sprayer except that the pump barrel is mounted on a steel frame and the pump is operated by a foot lever. This also can reach a height of 10 m.

Power sprayers: Power sprayers are used for high discharge and high pressure applications to cover large area. Portable power sprayers use petrol engine and can be easily taken to the field by attaching to shoulder of operator. In still bigger models, the entire unit is fixed to a trolley and taken to the field by manual towing. Still higher capacity models are attached either to a power tiller or a tractor. The motorized knap sack sprayer having an overall weight of 11 kg, operated by a 1.2 hp motor can have a horizontal spraying range of 10-15 m. and a vertical range of 6-8 m. The field capacity of the unit is 3 ha/hr. Motorized tree sprayer, operated by a 3 hp petrol engine which can be carried by two persons in a stretcher type frame is able to spray upto 20-25 m. vertically with a field capacity of 3-4 ha/day depending up on the plantation. Power tiller mounted orchard sprayer consists of a pump, trailed type main chasis, chemical tank with hydraulic agitation system, cut off device and boom equipped with turbo nozzles. Depending up on plant size and raw spacing, orientation of boom can be adjusted. It has a capacity to cover 0.4-0.7 ha/h. Blower sprayer attached to a tractor is suitable for spraying plantation crops, which consist of high capacity tank, axial or centrifugal fan to generate air flow and adjustable spray heads mounted on the three point linkage of a tractor. The pump is operated by the tractor power. The spray heads are arranged to cover two rows of trees and can be adjusted to cover tall trees depending up on requirement. The pump sucks the chemical and atomizes and at the same time the droplets meet the air stream to take it to the target. Tractor mounted aero blast sprayer is also a similar unit operated by tractor P.T.O, but having a manual adjustment of the blast outlet duct to direct the blast across 180 degrees. It requires a tractor of 35 hp or above power and the machine has a field capacity of 1.5-2 ha/h (Bharadwaj *et al.,* 2004).

Aerial sprayer: Aerial spraying is accomplished through air planes to cover a larger area. The chemical is either dropped down by gravity or pumped through nozzles using a centrifugal or gear pump, which gets driver from a wind driven propeller. Aluminium tanks are commonly used to reduce load.

Dusters: The duster basically consists of a hopper with agitator, a metering device, a blower unit and a discharge pipe. The chemical powder is stored in the hopper and the agitator prevents clogging. The agitator may be mechanical stirrer or a pneumatic device. Metering device controls flow of chemical out from hopper to discharge pipe. The metered chemical is put to an air stream, the flow rate and velocity of which decide the throw range of chemicals. The deflector provided at the end of discharge pipe spread the chemical uniformly. Manually operated and power operated dusters of different capacity are available.

Spraying practical considerations

The uniformity of spraying in a field is important to ensure effective application of chemicals on plant surface without wastage, for which following points need to be observed:

- The distance from nozzle to the plant surface should be constant.
- The angle of spray should be uniform.
- The nozzle spacing should be constant in case of multiple nozzle sprayers.
- Worn out nozzles should be replaced in time.
- The spray boom should be level and parallel to the plant surface.
- Periodical cleanings of the nozzle need to be ensured.

Nozzles are important components of the sprayer, which determine the spraying pattern and flow rate. Different spraying patterns like hollow cone spray, solid cone spray, flat spray, even spray and flooding spray can be achieved by selecting suitable sprayer nozzles. In high capacity sprayers wherein multiple nozzles are used for spreading the chemical, the uniformity of discharge should be checked periodically by calibration. This involves filling the tank with water and collecting the sprayed water in individual containers from each nozzles. If the quantity is uneven, such nozzles need to be replaced to ensure uniform spraying in the field.

Disaster Management/ Reclamation

The owners of plantations, which are prone to disasters like cyclones, tsunami and floods should be vigilant and suitable protection and reclamation techniques need to be planned. The technology for rejuvenation of cyclone affected oil palm plantations was standardized and replanting was economical comparing to abandoning the uprooted palms. (Rethinam, 2001). Among the machineries tried for replanting like earth movers, lever cranes and tractors with iron ropes. JCB make earth mover was efficient and economical for pit making and lifting of the palms.

8. Irrigation

Irrigation is the artificial application of water to the soil for assisting in growing crops by replacing missing rainfall in periods of drought. Controlled and judicious application of irrigation water is essential since natural water resources are getting scarcer and also there is need to conserve the energy sources. Controlled water application improves efficiency of water intake and reduces losses.

Water pump is the most important component of any irrigation system, operated by a prime mover either electrically operated or fuel operated. Solar and wind powered irrigation pumps are also in practice particularly where there is shortage of electricity. Selection of a pump and prime mover the best suiting to the required irrigation system will not only reduce investment and operating costs but also improve the efficiency of the system. Tractor or power tiller operated mobile pumps are also popular, which can be shifted from field to field. Hydraulic rams can be used for lifting irrigation water in plantations in remote hilly areas. Plantation crops being covered by large holdings in remote locations in general may have limited access to conventional power sources and hence harnessing non-conventional energy sources for irrigation water lifting is worth considering.

Non-Conventional energy for water lifting

Wind mills: Wind mill essentially consists of a vane rotating in an elevated tower with piston of a piston pump connected eccentrically to the vane wheel. The pump cylinder is connected to water source through a piping. When a sufficiently strong wind blows, the piston pump works and the lifted water is stored in an elevated tank for further use whenever required. The operation of windmill would be seasonal and occasional. The capacity depends on frequency and velocity of wind. Operational cost of the unit is practically nil and it is suitable for lifting water from shallow and moderately deep wells. A wind velocity of over 6.5km/h is necessary for satisfactory operation of wind mills (Michael, 1999).

Solar pumps: Solar pump system runs with help of sun energy through photo-voltaic module. The system consists of photo voltaic (PV) module, D.C. Pump, module mounting structure with hardware and cables and piping. While installation of solar pumps, it must be ensured that neighboring buildings and trees do not cause shadows on the PV module during any time in the day. While installing the system, tracking should be done by adjusting angle of PV module in a location where maximum delivery of water is found. Generally the panels should be placed towards North-South direction. Daily cleaning of the PV module surface by pouring water should also be ensured. To ensure more delivery of water, the module mounting structure may be positioned three times in a day such that the modules are directly facing the sun. Even though, the initial investment of solar pump is high, it has low maintenance cost and practically no operating costs.

Types of pump

Positive Displacement pumps: These pumps are suitable for moving very thick liquids, or creating very high pressures. They are mainly used in fertilizer injectors and spray pumps. They are used for high head applications, but their capacity is small compared to other types of pumps. Reciprocating pump is a kind of displacement pump in which an air tight piston or plunger moves up and down in a cylinder to suck and displace water. Size of the cylinder and length and number of strokes determine capacity of the pump. Another type of displacement pump is rotary pump, which mainly finds its application in boosting pressure in existing water lines to suit the operating pressure of irrigation system used. Rotary pump consists of two counter rotating metallic gears enclosed in a metallic casing, which squeeze water between the gear teeths and force water through discharge opening.

Variable displacement pumps: The variable displacement pump pumps more water at low heads and less water at high heads. Hence it requires more power if water head is low. Different types of variable displacement pumps like impeller, centrifugal, mixed flow and propeller pumps are largely used in irrigation pumping.

Centrifugal pump is the most common irrigation pump among these, which consists of an impeller with rotary vanes rotating inside a casing. The water is drawn inside through the center of casing due to creating of low pressure at inlet as the impeller rotates. The drawn in water is forced through the outlet. Depending upon the configuration of the vanes in the impeller, centrifugal pumps are classified into volute and diffuser/turbine pumps. Volute pumps are suitable for large capacity low head applications and diffuser pumps the best suited for high head conditions. Multistage centrifugal pumps are equipped with more than one impeller wherein the discharge from one impeller is directed to the suction of the next impeller. This is used for high head applications. The discharge capacity and efficiency are almost same. In all these types, there are different types of impellers like open, semi-open and enclosed. depending up on solid content in irrigation water. Open impellers are suited for water having considerable amounts of solids. Impellers may be made of bronze or cast iron for pumping ordinary water and gunmetal impellers are commonly used for brackish or salt water. Horizontal centrifugal pumps are mounted horizontally with their impeller fixed vertically in a horizontal shaft. This type of pumps is easily maintainable. It should be installed always above the water as close as possible to water table. The suction lift of such type pump should not exceed 4.5 to 6.0 m. (Michael, 1999). Horizontal centrifugal pumps are the most common in irrigation. Vertical centrifugal pumps are mounted vertically with their impeller fixed horizontally in a vertical shaft. It can be lowered near to the water level or even can keep submerged in water. The cooling of submersible pumps would be effective and it improves pump efficiency. The irrigation pumps may be unibuilt, direct connected or belt-driven. The pump and the driver are mounted in a common shaft for unibuilt pumps and both are mounted on a common base plate and connected through a flexible coupling in direct connected and both are connected through a pulley and belt mechanism in belt driven. Belt driven is suitable when both pump and prime mover are located away.

Propeller and mixed flow pumps are suited for low head high discharge operations. They are suitable for pumping water from rivers, canals or streams. Propeller pumps develop most of the head by propelling or lifting action of impeller vanes on the liquid, where as mixed flow pump develops head partly by centrifugal force and partly by lift of impeller vanes. Jet pumps, suitable when high suction lifts (12-18 m.) are required is a combination of centrifugal pump and a jet mechanism for providing additional lift.

Hydraulic Ram pump: Hydraulic ram pumps are time-tested technology(s) that use the energy of a large amount of water falling a small height to lift a small amount of that water to a much greater height. In this way, water from a spring or stream in a valley can be pumped to an irrigation scheme on hillside. Depending on the difference in heights between inlet pipe and outlet pipe, these water pumps will lift 1-20 % of water that flows into it. In general, a ram can pump approximately one tenth of received water volume to a height ten times greater than the intake. A hydraulic ram pump is useful where the water source flows constantly and usable fall from the water source to the pump location is at least 91 cm (3 ft). The advantages of hydraulic ram pumps are that it is inexpensive, very simple to construct and easy to install, does not consume petrol, diesel or electricity, minimum maintenance and pollution free operation.

Selection of pumps

Different types and models of pumps are available in the market to suit various working environments, different quality of water, varying head and capacity required. The pump selection should be done in consultation with experts. Most of the pump dealers would be aware of pump selection. A basic idea is required for the consumer also.

The basic factors to be considered in a pump selection are safe yield of water source, water requirement of crop to be irrigated and power source available. Well yield is the volume of water discharged per unit time. The well characteristic curve and pump characteristic curve should be matched. While calculating total pumping head friction head due to pipe length, curves and bends also should be considered. Standard formulae are available for such calculation depending up on the material and size of pipes. Each pump manufacturer would provide a performance curve for their pump, which include information on speed of pump, power of motor, impeller size and pressure-flow relationship. Pump characteristic curve should be referred while selection.

Mechanized Irrigation Systems

Drip Irrigation: In this method, water is delivered at or near the root zone of plants, drop by drop. This type of system can be the most water-efficient method of irrigation, if managed properly, since evaporation and runoff are minimized. The process of mixing fertilizer in irrigation water and feeding through drippers is also possible, which is popularly known as fertigation. Major parts of a drip irrigation system consist of a water source, pumping mechanism/overhead tank, main pipeline, back flow preventer, pressure regulator, filter, lateral tubing, emiters and end caps. Drip irrigation methods range from very high-tech and computerized to low-tech and relatively labor-intensive. Success of a drip system basically lies on care and attention given by periodical cleaning, using filters to prevent clogging. Drip system is suitable for crops like oil palm, coconut, cocoa and mangoes.

Sprinkler Irrigation: In sprinkler irrigation, water is distributed within the field by overhead high-pressure sprinklers or guns, wherein the water spray somewhat resembles rain. The sprinklers or guns undergo a full-circle or semi-circle rotating action over the risers there by distributing the water spray uniformley around. Higher pressure sprinklers that rotate are called *rotors* and are driven by a ball drive, gear drive, or impact mechanism. Guns are similar to rotors, except that they generally operate at very high pressures. By carefully selecting the nozzle size, water pressure and riser spacing, the amount of water required for plants in a particular area can be supplied nearly uniformely at a rate suiting the infiltration rate of soil by sprinkler system. Sprinkler irrigation is not suitable for soils of low infitration capacity. It is the best suited for sandy soils. The sprinkler equipment is flexible that it can be layed in steep slopes and irregular topography also. Sprinkler irrigation is also suitable for saving plants from frosts and also from high temperatures. It is being increasigly used for irrigating high value crops like coffee, tea and cardamom. The distortion of water spray by wind and the impact of spray on ripening fruits are the main disadvantages of this system, which should be taken care accordingly.

Sprinklers may also be mounted on moving platforms connected to water source by a hose. Automatically moving wheeled systems known as *traveling sprinklers*. Most of these utilize a length of polyethylene tubing wound on a steel drum. As the tubing is wound on the drum powered by irrigation water or a small engine, the sprinkler is pulled across the field. When the sprinkler arrives back at the reel, the system shuts off.

Centre pivot irrigation: It is a method of irrigation in which the irrigation equipment rotates around a pivot. A circular area centred on the pivot is irrigated. Centre pivot irrigation is a form of sprinkler irrigation consisting of several segments of pipe (usually galvanized steel or aluminum) joined together and supported by trusses, mounted on wheeled towers with sprinklers positioned along its length. There are two systems of centre pivot irrigation, viz. fixed centre point and towable centre point. The later enables to minimise costs by enabling to irrigate more than one circle of different crops in the one season. The above mentioned equipment can also be configured to move in a straight line, where the water is pulled from a central ditch. In this scenario, the system is called a *linear move* irrigation system. Centre pivot irrigation system is suited for nurseries of plantation crops in large area.

Irrigation water filtration

The most important pre-requisite for any mechanized irrigation system is that the irrigation water should be clean, free from debris and sand to ensure clog-free functioning of pipelines and nozzles. Knowledge on the filtration system is essential for its smooth functioning. Even if small sand particles can pass through the system without clogging it, they cause wear on the equipment. Automatic valves contain very small water passageways in them which can become plugged resulting in the valve failing to either open or close. A small grain of sand caught in a spray nozzle results in a dry, dead spot in the field.

There are different types of filters. Screen filters are probably the most common filters and the least expensive. Screen filters are excellent for removing hard particulates from water, like sand. Screen filters are cleaned by flushing them with a stream of water or removing the screen and cleaning it by hand. The most effective method of flushing is the backwash method, in which the flush water is forced backwards through the screen for a very effective cleaning. Cartridge filters essentially contain a paper filter which works just like a screen filter. Most cartridge filters remove organics well because the paper texture is rough enough to snag the organic matter. Some cartridges can be washed, others need to be replaced when dirty. Media filters clean the water by forcing it through a container filled with a small, sharp edged, "media". In most cases, the media material is uniform sized crushed sand. The water passes through the small spaces between the media grains and the debris is stopped when it cannot fit through these spaces. Media filters are the best for removing organic material from the water. Disk filters are a cross between a screen filter and a media filter, with many of the advantages of both. Disk filters are good at removing both particulates like sand and organic matter. A disk filter consists of a stack of round disks. For automatic cleaning of the filter, the disks are separated from each other which frees the debris to be flushed out through a flush outlet. Centrifugal filters are primarily for removing particulates, like sand, water. They are good for situations where a lot of sand is present in the water as they do not clog up as quickly as other types of filters. The dirty water enters the filter where it is swirled around the inside of a cylinder. The centrifugal force causes the sand particles to move to outside edge of the cylinder where they gradually slide down the side to a holding tank at the bottom.

9. Manuring

Mechanical fertilizer application using ground vehicles in mature plantations meets most requirements in terms of even fertilizer application within fields, reduced labour requirements and rapid application. Tractor-mounted spreaders can be used conveniently on flat mineral soils. A tractor operated fertilizer spreader essentially consists of a container for holding the fertilizer, feed conveyor, beaters for breaking the flakes and a spreader with spiral auger like blades for distributing the fertilizer. The spreader is operated by the tractor P.T.O shaft. Ground driven spreaders are also available which are similar to the tractor operated ones, but the drive is from wheel of the spreader through suitable sprocket and chain. Fertilizer application is only effective if uniform coverage are ensured.

Field calibration is required to determine actual application rates. Calibration should take traveling speed, amount of fertilizer to be applied, width of spread, distribution pattern required and type of fertilizer used into consideration.

Organic manuring is getting more popularity in recent times as it reduces application of chemicals, which contaminate both soil and the produce. Disposing the wastes within the plantations by way of composting and applying in the field itself is a promising technique, the effectiveness of which, by and large, depends on use of shredding machines for size reduction of organic matter to ensure faster decomposition and easy transportation. There are many models of waste shredders operated by electrical motors and diesel engines. Power tiller operated and tractor operated waste shredders have the advantages of taking into the plantation where there is limited access to electrical power. It enables in-field shredding and composting of wastes. Quality of cutting blade to maintain sharpness for more time is the most desirable character of such waste shredders. Periodical removal of blades for sharpening is required for smooth running of waste shredders.

A mobile waste shredder for oil palm plantations consists of a chaff cutter with specially designed blades, fitted to trailer of a power tiller. The power from the power tiller flywheel was taken through a power take off mechanism and V belts. Discharge of the shredded waste is towards rear side of trailer while feeding is done from the side. This can as well be used for other plantation crops also. A tractor-mounted machine of bigger capacity designed for oil palm plantations takes power from a tractor P.T.O shaft. The machine can be attached to the three-point linkage for shifting the machine from one place to other and it has to be aligned in a location for connecting to the tractor P.T.O shaft. The alignment, connecting to P.T.O shaft requires 10-15 minutes.

10. Harvesting

Crops are harvested after normal maturity with the objective to take out the produces without much loss. It involves cutting / digging / picking, laying and gathering. In cereals like wheat and paddy, the plants are straight and smooth and ears containing grains are at the top whereas most of the plantation crops have branches and also a few have heights 1.5 to 1.75m, which create problems' in harvesting by manual or mechanical means. Harvesting is a crop specific operation as it largely depends on the characteristics of the crops. Light weight poles coupled with sharp sickles are generally used for harvesting tall trees by standing on the ground. There are wide variety of materials for both poles and sickles and also different designs of sickles. Hold on and twist type tools attached in long metallic poles can harvest individual fruits with care. Elevated platforms of different designs are also available to take the labour to top of tree for manual or mechanical harvesting. Tree shakers are used for applying impact force to detach fruit from the tree. The fruits harvested in this method undergo mechanical damage.

Pole and sickle method: Harvesting using pole and sickle by standing on the ground cause damage to the fruit due to impact of falling in most cases. Hence fruits harvested by this method should go for immediate processing or in other words fruits, which require immediate processing, only should be harvested by this method. The pole and sickle method is suitable for heights up to 12 meters (40 feet) as balancing the pole beyond such heights is difficult. Aluminium poles, alloy poles and ultra light weight poles of diameters 31mm, 37mm and 44mm are available in different lengths of 6m, 7.5m and 9 m. The most common method is to join 2-3 pieces together to form a long pole either by male-female screw joints or telescopic arrangement. Poles of height in the range of 15 m. are available in such combination. Length of screw should not be less than 75 mm (3 inches) in the male-female screw type joint to prevent failing the joint in long run. Efficiency of pole-sickle combination too depends on sharpness of sickle and experience of operator. Oil palm is commonly harvested by this method as it is difficult to climb over the tree due to presence of remnants of frond base in the trunk and thorns in fronds. Due to lack of alternative methods and difficulty in harvesting beyond a height of 12-15 m, such palms are abandoned.

Sickles consist of curved steel edged blade fixed to wooden handle. The cutting edge may be plain or serrated depending upon the design. The curvature of blade differs widely in plain edged and serrated sickles. The blade is fabricated from medium carbon or high carbon steel and hardened to resist wear. Manganese steel, tool steel and alloy steel can also be used to produce better quality sickles which have more life than sickles fabricated from plain carbon steel. The blade has a knife section and thickness reduces towards the cutting edge. Tang of the blade is inserted in to the wooden handle for joining the blade to the handle or attached to a small piece of pole to rivet to top end of the pole.

Tree climbers: Simple, safe and easy to use device is available for climbing coconut and arecanut for various operations like harvesting and spraying using manual power. The device is made up of MS steel, steel rope wire (steel wires of 19 strings – as a rope) and rubber pad, constituting a total weight of 8.5 kg. The climbing time is 1-2 minutes for a 40 m tree as against 4-5 minutes traditionally required.

Fruit harvesters: Fruit harvesters are operated manually, which consist of one fixed blade and another moving blade actuated by a spring. There is a net basket attached to it to collect plucked fruits. The long handle facilitates reaching fruits from ground. During harvesting, the fruit pedicel is adjusted to rest on fixed blade and pressing the lever at grip end of handle actuates the moving blade. The device is used for harvesting mangoes, oranges apple and sapota.

In another model, individual fruit is first held between two jaws and then twisted to shear off the stock. The jaws are made of 14 gauge mild steel sheet. These are held together by a tension spring on a pivot fitted on 10 mm mild steel rod. A handle can be fitted to the tool. One of the jaws has a fever bracket and rope arrangement for operating the jaw. Three mm thick rubber sheet padding is provided on inside of the to avoid any skin damage while holding fruits. After its detachment, fruit is released by pulling the cord into a ring. A cloth conveyor or net is provided below the jaws for collection of harvested fruits at ground level without any damage. The tool is suitable for harvesting peach, pear and orange.

A manual harvesting device developed and commercialized for harvesting mango fruit consists of oval shaped rings. The bottom ring is meant for fastening nylon net. A cutting mechanism is provided at top of ring and it consists of double bladed triangular plate together with toothed wheel. This toothed wheel is riveted at center of two fixed cutting blades. The wheel rotates freely about its central rivet and acts as a conveyor of a mango stock. For fixing a bamboo handle of desired length, a holder is provided to the harvester opposite to cutting mechanism. A plastic divider rod bisecting the cutting mechanism is provided in the ring to guide stalk of fruits either to left or to right side of cutting blade. For harvesting mango, harvester is raised and fruit is taken in ring by pulling the harvester. Pedicels of fruit are taken in between the toothed wheel and blade. On rotation of a toothed wheel, the pedicle is guided over the sharp edge of blade where it is sheared. Field capacity of the device is 140 fruits/h (Bharadwaj *et al,* 2004).

A manually operated sapota harvester consists of a main body of PVC having cylindrical shape. Upper end of the body is closed while bottom end is open to which nylon net for collecting the fruits is tied. A stretched string closes the other end of net. A gate is made on the body for entry of fruits to be harvested. On the lower surface of the body, a metal holder is fixed to hold the bamboo of required length. Two fingers cut in V -shape and with small sharp blades are provided at the closed end of the body of the harvester. Fingers help to select and hold the fruit to be harvested from the bunch. By pulling the harvester, fruit is detached from the bunch, which falls in the body and rolls into the net. To unload the harvested fruits in the net a stretched string at the closed end of the net is loosened.

Lifting platforms: Climbing over the tree is more laborious than picking/cutting the fruit in most of the cases, which can be solved by a mobile lifting platform of suitable working height. Many research has gone into this topic and different models of lifting platforms were developed suiting to various crops. The economic feasibility of many such models are still under validation.

Tractor mounted telescopic hoist described by Bharadwaj *et al.* (2004) can be used in orchards and plantation crops for trimming, pruning, plant protection and harvesting operations. It consists of two square aluminum ladders, each made of U-sections as frames and round hollow pipes as cross members. The U-sections of the outer ladder are inward facing while U-sections of the inner ladder face outward, sliding one over the other. Two wire ropes are provided, driven with a hydraulic motor. The motor while running in clockwise direction helps in lifting the inner ladder and while running in anticlockwise direction lowers inner ladder down. A support frame fitted to the platform on the top end of inner ladder helps as a safety frame. Maximum height of platform is 9.5 m and minimum height of platform is 5.5m. Field capacity is 30 minutes / tree.

A towable hydraulic lifting platform developed for harvesting oil palm trees at National Research Center for Oil Palm, Pedavegi, requires a prime mover for shifting the machine from one location to other. The machine can be stabilized in each location with four arms spread over the ground. The platform is lifted by employing hydraulic system operated by separate 5 hp diesel engine by either controlling from ground or from the platform. It's 360 degree slewing action helps harvesting all the palms around when kept at center. Field capacity of this machine for harvesting oil palm plantation is 9.5hr/ha.

11. Collection and Transportation

The collection of the harvested produce and transportation from the plantation up to mainline transport system and through mainline transportation system require attention to reduce drudgery, early processing or marketing of the produce and ensuring quality produce. Transportation system should be selected based on plantation size, type of produce, quality parameters and post harvest loss factor.

Tractor or power tiller with trailer are the most common in-field transportation devices. Different capacity tractors are available. The mini-tractor mechanical loading system described by Aribi (1995) for oil palm fresh fruit bunch loading consists of a 20-25 hp 4-wheel drive tractor complete with an attachment of one tonne payload, low profile and rear tipping trailer. A hydraulic crane is mounted on to the rear of the tractor. A grabber unit is fixed to the crane. The crane has a maximum reach of 2.25 m with a maximum lifting capacity of 25 kg. The crane can be rotated 180° which allows picking up from both sides of tractor. The operation is carried out by one operator and assisted by a helper, when necessary to collect loose fruits. Another mini-tractor with manual loading consists of a 16-18 hp, 4-wheel drive tractor complete with an attachment of 1 tone pay load, low profile and rear tipping trailer. The system operates by traveling along harvester's path proceeded by the loader who throws the FFB on to the trailer as the vehicle moves along.

In terrace, areas a range of three wheelers are available for in-field transporting of oil palm fresh fruit bunches (D'Souza, 2003). Three wheelers fitted with hydrostatic drives and hi-lift capability work more efficiently with higher capacity. Transport devices for plantations wherein the soil has low bearing capacity are challenging problems. All Terrain Vehicles(ATV) and commercial use of hovercrafts are under active consideration. Transport devices using LGP tyres (low ground pressure) have been tested and found that the tyre design reduced the ground pressure and allowed better floatation. Tractor with dual tyre at rear allows large area and hence reduces ground pressure thereby gives the tractor additional stability. Jackpack container system is used for collecting oil palm FFB, which runs on three wide, low pressure types and is driven by a 5hp diesel engine and lifts the container hydraulically to a height of about 18 cm, carrying them at speeds up to 12 miles / h. It is very suitable for use on flat or gently undulating land. The handling capacity of this vehicle is approximately 25 t / day and this would be equivalent to one machine / 275-300ha of mature oil palms (Zakaria, 1980).

Webb (1977) described a system of in-field transportation in banana plantations, which uses a wire system in hilly areas. The system involves the installation of upper end of wire system fixed to a point up the hill and the lower end is at the bottom of hill at dispatch point. The load is enclosed in a net attached to a pulley, which runs on a fixed wire, but is roped to an end-less wire running around grooved wheels at each end of the system. When the net is off loaded at the dispatch point, the pulley is transferred to the second fixed wire, hitched to endless wire and returned to rows for loading. This system minimizes soil compaction due to movement of vehicles in plantation.

12. Integrated Mechanization

Concept of a new integrated mechanization system for oil palm plantation industry in Malaysia is reported by Pebrian and Yahya (2007). The system consists of a four-wheeled driven multi purpose prime mover that is equipped with the respective attachments for conducting various tree crop field operations in both hilly and undulating terrains. The prime mover is a single chassis unit with a hydrostatic drive, oscillating drive axles and low ground traction tires. A diesel engine of 50.5 kW (67.7 HP) with rated speed of 2600 rpm is used to drive the main hydrostatic pump having a displacement of 46 cc/rev at a continuous pressure of 210 bar. The pump then runs two hydraulic motors having a displacement of 245 cc/rev at a continuous pressure of 205 bar with either series and parallel connection options to provide the required torque and speed at drive wheels of the prime mover. Proper prime mover wheelbase and ground clearance height are employed to give better stability and maneuverability to prime mover for the typical plantation terrain conditions. Mounting provisions for the seedling transplanting, fertilizer application, spraying application and the infield fruit collection-transportation were made on the prime mover. The available implement attachments are designed to have a payload capacity of 20 seedlings for field transplanting of tree seedlings, 600 kg load for crop fertilizing, 450-liter tank capacity for crop spraying, and 1500 kg load for the infield fresh fruit bunches (FFB) collection-transportation.

Mechanization in Plantations-basic requirements

A key element of a successful mechanization system is the quality and level of maintenance that is put into the machinery. Mechanization has to be supported with availability of a competent mechanic, basic workshop set-up like washing bays and work benches additional set of tools/related equipment/air compressor structured training programme for the staff.

References

Aribi, K.H. (1995) Mechanization in oil palm plantations. *The planter*, 71(834):417-426.

Bharadwaj, K.C., M.M. Pandey and Singh.G. (2004) Horticultural Tools and Equipment, *Technical bulletin* No.CIAE/2004/110, Central Institute of Agricultural Engineering, Nabi Bagh, Bhopal, India.

Denich M., K. Vielhauer, M.S. de A. Kato, A. Block, O.R. Kato, T.D. de Abreu Sá, W. Lücke and Vlek. P.L.G. (2004) Mechanized land preparation in forest-based fallow systems: The experience from Eastern Amazonia, *Agroforestry Systems*, 61-62(1-3): 91-106.

D'Souza (2003) Basics of an integrated FFB evacuation system. *The Planter*, 79(925):235-249.

FAO (1986) Watershed management field manual Slope treatment measures and practices, FAO Conservation Guide 13/3.

Gers, C and Schmidt. E.J, (2001) Using spot 4 satellite imagery to monitor area harvested by small scale sugarcane farmers at Umfolozi, *Proceedings South African Sugar Technologists' Association*, SASA (South Africa).

Kerpauskas, P., Srivydas, A.P., Lazauskas, P., Vasinauskiene, R. and Tamosiunas, A. (2006) Possibilities of weed control by water steam, *Agronomy Research*, 4(Special issue): 221-225.

Krishna Rao PV, Venkateswara, I. and Venkateswara Rao. V. (1999) Remote sensing approach for acreage estimation of sugarcane crop in part of Krishna district Andhra Pradesh. *Proc STAI* 61:79-87.

Lennartsson, E.K.M. *et al.* (1990) The use of light exclusion techniques for clearing grass pasture in organic horticultural systems. (In) *Proceedings of the Third international Conference of Non-chemical Weed Control held in Linz*, Austria, October 1989 (G. Plakolm, ea.) (In print).

McKillop, G., Pepper, H.W., Butt, R. and Poole. D.W. (1995) Electric Fence Reference Manual, Research and Development Surveillance Report 607, Department for Environment, Food and Rural affairs, London.

Michael J.B, Wills, J.B., Gill, W.W. and Lane. C.D. (1990) Planning and Building Fences on the Farm, University of Tennesse Agriculture Extension Series Publication No.PB1541.

Michael, A.M. (1999) Irrigation-Theory and Practice,Vikas Publishing House Pvt. Ltd., New Delh 110 014.

Narciso G and Schmidt. E.J. (1999) Identification and classification of sugarcane based on satellite remote sensing. *Proc S Afr Sug Technol Ass* 73:189-194.

Noonan M. (1999) Classification of fallow and yields using Landsat TM data in the sugarcane lands of the Herbert River catchment. Herbert Resource Information Centre. http://hric.tag.csiro.au. Queensland, 4850.

Ojah, T.P. and Michael. A.M. (1998) Principles of agricultural engineering, Volume I, M/s Jain Brothers,, New Delhi 873, East park Road, Karol bagh, New Delhi.

Pebrian, D.E. and Yahya, A. (2003) Design and development of a prototype trailed type oil palm seedling transplanter. *Journal of Oil Palm Research* 15(1):32-40.

Pebrian, D.E. and Yahya, A. (2007) A New Concept of Integrated Mechanization System for the Oil Palm Plantation Industry in Malaysia. *Proceedings of the International Conference on Crop Harvesting and Processing*, American Society of Agricultural and Biological Engineers, St. Joseph, Michigan

Rethinam, P. (2001) Rejuvenation of Oil Palm Plantations Damaged by Cyclones-a Success Story. *International Journal of Oil Palm*, 2(1):7-10.

Sirvastva, O.P., Kumar, A. and Nagar. Y. (1999) Application of satellite technology in management and planning of sugarcane cultivation in the Saraswati sugar mills, Yamuna Nagar. *Proc STAI* 61: 145-156.

Slaughter, D.C., Giles, D.K., Lamm, R.D. and Lee. W.S. (2000) Robotic Weed Control Systems for California Row Crops, Paper number 00-PA-018, Biological and agricultural Engineering, University of California, Davis, Davis CA. http://baesil.engineering.ucdavis.edu/files/ageng2000.pdf

Slusher, J.P. (1993) Mechanical tree planters. University of Missouri Extension Publications No. G-5005.

Steve Diver (2002) Flame weeding for vegetable crops, Current topic of 'Appropriate Technology Transfer for Rural Areas'(ATTARA), www.attra.ncat.org.

Webb, B.J. (1977) The development of suitable in-field FFB harvesting system to improve the efficiency of oil palm production. (In): International development in oil palm. (Ed): D.A. Earp and W.Newell, pp:741-759, Kuala Lumpur: Incorporated Society of Planters.

Zakaria, Z.Z. (1980) Problems and possible solutions of oil palm harvesting and infield transportation. In: *Proceedings of Workshop on Farm Mechanization in the Oil Palm Industry*, 3rd-4th December 1980, Kuala Lampur, Malaysia.

Chapter – 7

Postharvest Management of Fruits and Vegetables

K.B. Sheela

India is a horticultural flora rich country. It has wide variety of climate and soil on which a large number of horticultural crops such as fruits, vegetables, spices, ornamental and medicinal plants are grown. In India, horticulture crops are grown in an area of 21.8 million hectares with an annual production of 240.5 million tones (National Horticulture Board Database). Horticultural crops are high value commercial crops and play a unique role in country's economy, nutritional security, poverty alleviation and employment generation. The horticultural crops, because of their inherent high moisture content are more liable to deteriorate especially under tropical conditions. Moreover, they are biologically active and carry out transpiration, respiration, ripening and other biochemical activities, which deteriorate the quality of the produce. Due to inadequate postharvest handling and processing facilities a significant portion of production is lost after harvest. It is estimated that 20 to 30% of the horticultural produce is lost before consumption which accounts for Rs.50000 crores, because of poor harvesting, handling, storage, transportation and marketing practices.

India is the largest producer of fruits and second largest producer of vegetables in the world. The total fruit and vegetable production is about 258.01 million tones per year is 91.44 million tones of Fruits and 166.57 million tones of vegetables. Fruits and vegetables are an important supplement to the human diet as they provide the essential minerals, vitamins and fibre required for maintaining health. Fruits and vegetables are highly perishable commodities and the high ambient temperature prevalent in a tropical country like India makes them more susceptible for rapid development of senescence, decay and rotting.

To achieve our target of feeding the growing population as well as meeting the requirements of processing industry and export trade, increasing the production and productivity will not be enough. Postharvest loss reduction is essential to make available more food from the existing level of production. The cost involved in preventing the losses is always cheaper than the cost of production; hence post harvest management attains greater significance. Postharvest technology in horticulture is the systematic study of harvested crops and their response to postharvest environment. It includes design of appropriate handling and storage systems for marketing these crops in a living state.

1. Postharvest Losses of Fruits and Vegetables

Postharvest losses are defined as losses that occur after harvesting and till the produce reaches the consumer. Postharvest losses occur at different stages in the chain of harvesting, packaging, storage, transport and marketing. It can be a loss in quantity or quality or both or merely a loss in market value. Postharvest losses are much more painful and costlier than pre harvest loss both in terms of money and man hours.

Fruits and vegetables are highly perishable. They have high moisture content (80-90%). They are live commodities and continue their life process like respiration and transpiration even after harvest. As long as the fruit is attached to the parent plant water and photosynthesis are continuously supplied to it. These losses are not replaced in the postharvest environment and hence the commodity has to depend on its own food reserve and moisture content, with the result they perish fast. Moisture is lost from the commodity due to transpiration and food reserves depleted by respiration.

High moisture content in the commodity is conducive to growth of the microorganisms. Plant has some resistance mechanism to fight against microorganisms. So when the fruit is attached to the plant it may be protected from microorganisms, when the fruit is harvested, the resistance potential is lost.

Postharvest losses can be categorized under three classes viz: physical, physiological and biological.

Physical losses are losses occurring due to injuries during harvesting, handling and transportation. These injuries serve as entry point for microorganisms and also enhance moisture loss and gaseous exchange.

Physiological losses are losses occurring due to the physiological processes of respiration and transpiration. Loss of moisture due to transpiration will result in wilted or shriveled appearance of the commodities. Fresh mass loss occurs due to respiration and transpiration.

Biological losses are losses caused by microbiological and macro biological agents. Fruits and vegetables are easily spoilt by microorganisms like yeast, bacteria and fungi. Rodents and pests cause damage to commodities during storage.

Postharvest losses in the field, during transportation and storage

The important sites where postharvest losses occur in India are 15-20% at farmers field 15-20% in packaging, 30-40% in transportation and 30-40% in marketing. Harvesting at improper stage of maturity, mechanical damage by harvesting implements, rough handling and excessive exposure to sunshine, leads to losses at the field level. During transportation, losses can occur due to use of unsuitable transport containers, heat accumulation or poor ventilation within the transport vehicle, irresponsible driving and rough roads. Inadequate cold storage facilities and manifestation of damage by pests and diseases contribute to storage losses.

2. Minimization of Postharvest Losses

Two approaches are there for minimisation of postharvest losses of fruits and vegetables. The first approach is to following scientific postharvest management of fruits and vegetables. Another approach for loss reduction is processing into value added products. Postharvest technology of horticulture crops envisages the developing of appropriate techniques to reduce postharvest losses, to prevent spoilage and help utilize maximum crops in a nutritious and safe manner.

3. Postharvest Management Practices

Postharvest losses can be reduced to some extent by breeding fruits and vegetables for longer shelf life, improvement of pre harvest factors and harvesting techniques, proper methods of handling, marketing, packaging, transportation and storage, development of appropriate processing technology and utilization of waste from fruits and vegetables. Some of these points are discussed below.

4. Selection of Varieties

Varieties with better keeping and processing quality and lesser handling susecptability should be bred and selected for different fruits and vegetables. A few examples of varieties with long shelf life are Banglora (mango), Arka Vishal, Pusa Gaurav (tomato), Arka Nidhi and Arka Neelakandh (brinjal).

5. Harvesting

The goal of harvesting is to gather a commodity from the field at the proper level of maturity with a minimum damage and loss, as rapidly as possible and at minimum cost. Harvesting is done either during early morning or late evening hours. A temperature of above 27° C during harvesting should be avoided. The commodities that are to be send to distant markets are harvested in the evening and transported in the cool hours of night whereas commodities for local markets are harvested early morning. Harvesting should not be done immediately after rain or irrigation. Harvesting at optimum stage of maturity ensures maximum quality and yield. Maturity standards have been worked out for important fruits and vegetables. Care during harvesting should be taken to avoid mechanical injury to fruits and vegetables.

Hand harvesting and mechanical harvesting are two methods of harvest of horticulture commodities. Mechanical harvesting is common in developed countries, but can result in considerable damage to the produce. Hand harvesting is adopted for commodities meant for fresh market. Harvesting aids have been developed for harvest of fruits like mango, sapota, citrus etc. with better harvesting efficiency and cause less damage to harvested fruit.

6. Sorting /Grading

Sorting of the harvested produce is done to remove rotten, damaged, diseased, overripe, misshaped and insect attacked fruits and vegetables. Diseased/insect infested commodity should be immediately discarded as it may spread infection to healthy commodity.

Grading of fruits and vegetables means grouping them according to set standards. Grades are based on soundness, firmness, cleanliness, size, weight colour, shape and maturity. Most countries have their own set of standards for domestic trade. International standard grades are to be followed for the export of these commodities. Grading facilitates trading and marketing.

7. Washing

The produce is cleaned/washed to remove adhering dirt, dust, insects, mould and sprays residues and to improve appearance. Washing in cold water can also reduce the field heat and thereby extend shelf life of fruits and vegetables. Onion, garlic, okra, mushrooms etc. are not washed after harvest. Washing under a shower with pressurized water is more helpful in removing soil from root vegetables like carrot, radish etc. Rotary drum or barrel washer is used for potato and other root vegetables. Brush washers are used to clean individual fruit or vegetable.

To the wash water, mild detergents (soap solution), glacial acetic acid or NaCl (1%) can be added for surface decontamination. Chlorinated water (100 ppm chlorine) is also effective in surface decontamination. Fruits and vegetables are to be rinsed again with clean water and excess water allowed to dry before packing.

8. Trimming

Trimming is done especially in vegetables like cabbage, lettuce etc to remove unwanted, discoloured, rotten and damaged parts. Trimming enhances visual quality, minimizes deteriorative processes, facilitate handling, packaging and transport.

9. Curing

The term curing is applied to the measures used to prepare starchy staple root crops, garlic and onions for long term storage. Curing means, holding harvested commodities for a specified period, under defined conditions of temperature and relative humidity. Curing is a natural wound healing process in which harvested root and tuber crop replace and strengthen damaged areas by forming a corky layer which protects against water loss and infection by decay organisms. In contrast, curing of onions and garlic is mainly a drying process for toughening of outer skin and tightening of necks. Potato curing is most effective at about 20° C and 80% relative humidity; whereas sweet potatoes are effectively cured at 29° C.

10. Delatexing or De-sapping

This process involves allowing latex to fall out without falling on the fruits or removing latex exuded during harvesting. The latex if fallen on fruits cause sap burn/ discolouration reducing the marketability of the commodity. Desapping is an inevitable operation for fruits like mango, banana, jackfruit etc.

11. Waxing

Waxing of fruits and vegetables are done primarily to reduce moisture loss and reduce shriveling and wilting and to extend the storage life. Wax seal off the stem near the petiole and the pores on the surface of fruits which are the main routes of transpiration. Waxing also improves appearance of the produce.

Paraffin wax, carnuabawax and various resins are common types of waxes used for the preparation of wax – emulsion. Recently sucrose poly ester waxes are gaining popularity due to their water solubility and biodegradability. Waxes are generally applied by foaming, spraying and brushing of which foaming is the best, since it leaves a very thin coating. Some of the common coating materials are Semperfresh, Prolong, Waxol etc.

12. Pre-cooling

Pre-cooling is the process of removal of field heat from the harvested commodity, particularly when harvested during hot weather. Rapidly respiring commodities which have a short shelf life should be rapidly cooled immediately after harvest. Pre cooling helps in decreasing rate of transpiration, respiration, delay ripening and easing the load on the cooling system of transport or storage chambers. Various methods like room cooling, forced air cooling, hydro cooling, contact icing and vacuum cooling are used for pre cooling of fruits and vegetables.

13. Hot Water Treatment

Hot water treatment successfully eradicates incipient infections on several fruits. Dip in hot water at 50 ± 2° C for 1 – 2 minutes inactivates infections of *Phytopthora* sp. in tomato and oranges, *Colletotrichum* in papaya, mango and crown rot in banana. Hot water treatment of mango results in uniform ripening and reduction in spoilage during storage.

Vapour heat treatment

Vapour heat treatment is an eco friendly treatment for controlling infection of fruit flies in fruits. The temperature and exposure time are adjusted to kill all stages of the insects (egg, larvae, pupae and adult), but not damage the fruit. Vapour heat treatment in addition to disinfecting fruit, controls storage decay, but it is relatively expensive operation in terms of initial investment for equipment and operational cost. It consists of stalking the boxes of fruits in a room which is heated and humidified by injection of steam.

Heated air at 95% RH is circulated through crates of freshly harvested fruits. Total treatment time for mangoes is 195 minutes. This consists of 125 minutes approach period, 10 minutes holding time at 50°C, 40 minutes air cooling and 20 minutes cooling period. The time and temperature vary from fruit to fruit.

Hot air treatment

Hot air can be applied by placing fruits or vegetables in a heated chamber with a ventilating fan, or by applying forced hot air, during which the speed of air circulation is precisely controlled. This method heats more slowly than hot water or vapour heat. High humidity in vapour heat may sometimes damage the fruit being treated, but the slower heating time and lower humidity with forced hot air causes only less damage.

14. Postharvest Disease Management

A significant portion of losses of fruits and vegetables during post harvest is attributed to diseases caused by fungi and bacteria. Succulent nature of fruits and vegetables make them easily invadable by microorganisms. Mechanical injuries, contamination by diseased fruits and vegetables, heat and other environment agencies predispose produce to diseases. Postharvest diseases can controlled by use of fungicides as sprays or dips, incorporated in wax or impregnated in packaging materials.

15. Regulation of Ripening

Hastening of ripening in fruits like banana, mangoes, citrus etc is carried out in special treating rooms with controlled temperature in which low concentrations of ethylene are applied. Ethylene is generated in ethylene generator where ethanol is heated in presence of a catalyst. Ethylene is also released when fruits are placed in a closed chamber along with alkaline solution of ethrel. The ratio of caustic soda to ethrel is approximately 3g/20 ml of ethrel for effective release of ethylene. Postharvest dip in ethrel solution also result in degreening and ripening. The best degreening temperature is 26°C and relative humidity 85-92%.

The ripening process can be delayed and shelf life extended by removing ethylene from storage atmosphere using ethylene absorbent, potassium permanganate ($KMnO_4$). 'Purafil" is a commercial preparation of ethylene absorbent.

16. Sprout Inhibition

Tuber and bulb crops like onion and potato enter a dormant stage at maturity. Sprouting commences at the end of dormancy or rest period. Sprouting is a growth resumption process. Sprouting results in huge postharvest losses. Quality of sprouted potatoes become poor due to high respiratory utilisation of substrates Maleic hydrazide (MH-40), 3- chloroisopropyl –N- phenyl carbamate (CIPC), Methyl napthaleneacetic acid

(MENA) and 2,3,4,6 tetranitrobenzene (TCNB) are commonly used as sprout inhibitors. Gamma irradiation at 0.02 – 0.15 KGy has been widely accepted by many countries for successful sprout inhibition of onion and potato, without affecting the other quality attributes.

17. Packaging

Packaging is a fundamental tool for management of highly perishable commodities. The two main functions of packaging are to assemble the produce into convenient units for handling and to protect the produce during distribution, storage and marketing. Packaging materials are selected on the basis of product characteristics. It improves storage life and provides greater appeal to the produce. Good package protects commodity from physical, physiological and pathological deterioration throughout storage, transport and marketing. Packaging and mode of transport selected should be such that it causes no or minimum vibration and bruising injury to the product. To avoid this cushioning materials should be used (paper shreds, paddy straw, thermocole etc). Packaging for fresh produce is of several types like bamboo baskets, sacks (made of plastic or jute), wooden crates, corrugated fibre board boxes, plastic crates, biodegradable plastic etc. Corrugated fibre board (CFB) cartons are becoming popular for transport of fruits because of their lighter weight and cheaper cost. Sometimes the produce is enclosed in smaller individual packages of consumer units which is called prepackaging and then the individual units are put into bigger containers. Materials used for prepackaging are plastic films like poly ethylene, polypropylene, polystyrene, nylon net bag, pulp tray etc. Prepackages are packed in large containers like wooden boxes, plastic crates, corrugated fibre board boxes, for transportation and distribution. The new innovations in packaging are aseptic packaging, controlled and modified atmosphere packaging.

18. Transportation

Transport is an important linkage in postharvest handling, storage and distribution. Transportation of the horticultural produce from field to the distribution markets is done by rail, truck, airplane, ship etc. Serious losses occur due to rough handling, careless loading and unloading and use of improper containers. Transport of commodities during cool hours of the night, use of ventilated, insulated, evaporative cooled or refrigerated vehicles ensures preservation of quality. Pallets are used in many developed countries for trading of horticultural produce. It is also essential to introduce mechanical loading and unloading particularly with use of fork lift trucks. In advanced countries refrigerated containers known as reefer containers are used for shipment of fruits, vegetables and flowers. In India we can use containers working on evaporative cooling technique.

For effective preservation of quality from harvest to final consumption introduction of cold chain is important. The harvested produce is precooled to remove field heat and then transported in refrigerated trucks to the cold storage or wholesale/ retail markets with cold storage facility. This type of cold chain linkage helps in introducing a systematic approach and result in reduction of wastage at farm level, transport and storage.

19. Marketing

Marketing problems are more in fruits due to their high degree of perishability, season bound availability and bulky nature. Development of marketing systems is essential to control activities of middlemen and to help regulation of trade. National Diary Development Board in Delhi, HOPCOM in Bangalore, Fresh in Hyderabad, Maxworth Harvest in Chennai are some of the marketing organizations formed to benefit growers and consumers. National Horticultural Board with its well spread net work centres is providing good market intelligence service for horticultural commodities.

20. Storage

Storage is essential to enhance usefulness of commodities, avoid market glut, ensure supply throughout year and increase profit to the producers. The principal aim of storage is to reduce and control transpiration, respiration and disease infection at the same time maintaining life processes at the required level. The various method of storage of horticultural commodities are refrigerated storage, controlled / modified atmosphere storage, hypobaric storage, zero energy cool chamber etc.

Cold storage

Cold or low temperature storage is most reliable and extensively used for retention of freshness and extending shelf life of fruits and vegetables. Exposure to high temperature is the biggest factor in the deterioration of horticultural produce. Cold storage is recommended for perishable commodities since it retards the rate of metabolic change, moisture loss, respiratory heat production and spoilage caused by microorganisms. The ideal condition for storage of fresh fruits and vegetables is the lowest temperature which does not cause chilling injury to the produce. The prevalent practice is to use mechanical refrigeration to store horticultural produce. Cold storage is not only energy intensive and expensive but involves huge initial capital investments. The cool storage space available in our country is primarily used for storage of potato.

Controlled / Modified Atmosphere Storage

The main purpose of controlled atmosphere (CA) or modified atmosphere is to adjust the atmosphere composition of gases surrounding the commodity by removal or addition of gases thus resulting in an atmospheric composition different from that of normal air. Modified atmosphere does not differ in principle from controlled atmosphere storage expect that the control of gas concentration is less precise. Modification of atmosphere for packaging and storage of fresh horticultural commodities involve reduction in oxygen and elevation of CO_2 concentrations. Modified atmosphere can be created by packaging in plastic films wherein the change in gaseous composition occurs due to the metabolic process like respiration. The first requirement of controlled atmosphere is a sufficiently gas tight chamber and the second is some means of maintaining the concentration of CO_2 and O_2 at the desired level. This method in combination with refrigeration markedly enhances shelf life of fruits and vegetables. The beneficial effect of CA/MA treatments are retardation of ripening, senescence and reduction in chilling injury of various commodities.

Hypobaric storage

This is also known as reduced atmospheric pressure/low pressure storage /sub atmospheric storage. The principles of such storage is to remove ethylene gas from the storage atmosphere and lowering the partial pressure of 0_2 which leads to slow ripening, hence fruits can be kept for longer periods. It is a type of CA storage with emphasis on reducing the pressure of 0_2 and ethylene gas. Low pressure of 0_2 (102mm Hg) reduces the growth of pathogenic organisms. Hypobaric is a form of controlled atmosphere storage in which produce is stored in partial vacuum. The vacuum is vented by vacuum pump to the desired low pressure. The process of ripening and senescence is greatly retarded by decreasing the rate of respiration and removal of ethylene produced by the ripening fruit. Hypobaric storage is limited to high value crops like cut flowers due to the high initial investment required.

Evaporative cool storage

This is one of the simple and effective methods for short term storage of fruits and vegetables at farm level. This reduces shriveling, extend storage life and prevent losses during marketing and storage. This type of storage maintains high relative humidity with temperature nearer to wet bulb and this reduces the weight loss considerably in stored fruits and vegetables. It is one of the low cost technologies compared to cold storage which is highly energy extensive and requires huge capital investment.

Zero energy cool chamber

Zero energy cool chambers are based on the principle of evaporative cooling and developed for short term storage of fruits and vegetables. In a tropical country like India tremendous amount of quality deterioration takes place immediately after harvest due to lack of on farm storage facilities. In order to overcome this problem, low cost environment friendly zero energy cool chambers have been developed by IARI. These chambers are constructed using locally available materials like brick, sand, bamboo etc. The temperatures in these chambers are less than surrounding atmosphere. These chambers can be used for short term storage of products at the farmers field itself.

21. Conclusion

Postharvest loss is even more serious than production loss. The reduction of postharvest losses will significantly increase availability of fruits and vegetables without bringing additional land into production and without using additional inputs. It is not possible to reduce looses completely, but it can be minimized by following some of the modern cultural, harvesting, handling, marketing and processing techniques. Appropriate postharvest technology methods should be technically feasible, economically viable and acceptable to all sectors in the handling chain.

References

Champp, B.R, Highley, E and Johnson, G.I (1993) Postharvest handling of tropical fruits. *Proceedings of International Conference held at Chiang Mai,* Thailand 19-13 July 1993.

Dash, S.K; Chandra, P and Kar, A. (2006) Evaporativly cooled storage of horticultural produce: A review. *J Food Sci. Tech.* 43(2): 105-120.

Lal, G; Siddappa, G.S and Tandon, G.L. (1998) Preservation of Fruits and Vegetables. ICAR, New Delhi.

Mitra, S.K. (1997) Postharvest Physiology and Storage of Tropical and Subtropical Fruits CAB International New York.

Rao, S.D.V. (2004) Pre-storage treatment for minimisation of postharvest losses in fruits and vegetables. Training manual on Minimisation of Postharvest Losses in Fruits and Vegetables organized by IIHR, Bangalore.

Salunkhe, D.K and Desai, B.B. (1984) Post harvest Biotechnology of Fruits. Vol I&II. CRC Press, Florida.

Salunkhe, D.K and Desai, B.B. (1984) Post harvest Biotechnology of Vegetables. Vol I CRC Press, Florida.

Sandooja, J.K; Mangal, J.L and Lal, S. (2002) Postharvest Management of Vegetable Crops. - An Overview. *Proceedings of National Workshop on Postharvest Management of Horticultural Produce,* Organized by National Productivity Council Chandigarh from 7 – 8th February 2002 p. 31- 41.

Sethi, V. and Maini, S.B. (1989) Appropriate technology for reducing postharvest losses in fruits and vegetables. *Indian Food Packer* pp 43-55.

Sudheer, K.P. and Indira, V. (2007). Peter K.V. (Ed) Post Harvest Technology of Horticultural Crops, New India Publishing Agency, New Delhi. 291p.

Thompson, A.K (1996) Postharvest Technology of Fruits and Vegetables. Blackwell Science Inc.

Chapter – 8

Preservation of Fruits and Vegetables

K.B. Sheela and P.B. Pushpalatha

The act of preservation of food has been known to mankind since ancient times. Since food is a sensitive commodity and directly affects human health and safety, it needs special care in handling right from raw material through processing, packing and transport to the consumption point. When food spoils, they undergo chemical and physical changes that may render the food inedible and hazardous. Food preservation can be defined as the science which deals with methods of prevention of spoilage of food, thus allowing it to be stored in a fit condition for future use.

1. Factors Causing Spoilage of Foods

All foods undergo varying degree of deterioration during storage. Deterioration may include losses in organoleptic quality, nutritional value and safety. Food is subject to physical, chemical and biological deterioration. The principle cause of spoilage in foods are the growth of microorganisms, the action of naturally occurring enzymes in the foods, chemical reactions and physical degradation and desiccation. All foods deteriorate, some rapidly than others. The type of spoilage for a particular food item depends to a great extend on the composition, structure, types of microorganisms involved and conditions of storage of food. The factors that control microbial food spoilage are moisture content, temperature, oxygen concentration, nutrient availability, the degree of contamination with spoilage organisms, and the presence of growth inhibitors. Generally control of one or more of these factors controls microbial spoilage (Desrosior and Desrosior, 1977).

The enzymes present in the food may affect the flavour and colour adversely. eg: browning of banana/apple juices due to the activity of enzyme polyphenol oxidase. The chemicals present in the food may react with one another and spoil its taste and aroma. Non enzymatic browning is observed in foods due to interaction of sugars and pectins. Exposure to air for a short period can result in undesirable flavour and taste. Citrus juices turn bitter when kept open for some time.

Metallic contamination (from processing equipment, containers) would not only cause discolouration, but also spoil taste and aroma of products. Blackening of pickles is observed due to contamination by metal iron.

2. Methods of Preservation

Different methods of preservation adopted for preventing spoilage of food are dehydration, thermal processing, use of chemical preservatives, fermentation, irradiation and low temperature storage.

Principles underlying different methods of preservation

Prevention or delay of microbial decomposition

- By keeping out microorganisms (asepsis)
- By removal of micro organisms (filtration)
- By hindering growth and activity of microorganisms eg: by low temperature, dehydration, fermentation, chemical preservatives
- By killing the microorganisms eg: thermal processing, irradiation

Prevention or delay of self decomposition of food

- By destruction or inactivation of enzymes eg: by blanching.
- By prevention or delay of chemical reactions eg: prevention of oxidation by use of antioxidants

Prevention of damage by insects, animals, mechanical causes

The general methods of preservation are

Asepsis (absence of infection)

Cleanliness is one of the most important deterrents of food spoilage. Microbiological deterioration of a food involves multiplication and growth of billions of cells. The larger the initial population of microorganisms, the more rapidly the food is spoilt. Absence of infection through general cleanliness during handling and processing is asepsis. Packaging of foods is a widely used application of asepsis.

Preservation by filtration

This is a cold process, involves removal of microorganisms by ultra filtration. Removal of microorganisms may be accomplished by means of filtration, centrifugation, washing or trimming. Filtration is the only successful method for the complete removal of organisms, and its use is limited to clear liquids. In this method microorganisms are removed by filtration using various types of germ proof filters. Juices are clarified by using ordinary filters and then passed through special filter which is capable of retaining yeast and bacteria. The liquid is filtered through a previously sterilized 'bacteria proof' filter made of sintered glass, diatomaceous earth, unglazed porcelain, membrane pads or similar material, and the liquid is forced through positive or negative pressure. This method has been used successfully with fruit juices, beer, soft drinks, wine and water. Centrifugation and washing are not generally very effective, in that some but not all the microorganisms are removed.

Preservation by Low temperature

Low temperatures are used to retard chemical reactions and action of food enzymes and to slow down or stop growth and activity of microorganisms in food. The lower the temperature, slower will be the chemical reaction, enzyme action and microbial growth and a low enough temperature will prevent growth of any microorganisms.

Preservation by removal of water

Microorganisms need moisture for their survival. When moisture level in food is brought down below a certain level, they are unable to grow. Drying can be done using solar energy (sun drying) on by artificially produced heat under controlled conditions of temperature, relative humidity and air flow (dehydration).

Preservation by chemicals

Microbial spoilage in food can be controlled by use of preservatives. A chemical preservative is a substance capable of inhibiting, retarding or arresting the process of fermentation, acidification or other decomposition of food. The preservatives are broadly classified as class I (natural preservatives) eg: common salt, sugar, honey, dextrose etc and class II (chemical preservatives) eg: sulphur dioxide and benzoic acid.

Preservation by fermentation

Fermentation is the decomposition of carbohydrates by microorganisms or their enzymes. Alcoholic, acetic and lactic fermentation are used for preservation of foods. Foods are preserved by virtue of alcohol or organic acid formed by microbial action.

Thermal processing

The process of preserving food by application of heat is known as thermal processing. The food in closed containers is subjected to varying degrees of heat to inactivate or destroy microorganisms. Coagulation of proteins and inactivation of their metabolic enzymes by application of heat, leads to the destruction of microorganisms present in the food. Heating can also inactivate the enzymes present in the food. The temperature and period of treatment depends on the organisms that have to be killed, nature of food to be preserved and other methods of preservation that may be used in addition to high temperature. High temperatures generally for preservation are sterilization (100^0C and above) and pasteurization (below 100^0C)

Irradiation

The utilization of ionising radiation for stabilization of foods offers a method of cold sterilization, wherein foods are preserved without marked change in their natural characters. Food irradiation consists of exposing the food to ionising radiation, emitting from radio active isotopes Co -60 and Ce-137 or from electrical machines, generating electrons or X – rays.

Hurdle technology stabilization of foods

Hurdle technology stabilization of foods is another emerging area of preservation of fruits and vegetables. Hurdle technology has been defined by Leistner (2000) as an intelligent combination of hurdles which secures the microbial safety and stability as well as organoleptic and nutritional quality and the economic viability of food products. Hurdle technology processing is a combination of processing techniques in which several technological approaches and unit operations such as freezing, refrigeration, dehydration, irradiation, use of chemical preservatives are used in a synergistic manner to bring about desired microbiological safety and sensory acceptability. Microorganisms must overcome these hurdles to cause deterioration or spoilage of food. No single hurdle can effectively prevent spoilage unless employed at their extremes at which the sensory and nutritive quality of foods is often compromised.

High hydrostatic pressure

High pressure processing is a very promising non thermal food preservation method. High pressure can be used not only to preserve foods, but also to improve heological and functional properties of foods and is environmental friendly causing less effluent problems. An important aspect of high pressure technology is the inactivation of enzymes, at the same time retaining nutrients and flavours in the food, giving a fresh like flavour and texture to pressure processed foods. At a pressure of 4000-9000 atmosphere, enzyme and bacteria are inactivated but taste and flavour are unaffected. HPP subjects liquid and solid foods with or without packaging to pressures between 100 and 800 M pa. Process temperature during pressure treatment can be from below 0^0c to above 100^0c. Exposure times can range from a few seconds to over 20 minutes. High pressure with 4000 – 9000 bars extends refrigerated shelf life from two months to six months at 4°C.

Pulsed light processing

This technology uses intense flashes of broad spectrum white light to kill microorganisms on food surface and in air and water. Flash duration is typically about 3000 ìs and the light intensity is greater, 20,000 times that of sunlight. A few flashes applied in a fraction of a second kills all exposed microorganisms, spores and viruses. Antimicrobial mechanism is based on the sensitivity of microorganisms to UV wave lengths that are readily absorbed by microbial nucleic acid and the very high power and intensity of the flash.

High Intensity pulsed electric field

This technology relies on the lethal effect of electric field on microorganisms. The electric field applied in the form of short pulse with pulse duration ranging between few microseconds and milliseconds rather than using electricity directly. Minimal heat is generated and processing generally remains nonthermal. Foods are processed in a short period of time and the energy lost due to heating is minimal. It partially inactivates the enzymes and spores avoiding the undesirable changes such as protein denaturation and vitamin destruction. The risk of dielectric break down limits this technology feasible for liquid foods like diary products and fruit juices. PEF holds potential as a type of low temperature alternative pasteurisation process for sterilizing food products. In PEF processing a substance is placed between two electrodes, then the pulsed electric field is applied. The electric field enlarges the pores of the cell membranes which kills the cells and releases their contents.

Oscillating magnetic field

The utilization of oscillating magnetic fields (OMF) to inactivate microorganisms has the potential to pasteurize food with an improvement in the quality and shelf life, compared to conventional pasteurization process. Strong static (SMF) or oscillating (OMF) magnetic fields (5-50 Telsa) have the potential to inactivate vegetative microorganisms.

The impulse duration is between 10 hrs and several milliseconds. The frequencies are maximally 500 MHZ, because above that value the items begin to warm up noticeably. Preservation of foods with OMF involves sealing foods in a plastic bag and subjecting it to 1-100 pulses in an OMF at temperature of 0° C to 50° C for a total exposure time ranging from 25 to 100ms. No special treatment of food is required before treatment of food by OMF. Magnetic field treatments are carried out at atmospheric pressure and at a temperature that stabilizes the food material. The temperature of food increases by 2^0 c to 5^0 c, and the organoleptic properties change very little after magnetic field treatment.

The technical advantage of inactivation of micro-organisms with oscillating magnetic field include minimal thermal denaturation of nutritional and organoleptic property, reduced energy, potential treatment of foods inside a flexible package to avoid post-process contamination. However the conditions necessary for magnetic fields to exhibit inhibitory stimulus on microorganisms are not clearly understood. Oscillating magnetic field treatment of foods in containers ensures no post process contamination unlike in thermal process. A magnetic field of 5 – 50 T with short pulses of 25 ms to 10 ms and frequency 5-50 HZ are required to inactive microorganisms.

Manothermo sonication

It represents a combination of low pressure (0.3Mpa), mild heat treatment and ultra sonic wave treatment that seems effective for inactivation of microorganisms. The combined heat treatment can be used in processing fruit juices and drinks to solve the problem of thermo stable enzymes.

Ohmic heating

Ohmic heating systems are now in commercial use in USA, Europe and Japan in aseptic processing plant. Problems in conventional heat transfer techniques have been overcome by the use of ohmic heating system thereby preventing the destruction of flavour and nutrients and mechanical damage to the particulate. The ohmic heating effect occurs when an electric current is passed through an electricity conducting product. Generally the product does not experience any large temperature gradient within itself as it heats both liquid and particulate simultaneously resulting in less heat damage to the liquid phase and avoiding overcooking of solid phase. As there are no heat transfer surfaces, possibility of deposit formation and transfer of burnt particles to the finished products is avoided. Commercial ohmic heating based plants are now available with fully automatic temperature control system.

Minimal processing

The consumer demand for minimally processed (fresh cut) fruits and vegetables has increased considerably in the recent years. Consumers all over the world prefer convenient, ready to use and ready to eat fruit and vegetables with fresh like quality and containing natural ingredients. Fresh cut fruits and vegetables are cleaned, peeled, cut, sliced and packaged in polymeric films. Minimally processed vegetables/fruits are highly perishable unlike conventionally processed foods. Selection of appropriate cultivars, good sanitation, prestorage treatment with antioxidants, preservatives, edible coating, irradiation, low temperature and modified atmosphere storage can be effectively used to enhance shelf life of fresh cut products.

3. Preservation by Use of Preservatives

The most common method of preservation of foods is by use of preservatives. Preservatives which help in keeping food wholesome and appealing are essential tools in processing industry.

Definition

Preservatives are substances which when added to food is capable of inhibiting, retarding or arresting the process of fermentation, acidification or other decomposition of food.

Classification of preservatives

Preservatives are broadly classified into 2 categories viz. class I preservatives (natural preservatives) and Class II preservatives (chemical preservatives)

a) Class I Preservatives

Natural preservatives are substances obtained from plants, animals or microorganisms which when added to food tend to prevent or retard the microbial growth, food deterioration and thus increase the shelf life. The commonly used natural preservatives are common salt, sugar, honey, dextrose, vinegar, vegetable oil, organic acids, spices and bacteriosins.

Mode of action of natural preservatives

Common salt (NaCl) acts as a preservative at a concentration of 15 to 20 %. It is used as a preservative in pickle, dehydrated products either alone or combination with other preservatives. Common salt is known to decrease the water activity of food products, which restrict the growth of microorganisms. It also decreases the solubility of O_2 in water, sensitizes cells to CO_2 and interferes with action of proteolytic enzymes. Furthermore, the increased osmotic pressure produced by salt causes plasmolysis of microbial cells.

Sugars (glucose, sucrose, lactose, dextrose) act as preservative when used at higher concentration (>68%). Sugar is used for preserving jams, jellies, syrups, candy etc. Sugars reduce the water activity of the food to the point where microbial growth is impossible. It also raises the osmotic pressure of the solution and thus causes plasmolysis of microbial cells.

Acid conditions inhibit growth of many microorganisms and hence organic acids are added or allowed to form in foods to preserve them. Acetic acid (vinegar), citric acid and lactic acids are commonly used for preservation.

Edible oils like gingelly oil and mustard oil have antimicrobial effect. A layer of oil at top of pickles creates anaerobic condition, which prevent growth of mould and yeast.

Many spices like turmeric, mustard, pepper, asafetida, cinnamon, clove, garlic contain antimicrobial substances and their addition to food not only provide the desired flavour but also imparts antimicrobial properties.

Bio-preservatives

Bacteriosin constitutes an important segment of bio preservatives. Bacteriosins are low molecular weight proteins produced by many starter bacteria. eg: Nisin, subtilin, pediocin etc. They are bactericidal to closely related strains and species. Many genera of starter bacteria like *Lactococcus, Leuconostoc, Pediococcus, Lactobacillus* have been identified as bacteriosin producing strains. Nisin is a bacteriosin produced by *Lactococcus lactis*. It has been used in processed teas, pasteurised milk and alcoholic beverages.

b) Class II Preservatives

According to FSSAI regulations, the only chemical preservatives permitted in fruit and vegetable products are

1. Sulphurous acid and its salts
2. Benzoic acid and its salts

Mode of action

- By interfering with the genetic mechanism of cell division of microbes
- Interfering with the enzyme system of the microbes.
- By inhibition of biosynthesis of vitamins
- By formation of complexes with heavy metal ions.

Sulphur dioxide

Because of its versatile properties and cheapness SO_2, sulphites and bisuphites are largely used by food industry. It is used widely throughout the world for treating foods of plant origin. Sulphur dioxide is a broad spectrum antimicrobial agent inhibiting yeast, mould and bacteria. Sulphur dioxide is more effective against bacteria and mold. It is less effective against yeast and hence used in fermentation industries such as wine making. Sulphur dioxide acts as an antioxidant and hence helps in retention of ascorbic acid, carotene and other oxidisable compounds. It is used for controlling both enzymatic and non enzymatic browning. It cannot be used in naturally coloured juices like phalsa, jamun, pomegranate, strawberry on account of its bleaching action on anthocyanin.

Sulphur dioxide can be incorporated in foods either as gaseous SO_2 or as sulphite salts. Gaseous SO_2 is incorporated by a process called sulphuring in which fruits are exposed to fumes of burning sulphur. Sulphuring prior to dehydration not only protects certain nutrients and control discolouration, but also controls microbial and insect activity effectively. The sulphite salts commonly used by the food industry are potassium and sodium metabisulphite.

Preservative action of sulphur dioxide

Potassium metabisulphite is used as a stable source of SO_2. It is a crystalline salt and fairly stable in acid and alkaline media. When added to the fruit juices, the potassium radical reacts with the acid of the fruit juices forming the corresponding potassium salt and SO_2 which is liberated forms sulphurous acid with the water of the juices. The preservative effect of H_2SO_3 depends not on its total quantity, but on the available amount of SO_2. Combined So_2 has very little preservative action. Free sulphurous acid is 120 times more effective than combined sulphurous acid. Pure sucrose does not combine with SO_2 but glucose, unidentified aldehydes, ketones, pectin have property of combing with SO_2 resulting in low effectiveness of SO_2. Quantity of SO_2 that is active in foods is far less than quantity added this may be due to loss of So_2 during processing and storage and formation of additive compounds with carbonyl groups thus making it unavailable Optimum microbial inhibition is at pH 2.5 -3. In fruit juices with low acidity, acid should be added to lower the pH value and thereby increase the preservative action of sulphur dioxide. Sulphur dioxide is heat unstable. Excess of sulphur dioxide can be removed by heating or subjecting to vacuum. (The maximum permitted level of SO_2 in processed products is furnished in Table 1)

Benzoic acid

Occurs naturally in strawberries. Sodium benzoate is widely used in preserving acid foods. It is more effective against yeast and mould than bacteria. The acidity of the substrate in which benzoates are added influence the effectiveness of the chemical preservative. In a highly acid food, the germicidal action is in the order of 100 times more effective than in slightly alkaline food. Optimum microbial inhibition is at pH range of 2.5-4. Hence it is used in acidic foods such as carbonated beverages, fruit juices, cider, pickles etc. Benzoic acid is sparingly soluble in water and hence used in the form of sodium benzoate which is highly soluble in water. Sodium chloride is believed to have a synergistic effect with sodium benzoate. Permitted levels are 0.2 – 0.3% but in practice only 0.05 – 0.1% sodium benzoate is added in foods and beverages. Addition of sodium benzoate and mild heating is complementary. Benzoates in foods at a concentration of 0.1% and above may be noticeable and can impart a disagreeable peppery or burning taste to the food.

Table 1. Permitted levels of chemical preservatives in processed products

Sl.No.	Products	Maximum permitted level in ppm	
		Sulphur dioxide	Benzoic acid
1.	Canned fruits	Nil	Nil
2.	Fruit pulp	1000	Nil
3.	Fruit juice concentrates	1500	Nil
4.	Squash, crush, cordial, syrup	350	600
5.	Jam, jelly, preserves	40	200
6.	Ready to serve beverages	70	120
7.	Pickles/chutneys	100	250
8.	Tomato and other sauces	Nil	750
9.	Dehydrated vegetables	2000	Nil
10.	Alcoholic wines	150	Nil
11.	Dehydrated fruits	750	Nil

4. Preservation by Irradiation

Food irradiation is an important application of nuclear energy for the benefit of food preservation. Radiation processing of food involves exposure of food to short wave energy to achieve a specific purpose such as extension of shelf life, insect disinfestation and elimination of food born pathogens and parasites. The food is freed of microorganisms without raising temperature and hence this method of preservation is also known as 'cold sterilization'. Food can be irradiated wet, dry, thawed or frozen. It does not cause change in texture and freshness of food unlike heat. Radiation technique is very effective and due to its highly penetrating nature, it can be used on packed food commodities.

Radiation preservation of food was first carried out at the Masachusetts Institute of Technology in 1943. The use of ionizing radiations for food preservation has not been adopted very widely. This is largely due to the unacceptable flavour of some foods irradiated to sterilization stage. Also, the fear that radio activity might be induced in a food has come in the way of its extended use. However recent reappraisal by the WHO and the International Atomic Energy Authority (IAEA) has resulted in the recommendation that irradiation upto 10 KGy is safe for human consumption. Independent technical and scientific commodities constituted by several countries including Canada, Denmark, France, India, Sweden, UK, European Economic Community (EEC) and the USFDA have declared that radiation processed food is safe for human consumption.

Radiation processing

Radiation processing of food involves the controlled application of energy from ionizing radiations such as gamma rays, electrons and X rays for food preservation. Gamma rays and X rays are short wave length radiations of the electro magnetic spectrum which include radio waves, micro waves, infra red, visible and ultra violet light. Gamma rays are emitted by radio isotopes such as Cobalt – 60 and Caesium – 137 while electrons and X - rays are generated by machines using electricity. X rays are photons and have much better penetration and are able to penetrate through whole carton of food products.

Mode of action of radiation

When gamma rays and electron beams pass through foods, there are collisions between ionizing radiations and food particles at the atomic and molecular levels, resulting in production of ion pairs and free radicals. The reactions of these products among themselves and with other molecules result in physical and chemical phenomenon inactivating microorganisms in foods. Irradiation works by disrupting the biological process that lead to decay. In their interaction with water and other molecules that make up food and living organisms, radiation energy is absorbed by the molecules they contact. The reactions with DNA cause the death of microorganisms and insects and impair the ability of potato and onion to sprout.

Use of radiations

Ionsing radiations can be used for

1. Disinfestation of insect pests in stored products
2. Inhibition of sprouting of tubers and bulbs and delay in ripening of fruits
3. Destruction of microbes responsible for spoilage
4. Elimination of pathogens and parasites of public health importance

Radiation dose

The unit used to measure the effects of radiation are gray and sievert in accordance with recommendations of International Organization for Standardization (ISO). The gray

(Gy) is the unit used to measure absorbed dose of radiation and is equal to one joule of energy absorbed per kg of matter being irradiated. 1 Gy corresponds to 100 rad.

The radiation dose must be carefully controlled to destroy the microorganisms and inactivate many foods enzymes. It should be sufficient to destroy the pathogenic and spoilage causing organisms. Apart from the intensity of radiation, the amount of radiation absorbed and length of time of radiation are to be controlled. On the basis of radiation dose, application of radiations can classified into

Low dose application upto 1 KGy.

- Sprout inhibition in bulbs and tubers 0.03 – 0.15 KGy.
- Delay of fruit ripening 0.25 – 0.75 KGy.
- Insect disinfestation including quarantine treatment and elimination of food born parasites 0.25 – 1 KGy.

Medium Dose Application

- Reduction of number of microorganisms in spices to improve hygienic quality-10 KGy.

High dose Applications

- Sterilization of packaged meat and their products which are shelf stable without refrigeration 25 -70 KGy.

Process of gamma irradiation

In a gamma irradiation facility, irradiation is carried out inside an irradiation chamber. The chamber is shielded by 1.5 – 1.8 meter thick wall. Radiation source Co-60 is contained in slender pencil like stainless casings which in turn are enclosed in lead shield. Food products, pre packed are placed in suitable containers and carried into the irradiation chamber with the help of automatic conveyer. The conveyer goes through a wall labyrinth. The concrete wall serves as a shield and prevents radiation from reaching the surrounding area. For irradiation of food products, the radiation source is raised above the water shield, after activating all safety devices. The food containers move around the source in the cell and get exposed for a specific length of time that allows them to absorb a defined radiation dose. Gamma rays pass into the food, affect the food or target organisms present in food and leave the food. The effect of irradiation depends on type of food products, radiation dose and organisms present in food. Irradiated food should be properly handled and stored, because irradiation connot prevent further contamination of food.

Irradiators are designed with several layers of overlapping protection systems/ safety devices to detect any malfunctioning of the equipment and to protect working personnel from accidental exposure to the radiations. All irradiation facilitates are under strict control of regulatory agencies. In India, irradiation facilities operate under direct control of AERB (Atomic Energy Regulatory Board), Bhabha Atomic Research Centre, Mumbai.

Influence of radiation on nutritive and organoleptic properties of foods

Irradiation does not very much affect nutritional properties of foods. The destruction of various amounts of nutrients is of the same degree as in heat processing. Irradiation of food can also result in certain undesirable effects. Ionising radiation in excessive dose can alter the structure of organic and biochemical compounds in food resulting in food damage. Degradation of carbohydrates may result in loss of texture and flavour. Protein degradation results in undesirable changes in colour and odour. Irradiation of fats results in development of off flavour and loss of the natural antioxidants.

In India the use of ionzing radiation for food preservation has been adopted very slowly only. In 1994, Government of India approved irradiation of onion, potato and spices for internal marketing and consumption. Even if ultimately proven to be quite safe, food irradiation will have its limitations as do all other methods of preservation. Sterilizing dose of irradiation can alter various components of foods, causing a wide range of organoleptic changes. Lower doses of irradiation if coupled with refrigerated storage would prove technically and economically feasible. Improved keeping quality and microbiological safety from irradiation could play a substantial role in international food exports and imports of the future.

5. Preservation by Low Temperature

Freezing and cold storage are among the oldest methods of food preservation. Temperature management is crucial in preventing spoilage and extending shelf life of perishable commodities. Microbial growth, enzyme and chemical reactions are retarded in foods stored at low temperature. The lower the temperature, the greater is the retardation. Fruits and vegetables if stored at low temperature remain fresh and nutritious for a longer time. In fresh produce, low temperature reduces physiological activity, rate of respiration, transpiration, ethylene production and other biochemical reactions responsible for rapid ripening and senescence. Cold preservation may be categorized into two groups, refrigeration and freezing.

Refrigeration

Refrigeration is the most common method of chilling or low temperature storage of fruits and vegetables. Refrigerated storage helps to retard the spoilage in perishable crops in the following ways.

- Retard aging due to ripening, softening, textural and colour changes.
- Minimizes undesirable metabolic changes and respiratory heat production

- Decrease moisture loss and wilting
- Reduce spoilage due to invasion by bacteria, fungi, and yeast.
- Inhibits undesirable growth such as sprouting of potatoes.

One of the important functions of refrigeration is to control the respiration of crops. The higher the storage temperature, the higher will be the respiration rate and shorter the shelf life.

Cold storage is useful as the principle method of preservation and as an adjunct to other methods of preservation. By cold storage we generally mean storage at temperature above freezing (-2.2-16°C). Commercial and house hold refrigerators are usually run at 4.4-7.2°C. While pure water will freeze at 0°C (32°F), most foods will not begin to freeze until about – 2.2°C (28°F) or lower is reached. Refrigerated storage will preserve perishable food for days or weeks, depending open the food.

Refrigeration system

Cold storage requires low temperature which is achieved by taking heat away from the storage area by the principle of latent heat of vaporisation. If the state is changed from liquid to gas or from gas to liquid without affecting its temperature an amount of heat must be added or removed. Liquids like ammonia and Freon (Cl_2F_2) are used as refrigerants. . A good refrigerant changes into vapour phase upon heating and to liquid state upon cooling. The refrigeration system consists of 4 important units namely compressor, condenser, cooling coil and expansion valves (Fig.1). Liquid refrigerant is circulated through evaporator, within the storage area. In the evaporator, the refrigerant evaporates and absorbs heat from the storage area. The heat comes from the warm moist air blown across the evaporator coil. As it passes over the cool coil, it gives up some of its heat and moisture may condense from it. The cooler drier room air is recirculated by a blower into the space to be cooled. The vapour refrigerant next moves into the compressor, which is basically a pump that raises the pressure so it will move through the system. The increased pressure from the compressor causes the temperature of the refrigerant to rise. In the compressor the refrigerant is converted from the low pressure cold condition to a high pressure hot one. As it leaves the compressor, the refrigerant is a hot vapour, about 120 - 140° F. It then flows into the condenser, where the refrigerant condenses into liquid form. Heat is extracted from the refrigerant, air/water or a combination of air and evaporating water is used to cool the condenser. As the refrigerant leaves the condenser, it is cooler, but still under pressure provided by the compressor. It then reaches the expansion valve. The expansion valve converts the high pressure refrigerant to a low pressure cool liquid. This pressure drop causes expansion followed by evaporation. Evaporation further causes heat absorption making refrigerant cool. The cycle is complete as the cool liquid refrigerant reenters the evaporator to pick up heat from storage area. The temperature of vaporization of the refrigerant is 5 - 6°C below the storage temperature.

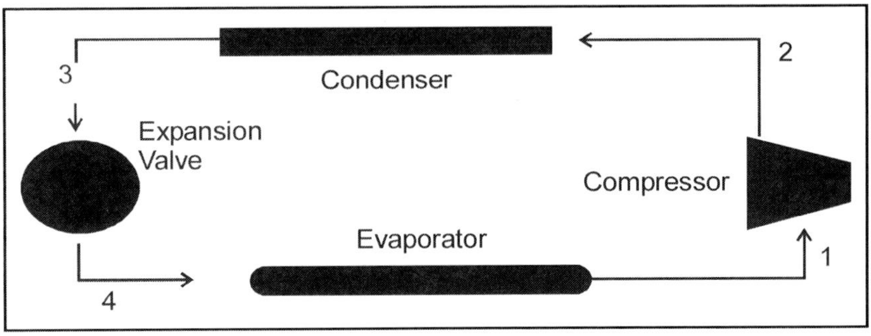

Fig 1. Refrigeration cycle

For the construction of commercial storage rooms, refrigeration load is taken into consideration. This is the quantity of heat that must be removed from the product and the storage area in order to go from a initial temperature to the selected final temperature and then maintain the temperature for a specific time. This depends upon the storage area and other factors that may generate heat within the storage area or influence the removal of heat from the area. Also the specific heat of food and the rate of respiration of such foods as fruits and vegetables have a bearing on the extent of refrigeration required. The heat produced due to respiration in the vicinity of the food surface in cold storage is to be transferred towards the refrigerator. This is achieved by air circulation. Air that is circulated within a cold storage must not be too moist or too dry. If it is high in humidity, moisture will condense on surface of cold foods and moulds will grow on these surfaces. On the other hand, if air is too dry, it will cause excessive drying. Foods can be protected from losing excessive moisture by using several packaging methods. The humidity of air is to be controlled, since it influences water loss in the produce. Most foods store best at a refrigeration temperature when the relative humidity of air is between 80 and 95 per cent.

Frozen storage

Frozen storage as the name implies, refer to storage at temperature where the food is maintained in frozen condition. Good frozen storage generally means storage at -18°C (0°F) or below. Frozen storage will preserve perishable foods for months or even years. Most food spoilage microorganisms grow rapidly at temperature above 10°C. Psychrotropic organisms will grow slowly within the range of 4.4-9.4°C (40-65°F). Below -9.4°C (15°F) there is no significant growth of microorganisms in food, but there is a gradual decrease in the numbers of living organisms. Destruction of microorganisms by cold is not complete; when the food is thawed there can be rapid microorganism multiplication and spoilage.

Foods can be quick frozen in about 90 minutes or less by (i) placing them in contact with the coil through which the refrigerant flows (ii) Blast freezing in which cold air is blown across the food and (iii) dipping in liquid nitrogen.

Quick frozen foods maintain their quality and freshness when they are thawed became only very small ice crystals are formed when foods are frozen in this manner. Frozen foods should always be kept at temperatures below -5°C. Vegetable are blanched at 80°C before they are frozen to inactivate certain enzymes which continue to be active even after quick freezing. Freeze preservation is a best means of preserving pure fruit juice, since the colour, aroma and freshness is retained for a longer period of time. This method is particularly useful in case of juices whose flavour is adversely affected by heating.

6. Methods of Food Freezing

Quick fast freezing

It is the process by which the temperature of the foods is lowered to about -20°C within 30 minutes. This treatment may be achieved by direct immersion or indirect contact of foods with the refrigerant, and the use of air blasts of frigid air blown across the foods being frozen.

Slow freezing

It refers to the process whereby the desired temperature is achieved within 3 to 72 hours. This type of freezing is utilised in the home freezer. Quick freezing process gives more advantage than slow freezing from the stand point of overall product quality.

Air freezing

Air freezing is the oldest and least expensive method of freezing. In this method, food packs are placed in the freezer in such a way that air can circulate between the packages. The freezer is maintained in the range of -23 – -28°C. In air blast freezing food packages are carried in air on an open mesh belt through a tunnel through which air at temperature -20 – -46°C is forced at velocities of 10 to 15 m/s. under or through the product. Under these conditions a food packet that takes 72 hours to freeze in still air can be frozen in about 12 to 18 hours.

Indirect contact freezing

Solid foods in consumer size flat packets are placed on hollow metal plates chilled by circulating refrigerant, so that the foods are in direct contact with cold metal wall but in indirect contact with the refrigerant. The efficiency of freezing depends upon the extent of contact between the plate and the food. The freezing of liquids and purees by indirect freezing is carried out by pumping them through the tubes on the outside of which the refrigerant flows.

Immersion freezing

In immersion freezing there is direct contact between the food or package and the refrigerant. This is particularly important in freezing irregularly shaped food pieces and immersion freezing minimizes the contact of food with air during freezing which is desirable for foods sensitive to oxidation. The refrigerants used for immersion freezing should be nontoxic, pure, free from taste, odour, colour or bleaching action. Low freezing point liquids by indirect contact with another refrigerant or cryogenic liquids like liquid nitrogen, carbondioxide or freon are used as refrigerants.

The low freezing point liquids used are solutions of sugar, salt or glycerol. A temperature as low as -21°C can be reached with a 23 per cent sodium chloride or 62 per cent sucrose solution. Cryogenic liquids are liquefied gases of extremely low boiling point such as liquid nitrogen (BP -196°C), liquid carbon dioxide (BP -79°C) and freon 12 (BP - 30°C). The most commonly used cryogenic liquid is liquid nitrogen. Its low boiling point helps very quick freezing of even large food pieces and it can give a quality unattainable by other non cryogenic freezing methods. Further, since the cold temperature results from evaporation of liquid nitrogen, freezing by liquid nitrogen is nontoxic and inert to food constituents.

Evaporative cooling / zero energy cool chamber

Evaporative cooling is nature's very own method / Evaporation of water produces considerable cooling effect and the faster the evaporation, the greater is the cooling. Evaporative cooling occurs when air, that is not already saturated with water vapour, is blown across any wet surface. The principle of evaporative cooling can be gainfully utilised for storage of fresh produce.

Zero energy cool chamber is a low cost storage structure based on the principle of evaporative cooling. The most important advantage of this low cost technology is that it requires very minimal electricity or power to operate. The raw material required for the construction of cool chamber are bricks, river bed sand, bamboo, khaskhas, rice straw, gunny bags etc. The floor of the chamber is made with single layer of bricks and the side walls are made with a double layer of bricks leaving a cavity of about 7.6 cm space in between two layers of bricks. This cavity is filled with river sand and chamber cover is made with straw or gunny bags on a bamboo frame structure. After construction, wall, floor sand and top cover should be wetted fully by water. Sprinkling of water once in the morning and once in the evening is enough to maintain the temperature and humidity. These cool chambers can reduce the temperature by at least 10°C than the atmosphere and maintain very high humidity throughout the year.

The chamber should be constructed under the shade and the site should not be prone to water logging condition. The shelf life of fruits and vegetables could be increased from a few days to few weeks in these chambers. Increased shelf life of fruits and vegetables in the cool chamber is not only because of low temperature, but also due to uniformity high humidity.

7. Principles of Preservation by Fermentation

Fermentation as a method of preservation has been known to mankind since time immemorial. Microorganisms play beneficial and detrimental roles in preservation. Beneficial role of microorganisms are made use of in preservation of foods by fermentation. Acetic, lactic and alcoholic are the three important kinds of preservation involved in fruit and vegetable preservation. Wine and other alcoholic beverages, cheese and other fermented milk products and bread are some examples of fermented products.

Definition

Decomposition of carbohydrates by microorganisms or enzymes is known as fermentation. Preservation is by virtue of ethyl alcohol, acetic acid and lactic acid formed in alcoholic, acetic and lactic fermentation respectively. Distinct new food products like wine, vinegar and fermented pickles are produced by fermentation.

Alcoholic fermentation

Ethyl alcohol is produced by fermentation of any carbohydrate containing fermentable sugars. The most important product of alcoholic fermentation is wines. The wine is a product obtained by alcoholic fermentation of grape or grape juice by yeast (*Saccharomyces* sp.). The alcoholic product obtained from other fruits is known as fruit wines. Wine represents a non toxic healthful beverage which provides calories, vitamins, minerals and other nutrients. It is not distilled before consumption. It differs from the spirits which are obtained by distillation of yeast fermented grains, grain products eg. whisky, brandy.

The equation that describes the net result of alcoholic fermentation by yeast is as follows.

$$G_6H_{12}O_6 \xrightarrow[\text{Yeast (anaerobic conditions)}]{\text{Yeast}} \text{-°\%}2\ C_2H_5OH + 2\ CO_2$$

Fermentation is carried out under anaerobic conditions employing true wine yeast or *Saccharomyces cervisia.* Theoretically 100g of a hexose sugar should yield 51.1 g of ethyl alcohol and 48.9 g of carbon dioxide. The alcohol content of wine is expressed as volume per cent. ie. cc of alcohol per 100 cc of wine. Wine usually contains 8 – 10 % alcohol.

Classification of wines

Wines are classified based on colour, alcohol content, sugar content, presence of dissolved CO_2 etc.

Based on colour, there are mainly red and white wines. Red wine is prepared from deep purple grapes and skin is retained during wine preparation. The alcohol act as a solvent for anthocyanin pigment present in the skin, yielding a red wine. White wine is prepared from white on greenish grapes or from the juice of grapes from which the skin has been removed.

Wines are classified into dry wines (that does not contain residual sugar) semidry wine (with some amount of residual sugar) or sweet wines (that contains sugar sufficient enough to leave a distinct sweet taste).

Based on the content of alcohol wines are classified into strong (16 – 21% alcohol) medium (16 – 19% alcohol) and light (6 – 9% alcohol). Wines may be sparkling (contain CO_2 dissolved under pressure) or still (does not contain CO_2).

Fortified wines contain added alcohol in the form of brandy. Vermouth is a flavoured wine in which different varieties of herbs and spices are added.

Steps in preparation of wine

Flow chart- wine making

Grapes → Washing → Crushing, destemming, SO_2addition → Sugar acid adjustment → Addition of starter culture → Fermentation → Racking → Fining → Aging and maturation → Blending, pasteurization and bottling

Grapes form the important raw material used in wine preparation. Fruit variety, sugar content, juice yield, yeast strain for fermentation, all affect the final qualities of the wine. *Saccharomysis cervisiae* is the most important strain of yeast for fermentation. It is tolerant to high ethanol concentration, grows even at low pH, ferments 'must' at low temperature, induces desirable aroma to wines and also is resistant to sulphur dioxide level which suppresses the growth of other microorganisms. Selected ripe grapes are washed, berries separated from rachis, crushed and the liberated juice or 'must' is transferred to the vats or closed fermenters. Stalk of berries contain tannins that would give an undesirable colour and flavour to the wine and hence should be eliminated.

For white wine preparation, skin is removed and the juice alone is kept for fermentation. For red wine, skin is retained during early stages of fermentation so as to extract anthocyanin pigments. Sulphur dioxide is added to the must @ 100 ppm to control wild yeast and other microorganisms. TSS of the 'must' is adjusted to 20 -40° Bx by adding cane sugar. The optimum pH for fermentation is 3 – 3.9

Inoculum of wine yeast (*Saccharomyces cervevisiae*) at 2 -3% is added, stirred and mixed evenly. A temperature below 30°C is usually favoured for fermentation and it is found that aroma develops best below 24°C. Fermentation is completed in 3 – 5 days. The layer of grape tissue, skin etc present on the surface of the fermented liquid are removed. The partly fermented wine is decanted from the pomace into storage tanks, allowed to ferment slowly during the next 7 – 10 days. When fermentation is complete, the new wine is decanted or racked from the sediment of yeast and particles of grape tissues to facilitate clarification. To remove the cloudy appearance of wine bentonite/casein/egg white is used as clarifying or fining agent. The wine is allowed to mature/age in cooler rooms. Wine should be matured at least for 8 to 10 months before consumption and flavour continues to improve over several years, especially for red wines. Wines are aged to achieve their final bouquet and essence of flavour.

Acetic fermentation

The production of vinegar (acetic acid) from fruit juice is one of the oldest organic acid fermentation known. Acetic acid is produced by oxidation of ethyl alcohol by bacteria such as *Acetobacter aceti, A. orleansis, A. schutzenbachi* etc. Two distinct processes are involved in the preparation of vinegar.

1. Transformation of sugars of fruits or any sugary materials to ethyl alcohol by yeast anerobically.
2. Oxidation of alcohol into acetic acid by acetic acid producing bacteria in the presence of air.

The biochemical reaction by which acetic acid is produced is as follows

$$\underset{\text{Glucose}}{C6H12O6} \xrightarrow[\text{Ethyl alcohol}]{\text{Yeast (anaerobic conditions)}} 2\ C2H5OH + \underset{\text{Carbon dioxide}}{2\ CO2}$$

$$\underset{\text{Ethyl alcohol}}{2\ C2H5OH + 2O2} \xrightarrow[\text{Oxygen}]{\text{Acetic acid bacteria (anaerobic conditions)}} \underset{\text{Acetic acid}}{2\ CH3COOH} + \underset{\text{Carbon dioxide}}{2CO2}$$

Theoretically 100 parts of sugar (sucrose) should yield 51 parts of ethyl alcohol and 67 parts of acetic acid.

Vinegar is a liquid obtained by alcoholic and acetic fermentation. It contains 5 per cent acetic acid and has germicidal and antiseptic properties. The amount of acid in vinegar is expressed as grain strength which is ten times the per cent of acetic acid present in it *ie.* vinegar having 5 per cent acetic acid is termed as vinegar of 50 grain strength.

Vinegar is prepared from fruits, starchy materials (potato) and sugar containing substances (molasses, honey).

Acetic acid fermentation is brought about by acetic acid bacteria (*Acetobacter* sp.) which are strongly aerobic, but whose activity is greatly reduced by light. So acetic acid fermentation is carried out in the dark. For acetic acid fermentation, the alcohol content of fermented mass is adjusted to 7 -8 per cent by dilution with water, since acetic acid bacteria cannot grow well at higher concentration of alcohol. Mother vinegar containing acetic acid bacteria is added at the rate of one part to ten parts of fermented mass is order to check growth of undesirable microorganisms and to hasten the fermentation process. Vinegar can be produced by different processes like slow process, Orleans slow process, quick process, fringe process and submerged culture process.

Lactic acid fermentation

Lactic acid fermentation is an anaerobic intra-molecular oxidation reduction process. Vegetables like cabbage, cauliflower, carrot, turnip, radish, cucumber, tomato, onion, garlic and fruits like mango, apple, lime, pear etc are suitable for lactic fermentation. Fermented vegetables are common ingredient of European diet.

Pickling is the result of fermentation by lactic acid forming bacteria, which are generally present in large numbers on the surface of fresh vegetables and fruits. Generally bacteria prefer low or no acid medium for their growth. The lactic acid bacteria however can grow in acid medium and also produce acid through their action on the substrate. They can grow in the presence of 8 – 10 % salt. Advantage is taken of these two factors in pickling. The growth of undesirable microorganisms is inhibited by adding salt, while allowing the lactic acid fermentation to proceed.

The reaction involved is as follows:

$$\underset{\text{Glucose}}{C_6H_{12}O_6} \xrightarrow{\text{Lactic acid bacteria}} \underset{\text{Lactic acid}}{2\ CH_3\ CHOH\ COOH}$$

Sauerkraut is an important product obtained by lactic acid fermentation, popular in European countries.

Steps in preparation of sauerkraut

Fresh cabbage → trimming, cleaning and shredding (2.5 mm size) → salting at 2.25% (w/w) → fill in vats or containers → cover and seal with proper weight → fermentation (30 days) → pasteurization (at 74°C for 3 minutes) → filling in containers

Fermentation takes place fairly well in a brine containing 5% salt, but proceeds slowly with 10% salt. It is therefore customary to place the vegetables in a 10 per cent salt solution to allow lactic acid fermentation to take place and then increase the concentration of salt gradually to 15%. When vegetables are placed in brine (dry salt or brine solution) soluble material present in them diffuse into salt water due to osmosis, this soluble material contains fermentable sugar. Lactic acid bacteria present on the surface of fruits and vegetables convert these sugars into lactic acid. When sufficient lactic acid is formed, the lactic acid bacteria cease to function and any further change in the composition of the material is prevented. Materials that have undergone lactic acid fermentation must be kept sealed from air to prevent growth of acid destroying bacteria.

8. Preservation by dehydration

Drying and dehydration of foodstuffs are very age old practice. In each case some amount of moisture is removed and concentration of food constituents occur. Drying is the simplest method of food preservation, achieved under the influence of non-conventional energy sources like sun and wind. In this method there is no possibility of temperature and humidity control and it is dependent upon the elements which are beyond strict control.

Dehydration technique requires standardization of many factors viz., time, temperature, relative humidity and air velocity depending upon the products to be dehydrated.

It is more rapid than sun drying and can be done under very hygienic conditions. Dehydrated products are generally free of contamination and are of good quality and hence can command a sufficiently higher price.

Drying and dehydration have several advantages as well reduction in weight of the products is achieved to 1/4th to 1/9th of their original or fresh weight, reducing the cost of transport. Due to reduction in bulk of the products, less storage space is required. Cost of processing is also very low compared to other methods as less labour and ingredients are required, Food dehydration causes minimum or no change to food value. Natural flavor as well as characteristic cooking quality of the fresh material is also retained. It leaves the mineral contents virtually unchanged. Therefore, the process is helpful in preserving the nutritive content of final products. They provide consistent product which is an important characteristic for marketing.

Factors affecting dehydration

Whatever method we adopt for dehydration, it basically involve two processes (1) Entry of heat into the product (2) Removal of moisture out of the product. To operate these two occurrences in a congenial way the following factors are to be taken into account.

Surface area

If the product to be dehydrated are made to small pieces or thin layers the drying rate will be more. This is due to the fact that larger surface area of the product come in contact with heat. Secondly smaller pieces or thinner layers reduce the distance that the heat must travel to the center of food and reduce the distance through which the moisture in the center of the food must travel through the surface and escape.

Temperature

Temperature is the driving force for removal of moisture. The greater, the temperature difference between heating medium and the food, the greater will be the rate of heat transfer into the food. If air is the heating medium, the velocity of dry air and volume of air that come into contact with the product to be dehydrated is important.

Air velocity

If the velocity of hot air is more, the sweeping of moisture from product will be also more. Air in motion will prevent the held up of moisture in the atmosphere which would slow down subsequent moisture removal.

Dryness of air

The drier the air, the more rapid is the rate of drying. Dry air is capable of absorbing and containing more moisture while the moist air is close to saturation and hence can absorb less moisture. The extent of dryness of air also determines how a low moisture content food product can be dried to further low moisture content. Dehydrated food is hygroscopic and each food has its own, equilibrium relative humidity (ERH), which is the humidity at a given temperature where the food will neither loose moisture to the atmosphere nor pick up moisture from the atmosphere.

Time

The duration of temperature treatment given to the product to achieve moisture removal is very important. Generally drying process which employ high temperature for short time do less damage to food than drying processes employing lower temperature for longer times.The above factors affecting heat and mass transfer such as temperature, humidity, airvelocity, maximum surface area and the factors alike are easy to optimize and control.

Drying rate factor curves

The drying rate of the product is dependent on the time, and moisture content of the products. The relation between these factors and drying rate can be plotted on graph (Fig 1).

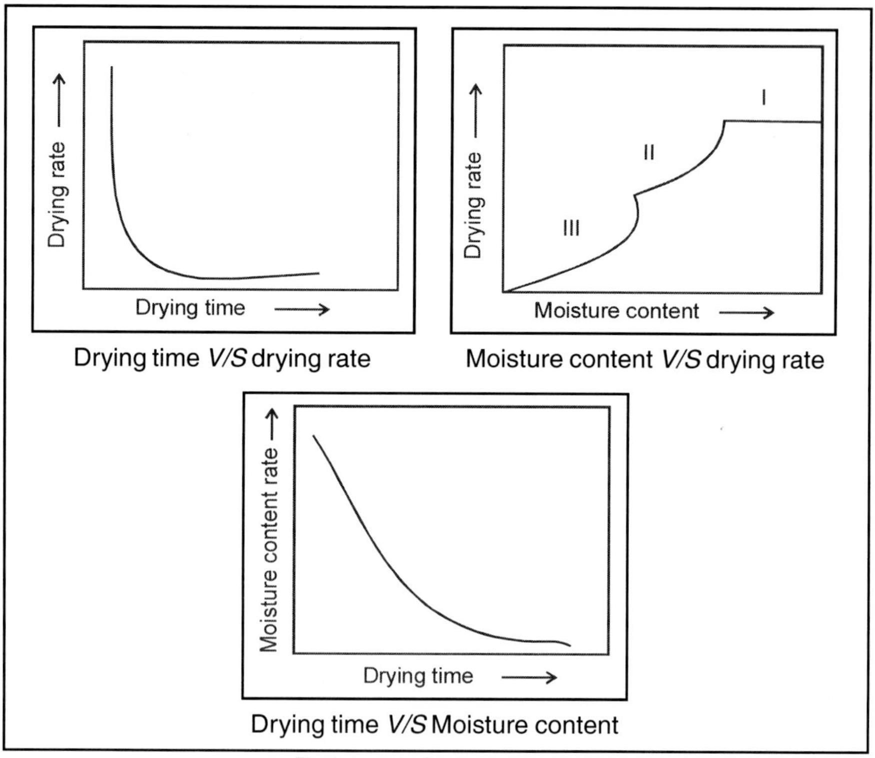

Drying rate factor curves

Drying time *V/s* drying rate

The foods subjected to drying do not loss water at constant rate and the rate of water removed drop off as the drying progress. We can remove 90% of water in 2 hrs, but it may require more than 2 hrs to remove the remaining 10% water. This relation can be graphically represented as follows.

Moisture content *V/s* drying rate

Generally drying of most fruits and vegetables shows constant and falling rate periods. This relationship is given in moisture content Vs drying rate. However shape of curve may vary with different food materials, different types of dryers and drying conditions.

As the moisture content decreases, drying rate changes from constant rate of drying to falling rate of drying. During falling rate period, rate of water movement from interior of foods to surface falls below the rate at which water evaporates to surrounding air.

Drying time *V/s* moisture content

The moisture content of the food stuff register a drastic reduction in the initial phase of drying and reduces to a minimum as the drying progresses. The removal of moisture below about 2% without product damage is extremely difficult in dehydration.

Pre-treatment for drying and dehydration

To obtain the excellent quality of a dried product the raw produce needs to be harvested and handled properly. For better quality of dried products following pre-treatments are required.

Harvesting

Harvesting the product at optimum maturity according the product to be prepared out of it is very important.

Washing

Fruits and vegetables are usually rinsed in cold water to remove surface dust particles and pesticides spray residues. A residual chlorine content of 10-20 ppm of water is recommended to decrease the initial count of microorganisms on raw material to minimum.

Peeling and slicing

Some fruits and vegetables, apple and sometimes cling-stone peaches are peeled prior to drying. This is accomplished by various means: refractory, lye, hot brine peels, high pressure steam for root crops, mechanical knife peelers for apples and dilute lye for cling-stone peaches. Prunes, grapes, berries and cherries are dried as whole, although latter may be pitted.

Pricking

Pricking is given to increase the drying rate. Whole fruits, vegetables, or their slices are uniformly pricked with stainless steel prickers or forks. It also helps osmotic dehydration of products.

Soaking

The fruits and vegetable slices after pricking are soaked in plain water, lime water, brine or alum solution for few hours to few days before blanching and cooking

Curing

Curing is practiced in onion and garlic. It dehydrates raw material by sun drying or by hot air drying at a temperature of 30^0C.In garlic it is carried out until the segments can be easily separated.

Alkaline dip

Alkaline dip is used primarily for fruits that are dried as whole. It facilitates drying by forming fine cracks in the skin. A sodium carbonate or lye solution(0.5% or less) is used at a temperature ranging from 93-100^0C

Acid dip

Before sulphuring use of an acid pre-dip provides better color stability to a product. A dip of 1% ascorbic acid and 0.25% malic acid has been used to retard enzymatic browning in peaches.Steepinng of mushroom in 0.5% to 1.0% citric acid for 16 hrs results into better quality dried products in respect of color, appearance, rehydration properties and keeping quality.

Blanching

Blanching is partial pre-cooking treatment in which vegetables/fruits are usually heated in water or in live steam to inactivate the enzymes before processing.

Purpose of blanching

- Reduce drying time
- Removes inter-cellular air from the tissues
- Causes softening of texture
- Retards the development of objectionable odour and flavor during storage
- Retain carotene and ascorbic acid during storage
- Remove pungency(onion)

- Impart desired translucent appearance to the product

Sulphiting

Sulphitation is a process in which the product is dipped in SO_2 containing water

Purpose of sulphiting

- Maintains an attractive color
- Prevents spoilage
- Preserves certain nutritive attributes until marketing
- Act as antioxidants and protect the carotene and ascorbic acid content of dried fruits.

Post Drying treatments

Sweating

Sweating is a practice of storage of dried product in bins or boxes for equalization of moisture or re-addition of moisture to a desired level. It is used primarily with some dried fruits and some nuts (almonds and walnuts)

Packaging

Most products are packaged after drying for protection against moisture, contamination with micro-organism, and infestation with insect, although some dried foods may be held as long as a year before packaging.

Osmotic dehydration

Osmotic dehydration has received greater attention in recent years as an effective method for preservation of fruits and vegetables.Osmo dehydration is one of the complementary treatments and food preservation technique in the processing of dehydrated foods, since it presents some benefits such as reducing the damage of heat to the flavour, colour, inhibiting the browning of enzymes and decrease the energy costs. Osmotic dehydration is a common pre treatment used before air drying. The technique consists of immersing the fruit in a hypertonic solution to remove part of the water from the fruit. It involves dehydration of fruit slices in two stages, removal of water using an osmotic agent and subsequent dehydration in a drier where moisture content is further reduced to make the product shelf stable. The different types of solutes used as osmotic agent for osmo dehydration are fructose, corn syrup, glucose and sodium chloride.

Common types of driers

Solar driers

The matrix type solar air heater consists of an iron scrap as an absorber. The upper surface of this matrix is painted dull black. Air is made to flow through the matrix in the upward direction. Similarly in rock type solar air heater is connected with the cubic shaped dehydration chamber through an air duct of appropriate dimension. Dehydration chamber has provision for holding trays and thermometer a glass door and a chimney. Solar driers are different types like tent type drier, walk-through drier, air flow inside indirect drier cabinet, indirect chimney driers etc.

Air convection driers

All air convection driers have some sort of insulated enclosure, a means of círculating air through the enclosure, and a means of heating this air. They also will have various means of product support, special devices for dried product collection. Some will have air driers to lower drying air humidity, and will be equipped with various kinds of instrumentation and controls.

Movement of air is generally controlled by fans, blower and baffles. The air is in direct contact with a flame or combustion gases. In indirect heating the air is in contact with a hot surface such as pipes or fins heated by steam, flame or electricity.Klin cabinet ,tunnel , belt trough, airlift, fluidized bed and spray drier are the different versions of air convection driers.

Roller drier

Roller drier consists of one or two hollow drums which are fitted so that a heating medium usually steam, water or a special high temperature, heat transfer liquid can be directed through them. The drums are mounted to rotate about the symmetrical axis and are customarily driven with a variable speed mechanism. The feeding device applies a thin, uniform, layer of the material to be dried on the hot drum surface. A knife or doctor blade is fitted to the drum at an appropriate location. The material is dried as the heated drum rotates towards the doctor blade which scrapes the thin layer of dry material from the drum surface. Drum and foam-mat driers are common types of roller driers.

Vacuum dryers

Vacuum dehydration methods are capable of producing the highest quality dried products, but costs of vacuum drying are generally higher than other methods which do not employ vacuum. In vacuum drying, the temperature of the food and the rate of water removal are controlled by regulating the degree of vacuum and the intensity of heat input. Heat transfers to the food is largely by conduction and radiation. Vacuum drying methods usually can be controlled with a higher degree of accuracy than other methods depending upon air convection heating.

Freeze dryers

Freeze –drying can be used to dehydrate sensitive high value liquid foods such as coffee and juices, but it is especially suited to dry solid foods of high values such as stawberrries, whole shrimp, chicken dice, mashroom slices, and sometimes food pieces as large as steaks and chops. These types of foods, in addition to having delicate flavors and colors have textural and appearance attributes which cannot be well preserved by any current drying method except freeze-drying. In freeze dryers the material such as fruit juice concentrate, is first frozen on trays in the lower chamber of a freeze driers and frozen material dried in the upper chamber under high vacuum. The material dries directly by sublimation of ice without passing through the intermediates liquid stage. The dried product by this method is highly hygroscopic. It reconstitutes readily.

9. Preservation by thermal processing

The process of sealing food stuffs hermetically in containers and sterlise them by heat for long storage is known as canning. The process gained the name in honor of Nicholas Appert (1804) who invented the process of sealing food stuffs hermetically in containers and sterilizing them by heat.

Canned products have long shelf life since the process involve destruction of complete microorganism. They can be sold in off season fetching better returns to the entrepreneurs.

Selection of raw materials

For canning raw materials should be absolutely fresh. The fruit should be ripe, but firm and evenly matured. It should be free from the blemishes insect damage and malformation. Over ripe fruit is generally infected with microorganisms and would yield a pack of poor quality. Under ripened fruit will generally shrivel and toughen on canning. The vegetables should be tender and reasonably free from soil, dirt *etc*. Tomatoes should however, be firm, fully ripe and of deep red color .The food item should be properly prepared and tested for purity and quality.

Sorting and grading

After a preliminary sorting of for the undersize, over ripe, immature, insect and pest damaged ones, fruit and vegetables will be graded. It can be done based on weight, shape and color. It is done by hand or with the help of grading machines .Among several mechanical graders, screen rollers are most commonly employed.

Washing

A thorough wash of fruits and vegetables are to be done to achieve quality products. Fruits or vegetables are generally soaked or agitated in water, washed with cold, hot or salty water to disinfect them. Washing by means of high pressure water spray is the most satisfactory method.

Peeling, coring or pitting

Peeling of fruits and vegetables are done manually or mechanically through heat or lye treatment even though hand peeling is commonly practiced.Mechanicl peeling have been recently introduced for pears, apples root vegetables like carrots, turnips, potatoes etc.The heat treatment soften and loosen the skin of some fruits and vegetables, particularly certain varieties of peaches and potatoes and enable easy removal of skin.

Lye peeling is also practiced in the case of some fruits and vegetables like peaches, apricots,sweet orange,carrots,sweet potatoes etc.For this the raw materials are treated by dipping them in boiling caustic soda or lye solution of 1-2% strength for a short period ranging from 30 seconds-2 minutes.

Blanching is given to fruits and vegetables prior to filling them in cans by treating them with boiling water or steam for short periods.

Blanching helps to:

- Remove respiratory gases
- Inhibit enzymatic reactions,
- Promote shrinkage of products
- Hydrate the products
- Preheat the product to assist vacuum formation in cans.

In small scale canning, the fruits and vegetables to be blanched are placed on the wire of a perforated basket, which is first dipped in hot water for a short period and then in cold water. In large canneries, blanching is done on belt conveyors passing through boiling water or steam or in rotary horizontal cylinders.

Can filling

The prepared fruits or fruit slices are filled in cans, which are sterilized by various methods. In India washing empty can in open tanks containing hot water is the usual practice. However simple devices designed to sterilize the cans in steam also there in use. Automatic can filling machines are used in large canneries but to ensure a properly graded pack and to prevent bruising, it is better to adopt hand filling. In India filling by hand using rubber gloves is the common practice.

Syrup or Brine

The cans are filled with hot sugar syrup (usually sucrose) for fruits and with hot brine for vegetables. The objective of adding syrup or brine is to improve the taste of the canned product, to fill up the interspaces between fruits and vegetables and to facilitate further processing .In some canneries syruping or brining is done with the help of automatic machine. The syrup or brine prepared at 79-82^{0}C is added to the products leaving suitable headspace in the can

Lidding or clinching

The process of lidding involve covering the lid loosely on the cans, before passing the cans through the exhaust box. But now, instead of lidding, clinching is done, where the lid is partially seamed to cans by a double seamer.The lid is left sufficiently loose to allow the escape of dissolved as well as free air from the content

Exhausting

Before giving final sealing all the air present in the can is removed by the process of exhausting

Containers are exhausted either by heat treatment or mechanical means, the cans are passed to a tank of hot water at about 82-87^{0}C or on a moving belt through a covered steam box. The time of exhaust varies from 5-25minutes depending on nature of products. At the end of exhaust the temperature of the center of can should be 79^{0}C.

The advantages of exhausting are

1. It reduces the risk of corrosion of tin plate and pin holing during storage
2. Prevent the discoloration of the products due to oxidation
3. Removal of air also helps retention of vitamins especially vit C
4. It prevents bulging of cans when stored at high altitudes in hot climate, reduces the chemical reaction between the container and the contents ad prevents the excessive pressure as well as strain during sterilization.

Sealing

After exhausting, the cans are closed tightly with the help of a machine known as double seamer. Manual semiautomatic or fully automatic double seamers are available. Glass containers for processed products are generally vacuum sealed.

Processing

The term processing in canning refers to the sterilisation of canned foods by heating or cooling to inactivate the bacteria. Absolute sterilization is difficult to attain, as many bacteria can form very minute and highly heat resistant bodies called spores, which can be killed only either by very high or very low temperature treatment or by prolonged coring. This is usually achieved by heating for an accurately predetermined time and temperature in an atmosphere of saturated steam, in hot water or occasionally in a steam air mixture.

The temperature at which processing is done depends upon the nature of food, mainly its acidity. Fruits and vegetables products can be classified into following four groups according to their pH values.

Low acid foods (75.0 pH)	:	Peas, cauliflower, potato, French bean, spinach etc.
Medium acid (4.5 to 5.0 pH)	:	Turnip, carrot, cabbage, green bean, okra etc.
Acid (3.7 to 4.5 pH)	:	Tomato, pear, banana, mango, pineapple, apple and other fruits
High acid (< 3.7 pH)	:	Citrus juice, sauerkraut, pickle, chutney etc.

The bacterial spores won't be in active phase below pH 4.5.So the canned products (generally fruits) with less than pH 4.5 need to be processed only in boiling water. Products with pH above 4.5 (Non acidic vegetables) require processing at 0.70 to 1.05kg/cm^2. It should be ensured that the high temperature at the centre of the can needs to be maintained for sufficient period to destroy the spores of the more resistant bacteria.

Time and temperature of processing vary with size of can as well. Larger the can, greater is processing time. Processing may be carried out in batch or continuous retorts which are equipped with specially constructed hydrostatic seals, through which the containers enter and leave the pressure zone. Fruits and acid vegetables are usually processed in open type cookers or continuous nonagitating cookers, while vegetables are processed under steam pressure in cold containers.

Effect of altitude on processing time

The boiling point of water decreases with an increase in altitude. The processing time for different kind of fruits and vegetables, as standardized at places about 152meters above mean sea level, has to be relatively increased with further increase in the altitude. As a rule, for every increase of 152 meters in altitude, the boiling point of water decreases by about 10^{0}C and normal processing time has to be increased by about two minutes.

Thus, an A 2$^1/_2$ size can of pears, which requires processing for 30 minutes at sea level, will require a total processing time of 30+25= 55minutes at an altitude of about 1830 meters above mean sea level.

Effect of altitude on processing pressure and temperature

As the altitude above sea level increases, the pressure required to maintain the specified processing temperature, say 115^{0}C or 121^{0}C,will also increase. Since process-time schedule are based on the exposure of the canned food to the specified temperature in steam, it may be necessary to maintain higher gauge pressure if the processing is done at an altitude substantially grater than 305 meters above mean sea level.

Cooling

Immediately after processing, the cans are cooled to stop the cooking process. This will help to prevent stack burning, especially in cans at the center of large stacks where they may remain hot for several hours, sometimes for several days. Prolonged heating results in an inferior or uneven pack.

In extreme cases peaches and pears become dark in color, tomatoes turn brownish and become bitter in taste while peas become mushy with a cooked taste. Many vegetables develop flat sour on prolonged heating followed by improper cooling due to heat resistant spore forming bacteria which survive ordinary commercial processing.

The methods commonly adopted for cooling are

a. Immersing the hot cans in tanks containing cold water

b. Spray with jets of cold water

c. Turning in cold water into the pressure cooker in the case of canned vegetables

d. Exposing the cans to air in small lots when water supply is scarce.

Storage

After labeling cans, they should be placed in storing wooden cases or corrugated cardboard cartons and stored in a cool and dry place. Storage of cans at high temperature should be avoided as it shortens the shelf life of products.

India has the ideal agro climatic conditions suitable for cultivation of a wide range of fruits and vegetables. There is rich diversity of indigenous fruits and vegetables that are underexploited. Many of these fruits and vegetables are well known for their nutritive and therapeutic values, but may not be palatable in raw form eg. Karonda and Kokkum. These fruits have the potential for processing into value added products.

Most of the tropical fruits like mango, jackfruit and cashew apple are seasonal in nature. During peak harvesting season ,a good amount of quality fruits and vegetables are available in abundant quantity resulting in market glut and all of these quality produces cannot be marketed in fresh form. As a result, there is huge spoilage and wastage of fresh fruits and vegetables. This wastage can be reduced if they are timely processed into different value added products or preserved by different methods during harvesting season.

Value Addition

Value addition is the process of increasing economic value of an agricultural commodity by means of processing, packaging or diversification.

10. Advantages of Processing

- Minimization of postharvest losses
- Extension of shelf life of foods
- Value addition
- Ensuring availability throughout the year
- Ensuring availability in areas other than places of production

- Employment generation
- Adding variety to diet.

The fruit and vegetable processing industry is generating waste to the extent of 20 – 50% of raw materials utilised. Many of these wastes are rich sources of vital constituents of commercial use like pectin, starch, pigments and oils. Effective utilisation of waste not only serves to alleviate problem of environmental pollution, but also reduces price of primary product.

11. Demand for Processed Products

In recent years, scope for processing industry has increased due to following reasons.

- Urbanization and growth of small and nuclear families and increasing number of working women have increased the demand for processed and convenient foods (ready to eat, half cooked, full cooked etc).
- Increase in purchasing power, change in food habits and life style due to modernisation and effect of mass media.
- Fluctuation in availability and price of fresh commodities. Food industry provides foods at reasonable and steady prices throughout the year.

Consumers' desire today is for foods, more convenient to store and prepare for consumption, higher in quality and freshness, more natural, nutritively healthier and medicinally enriched.

Across the world, food industry plays a vital role in growth of economy. In spite of tremendous potential, only less than 2% of production of fruits and vegetables in India is utilised for processing. This is very low compared to USA, Brazil, Malaysia, Europe and Thailand where fruits and vegetables are utilised for processing. Range of products processed by the industry is also narrow, mainly based on fruits, especially mango.

India's share in the world trade of processed fruits and vegetables is still less than 2%. There is great demand for fresh and processed fruits and vegetables in the world market which will not only earn the much needed foreign exchange for the country but also create job opportunities. The export of processed products increased from Rs. 51 to 60 crores in 1988-89 to Rs. 1265 crores in 2001-02. There are about 5500 fruit and vegetable processing units in the country of which only 13% are large scale units. Considerable processing of fruits and vegetables is done in home and cottage scale sector. Large quantities of pickle, chutneys, jam and jellies, are prepared in house hold sector. The installed capacity of processing units is 21.1 lakh t. The capacity utilization continues to remain low at 45-50 % level for several years.

Value added products

Major processed products from fruits and vegetables are fruit pulps and juices, ready to serve (RTS) beverages, canned products, jams, squashes, preserves, ketchups, pickles, dehydrated products and fruit juice concentrates. The more common processing methods to ensure stability of products are use of chemical preservatives and thermal processing. There is always an intense need for improved preservation technology that effectively inactivates undesirable microorganisms and enzymes in foods and yet maintain quality and meet the changing requirements of consumers.

Fruit juice concentrates

Fruit juices and pulp are preserved as concentrates by using units operations like vacuum evaporation, freeze concentration and membrane processing techniques like osmosis, reverse osmosis and ultra filtration. Most of the fruits are very high in water content. Decreasing water content brings desired microbial stability. The developed product is light in weight and volume and permits economy in packaging transportation and distribution.

Intermediate moisture foods

Intermediate moisture foods (IMF) are products having a moisture content of 30–40 %. In these products, instead of removing bulk of water from the fruit as in traditional drying, just enough water is removed or bound through addition of humectants to retard or greatly reduce microbial growth, with improved retention of sensory and nutritional attributes. Technology(s) for preparation of IM fruit slices and fruit bars of guava, pineapple, mango, banana, jackfruit, apple and sapota are available.

Minimally processed products

Minimally processed (fresh cut) fruits and vegetables are ready to eat products which include fresh, washed and chopped ones, ready for use and packaged in sealed polymeric films or trays. Minimally processed products offer a number of advantages like convenience, time saving and reduction in accumulation of solid waste. They retain flavour, aroma and nutrition better than conventionally processed fruits and vegetables.

H T stabilized products

H T preservation is a combination of preserving techniques in which several technological approaches like freezing, refrigeration, dehydration, irradiation, use of chemical preservatives are used in unison in a synergistic way to bring about desired microbial safety and sensory acceptability. Two or more hurdles (temperature, water activity, pH, redox potential and preservatives) are used in combination to prevent spoilage. Microorganisms have to overcome these hurdles to cause spoilage of food.

Value added products from fruit and vegetable waste

Fruit waste, generated in large quantities, is a problem to the processing industries and pollution control agencies. Utilization of waste not only economizes cost of finished products but also reduces pollution level. Waste could be used either as animal feed or some value added products be obtained by chemical or fermentation processes. Some of the important products like fibre, starch, protein, oil, flavour and wax can be extracted from waste materials. Many important metabolites can be obtained by fermenting solid or liquid wastes with yeast, fungi or bacteria. Major products obtained by fermentation are enzymes, protein, organic acids, ethyl alcohol and biogas. The organic waste decomposes over a period of time to produce manure. A variety of processed products also can be prepared from fruit waste.

Value added products from fruits

Fruits	Products
Mango	Pickle (tender and mature mango), amchur, raw and ripe mango beverage, jam, sauce, fruit bar, cereal flakes, chutney, canned products, frozen product
Banana	Puree, fig, raw and ripe fruit powder, flakes, jam, chips, jelly, RTS beverage, wine, pickle, fruit bar
Pineapple	RTS beverage, jam, jelly, fruit bar, sauce, sauce
Apple	RTS beverage, jam, jelly, fruit bar, sauce, apple butter
Papaya	Papain, tutty fruity, jam, beverages, jelly, nectar, pickle, sauce, fruit bar
Grapes	Raisins, juice, wine, concentrates
Guava	Jelly, jam, beverage, cheese, puree
Sapota	Jam, fruit bar, chunks, squash
Anola	Candy, pickle, beverage
Jack fruit	Jelly, candy, slices in syrup, pickle, chips
Litchi	Jam, jelly, beverage, wine
Bael	Nectar, squash, fruit bar, candy
Ber	Powder, jam, jelly, candy, wine
Karonda	Preserve, pickle
Acid lime	Pickle, beverage, dehydrated products

Value added products from vegetables

Products	Ideal vegetable
Pickle	Garlic, cabbage, cauliflower, ivy gourd, bitter gourd, carrot
Jam	Water melon, tomato, pumpkin
Jelly	Pumpkin, rozelle
Sauce	Tomato, chilli, pumpkin
Beverages	Tomato, water melon, carrot
Dehydrated products	Bitter gourd, cowpea, carrot, onion, garlic, peas, cauliflower
Candy	Ashgourd, pumpkin
Slice in syrup	Musk melon, water melon
Slice in brine	Cucumber, gherkin, shelled beans, pea
Puree	Tomato

Contd.

Products	Ideal vegetable
Powder/Flakes	Onion, Garlic, Potato, Chilli
Chips	Potato, cassava, yams
Canned Products	Tomato, okra, shelled peas
Frozen dry	Peas, cauliflower, carrot, okra
Soup	Asparagus, sweet corn, spinach, tomato
Chutneys	Chillies, tomato, gogu

Processed products / by products from fruit and vegetable waste

Products	Fruit/vegetable waste used
Jelly	Watermelon rind, jackfruit rind and perigones, passion fruit rind
Jam	Watermelon rind, tomato trimmings
Dehydrated products	Rind of cucurbitaceous vegetables (ash gourd, pumpkin, watermelon)
Candy	Pineapple core, watermelon rind, citrus peel
Cheese	Guava pomace
Pickle	Banana rind, watermelon rind
Wine	Pineapple peel
Pectin	Citrus peel, apple peel, jackfruit rind
Vinegar	Pineapple peel
Oil	Citrus peel, grape seed, tomato seed
Enzymes (pectinase)	Apple peel, pineapple peel
Starch	Mango kernel, jackfruit seed
Fibre	Pineapple peel, jackfruit rind, banana peel
Natural colour	Grape skin, kokum, jamun
Citric acid	Pineapple waste
Cattle feed	Jackfruit seed, pineapple peel, grape skin

India is endowed with ideal agro-climate conditions to attain higher levels of productivity with diversity. To substain the agro based economy of the country and to prevent losses of horticultural produces, processing of fruits and vegetables is inevitable. The processed sector is gaining importance due to its vast potential for wealth and employment generation.

References

Desrosier, W and Desrosier, J.N (1987) The Technology of Food Preservation. CBS Publishers and Distributors, New Delhi.

Fellows, P.J. (2005) Processing Technology: Principles and Practices. Woodhead Publishing Limited, Cambridge, England.

Jood, S and Paul, N.K (2002) Food Preservation. Agrotech Publishing Company Udaipur.

Khader, (2001) Preservation of Fruits and Vegetables, Kalyani Publishers, New York.

Khader, (2001) Food Science and Technology. ICAR, New Delhi.

Khatkar, B.S (2007) Food Science and Technology. Daya Publishing House New Delhi.

Manoranjankalia and Sood, S. (2002) Food Preservation and Processing. Kalyani Publishers, Ludhiana.

Potter, N (1987) Food Science. CBS Publishers and Distributors, New Delhi.

Potter, N and Hotchkiss, J.H (1996). Food Science. CBS Publishers and Distributors, New Delhi.

Ramaswamy, H and Marcotte, M (2005) Food Processing: Principles and Applications. Taylor & Francis, New York.

Srivastava, R.P and Sanjeevkumar, (2006) Fruit and Vegetable Preservation - Principles and Practices. International Book Distributing Company, Lucknow.

Sudheer, K.P. and Indira, V. 2007. K.V. Peter (Ed.) Post Harvest Technology of Horticultural Cros. New India Publishing Agency, New Delhi, 291p.

Subbulakshmi G, Udippi, S.A (2001) Food Processing and Preservation. New Age International (P) Limited Publishers, New Delhi.

Chapter – 9

Apomixis in Plants

N.K. Parameswaran

Apomixis is the term first introduced by Winkler in 1908 to denote the form of reproduction in plants which does not involve meiosis and syngamy (Maheswari, 1950). It is a substitution for the normal and usual sexual reproduction and results from production of a zygote which bypasses the sexual process of meiosis and fertilization. The genotype of the zygote and resulting embryo will be the same as the seed parent. The hall mark components of apomixis include female gamete formation without meiosis (apomeiosis), fertilization, independent embryo development (parthenogenesis) and developmental adaptations to ensure functional endosperm formation. Tracing the molecular mechanisms underlying apomixis, a developmentally fascinating phenomenon is critical for the successful induction and utilization in crop plants (Promila Gupta *et al.*, 1996).

Clonal seedlings coming out of apomixis are known as apomicts and there are two types of apomicts *viz.,* obligate apomicts and facultative apomicts. Obligate apomicts include species or individuals producing only apomictic embryos and facultative apomicts include species which produce both apomictic and sexual embryo (Singh *et al.*, 2011).

1. Groups of Apomixis

There are two main groups of apomixis:

Vegetative reproduction

Plants are propagated by a part of the body other than the seed. The structural units used for this purpose are termed as propagules.

Agamospermy

Plants belonging to this category have retained seed as the agent of propagation but the embryo is formed by some process in which normal meiosis and syngamy are eliminated.

There are three different types of agamospermy.

Adventive embryony: Embryogenesis takes place directly from diploid sporophytic cells (nucellar or integuments). Development of sexual embryosac takes place normally but zygotic embryo either degenerates or competes with apomictic embryos.

Diplospory: An archesporium differentiates, but the megaspore mother cell (MMC) develops to an unreduced embryo sac. The embryos are formed by the unfertilized egg (parthenogenesis) or some other cells of the embryosac (apogamety).

Apospory

A somatic cell in the nucellus directly forms an undivided embryosac and the diploid egg parthenogenetically develops in the embryo. The apomictic embryo sac may develop in addition to the haploid embryo sac derived from a functional MMC. The latter may either degenerate at an early stage or form a sexual embryo after fertilization.

Maheswary (1950) classified ***apomixis*** in to:

- **Recurrent apomixis**
- **Non recurrent apomixis and**
- **Vegetative apomixis.**

2. Recurrent Apomixis

An embryosac (female gametophyte) develops from the egg mother cell (or from some adjoining cell, the egg mother cell disintegrating) but complete meiosis does not occur. Consequently the egg has the normal diploid number of chromosomes, the same in the mother plant. The embryo subsequently develops directly from the egg numbers without fertilization.

Recurrent apomixis occurs in a few species of *Crepis, Taraxacum, Dandelion, Poa* (blue grasses) and *Allium* (onion) without the stimulation of pollination. In others eg. species of *Parthenium* (guayule), *Rubus* (raspberry), *Malus* (apple), some *Poa* species and *Rudbeckia*, pollination is necessary either to stimulate embryo development or to produce a viable endosperm.

3. Non-reccurrent Apomixis

An embryo arises directly from the egg without fertilization. Since the egg is haploid, the resulting embryo also will be haploid. Such cases are rare and primarily of genetic interest. It does not consistently occur in any particular type of plant as does recurrent apomixis.

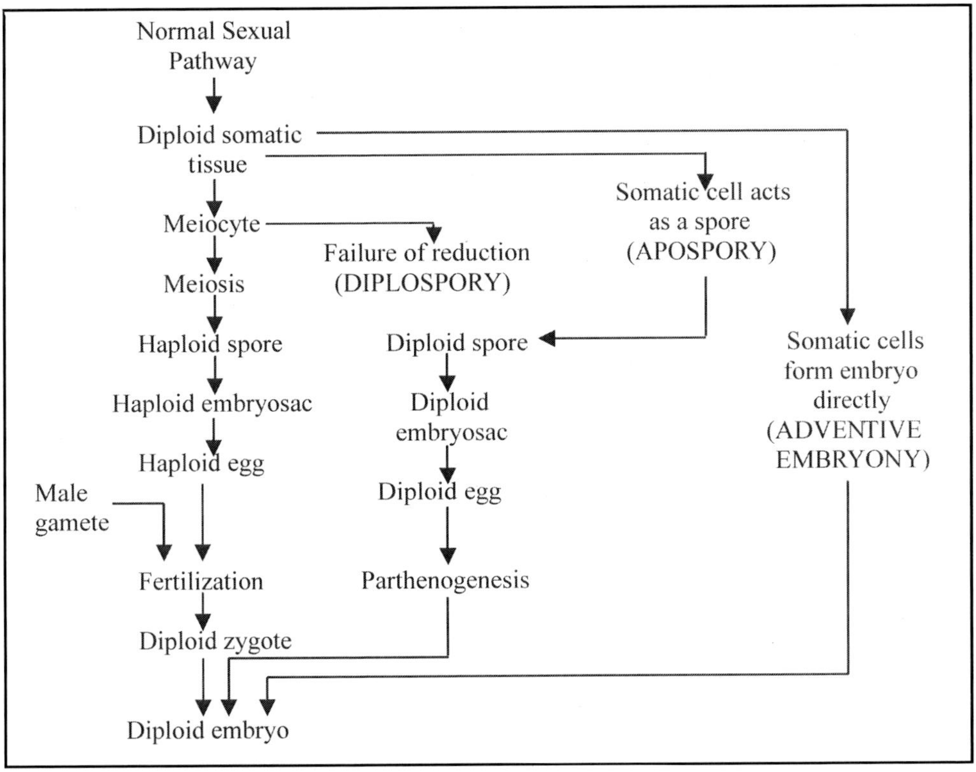

Developmental pathways of various types of agamospermy
(Modified after Bhojwani and Bhatnagar, 1974)

4. Vegetative Apomixis

In some cases, vegetative buds or bulbs are produced in the inflorescence in place of flowers. This occurs in *Poa bulbosa*, and some Allium, Agave and grass species.

5. Genetic Control of Apomixis

Apomixis is known in over 40 plant families. The pioneering studies by Nogler (1995) and Savidan (2001) proved that apomixis is under genetic control, although its expressivity may be affected by genetic modifiers or environmental conditions. In genetic crosses, the apomict is used as the pollen donor to a sexual female. Genetic analysis of apomixis is complicated by the polyploidy nature of most apomicts, the suitable availability of compatible sexual plants, and difficulties in scoring the progeny, particularly as the components of apomixes may segregate.

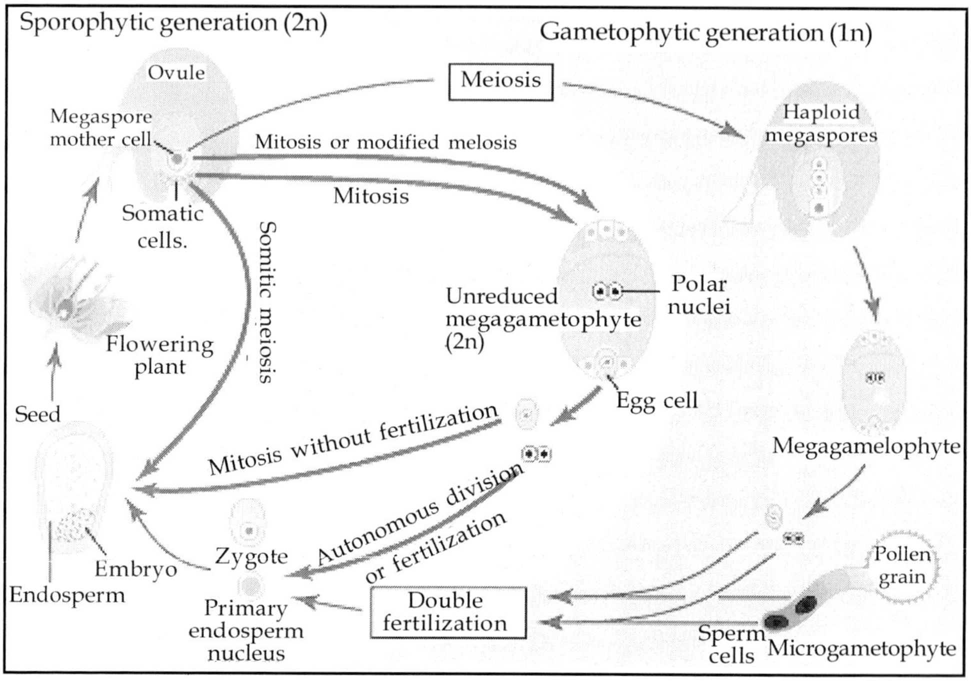

Apomixis in angiosperms

In apomictic *Panicum* spp., *Ranunculus* spp. and *Hieracium* spp., the dominant locus controlling apospory co-segregates with parthenogenesis, suggesting that a single locus, either simple or complex, controls apomixis. In other apomicts, the genetic loci controlling apomeiosis, parthenogenesis, and functional endosperm formation can be separated from each other. Thus, in these species at least three distinct loci are involved in apomixis. Moreover, the genetic control of each apomixis component may be complex and involves more than one gene.

Most of the polyploidy forms of guayule (*Parthenium argentatum*) are obligate apomicts, and the diploid individuals are sexually reproducing. Experimental studies indicated that three pairs of genes determine the breeding behaviour of individual plants of this species. In homozygous condition, gene ***a*** leads to the formation of unreduced egg, gene ***b*** prevents fertilization and gene ***c*** causes the egg to develop without fertilization. Plants with the genetic constitution *aaBBCC* form unreduced eggs but these cannot develop into embryos without fertilization. Plants with the genotype *AAbbCC* produce reduced eggs but embryos are not formed because fertilization is prevented. Those with *AABBcc* genotype have a normal sexual behaviour. The gene *c* has no effect in the presence of *A* and *B* since the eggs are reduced and fertilization takes place. Only those plants with an *aabbcc* genetic constitution would be apomictic. In a population of mixed genotypes, therefore, segregation would occur and fully sexual plants could give rise to apomictic offspring.

In facultative apomicts, environmental conditions seem to play a key role in the shift from sexual mode of reproduction to apomixis.

6. Mechanism of Apomixis

Koltunow and Grossniklous (2003) have critically gone through the phenomenon of apomixis in plants on a developmental perspective and narrated the events during the process with reference to the normal sexual reproductive pathway.

7. Ovule, Seed and Fruit Development in Sexual Plants

In sexual species, a strictly ordered and defined sequence of events is required to produce viable seeds and associated fruit structures that facilitate seeds dispersal. By contrast, apomixis results in seed and fruit formation despite omission of meiosis and double fertilization. Analysis of apomictic development in plants requires a fundamental understanding of sexual reproduction. Before examining how apomixis occurs and is accommodated within the ovule, an overview of ovule morphogenesis and early seed development during sexual reproduction is provided in a concise format.

Ovule Development

Chronologicaly, the ovule is the last organ to form in the angiosperm flower. Since it is the progenitor of seed, the ovule is the site of most processes relevant to sexual and apomictic reproduction. Ovules vary in shape, which is governed in part by the extent of their curvature and the elaboration of their epidermal cells, but all ovules share a number of structural elements. The embryo sac, or female gametophyte, is located in the core of the mature ovule, surrounded by one or two integuments that eventually form the seed coat. The integuments are not fused at the tip of the ovule where they form an opening called the micropyle, a canal that allows entry of pollen tube during fertilization. In some species, the endothelium, a single layer of cells, differentiates from the innermost cell layer of integument. The endothelium forms a tube, open at both ends, which immediately adjoins the embryo sac. The ovule is attached to the placenta by a vascularized stalk called the funiculus. The development of the female gametophyte is coordinated with that of the sporophytic tissues of the ovule.

The Female Gametophyte Originates from Nucellar Tissue Derived from the LII Layer

Ovule development initiates with formation of a primordium, comprising three layers, LI, LII and LIII. The ovule primordium initiates as an outgrowth from the carpel placental tissue with cells dividing in the innermost LIII layer. Soon after initiation, LIII cell division slows down and the cells of the sub-epidermal layer LII begin to multiply rapidly. These cells give rise to the nucellus, from which the sporogenous tissue differentiates. The ovule epidermis is derived from the outermost LI layer.

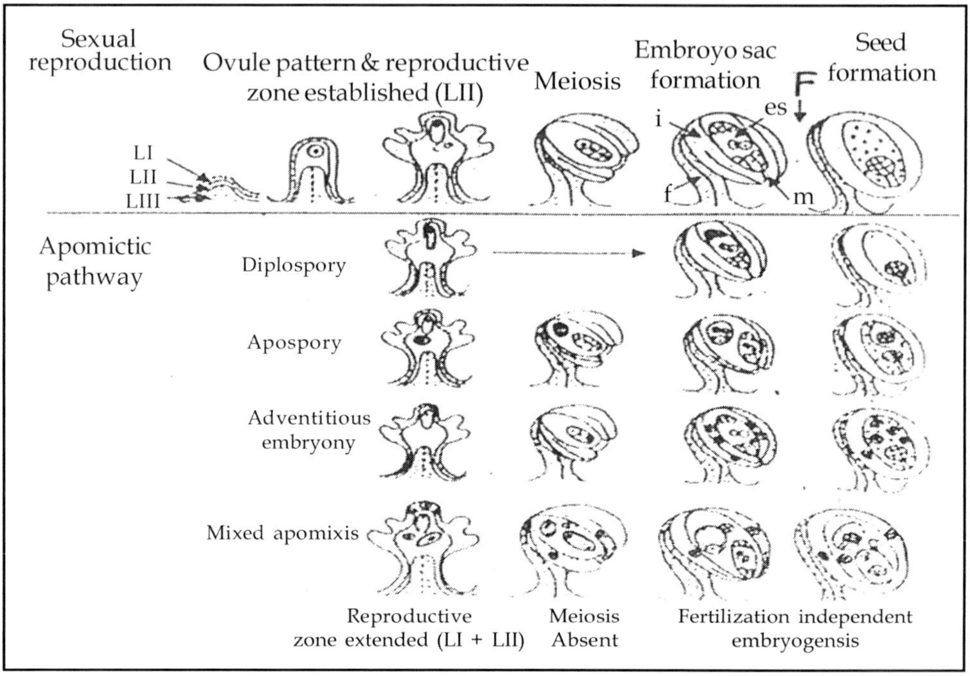

The diagram summarizes the ovule development beginning with the formation of the primordlum made up of the LI, LII, LIII layers. Events in sexual reproductive pathway are represented in the top panel above the line. Apomictic processes are shown below the line and cells and structures involved in apomixis are shaded and while those pertaining to sexual reproduction remain white. (Modified after Koltunow and Grossniklaus 2003)

Abbreviations : a - archesporial cell ; e-embryo; en - endosperm ; es - embryo sac; f - funiculus; F - double fertilization; i - integuments; m - micropyle; mmc - Megaspore mother cell

Sexual and apoimictic reproduction in the ovule of flowering plants - A schematic presentation

Megasporogenesis begins with the differentiation of an archesporial cell in the sub-epidermal layer of the ovule. Differentiation of more than one archesporial cell is not uncommon in angiosperms. Thus, reproductive potential is not confined to a single cell in the sub-epidermal layer but to several cells possibly derived from other cell layers of the nucellus. In general, only one cell develops into the megaspore mother cell (MMC) that undergoes meiotic reduction. Differentiation of more than one MMC has been noted in sexual species such as *Arabidopsis, Datura* and soybean. The infrequency of multiple MMC formation suggests presence of a mechanism to restrict reproductive potential in the nucellus to a single cell.

Integument formation begins with the onset of archesporial cell differentiation in most plants. The integument(s) initiates just below the base of the archesporial cell and develops from the outermost layer, LI and grows to envelope the differentiating MMC. The sequential growth of cells in the three layers means that the young linear ovule primordium can be divided roughly into three portions, with the funiculus derived from the LIII, and the nucellus and integuments derived from the LII and LI, respectively.

8. Double Fertilization and Control of Embryo Initiation

The egg cell and the central cell of the embryo sac are progenitors of the embryo and endosperm in the seed, respectively. Accessory cells in the embryo sac are crucial to double fertilization process. The synergids, flanks the egg cell, produce a pollen tube attractant, which guides the pollen tube through the micropyle into the embryo sac. Generally only one of the synergids starts to degenerate prior to the arrival of the pollen tube, although the exact time-point, at which synergid degeneration starts, varies from species to species. The pollen tube penetrates the degenerating synergid and upon entry, it bursts and releases the two sperm cells. Sperm cell release requires communication between the pollen tube and the synergid. If this signaling process is disrupted, the pollen tube does not burst and the sperm cells are not released. One sperm cell fuses with the egg cell to form the zygote, the other fuses with the central cell to initiate endosperm development. In the absence of fusion with a sperm cell, the egg and central cell remain in a quiescent state and eventually degrade as the flower undergoes senescence. This suggests the requirements for signaling processes to activate development of the fertilization products. The mechanisms that prevent the egg cell in plants from initiating embryogenesis without fertilization are unknown.

9. Sequential Events in Apomictic Pathway

The apomictic mode of reproduction omits critical events observed in the sexual pathway. Meiosis is avoided prior to embryo sac formation (apomeiosis), embryo development is autonomous (parthenogenesis) and endosperm formation requires certain developmental adaptations. These three factors are often referred to as components of apomixis. Sexuality and apomixis are not mutually exclusive modes of reproduction. With a few exceptions, most apomicts retain the capacity for sexual reproduction and are thus facultative in nature. Depending on the mode of apomixis, the events of sexual reproduction might occur in the same ovule or in different ovules of the same apomictic plant.

Apomixis initiates from Cells Derived from Both LI and LII layers

The course of apomictic reproduction varies in terms of origin of cells initiating apomixis and whether their immediate fate is the formation of an embryo sac or an embryo. The means by which meiosis avoided, the degree to which the sexual process continues in parallel, and the requirement of fertilization and specific adaptation for endosperm formation are all factors contributing to apomictic processes.

Collectively, apomixis can initiate at various stages during ovule development. The often concurrent events of sexual reproduction in the same ovule, or in adjacent ovules, are used as reference points for the timing of initiation of apomixis and its programme. A survey of described apomictic processes indicates that cells initiating apomixis can initiate from nucellum, the integument, or the epidermal cells in the ovule. Therefore in contrast to sexual plants where the cells initiating reproductive events originate from the LII layer, the zone of reproductive competency in ovules of apomicts has been significantly extended to include cells originating from both L and LII layers.

10. Gametophytic Apomixis

Diplospory and apospory

Cells initiating apomixis early in ovule development usually undergo gametophytic apomixis, which is the mitotic formation of unreduced embryo sacs. It is subdivided into two forms based on the origin of cell that initiates unreduced embryo sac formation. If the progenitor cells differentiate in the position usually occupied by the MMC, the mode of embryo sac formation is called diplospory. Entry into meiosis is observed cytologically in some Diplosporous species, but the process aborts, and after chromosome re-organisation, the cell undergoes mitosis to form an embryo sac. Initiation of meiosis is not observed in all species classified as diplosporus.

The mitotic formation of an unreduced embyosac from a cell in a position other than that occupied by a MMC is called apospory. Diplosporous and aposporous embryo sacs may or may not resemble the reduced embryo sac observed in related sexual plants. Embryogenesis occurs autonomously in both diplosporus and aposporus embryo sacs and endosperm formation may be autonomous or may commonly require fertilization (referred to as pseudogamy).

Multiple aposporous initial cells can form an individual ovule giving rise to multiple embryo sacs.

Adventitious Embryony and Mixed All Sorts of Apomixis

If the cell initiating apomixis directly develops into an embryo, this process is called as adventitious embryony. This occurs late in ovule development from cells of either the integument or the nucellus. Usually embryo survival depends on successful fertilization of the adjacent meiotically derived embryo sac and on the ability of the adventitious embryo to grow sufficiently to gain access to the nutrient endosperm. More than one embryo can initiate and the zygotic embryo may or may not survive. If fertilization does not occur, the adventitious embryos grow by obtaining nutrients from the degrading nucellar and integumentory cells. These embryos get arrested in development in early stage itself.

It is noteworthy that diplospory, apospory and adventitious embryony can and do exist in apomictic species of Cruciferae and Rosaceae. In some aposporous grasses, citrus and certain *Heiracium* spp., apospory and adventitious embryony co-exist.

11. Significance and Scope of Apomixis in Crop Plants

Apomixis is significant in agriculture and horticulture because the seedling plants have the same genotype. The asexual processes automatically eliminating variability, **'fixes'** the characteristics of any cultivar immediately. The apomictic life cycle has the same juvenile, transitional and mature phases as does a seedling cycle. It differs in having no reduction division and fertilization occurring to produce the embryo. The value of apomictic cultivar depends on the use of the particular cultivar.

Citrus is a noted example of apomixis. For cultivar propagation, apomictic seeds are not used because juvenile apomictic seedlings have many undesirable characteristics for fruit production like excessive vigour, slowness to come to bear, thorniness and poor quality. On the other hand, viruses are screened out during the process of embryogenesis and apomictic seedlings are used to produce new "clean" version of citrus cultivars.

Apomixis is beneficial in plants where vegetative growth predominates. In grasses a number of apomictic species and cultivars exist. Kentucky blue grass (*Poa pratensis*) plants are facultative apomicts. Certain apomictic pasture grass cultivars were developed, including 'King Ranch', blue stem Andropogon and Argentine (*Paspalum notatam*).

References

Bhojwani, S.S. and Bhatnagar, S.P. (1974). The Embryology of Angiosperms, Vikas Pub. House Pvt. Ltd., New Delhi, 369p.

Hartmann, H.T., Kester, D.E. and Davies, F.T. (1993). Plant Propagation Principles and Practices. 6th Ed. Prentice Hall of India Pvt. Ltd., New Delhi, 647 p.

Koltunow, A.M. and Grossniklous, U. (2003). Apomixis: A Developmental Perspective. *Annu. Rev. Plant Biol.* 54: 547-74.

Maheshwari, P. (1950). An Introduction to the Embryology of Angiosperms. Tata Mc-Graw Hill Publishing Co. Ltd., New York, 453p.

Nygren, A. (1967). "Apomixis in Angiosperms". (In) Ruhland, W. (Ed.), Encyclopaedia of Plant Physiology, Springer-Verlag, *Berlin*, 18: 551-96.

Promila Gupta, Shivanna, K.R. and Mohan Ram, H.Y. (1996). Apomix is and Polyembroyoning in gergal plant *commiphora wightii Annals of Botany*. 78:67-72.

Radha, T and Lila Mathew. (2005). Peter, K.V. (Ed) Fruit Crops. New India Publishing Agency, New Delhi, 435 p.

Singh, S., Dhyani, D. and Ahuja Paramvir Singh (2011). Apomixis in plants-embryology genetics and molecular basis. *J. Cell and Plant Sciences*. 2:24-31.

Chapter – 10

Polyembryony in Horticultural Crops

N.K. Parameswaran and K. Dhinesh Babu

The emergence of multiple seedlings from a single seed was first observed by Leeuwenhock as early as 1719 itself (Maheswari, 1950; Bhojwani and Bhatnagar, 1974). This condition connotated as polyembryony (PE) is widely prevalent among angiosperms. In most of the cases, additional embryos do not always mature and remain arrested at very early stages or may degenerate in the course of seed development. Thus if mature seeds are taken into account, the percentage of PE in a spp. is far lesser than its actual frequency. Therefore in strict terms, PE includes all those instances in which there is a clear indication of actual occurrence of two or more pro-embryos or embryos in a developing ovule.

Additional embryos (AE) in PE result from differentiation and development of various maternal and zygotic tissues associated with embryosac; the maternal tissues include nucellus, integument, antipodals and synergids, while the embryonic tissues include the suspensor and zygote per se. The triploid tissue, endosperm, comprising one paternal and two maternal genomes also contribute to the formation of PE.

About 244 species from 140 genera belonging to 50 angiosperm families are reported to exhibit either AE or PE.

Species showing various types of polyembryony

Tissues of embryosac originating additional embryos	No. of species
Nucellar	122
Suspensor	22
Synergids	34
Embryo	21
Integument	21
Endosperm	7
Antipodals	6

Frequency of maternal, endosperm and embryo structures originating additional embryos

Tissues	Number of species	(%)
Maternal	183	78
Endosperm	7	3
Embryo	43	19

Maternal structures include nucellar, synergids, antipodals and integuments.

1. Groups of Polyembryony (PE)

The very early classification of PE includes

1. Cleavage polyembryony
2. PE resulting from cells of embryosac other than the egg
3. PE resulting from occurrence of more than one embryosac in the same ovule
4. PE resulting from some sporophytic cells of the ovule

a. Cleavage PE

It is less frequent in angiosperms and more in gymnosperms. In this situation, cleavage and proliferation of zygote or its derivatives result in establishment of separate embryonal primordia (Durzan, 2008). Among angiosperms, cleavage polyembryony is commonly observed in orchids. Swamy (1943) recorded three different modes of super numerary embryo formation.

- The zygote divides irregularly to form a mass of cells of which those lying towards the chalazal end grow simultaneously and give rise to many embryos.
- The proembryo gives out small buds or outgrowths which may themselves function as embryos.
- The filamentous embryo becomes branched and each branch gives rise to an embryo. Unlike the cleavage embryony, arising during seed development very common in orchids, the formation of plural embryos during seed germination is exclusively observed in Vandas only. The apical promeristem of embryo in this genera divides into a number of primordia (3-9) and each of which further organizes into an embryo.

b. PE resulting from cells of embryosac other than the egg

Synergids form common source of additional embryos in this situation. Depending on whether it arises from fertilized synergid or unfertilized synergid, the embryo may be diploid or haploid.

In *Aristolochic bracteata, Poa alpinia* and *Sagittaria graminea* besides the egg and the polars, one or both the synergids may get fertilized. This can be brought about either by the entry of more than one pollen tube into the embryosac or by the presence of additional sperms in the same pollen tube. In such a situation, the zygotic as well as the synergid embryos are diploid.

Formation of embryos from antipodals as well as from endosperm is rather rare and no authentic reports are available on this.

c. PE resulting from occurrence of more than one embryosac in the same ovule

Multiple embryosacs in an ovule emerge from:

- Derivatives of some megaspore mother cells
- Derivatives of two or more megaspore mother cells
- Nucellar cells

Formation of twin embryosacs within an ovule is common in *Casuarina equisetifolia* and citrus. In *Pennisetum ciliare,* 22% seeds contain twin embryos. The normal embryosac develops only upto the 4 nucleate stage and multiple embryos are formed by aposporous embryo sacs.

d. PE resulting from sporophytic cells of the ovule

The process of development of embryos from tissues other than the zygote or endosperm is generally referred to as adventitious embryony. Maternal sporophytic tissues outside the embryosac. i.e., nucellus and integuments form the resource tissues in this situation. In addition to the well known examples of *Mangifera* and *Citrus* genera, nucellar polyembryony is observed in a few species under the genera *Opuntia* and *Trillium*. Species variation with respect to manifestation of polyembryony is noted in *Citrus* in which some are monoembryonate (*C. grandis, C. limon*) while others are polyembryonate (*C. microcarpa, C. reticulata* and *C. jambhiri*) (Kundan Kishore *et al*., 2012). Seeds with as many as 40 embryos are observed in *C. unshiu*. In polyembryonate species, the adventive embryos arise by proliferation of nucellar cells with rare exceptions. Nucellar embryos arise from the micropylar half of the nucellus (Batyjina and Vinogradova 2007). In *Mangifera*, the nucellar cells destined to form adventive embryos can be distinguished from other cells of the nucellus by their dense cytoplasm and starchy contents. The inception of nucellar embryos takes place outside the embryosac, but they are gradually pushed into the embryosac cavity where they divide and differentiate into mature embryos. The adventive embryos do not show synchronous development. A single seed may show embryos at various stages of development. Nucellar embryos can be differentiated from the zygotic

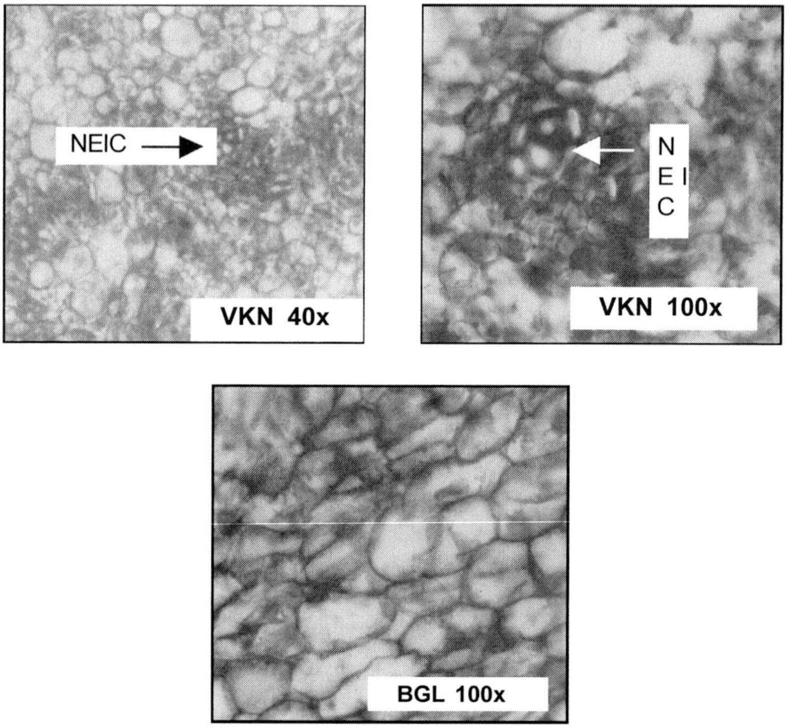

Photomicrographs showing nucellar buds of Muvandan Vellaikolumban and nucellus of Bangalora

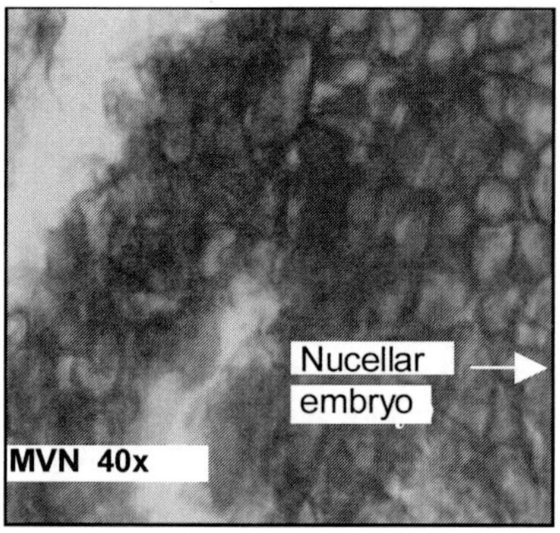

Photomicrograph showing nucellar embryo of Muvandan with prominent nucleus and dense cytoplasm

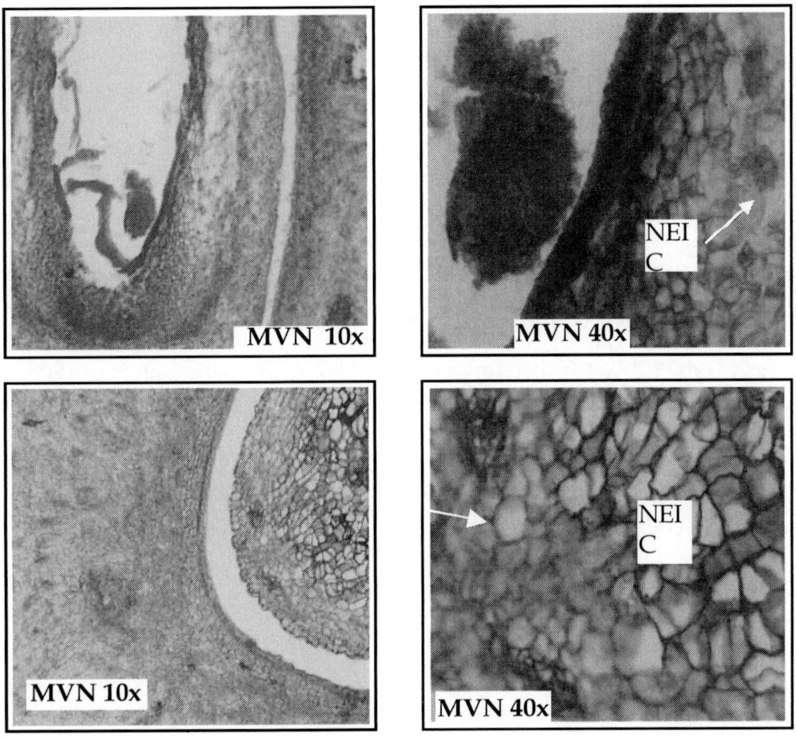

Photomicrographs showing adventive embryos arising outside embryosac in Muvandan

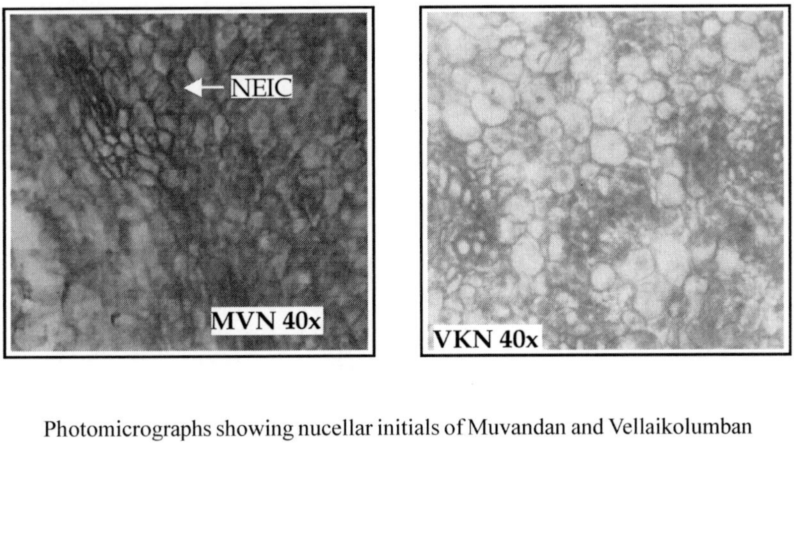

Photomicrographs showing nucellar initials of Muvandan and Vellaikolumban

embryos by their lateral position on the embryosac, irregular shape and lack of suspensor. But in *Citrus microcarpa*, the nucellar embryos possess a distinct suspensor from very early stages. Mostly, for the initiation of nucellar cells, the pollination stimulus is essential.

Suspensor polyembryony

It is common in some of the genera belonging to family Santalaceae, where, as many as six embryos may develop simultaneously in an ovule by the proliferation of the suspensor cells. But only one of them fully develops into full maturity.

Sequential Events During Adventive PE

Mango

Dhinesh Babu (2005) while studying the polyembryony in mango observed that the adventive embryos originated from the epidermal cells of the nucellus and they were located just adjacent to embryosac near the micropylar region. Since the adventive embryos originated only from the nucellar region, it was referred as nucellar embryony. The inception of nucellar embryos occurred after a definite time gap of 10-20 days after fertilization during which the zygotic embryos attained the stage of a near heart shaped embryo.

The inception of nucellar embryos was found to occur after a definite time gap of 10-20 days after fertilization during which the zygotic embryo attained the stage of a near heart shaped embryo.

These nucellar embryos developed only after the zygote gave rise to a sexual proembryo. The precursor cells in the nucellus giving rise to adventive embryos were characterized by their prominent nucleus and dense cytoplasm. They were meristematic and rich in metabolic contents like insoluble polysaccharides, RNA and proteins and lateral in position. The role of fertilization in providing stimulus for nucellar embryogenesis as reported by Sachar and Chopra (1957) was very much observed in the present study also. In both the polyembryonic cultivars studied, the inception of nucellar embryonic initial cells (NEICs) occurred only after fertilization.

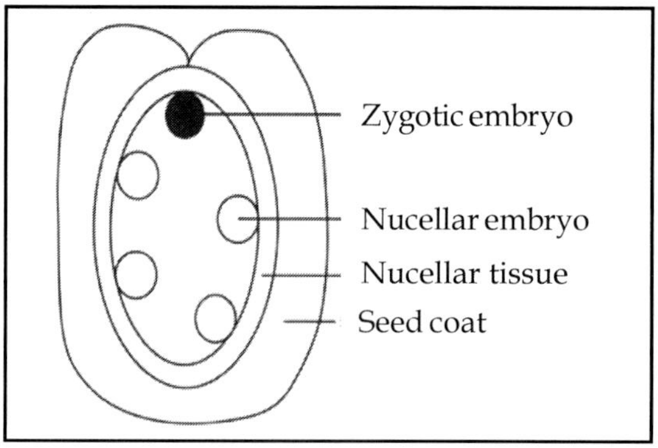

Nucellar Embryogenesis in Mango Cultivars Muvandan and Vellaikolumban (Polyembryonic) with Reference to Monoembryonic Cultivar 'Banglora'.

The most pertinent hallmark of nucellar embryogenesis, as it could be traced through the investigations, was the close dependence of the adventive embryogenesis on sexual zygotic embryogenesis. Inception of NEICS was observed to be an event just succeeding the fertilization process and zygote formation. Any of the fruitlets of the polyembryonic cultivars examined at different stages during the studies did show a bypassing of sexual fertilization and zygote formation and just producing adventive embryos only. In all the cases, it took place as one event leading to the occurrence of yet another event upon receiving some stimulus from the former. What is the nature of this stimulus and how exactly it triggers the processes leading to adventive embryo formation remained as puzzles.

Citrus

Adventive embryogenesis in *Citrus* can be divided into four steps: (1) formation of Adventive Embryo Initial Cells (AEICs), (2) differentiation of AEICs to acquire appearance of a zygote, (3) division of AEICs, and (4) development of adventive embryos. AEICs generally appear before anthesis and are characterized by their homogeneous cytoplasm, a large and spherical nucleus, irregular plastids with single thylakoids, numerous oval mitochondria (0.1-0.2 μm in length) with short cristae, some single strands of rough ER together with many free ribosomes and polysomes, some dictyosomes with 4-5 cisternae and associated vesicles. The walls of these cells stain intensely and show the presence of granular, heterogeneously dispersed, electron dense material between the plasma membrane and the cell wall. Before dividing, the AEICs show complete change in their shape, composition and ultrastructure, and attain zygote-like appearance. The initials become spherical and are isolated from the surrounding degenerating nucellar cells by blocking of plasmodesmata, formation of a completely new cell wall, and ultimate breakdown of original cell wall. After pollination, their cytoplasm shows strong reduction in the number of proplastids and increase in the number of ER, free polysomes, dictyosomes and their associated vesicles, and the amount of lipid. In fertilized seeds, the division of the AEICs generally occurs before the division of zygote but after free nuclear endosperm has developed to some extent. In unfertilized seeds the AEICs may divide 40-50 days after anthesis.

The studies have clearly shown that the initiation of adventive embryos in *Citrus* occurs autonomously all around the embryo sac; it is not affected by pollination, fertilization or the development of zygotic embryo or endosperm (Kobayashi *et al.*, 1978, 1981; Wilms *et al.*, 1983; Wakana and Uemoto, 1987a, b). The development of adventive embryos is strongly influenced by endosperm development. Lack or poor development of endosperm results in poor development of adventive embryos. Therefore, in fertilized seeds, the nucellar embryos fail to develop beyond the globular or early dicotyledonous stage and are incapable of normal germination under natural conditions. Presence of endosperm promotes development of adventive embryos at the micropylar end, but represses their development towards chalazal end; the degree of suppression is directly related to distance of embryos from micropylar end. In normal seeds, the AEICs at the chalazal end generally do not develop beyond the initial-celled stage. The suppressive effect of endosperm on the development of adventive embryos towards the chalazal end is supported by the observations that in abnormal seeds with 3n embryo and 5n endosperm or where endosperm either aborts or fails to develop altogether the adventive embryos at the chalazal end may develop up to the early cotyledonary stage. Moreover, the AEICs from normal seeds if removed from the inhibitory effect of endosperm and cultured on simple, hormone-free culture medium give rise to normal embryos. This differential effect of endosperm on the adventive embryos at the micropylar end and at the chalazal end awaits explanation.

2. Mechanism of Polyembryony and the Genetic Bases

Attempts are made from very early times to trace the mechanism under lying the occurrence of polyembryony as observed in different plant species. Accordingly many theories are advanced in this line but none of this early explanations is sufficiently validated. In these different theories, polyembryony is ascribed to hybridization, necrohormones and the effect of recessive genes.

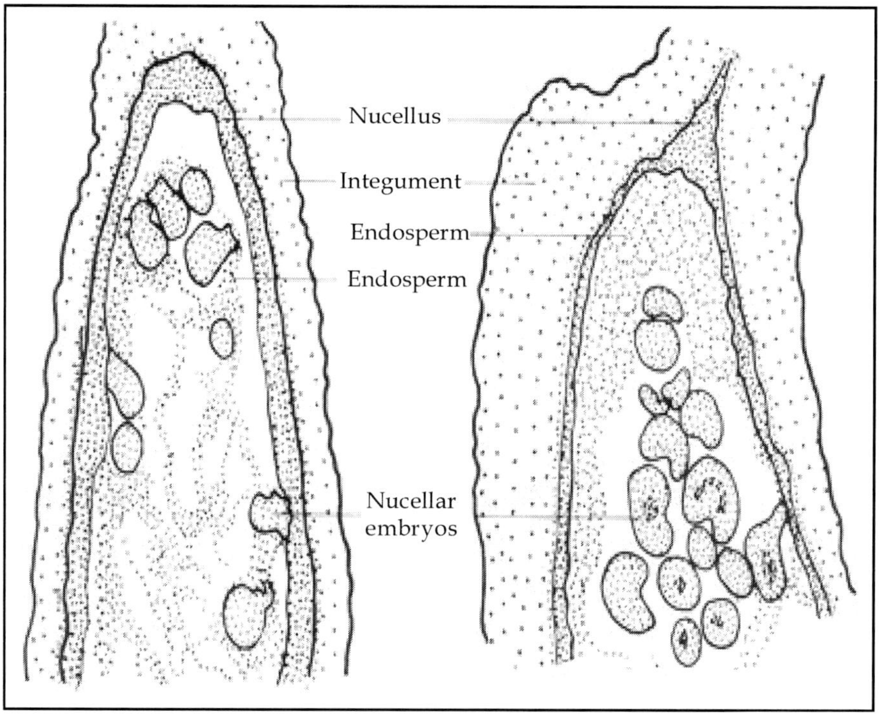

Nucellar Polyembrony in citrus

Haberlandt (1921, 1922) proposed the "necrohormone theory". He regarded, the degenerating cells of the nucellus as source of stimulus for adjacent cells to divide and form adventive embryos. Haberlandt attempted to induce adventive polyembryony in *Oenothera* by damaging cells by pricking the ovules with a fine needle and by gently squeezing the ovary. In one ovule, he got two embryos which he considered to be of nucellar origin.

Attempts by subsequent workers to induce adventive polyembryony using Haberlandt's method yielded only negative results.

Leroy (1947) suggested that polyembryony in mango was caused by one or more recessive genes. He stated that while primary centre of origin (Eastern India) contains only monoembryonate forms carrying dominant genes, the secondary centres of origin (China, Philippines, Sudan) have forms with necessive genes and polyembryonate seeds. But it is not true however, since many number of varieties especially in the South West Coast of India show polyembryonate seeds. Frusato *et al.* (1957) showed the embryo number in *Citrus* seeds may be influenced by the following factors:

1. Age of the tree; increased incidence in older trees.
2. Fruit-set: being higher in years of higher fruit-set.
3. Nutritional status of the plant – direct relationship.
4. Orientation of the branch of the tree; being higher on northern than on southern branches.

The monoembryonate condition in some species of *Citrus* is ascribed to the synthesis and release of certain volatile and non-volatile embryogenic inhibitors in their ovules which do not occur in the ovules of polyembryonate species.

Ganeshaiah *et al.* (1991) dealt evolution of polyembryony at length by taking examples from that of *Citrus* and developed genetic and inclusive models to derive the conditions that permit evolution of polyembryony under maternal and off spring control. They also derived expressions for optimal allocation (evolutionarily stable strategy, ESS) of resources between zygotic and nucellar embryos, and concluded that nucellar polyembryony has evolved as a counter strategy by the maternal parent to mitigate the consequences of sibling rivalary. This can be elucidated by taking examples from the situations favouring polyembryony in *Citrus* spp. Most species of *Citrus* exhibit polyembryony. They produce very large number of ovules, of which invariably only a few develop to maturity. Hence the number of seeds set in a fruit provide an inverse index of the extent of brood reduction. If polyembryony is a maternal counter strategy to over come the loss in her fitness due to brood reduction, then one can predict that the average number of polyembryos seed should be negatively related to seeds/fruit i.e., a small number of seeds implies a higher level of brood reduction; which should be countered by a higher extent of polyembryony, obtained by producing more nucellar embryos. Upon analyzing the situation prevailing in 27 cultivars of 8 species of *Citrus* it was found that the polyembryos/seed was negatively correlated with seeds/fruit. Such a brood reduction driven by sibling rivalry is observed to be intense in species where the indehiscent fruits are dispersed as a single unit through wind, water or animals.

3. Classification of Polyembryony

In broader terms, polyembryony can be classified into two types:

1. Spontaneous: Includes instances of naturally occurring polyembryony

2. Induced: Includes instances of experimentally induced polyembryony.

Spontaneous polyembryony is sub-divided into two categories:

1. *True polyembryony:* Two or more embryos arising in the same embryo sac; from the zygote or embryo (*Eulophia, Vanda*), from synergid (*Sagittaria*), from antipodal cell (*Ulmus*), or from nucellus or integument (*Citrus, Spiranthes*).
2. *False Polyembryony:* Development of embryos in more than one embryo sac in the same ovule (*Fragaria*) or placenta (Loranthaceae).

Yakovlev (1967) proposed a classification of polyembryony on genetic basis. He distinguished two types of spontaneous polyembryony.

1. *Gametophytic:* Arising from any gametic cell of the embryo sac after or without fertilization.
2. *Sporophytic:* Arising from the zygote, proembryo or the initial sporophytic cells of the ovule (nucellus, integuments).

Bouman and Boesewinkel (1969) put forth another classification based on origin of additional embryos. According to them, spontaneous polyembryony should be split into four categories:

1. Supernumerary embryos arising from sporophytic cells of parental generation (integuments and nucellus).
2. Supernumerary embryos arising from cells of the gametophyte (by one or more embryo sacs in the same ovule).
3. Supernumerary embryos arising from the new sporophyte (fertilized egg or proembryo).
4. Supernumerary embryos arising from the male gametophyte (disputed group).

4. Polyembryony – Its Practical Implications

Nucellar adventive polyembryony is of great significance in horticulture. The adventive embryos serve as source of uniform seedlings of the parental type, akin to vegetative propagation by cutting. However, nucellar seedlings of *Citrus* furnish better clones of orchard stock than cuttings. It is due to the following reasons:

1. The nucellar seedlings have a tap root and, therefore, develop a better root system than do the cuttings. The latter have only a small lateral root system.
2. The nucellar seedlings show a restoration of the vigour lost after repeated propagation by cuttings.
3. The nucellar embryos are free from diseases. So far nucellar polyembryony is the only practical approach to raise virus-free clones of polyembryonate *Citrus* varieties in nature. For monoembryonate cultivars of *Citrus*, there is no *in vivo* method for

raising virus-free clones. However, it can be achieved by culturing their nucellus and inducing embryoid formation.

In vitro embryogenesis is looked upon as a practical approach to large scale propagation of selected genotypes. In tissue cultures, plant generation occurs via organogenesis or somatic embryogenesis. The latter process offers the possibility of using bioreactors to automate large scale production of somatic embryos. Moreover, somatic embryos directly germinate to form plantlets because of the presence of root and shoot primordial. Of late, there is considerable interest to convert somatic embryos into "artificial seeds" which can be sown in the field using suitable mechanical devices similar to those used for normal seeds. Synthetic seeds are produced by encapsulating individual somatic embryos with a hydrogel. The major technological problems which need to be overcome before artificial seeds can be commercialized are: (1) occurrence of somaclonal variations in the plants produced through tissue culture, (2) low rates of conversion of somatic embryos into plantlets due to poor quality of the embryos, (3) Inadequate information on somatic embryogenesis in many-a-crop plant, and (4) a little know how on the regulation of embryo development.

Information on the intricacies about the molecular biology of embryogenesis is still very superfluous and one of the difficulties is that, the zygotic embryo is embedded in the seed and hence not easily accessible to experimental manipulations. Somatic embryogenesis serves as an excellent source in such a situation for analyzing the 'how and why' of the molecular biology of the process.

References

Batygina, T.B. and Vinogradova, G.Y. (2007) Phenomenon of polyembryony: genetic heterogenity of seeds. *Russel. J. Dev. Biol.* 38:126-151.

Bhojwani, S.S. and Bhatnagar, S.P. (1974) The embryology of angiosperms, Vikas Publishing House Pvt. Ltd., New Delhi, 357 p.

Dhinesh Babu, K. (2005) Histological and Biochemical Characterisation of Polyembryony in Muvandan and Vellaikolumban Mangoes. Ph.D. thesis submitted to KAU (Unpublished).

Durzan D.J. (2008) Monozygotic cleavage polyembry genesis and conifer tree improvement. *Cytology and Genetics* 42:159-173.

Ganeshaiah, K.N., Shanker, R.U. and Joshi, N.Y. (1991) Evolution of polyembryony: Consequences to the fitness of mother and offspring. *J. Genet.* 70: 103-27.

Kundan Kishore, Monika, N., Rinchen, D., Boniface Lepcha and Brijesh Pandey (2012) Polyembryony and seeding emergence traits in apometic citrus. *Scientia Horticulture.* 138:101-107.

Maheshwari, P. (1950). An Introduction to the Embryology of Angiosperms. Tata Mc-Graw-Hill Publishing Co. Ltd., New York, pp.453

Wakana, A. and Uemoto, S. (1987a) Adventive Embryogenesis in Citrus. The Occurrence of Adventive Embryos without Pollination or Fertilization. *Am. J. Bot.* 74: 517-30

Wakana, A. and Uemoto, S. (1987b) Adventive Embryogenesis in Citrus (Rutaceae) II Post Fertilization Development. *Am. J. Bot.* 75: 1033-47.

Zhang, X.P., Rhodes, B.B., Baird, W.V., Skorupska, H.T. and Bridges, W.C. (1996) Development of genetic male sterile watermelon lines with delayed-green seedling marker. *HortScience* 31:123-126.

Zhiyuan, F., Wang, X., Dongyu, Q. and Guangshu, L. (2000) Hybrid seed production in cabbage. (In) A.S. Basra (ed.). Hybrid Seed Production in Vegetables : Rationale and Methods in Selected Crops. Food Products Press. pp 109-129.

Zhou Changjiu and Zhang Youliang (1994) Studies on several properties of radish male sterility. *Acta Horticulturae Sincia* 21:65-70.

Chapter – 11

Male Sterility and Self-Incompatibility in Selected Vegetable Crops

Hari Har Ram and M.N. Badukla

Today hybrid development breeding technology is the most widely used by vegetable breeders all over the world. In the hybrid technique, an F_1 hybrid is produced by crossing two distinct and usually contrasting lines. In a review on heterosis in vegetable crops, Pearson (1983) cites a report by Hayes and Jones (1916) on first generation crosses in cucumber as probably the first suggestion that hybrid vigour can be exploited in vegetable breeding as reviewed by Riggs (1988). Presently, hybrids predominate in several vegetable crops like sweet corn, onion, tomato, hot pepper, sweet pepper, eggplant, summer squash, cabbage, cucumber, watermelon, muskmelon, asparagus, spinach, broccoli, brussels sprouts, radish, carrot, beet root, celery, bottle gourd, bitter gourd, ridge gourd, smooth gourd, pumpkin, etc. The advantages of hybrid vegetables include uniformity, increased vigour, earliness, higher yield, and resistance to specific pests and pathogens. It must be kept in mind that these advantages may not be always present in any one hybrid cultivar of a vegetable species. Besides, hybrids provide opportunities for rapid deployment of desirable dominant genes in the F_1 itself and built-in exclusivity conferred by the ownership and maintenance of the parental lines of the superior hybrids. In India, hybrid breeding in vegetable crops got boost 1988 onwards when Govt. of India, Ministry of Agriculture, Department of Agriculture and Cooperation announced a new policy on import of seed/ planting material on Sep. 16, 1988 for the benefit of all those concerned with seeds. This new policy came into effect from 1st October, 1988 and gave opportunities to private sector seed companies to expand their R&D and business with focus on development of hybrids and their seed production including import of hybrid seeds in bulk in several crops like, cucumber, watermelon, muskmelon, cabbage, temperate cauliflower, , tomato and hot pepper, *etc.* (Ram, 2012). This trend is still continuing with increasing focus on development of hybrids in India itself.

1. Breeding Systems Utilized in Production of Vegetable Hybrids

Singh and Swarup (1971) highlighted the importance and potential of exploitation of hybrid vigour in vegetable crops through a technical bulletin. This was followed by several comprehensive reviews and text books where the techniques of hybrid seed production in vegetable crops were described in detail along with scope of hybrid breeding in vegetable crops (Swarup, 1977; Frankel and Galun, 1977; Wills and North, 1978; Kalloo, 1988; George, 1999; Hazra and Som, 2005; Ram, 2012 and Swarup, 2006). The main methods applied for hybrid seed production in vegetable crops include:

- Genetic male sterility
- Cytoplasmic genetic male sterility
- Self-incompatibility
- Gynoecious lines
- Dioecy
- Monoecy
- Chemical suppressants
- Chemical hybridizing agents
- Manual methods

The present chapter deals with basics of male sterility and incompatibility in selected vegetable crops with focus on production of hybrid seed on commercial scale using these two systems along with relevant basic information as the background.

2. Male Sterility: Theory

Male sterility is defined as absence or nonfunctioning of pollen in plants. It could be of three types:

- **Genetic male sterility**
- **Cytoplasmic male sterility**
- **Cytoplasmic-genetic male sterility**

Genetic male sterility

Genetic male sterility which has been reported in many cultivated plant species, such as wheat, maize, sorghum, sugar beet, pearl millet, tomato, brinjal, cauliflower, cabbage, broccoli, watermelon, muskmelon, cucumber, summer squash, pumpkin, chive and soybean, etc. is usually inherited as a recessive nuclear gene designated as st or ms. Detailed information has been provided by Kaul (1988), Kuckuck *et al.,* 1991) and in a review on male sterility in vegetable crops by Kumar *et al.* (2000). An overview of information on genetic male sterility (GMS) as reviewed by Kumar *et al.* (2000) in vegetable crops is given in the following Table 1

Table 1: An overview of important reports on genetic male sterility in some vegetable crops (Kumar *et al.*, 2000).

Vegetable crops	Chief features of GMS	References
Tomato	More than 40 male sterility alleles reported, functional sterility also reported, sl-2, ms-13 and ms-15 temperature sensitive, several ms alleles linked with seedling marker traits	Rick, 1944; 1948; Georgiev, 1991;Karbinskaya *et al.*, 1985; Kaul, 1988; Gorman *et al.*, 1996, Gorman and McCormick, 1997; Sawhney, 1997; Masuda *et al.*, 1998
Brinjal	Two recessive genes controlling male sterility reported, functional sterility available	Jasmin, 1954; Nuttall, 1963; Chauhan, 1986; Kaul, 1988; Phatak and Jaworski, 1989
Chilli/bell pepper	More than 12 recessive male sterility alleles reported, MS 12 (ms-509) and ms-3 commercially being used in India and Hungary respectively	Shifriss and Frankel, 1969; Daskalov, 1972; Shifriss, 1973; Shifriss, 1997
Cauliflower	Dominant allele for male sterility also reported	Nieuwhof, 1961
Cabbage	Dominant allele for male sterility also reported	Nieuwhof, 1961; Fang *et al.*, 1997
Broccoli	Six recessive male sterility alleles reported with linkage of ms gene with bright green hypocotyle	Cole, 1957; Dickson, 1970
Watermelon	Linkage of ms gene with delayed green seedling marker gene (dg) reported	Watts, 1962; Zhang *et al.*, 1996
Muskmelon	Five recessive male sterility alleles reported, ms-1 being commercially utilized in India	Bohn and Principe, 1962; Pilrat, 1990; McCreight *et al.*, 1993
Cucumber	Limited scope to use male sterility due to availability of gynoecious lines	Barnes, 1961
Summer squash (*C. pepo*)	-	Eisa and Munger, 1968
Pumpkin (*C. maxima*)	-	Scot and Riner, 1946; Singh and Rhodes, 1961

As per extensive review by Kumar *et al.* (2000), the genetic male sterility has been reported in about 175 plant species including major vegetable crops (Kaul, 1988). As the name itself indicates, this type of sterility is controlled by nuclear gene(s) which are usually recessive with a few exceptions in cole crops and genetically transformed male sterile lines (Williams *et al.*, 1997). All the transgenic male sterile lines developed are genetic male sterile (GMS) as they have been developed through transformation of male sterility causing gene construct inside the nuclear gene. There are certain mutants, which produce functional pollen grains but fail to self-fertilize either due to non-dehiscence of pollen or special floral morphology and these are termed as functionally male sterile. Certain mutants with exerted stigma in tomato, brinjal, etc. come under this category.

There are certain GMS lines which retain their male sterility under certain environmental conditions and turn into fertile lines once the critical environmental conditions (usually temperature and/or photoperiods) are changed. This type of male sterility is called as environmental sensitive genetic male sterility (EGMS). The critical factor to affect such genetic male sterility is temperature. Initially, this type of GMS was considered to be of less practical value but lately it is being recognized as an effective system of hybrid seed production and it has been well established in rice as a two line system of hybrid seed production on commercial scale. This has been called as two line system in contrast to three line system based on cytoplasmic genetic male sterility where three lines A, B and C or R are used. Environmental sensitive male sterility mutants have been reported in cabbage, Brussels sprouts, broccoli, pepper, carrot and tomato.

Utilization of genetic male sterility for hybrid seed production

The genetic male sterility thought to be of limited value initially for commercial hybrid seed production is now being used commercially in India by the private sector seed companies particularly in hot pepper and muskmelon. The stock seed of the female parent always segregates into a ratio of 1 fertile : 1 sterile. This stock seed is developed as follows:

Generation	Genotypes			Allele frequency in male gametes	
Parents	Female msms	X	Male MsMs	1.0 Ms	0.0 ms
F_1	Msms (self pollinated)			0.5 Ms	0.5 ms
F_2	25 % msms	X	25 % MsMs and 50 % Msms	0.66 Ms	0.33 ms
F_3	33.3 % msms	X	66.6 % Msms	0.50 Ms	0.50 ms
F_4	50 % msms	X	50 % Msms	0.50 Ms	0.50 ms
F_5	50 % msms	X	50 % Msms	0.50 Ms	0.50 ms

This shows that stock seed in F4 and onwards shall segregate into a ratio of 1 fertile : 1 male sterile if the seeds set on sterile plants are harvested and those set on fertile plants are discarded. This stock seed can be used to maintain the female parent by planting in a separate block and harvesting the seeds set only on sterile plants. For hybrid seed production, the stock female seed and the desirable male parent seed are planted in another block in a certain ratio where the fertile plants (50%) from female rows are rogued out (before pollination in cross pollinated crops) and the remaining 50% sterile plants are used as female for pollination by the adjoining male parent. The detection of the sterile plants before flowering is desirable for effective roguing. However, this is still a problem in most of the cases as male sterile plants are morphologically not distinguishable from the counterpart fertile plants except in few cases where male sterile flower size is reported to be smaller than that of fertile flowers, as in tomato (Sawhney, 1997).

In certain cases, ms gene in genetically male sterile lines linked with the recessive phenotypic marker gene. Such marker genes which can express at the seedling stage/ before flowering, can be utilized for identification of male fertile plants before flowering and consequently for roguing of fertile plants from the plot of female parent. This avoids the complicated and time consuming process of identifying the fertile plants at flowering and removing them from the seed production plot. Further, if the fertile plants can be identified at the seedling stage, they can be removed before transplanting the female parent block and all the plants in the female parent block will be male sterile and hybrid seed production will not suffer adversely. The linkages of ms gene with the marker gene reported in vegetables are bright green hypocotyle in broccoli (Sampson, 1966), green stem colour in tomato (Kaul, 1988) and delayed green seedling in watermelon (Zhang, *et al*.,1996). Tanksley and Zamir (1988) reported two genes, Prx-2 (Peroxidase-2) and aa (anthocynanin absent) which are linked to ms-10 gene in tomato.

Use of genetic male sterility in hybrid seed production on commercial Scale

In India, this system is in use on commercial scale in two crops, namely, hot pepper and muskmelon. The details are as follows:

3. Hybrid Seed Production in Chilli Using GMS

Production of foundation seed

The crop requires slightly humid warm climate for flourishing growth and flowering. The crop can withstand the heat but below 25 degree centigrade temperature, flowering may be hampered. As regards the seed production, the *Kharif* season planting i.e. nursery sowing in May-June is ideal for the crop growth, flowering and ultimately higher seed yield in Maharashtra and Karnataka regions where bulk of the commercial hybrid seed of hot-pepper is produced. It can be successfully grown on medium to heavy soils with good drainage. 20 tones of FYM and 200:100:100 kg of NPK / ha will help to achieve good crop. For healthy seedlings adopt the seed dressing by Bavistin 3 g/kg of seed. Protect the seedlings from damping off by adopting thin sowing and spraying with Dithane M-45 (3 g/l of water). In the main field, protect the crop from powdery mildew by spraying Sulfex 2 mg/ 1 of water. During crop growth, it is very necessary to protect the crop from sucking pests and Spodoptera which mainly create the nuisance in seed production under net house. For sucking pests, intermittent sprayings of Trizophos and Confidor (0.5 ml/ 1 of water) may be used. For controlling Spodoptera, the crop should be initially sprayed with the Thiodon, Monocrotophos, etc. and finally by any synthetic pyrethroids specially for seed production in net houses. Spray the nylon net area from the out side up to the crop height level as the adults lay the eggs on the outside surface of the net. Seed rate is 250-300 g/ha. The seeds need to be planted on 25 to 30 cm raised beds of 5 x 0.5 m size which will provide seedlings sufficient for planting one hectare land. The required isolation distance is 1000 meters. In case of seed production in the net house, there is no need of isolation distance. The required spacing is 75 x 45 cm. The crop should be regularly inspected at various stages. At least 2 pre-flowering inspections are required. Check and rogue out every time the volunteers and any morphological variant.

At least two inspections are required for complete flower checking (with the consideration of non-availability of sterility marker) for sterile/ fertile plants. At this time, remove male fertile plants from 75 % of planted area. Remove sterile plants from the remaining 25 % of the planted area from one particular side of the plot. Fertile plants from this area will act as a pollinator (B line) counterpart (male) for seed production of the female line which shall be used for commercial hybrid seed production next year by planting and crossing with a well recognized male parent after removal of 50 % of fertile plants.

At least one inspection is required at fruit development stage. Do not start pollination work until at least one fruit with complete skin colour and fruit shape developed is seen on each pollinator plant. After careful examination, all the variant plants are removed. In this process, there may be a delay of 10-15 days for pollination work. In order to avoid this, male seedlings can be nursery sown and transplanted 15 days earlier than the female. Two inspections are required during pollination. First, just to explain the procedure for pollen extraction from male flowers and actual pollination work and second inspection is just to confirm that the work is being done with the prescribed procedure. Prior to harvest, one inspection is required to instruct to harvest only pollinated and completely red ripe fruits. One can also harvest the seed of B line. Allow the fruit set on B line openly after completing the pollination operation. We can permit the harvest of red ripe fruits from the B line after completing the harvesting of crossed fruits on the A line. Care should be taken that there are not any admixture with the crossed fruits.

The pollination work is carried out using following steps. One day before starting the actual pollination work in the morning, pluck all the open-pollinated set fruits and open flowers from the female and male plants. Then in the evening generally at 4 pm pluck all the flower buds from male parent which are about to open in next day morning in a plastic bowl. Then separate the anthers from the flowers and keep them in a plastic bowl in the room or at a protected area. Next day morning take the bowl with the anthers in the field and expose to sunlight for 15 to 30 minutes or till the anthers start bursting. Alternatively, take two plastic or steel containers like *Katories* (of about 300-500 ml) with equal mouth diameter. Put some of the anthers in one container and cover its mouth with muslin cloth. Put other container over the mouth of the container containing the anthers in an inverted position. Hold these two containers in an inverted position and shake them vigorously so that the pollen grains will accumulate in the lower empty container. Collect the pollen grain powder and fill in the small plastic cups specially designed for filling the pollen grains and bearing them like rings for pollination purpose. For 10 R (in Maharastra State 1 R=100 sq. m) area, 10 to 20 labors (specially ladies) are sufficient for pollination work, depending up on the stage of the plant and flush of the flower.

After filling the pipes (50 % blocked by cotton lint) by pollen grains insert the ring in a finger. Form the pairs of ladies. One lady in a pair will tie the thread to pedicel of the flower and the other will pollinate the stigma by inserting it in a pipe attached to ring and filled with pollen grains. Pollinate accordingly the flowers which are completely opened with receptive stigma. Threads need to be tied prior to pollinating the flowers to avoid further immediate handling of the pollinated flower. To achieve the desired load of the fruits on the plant, in general, 30 day- pollination work is sufficient.

Pluck the fully red ripe fruits (pedicel tied with the thread) and dry them in sunlight. For emergency, power driven dryers can also be used.

As regards the small scale production, after complete drying of the fruits the seeds are separated by beating the fruits followed by winowing. Here also power driven machines are in use for seed extraction. After the extraction the seed should be well cleaned, graded, dried and stored after proper labeling. Prior to use further, the seed should be tested as per the seed standards. Under normal conditions one can get average yield of 10 kg in 10 R area (1R=100 square meter). In case of production under net house, depending upon the variety, the seed yield can be more.

Hybrid seed production

Here any proven and desirable male fertile line other than the maintainer can be used for F1 seed production. Female and male ratio can be used as 3 : 1. Thus, we can say that this is a two line theory for F1 seed production. All other activities for hybrid seed production remain the same as in foundation seed production.

4. Hybrid Seed Production in Muskmelon Using GMS

Genetic male sterility has huge potential for hybrid seed production in musk melon also. Five recessive nuclear genes (ms1, ms2, ms3, ms4 and ms5) are known. The genetic male sterility gene ms1 was introduced in India more than 30 years ago and is in use in hybrid seed production (Nandpuri *et al.*,1982). The development of hybrid muskmelon cultivar Punjab Hybrid (ms1 x Hara Madhu) is one such example. The male sterile line (ms1ms1) is maintained by crossing to the isogenic maintainer heterozygous for recessive sterility allele. Further the seed stock which segregates for sterility and fertility in a ratio of 1:1 is used as seed/female parent in the hybridization block to produce F1 seed with a standard male parent, for example, Hara Madhu is the standard male parent for Punjab Hybrid. However, one should keep in mind that in the seed production block, 50 % of the fertile segregates (Ms1ms1) will have to be removed from the female parent block. For this purpose, anthers from 3 freshly opened flowers from each of the female parent are pierced with the needle in early morning. The male fertile plants give light yellow powdery mass of the pollen grains and those from the sterile plants are sticky, green and immature. Identified male sterile plants are tagged and the confirmation test is once again done on the last day of rouging period and in this process all the fertile plants are removed and the sterile plants are used as female for hybrid seed production using Hara Madhu as the male parent. Genetically, Hara Madhu is Ms1Ms1 and the sterile plants are ms1ms1 and thus the hybrid seed produced is Ms1ms1 which is fully fertile. Pusa Rasraj (ms3 x Durgapura Madhu) is another good hybrid based on this system.

5. Cytoplasmic Genetic Male Sterility

This is a type of sterility where the offspring of male sterile plants are not necessarily male sterile but can be male fertile when certain stocks are used as pollinators. The situations where certain male plants when crossed with the male sterile females, give rise to fertile progenies are supposed to carry the restorer gene(s) having power to restore the pollen producing ability of the male sterility cytoplasm. Here both nuclear and cytoplasmic factors interact and this system has been historically discovered in onion by Jones and Clarke (1943). The mode of operation of this nuclear-cytoplasmic system can be understood by considering the evidence on inheritance of male sterility in onions. The recessive gene ms converts the normal hermaphrodite onion into a potentially male-sterile form but the homozygous recessive genotype, msms can only express itself in one particular type of cytoplasm designated as S type. In normal cytoplasm designated as N, the msms genotypes produce normal male fertile plants. Plants of the genetic constitution MsMs or Msms are male fertile irrespective of whether the cytoplasm is S or N type. The genotypes and phenotypes based on this system are as follows (Williams, 1964).

Male fertile:

Cytoplasm	Nuclear genotype
N	MsMs, Msms, msms
S	MsMs, Msms

Male sterile:

Cytoplasm	Nuclear genotype
S	msms

Cytoplasmic genetic male often abbreviated as CMS is reported in a large number of vegetable crops as reviewed by Kumar *et al.* (2000) and given in the following Table 2.

In principle, the system of hybrid seed production using CMS is typically called as three line hybrid breeding system where the three lines used are as follows:

Line A	Sterile cytoplasm as S with nuclear gene as msms
Line B	Normal fertile cytoplasm as N with nuclear gene as msms
Line C or R	Normal fertile cytoplasm as N with nuclear gene as MsMs

Line A is normally called as the female for hybrid seed production, line B is the maintainer line to maintain the A line and the C or R line is the restorer line used as male parent for commercial hybrid seed production with line A as the female parent.

Table 2: Cytoplasmic genetic male sterility reported in vegetable crops

Vegetable crops	Chief features of CMS	References
Tomato	Restorer allele not reported, sterile cytoplasm derived from distinct species	Melchers *et al.*, 1992;Petrova *et al.,* 1999
Chilli/bell pepper	Identical sterile cytoplasm (mostly temperature sensitive) derived from three sources, occurrence of R allele common in hot pepper and that of r allele in sweet pepper	Peterson, 1958; Shifriss and Frankel, 1971; Kaul, 1988; Yoo Woong, 1990; Shifriss, 1997
Cole crops	Sterile cytoplasm incorporated from *B. nigra* and Ogura of radish, cybrid CMS being utilized in France to produce hybrid seed	Pearson, 1972; Dickson, 1975; Hoser-Krause and Antosik, 1987; Kaul, 1988; Crisp and Tapsell, 1993;Sigareva and Earle, 1997;Cardi and Earle, 1997
Onion	CMS widely used to produce hybrid seed	Jones and Emsweller, 1936; Jones and Clarke, 1943; Davis, 1957; Cohan and Weigle, 1966; Van Der Meer and Bennekom, 1971
Bunching onion	-	Zhang *et al.*, 1995
Carrot	Two types of CMS, namely, petaloid and brown anthers reported with complex genetics of restoration	Banga *et al.*, 1964; Erickson and Peterson, 1979; Morelock *et al.*, 1996
Radish	Sterile cytoplasm widely distributed in wild radish, restorer gene R available in European and Chinese cultivars in contrast to presence of r allele in Japanese cultivars.	Ogura, 1968; Zhou and Zhang, 1994; Hawaldar *et al.*, 1997;Yamagishi, 1998

6. Hybrid Seed Production in Onion Using CMS

Onion is a historic crop for discovery and use of CMS in hybrid seed production. Male sterility in onion was discovered for the first time in 1925 in the cultivar Italian Red grown in California, USA as reviewed by Jones and Mann (1963). This sterility is determined by an interaction nuclear gene and cytoplasm. In the nucleus, there is a restorer gene with two alleles Ms and ms. On the cytoplasm side, therer are two types of cytoplasm, N (normal) and S (sterile). The first discovered CMS cytoplasm is known as CMS-S and the second source discovered by Berninger (1965) in French cultivar Jaune paille de Vertus is known as CMS-T. Male sterility with new cytoplasm has been reported to be governed by a dominant allele at locus A or dominant alleles at two complementary functioning loci, the so called B and C loci (Shigyo and Kik, 2008). CMS-S is most commonly used worldwide these days. CMS-T is less used as it has been discovered only in Dutch and Japanese cultivars (Havey, 2000). Development of molecular markers to differentiate fertile and sterile cytoplasm has been found to be difficult because of too much variability within groups. There are reports of variation in frequency of restore allelels for CMS-S among populations. In the tropical populations, the frequency of M or Ms allele is close to 100 % and because fo this, discovery of maintainers becomes extremely difficult in the tropical onion populations (Van der Meer, 1994; Havey, 1995). Restorer locus for CMS-S has been is present on chromosome 2 by Martin *et al.* (2005). A tightly linked (0.9 cM) molecular marker to this restorer locus has been reported (Shigyo and Kik, 2008). This may facilitate breeding process of new onion cultivars. Pathak (2000) has given a detailed account of this system in onion. Onion is highly cross-pollinated crop and is sensitive to temperature, photoperiod and stages of bulb development. Cool temperature plays an important role in flower induction. Optimum temperature required for vernalization is 7-12 degree C for about 4 to 6 weeks. Larger the bulb size, more easily it can be induced for flowering. Appearance of scape (flower stalk) signals the initiation of flowering. Each flower stalk bears umbels containing about 200 to 600 flowers. Individual flower consists of 6 perianth lobes, 6 stamens, and 3 carpels united in a pistil. Each carpel has 2 ovules. The flowers are protandrous favouring cross-pollination. Inbred development is essential for heterosis breeding.

Production of onion hybrids

Onion is highly cross-pollinated crop and inbred lines are developed through inbreeding. However, inbreeding depression in onion is so high that normally 2-3 generations of inbreeding is used to develop partially inbred populations. Cytoplasmic male sterility in onion results from an interaction between the cytoplasm and the nuclear genes. The male sterile line used as female in the hybrid seed production is called as the A line and has the genetic constitution as Smsms, that is a combination of sterile cytoplasm and the sterile nuclear gene. Plants with N cytoplasm are always fertile irrespective of the nuclear gene constitution. The maintainer lines with the genetic composition Nmsms are identified in the field. For this purpose, the male sterile plants are crossed with several fertile plants originating from different lines and the resultant F_1 hybrids are examined for pollen fertility.

The crosses producing sterile pollen are selected and the male parent of this cross is identified for developing maintainer line. Backcrossing is done with the sterile (A line) line to develop an A line with the nuclear genetic make-up of the maintainer, but with sterile cytoplasm.

For the production of the hybrid seed, three parental lines (A, B and C or R) are used. The A line which is male sterile is used as the female parent. This is crossed with the male parent C and the seeds obtained are the commercial F_1 hybrid seed. The B line is used as the maintainer to maintain the A line. In this system, the male sterile line A is used as the female and the maintainer line B is used as the male. The C line is male fertile line and is genetically diverse from the A line. It is produced by selfing the plants in the selected lines for 1-2 generations and then maintained by sib-mating. To identify a suitable commercially viable F_1 hybrid, an A line is crossed with several C lines. The best combination giving maximum heterosis for yield and other desirable traits is identified for commercial production. Maintenance of A, B and C lines in pure form is the most important part of any onion hybrid seed production programme.

The first commercial F_1 hybrids of onion were marketed in USA in early fifties. Since then, there has been increase of hybrid onion seeds globally particularly in temperate onion where about 90% of seed market is of hybrid onion. Seed market for tropical onion hybrids is much less. F_1 hybrids of any crop including onion are advantageous in terms of heterosis for yield (greater yield), increased uniformity, relative ease to combine traits from the parents and built-in variety protection as the hybrid owners have control over the parents. However, in case of onion, the advantages in terms of greater yield (yield heterosis) and increased uniformity are not strong enough because the parental lines used in hybrid production in onion are mostly at I2 or I3 stage and still have considerable within line genetic variability This is the situation with long day onion cultivars and the matter is still worse in short day onion cultivars where inbreeding is mostly applied only once in parental lines.

Therefore one can understand that uniformity in onion F_1 hybrids is not necessarily higher as compared to OP cultivars which are normally developed through strong positive mass selection. However, sometimes it is argued that exclusion of selfing in F_1 hybrids may lead to increased uniformity in the hybrids over the Ops. There is overwhelming opinion that heterosis in onion is not strong and obvious. It is worth mentioning that heterosis in onion has been explained primarily due to dispersion of partially dominant alleles in the parents rather than based on overdominance (Werner *et al.*, 1998, 1990) Thus, it seems that the strongest factor in favour of hybrid onions is in terms of built-in variety protection and that is a kind of safe guard to the organizations/companies who are putting their money in the R&D.

Seed increase of parental lines in onion

The initial seed increase of the three lines A, B and C should be done in isolation or in screen cages. The A and B lines are generally planted in the same cage while line C is planted in a separate cage. The size of the cage depends upon the quantity of the seed to be produced. Equal numbers of bulbs of the a and B lines are planted in separate rows in the same cage. At anthesis, a net cage is put over each group of plants. It is essential to stake and tie up the seed stems on the outside rows of the cages to prevent them from touching the net and becoming contaminated with foreign pollen carried by insects. At the initiation of anthesis, each plant of male fertile A line is checked for production of fertile pollen and any plant which is found to be male fertile, is rogued out. Likewise, any pollen sterile plant found in the B line is removed. Once all the plants are checked, a beehive is placed inside the cage to ensure sufficient cross-pollination. Seeds of the A and the B lines are harvested carefully to avoid mechanical mixing. Line C is also maintained in isolation or under the cage. For large scale seed increase, the parental lines, A, B and C are planted in the open field under proper isolation which should not be less than 3000 m (Pathak, 2000). The stock seed of the A line is produced by planting alternate rows of A and B lines. Seeds of the C line are also planted and multiplied in isolation. It is recommended to keep honey bee hives in the seed field to increase cross-pollination and seed yield. The parental seed increase of the lines A, B and C is known as foundation seed production.

Production of F_1 hybrid seed in onion

The hybrid seed of onion is produced in the areas which have low humidity, mild cool temperatures during initial crop growth followed by increasing temperature at the later stages of the growth. Long rainy periods or heavy dew and fogs are conducive for the diseases like downy mildew, purple blotch and stemphyllium leaf blight. During flowering stage in crop, bright sunny days are required to ensure insect activity and the cross-pollination. Hot, dry weather is required during harvesting, curing and threshing of seed. There are specific areas for seed production of long day and short day onion hybrids. Long day cultivar seeds of onion in USA are produced in Idaho and Oregon while the seeds of the short day onion cultivars are produced in Imperial Valley of California. Italy, southern France and Spain are the ideal locations for onion seed production in Europe. In India, the major onion seed is produced in the State of Maharashtra which also happens to be the major onion producing State and very appropriately, there is Directorate of Research on Onion and Garlic at Pune in Maharashtra. Heavier soil should be preferred for onion seed production because of higher water retention capacity. Soil should be free from soil borne pathogens like pink root, basal rot and nematodes.

Mother bulb production is same as in case of commercial bulb production. However, sometimes higher seed rate is used in case of mother bulb production for seed crop to have little smaller bulbs which have better storage life than the larger bulbs. Seedlings of the lines A, B and C are generally planted in the same field with proper labeling to avoid mechanical mixture. Bulbs of 4 to 6 cm diameter are preferred. Mother bulbs are harvested when the tops have fallen. Bulbs are cured for about one week to increase storage life. Tops are pruned after the necks have dried. Bulbs are stored until planted. Mother bulbs are stored as market bulbs. In tropical conditions, the bulbs are stored at ambient high temperature before planting in cool season. No vernalization temperature is required for tropically adapted lines as is the case in India otherwise 3 month storage of bulbs at 10 degree C is sufficient to induce flowering in most other cultivars.

In warm climates like India, onion bulbs for seed are planted in autumn, generally in rows 50-100 cm apart to facilitate cultural operations. In case of inbred lines, it is better to keep higher plant density to have increased seed yield.

Selected bulbs of male sterile line A and pollen parent C are planted in alternate rows in field. Usually 8 rows of sterile female are alternated with 2 rows of pollinator parent that is the C line. Commercial F_1 hybrid seeds are produced in isolation during the optimum season for seed production. Onion is insect-pollinated and hence more isolation distance is required from other seed producing fields and even from chance bolters from commercial onion fields. Generally isolation distance should not be less than 2 km, however, higher isolation distance than 2 km is preferable. The best isolation distance for onion seed production is 5 km (Pathak, 2000).

Fertilizer requirements of a seed crop are as applicable to commercial crop. Availability of high soil moisture along with high soil fertility is essential for good seed yield. All phosphorous should be applied as basal dose before bulb planting. Nitrogen and potassium are applied as side dressing. Adequate care should be taken to control downy mildew, stemphyllium blight and purple blotch. Roguing should be carried out at the bulb stage and at the time of flowering. Bulb colour, shape, foliage type and seed stalk height are the some of the characters used in off type plants. In hybrid seed production, fertile plants in the female lines are immediately rogued out to avoid cross-pollination and contamination. The flowering period of A and C lines must be synchronized to have good cross pollination and seed yield. For good seed yield, 8-10 bee hives should be placed per hectare.

Seed reaches to maturity approximately 30-50 days after anthesis. Seed stalks of A and C lines are harvested separately and C line should be harvested before A line. Hybrid seed yield of 5-8 q/ha are obtained.

Viability of onion seed is short lived and depends on the moisture content of the seed and the seed storage temperature. At high temperature and humidity, the viability of onion seeds is lost within 1 year. The seed with moisture content of about 6 % stored in moisture-proof cans or foil packets can retain viability for at least 3 years.

7. Hybrid Seed Production in Hot Pepper Using CMS

Foundation seed production

In this method the sterility of the female is 100 % and there is no need to rouge out fertile plants (except volunteers) from the female population. There is a separate maintainer line called as 'B' line which maintains the sterility of the female. Second advantage of this system over the GMS system is that the field does not remain gappy due to roguing of male fertile plants from the female area under seed production. Female and male ratio is 3:1 and the rest of the activities are same as in the foundation seed production.

Hybrid seed production

Here any proven and desirable male fertile counterpart line having the gene for fertility restoration can be used for F_1 seed production. Female and male ratio can be used as 3:1, to have sufficient male flowers. Thus, we can say that this is a three line theory for F_1 seed production where A, B and R lines are used. All other activities for hybrid seed production remain the same as in case of foundation seed production.

There are reports that fertility is at times unstable in hot-pepper. Fertility is mostly restored at relatively cooler temperatures. Further, sterility is highly influenced by the genetic background of the CMS line and maintainer also influences the expression of the sterility (Shifriss and Guri, 1979). Lee *et al.* (2005) described a gene *St* which affects the stability of CMS. Kim and Kim (2005) have developed and reported molecular markers which can discriminate between sterile and fertile cytoplasms.

8. Self-Incompatibility Systems

Basic principles

Self incompatibility (SI) is a general name for several genetic mechanisms in angiosperms, which prevent self-fertilization and thus encourage outcrossing. In plants with SI, when a pollen grain produced in an anther reaches a stigma of the same or another plant with a similar genotype, the process of pollen germination, pollen tube growth, ovule fertilization, and embryo development is halted at one of its stages, and consequentaly no seeds are produced. The term incompatibility is used to describe the failure of pollen tubes to penetrate the full length of the style and to effect fertilization. SI is one of the most important means to prevent selfing and promote the generation of new genotypes in plants and it is considered as one of the causes for spread and success of the angiosperms on our planet. System of incompatibility is different from sterility where failure to produce viable seed is due to chromosomal abnormalities or by some form of gross physiological derangement affecting gamete formation or embryo development. Major vegetable crop families where incompatibility has been reported are Leguminosae, Solanaceae and Cruciferae. The best studied mechanisms of SI act by inhibiting the germination of pollen on stigmas or by arresting the growth of pollen tube in the style. These mechanisms are based on primarily protein-protein interaction where each mechanism is controlled by a single locus known as S which has many different alleles. Despite similar morphological manifestations, these mechanisms have evolved independently and are based on different cellular components (Charlesworth *et al.*, 2005).

The S locus has 2 basic SI genes. One is expressed in the pistil and the other in the pollen. These are referred to as female and male determinants respectively. Because of their physical proximity, these are genetically linked and are regarded as a single allele. These alleles produce two proteins which when interact, produce an inhibitory protein which is responsible for pollen germination and / pollen tube growth which ultimately leads to failure of fertilization and seed setting on selfing. However, when a female determinant interacts with a male determinant of different allele, no SI is created and a successful fertilization takes place.

9. Types of Incompatibility

Genetically, it is of two types.

Gametophytic incompatibility

This was first reported in *Nicotiana sanderae* by East and Mangelsdorf (1925). This system originally called as oppositional factor system is characterized by the independent action in both pollen and style of the two alleles of the incompatibility locus (S) present in any one diploid individual. For example, the pollen of an S12 plant behaves either as S1 or S2 according to which of the two alleles is borne in the pollen. Neither allele shows dominance or any other form of inter-allelic interaction. Similar action is seen in the style also and ultimately it is the autonomous action of each allele in the gametophyte that has prompted the use of the term gametophytic incompatibility to describe this system. Further, there are a large number of multiple alleles at the S locus. In gametophytic system, the incompatibility reaction takes place in the style when the pollen grain carrying a particular S allele meets the same allele in the style (Williams, 1964). Depending upon the allelic constitution of the parents, the gametophytic system gives rise to three types of pollinations:

- Fully incompatible-S1S2 x S1S2
- Half compatible-S1S2 x S1S3
- Fully compatible-S1S2 x S3S4

Sporophytic self-incompatibility (SSI)

In sporophytic self-incompatibility, the SI phenotype of the pollen is determined by the diploid genotype of the anther (the sporophyte). This form of SI has been identified in the families Brassicaceae, Asteraceae, Convolvulaceae, Betulaceae, Caryophyllaceae, Sterculiaceae and Polemoniaceae. Since SSI is determined by a diploid genotype, the pollen and the pistil each expresses the translation products of two different alleles i.e. two male and two female determinants. Dominance relationships often exist between pairs of alleles, resulting in complicated patterns of compatibility/self-incompatibility. Because of these dominance relationships, homozygous individuals are part of the system.

In *Brassica,* the SI phenotype of the pollen is determined by the diploid genotype of the anther where the pollen coat derived from the anther-tapetum tissue, carries the translation product of two alleles. These have been found to be small, cysteine-rich proteins. The gene responsible for coding of these proteins is termed as SCR or SP11 and is expressed in the anther tapetum as well as in the microspore and the pollen (Takayama *et al,* 2000). The female determinant of the SI response in this genus is a transmembrane protein called as SRK which has an intracellular kinase domain and a variable extracellular domain. SRK is expressed in the stigma and probably functions as a receptor for SCR/SP11 protein in the pollen coat. Another stigmatic protein termed as SLG is highly similar in sequence to the SRK protein and seems to function as a co-receptor for the male determinant amplifying the SI response (Takasaki *et al.,* 2000). The interaction between the SRK and SCR/SP 11 proteins results in autophosporylation of the intracellular kinase domain of SRK and a signal is transmitted into the papilla cell of the stigma. Another protein essential for the SI response is MLPK, a serine-threonine kinase which is anchored to the plasma membrane from its intracellular side.

Incompatibility is expressed in one of three ways. Germination of the pollen may be decreased or inhibited on the stigma, for example, in broccoli. In these cases, removal of stigma allows normal pollen germination. This is the basis of selfing in such crops through bud-pollination. In the second case, pollen germination may be normal, but the pollen tube growth is inhibited in the style, for example, in tobacco. In the third situation, incompatibility reaction may occur after fertilization, for example, in *Gesteria.* This system is rare.

Use of self-incompatibility in hybrid seed production in cole crops

The major cole crops where considerable heterosis has been reported and hybrid seed production is in abundance are broccoli, Brussels sprout and cabbage. In cauliflower, the tropical Asiatic types are self incompatible as is the case with broccoli, Brussels sprout and cabbage and therefore, they show considerable heterosis and accordingly, attempts have been made to produce hybrid cultivars in them. The summer type of European cauliflower also known as Snow Ball group of cauliflowers are self compatible and behave like typical self- pollinating crops. Therefore, the Snow Ball group of cauliflower has not been subjected to heterosis breeding. The typical system applicable for hybrid seed production in cole crops is based on the use of sporophytic system of self-incompatibility. Through continuous self pollination and selection, incompatible lines with very low seed set or even no seed set when pollinating between plants within the same line can be developed. Furthermore, when two self incompatible lines are used as parents to produce hybrid, the reciprocal cross seeds can be harvested as hybrids. In 1950, Nagaoka No.1 which was the first cabbage hybrids in the world was developed through self- incompatible lines in Japan. In 1960s to 1970s, Self incompatible lines were used extensively in the production of hybrid seed in various countries in Europe, America and Asia. The general system applicable to cole crops described here is specifically for hybrid seed production in cabbage (Zhiyuan *et al.*, 2000).

The superior self- incompatible lines for hybrid seed production should possess the following characters.

- Stable self- incompatibility
- High seed set after self- pollination at bud stage
- Favourable economic characters
- Desirable combining ability

Development of inbred lines through selfing following bud pollination

It is difficult to obtain seeds of self- incompatible lines by self pollination and sister pollination at the flowering stage. The basic seeds can be propagated by hand pollination at bud stage. The bud top is removed by tweezers and strippers to expose stigma which is then pollinated with pollen collected from the same line. Tweezers and hands should be sterilized with alcohol when pollination at one line is finished. The seed production plot must be covered with net to avoid contamination by bees or other insects. Pollen must be freshly collected from the opened flowers on the same day or the day before. Mixed pollen collected from the same line should be used for pollination to avoid viability depression from continuous selfing. If the bagging isolation is applied to multiply the basic seeds, the flowering branch has to be covered with paraffin bag or muslin cloth bag before the buds open. After pollination, the flowering branch has to be covered with paraffin/muslin cloth bag immediately. The bud pollination conducted to 2-4 days prior to flowering gives higher seed set.

Propagation of basic seeds of incompatible line by bud pollination is labour intensive and costly. In consideration of this advantage, the electricity aided pollination, wire brush pollination, thermal aided pollination, carbon dioxide enrichment, etc. have been used. However, each one of them has its limitation and has not yet been used on commercial scale. In recent years, spraying a solution of 5% common salt has been reported to be effective to overcome self-incompatibility and increase the seed set by scientists in China (Zhiyuan *et al.,* 2000).

Production of F_1 Hybrid Seeds

The planting ratio of female to male is maintained as 1:1 and the parental lines are planted in alternate rows. If the plant vigour of both parents differs from each other greatly, ratio might be 2:2 which is in favour of insect pollination. Self incompatible lines of cabbage are easily pollinated by other brassica lines. Therefore, hybrid production plot should be isolated by at least 2000 m where no cauliflower, kohlrabi, broccoli, Brussels sprout, etc. are planted. Early bolting plants of early cabbage, cauliflower and kohlrabi in the commercial production plot around F_1 hybrid seed production plot should be prohibited.

To improve seed set, it is necessary to have sufficient bee population for pollination. Fifteen boxes of bee must be placed in one hectare area for pollination which is helpful to increase hybrid seed yield.

The cabbage stock plants get easily lodged during flowering and pod setting. The lodging affects insect pollination and also causes pod mildew which decreases seed yield and germination. Therefore, at early flowering stage, a framework should be built up with bamboo sticks and straw rope or stick stands beside each plant to prevent lodging.

Major pests are aphids, cabbage worm and major diseases are black rot, virus and damping of at seedling stage and powdery mildew at flowering stage. The control measures should include spraying chemicals besides using resistant parent and avoiding continuous cropping with cabbage crop.

Hybrid breeding has been primarily applied to Brussels sprout, cabbage and a few other cole crops despite difficulties with seed production. All commercial F_1 hybrids are produced by utilizing the incompatibility system to ensure crossing between parental lines but in practice, some self-or sister-brother fertilization within one or both parent lines occurs almost invariability, producing the so called 'sib' seed contaminants. Depending upon the parental lines and conditions during seed production, the proportion of such sib seed may vary from nil to as much as 80% (Wills *et al.*, 1979). It is essential therefore, for seeds-man to estimate sib proportions actually in order to avoid the sale of hybrid seed with unacceptable levels. The usual method of assessment is to conduct a grow out test (GOT) where plants from a sample are grown until they are sufficiently developed to judge phenotypic differences between sibs and hybrids. This is time consuming depending upon at what stage sibs could be distinguished from the hybrids. A biochemical test of seed extract as an alternative to the grow out test was refined by Wills and Fyfe (1974). It utilizes the variability of acid phosphate isoenzyme constitution revealed by appropriate staining of individual seed extracts following separation by electrophoresis on polyacrylamide gels and it enables objective comparison to be made between sibs and hybrids of suitable genotypes. Wills *et al.* (1979) demonstrated the utility of electrophoretic detection of sib vs hybrid seeds in *Brassica oleracea* using acid phosphate as the target isoenzyme. Fitzgerald *et al.* (1997) have described the technique for determining aberrant seed proportion characterization in hybrid seed using image analysis.

10. Hybrid Cultivars in Cauliflower and Cabbage

There are a large number of hybrids of cabbage in private seed establishments in comparison to those in cauliflower. However, heterosis breeding has been underway at the Division of Vegetable Crops, Indian Agricultural Research Institute, New Delhi with main emphasis on the exploitation and development of self- incompatible lines. This programme has been pursued at Pantnagar also. At IARI, a large number of test crosses were made using self- incompatible lines as female parent. On the basis of their performance for two years in 1985-86 and 1986-87, the best cross combination ccX Sel. 1-3-18-19 was put in all India coordinated trial under November maturity group. On the basis of its performance, Pusa Hybrid-2 was identified by the All India Coordinated Vegetable Improvement Workshop, 1992 held at Acharya N.G. Ranga Agricultural University, Hyderabad. It was later on released by the central sub-committee on Crop Standards, Notification and Release of Varieties in 1993 for cultivation in zones II and IV of the country comprising humid Bengal-Assam basin and sub humid Sutlej-Ganga alluvial plains.

Pusa Hybrid-2 is the first F_1 hybrid in cauliflower developed in India using self-incompatible lines in November maturity group. Plants of this variety are semi-erect with bluish green upright leaves. Curds are creamy white and compact. Average curd weight is 900 g. Curd to plant ratio is 10% higher in this hybrid because of its very compact curd and less area of leaves. Average curd yield is 150 q/ha (Singh *et al.,* 1994). Somehow, the parental line seeds could not be maintained and the hybrid could not reach to commercial production stage.

The lack of development of F_1 cultivars of cauliflower is in marked contrast to other *Brassica* vegetable crops and has been due to following factors (Crips and Tapsell, 1993).

- Selfing of parental plants or sib-mating within the parental lines giving rise to off types after planting of commercial hybrid seed.
- Less effective self-incompatibility in cauliflower than in other brassicas.
- Non-synchronous flowering of male and female parents leading to increased proportion of sibs.
- The inflorescence of cauliflower tends to be cymose than racemose as in other brassica, which results in a flush of flower production over a shorter length of time, leading to greater tendency for non-synchrony of flowering of the parents of a cauliflower hybrid.
- Minor heterosis for curd size in contrast to substantial heterosis in other brassicas (cabbage, Chinese cabbage, Brussels sprout, etc.) which are grown primarily for their vegetative leafy tissues.

11. Male Sterility in Tomato and its Use in Hybrid Seed Production

Three types of male sterility have been summarized by Diez and Nuez (2008) in tomato. All are governed by single recessive genes. i) A long series of *ms* genes produce sterile pollen. All these genes are recessive except *MS* 48 which is dominant. As ususal this type of sterility as explained in muskmelon is maintained by backcrossing (*msms* x *Msms*). ii) Stamenless gene (*sl*) produce flowers with only a rudimentary development of stamens which normally produce no anthers or pollen. iii) Functional male sterility mutants which include *positional sterile* (*ps*), *positional sterile 2* (*ps-2*), *cleistogamous 2 (cl-2)*, *dialytic* (*dl*) and *exerted stigma* (*ex*). However, in spite of significant amount of research work done on these issues, male sterile lines in tomato have not been used on commercial scale in hybrid seed production.

Cytoplasmic male sterility in carrot

Cytoplasmic male sterility in carrot is available in two forms, namely, "brown anther" and "petaloid" types. The details are available in Simon *et al.* (2008). The brown anther male sterility was first discovered in cultivar Tendersweet and has been reported in several other cultivars. These plant start with formation of normal anthers but anthers fail to continue through to mature pollen production, remain rudimentary and turn brown. This sterility is usually unstable. The petaloid male sterility discovered in North American wild carrots is characterized by second whorl of petals in place of anthers. This is most commonly used form of male sterility for production of commercial carrot hybrids in North America. It is stable across a wide range of environments throughout flowering and seed production, although there are reports of break-down in late season umbels in some genetic backgrounds.

Several other forms and sources of CMS have been reported in carrot. The incorporation of CMS into carrot inbreds is almost similar to that in other crops. Generally this process begins at the F_2 or F_3 generation with intercrossing of individual selected fertile plants to a male sterile plant. In the next generation, there is an examination of progenies from male sterile plants for male fertility. This is followed by backcrossing for sterile line development. Presence of male fertile plants indicates the presence of restorer gene from the original male fertile parent.

References

Banga, O., Pitiet, J.and Van Bennekom, J.L. (1964). Genetic analysis of male sterility in carrot, *Daucus carota L. Euphytica* 13:75-93.

Barnes, W.C. (1961). A male sterile cucumber. *Proc. Amer. Soc. Hort.Sci.* 77:415.

Berninger, E. (1965). Contribution a l'etude de la sterilite de male de l'onignon (*Allium cepa* L.). *Ann Amelior Plan*t (Paris) 23: 183-199.

Bohn, G. W. and Principe, J. A. (1962) A second male sterility gene in the muskmelon. *J. Hered.* 55:211-215.

Cardi, T. and Earle, E.D. (1997). Production of new CMS *Brassica oleracea* by transfer of Anand cytoplasm from *B. rapa* through protoplast fusion. *Theor. Appl. Genet.* 94:204-212.

Charlesworth, D., Vekemanas, X., Castric, V. and Glemin, S. (2005). Plant self-incompatibility systems: a molecular evolutionary perspective. *New Phytologist* 168:61-69.

Chauhan, S. V. S. (1986). Studies in genetic male sterile *Solanum melongena* L. *Indian J. Genet. Plant Breed.* 46:367-371.

Cohan, S. M. and Weigle, J. L. (1966). Chemically induced male sterility in onion (*Allium cepa*). *Hort Science* 1:6.

Cole, K. (1957). A genetical investigation of male sterility in green sprouting broccoli. *Proc. Genet. Soc. Canada* 2:44.

Crisp, P. and Tapsell, C.R. (1993). Cauliflower – *Brassica oleracea* L. (In) G. Kalloo and B. O. Bergh (Eds.). Genetic Improvement of Vegetable Crops, p. 157 – 178. Pergamon Press, Oxford, New York.

Crisp,P. and Tapsell, C.R. (1993). Cauliflower. (In) G. Kalloo and B. O. Bergh (eds.). Genetic Improvement of Vegetable Crops. Pergamon Press, UK, pp 157-177.

Daskalov, S. (1972). Male sterile pepper (*C. annuum* L.) mutants and their utilization in heterosis breeding. *Proc. Eucarpia Meet. Capsicum* 7:202-210.

Davis, E. W. (1957). The distribution of the male sterility gene in onion. *Proc. Amer. Soc. Hort. Sci.* 70:316-318.

Diez, M. J. and Nuez, F. (2008) Tomato. (In) J. Prohens and F. Nuez (eds.). Handbook of Plant Breeding-Vegetables II. Pp 249-323. Springer.

Dickson, M. H. (1970) A temperature sensitive male sterile gene in broccoli, *Brassica oleracea* L. var. *italica*. *J. Amer. Soc. Hort. Sci.* 95:13-14.

Dickson, M.H. (1975). G 1117A, G 1102A and G 1106A cytosterile broccoli inbreds. *Hort Science* 10:535.

East, E. M. and Mangelsdorf, P. (1925). A new interpretation of the hereditary behaviour of self-sterile plants. *Proc. Nat. Acad. Sci. U.S.* 11:166.

Eisa, H. M. and Munger, M. H. (1968). Male sterility in *Cucurbita pepo. Proc. Amer. Soc. Hort. Sci.* 92:473-479.

Erickson, E. H. and Peterson, C. E. (1979). Asynchrony of floral events and other differences in pollinator foraging stimuli between fertile and male sterile carrot inbreds. *J. Amer. Soc. Hort. Sci.* 104:639-643.

Fang, Z., Sun, P., Liu, Y., Yang, L., Wang, X., Hou, A., and Bian, C. (1997). A male sterile with dominant gene (Ms) in cabbage (*Brassica oleracea* var. *capitata*) and its utilization for hybrid seed production. *Euphytica* 97:265-268.

Fitzgerald, D. M., Barry, D., Dawson, P. R. and Callels, A.C. (1997). The application of image analysis in determining sib proportion and aberrant characterization in F_1 hybrid *Brassica* populations. *Seed Science and Technology* 25:503-509.

Frankel, R. and Galun, E. (1977). Pollination Mechanisms, Reproduction and Plant Breeding. Springer-Verlag, Berlin.

George, R. A. T. (1999). Vegetable Seed Production, 2nd Edn. CAB International.

Georgiev, H. (1991). Heterosis in tomato breeding. (In) G. Kalloo(ed.). Genetic Improvement of Tomato. Monographs on Theoretical and Applied Genetics. 14. Springer-Verlag, Berlin, pp 83-98.

Gorman, S. W. and McCormick, S. (1997). Male sterility in tomato. *Critical Rev. Plant Sci.* 16:31-53.

Gorman, S. W.,Banasiak, D., Fairley, C. and McCormick, S. (1996). A 610 kb YAC clone harbors 7 cm of tomato (*Lycopersicon esculentum*) DNA that includes the male sterile 14 gene and a hotspot for recombination. *Mol. Gen. Genet.* 251:52-59.

Havey, M. J. (1995). Identification of cytoplasm using polymerase chain reaction to aid in the extraction of maintainer lines from open pollinated populations of onion. *Theor. Appl. Genet.* 90: 263-268.

Havey, M. J. (2000). Diversity among male sterility inducing and male fertile cytoplasms of onion. *Theor. Appl. Genet.* 101: 778-782.

Hawaldar, M. S. H., Mian, M.A.K. and Ali, M. (1997). Identification of male sterility maintainer lines for Ogura radish (*Raphanus sativus* L.). *Euphytica* 96:299-300.

Hayes, H.K. and Jones, D.F. (1916). First generation crosses in cucumber. Report of the Connecticut Agricultural Experiment Station, 319-322.

Hazra, P. and Som, M.G. (2005). Vegetable Science. Kalyani Publishers, Ludhiana.

Hoser-Krause, J. and Antosik, J. (1987). Horticultural value and seed setting of cytoplasmic male sterile cauliflower line with *Raphanus sativus* CMS (Bannerot). *Eucarpia Cruciferae Newsl.* 12:34.

Jasmin, J. J. (1954). Male sterility in *Solanum melongena* L.:Preliminary report on functional type of male sterility in eggplant. *Proc. Amer. Soc. Hort. Sci.* 63:443.

Jones, A. M. and Emsweller, S. L. (1936). A male sterile onion. *Proc. Amer. Soc. Hort. Sci.* 34:582-585.

Jones, H. A. and Mann, L. K. (1963). Onions and their Allies. Leonar Hill Ltd., London, pp 286.

Jones, H. A. and Clarke, A. E. (1943). Inheritance of male sterility in the onion and the production of hybrid seed. *Proc. Amer. Soc. Hort. Sci.* 43:189.

Kalloo,G. (1988). Vegetable Breeding, Vol. 1. CRC Press, Inc., Boca Raton, Florida.

Karbinskaya, E. N., Kosova, A. I. and Zaginailo, N.N. (1985). Stamenless sterility in tomato and its use in breeding. *Ref.Zh.* 7:65-289.

Kim, B. and Kim, D. H. (2005). Devedlopment of SCAR markers for early identification of cytoplasmic male sterility genotype in chilli pepper (*Capsicum annuum*). *Mol Cells* 20:416-422.

Kaul, M. L. H. (1988). Male Sterility in Higher Plants. Monographs on Theoretical and Applied Genetics. 10. Springer-Verlag, Berlin.

Kuckuck, H., Kobabe, G. and Wenzel, G. (1991). Fundamentals of Plant Breeding. Springer –Verlag, Berlin.

Kumar, Sanjeet., Banerjee, M. K. and Kalloo, G. (2000). Male sterility:Mechanisms and current status on identification, characterization and utilization in vegetables. *Veg. Sci.* 27:1-24.

Lee, J., Lee, D.H. and Park, H. G. (2005). Mapping of St locus flanking region related to incomplete phenotype of cytoplasmic genic male sterility in chilli pepper (*Capsicum annuum*). Plants and Animal Genome XIII. P 194.

Martin, W.J., McCallum, J., Shigyo, M., Jakse, J., Kuhl, J.C., Yamane, N., Pither-Joyce, M., Gokce, A. F., Sink, K.C., Town, C.D. and Havey, M.J. (2005). Genetic mapping of expressed sequences in onion and in silico comparisons with rice show scant colinearity. *Molecular Genetics and Genomics* 274: 197-204.

Masuda, M., Furuichi, T., Ma, Y. and Kato, K. (1998). Pollen degradation and inheritance of male sterility in three mutants in tomato (*Lycopersicon esculentum* Mill. cv. First). *J. Jap. Soc. Hort. Sci.* 67:583-588.

McCreight, J. D., Nerson, H. and Grumet, R. (1993). Melon. (In) G. Kalloo and B. O. Bergh (eds.). Genetic Improvement of Vegetable Crops. Pergamon Press, UK. Pp 268-270.

Melchers, G., Mohri, Y., Watanabe, K.,Wakabayashi, S. and Harada, K. (1992). One step generation of cytoplasmic male sterility by fusion of mitochondrial inactivated protoplasts with nuclear inactivated *Solanum* protoplasts. *Proc. Nat. Acad.Sci.* (USA) 89:6832-6836.

Morelock, T. E., Simon, P. W. and Peterson, C. E. (1996). Wisconsin Wild: another petaloid male sterile cytoplasm for carrot. *Hort Science* 31:887-888.

Nandpuri, K. S., Lal, T., Singh, S., and Singh, H. (1982). Seed production of muskmelon variety Punjab Hybrid on commercial scale. *The Punjab Vegetable Grower* 13: 51-53.

Nieuwhof, M. (1961). Male sterility in some cole crops. *Euphytica* 10: 351-356.

Nuttall, V. W. (1963). The inheritance and possible usefulness of functional male sterility in *S. melongena. Can. J. Genet. Cytol.* 5:197-199.

Ogura, H. (1968). Studies on the new male sterility in Japanese radish with special reference to the utilization of this sterility towards the practical raising of hybrid seed. *Mem. Fac. Agric.*, Kagoshina Univ. 6:39.

Pathak, C. S. (2000). Hybrid seed production in onion. (In) A. S. Basra (ed.). Hybrid Seed Production in Vegetables: Rationale and Methods in Selected Crops. Food Product Press, New York, London, Oxford. pp 89-129.

Pearson, O. H. (1972). Cytoplasmically inherited male sterility characters and flavour components from the species crosses *Brassica nigra* (L.) Koch x *B. oleracea* L. *J. Amer. Soc. Hort. Sci.* 97:397-402.

Pearson, O. H. (1983). Heterosis in vegetable crops. (In) R. Frankel (ed.). Monographs of Theoretical and Applied Genetics, Vol. 6 Heterosis, p 138-188, Springer-Verlag, Berlin.

Peterson, (1958). Cytoplasmically inherited male sterility in *Capsicum. Amer. Nat.* 92:111-119.

Petrova, M., Yulkova, Z., Gorinova, N., Izhar, S., Firon, N., Jacquemin, J. M., Attanassov, A. and Stoeva, P. (1999). Characterization of a cytoplasmic male sterile hybrid between *Lycopersicon peruvianum* Mill x *Lycopersicon pennellii* Corr and its crosses with tomato. *Theor. Appl. Genet.* 98:

Phatak, S.C. and Jaworski, C. A. (1989). UGA 1-MS Male sterile eggplant Germplasm. *Hort Science* 24:1050.

Pilrat, M. (1990). Gene list for *Cucumis melo* L. Cucurbit Genet. *Coop. Rep.* 13:58.

Ram, Hari Har. (2012). Vegetable Breeding-Principles and Practices, Third Revised Edn. Kalyani Publishers, Ludhiana, pp 791.

Rick, C. M. (1944). A new male sterile mutant in tomato. *Science* 99:543.

Rick, C. M. (1948). Genetics and development of nine male sterile tomato mutants. *Hilgardia* 18:599-633.

Riggs, T. J. (1988). Breeding F1 hybrid varieties of vegetables-Review article. *Journal of Horticultural Science* 63:369-382.

Sampson, D. R. (1966). Linkage of genetic male sterility with a seedling marker and its use in producing F1 hybrid seed of *Brassica oleracea* (cabbage, broccoli, kale, etc.). *Can. J. Plant Sci*. 46:703.

Sawhney, V. K. (1997). Genetic male sterility. *In* K.R. Shivanna and V.K. Sawhney (eds.). Pollen Biotechnology for Crop Production and Improvement. Cambridge Univ. Press, pp 183-198.

Scott, D. H. and Riner, M. E. (1946). Inheritance of male sterility in winter squash. *Proc. Amer. Soc. Hort. Sci.* 47:375.

Shifriss, C and Frankel, R. (1969). A new male sterility gene in *Capsicum annuum. J. Amer. Soc. Hort. Sci.* 94:385-387.

Shifriss, C. (1973). Additional male sterile mutant in *Capsicum annuum L. Euphytica* 22:527-529.

Shifriss, C. (1997). Male sterility in pepper (*Capsicum annuum* L.). *Euphytica* 93:83-88.

Shifriss, C. (1997). Male sterility in pepper (*Capsicum annuum* L.). *Euphytica* 93: 83-88.

Shifriss, C. and Guri, A. (1979). Variation in stability of cytoplasmic genic male sterility in *Capsicum annuum*. *J. Amer. Soc. Hort. Sci.* 104: 94-96.

Shigyo, M. and Kik, C. (2008). Onion. (In) J. Prohens and F. Nuez (eds.). Handbook of Plant Breeding – Vegetables II. Springer. Pp 121-159.

Sigareva, M. A. and Earle, E. D. (1997). Direct transfer of a cold tolerant Ogura male sterile cytoplasm into cabbage (*Brassica oleracea* ssp. *capitata*) via protoplast fusion. *Theor. Appl. Genet.* 94:213-220.

Simon, P. W., Freeman, R. E., Vieira, J. V., Boiteux, L. S., Briard, M., Nothnagel, T., Michalik, B. and Kwon, Y. (2008). Carrot. (In) J. Prohens and F. Nuez (eds.). Handbook of Plant Breeding-Vegetables II. Pp 327-357. Springer.

Singh, H. B. and Swarup, V. (1971). Exploitation of hybrid vigour in vegetables. ICAR Tech. Bulletin No. 33. Indian Council of Agricultural Research, New Delhi.

Singh, S. P. and Rhodes, A. M. (1961). A morphological and cytological study of male sterility in *Cucurbita maxima*. *Proc. Amer. Soc. Hort.Sci.* 78:375-378.

Swarup, V. (1977). Breeding Procedures for Cross Pollinated Vegetable Crops. Indian Council of Agricultural Research, New Delhi.

Swarup, V. (2006). Vegetable Science and Technology in India. Kalyani Publishers, Ludhiana.

Takasaki, T., Hatakeyama, K., Suzuki, G., Watanabe, M., Isogai, A. and Hinata, K. (2000).The S receptor kinase determines self-incompatibility in *Brassica* stigma. *Nature* 403:913-916.

Takayama, S., Shiba, H., Iwano, M., Shimosato, H., Che, F. S., Kai, N., Watanabe, M., Suzuki, G., Hinata, K. and Isogai, A. (2000). The pollen determinant of sel-incompatibility in *Brassica campestris*. *Proc. Natl. Acad. Sci.* USA 97:1920-1925.

Tanksley, S. D. and Zamir, D. (1988). Double tagging of male sterile gene in tomato using morphological and enzymatic marker gene. *HortScience* 23:387-388.

Van der Meer, Q. P. (1994). Onion hybrids: evaluation, prospects, limitations and methods. *Acta Hort.* 358: 243-248.

Van Der Meer, Q.P. and Van Bennekom, J.L. (1971). Frequency of genetical factors determining male sterility in onion (*Allium cepa* L.) and their significance for the breeding of hybrids. *Euphytica* 20:51-56.

Watts, V. M. (1962). A marked male sterile mutant in water-melon. *Proc. Amer. Soc. Hort. Sci.* 81:498-505.

Werner, C. P., Dowker, B. D., DeSouza, D.C., Setter, A.P., Crowther, T.C. and Horobin, J. F. (1998). Triple test cross predictions of the performance of recombinant inbred lines from a wide cross in onions. *Ann. Appl. Biol.* 112: 525-535.

Werner, C.P., Kearsey, M. J., Crowther, T. C. and Dowker, B. D. (1990). Prediction of performance of inbred lines derived from a population cross in autumn-sown onions (*Allium cepa* L.). *Theor. Appl. Genet.* 79: 507-512.

Williams, M. E., Leemans, J., and Michiels, F. (1997). Male sterility through recombinant DNA technology. (In) K. R. Shivanna and V. K. Sawhney (eds.). Pollen Biotechnology for Crop Production and Improvement. Cambridge Univ., Press. pp 237-257.

Williams, W. (1964). Genetical Principles and Plant Breeding. Blackwell Scientific Publications, Oxford. pp 504.

Wills, A. B. and North, C. Problems of hybrid seed production. *Acta Horticulturae* 83:31-36.

Wills, A. B. and Fyfe, S. K. (1974). Practical application of seed isoenzyme analysis. Proceedings of Eucarpia Meeting – Cruciferae, pp. 74-79.

Wills, A.B., Fyfe, S.K. and Wiseman, P. M. (1979). Testing F_1 hybrids of *Brassica oleracea* for sibs by seed isozyme analysis. *Ann. of Appl. Biol.* 91 : 263-270.

Yamagishi, H. (1998). Distribution and allelism of restorer genes for Ogura cytoplasmic male sterility in wild and cultivated radishes. *Genes and Genetic Systems* 73:79-83.

Yoo Woong, I.L. (1990). The inheritance of male sterility and its utilization for breeding in pepper (*Capsicum* spp.). Ph. D. Thesis, Kyung Hee Univ., South Korea.

Zhang, Qipei, Wei You Ying and Zhang Song (1995). Breeding of a male sterile line on the welsh onion (*Allium fistulosum* L. var *giganteum* Makino) and preliminary study of its heterosis. *Acta Horticulturae* 40:273-277.

Zhang, X. P., Rhodes, B.B., Baird, W. V., Skorupska, H. T. and Bridges, W.C. (1996). Development of genetic male sterile watermelon lines with delayed-green seedling marker. *Hort Science* 31: 123-126.

Zhiyuan, F., Wang, X., Dongyu, Q. and Guangshu, L. (2000). Hybrid seed production in cabbage. *In* A.S. Basra (ed.). Hybrid Seed Production in Vegetables : Rationale and Methods in Selected Crops. Food Products Press. pp 109-129.

Zhou Changjiu and Zhang Youliang (1994). Studies on several properties of radish male sterility. *Acta Horticulturae Sincia* 21:65-70.

Chapter – 12

Pests Management in Horticultural Crops

A.M. Ranjith

When we grow crops, we also provide food for pests attacking such crops. Pests would mean insect and non-insect pests of animal kingdom, diseases and weeds. Weed management is dealt with separately. Pest management is a strategy to reduce the conditions for multiplication of pests and to provide situations that maximize growth of crops. This, in turn would increase the profit of farmer. In modern horticulture even though it is not expressly written, this also means environmentally safe, ecologically sound and economically sustainable crop production.

Pest management involves a total system that considers agro-eco system as its base. Additional interactions are sourced from ecological perspective and also human perspective. Individual pest management, thus will have to be limited to the very specific pest situation only and when the pest in question is the most important. Other wise the agro- ecosystem encompassing the environment, the crop and the pest complex is a total unit for pest management.

1. What is IPM?

Integrated Pest Management (IPM) is the total strategy or totality of actions in the crop against individual or a group of pests that involve tactics based on cost : benefit analysis involving impacts on crops, pests, environment and end consumers.

2. What is a Pest?

Any organism can be a pest only under specific crop situations and under increasing populations that cause the damage. Damage can also be quantitative or qualitative. In a few cases, even a small blemish on a crop part is not accepted, but in most situations, a small amount of damage is tolerated. In nature, many biotic and abiotic (mostly climatic) factors cause natural population regulation to limit organisms from reaching pest status.

3. Importance of Crops in IPM

Specific value of the crop also decides the IPM strategy. High value crops require interventions at a much lower damage level. The specific crop part that is affected also defines the IPM strategy. The fruit borer in Apple even at very low population cannot be tolerated while leaf eating caterpillars at low levels can be left to be suppressed by natural enemies. The fruit borer in this case is a direct pest, while the leaf eating caterpillar is an indirect pest. (Direct pests are those that affect the economic produces like fruits, seeds, leaves or tubers. Indirect pests, on the other hand, feed on other non-economic parts like leaves, stem or roots.)

4. Impact of Environment in IPM

Within normal ranges of temperature and humidity, rainy season with a lot of humidity is ideal for increase in diseases and decrease in insect pests. During summer, the reverse is true.

Pest specific IPM- a case study of Helicoverpa

Crop surveys indicate that before 1975 only 20% of pigeon pea farmers in India were using insecticides; but by 1993, all farmers used chemical control in India. There are no resistant varieties and even six chemical sprays between 127 and 160 days after sowing did not give satisfactory results. Shaking and removing when 1-2 larvae/plant are noticed, a 85% reduction in insect population was achieved.

Impact of a bollworm infestation in cotton depends on number of larvae present, age of larvae, and timing of damage relative to Biological and cultural controls. *Bacillus thuringiensis* applications are acceptable for use on organically grown cotton.

Host plant resistance

Agricultural research programmes of countries where cotton is important, have in their germplasm, collections of cotton genotypes with partial resistance to *Helicoverpa* and other pests. Scientists made large investments into seeking resistance to *Helicoverpa* but conventional breeding programs have not repaid this investment in the form of plants performance in farmers fields.

Genotypes with only partial resistance are identified and the characteristics include antibiosis-genotypes with high levels of gossypol and other terpenoids,'Okra' leaves and glaucous (shiny) leaves combined with absence of nectaries (nectariless) and short duration with determinant growth habit are associated desirable traits.

5. Crop Stage Specific IPM

A single pest alone does not attack a crop. Pest attack is based on crop stage. Application of management strategies including crop protection chemicals affects all the pests in a particular growth stage. It is thus better to plan IPM, based on crop stage. A good illustration of this can be seen in classical guides by IRRI on IPM in Rice. Pests at different growth stages and strategies for management are considered while treating individual pests as a subgroup.

6. Management Practices

General instructions in IPM

- Apply insecticides only when the target insect population is above economic threshold level. Do not apply insecticides as a matter of routine.
- Do not use insecticides which induce vegetative growth, resurgence of sucking pests and development of resistance.
- Do not mix two insecticides as they hasten development of resistance in insects.
- Do not apply the same insecticide repeatedly.
- When a dose indicated in acre, multiply by 2.5 for getting per ha dose.
- Adopt suggested waiting period between application of pesticides and harvest.
- Ensure safe period between application of pesticides and release of parasitoids and predators.

Potato

Aphids: *Myzus persicae, Aphis gossypii* - Aphididae

Aphids suck the sap from potato foliage, causing wrinkling, cupping and downward curling of young leaves. The older leaves become yellow, wilt and fall off. Major damage by *M.persicae* and *A. gossypii* are through transmission of potato leaf roll virus

Leafhoppers: *Amrasca devastans(biguttula biguttula)* - Cicadellidae

The leafhoppers suck the leaf in large numbers and desapping causes leaves to become brown, curl upwards and dry up.

Mites: *Hemitarsonemus latus*-Tarsonemidae

Damage starts from apex leaf and goes downward. Leaves roll inward, showing oily spots on upper surfaces. Lower surface gets sooty appearance. The leaves ultimately become bronze in colour and wither. Plants become stunted and finally die. Heavy mite infestation might cause 50% loss in yield. It is a major pest in Deccan plateau.

Tuber moth: *Pthorimaea operculella* - Gelechidae

This is a major pest of potato in the field and in storage. Larvae of insect make mines in tender leaves, petiole and stem. They also bore into tubers and form irregular silk lined galleries filled with excreta. Saprophytes grow in such damaged tubers and cause rotting of potato tubers.

Cut worms: *Agrotis segetum, Agrotis ipsilon* - Noctuidae

The youngest larvae are leaf miners. Later they feed on leaves and grown up larvae cut young plants at ground level and feeding on shoot resulting in stunted plant growth. After tuberization, they may also feed by boring and nibbling into tubers.

Bihar Hairy caterpillar: *Spilosoma (Diacrisia) obliqua* - Arctiidae

The young larvae scrape leaves in a group and skeletonise them. Later instars feed fully on leaves.

Golden cyst nematode: *Globodera pallida, G.rostochiensis* - Heteroderidae

Initially, growth of plants is retarded and leaves wilt during hot day and recover in evenings. Plants remain stunted; leaves become yellow and plants wilt in patches. The underground symptoms include poor development of roots and reduction in number and size of tubers. This is one of the most important pests of potato.

IPM in potato

IPM starts in potato from planting itself. Avoid growing potato and brinjal together in nematode infested field and follow a rotation of once in three years. Seeds should be from nematode free areas and avoid irrigation water from infested field to new fields. Sprinkler system during day time would cause cut worms to come out and to be eaten by birds during day time. Aldicarb 10 G @ (20kg/ha) can be applied in heavily infested fields.

Summer ploughing of soil exposes larvae and pupae of cutworms to be eaten by birds. This also exposes grubs of cockchafer beetles for predation. Seed tubers should be free from life stages of potato tuber moth and should be planted at 15cm depth to avoid fresh oviposition. Cover top layers of seed tubers with lantana or eucalyptus branches to repel ovipositing adults.

From nursery stage itself, set up monitoring traps for different pests. Pheromone traps should be installed at the rate against potato tuber moth and cutworms 12/20/ha. Against white flies, aphids and leaf hoppers, set up yellow sticky traps for monitoring. Remove mechanically the nucleus infestations by whitefly, aphid, leafhoppers, Epilachna beetles and leaf eating caterpillars. Apply insecticides as spot application at the foci of the infestation to prevent spread to other areas. Pest monitoring should be continued until harvest of the crops. Set up light traps or bonfires during dusk to attract and kill adults of cut worms and moths.

Release egg larval parasitoid ***Chelonus blackburnii***@(25,000/ha) twice, on 40 and 70 days after planting.

Pest control using insecticides

Cut worms

Apply (Neem Seed Kernal Extract), NSKE (5%), or neem oil 5 lit/ha @500 litres of spray fluid/ha); use high volume sprayer and focus the nozzle at collar region. Apply insecticides during evening hours.

Whitefly

Spray Quinalphos 400ml or monocrotophos 350 ml or chlorpyriphos 500 ml/0.4ha.

Epilachna beetles

Carbaryl 50WP 800 g, endosulfan 35 EC 600 ml/0.4ha (450 litres of Water)

Leaf eating caterpillars

Endosulfan 35EC 600 ml/0.4ha (450 litres water). Against tuber moth, a dose of 1000ml quinalphos or endosulfan may be needed.

7. Tropical Tuber Crops

Aroids and Yams

Aphids: *Aphis gossypii* - Aphididae

Aphids congregate on growing points and suck the sap. Honey dew secreted by the aphids causes sooty mold in severe infestation. Aphids attack leads to yellowish speckling and curling of leaves.

Thrips: *Caliothrips indicus* - Thripidae

Characteristic silvery white specks with black dots, mostly faecal pellets develop on leaves. Drying of leaves takes place.

Leaf-eating caterpillars: *Spodoptera litura* - Noctuidae

Young caterpillars feed gregariously by scraping on leaves. Older caterpillars feed on entire leaf.

Scale insect: *Aspidiotus (Aspidiella) hartii* - Diaspididae

These are whitish yellow, hard scales that desap aerial parts and tubers. In severe infestation, the plant shows withering and drying, reducing vigour and vitality of plants. The infested tubers lose quality and viability.

Arecanut beetle: *Araecerus fasciculatus* - Anthribidae

Basically a storage pest, the grubs bore into tubers and convert to powder, causing loss in weight and quality.

Mealybugs: *Pseudococcus citriculus, Rhizoecus* sp.

Nymphs and adults have a powdery coating and they cover the tuber surface also. Sap removal leads to shrinkage, poor germination and loss in cooking quality.

Cassava or Tapioca

Cassava scale: *Aonidomytilus albus* - Diaspididae

White elongated scales appear in colonies on stems. Plants die in the field and under storage. Stem becomes weak and dry, causing side branching to give bushy appearance. Viability of planting material is reduced, resulting in poor establishment.

Termite: *Odontotermes obesus* - Termitidae

They eat the stems and roots and cause poor establishment and drying of plants.

White grub: *Leucopholis coneophora* - Melolonthidae

The grubs feed on developing roots and tubers. Attacked plants become pale-yellowish, wither and dry up or have reduced yield, prevalent in sandy or loose soils.

Spidermites: *Tetranychus cinnabarinus, T. neocaledonicus* - Tetranychidae

These are spider mites that colonize on lower surface, causing elongate yellowish streaks, chlorosis, withering and drying of leaves. The photosynthetic efficiency and yield are reduced. The pest is ubiquitous in occurrence.

Eutetranychus orientalis - Tetranychidae,

Oligonychus biharensis - Oligonychidae

Both the species infest upper surface of leaves, causing characteristic bronzing and typical leathery appearance. Such leaves curl from margins and fold into a roll. Heavy incidence causes depletion of chlorophyll content of leaves, photosynthesis and tuber yield.

Thrips: *Retithrips syriacus* - Thripidae

The reddish nymphs and adults lacerate and suck sap from leaves. The attacked leaves become pale, and show dirty brown excreta. In severe infestation, the leaves roll and dry up. Damage reduces tuber yield.

Spiralling whitefly: *Aleurodicus disperses* - Aleyrodidae

Adults and nymphs are covered with white waxy and powdery scale. They remain in large numbers on undersurface of leaves. Damage causes yellowish speckles with crinkling and curling of leaves. Honeydew secretion leads to development of black sooty-mould.

Cotton whitefly: *Bemisia tabaci* - Aleyrodidae

This insect is important as a vector of Indian cassava mosaic disease (ICMD).

Chips borers: *Araecerus fasciculatus* - Anthribidae, *Lasioderma serricorne* - Anobidae,

Rhyzopertha dominica - Bostrychidae, *Sitophilus oryzae* - Curculionidae

These are storage pests that bore into dried tapioca and reduce them into a powdery mass in due course. Grubs of these beetles bore dry bits and consume them. It causes loss in weight and quality.

Angumois grain moth: *Sitotroga cerealella* - Gelechidae, Larvae webbed together the flour, and feed within, causing weight and quality loss.

Red Flour beetle: *Tribolium castaneum* - Tenebrionidae. Grubs feed on dry cassava, causing weight and quality loss.

IPM in Tapioca

Tapioca tolerates a lot of insect pests. The management practices are mostly for managing whiteflies, which are mosaic vectors. Light traps and yellow sticky traps during dusk and dawn are to be used to attract and kill adults. The sticky traps should be cleaned and fresh sticky materials like oil or grease should be applied everyday morning. For light traps, use CFL tubes which have more rays in the UV range. If pesticide is needed, apply dichlorvos (1ml/litre) along with 5ml teepol. Apply in strips to conserve parasitoids like *Encarsia* spp.

Chinese Potato or 'Koorka'

Leaf-and shoot folders: *Hymenia recurvalis* - Pyraustidae

Pycnarmon cribrata

Phostria piasusalis

The larvae web together the leaves and feed within. Folding and webbing of leaves are noticed. In heavy attack, the leaf area is reduced considerably

Root-knot nematode: *Meloidogyne incognita* - Heteroderidae

An important pest of Coleus (Chinese potato). Heavy incidence by nematodes suppresses root growth and cause gall formation and poor growth. Tubers are malformed, hypertrophied and undersized. Both yield and quality of tubers are reduced.

Sweet Potato

Sweet potato weevil: *Cylas formicarius* - Curculionidae

Adult weevil punctures on vines and tubers. Damage is mainly by grubs that bore tubers and feed by making tunnels. Feeding holes and adult emergence holes can be seen on vines and tubers. In severe infestation, the collar region of plants shows unusual thickening due to proliferation and hypertrophy of tissues emitting a characteristic terpenoid odour. Even the slightly damaged tubers are unsuitable for cooking and consumpton. Yield loss in practice may come to 100% in severe cases.

Vine-borer: *Omphisa anastomosalis* - Pyraustidae

The caterpillars bore into vines and kill the vines. Wilting of vines and terminal drying are noticed.

Tortoise beetles: *Aspidomorpha miliaris, Cassida (Metriona) circumdata, Chirida bipunctata* - Cassididae

Grubs and adults of these beetles bite holes on leaves and feed on contents.

Spiny beetls: *Oncocephala tuberculata* - Hispidae

Damage is mainly by grubs that mine leaves. The adults feed on leaves making small holes.

IPM

- After every cropping season, remove all crop residues and also destroy left over in nursery.
- Remove and destroy alternate hosts of sweet potato (*Ipomoea* spp.) weevil.
- Use planting materials without any life stages of the pests.
- Seed tubers or vines should be dipped in fenthion (0.05%), fenitrothion (0.05%) or monocrotophos (0.05%).
- At 50 days after planting, earthup the ridges to prevent oviposition by adults and for removal of weeds.
- Tuber traps using 100G bits of sweet potato can be kept at 5m intervals in the evening at 50 to 80 DAP. Collect and destroy adult weevils every next day morning.
- Do not delay harvesting.
- In heavy infestation, drench the soil with endosulfan (0.05%) at 65DAP.
- Collect and destroy adults of tortoise beetles with sweep nets.
- Sphingid caterpillars can be handpicked and destroyed.

Mushroom

Sciarid (mushroom flies): *Bradysia paupera, Lycoriella auripila* - Sciaridae

The larvae feed on mycelial attachments and restrict growth of spawn. The maggots also feed on developing mushrooms or pinheads and thus cause loss in yield. They also bore developing mushrooms and render them hollow.

Phorid (dung flies): *Megaselia halterta,M.agarica,Megaselia* spp. - Phoridae

The female lays eggs in the base growing medium or directly in the gills of opened mushrooms. The larvae feed through bottom or directly into the cap from gills. The infested mushrooms become yellowish brown and they stop growing.

Staphylinid beetle: *Staphylinus* spp. - Staphylinidae The grubs of this beetle feed on soft gills of mushroom

Mite (mushroom mite):*Tyrophagus putrescentiae* - Acaridae

These are translucent white with long hairs on body. They feed on mycelium and damage the sporophoras. The caps shrink and develop brown rusted spots and can also kill young buttons.

Mycetophagous mites: *Pygmephorus* spp. - Pyemotidae. The mites feed on the mycelium.

Nematodes: *Aphelenchoides composticola, Ditylenchus myceliophagus* - Tylenchiae

These nematodes feed on the mycelium, which becomes sparse and patchy and the compost colour turns brown. Flushes become poor and delayed with decline in the yield.

IPM in mushroom

Sciarid flies, Phorid flies and Cecid flies

- Do not keep the spent compost and casing material at moist and shady places as they serve as ideal substratum for fly breeding during non cropping period.
- Dispose the material in the manure pit and cover it with at least 10cm thick layer of manure to check fly breeding in surrounding areas of mushroom house.
- Cook-out of the spent compost and casing material at 71°C with steam for 2 hours to kill immature stages of fly in spent material.
- Put screen for doors and ventilating openings of mushroom house with nylon wire nets of 14-16 mesh/cm to check entry of flies into mushroom house.
- Incorporate chlorpyriphos (50 ppm) in the compost and casing materials to reduce larval population in beds and fly population in mushroom house.

8. Ornamental Plants

Anthurium

Whitefly: *Aleurotulus anthuricola* - Aleyrodidae

Adults and nymphs with a white waxy coating colonize underside of leaves and petiole sheath. They suck sap from these parts causing yellowing, wilting and drying of affected parts.

Thrips: *Chaetanaphorothrips orchidii* - Thripidae

Thrips are hot weather insects whose damage starts in March and reaches to a peak in September. Thrips suck cell sap and produce white streaks and distortion of spathe.

Mite: *Brevipalpus* spp. - Tenuipalpidae

Mite colonies are seen on lower surface of leaves. Desapping and feeding by nymphs and adults cause speckling, mottling and wilting of affected leaves.

Carnation

Carnation tortrix moth : *Tortrix pronubata* - Tortricidae

The larvae of this tortrix attack the region around the growing points of shoots. They fasten the leaves together with silken threads and feed within.

Bud-borer: *Helicoverpa armigera*-Noctuidae

The adults lay eggs singly on growing buds. Larvae bore into buds and feed on developing petals. They also attack open flowers resulting in dropping of petals.

Thrips*:* *Haplothrips spp.* - Thripidae

Both nymphs and adults desap buds and flowers. The damaged buds show white streaks and get distorted in severe cases.

Two spotted spider mite: *Tetranychus urticae* - Tetranychidae

The mites are seen in colonies on lower surface of leaves, covered by dirty webs. The mite is serious on carnation under polyhouse conditions. Infestation spreads to buds, flowers and plants in severe cases. Affected leaves get discoloured and dry. Attacked plants lose vigour, get stunted in growth, wilt and dry, if damage is severe.

Chrysanthemum

Aphid: *Macrosiphoniella sanborni* - Aphididae

Sap sucking from tender shoots causes yellowing and drying of tender shoots. Greenish black nymphs and dark brown adults suck the cell sap from growing shoots and lower surface of leaves. Damage results in loss of vigour, yellowing and premature leaf fall and stunted growth. Honey- dew falling on leaves that are lower to the aphid attack favours development of sooty mould.

Thrips: *Microcephalothrips abdominalis, Frankliniella* spp. - Aphididae

Slender white nymphs and black adults feed on tender leaves causing silvering, mottling and distortion of leaves. Damaged flowers discoloured, withered and dried due to scorching.

Leaf folder: *Hedylepta indicata* - Pyraustidae

Pale white eggs are laid singly or in small groups on lower side of leaves. Green larva with white stripes fold leaves together and feed on chlorophyll. Affected leaves become skeletonized and dry. The larvae also feed on petals of flowers.

Serpentine Leaf miner: *Liriomyza trifolii* - Agromyzidae

The incidence of leaf-miner is severe in March-June. Eggs are deposited in leaf tissue and larvae mine leaves by feeding on mesophyll. Severely damaged leaves show extensive serpentine mines and may dry up.

Bud-borer: *Helicoverpa armigera* - Noctuidae

Round cream coloured eggs are deposited singly on bracts and petals of buds. Larvae feed on growing flowers causing loss in marketable flowers.

Hairy caterpillar: *Spilosoma obliqua* - Arctiidae

Pest attack starts during rainy season and continues till winter. Eggs are laid in clusters on lower side of leaves. Early instars are gregarious in habit and skeletonize leaves by scraping. Mature larvae eat leaves completely.

Thrips: *Microcephalothrips abdominalis, Haplothrips ramakrishnae* - Thripidae

The thrips desap the leaves and cause bronzing and drying of leaves and flowers

Spider mite: *Bryobis sp., Tetranychus sp.* - Tetranychidae

Mites are usually seen on upper surface of leaves. Damage starts in March and reaches a peak during April- July. Affected leaves become discoloured, wilt and fall off.

China Aster

Chrysomelid beetle*:* *Anlacophora foveicollis* - Chrysomelidae

Adults cut tender shoots and feed on leaves resulting in holes and loss of leaf. Grubs feed on roots and underground stems leading to yellowing,wilting and drying of attacked plants.

Semilooper: *Clenoplusia albostriata* - Geometridae

These are deep green semiloopers, with wavy lines and they feed extensively on leaves.

Flower caterpillars: *Helicoverpa armigera* – Noctuidae.

The caterpillars feed on flower heads in large numbers and make them unfit for sale. Caterpillars of *Phycita* sp. are green in colour with brown head. *H. armigera* incidence is seen in cooler months. Larvae damage flowers by making groove in central receptacle and feed on ovaries and stamens of flower buds. Flower is lost.

Stem borer: *Platyptilia molopias* - Pterophoridae

The adult plume moth lays white elongated eggs on tips of shoots. Larvae bore into stem and side branches. Cream coloured larvae with black head feed inside shoots resulting in hollowing and wilting of affected shoots. Plants may die due to the attack.

Crossandra

Spike borer: *Helicoverpa armigera* - Noctuidae

Larvae feed on spikelets, flowers and developing seeds. Early instar larvae bite holes into spikelets, whereas full grown ones bore into mature spikes.

Brown scale: *Saissetia nigra* - Coccidae; **White scale:** *Orthezia insignis* - Orthezidae

Yellowish nymphs and dark brown adult scales are seen in large numbers under leaves and on petioles and stem. Severe infestation results in stunting of growth and cause leaves to turn yellow and drop. Honey dew secreted by *Orthezia* sp. favours development of sooty mould on plant parts on which the dew falls.

Whitefly: *Lipaleyrodes* sp.- Aleyrodidae

It occurs in large numbers on undersurface of leaves. They desap from undersurface of leaves and infestation leads to chlorosis and development of sooty mould.

Bug: *Gynenciaa affinis*

Nymphs and adults of crossandra bug suck sap resulting in brown lesions on leaves and flowers. Severely infested plants fail to produce flowers. High population is seen during winter.

Midge: *Contarinia* sp.

Black coloured adult fly lays eggs singly within buds of spikes. Orange maggots feed on ovaries of developing buds and pupate inside. Due to damage, the buds fail to develop and rot in severe cases.

Gladiolus

Thrips: *Taeniothrips simplex* - Thripidae

Yellow coloured nymphs and black adults damage leaves and spikes by rasping tissues and sucking sap. Affected leaves and spikes develop silver streaks, turn brown, get deformed and dry when attack is severe. Corms in storage are also attacked by thrips. Infested corms are sticky, get shriveled and produce weak plants.

Cut worm: *Agrotis segetum* - Noctuidae

The female moth lays eggs near ground level on plant parts. Emerging larvae feed on developing shoots at night. The clay coloured larvae eat shoots and grown up larvae cut the plants at ground level. Young plants upto around 3-4 leaf stage are the most susceptible. The larvae also damage underground corms and developing spikes.

Leaf caterpillar: *Spodoptera litura* - Noctuidae

The adults lay eggs in groups on lower side of leaves. Early instar larvae feed on this lower side by scraping and skeletonizing them. Blackish brown mature larvae feed voraciously on leaves.

Mite: *Tetranychus equatorious* - Tetranychidae

Attack begins at a very early stage when plants are young. Green coloured nymphs and carmine coloured adults colonize on leaves and suck the sap. The damage results in discolouration, bronzing and shedding of affected leaves.

Mealy bug: *Ferrisia virgata* - Pseudococcidae

The infestation begins in field on underground corms during dry conditions and is carried on to storage. Nymphs and adults damage corms by sucking sap resulting in shriveling and drying of affected corms.

Jasmine

Bud-worm: *Hendecasis duplifascialis* - Pyraustidae

This pest is serious on *Jasminum sambac*. The larvae are dark green with black head which bores into immature buds. They make a hole on buds, and feed inside, leaving behind faecal pellets entangled in a few silk strands. Webbing is seen in severe infestation and pupation takes place in soil.

Leaf web worm: *Nausinoe geometralis*- Pyralidae

The larvae is green with dark warts. Young larvae scrape the tissue, and mature larvae web leaves together with loose silken threads. Severely affected bushes show burnt appearance due to leaf drying. Pupation takes place inside the web. Damage is more in irrigated fields during summer.

Gallery worm: *Elasmopalpus jasminophagusi*- Pyralidae

This is a serious pest on *J. auriculatum* and causes webbing on terminal leaves, shoots and flower heads. The larvae is green with a red head. It builds galleries of faecal matter and silk and feeds inside. Pupation takes place in galleries.

Blossom midge: *Contarinia maculipennis* - Cecidomyidae

It is a pest on *J. sambac* and *J. auriculatum* .The attack causes the bud to show purple discolouration 4-5 days after oviposition. Maggots are dull white and feed on inner 1-4 petals, anthers and stigma. Mature yellowish orange maggots fall down to soil and pupate.

Shoot borer: *Sycophila* sp. - Eurytomidae

This is observed on *J.grandiflorum*. Larvae tunnel inside shoots and cause a series of grooves and pupate inside the shoots. Series of entry holes are seen on damaged shoots.

Gall mite: *Aceria jasmini* - Eriophyidae

It is a major pest of *J. auriculatum* and *J. pubescens*. The mites feed on leaves and flowers and cause malformation and formation of felt outgrowths in vegetative and floral parts and consequent reduction in growth and yield.

Flower thrips: *Thrips orientalis* - Thripidae

The nymphs and adults of thrips suck sap from flowers and buds. The petals show brown streaks and buds fall.

Marigold

Bud caterpillars: *Helicoverpa armigera,* - Noctuidae

Eggs are laid singly on young buds. Larvae feed on developing flowers by damaging florets. They can also eat the leaves.

Leaf webber: *Phycita* sp. - Phycitidae.

Larvae of *Phycita* sp. feed on heads of buds and flowers.

Aphids: *Aphis gossypii,* - Aphididae

Aphids mainly infest lower surface of flowers and base of petals. Nymphs and adults suck sap from flowers causing discolouration and withering. Congregation is also seen on the fresh shoots.

Thrips *Neohydatothrips samayunkur*: - Thripidae

Thrips infest young leaves, buds and flowers and suck the sap. Affected leaves get distorted, while petals of flowers turn brown and dry.

Orchids

Biosduval Scale/Orchid scale: *Diaspis boisduvalii* - Diaspididae

It is round flat and yellow in colour. Only the first instar crawlers move. Others are sedentary. The young and mature scales suck juice from leaves, petioles and shoots and also inject toxins into the plant. Affected plants turn yellow in patches, lose vigour and get stunted.

Mealybug: *Pseudococcus maritimus* - Pseudococcidae

These are commonly referred as grape mealy bugs, yellow coloured with a covering of white powdery wax. They suck sap from plant parts and secrete honey dew. Sooty mould develops on leaves in severe infestation. Attacked plants turn yellow, withered and deformed.

Thrips: *Anaphothrips orchidaceous* - Thripidae

Both adults and nymphs suck cell sap from leaves and growing buds. Affected leaves curl, while petals develop streaks and margins of leaves turn brown and dry up.

Giant African Snail: *Achatina fulica* - Achatinidae

Adult and young ones of snail eat plants during night. Snails attack is heavy during monsoon. Grown up snails also cut tender shoots especially.

Rose

Hard Red scale: *Aonidiella aurantii* - Diaspididae

Both adults and young scales suck sap from the stem and cause spot yellowing and loss of vigour of plant. Infested shoots bear reddish brown encrustations of scale and in case of severe damage, the shoots dry up.

Blackfly: *Aleurocanthus spiniferus* - Aleyrodidae

The puparia are black oval and seen on underside of leaves. The nymphs and adults drain sap, causing yellowing and crinkling of leaves.

Thrips: *Rhipiphorothrips cruentatus; Scirtothrips dorsalis* - Thripidae

These are pests that attack new flush after pruning. Blackish brown adults and reddish nymphs of *Rhipiphorothrips* suck cell sap from tender leaves, buds and flowers. *Scirtothrips* is cream coloured and is a major problem in polyhouse. Damage is seen as curled leaves with silvery yellow patches, brown marks and deformed buds with burnt margins.

Aphid: *Macrosiphum rosaeformis/rosae* - Aphididae

Blackish green coloured aphids cluster around growing shoots and buds. Both nymphs and adults, suck cell sap and devitalize plants. High humidity and cloudy weather favour rapid build up of aphids. Affected leaves get deformed while buds retard in growth.

Leaf cutter bee: *Megachile anthracina* - Megachilidae

The adult cuts leaves in a characteristic ring like fashion and almost fully removes it. The leaf is used in the nest by the bee.

Hairy caterpillar: *Euproctis fraterna* - Lymantriidae

Eggs are laid in clusters on lower side of leaves. Early instars are gregarious in habit and feed from surface causing skeletonization and drying of leaves. Black coloured mature larvae, feed voraciously on leaves, buds and flowers, resulting in severe defoliation and flower loss.

Bud borer: *Helicoverpa armigera* - Noctuidae

Eggs are laid singly on growing buds. Larvae bore into buds and feed on developing petals. Open flowers are also attacked by larvae resulting in dropping of petals.

Chafer beetle: *Adoretus* spp. - Rutelidae

Damage starts with of onset of monsoon. Black and brown coloured adult beetles emerge from soil in large numbers after dusk and feed on leaves, buds and tender shoots. Damage is mainly due to grubs that feed on roots and cause plants to lose vigour and die.

Ash weevil: *Myllocerus* spp. - Curculionidae

Eggs are laid in soil. The grubs feed on roots, while the grey coloured adults feed on leaves from margins in a regular fashion and cause severe foliage damage.

Termite: *Microtermes obesi* - Termitidae

Termite attack starts under dry soil conditions. The termites feed on roots and may attack stem and bark in severe infestations. Affected plants wilt, dry and die consequently.

Two spotted spider mite: *Tetranychus urticae* - Tetranychidae

The mite is found in colonies in lower surface of mature leaves under white-silky webs. In severe infestation, population spreads to entire foliage including flowers and buds. Hot and dry conditions favour mite multiplication. Cream coloured nymphs and red coloured adults with two black spots scrape and suck sap resulting in discolouration, bronzing and drying of affected leaves. Damage on tender buds leads to discolouration, retardation and drying of leaves and buds.

Tuberose

Thrips: *Thrips tabaci* - Thripidae

These minute insects, preferring hot humid weather, rasp and suck the cell sap, causing bronzing, withering and drying up of leaves and flowers. This is the most serious pest on tuberose. Damage to young plants results in stunted growth.

Bud borer: *Helicoverpa armigera* - Noctuidae

Eggs are deposited singly on growing spikes. Larvae bore into buds and flowers and feed on them by making holes, causing loss of flowers and fresh growth

Weevils: *Myllocerus* spp. - Curculionidae

Ash coloured adults feed on leaves by making characteristic cuts from margins during night.

Pentatomid bug:*Nezara viridula* - Pentatomidae

Eggs are laid in groups on leaves. Green coloured nymphs and adults suck the sap from basal portions of growing buds and flowers. Affected buds wither and drop off.

Aphids: *Myzus persicae* - Aphididae

Small green coloured aphids suck sap from base of flower buds and growing spikes. Damaged buds and flowers fall off while affected spikes look withered and droop.

IPM in ornamental plants

Ornamental and flowering plants grown in the open are the best protected during rainy season by giving rain cover and during summer by giving rain type irrigation. Summer pests like thrips, mites and whiteflies are regulated by decreasing temperature in the microenvironment around plants. Monitor pests everyday and mechanically, remove and destroy affected parts. When insecticides are sprayed, do not repeat the same chemical consecutively.

Grubs of chafer beetles can be managed by drenching soil with chlorpyriphos (2.5ml/ litre of solution). When population of aphids, thrips and leaf eating caterpillars increase, apply Methyl Demeton (2ml/litre) or Dimethoate (1ml/litre) or quinalphos (2ml/litre) in need based manner. Against budworms in jasmine, profenofos (1.5 ml/l) or Lamda cyhalothrin (1.5 ml/l) or Malathion (1ml/l) can be alternated, based on requirement.

Pest management in green house

Mites, Thrips, Whiteflies and Aphids predominate in greenhouses due to greenhouse effect. Providing cross ventilation at plant canopy height reduces the pest population substantially. Green house pest management should also use natural enemies like coccinellids, lacewings (chrysopa) and predatory mites. A simple way to increase population of predatory mites is to grow pollen rich flowering plants. Use low residual chemicals like Dichlorvos, or Malathion and do not use synthetic pyrethroids repeatedly against caterpillars.

9. Medicinal Plants

Aswagandha

Grasshoppers: *Trilophida annulata* - Acrididae

The nymphs and adults of this grasshopper feed on margins of leaves

Long horned grasshoppers: *Phenoropiara gracillis* The adults and nymphs feed on leaves in a shot hole manner.

Mealy bug: *Coccidohystrix insolitus* - Pseudococcidae

The nymphs and adults remain on under surface of leaves and suck the sap. Affected plants turn yellow, wilt and dry. Honey dew secreted by these bugs favours development of sooty mould on areas where the honeydew falls.

Leaf hoppers: *Penthima* sp. - Cicadellidae

Nymphs and adults congregate on leaves and suck the sap, causing yellow patches

Cow bug: *Gargara mixta* - Membracidae

Nymphs and adults of this bug are seen feeding on shoots, causing yellowing and devitalisation.

Sphingid larvae: *Dilephila nerii* - Sphingidae

Larvae feed voraciously on leaves resulting in severe defoliation.

Epilachna beetle: *Henosepilachna vigintiopunctata* - Coccinellidae

Eggs are laid on under surface of leaves. Grubs and adults feed on leaves by scraping. Damage results in skeletonization and drying of leaves.

Ash weevil: *Myllocerus viridanus* - Curculionidae

Adult weevils cause damage by feeding on leaves from edges. Grubs feed on roots and cause wilting and drying of plants.

Weevil: *Blocyrus inaeagualis* - Curculionidae

The adult weevils eat the leaves and thus make perforations in leaves.

Coleus aromaticus

Lace wing bug: *Monantha globulifera* - Tingidae

Eggs are inserted slantingly into leaves and other tender foliage. Black spiny nymphs and lace winged adults suck sap from leaves. Attacked leaves develop yellow patches, become brown, shrivel and dry up in severe cases.

Spike borer: *Helicoverpa armigera* - Noctuidae

Female moth lays eggs on tender foliage. Larvae initially feed on foliage and later damage developing spikes.

Isabgol

Aphid: *Aphis gossypii* - Aphididae

Yellowish green or brown aphids suck sap in colonies from tender shoots and lower side of leaves. Attacked leaves curl up, wither and fall.

Opium

Cut worms: *Agrotis sulfusa* - Noctuidae

Adult moth lays eggs in moist soil or on lower side of leaves. Dark brown larvae cut younger plants at ground level and shoots of grown up plants. Damage results in growth retardation of plants and reduced yield.

Capsule borer: *Helicoverpa armigera* - Noctuidae

Spherical yellowish eggs are laid on tender foliage. Larvae feed on tender foliage and later bore into developing spikes and capsules.

Aphid: *Myzus* sp. - Aphididae

Aphids damage growing tender shoots by sucking sap. Affected shoots wither, get deformed and retard in growth.

Periwinkle/Lavender Blue

Sphingid larva: *Dilephila nerii* - Sphingidae

The large larvae cause severe damage to plants by feeding extensively on leaves, buds and flowers.

Leaf eating caterpillar: *Anomis flava* - Noctuidae

Larvae feed on leaves resulting in considerable foliage loss.

Grass hopper: *Acrida exaltata* - Acrididae

Both young and adult stages of this grasshopper damage the foliage by making holes on leaves and cutting tender shoots.

Sarpagandha

Scale: *Saissetia coffeae* - Coccidae

Brownish black scale infests lower surface of leaves along with mid ribs and mid veins. They also cover petioles and tender shoots. Due to draining of sap, leaves turn yellow and drop, plants become stunted.

Grass hopper: *Orthacris simulans* - Acrididae

Both nymph and adult stages feed on leaves and cut tender shoots.

Leaf folder: *Glyphodes vertumalis* - Pyralidae

Larvae fold leaves together and feed on green tissue causing drying of shoot tip and stunted growth of plant.

Chafer beetle: *Anomala polita* - Scarabidae

The young larvae cut seedlings about 2cm above the ground. Adults emerge with onset of first showers and feed on foliage, buds and flowers at dusk.

10. Aromatic Plants

Ambrette

Spotted boll worms: *Earias* spp. - Noctuidae

Moths lay eggs singly in leaf axils, bracts and leaf veins and on under surface. Larvae bore into tender shoots, buds, flowers and fruits and feed on internal contents. Damage results in drooping of shoots and shedding of buds and flowers.

Aphid : *Aphis gossypii* - Aphididae

Small greenish brown aphids are found on tender shoots and lower surface of leaves and suck sap. Affected leaves fade, curl and dry up. Black sooty mould develops on leaves where honey dew secreted by the pest has fallen.

Red spider mite: *Tetranychus telarius* - Tetranychidae

Mite colonies are found on lower side of leaves. They suck sap from leaves and produce white or yellow specks. Severe infestation causes bronzing and drying of leaves.

Davana

Semilooper: *Thysanoplusia orichalcea* - Noctuidae

The semilooper larvae feed voraciously on leaves and cause defoliation of plants.

Pentatomid bug: *Dolycoris indicus* - Pentatomidae

Both nymph and adult stages suck the sap from leaves, resulting in yellowing, curling and falling of leaves.

Fennel

Thrips: *Thrips flavus* - Thripidae

Infestation of thrips is generally during hot months with high humidity. The attack by thrips leads to discolouration, bronzing and drying of affected leaves.

Semilooper: *Thysanoplusia orichalcea* - Noctuidae

Bluish green or grey coloured eggs are laid singly on lower surface of leaves. Larvae feed voraciously on leaves and defoliate plants.

Pentatomid bug: *Dolycoris indicus*-Pentatomidae

A summer pest, maximum population of bug was found during May. Both nymphs and adults suck sap from growing spikes and cause flower drop.

French Basil

Aphid: *Aphis gossypii* - Aphididae

The aphid is small yellowish green or greenish brown and is found in colonies on shoots, leaves and inflorescence. Due to sucking of sap, leaves become discoloured and deformed. Affected buds and flowers fall off. Sooty mould develops in case of heavy dropping of honey dew.

Leaf eating caterpillar: *Pycharmon caberalis*

Larvae feed on leaves and cause severe foliage loss.

Whitefly: *Dialeurodes* sp. - Aleyrodidae

Eggs are laid singly on lower surface of leaves. Nymphs and adults suck sap from leaves. Affected leaves turn pale, curl and dry, sooty mould develops on honey dew secreted by the pest.

Tingid bug: *Monanthia globulifera* - Tingidae

Female bug inserts eggs into the tissue of leaves. Black spiny nymphs and lace winged adults suck sap from leaves causing yellowing, curling and drying of infested leaves.

Lemon Grass

Shoot borer: *Chilo infuscatellus* - Pyralidae

More commonly known as a pest of sugarcane. Eggs are laid in batches on lower side of leaves. Emerging larvae move down and cut a hole at ground level and enters stem. Due to feeding by larvae central shoot begins to wilt and dry causing dead heart.

Spittle bug: *Clovia bipunctata* - Cercopidae

Nymphs and adults suck sap from young leaves causing yellowing of leaves.

Aphid: *Macrosiphum miscanthi* - Aphididae

Aphids suck sap from leaves and tender shoots resulting in yellowing of leaves and shriveling of tender parts.

Termite: *Microtermes obesi* - Termitidae

Usually a dry season pest. The termites feed on roots and also on stem portion of plant. Affected plants show wilting and drying of leaves followed by death of the plant.

Mentha (Mint)

Cabbage semilooper: *Thysanoplusia orichalcea* - Noctuidae

Eggs are laid singly on lower surface of leaves. The emerging larvae feed as a scraper initially and then feed voraciously on leaves.

Leaf webber: *Syngamia abruptalis* - Pyralidae

Eggs are laid on tender foliage. Larvae web leaves together and feed on them. When fully grown, they pupate inside leaf fold. Damaged leaves become skeletonized and dry up in heavy attack.

Gram caterpillar:*Helicoverpa armigera* - Noctuidae

Eggs are laid singly on tender foliage. Larvae feed on leaves and cause defoliation.

Aphid: *Aphis affinis* - Aphididae

The pest is active during March-April. Colonies of nymphs and adults are found on tender shoots and lower surface of leaves and suck the sap. Attacked leaves fade, curl up and fall. Tender shoots dry. Black sooty mould is seen on infested parts due to secretion of honey dew.

Patchouli

Leaf webber: *Syngamia abruptalis* - Pyralidae

Caterpillars feed extensively on tender foliage leaving veins alone and growth of plants are affected.

Leaf roller: *Psara stullalis* - Pyralidae

Larvae are leaf rollers, and rolls the leaves and feed inside causing browning and drying of affected leaves.

Pentatomid bug: *Nezara viridula* - Pentatomidae

Green coloured bugs suck sap from tender leaves and shoots. As a result, leaves get deformed and growth of plant is stunted.

IPM in aromatic and medicinal plants

Insecticide application in these crops is to be avoided. Most of these have repellency properties against pests. Use a crop mix that ensures deterrency to all pests of crops. This is ensured by planting specific crops in strips, so that volatiles from one strip deter pests of other strip. Crops that are the least preferred by pests, should be grown as border crops. *Vitex nigundo, Nerium oleander* and lemongrass can be used for this purpose.

11. Plantation Crops

Arecanut

Spindle bug: *Carvalhoia arecae* - Miridae

Adults and nymphs suck sap from emerging spindle and tender leaves. Leaves develop linear brown and black necrotic streaks, and become stunted, twisted and malformed. The leaves dry up later and drop off in the wind.

Root grub: *Leucopholis burmeisteri* - Melolonthidae

These are serious pests in loose sandy soil, and when the attack is severe, cause yellowing of leaves, leaf or inflorescence drop, tapering of stem, reduction in yield and even death of young palms in 2-3 years. The grub feeds on roots of any vegetation, including that of areca nut and in severe cases, the entire bole region of palm is destroyed.

Inflorescence caterpillar: *Tirathaba mundella* - Pyralidae

The caterpillars are seen inside unopened spadix where they feed voraciously within silken webbings. The webbed flower stalks has also the excreta in between. When the flower has opened, these webbers can be seen actively feeding within the webbings and amidst excreta. Due to this, the flower fails to open in time as the inner floral parts are webbed together by caterpillars inside it. Drying up of inflorescence is another symptom.

Pentatomid bug: *Halyomorpha marmorea* - Pentatomidae

The nymphs and adults pierce tender nut and suck kernel sap. Due to the feeding, tender nuts are shed. The shed nuts show a pin prick like puncture mark below the perianth or on general nut surface where the nut has shrunk.

Scale insect: *Aonidiella orientalis* - Diaspididae

The scale insects colonize leaves, spathe, bunches and nuts and suck sap from soft parts. Feeding on leaves and tender nuts result in yellowing of tissues. In severe cases, kernel may not develop and turns black. Nuts fall.

Foliage mites: *Raoiella indica,* - Tenuipalpidae; *Oligonychus indicus* - Tetranychidae

Nymphs and adults of these mites congregate in colonies on undersurface of leaves and suck plant sap. Mostly, damage is appreciable in young palms. As a result of their feeding yellow speckles are formed on upper lamina. Later they coalesce becoming bronze coloured and finally the entire lamina dries up.

Calyx mite: *Dolichotetranychus* sp. - Tenuipalpidae

Colonies of mites are seen inside perianth and due to their feeding, tender nuts are shed.

IPM in arecanut

Root grubs can be managed by applying chlorpyriphos (2.5 ml/l) to root zone just after pre-monsoon showers. Set up light traps during dusk in periphery of garden during pre-monsoon showers when adults emerge. Burn trash, old dried leaves, petioles and leaf bases during dusk to attract and kill adults. Mites can be managed with Dicofol (2ml/l), or Dimethoate (1.5ml/l). Keep perforated sachets containing phorate 10G in the innermost leaf axils to deter spindle bugs. If inflorescence caterpillars are prevalent, force open the spadix and pour Malathion at (2ml/litre) into the spadix.

Cashew

Stem and root borers: *Plocaederus ferrugineus, Bactocera rufomaculata* - Cerambycidae

Grubs of this beetle are seen tunneling roots and also trunk of cashew. Eggs are laid in the bark or root and the first instars eat outer layers first and then bore deep inside. Presence of small holes in collar region, gummosis, extrusion of frass through the holes,yellowing and shedding of leaves and drying of twigs are symptoms of pest infestation.The tree eventually dies.

Tea mosquito bug: *Helopeltis antonii* - Miridae

One of the most important pests of cashew. Both nymphs and adults suck sap from shoots, inflorescence, nuts and apples. The feeding results in formation of a necrotic area around point of stylet entry. These lesions coalesce and ultimately result in shoot blight or inflorescence blight. The developing nuts are shriveled and dry up. A combined attack by the fungus *Colletotrichum gloeosporioides* is also seen in recent years.

Cashew Leaf miner: *Acrocercops (Conopomorpha) syngramma* -

The leaf miner causes heavy damages on tender foliage, especially younger plants. Mining in between epidermal layers of leaves results in formation of blisters and when the leaves mature, they appear as holes due to drying up of damaged tissues.

Leaf blossom webber: *Lamida (Macalla) monocusalis* - Pyralidae

The larvae web the apical portion of new tender leaves and inflorescence causing drying up of webbed portion.

Apple and nut borers: *Thylocoptila panrosema, Nephopteryx* sp.

Several species cause damage on apple and developing nuts.

Flower thrips: *Solenopsis rubrocinctus, Rhynchothrips rauensis, Scirtothrips dorsalis*- Thripidae

Thrips are seen in large numbers on inflorescences, devouring on male and female flowers. The desapping and shrinkage of female flowers cause reduced yield. They also cause shedding of flowers. Scabs on floral branches, leaves, apples and nuts are also seen.

Leaf beetles/red beetle: *Monolepta longitarus*

The adult beetles scrape green matter of leaves from lower surface ultimately resulting in drying up of leaves. Seedlings and young trees are severely affected.

IPM in cashew

The most important yield limiting factor in cashew is the attack by tea mosquito bug in endemic areas; cashew has to be sprayed from flushing time itself. Apply carbaryl 2gm/l during flushing or endosulfan (1.5 ml/l) during flowering and quinalphos (2ml/l) during initiation of fruit set. Add the fungicide copper oxychloride (2 g/l) or Mancozeb (2g/l) with the first and second spray in areas where anthracnose is a problem. The above insecticides will also manage leaf miner, and leaf eating caterpillars.

Remove and destroy roots and branches with attack by root and stem borer. The grubs from main trunk may be scooped out, if possible. Swab the stem with a suspension of mud slurry + carbaryl (0.2%) twice a year in areas of regular attack. Collect and destroy adults whenever they are spotted.

Cocoa

Mealy bug : *Planococcus lilacinus, P. citri* - Pseudococcidae

Nymphs and adults desap the tender shoots, terminal buds, flower cushions, cherelles and mature pods. Heavy colonization by the pest on shoot tip causes stunting of apical buds. Cushion abortion and cherelle wilt are also experienced on its infestation on flower cushions and cherelles.

Black aphid: *Toxoptera aurantii* - Aphididae

Adults and nymphs accumulate on tender leaves, shoots, flower cushions and developing cherelles and suck sap. Feeding by the pest results in crinkling of leaves, shedding of flowers and stunting of terminal buds. Attack by this pest is high during flushing season.

Stem borer: *Zeuzera coffeae* - Cossidae

The caterpillars of this moth make hollow tunnels inside the shoots and tender stem. They cut a round hole on stem, go inside and cause drying of portion above the hole. Presence of red excreta coming out of hole is seen. Rarely the pest is also observed to affects the main stem and then the plant dies.

IPM in cocoa

Stem girdler is serious in young seedlings and branches of young trees. Prune the affected branches, extract the grubs and destroy them. Apply lindane or carbaryl paste at pruned area. This practice should be done for stem borer also. Mealy bugs, aphids and leaf eating caterpillars can be managed with fenthion (0.5ml/l) Dimethoate (1.5 ml/l). If carbaryl has been sprayed against stem borer, or stem girdler, a second application against these pests will not be necessary.

Oilpalm

Spindle bug: *Carvalhoia arecae*

Nymphs and adult bugs suck plant sap from spindle leaf and tender leaves. Feeding marks are seen as necrotic lesions in spindle and fully opened leaves. The pest is seen in the nursery plants round the year but incidence is high during rainy season.

Tussock caterpillar: *Dasychira mendosa*

Caterpillars of this polyphagous pest feed on young and mature leaves of seedlings, causing considerable defoliation. Incidence of the pest is more during June and July after a few heavy rains.

Rhinoceros beetle: *Oryctes rhinoceros*

The adult beetle bores through into spear leaf, resulting in snapping of fronds at feeding sites. Beetles also bore into the male and female inflorescence even when they are within spathe.

Bagworm: *Manatha albipes, Metisa plana*

The caterpillars of these species of bagworms cause severe defoliation occasionally on oil palm.

Vertebrate pests: *Crows, parrots and babbler*

They feed on ripe fruits and destroy them.

Rats: *Bandicoota indica*

Rattus rattus wroughtoni

Rats tunnel into bole of seedlings causing their death. They also damage ripe bunches and spikelets of male inflorescence.

Coconut

Rhinoceros beetle: *Oryctes rhinoceros* - Scarabaeidae

The damage is mostly by adult beetles, which bore and chew up the unopened fronds and rarely the spathe. When fully opened, the affected frond shows characteristic V shaped cuts. Attack by the adult on spathe causes wilting of inflorescence, amounting to direct crop loss. The grubs are seen in manure pits and decaying organic matter.

Red Palm weevil: *Rhynchophorus ferrugineus* - Curculionidae

The adults lay eggs on cabbage region of crown, on cuts in stem, or in boll region. The grubs' tunnel extensively inside and sometimes may push out frass to outside. Tunneling causes the plant to weaken, and eventually death of tree. The attack is seen more prevalent in young palms in 15-20 age groups.

Black headed caterpillar: *Opisina arenosellai*-Cryptophasidae

Larvae with a black head and pink body feed on undersurface of leaflets protected in silken galleries of frass. Presence of dried leaf lamina on upper surface, larval galleries with moults and pupal cases on lower lamina are its symptoms. This hot weather pest causes extensive leaf damage on lower leaves and great reduction in yield.

White grub: *Leucopholis coneophora* - Melolonthidae

Pest damage occurs on roots and bole region of palm. The pest has an annual life cycle. The adults emerge with pre-monsoon showers and lay eggs in soil. The grubs feed on almost any root, including coconut roots and boll region. Continuous infestation by the pest results in yellowing of leaves, tapering of crown, premature nut fall, delayed flowering, retarded growth and reduction in yield. Seedlings and young palms are the most affected.

Coreid bug: *Paradasynus rostratus* - Coreidae

Both adults and nymphs feed on buttons and developing nuts. The feeding marks appear initially as brownish lesions and later these result in formation of cracks, shrinkage and malformation with gummosis on affected side of nut surface. Tender nut drop is another symptom. Pest incidence reaches a peak in post-monsoon period from September to March.

Mealy bug: *Pseudococcus longispinus, P.cocotis, Pulmiculator palmarum* - Pseudococcidae

These attack spear leaf and inflorescence, causing characteristic crinkling and failure to open. Severe infestation on spearleaf results in suppression of plant growth and wilting.

Scale insects: *Aspidiotus destructor, Aonidiella orientalis, Lepidosaphes mcgregori* -Diaspididae

These scales infest leaf lamina, buttons, nuts and inflorescence. They cause yellowing of leaves in specks in initial stage and may also contribute to button shedding. Highest pest population is seen in summer.

Defoliating caterpillars: *Latoia lepida, Contheyla rotunda, Macroplectra nararia* -Cochlididae

The slug caterpillars with grotesque shapes and poisonous or itch causing spiny scoli eat away leaves.

Nut borers: *Cyclodes omma, Tirathabac* sp

The caterpillars of these moths bore into developing buttons through perianth and feed on inner tissues causing nut fall.

Termite: *Odontotermes obesus*

Colonies of termite feed on husk portion of nuts and collar region of seedlings in nursery. Severe infestation results in wilting of central shoot.

Mites: *Raoiella indica, R. vandergooti, Aceria guerreronis, Oligonychus iseilemae, Tetranychus ludeni*

The foliage mites colonize foliage and suck leaf sap. Seedlings are more prone to infestation. The orange mite infests perianth region on nuts causing severe nut fall. Of late, the eriophyid mite has extensive damage to coconut causing tender nut fall and malformation of nuts.

Rats: *Rattus rattus wroughtoni, Bandicota bengalensis, B.indica*

Rats cause extensive damage to tender nuts. Living on the crown of palms, they make entry hole in developing nuts and feed on internal contents and live inside burrows in coconut plantations.

IPM in coconut

IPM in coconut stresses need for clean cultivation. The premises should not contain any decaying debris or cowdung or cut and decaying stem. These harbour grubs of rhinoceros beetle, pupae of black headed caterpillar and other life stages of pests. Compost pits and manure pits should be surface treated with dusts of carbaryl or lindane (0.01% v/v). This should be repeated once in three months. After every 3 months, the innermost leaf axils of trees have to be filled with a mixture of lindane/carbaryl dust with sand to ward off attack by rhinoceros beetle and coreid bug. Cut holes on stem should be sealed with mud slurry containing carbaryl, to prevent egg laying by red palm weevil. Against eriophyid mite attacking buttons and mites, apply Dicofol (5 ml/l) or Dimethoate (1.5 ml/l) or Neemoil/garlic extract/soap suspension (20:20:5g for 1 litre) once in three months on first three bunches after pollination. Hook out adult rhinoceros beetles immediately on spotting attack.

Dried and old leaves in garden should be removed as firewood or burnt, to destroy pupae of leaf eating caterpillar. Coconut log traps can be set up longitudinally splitting cut stumps (50cm long) and pouring toddy or fermented coconut water into central pith. Adults of rhinoceros beetle or red palm weevil collecting on them should be destroyed every day. Apply 2 carbaryl solution in one litre to the pith once a fortnight to kill grubs of scarabeids /weevils developing on pith. Set up pheromone traps 50m away from the garden in non-coconut area or at least in outermost periphery of garden. Collect and kill the adults everyday, provide pineapple waste or fermented coconut water or sugarcane splits in the trap, with Malathion 1ml. This bait should be removed and fresh bait provided every week.

Release parasitoids like *Bracon brevicornis* @ 1800/0.4 ha or *Goniozus* (*Parasierola) nephantidis* @1200/0.4ha once every month during summer months when the pest attack is heavy.

Root grubs can be managed by applying chlorpyriphos @ 2.5 ml/l to the root zone just after the pre-monsoon showers. Set up light traps during dusk in the periphery of the garden during pre-monsoon showers when the adults emerge. Burn trash, old dried leaves, petioles and leaf bases during dusk to attract and kill adults.

Coffee

White stem-borer: *Xylotrechus quadripes*

Serious pest of *arabica* coffee. Infested plants show external ridges around the stem. Affected plants also show yellowing and wilting of leaves.

Coffee berry-borer: *Hypothenemus hampei*

Pin hole at the tip of berries indicates presence of the pest. In severe cases of infestation two or more holes may be seen. Infested berries may fall due to injury or secondary infection. Severe infestation may result in heavy crop loss.

Shot hole borer: *Xylosandrus compactus*

Withered or dried branches with shot holes indicate presence of the pest. Attacked branches dry up fast. Leaves distal to the point of attack fall prematurely. The terminal leaves wilt, droop and dry up.

Mealy bug: *Planococcus citri, P.lilacinus*

Attack nodes, spikes, berries, tendr branches,leaves and roots leading to debilitation of the plant and crop loss. Ant activity noticeable on the plants affected by mealy bugs.

Green Scale: *Coccus viridis*

Debilitation of older plants and death of nursery plants. Honey dew secreted by the scale forms a film on leaves and a black fungus (sooty mould) grows on it. Ant activity noticed.

Root-lesion nematode: *Pratylenchus coffeae*

Affected young plants are lean and lanky. Older leaves become yellow and drop, leaving a very few undersized, chlorotic and crinkled leaves at tip of stem giving a 'tufted' appearance. Affected bearing plants show thinner stem and have adequate foliage to support the crop. Affected plants are left with two or three pairs of leaves at tip producing a 'goose-neck' condition after receipt of rains.

Brown scale: *Saissetia coffeae*

Damage similar to that caused by mealy bugs and green scale.

Cockchafers or white grubs: *Holotrichia* spp.

Yellowing of leaves and stunted growth. Affected plants can be easily pulled out and may wilt and dry.

Coffee Bean: *Areaecerus fasciculatus*

Though the pest attacks berries in field, severe infestation occurs only in storage. Infested beans show a circular hole. Attacked fruits in field shrink and become black. The adults and larvae make feeding tunnels in all directions in the beans.

Hairy caterpillars: *Eupterote* spp.

Caterpillars feed on leaves and cause severe damage by denuding plants.

Red borer: *Zeuzera coffeae* - Cossidae

In the early stages of attack, young plants/branches show sign of wilting. Infested plants usually bear one or two holes through which pellet like excrement of the larvae hangs out, and accumulates at the base of plant. In advanced cases, branch of the whole plant dries up.

Snail: *Ariophanta solata*

Feeds on leaves of Arabica coffee and dadap, and occasionally on bark of tender branches and the skin of fruits.

Termites or white ants

Cause poor establishment of dadap stakes in coffee plantations.

Thrips: *Heliothrips haemorrhoidalis, Haemorrhoidalis* sp; *Retithrips syriacus; Scirtopthrips bispinosus; Thrips nilgiriensis*

Lacerations appear as brownish irregular streaks or silvery patches on leaf surface. Infested leaves become distorted, crinkled or mottled. Damage is more pronounced in exposed areas during dry weather.

IPM in coffee

Berry borer, cockchafers, mealy bugs and scale insects are main pests in coffee.

Coffee berry borer management requires complete avoidance of left over berries that would allow residual population to develop. Pick the main crop completely and remove and destroy all unmarketable berries. This is a must in Coffee IPM. Avoid excess shade, dry the berries to 10% moisture, and set up alcohol traps (Ethyl: Methyl 1:1) to attract adults and kill them. In heavy incidence areas, spray endosulfan 35 EC (350ml/0.4ha) at 120- 150 days after flowering. This would also kill leaf and flower feeding caterpillars, cockchafer adults, and other surface feeders. Add sandovit or other wetting agent (100ml or one kg fish oil rosin soap/every 200 l spray fluid). This ensures management of scale insects and mealy bugs. Specific chemicals for mealy bugs include Quinalphos (300ml), fenitrothion (300ml) or fenthion (150ml/200 l) spray fluid. Mealy bug and scale management require ant control using dusts of quinalphos or malathion or methyl parathion, which also kill surface feeders. Release ladybird beetle (*Cryptolaemus montrouzieri)* against mealy bugs and scales and also spray *Verticillium lecani* (6x106 spores/ml). Cockchafer grubs can be killed by drenching with lindane 20 EC (750 ml in 200 l) of water per 0.4 ha, one to two months after pre monsoon showers where first instar grubs are seen as surface soil root feeders.

Tea

Pink, purple, scarlet and red spider mites: Discolouration of leaves.

Thrips: Stunted shoot growth, corky patches on leaves.

Tea mosquito: Brown punctures on leaves. Curling of leaves; shoot growth arrested.

Jassids: Curling of young leaves; they turn brown and dry up.

Looper caterpillars: Nibbling of leaf margins. Hole on leaves.

Red slug caterpillar: Feeding on epidermis of leaves. Presence of skeletonised leaves on small branches.

Bunch caterpillar: Feeding on leaves. Caterpillars remain as bunches on small branches.

Flushworms and leaf rollers: Characteristically rolled apical shoots; Rolled leaves.

Termites: Damage to bark and sometime to live wood and roots depending on species involved.

Shot hole borer: Construction of galleries in stems; branch breakage.

Root-knot nematode : Galls on roots of young nursery plants.

IPM in tea

Mites and flush worms are important pests in tea that require regular monitoring and management. The volume of spray fluid required for various pesticide operations vary in Tea. Low volume and high volume sprays are employed. After pruning and in first year, volume of 250 litre for motorized spray and 500 litre for high volume spray are needed. It increases to 350/650 litre as the crop canopy thickens. The spray should wet the leaf surface fully at any spraying, to be effective. Insecticide treatment will have to be repeated after two weeks, to control flush worms, leaf rollers and other leaf feeders. Against tea mites, Dicofol or Ethion or Phosalone (400ml per 0.2ha) can be sprayed in non repetititive cycles, based on intensity of attack. This also prevents scales and mealy bugs along with flush worms, nettle grubs, loopers, lobster caterpillars and other surface feeders.

12. Masticatory

Betelvine

Scale insect: *Lepidosaphes* sp.

Dark brown insect occurs on leaves, petioles and main vines, causing shriveled leaves from sucking of plant sap. In severe infestation, leaves dry up and give sticky look and gradually vine wilts.

Betelvine bug: *Disphinctus politus*

It is a serious pest in many parts of betelvine areas. Tender leaves are punctured by insect while feeding on leaf and this results in a number of minute black spots all over leaves. When such leaves matured, these spots result in small holes. Very badly affected leaves curl, shrivel and give a puckery appearance. The bug is active during June- October and multiply rapidly under warm and humid conditions. They take shelter in dark and protected places and thrive by feeding on plant sap.

Mealy bug: *Ferrisia virgata*

White cottony mealy bug causes considerable damage to tender leaf petioles and apical shoots mainly during rainy season. Insects sucks sap from leaves and twigs. White patches are seen from a distance.

White fly: *Dialeurodes pallida*

This is a serious pest in the Eastern and North Eastern regions of India. Both adults and nymphs of the insect, suck sap from lower surface of leaves resulting in pronounced chlorosis. Sooty mould develops on honey dew excreted by insects.

Red spider mite: *Tetranychus* sp.

All the stages of the pest feed on under-surface of leaves by sucking leaf sap. Characteristic blotching of leaves develops due to removal of chlorophyll and other plant pigments and leaves present a pale yellow look which after some time dries off.

13. Spices

Black pepper

Pollu beetle: *Longitarsus nigripennis*

Feeding holes on tender leaves; black patches on spikes;black berries that crumble when pressed. Adults feed on tender shoots, leaves and spikes. Larvae bore into tender spikes and berries and feed on internal tissues.

Top shootborer: *Cydia hemidoxa*

Decayed and black terminal shoots. Larvae bore into tender terminal shoots and feed on internal tissues and retard growth of young vines.

Scale insects: *Lepidosaphes piperis, Aspidiotus destructor*

Chlorotic patches on leaves; wilted and dried leaves and stems. Adults and larvae suck plant sap from leaves, stems and berries resulting in wilting and drying.

Leaf gall thrips: *Liothrips karnyi*

Marginal tubular galls on leaves. Adults and larvae feed on tender leaves inducing formation of marginal tubular galls. Leaves are crinkled and reduced in size.

IPM in Black pepper

Pollu beetle, top shoot borer and scale insects are the major pests in Black pepper. Pollu beetle is predominant in gardens with heavy shade. Prune shade trees during onset of monsoons and lightly rake up base of plants to expose pupae of beetle to predators or to be dried by sun. Collect and destroy all new flushes affected by top shoot borer in the pre monsoon flushing. When needed, apply first insecticide spray of Quinalphos (2ml/litre) against all pests at this flushing stage. Give a second spray at spike initiation with Endosulfan or Dimethoate (2ml/litre). Give a third non repetitive spray after one month. Add sandovit or teepol (5ml/litre) to enhance effectiveness against mealy bugs and scale insects.

Cardamom

Grasshoppers: *Mecopoda elongate, Oxya velox*

These pests occur both in nurseries and plantations. Both adults and nymphs feed on leaves by scrapping and eating outer tissues and left margins. The emerging leaves are nibbled and the spot holes appear when the leaf unfurls. In severe conditions, the entire leaf is eaten leaving behind only the midrib.

Hairy caterpillars: *Clelea plumbiola*

The pests appears during July-October. The caterpillars feed on chlorophyll content in gregarious forms leaving behind only the skeleton of midrib and veins. The infestation sometimes even extends to young flower and sucker buds. The fully -grown larva usually migrates and pupates in soil in membraneous cocoon. The adult is small, measuring 1.54-2.0 cm wingspan, light grayish forewings and black hind wing. The adult lays eggs on undersurface of leaves.

Stem borers: *Glyphipterix* sp.

The tiny caterpillar bores into leafy stem at a height of about 5-10 cm above the collar region, which feeds on central leaf core, causing dead heart. The pest is active during July-September. This is a very common insect causing substantial losses and hence often requires strict control measures.

Capsule borers: *Synoxy* sp.

This is a serious pest attacking immature capsules. The grub measures about 1.5-2.0 cm in length, bores through the capsule wall and eats away the young soft seed mass, making capsule hollow.

Leaf hoppers: *Kolla mimica and K. opponense*

Thrips: *Rhipiphorothrips cruentatus*

Mealy bugs and scale insects: *Paraputo group*

These are subterranean inhabiting pests, sucking sap from developing capsules, actively growing root tips and differentiating buds on mature rhizomes. Their incidence is more severe in drier zones, in plantations with scarce shade and those areas nearby having ginger and turmeric fields.

Aphids: *Pentalonia nigranervosa f. galadii; Micromyzus* sp*; Kalimpongensis* sp*;* and *Rhopalosiphum maidis*

These pests are vectors of 'foorkey' virus disease. They suck sap from tender portions and differentiating buds at collar region. If abundant, they can be seen on leafy stems. The affected clumps show foorkey symptoms. The differentiating spikes and floral parts are very susceptible to aphids. The infestations occur mainly during May- November.

Corn aphids are seen on lower surface of leaves congregating near midrib and veins. These are known to be vectors of mosaic streak or 'chirkey' virus disease. They suck the sap from leaves and secrete honey dew, which encourages sooty mould growth on leaves. This is a major pest causing outbreak of 'chirkey' viral disease.

Moth and Psocids

The reddish brown caterpillar measuring 1.5cm in length and 1mm thickness bores into the capsule wall and eats away mucilaginous seed coating. Hard seeds are unaffected but appearance of the capsules and seed quality are adversely affected. The adult is a small greyish coloured moth having about 1cm wingspan. Exposed frass material is later on fed by psocid insects. The incidence is common in flue cured cardamom which is kept under open conditions or with improper packaging.

Thrips: *Sciothrips cardamomi*

Wilting of panicles; scabbing and splitting of capsules. Adults and larvae lacerate and feed on plant sap from tender shoots, panicles and capsules.

Shoots-and capsule-borer: *Conogethes punctiferalis*

The pest bores hole on capsules and pseudostems with extrusion of frass. Larvae bore into panicles, capsules and pseudostems and feed on internal tissues.

Root grub: *Basilepta fulvicorne*

Stunted plants with yellow leaves. Larvae feed on roots and rhizomes and retard plant growth.

Hairy caterpillars: *Eupterote* spp, *Lenodera vittata* .Larvae defoliate plants.

IPM in cardamom

Thrips, shoots and capsule borer and the whiteflies cause appreciable damage to the crop. Regulate shade to reduce thrips incidence.

Quinalphos (2ml/l), fenthion (1ml/l), Dimethoate (2ml/l), or phosalone (2ml/l) can be applied to manage them. Against shoot and capsule borer, set up pheromone traps for monitoring and mass trapping. Collect and destroy locusts, sphingids and hairy caterpillars at the first occurrence itself.

Cinnamon

Cinnamon butterfly: *Chilasa clytie*

Larvae defoliate plants.

Leaf miner: *Conopomorpha civica*

Leaves with blister like patches. Larvae mine tender leaves resulting in drying of infested portions.

Clove

Stem-borer:*Sahyadrassus malabaricus*

Mat-like frass at basal region of stem; bore and gridle on stem. Larvae gridle main stem and bore downward killing the tree.

Scale insects: *Ceroplastes floridensis, Kilifia accuminata*

Wilting of seedlings and young plants in the nursery. Adults and larvae feed on plant sap and cause wilting and drying of seedlings and young plants.

Coriander, Cumin, Fennel and Fenugreek

Aphids: *Hyadaphis coriandri, Myzus persicae*

Wilting of tender shoots, leaves and flowers and shriveling of fruits. Adults and nymphs feed on plant sap and cause wilting of tender shoots, leaves and flowers and shriveling of fruits. Pest infestation leads to formation of sooty mold.

Seed midge: *Systole albipennis*

Larvae feed on developing embryo and endosperm of fruits.

Ginger and Turmeric

Shoot borer: *Conogethes punctiferalis*

Yellow/withered central leaf; bore hole on pseudostem with extrusion of frass. Larvae bore into pseudostems and feed on internal tisssues.

Rhizome scale: *Aspidiella hartii*

Adults and larvae feed on plant sap leading to shriveling of rhizomes and buds resulting in poor germination.

Nutmeg

Scale insects: *Pseudaulacaspis cockerelli; Parasaisettia nigra*

Wilting of seedlings and young plants in nursery. Adults and larvae feed on plant sap and cause wilting and drying of seedlings and young plants.

14. Fruit Crops

Almond

Almond Weevil: *Myllocerus laetivirens, M. undecimpustulatus, M. discolor, Amylyrrhinus poricollis,* Curculionidae - Coleoptera

The adults and grubs cause damage. Young larvae (grubs) feed on roots of plants. The adult beetles cut holes on leaf or eat margins of leaves.

Aphids

Brachycaudus helichrysi See under peach

Myzus persicae See under citrus

Pterochlorus persicae - Aphididae: Homoptera. Aphids are sap suckers from plant parts causing shriveling of affected parts

Almond beetle/ chrysomelids

Mimastra cyanura, Altica sp., *Haplosoma* sp., *Merista* sp., *Monolepta* sp., *Phyllotreta* sp.: Chrysomelidae

These chrysomelids are small coloured beetles with black spots at anal end of elytra. Adult beetles of all species appear in swarms during May, defoliate trees, causing yield losses; peak activity is seen during July- August

Flat- headed tree borer: *Sphenoptera lafertei:* Buprestidae - Coleoptera

See under peach

San Jose Scale: *Quadraspidiotus perniciosus* - Diaspididae

See under apple

Aonla

Bark- eating caterpillars- *Indarbela quadrinotata, I. tertraonis* - Metarbelidae

The larvae of this species make Zigzag galleries and feed on bark under silken ribbon-shaped webs. They are stout and dirty brown in colour.

Fruit- borer : *Deudorix isocrates,* Lycaenidae - Lepidoptera

The larvae of this beautiful butterfly cause considerable damage (up to 42%) to fruits.

Shoot-gall maker : *Betanosa stylophora* - Lepidoptera

The young caterpillars bore into tender shoots and cause gall formation. All varieties are susceptible. Growth of severely infested shoots is suppressed

Aphid: *Schoutedenia emblica* - Aphididae

The nymphs and adults suck sap from leaves and new flushes.

Apple

San Jose Scale: *Quadraspidiotus perniciosus* - (Diaspididae: Hemiptera)

Nymphs and adults of this insect congregate in large numbers on bark and desap trees. Later infestation spreads to the new flushes, blossoms and fruits. These parts are also shrunken, blemished and damaged due to desapping. The infested region in bark becomes reddish pink and fruits purple colour. The adult female is round, slightly convex with a black pustule. The adult males are small and winged.

Woolly apple aphid: *Eriosoma lanigerum* - (Pemphigidae: Hemiptera)

Nymphs and adults of these aphids live in colonies, clustered on roots or branches of apple, pear, elm etc. On aerial parts, it is seen as a woolly mass. Active from April to December, but found in large numbers on aerial parts after rains. Damage is caused by sucking of sap from aerial parts and roots, resulting in gall formation, stunting, and retardation of growth and proliferation of fibrous roots

Cottony cushion scale: *Icerya purchasi* - (Margarodidae =Coccidae: Hemiptera)

The nymph is a pinkish crawler with long antenna and group of hairs. The female has a cottony ovisac. The desapping by nymphs and adults causes yellowing of leaves.

Blossom thrips- *Thrips rhopalantennalis*, T. *flavus, T. florum, Haplothrips tenuipennis* - Thripidae

These are tiny, yellow, pale-brown and black slender insects within flowers and flower buds. Adults and nymphs lacerate floral parts and suck sap resulting in brown spots. Infested flowers give sickly and faded look. Lesions cause flower fall, inferior fruits and loss in yield. Attack on tender leaves causes necrotic silvering, curling, deformation and withering.

Defoliating and fruit eating beetles (Cockchafers or May-June beetles)

Adoretus duvauceli,A. versutus,Anomala dimidiata, A. flavipes, A. lineatopennis, A. rufiventris, Holotrichia longipennis, Hilyotrogus holosericeous, Lachnosterna coriacea, L. crinicollis, L. longipennis,Mimela passerine, M. pectoralis, Mylabris mecilenta - Melolonthidae

These are pests in adult stage, where the different coloured beetles (Reddish brown, blue-black, and green), feed on foliage and developing fruits. The adults appear in May-June and lay eggs in soil at dusk during rainy season. They attack all temperate fruit plants whose leaves get skeltonized and small fruits which are eaten falling prematurely. Their immature stages (white grub) feed on almost any root; crops, weeds, potato, vegetables and grasses.

Leaf- roller and fruit- scrapper : *Archips fuscocuprianus* - Tortricidae

The black- headed green caterpillars of this pest fold leaves of apple from April-May. Attack is seen on foliage during summer and last generation of caterpillars attacks fruits in field and also in storage and damage them by scrapping. Sometimes 100% damage is done to stored apples.

Leopard moth: *Zeuzera multistrigata* Cossidae - Lepidoptera

The adults are white moths with attractive patterns, seen at dusk during May-June. The eggs are laid singly in cracks of barks. Within about a month, the emerging pinkish-white young caterpillars bore into branches and stems. The early instars scrap off the cambium and induce gummosis and later instars enter into the hardwood and make tunnels up to 25cm in the terminal shoots, secondary branches and main stem. This results in drying of shoots within 1-2 years and of the whole plant within 3-4 years.

Apple fruit moth: *Argyresthia conjugella* Argyresthidae - Lepidoptera

These are small moths whose larvae are greyish in colour. The larvae bore into fruits and feed on seeds. The infested fruits show exit holes. Fully-grown larvae emerge out after tunneling through pulp of fruit and render it unfit for use.

Apple zygaenid: *Aglaope phylina* - Zygaenidae

Moths are white in colour and the slug –like larvae mimic plant colour. Caterpillars feed on bark, leaf and flower spurs resulting in defoliation, reduced flowering and less yield.

Apple Root borer: *Dorysthenes hugelli* - Cerambycidae

The chestnut red beetles lay eggs in soil during July- August. Emerging grubs feed on thick roots and other organic matter for long periods. Their longevity is around 3.5 years and they have preference for sandy soil. Root borer attack makes plants weak and shaky and can cause death of plants.

Tent caterpillar : *Malacosoma indica* - Lasiocampidae - Lepidoptera

Brown coloured moths lay eggs in bands around small twigs in May. The larvae are dark brown to black, 40-50mm in length. These larvae spin triangular or quadrangular tent of silk at forking of twigs and remain there during day. Caterpillars feed gregariously on leaves at night, tracing their path through silken threads.

Apple hawk moth: *Langia zenzeroides* - Sphingidae

These are large moths. The huge larvae feed on large quantity of leaves.

Hairy caterpillars: *Euproctis signata* - Lymantridae

These occur sporadically; caterpillars are profusely hairy. They feed voraciously and defoliate trees.

Indian gypsy moth: *Lymantria obfuscate*

See under Apricot.

Apple leaf miner: *Gracillaria zachrysa* - Gracillaridae

Mines are seen on leaves. Young caterpillars make several mines on leaf surface; later they leave mines, roll young leaves longitudinally into tubular or cone pouch and feed within; the maximum damage is done during summer (April- May) and in autumn (September-October).

Apple stem borer: *Apriona cinerea, A. sarta* - Lamiidae

These are ashy-grey to greyish –yellow, long horned beetles with conspicuous blackish tubercles at base of elytra. The grubs damage the stem and shoots as a result of which plants may become weak and even die.

Stem borer: *B. rufomaculata* - Cerambycidae

These are dark brown beetles that could be above ground borers or root borers. Attack is evidenced by frass and faecal pellets or wilting of leaves and drying of terminal shoots.

Shot hole-borer : *Scolytus amygda, S.faciatus,Scolytoplatypus raja,Pitygenes* sp., *Xyleborus* sp. - Scolytidae

These are small dark coloured beetles with a deflexed head. The grubs of these bore into the trunk. There is an exudation of a dark fluid. The trees become weak and the branches dry up.

European red mites : *Panonychus ulmi* - Tetranychidae: Acarina

Two- Spotted spider mite: *Tetranychus urticae*

Mites can cause light infestation which lead to speckling of foliage, or heavy infestation when the foliage becomes pale, sickly or bronzed. They feed on the foliage causing loss of chlorophyll and bronzing followed by leaf drop and weakening of buds. This results in poor size and quality of fruits. Excessive feeding results in fruit drop during the season and reduced blossom and fruit set in subsequent years.

IPM in apple orchards

Integrated Pest Management (IPM) in apple orchards involves three main components – monitoring, damage assessment and preventive action. Year round management strategy is based on six seasons crop stage periods. These are the dormant winter, spring, blooming, post blooming, summer and pre-winter harvest stages.

Dormant winter: Monitor and assess mite and aphid egg population and San Jose scale. Red mites are the best controlled by the application of dormant oil. Allow low populations of rust mites to survive, so that predatory mites feed on them and control red mites later in the season. Normally, natural enemies often control the green aphids. A delayed dormant oil spray will take care of early damage. Only if more than 60% of the trees terminals are affected, especially in young orchards, a chemical like Imidachloprid (2.5ml / l or Diazinon 3ml/l) can be used. Do not repeat the spray with same insecticide. Dormant oil spray has to be done annually to manage San Jose Scale in orchards where the scale is endemic. There are ladybird beetles which are good predators, but insecticides during growing season kill them.

Spring: During spring, the aphids will be a problem if IPM was not done earlier. Caterpillar pests like leaf rollers and orange tortrix are mainly seen during this period.

Blooming: Codling moth monitoring and mating disruption with pheromone traps have to be done. Monitoring has to be continued for aphid and caterpillar pests.

Post blooming: Fruit damage by codling moth occurs only, if pheromone trapping and mating disruption were ineffective. Monitor also for aphids, leaf rollers and San Jose scale crawlers. If crawlers are abundant, as evidenced from sticky trap catches, a chemical control using chlopyriphos may be needed.

Summer: Trapping and mating disruption control codling moth. Previous IPM could have controlled leaf roller, tortrix, leaf miners and lygus bugs. Use a contact spray only if the populations are heavy.

Pre-winter harvest stage: Check fruit for codling moth larvae. Remove apple, pear and walnut trees in nearby abandoned orchards to destroy reservoirs of codling moth. Remove affected fruits and harvesting tools and containers from orchard and destroy affected fruits. Destroy dropped fruits regularly.

Apricot

Indian gypsy moth: *Lymantria obfuscata* - Lymantridae

Eggs are laid by bulky adult female moth, which has only rudimentary wings and hence cannot fly. Eggs are laid under bark on tree trunks or on fallen leaves. They feed on many shade trees and other trees also. Feeding is gregarious and done during night. The trees are almost completely defoliated.

Apricot chalcid: *Eurytoma samsonowi,* - Chalcididae: - Hymenoptera

Eggs of this very small chalcid are laid inside young fruits. The grubs feed on the developing seeds. Fruit growth is stunted and fruits fall off prematurely. Pupation takes place inside the fruit.

Apricot weevil: *Emperorhinus defoliator,* - Curculionidae

Symptoms are same to those in almond. Adults defoliate trees during summer-

Grey weevil: *Myllocerus discolor, M. undecimpustulatus, M. laetivirens* - Curculionidae

Nature of attack by this weevil is described under pests of almond. Adults defoliate trees during summer.

Chafer beetle: *Anomala polita*

Cockchafer beetles are pests in adult and grub stage. See under **apple**

Flat-headed borer: *Sphenoptera lafertei*

See under peach

Leaf rollers: *Cocoecia epicryta*

See under apple

Scale insects: *Eulecanium sp.* - Coccidae

Parlatoria oleae

Quadraspidiotus perniciosus - Diaspididae

See under **apple**.

Apricot fruit moth: *Cydia funebrana* - Tortricidae

The caterpillars feed on mesocarp of developing fruits and render them useless. The fruits become filled with frass and excreta.

Ber

Fruit Flies: *Bactrocera zonata, Carpomyia vesuviana* - Tephritidae; Diptera

The adults lay eggs on fruit. Maggots in the infested fruits cause browning, rotting and fruit drop. In North India, up to 80% fruits are damaged. Thin skinned, soft and sweet cultivars are more susceptible.

Fruit borer: *Meridarchis scyrodes* - Carposinidae : Lepidoptera

Very serious pest in South India. Eggs are laid on fruits at pea stage. On hatching, the reddish brown larvae bore into fruits and feed on pulp near seed. Extent of damage may vary from 20 to 80%.

Leaf eating caterpillars: *Euproctis fraterna,Thiacidas postica* - Lymantridae, Lepidoptera

The caterpillars in the initial instars remain gregarious and scrape leaves and tender fruits. Later instar caterpillars disperse and devour leaves, fruits and tender shoots.

Psorostichia (Tonica)zizyphi - Oecophoridae:Lepidoptera

See under **citrus**

Leaf butterfly, *Tarucus (Theophrastus)indica* - Lycaenidae:Lepidoptera

The larvae of this species cause defoliation .They are spindle shaped, and fleshy green. The adult is a blue butterfly with several white spots.

Mealybugs and scales: *Drepanococcus chiton,Nipaecoccus viridis* - Coccidae

Planococcus lilacinus - Pseudococidae

Kerria lacca - Lacciferidae

The waxy scale, *D. chiton*, causes extensive damage to *ber* in South India. They are found in large colonies, sucking sap from young shoots and tender leaves. Removal of sap leads to drying of shoots. *Kerria lacca,* the lac insect damages in a similar way except that the colour of shoots becomes dark brown due to encrustation by scales.

Thrips: *Florithrips traegardhi* - Thripidae

These thrips, cause serious damage on *ber* leaves. Excessive feeding injury results in curling and drying of leaves

Cherry

Flat- headed borer: *Sphenoptera lafertei* - Buprestidae

See under **peach**

Root borer: *Dorysthenus hugelli*

See under **apple**

Stem borer: *Aeolesthes holosericea*

See under **apple**

Plum Scale: *Eulecanium tiliae*

See under **plum**.

Citrus

Psylla: *Diaphorina citri* - Psyllidae: Hemiptera

The citrus psyllid is a serious pest of citrus and curry leaf throughout the country. Orange coloured eggs are laid on leaf axils of tender shoots. Nymphs are flat, orange coloured and adults are grey coloured. The nymphs and adults suck sap from buds and tender leaves. Affected leaves curl and shrivel while shoots get dried. It also acts as a vector of greening disease in citrus.

Whiteflies: *Dialeurodes citri* - Aleyrodidae

Black flies: *Aleurocanthus woglumi* - Aleyrodidae

Nymphs and adults of both these species suck sap from ventral side of tender leaves, excrete honey dew and encourage development of sooty mould. The infested plants lose vigour and fruit production is adversely affected.

Red scale : *Aonidiella aurantii* - Diaspididae

Brown scale:*Chrysomphallus aonidum*

Green scale :*Coccus viridis* - Coccidae

The red scale infests shoots, leaves and fruits. They suck the sap and affect the plant vigour. *C. aonidum* is found along with *A.* aurantii. The green scale infests shoots and leaves on the ventral side. Honeydew excreted by the scales and other sucking pests generally encourages development of sooty mould on those parts.

Mealy bugs: *Planococcus citri, P. pacificus* - Pseudococidae; *Icerya purchasi* - Margarodidae

The mealy bugs infest tender shoots and fruits. Severe infestation causes premature fruit drop and affects plant growth. Presence of sooty mold and white mealy bugs on fruits and shoots are symptoms.

Aphids: *A.gossypii, A. craccivora,Myzus persicae,Toxoptera aurantii,Toxoptera citricidus* - Aphididae

Normally, these aphids attack during flowering period but occasionally severe outbreak occurs when rainy season is followed by dry weather. Nymphs and adults suck sap from tender inflorescences and shoot and transmit the tristeza virus.

Thrips: *Thrips nilgiriensis* - Thripidae: Thysanoptera

Desapping by nymphs and adults causes ring like scarred area encircling the fruit stalk and irregular mottled patches on the rind. Adults are yellowish with fringed wings.

Mites: *Eutetranychus orientalis, E. banksi,Panonychus citri, Phyllocoptruta oleivorai* - Tetranychidae

Generally, mites are pests in hot weather. They damage leaves and fruits. Infested leaves turn grey or yellow from upper side. Infested fruits remain often small in size and drop prematurely. *P. olevora* causes rusting and bronzing of fruits. The damage is increased by drought conditions.

Leaf miner: *Phyllocnistis citrella* - Gracillaridae: Lepidoptera

The leaf miner is another serious pest of citrus in India. Adults of leaf miners are small white with fringed wings and a black spot at tip of forewing. The eggs are laid on tender leaves. The minute reddish or yellowish, apodous larvae mine the leaves with serpentine glistening mines causing distortion and drying up of leaf lamina. They feed on epidermis of tender leaves making serpentine mines which are silvery in colour. Affected leaves become papery white in patches and later dry up.

Leaf roller, *Psorostichia (=Tonica) zizyphi* - Oecophoridae: Lepidoptera

The Larva is pale yellowish green with black head. They roll up the leaves,causing crinkling and defoliation. The adult is a small, brown moth.

Blue butterfly, *Chilades laius, Tarucus (Theophrastus) indica*

The greenish larvae cause defoliation. The adult is a blue butterfly.

Lemon butterfly/Citrus butterfly, *Papilio demoleus* - Papilionidae: Lepidoptera

Early stage larvae of this pest resemble bird dropping. The grown up larvae are cylindrical, stout and green with brown lateral oblique bands. The caterpillars feed on tender leaves and defoliate the entire trees. Young plants in nursery are seriously affected. The adult is a dark brown swallow tail butterfly with numerous pale yellow markings.

P. polytes - Papilionidae: Lepidoptera

Larvae defoliate tender leaves. Adult is a black butterfly with white markings.

Shoot-borer/trunk-borer: *Oberea latteapicalis*

These are serious pests of citrus in Southern and North Eastern hilly regions. Incidence is high in citrus trees interplanted with coffee. Freshly hatched larvae bore into young twigs and feed on soft tissues. Affected twigs break and dry.

Orange trunkborer: *Anoplophora versteegi* - Lamiidae: Coleoptera

Damage is due to grub and adult. Adult feeds on leaf tissue along the midrib, keeping the margins intact. Grub, which is cream coloured, elongate and with brown head tunnels the trunk killing the tree.

Orange borer: *Chelidonium cinctum, Chloridolum alcamene* - Cerambycidae: Coleoptera

Grub is creamy white with flat head. They bore the trunk causing drying of terminal shoots in early stages, followed by wilting of thicker branches or whole tree if damage occurs in the main stem. Adult of *C.cinctum* is dull metallic green to dark violet with an yellow band across the middle of elytra. *C.alcamene* is a shiny blue beetle.

Bark borer: *Indarbela tetraonis*

See under drumstick.

Fruit sucking moths, *Othreis fullonica, O.materna, O. ancilla* - Noctuidae: Lepidoptera

Adults pierce the fruit and suck the juice causing rotting at the feeding site and fruit dropping. The larvae are semiloopers with orange blue and yellow spots on its velvety dark speckled body. They feed on weed hosts. Adults are stout built moths with grey and orange coloured wings.

O. fullonica: Tripod black mark in the fore wing & curved marking in hind wing

O. materna: Three black spots on the fore wings.

O. ancilla: White band in the middle fore wing.

IPM in citrus

Citrus has many scales and mealy bugs. The farmer has to monitor the population of these and small pests like mites and thrips using sticky traps or yellow sticky traps. Differentiate between males of scales, parasitoids and other small insects and monitor for increase in pest population to manage the rising populations. Soft scales, black flies and whiteflies can be managed from damaging populations by initial spot applications of Methyl parathion 50 EC (1.5 ml/l), or chlorfenvinphos 20 EC (2.5 ml/l) or quinalphos 25EC (2.0ml/l). Against mealy bugs, provide a ring band of insecticide paste (methyl parathion).Crawlers are caught on sticky traps. Place them on shoot bearing the fruits at a length of 5cm. Use dichlorvos (0.20%) in combination with fish oil resin soap (25g/l) as spray or for dipping the fruits for two minutes. In endemic areas, a single soil application of aldicarb 10 G (50 g/tree) around the base at the time of pruning can be done. Then, release the predator, Cryptolaemus montrouzieri

a. Mark the mealy bug infested plants early in the season.
b. Put a band of quinalphos or methyl parathion dust (8kg/0.4ha) around the tree trunk leaving about 30cm from the main stem to kill the phoretic ants. Follow the ant control methods like destruction of ant holes, red ant nests and skirting of citrus trees after fruit harvest which prevents the ant migration through side branches.
c. Release the beetles (10/tree) once the patrolling of ants on the trunk is stopped.
d. Make periodic check during first fortnight and put dried leaf mulch around the tree trunk 20 days after release of beetles to facilitate pupation of full grown grubs.
e. Make one to three releases/annum depending on mealy bug populations:

Monitor for cottony cushion scale build up and population of vedalia beetles. Shift vedalia to sites of build up and give augmentative releases. In the case of leaf miner attack, or caterpillars, hand pick the larvae and destroy. Fruit sucking moths develop on weeds and hence destroy the weed host *Tinospora*. Use poison bait for adults (Dilute suspension of the fermented molasses + Malathion 0.05% (50EC1ml/l)). Set up light traps or food lure (pieces of poisoned fruits) to attract adult moths. Leaf miners may be managed, with chemicals sprays like Dichlorvos 76 WSC (1ml/l), fenthion 80 EC 1ml,monocrotophos 36 WSC (1.5 ml/1). NSKE (5%), Neem formulations (2ml/1), Neem cake extract (5%) if not managed other wise. If leaf caterpillar becomes serious, spray endosulfan 35 EC (2ml/l). Chemicals that can be used in rotation are Methyl demeton 25EC (1ml/l) for aphids, or endosulfan 35EC (1.5 ml/l), carbaryl 50 WP (4g/l) against shoot borer/stem borers. Rust mite populations can be managed by Dicofol 18.5 EC (2.5 ml/l) or wettable sulphur 50 WP (2g/l). Nematodes can be managed with application of Aldicarb 10 G (250g), or carbofuran 3 G (750 g/tree).

Custard-Apple

Mealy bugs; *Ferrisia virgata, Planococcous lilacinus,P. pacificus, Maconellicoccus hirsutus, Perisopneumon ferox - Pseudococcidae*

Mealy coating over the fruit is seen due to colonies of small reddish-white mealy bugs. They also infest leaves,shoots and buds. Due to sucking of sap by nymphs and adult female mealy bugs,fruits remain small in size and become shrivelled. Large number of mealy bugs are seen in grooves of fruits epicarp along with sooty mould. This affects its market value. *Maconellicoccus hirsutus* has pinkish nymphs and adults

Fruit- borer: *Anonaepestis(Heterographis) bengalella* - Phycitidae: Lepidoptera

The fruit borer, *A. bengalella* is another sporadic pest of custard- apple in South India. The caterpillars bore into fruits making irregular tunnels, covered with black excreta and damaging mesocarp. Growth of infested fruits is affected and fruits drop. The adult is yellowish with a pale band on third tergite .

Fruit fly *Bactrocera (Dacus)zonata*

The attack causes fruit rotting and fruit drop.

See under **cucurbits**

IPM for custard apple

This crop is generally managed without pesticides. Lookout for mealy bug infestation in the beginning of the flowering season itself. Application of fish oil rosin soap (25g/lit) or teepol (25 ml/ l) will take care of crawlers. This has to be done on the main stem and bark of branches, before flowering season if the attack is endemic. For fruit fly attack, apply a bait spray of Malathion (2ml in a solution of 50 g jaggery plus 25 g fruit pulp/litre) from half ripe stage of fruit. The sprays can be repeated at need based intervals from this stage onwards. Monitor fruit borer attack and remove and destroy affected fruits immediately.

Fig

Stem-borer: *Batocera rufomaculata, B. rubus, Olenecamptus bilobus* - Cerambycidae: Coleoptera

These are large dark brown adult beetles that lay eggs in loose bark. Grubs bore into the trunk and make tunnels in branches and main trunks. This causes drying and wilting of branches. Adults of *O. bilobus* are pale whitish with long serrate antenna

Scale insects: *Parlatoria oleae*

The nymphs and adult females of scale insects suck sap from flushes and fruits. Affected leaves and shoots show yellowing and drying.

Leaf roller: *Phycodes minor, P. radiata* - Glyphiterygidae: Lepidoptera

Greenish caterpillars of leaf rollers feed on epidermis of leaves in spring, and later on fold the leaves and feed within infested leaves. Larvae of *P.radiate* are yellowish white with dark lateral stripes. Adult is small greyish brown. Larvae of *P. minor* are light green with yellow, shiny head and lateral stripes.

Gall makers (midges): *Udumbarie nainiensis,Anjeerodyplosis peshawarensis*

These are pests in North India and damage the flower buds by ovipositing in them. The maggot feeding causes discolouration, wrinkling and small protuberance. *A. peshawarensis* oviposits in tender fruits. On hatching, the maggots feed on fruit pulp. Fruits become elongated, shrivel and drop prematurely.

Leaf caterpillars: *Glyphodes* spp. - Pyraustidae: Lepidoptera

The larva is greenish and causes defoliation. Adults are green or black or white coloured moths.

Wild silk worm, *Ocinara varians* - Bombycidae: Lepidoptera

The larvae are pale coloured and humped anteriorly with a slight horn at the tail region and hence resemble twigs. The adult is reddish grey.

Leafworms *Hypsa ficus* - Hypsidae:Lepidoptera

The larvae cause defoliation. They are black with hairs arising from yellowish brown warts. Adults are yellow with black dots on wings.

Thrips, *Giganotothrips elegans* - Thripidae: Thysanoptera

Nymphs and adults of these large thrips desap from tender leaves causing curling and drying.

Grape Vine

Stem girdler

***Sthenias grisator* -** Cerambycidae: Coleoptera

Adults of the weevil oviposit in around fashion on stem and emerging grubs eat the internal contents. The above portions become dried up. Adults are medium sized, grey beetles with a white spot in the centre of elytra.

Defoliating beetles: *Adoretus lasiopygus, A. versutus, A. duvauceli, A. bengalensis* - Melolonthidae

The beetles feed on leaves from margins or as shot holes. In severe cases beetles may defoliate the vines of grapes completely.

Thrips: *Rhipiphorothrips cruentatus,Scirtothrips dorsalis* - Thripidae

These insects cause damage by sucking sap from foliage and flowers. Desapping causes crinkled leaves and corky lines over the berries.

Leaf roller: *Sylepta lunalis* - Pyraustidae

The caterpillars roll leaves and also feed on flowers.

Flat- headed borer: *Sphenoptera lafertei*

See under **peach**

Leaf miner *Phylocnistis toparcha* - Gracillaridae

The larvae mine the leaves and cause crinkling. Adult is a greyish brown moth.

Yellow and red wasp: Wasps cause damage by puncturing ripe fruits and feeding on them. Honeybees are also attracted to punctured fruits.

IPM in grapes

Grapes, being consumed raw, allow only minimal use of insecticides. As a general rule, even when insecticide use is warranted, use botanicals or chemicals with low residual and mammalian toxicity. Monitoring and managing at initial pest build up is the rule. Biocontrol of thrips, aphids and mealy bugs should be promoted. Monitoring of these in the blooming stage and immediate remedies based on surveillance reduces pesticide application. Release *Cryptolaemus montrouzieri* as a predator against mealy bugs and scales. In severe infestation, use Dichlorvos + fish oil rosin soap. Against stem girdler, deterrent chemicals/botanicals like neem oil/NSKE/garlic extract can be sprayed on the stem during incidence season. Remove and destroy loose bark and plant parts after every pruning.

Guava

Fruit flies: *Bactrocera dorsalis,B. zonata, B. diversa* - Tephritidae

Fruit flies are a problem in many susceptible varieties. The adults deposit eggs in soft skin of half mature fruits. On hatching, the maggots bore into fruit and feed on soft pulp. The infested fruits show dark greenish depression with maggots inside. The infested fruits may rot and fall.

Guava scale/Green shield scale: *Chloropulvinaria psidii*-Coccidae. These scales are flat, broadly oval in shape and yellowish –green in colour. They infest the ventral side of leaves, shoots and fruits.

Mealy bugs: *Ferrisia virgata,Plannococcus citri, P. lilacinus* - Pseudococcidae

Severely infested fruits are covered by white mealy bugs along with sooty mould and honeydew. Fruit quality as well as size is reduced.

Tea mosquito bug: *Helopeltis antonii,* - Coreidae

The tea mosquito bug is a major pest of guavas in South India. The nymphs and adults puncture fruits of all sizes, tender shoots and suck the sap. Pricking punctures on fruit result in brownish spots which develop into raised pustules. Severely infested fruits become hard and corky.

Fruit- sucking moths: *Othreis* spp.

The fruit sucking moths puncture ripe fruits and suck their sap during dusk. The area around feeding punctures on fruits becomes soft and rotting takes place. Infested fruits ultimately drop.

Bark-eating caterpillars: *Indarbela* spp.

The caterpillars feed on bark under webbed galleries of silk, chewed wood and excreta during night. The infested plants show presence of such galleries on main stem and branches.

Spiralling whitefly: *Aleurodicus disperses* - Aleurodidae

This is a recently introduced serious pest of guava in South India. Thousands of nymphs and adults congregate on under surface of leaves and suck the sap. Severe infestation initially gives silvery appearance on ventral side of guava leaves. Affected leaves turn yellow, crinkle and fall prematurely. Also cause sooty mould due to excretion of honey dew. Nymphs have numerous evenly spaced short silvery rods of wax along the sides of body; Eggs are laid in a spiral shape and covered with mealy coat and hence the name spiraling whitefly. The adults are powderry white, and active during early morning and late evening hours.

IPM in guava

Monitoring population of mealy bugs and whiteflies is to be done regularly. Plough the soil before every flowering period to expose pupae of fruit flies. Remove the fallen fruits and bury them deeply with lindane 1.3 D dusted over them. Release Cryptolaemus beetles @ 10/tree during beginning of season to prevent mealy bugs and whiteflies. Spray neem formulations (2ml/l) at monthly/bimonthly intervals at the time of flowering, if need be. Under heavy population, use (Triazophos 2ml+ neem oil 5ml/l) or phosalone (1.5 ml +

neem oil 5ml/l). Against tea mosquito outbreaks, use malathion 50EC (2ml/l), endosulfan 35EC (2ml/l) or fenthion 80EC (1ml/l).

Jackfruit

Shoot-and fruit- borer: *Diaphania (Margaronia) caesalis* - Pyraustidae: Lepidoptera

The shoot-and fruit-borer *D. caesalis* is a major pest of jackfruit in India. The larvae are reddish brown with pale brown head and thin hairs from black warts. The caterpillars bore into shoots, buds and fruits. Affected shoots wilt or dry up and fruits rot and drop. The adult is brown with grey elliptical patterns on wings.

Bud weevil/leaf-eating weevil: *Ochyromera artocarpi, Onychocnemis careyae, Telurops ballardi* - Curculionidae:Coleoptera

The small whitish grubs bore into tender flower buds,terminal shoots and fruits. This induces premature fruit drop. The adults are greyish – brown weevils.

Aphids: *Greenidia artocarpi, Toxoptera aurantii* - Aphididae: Hemiptera

Both these species of aphids feed in colonies on tender leaves and shoots which get distorted and devitalized. The adult is a pale green aphid.

Giant mealybug: *Drosicha mangiferae* - Pseudococcidae

These are flat,oval,waxy white bugs observed in clusters on tender shoots and inflorescence sucking the plant sap.

Leaf caterpillar : *Glyphodes bivitralis* - Pyraustidae:Lepidoptera

The larva is light yellow with black dots; adult is light yellow with wavy lines. The larvae eat away the leaves

Hairy caterpillar : *Perina nuda* - Lymantriidae: Lepidoptera

The adults are dimorphic moths; female with brownish white wings; male with half brownish and half transparent wings. Larvae cause defoliation

Lace wing bug : *Stephanitis chariesis* - Tingidae: Hemiptera

Bugs with lace like wings. They suck the sap from undersurface of leaves.

Spittle bug : *Clavia lineaticollis, Cosmoscrata relata* Cercopidae: Hemiptera

Yellowing and terminal drying of shoots due desapping by nymphs and adults..

Thrips : *Pseudodendrothrips dwivarna* Thripidae; Thysanoptera

Whitish patches are seen on leaves due to desapping by nymph and adult. These are bicoloured thrips.

Litchi

Fruit – borers: *Criptophlebia illepida,Rapala varuna,Deudorix epijarbas, D. isocrates* - Lycaenidae

Of the four species, *C.illepida* is more damaging the litchi fruits. The larvae bore into developing fruits and seeds, filling the bore-holes with excreta and the infested fruits start rotting.

Fruit- fly: *Bactrocera dorsalis,*- Tephritidae

Maggots of *B.dorsalis* feed on fruit pulp and induce rotting. It has been observed as a serious pest in Bihar.

Eriophyid mite: *Aceria litchi* - Eriophyidae

The eriophyid is a major pest of litchi in India. It attacks new leaves, tender shoots and fruits. Feeding of sap initially causes yellowish-grey area which later on turn to chocolate brown. The leaves become twisted, curled and thickened. Heavy attack results in low fruit-set and poor fruit yield.

Bark-eating caterpillars: *Indarbela tetraonis,I. quadrinotata*

See under **pomegranate**

Loquat

Fruit-fly: *Bactrocera dorsalis* - Tephritidae

The egg laying and ovipunctures make the fruits deformed. Feeding of maggots in the pulp induces rotting and fruit drop.

Bark-eating caterpillar : *Indarbela quadrinotata*

Severe damage on loquat due to *I.quadrinotata* was observed during October-November in Uttar Pradesh. Cultivar Thames Pride is less susceptible.

Loquat thrips: *Heliothrips* sp.

Infest flowers during winter and suck the sap. Affected flowers get dried and do not bear fruits.

Scales: *Eulecanium tiliai, Parlatoria oleae* - Coccidae

The scales suck sap from lower surface of leaves and shoot which affects the vitality and fruiting capacity of the plants.

Mango

Leafhoppers: *Amritodus atkinsoni, A. brevistylus,Idioscopus clypealis, I. nagpurensis, I.niveosparsus, Amrasca splendens* - Cicadellidae

The jassids or leafhoppers are pests of flushes and inflorescence. A grazing sound can be heard while walking under the infested tree, due to mass movement by the jassids. A lot of damage is caused by nymphs and adults sucking sap from tender foliage and inflorescence. Ovipunctures also cause damage to tender leaves and inflorescences. Severely infested leaves become curled and inflorescences get dried. Affected parts are covered with honeydew and sooty mould. *Amrasca splen*dens infestation causes typical hopper burn symptoms on tips of tender leaves. The extent of loss varies from 20 to 100% depending upon crop stage, time of attack and hopper population.

Mealybugs: *Drosicha mangiferae, D. stebbingi, Rastrococcus iceryoides* - Pseudococcidae

Greyish-white flattened, wingless nymphs and female bugs suck sap from inflorescence, tender leaves and fruit peduncles during December-April in North India. Infested shoots and inflorescence get dried. Infested parts are covered with honeydew and sooty mould. *R. iceryoides* is serious in South India.

Mango fruit flies: *Bactrocera dorsalis, B. zonata, B. diversa* - Tephritidae

Of the three species, *B. dorsalis* is the most serious. Light brown flies with hyaline wings lay eggs under fruit epicarp. The emerging maggots feed on pulp when fruit ripens. Extensive tunneling by maggots has a distracting effect and also induces fruit rot. Affected fruits show dark brown, rotten patches and fall-off prematurely. All commercial cultivars are susceptible; however, infestation is low in Langra, Dashehari and Bombay Green.

Stone/nut weevils: *Sternochetus mangiferae, S. gravis* - Curculionidae

Adult female weevils deposit eggs singly in the fruit skin in the marble stage causing fruit drop. Grubs after hatching enter the stone. They feed on the cotyledons in a zigzag fashion,destroying complete stone filled with black frass. Damaged stones lose their viability. Adults emerge from the stone during ripening, make an exit passage and thus render the fruit useless. Extent of damage in susceptible cultivars like Neelam, Totapuri and Baganapalli can go up to 60-65%.

Shoot-borer: *Chlumetia transversa* - Noctuidae

Larvae causes serious damage to new flushes of mango throughout the country. The affected shoots wilt, droop and finally get dried. The infested seedlings and branches become stunted and flowering is inhibited. Young trees up to the age of 8-10 years are more damaged.

Leaf Webbers: *Orthaga exvinacea,O.euadrusalis,O. mangiferae* - Pyralidae

The larvae of *Orthaga* spp web the terminal leaves and shoots into clusters which harbour several of them in initial stage. The larvae scrape leaf surface. Late instar larvae feed on whole leaf lamina leaving only midrib. Severe attack leads to drying of shoots/ branches and reduction in fruit yield. Dried up terminals are visible on tree long after attack is over

Shoot-gall psylla: *Apsylla cistellata*

Nymphs and adults of psyllids feed on apical and axillary buds are causing cone shaped green galls on mango shoots which get dried later on. Up to 90% shoots are infested in terai region of Uttar Pradesh and Bihar. Yield can also get reduced up to 90%.

Mango stem-borer: *Batocera rufomaculata, B. rubus* - Cerambycidae

These are large, long horned dark brown beetles. Their grubs feed on bark of branches and stem. Initially gum oozes out from a circular hole in trunk, but later on, chewed up frass is thrust out. Affected branches get dried. The yield of the trees is reduced.

IPM in mango

Fruit flies are major pests in Mango. Leaf and flower hoppers also cause yield reduction in Mango.

Set up Methyl eugenol traps to monitor and kill fruit fly adults. The traps must be cleaned every week. The converging adults can be killed by adding a little from a solution of 2ml Malathion and 20 g sugar dissolved in one litre water. This solution should also be sprayed in spots on underside of leaves of Mango and other vegetation to attract and kill adults. Collect all fallen fruits, and deep bury them after sprinkling lindane dust. Bait spray should be given at 10 days intervals from 50% maturity of fruits until harvest.

Mango hoppers, and other leaf and flower feeders like webbers, shoot weevils and mealy bugs are best managed by spraying Malathion (2ml / litre), Dimethoate (1ml / litre) or phosalone (1.5 ml / litre). For flush worms and flush weevils, dust with any insecticide dust, just one to two days from emergence of new flushes.

Olive

Olive-scale: *Metacernema japonica,*

Black scale: *Saissetia oleae* - Coccidae

Olive scales are tiny pale-greyish insects which settle on branches, leaves and fruits; active during April-November. *S. oleae* are round greenish brown scales. Sooty-mould appears on leaves and branches of infested plants as a consequence of honeydew secretion. Shoots get dried up and fruits remain undersized.

Olive psylla: *Euphyllura longinova* - Psyllidae

These psyllids have waxy strands covering body of nymphs, which give them a woolly appearance. They suck the sap from tender growing parts of plants and fruit formation is affected.

Black tingid (lacewing) bug: *Eteoneus sigillatus* - Tingidae

The bugs feed from the ventral side of leaves. Due to continuous sucking of sap, yellowing of leaves and sooty mould on upper surface is seen.

Papaya

Fruit flies: *Bactrocera cucurbitae,Toxotrypana curvicauda* - Tephritidae

The fruit flies lay eggs on maturing fruits. The emerging maggots feed in fruit pulp, causing rotting.

AK grasshopper: *Poekilocerus pictus* - Acrididae

Nymphs and adults of the AK grass hopper feed on papaya leaves and tender bark. Maximum damage is done during July-August in North India.

Aphids: *Aphis gossypii,Myzus persicae* - Aphididae

Nymphs and adult aphids suck sap from leaves. They are more important as a vector of papaya mosaic virus disease. The affected trees die in a year or two.

Red spider mite: *Tetranychus cinnabarinu*

Scarlet mite, *Brevipalpus phoenicis'* Tetranychidae

The red spider mite is an occasional serious pest. Nymphs and adults suck sap from leaves which turn yellow and ultimately dry. In case of severe infestation, fruits are also attacked.

Stem-borer: *Dasyses rugosellus* - Tineiidae

Noted as a minor pest in Bihar, Maharashtra and Tamil Nadu, its caterpillars bore into the stems and feed internally.

Grey weevil: *Myllocerus viridanus* - Curculionidae

Adults of this grey weevil feed and skeletonise leaves in South India. The grubs are root feeders of papaya and other nearby plants.

IPM in papaya

Generally papaya is not sprayed. A good monitoring mechanism and timely mechanical control will eradicate most of pest problems. Aphid build up has to be prevented by systematic monitoring and timely control by mechanical removal of colonies and application of botanicals. When mosaic appears, remove and destroy affected plants immediately.

Peach

Green peach aphid/peach leaf curl aphid: *Brachycaudus helichrysi* - Aphididae

These are green aphids that suck the sap from growing buds. Leaf buds become weak and result in poor setting and premature fruit fall.

Blossom thrips: *Thrips rhopalantennalis, T. flavus, T. florum, Haplothrips tenuipennis, Frankliniella dampfi* - Thripidae

See under **apple**

Peach fruit fly: *Bactrocera zonatus,* **Oriental fruit fly:** *B. (Dacus)dorsalis* - Tephritidae

The adults are brown to dark brown flies with ferruginous thorax; measure 7mm in length and 14mm across wings; females have tapering abdomen with a pointed ovipositor. Adults lay eggs inside fruits; the maggots feed on pulp; fruits get tunneled with active maggot feeding becomes soft, ferments and drops. The incidence reduces yield and quality of fruit.

Flat-headed borer: *Sphenoptera lafertei* - Buprestidae

Small coppery black adult beetles (10-13) feed on foliage; generally attack stems and branches exposed to sun. The plants become weak or die in severe infestation.

White Peach Scale: *Pseudoulacaspis pentagona* - Diaspididae

Nymphs suck sap from leaves during April-May; adults puncture ripe fruits during June-July; rot sets in at punctured spots.

Peach leaf roller: *Cacoecia epicryta* - Tortricidae

The larvae of leaf roller webb the leaves together. The adults are bell shaped moths while at rest.

Bark – eating caterpillar: *Indarbela quadrinotata, I. tetraonis* - Metarbelidae

The gregarious caterpillars feed upon bark as well as on stem under cover of webbing.

Termites: *Microtermes obesi,Odontotermes(Termes)obesus* - Termitidae

This is more a pest in dry weather. The termites feed on roots and stem bases near the ground level of almost all types of fruit trees in warm localities with sandy loam soil. The severely infested trees often dry and yield is reduced.

Pineapple

Mealy bug: *Dysmicoccus brevipus* - Pseudococcidae

Mealy bugs are major pests of pineapple in India. They can attack roots and aerial parts. Colonies of reddish –white mealy bugs infest roots, crown, and fruits and suck sap. The leaves change in colour from green to red or pink and the margins reflect inwards and leaf tip starts drying. Severely infested plants become stunted and bear only small fruits.

Fruit-eating beetle/Corn sapbeetle: *Carpophilus dimidiatus* - Nitidulidae: Coleoptera

The beetles feed inside ripen fruits. Damaged fruits rot due to infection of pathogens.

Termites: *Odontotermes* spp. - Termitidae

Termite infestation is noticed in plantations in sandy soil. This is more a pest in dry weather. The termites feed on roots and stem bases

Persimmon

Leaf roller: *Hypocala rostrata* - Noctuidae

Dark violaceous grey and orange moth lay eggs on leaves in early- July. Insects remain active throughout rainy season.

Caterpillars fold leaves, feed inside and pupate within the leaf folds.

Plum

Plum weevil: *Amblyrrhinus poricollis, Myllocerus laetivirens, M. sabulosus, M. undecimpustulatus* - Curculionidae

Adult beetles feed on epidermis of fresh and tender leaves leaving behind only the network of veins, affected leaves dry up and fall. They cause serious damage in young plantations.

Defoliating beetles

See under apple

Scale insects: *Eulecanium tiliac, Eriochiton theae* - Coccidae

The scale insects desap the plant parts and cause yellowing and wilting

San Jose Scale: *Quadraspidiotus perniciosus* - Diaspididae, *Hemiberlesia rapex, H. lataniae* - Diaspididae

Small nodule like brown female bodies of the armoured scales are observed on current growth of the tree. Large number of flies and ants get attracted due to honeydew secreted by them.

Adult females suck sap from shoots while nymphs thrive on leaves. Heavy attack of the pest results in poor fruit- set or undersized fruits.

Plum fruit moth: *Cydia funebrana* - Tortricidae

The larvae of these plum moths bore into developing fruit and feed on pulp. The fruits will ultimately rot and fall down.

Hairy caterpillars: *Euproctis fraternal, E. signata* - Lymantridae

See under **peach**

Leopard moth: *Zeuzera multistrigata* - Cossidae

See under **apple**.

Pomegranate

Fruit- borer (Anar butterfly): *Deudorix isocrates, D. epijarbas* - Lycaenidae

D. isocrates is more serious in India. Newly hatched caterpillars bore into fruits and feed on developing seeds. Hole made by caterpillars facilitates entry of pathogens and induce fruit rot and drop. The extent of loss varies from 20-80% depending upon area, variety and season.

Pomegranate Aphid: *Aphis punicae* - Aphididae

This aphid is serious on new flushes. The aphids infest tender shoots, leaves, flower buds and fruits. Severally infested leaves become curled.

Mealy bugs: *Planococcus* lilacinus, *Planococcus citri* - Psedococcidae

These mealy bugs infest flower buds and fruits. Severe infestation leads to dropping of flower buds and low fruit-set.

White fly: *Siphonimus phyllyreae* - Aleyrodidae

Nymphs and adults of whitefly infest leaves on ventral side, causing curling, yellowing and drying. Nymphs cause more damage due to desapping.

Fruit- sucking moths: *Othreis ancilla, O. cajecta, O. fullonica, O. materna, Achaea janata* - Noctuidae

The adults of these moths puncture mature fruits during dusk time by inserting proboscis and suck juice. Through feeding punctures secondary pathogens enter and cause fruit rotting. Fruit infestation varies from 40 to 90%. The semiloopers of *Achaea* also feed on seedlings.

Mites: *Oligonychus punicae* – Tetranychidae; *Aceria granati* - Eriophyidae

Adult and nymphs feed on the lower leaf surface, causing brown patches, curling and drying.

Bark-eating caterpillar: *Indarbela quadrinotata* - Metarbelidae

Its larvae feed on bark under webbed galleries of silk and excreta. The attacked plants show presence of such galleries on bark surface.

Strawberry

White grubs and cut worms: These insects cut roots and stems of young plants.

Hairy caterpillars: Defoliate plants during June-July

Walnut

Walnut weevil: *Conotrachelus juglandis* - Curculionidae

The adult weevil is dark brown with white markings and has a prominent snout. They excavate circular holes on fruit surface due to feeding. Grubs bore into fruits and reduce kernels into a black mass, and attacked fruits drop prematurely with grubs inside.

15. Vegetables

Amaranth and Spinach

Leaf-eating caterpillar: *Spoladea (Hymenia) recurvalis* - Pyralidae

Green slender caterpillars fold leaves on top of plant, feed inside and skeletonize them.

Bean leaf roller *Lamprosema indicata* - Pyralidae

The young larvae are gregarious green caterpillars that fold leaves, feed inside and skeletonise them.

Aphids: Pea aphid *Aphis craccivora,* Mustard aphid *Lipaphis erysimi* - Aphididae

These aphids sometimes attack young parts and suck plant sap thus reducing yield and appearance of produce.

Scale: *Coccus hesperidum* - Coccidae

Sometimes scales attack plants, suck plant sap and thus reduce marketability.

Stem weevil: *Hypolixus truncatulus* - Curculionidae

The grubs tunnel the stem and fill it with frass and excreta. These stems develop thickening, may split and cause drying of plants. Adults nibble leaves.

Blue beetle: *Altica caeruleus* - Chrysomelidae

The small grubs mine the cotyledonary leaves and pupate in the leaves.

Beet Root

Cut-worm: *Agrotis ipsilon* - Noctuidae

The adult moth has brown forewings and a dumb-bell mark in the centre of each wing, but specifically has a paler area around tips of the forewings. The hind wings are buff, darkening to margin, with dark veins. The moth has a wingspan of about 4 cm. The larvae which are cut worms cut seedlings in base at night.

Leaf-eating caterpillar: *Spodoptera litura* - Noctuidae

Eggs laid in clusters and young larvae gregariously feed and skeletonise leaves. The larger larvae bore into heads and cause loss of leaf area.

Brinjal

Shoot and fruit borer: *Leucinodes orbonalis* - Pyralidae

On plants in the pre flowering stage, the larvae bore into young shoots with holes plugged with excreta. In fruits, they bore holes and plug it with excreta. Larvae are pinkish in colour, active throughout year, but more during monsoon. Heavy losses up to 60% can occur.

Stem-borer: *Euzophera perticella* - Pyralidae

These bore into the plants and the infested plants start wilting. If stem is cut open, pale-gray caterpillars with hairs can be seen.

Ash-weevil: *Myllocerus subfasciatus* - Curculionidae

Adults and grubs cause damage. Adults nibble margins of foliage. The grubs feed on roots and also girdle the stem.

Midge: *Asphondylia* sp. – Cecidomyiidae: Diptera

The pest attacks flower buds or young fruits in the larval stage as maggots, causing drying of flowers or malformation of fruits. Loss is severe during cooler months. Up to 70% flowers are damaged and fail to set fruits.

Leaf- roller: *Antoba olivacea* - Noctuidae

The leaves are rolled longitudinally and with silken webbing. The larvae are seen inside the webbing inside. It is an occasional pest.

Horn moth: *Acherontia styx* - Sphingidae

These are sometimes seen. They are stout greenish caterpillar with yellow lateral bands and feed on feed on leaves voraciously. Whole plants can be eaten within a short time.

Bud- worm/ Potato tuber moth: *Phthorimoea operculella* - Gelechidae

Small pink larvae of these feed inside flower buds and reduce fruit formation.

Eggplant mealy bug: *Coccidohystrix insolitus* - Pseudococcidae

These occur on lower surface of leaves in large numbers, especially in older plants. They suck sap and weaken plants.

Aphids: *Aphis gossypii, Myzus persicae* - Aphididae

Colonies of aphids are found on tender leaves, flower buds, etc. Generally, aphid population can increase due to indiscriminate use of insecticides.

Leafhopper: *Amrasca biguttula biguttula* - Cicadellidae

The nymphs and adults suck the sap from underside of leaves. The Infested leaves curl upwards in margins.

Lace-wing bug: *Urentius hystricellus, U.sentis* - Tingidae

These are small black-brown insects with lace like wings that desap from under surface of leaves.

Red bug: *Aspongopus janus* - Pentatomidae

Medium sized reddish flat bugs are found sucking sap on tender parts.

Thrips: Onion thrips *Thrips tabaci,* Melon thrips*Thrips palmi* - Thripidae

Both nymphs and adults lacerate tender fruits and leaves. They usually congregate on the lower surface of older leaves.

Epilachna beetle(Hadda beetle): *Epilachna vigintioctopunctata, Epilachna dodecastigma* - Coccinellidae

Large number of adults with 28 or 12 spots and spiny yellow grubs of these beetles can be observed scraping on leaves. The eggs are cigar- shaped and bright yellowish, laid in clusters on the underside of leaves.

Termites: *Microtermes* sp., *Trinervitermes biformis* - Termitidae

These are usual in dry sandy soil. The roots are eaten and cause sudden wilting of seedlings.

Mite: *Tetranychus cinnabarinus* - Tetranychidae

The small red mites are seen on lower side of leaves. They spin small webs on undersurface and feed from beneath it.

Cabbage and Cauliflower

Diamond back moth : *Plutella xylostella* - Plutellidae

One of the most notorious pests of all cole crops throughout the world, the small green larvae feed on leaves, on leaf primordial, causing extensive defoliation. This causes prevention of head formation in both cabbage and cauliflower. In the plains of India, the damage can be very heavy.

Leaf- webber: *Crocidolomia binotalis* - Pyralidae

This is a common leaf Webber in cole crops, and the groups of hairy larvae feed on leaves and heads, which may rot.

Cabbage webworm /stem- borer: *Hellula undalis* - Pyralidae

They feed on midrib or veins of old plants. Main damage is due to larvae feeding into main stem immediately after transplanting and preventing head initiation and/or causing multiple heads.

Aphids: *Brevicoryne brassicae,Lipaphis erysimi* - Aphididae

Leaves with *B. brassicae* have white mealy covering. *H. erysimi* is lighter in colour. These feed and devitalize plants, affecting quality of curds and heads.

Gram- caterpillar: *Helicoverpa armigera* - Noctuidae

Most ubiquitous pest throughout the world. The initial larvae scrape the leaves, but the later instars cut big holes plugged with excreta on formed heads.

Tobacco caterpillar: *Spodoptera litura* - Noctuidae

Eggs laid in clusters and young larvae feed gregariously and skeletonise leaves. Larger instar larvae bore into heads.

Mustard saw-fly: *Athalia lugens proxima* - Tenthredinidae

Adults are orange bodied with smoky transparent wings. Larva is greenish black with wrinkled body and eight pairs of pro-legs. On slightest touch, the larva falls to round and feigns death. Initially the larva nibbles leaves, later it feeds from the margins towards the midrib. One of the a few pests in Hymenoptera.

Striped-flea beetle: *Phyllotreta striolata* - Chrysomelidae

The flea beetles are pests in adult stage. They feed on seedling and cause small holes.

Painted bug: *Bagrada cruciferarum* - Pentatomidae

Black and orange coloured adults and nymphs suck sap from leaves occasionally. They emit typical bug odour when disturbed. The damage causes leaves to wither.

Capsicum and Chilli

Thrips: *Scirtothrips dorsalis,Thrips palmi* - Thripidae

Thrips are very serious in chilli. They have fringed wings. Both adults and nymphs lacerate leaf tissues. The leaves curl inside, and are mostly a hot weather pest. The attack can be confused with chilli mosaic symptom. In capsicum and chilli, they also feed beneath calyx of flowers and fruits causing flower/fruit drop and reduction in yield and market value of capsicum.

Mites: *Polyphagotarsonemus latus, Tarsonemus translucens* – Tarsonemidae; *Tetranychus cinnabarinus* - Tetranychidae

These are small mites, and suck sap from underside of leaves. The leaves curl downwards. Very serious in chilli and causes up to 80% loss in chilli if left uncared for.

Aphids: *Aphis gossypii* - Aphididae

Colonies of aphids feed on tender leaves, flower and buds causing devitalisation and flower fall

Fruit-borer: *Helicoverpa armigera* - Noctuidae

Major pest in cotton. The larvae bore inside the fruits. In chilli, they cause serious loss, where chilli is grown as an intercrop along with cotton.

Fruit-borer: *Spodoptera litura* - Noctuidae

Larvae bore inside the fruits. In chilli, this pest causes serious loss by leaf eating and fruit boring.

Black cut-worm: *Agrotis ipsilon* - Noctuidae

Cut worms generally feed at night and hide during day. They cut seedlings at the base at night and hide in cracks, stubble,etc. during day.

White grub: *Anomala bengalensis, Holotrichia consanguienea, H. reynaudi* - Scarabaeidae

Eggs are laid in soil during pre-monsoon showers and grubs feed on roots. Adults feed on leaves.

Carrot

Many of the pests of cabbage and cauliflower are also pests of carrot.

Leaf hopper: *Empoasca punjabensis*

Both adults and nymphs suck sap and devitalize plants.

Cut-worm: *Agrotis ipsilon* - Noctuidae

The larvae cut seedlings in the base at night.

Cluster Bean

Aphid: *Aphis craccivora* - Aphididae

The aphids feed on young leaves, inflorescence and developing pods. They reduce yield and marketabilty.

Field bean pod-borer: *Adisura atkinsoni* - Noctuidae

The larvae are pod borers and cause damage if, left unprotected.

Bugs: *Megacopta(Coptosoma)cribraria* - Platasmidae

Both adults and nymphs suck sap and devitalize plants. They emit characteristic bug odour when disturbed.

Cowpea

Stem fly: *Ophiomyia phaseoli* - Agromyzidae

Seedlings are more affected. The maggots bore stems causing swollen petioles and dried leaves. White puncture marks on basal leaves are seen. Larvae are found at base of plant beneath the epidermal layer

Pod- borer: *Maruca testulalis* - Pyraustidae, *Lampides boeticus* - Lycaenidae

Caterpillars bore flower buds and pods. They cause heavy damage both in vegetables and grain cowpea.

Bruchid: *Callasobruchus chinensisi* - Bruchidae

Damage is mainly as a grain borer in storage. But infestation starts from field. Seeds are eaten by the grubs from inside and turned hollow. Most serious storage pest.

Bean aphid/Black cowpea aphid: *Aphis craccivora* - Aphididae

Colonies of aphids are noticed on young leaves and inflorescence stalk and young pods. Both nymphs and adults suck the sap from plant parts. Assumes seriousness only if left unchecked.

Pod bugs: *Riptortus pedestris, Clavigrella gibbosa, C.horrens* - Coreidae

The adults and nymphs drain sap from pods and cause them to be chaffy

Plant bugs: *Nezara viridula* - Pentatomidae

Both adults and nymphs suck sap and devitalize plants. They emit characteristic bug odour.

Leaf-eating caterpillar: *Spilosoma obliqua* (Bihar hairy caterpillar - Arctiidae) *Euproctis fraterna* - Lymantriidae

Large number of hairy caterpillars appear. They destroy leaves initially by scrapping and later by eating the whole lamina.

Cucumber, Gourds and Pumpkin

Red pumpkin beetle: *Aulacophora foveicollis, A. lewisii* - Galerucidae

Both adults and grubs cause damage. Leaves are eaten by different coloured adult beetles. The grubs feed on roots and plant parts touching ground.

Serpentine leafminer: *Liriomyza trifolii* - Agromyzidae

Introduced international pest of recent origin. The larvae mine leaves in a serpentine fashion

Fruit fly: *Bactrocera cucurbitae* - Tephritidae

Most important pest on fruits of cucurbits. Adults lay their eggs under skin of young developing fruits. The maggots bore inside, feed voraciously and cause rotting of fruits. Fruits drop, or are rendered useless. The pupae are seen in the soil, where the rotten fruits ultimately fall.

Blister beetle: *Mylabris pustulata* - Meloidae

The adults of these bright coloured beetles (3 black and 3 yellow bands) feed on flowers.

Stink bug: *Coridius (Aspongopus) brunneus, C. janus, C. observus* - Eupelmidae

Adults feed gregariously on leaves and tender parts. Bugs when touched emit characteristic smell.

Aphids: *Aphis gossypii, A. malvae* - Aphididae

Colony of aphids on tender leaves or flower buds and desap from there.

Mite: *Tetranychus neocaledonicus* - Tetranychidae

They web a colony on undersurface of leaves and feed on leaves.

Thrips: *Thrips tabaci, Frankliniella sulphurea* - Thripidae

Adults and nymphs of these thrips feed on tender leaves which later curl towards inside.

Plume moth: *Sphenarches caffer* - Pterophoridae

Small, green larvae with lateral brown stripes with hairs feed on leaves and cause holes. The adult is a small plume moth.

Leaf-eating caterpillar: *Diaphania (Margaronia) indica* - Pyraustidae

Long shining caterpillars feed on leaves. Pupae in silken cocoons can be seen.

Stem gall fly: *Lasioptera falcata* - Cecidomyidae

This is a serious pest of bitter gourd. The adults lay eggs on vines and maggots feed inside causing the plant to produce flat vine galls that cause stoppage of further growth.

French Bean

Stem-fly: *Ophiomyia phaseoli* - Agromyzidae

As soon as seeds germinate, adults of fly puncture leaf lamina to lay eggs. Larvae mine leaves, petiole and stem, causing extensive mining, stem galls and mortality of plants 20-30 days after sowing. Severe incidence observed during dry period of monsoon, causing 80-90% mortality of plants.

Leafhopper: *Empoasca kerri* - Cicadellidae

Large number of adults are seen feeding from undersurface of leaves, causing yellowing and crinkling in margins of leaves due to severe depletion of plant vitality and leaf drop.

Serpentine leaf-miner: *Liriomyza trifolii* - Agromyzidae

Mines are generally observed in basal leaves, rarely seen in trifoliate leaves. The leaves show the presence of large number of serpent shaped criss - cross mines with small larvae inside.

Thrips: *Calliothrips indicus, Megaleus distalis* - Thripidae

Leaves, flowers and pods are damaged. Occasionally serious and cause characteristic crinkling of leaves, flowers drop and streaks on pods due to laceration and removal of plant sap.

Red spider mite: *Tetranychus cinnabarinus* - Tetranychidae

Small red mites seen on lower side of leaves. When serious, cause webbing and feed within web. Web is seen generally on old leaves.

Garlic and Onion

Thrips: *Thrips tabaci* - Thripidae

The nymphs and adults of thrips desap from narrow spaces between inner leaf axils. Leaves are faded, with curled tips and shining silver streaks or brown spots. Many minute yellow or white insects moving, generally near base or middle of leaves. Plants faded and yield reduced.

Mite: *Aceria tulipae* - Eriophyidae

These mites desap leaves and cause malformation. Infested leaves show yellow patches. Plants get stunted.

Onion fly: *Delia antique* - Muscidae

This is coming under the family of houseflies. The maggots feed on developing bulb underground and cause rotting. Plants wither and attack predisposes to infection by fungus causing soft rot of onion.

Cut-worm: *Agrotis ipsilon* - Noctuidae

The young plants are cut at the base at night by cut-worms.

Ground nut earwig: *Anisolabis* (*Euborellia) annulipes* - Forficulidae

The nymphs and adults bore into onion bulbs.

Lablab Bean (Dolichos bean)

Pod-borer: *Helicoverpa armigera, Adisura atkinsoni* - Noctuidae; *Etiella zinckenella* - Phycitidae

Pod-borers are the most serious pests of lab lab. Their larvae bore flowers and pods, causing 10-90% losses, if left unprotected.

Aphid: *Aphis craccivora* - Aphididae

The colony of black aphids is found on young leaves and inflorescence, sometimes very serious in patches. They suck the sap from the shoots, flowers and pods.

Leaf-eating caterpillars: *Spilarctia(Spilosoma)obliqa* - Arctiidae; *Euproctis fraterna* - Lymantriidae.

The young ones are hairy caterpillars that appear in large numbers and destroy leaves.

Sap sucking bugs: *Nezara viridula* - Pentatomidae, *Anoplocnemis phasianai* - Coreidae, *Megacopta (Coptosoma) cribraria* - Plataspididae.

Both adults and nymphs suck sap and devitalize plants. They emit characteristic bug odour.

Weevils: *Alcidodes pictus,A.bubo* - Curculionidae

Galls with many grubs are observed on vines.

Red spider mite: *Tetrancychus cinnabarinus* - Tetranychidae

Small red mites are seen on lower side of leaves. When serious, they cause webbing and feed within the web. Web is seen generally on old leaves.

Leafhopper: *Empoasca kerri* - Cicadellidae

Nymphs and adults are seen on underside of leaves. Swarming of adults can be seen while walking in field, giving characteristic movement and sound, yellowing and crinkling in the margins of leaves and mottling of plants and loss of plant vitality.

Okra

Leafhopper: *Amrasca devastans(=biguttula biguttula)* - Cicadellidae

Small young nymphs of different colours are seen on underside of leaves. Infested leaves curl upward in margins. Plants are stunted bronzed and show hopper burn.

Shoot and fruit-borer: *Earias vittella, E.insulana* - Noctuidae

The larvae bore into flowers, flower buds and shoots, causing flower and fruit drop and drooping of affected shoots. Bored fruits have larvae inside or filled with excreta.

White fly*: Bemisia tabaci* - Aleyrodidae

More important as a vector that suck sap from leaves and transmit yellow-vein mosaic virus, particularly during summer. Losses may be even 80-90%.

Aphids : *Aphis gossypii* - Aphididae

Curling and crinkling of leaves,with stunting of plants. Colony of aphids on young leaves and fruits can be seen. Honey dew is secreted and causes development of sooty mould.

Mites: *Tetranychus cinnabarinus* - Tetranychidae

These are small red mites seen on lower side of leaves. They make webs and feed under it.

Blister beetle: *Mylabris pustulata* - Meloidae

Bright coloured adult beetles (with 3 black and 3 yellow bands) sometimes feed on flowers.

Petiole maggot: *Melanagromyza hibisci, M.obtusa* - Agromyzidae

The maggots feed on tender shoots and leaf petioles. This causes drying of shoots as also the leaves and petioles. Maggots are seen in dead plant base and on drying petioles.

Tomato fruit borer: *Helicoverpa armigera* - Noctuidae

The larvae are sometimes observed on flower buds and fruits. They feed from outside, as a fruit borer.

Cotton leaf roller: *Sylepta derogata* - Pyralidae

Young larvae feed on lower epidermis of leaves and after sometime, roll leaves and feed within. Pupation generally inside the roll.

Red cotton bug: *Dysdercus koenigii* - Pyrrhocoridae

Both nymphs and adults suck sap from fruits, and cause malformation of fruits, fruit drop and shrinking of seeds.

Dusky cotton bug: *Oxycarenus laetus* - Lygaeidae

Both nymphs and adults suck sap from fruits.

Semi-looper: *Anomis flava* - Noctuidae

Feed on leaves with characteristic semi-loop movement of the caterpillar.

Painted bug: *Bagrada cruciferarum* - Pentatomidae

Black and orange coloured adults and nymphs suck sap from leaves and cause withering .

Aphids: *Lipaphis erysimi, Myzus persicae* - Aphididae

Green to black nymphs and adults suck sap and devitalize plants.

Stripped flea beetle: *Phyllotreta striolata* - Chrysomelidae

Large number of adults feed on seedlings and cause small holes.

Tomato

Fruit-borer: *Helicoverpa armigera* - Noctuidae

Damage starts from flowering. Eggs are laid on young leaves damaged by young larvae. Later they migrate to developing fruits. They bore the fruits with the head inside and remaining parts of body outside fruits. One larva feed on many fruits, causing 5-50% losses.

Serpentine Leaf miner: *Liriomyza trifolii* - Agromyzidae

A recent introduction, it attacks several crops. The larvae mine leaves, particularly basal leaves, causing 10-15% loss in severe cases.

White flies: *Bemisia tabaci* - Aleyrodidae

Suck sap from leaves and transmit leaf curl virus, particularly during summer. Loss may be even 80-90%.

Mite: *Tetrancychus cinnabarinus* - Tetranychidae

Small red mites are seen generally on lower side of leaves. They cause webbing and feed from below.

Mealy bugs: *Ferrisia virgata* - Pseudococcidae

White cottony mealy bugs are seen on plants. The nymphs and adults suck plant sap and stunt growth of plants.

Fruit-sucking moth : *Othreis fullonica, O.materna,O.ancilla* - Noctuidae

Adult moths puncture ripening fruit and suck juice. Damaged area becomes soft and rots and close examination reveals pin hole.

Leaf-eating caterpillars: *Spodoptera litura, S. exigua* - Noctuidae

Eggs are laid in clusters. Young larvae gregariously feed and skeletonise leaves. Large larvae bore into fruits.

Bugs: *Nezara viridula* - Pentatomidae

Sometimes green flat bugs gregariously feed on tomato leaves.

Thrips: *Thrips tabaci, Calliothrips indicus* - Thripidae

Nymphs and adults lacerate leaves. Leaves may become pale and silvery shines appear on affected leaves.

IPM in vegetables

Reducing insecticide application to the lowest level is the motto of IPM in vegetables. Monitoring for the initial build up and timely mechanical removal of the first population is a must in vegetables. Always remove the first affected plants, by diseases like mosaic, and destroy them by burying or burning. Remove insect eggs, insect affected plant parts and adults and immature stages of insects, and destroy them every day. Do not throw them in the field. They will either come back, or pupate and attack later. During nursery stage, use wood ash, cow dung solution, and plant extracts to deter all pests. Use only low residual insecticides like Dichlorvos (1ml/l), Malathion (2ml/l), Endosulfan (2ml/l) or Quinalphos (2ml/l), in spots, at the beginning of attack itself. Use pheromone traps against fruit and shoot borer in brinjal, bhindi, diamond back moth in cabbage and cauliflower etc. Set up sticky yellow traps against aphids and whiteflies in cowpea, brinjal, bhindi, cabbage and cauliflower. Always grow resistant varieties and treat the seeds with seed treating dusts (STD) such as Imidachloprid, before planting. Release parasitoids wherever available, like *Chelonus blackburni* (4000/0.4ha) in bhendi, *Trichogramma chilonis* (20000/ 0.4ha)/week against Helicoverpa and *Cotesia plutellae* against Diamond Black Moth (DBM).

References

Atwal, A.S. and Dhaliwal, G.S. (2005) Agricultural Pests of South Asia and their management Kalyani Publishers, New Delhi India 505p.

Chadha, K.L. (2005) (Ed.) Handbook of Horticulture, ICAR, New Delhi 1031p.

Dhaliwal, G.S. and Arora, R (2006) Integrated Pest Management-Concepts and approaches- Kalyani Publishers, New Delhi 349 p.

Kerala Agricultural University (2007) Package of practices Recommendations: Crops13th Edition Kerala Agricultural University, Thrissur 334 p.

Nair, MRGK, (1999) A monograph on crop pests of Kerala and their control. Kerala Agricultural University 227p.

Norris, R F, Caswell-chen, E.P. and Kogan, M. (2003) Concepts in Integrated Pest Management, 586 p. Prentice Hall of India Private Limited, New Delhi.

Regupathy, A., Palanisamy, A, Chandramohan, N. and Gunathilakaraj, K. (2003) A guide on crop pests, Sheeba Printers, Coimbatore 276 p.

Chapter – 13

Nematodes Management in Horticultural Crops

M.S. Sheela, T. Jiji, K.R. Remya and T. Bharati Meena

Nematodes constitute a group of primitive microscopic invertebrate animals which occur in all conceivable habitats in amazingly large numbers. Besides bacteria, viruses, fungi and insects, the nematodes are regarded as important pests of agricultural and horticultural crops. Their success in all types of biosphere is related to their varied feeding mechanism. Approximately 3% of the expected nemic biodiversity have been identified. Nematodes cause damage directly or indirectly to food and fibre production and ultimately affect quality of life. Feeding on plant cells by nematodes deprives the host of nutrients and water for its growth. The conducting tissues responsible for translocation of nutrients and water are often blocked. Recent projections have estimated that in India, nematodes cause about 5% loss in oil seeds, 8% in cereals and pulses, 10% in sugarcane and fruit crops and 12% loss in vegetables. The price of crops lost is estimated Rs. 242 billion annually (Seshadri and Gaur 1999). Continuous cropping of any single crop leads to a gradual increase in nematode density. Five years of monoculture of sugarcane resulted in 20% increase in the initial population to final population year after year (Mehta, 1992).

Control of nematodes is far more difficult than any other kind of pests because they usually inhabit soil and usually attack underground parts of plants. Though, many methods for controlling nematodes are devised, only some have proved effective, others being particularly effective or ineffective. Chemical control of nematodes proved generally effective, but most of these chemicals are general biocides and is highly expensive, often causing health hazards and pollution problems. Generally, the nematode control measures are either corrective (e.g., chemical) or preventive (e.g., cultural).

In principle, nematode management is mainly applied ecology. A large number of abiotic and biotic factors play direct or indirect roles in regulating nematode population densities. Alteration of many of these is inherent in various agronomic practices whether or not intended to affect nematodes. Identification and suitable modification of such practices and exploitation of local manageable environmental factors is the art of an ecologically based nematode management strategy. Usually no single method gives sufficient control of nematodes. Combinations of several practices can be more effective. Supplementary use of drastic measures like nematicides may become economic, but their use is recommended on a limited scale such as for production of nematode – free planting materials and avoidance of nematode attack at early stages of crop. Consideration of ecological factors is essential while applying external components like chemicals and biocontrol agents. Nematode parasites are serious problems in most of the horticultural crops viz., fruits, spices, vegetables, fibres, plantations, medicinal plants and ornamentals.

1. Fruit Crops

Plant parasitic nematodes form an important limiting factor in production of fruit crops all over the world. Roots damaged by nematodes are not efficient in the absorption and utilization of moisture and nutrients in soil resulting in reduced functional metabolism

The investigations carried out during the last fifty years throughout the world have further revealed that these tiny organisms are not only the major limiting factors in our horticulture economy but they are also the basic cause in most plant diseases of complex nature. The nematodes have also been demonstrated to transmit a number of nepoviruses in fruit trees.

Banana

The most detrimental nematode involved in destruction of primary roots, disrupting the anchorage system and resulting in toppling of plant is the burrowing nematode, *Radopholus similis*. Some species of *Pratylenchus* and *Helicotylenchus multicinctus* are also seen in the rhizosphere. The sedentary endoparasites like *Heterodera oryzicola*, *Meloidogyne* spp. and *Rotylenchulus reniformis* also parasitize banana roots. In addition to above parasites, there are 146 species belonging to 43 other genera of nematodes associated with *Musa* spp. throughout the world. Nematode parasitism in banana plants is characterized by simultaneous infestation by several species.

Burrowing Nematode - *Radopholus similis* (Cobb, 1983) Thorne, 1949: The burrowing nematode, *R. similis* causes root rot of banana (Blake 1961, 1966) and is responsible for 30.76 to 41% yield loss in banana (Vadivelu *et al.,* 1987; Rajagopalan and Naganathan, 1977; Nair, 1979; Reddy *et al.,* 1996).

Physical methods

To avoid introducing inoculum into a new plantation, sets (suckers) may be disinfected by heat therapy. The pared suckers can be disinfected by dipping them in hot water at 55^0 C for 25 minutes without causing significant losses. Venkitesan and Charles (1983) observed that dipping of banana suckers in aldicarb (0.1%) solution effectively reduced burrowing nematode population, besides increasing plant height, bunch weight, fingers/bunch and inducing early flowering.

Paring and parlinage with carbofuran at (1.2 g a.i / sucker) recorded low incidence of *R.similis* and the highest yield (Rajagopalan and Naganathan, 1977). Soil application of fenamiphos at (3 kg a.i / ha) was effective in reducing nematode population besides, increasing plant height, plant girth, number of leaves and yield (Reddy *et al.*, 1991)

Certified seed material : Certified nematode-free (suckers) obtained by paring, heat treatment or selection is to be recommended while establishing new plantations.

Cultural methods

Number of nematodes in soil may be decreased either by fallowing, crop rotation or inter cropping. The banana biotype of *R. similis* has a narrow host range and limited survival period in soil in absence of host. Therefore, destruction of the infected area either mechanically with a bulldozer or by injection of a herbicide into the pseudostem and replanting the area with nematode free sets is a practical approach to rehabilitate infested plantations. After destroying infested plantations, planting of cover crops like *Panicum maximum* var. *trichoglume* and *Phaseolus atropurpureas* did not support *R. similis* population. Planting sugarcane in soil following destruction of bananas eliminated *R. similis* after 10 weeks (Loos, 1961).

Crop rotation

Rotation of banana with paddy, sugarcane, green gram, cotton or turmeric suppressed nematode population and increased yield (Rajendran *et al.,* 1979). Crop rotation with rice, sugarcane, green gram or cotton could suppress nematode population and increase yield of banana. Also, intercropping of banana with *Crotalaria juncea* reduces *R. similis* population besides better growth and yield of banana in India (Venkitesan and Charles, 1983).

Mulching

Mulching with black polythene at 20% moisture depletion recorded the highest yield with low population of *R. similis* (Vadivelu *et al.,* 1987; Sundararaju and Sathyamurthi, 2005).

Intercropping

Intercropping with sun hemp, coriander, marigold, or lucerne reduced nematode population and increased yield (Vadivelu *et al.,* 1987; Sundararaju, 2005). Banana intercropped with *Crotalaria* reduced *R. similis* population in root and increased plant growth and yield (Charles *et al.,* 1985).

Fallowing

Fallowing for a period of three months after banana effectively suppressed burrowing nematode population. Flood fallowing for about five months destroyed not only the *Fusarium* but also the burrowing nematodes (Rajendran *et al.,* 1979).

Organic amendments

Application of neem cake at 800g/plant (two split applications) or neem coated urea at 110g/plant reduced population of burrowing nematode and also increased yield of banana (Vadivelu *et al.*, 1987).

Biological methods

Generalo and Davide (1986) reported effect of *Pacilomyces lilacinus, P. anatolicum* and *Arthroboyrys cladodes* for biological control of *R.similis* .*A*pplication of arbuscular mycorrhizal Fungus (AMF), *Glomus fasciculatum* suppressed multiplication of *R.similis* in root and soil significantly. Banana plants inoculated with mycorrhiza also recorded higher N, P, K, Ca, Mg, Sugars, Phenols and Amino acids in their root system suggesting AMF as a potential biocontrol agent in management of *R.similis* (Reddy *et al.,* 2001). The effect of fungus *P. lilacinus* on *R. similis* management was reported by Devarajan and Rajendran, 2001.

Plant resistance

Banana Cvs / hybrids viz., Kadali, Octoman, Then Kunnan, Kunnan, Dulmal, Bhog, Malbhug, Tongat, Pedalimoongil, Pisang Lilin, Anil Vazhai, Velatothan, Dudhsagar, Chinachampa, Chinia Kurlan, Benid, Ayiramkapoovan, Pisang Seriby, Vennettukunnan, Beecha, Brilechinia Local Dwarf, Manick Champa, Singhlal, Pey Kunnan, Elakkibale Pakea, Annaikomban, Bodles Altafort and Hybrid – 74 (Mathi X Pisang Lilin) were tolerant / resistant to *R.similis* (Charles *et al.,* 1983; Reddy *et al.,* 1989; Ravichandra and Krishnappa 1985; Vadivelu *et al.,* 1987).

Integrated method

Eco friendly integrated management methods using biological agents and biocomponents were reported by many workers (Channabasappa *et al.,* 1995; Charles *et al.*, 1996). Paring + hot water treatment of suckers at 50^0 C for 15 minutes + application of neem cake significantly reduced nematode population and increased yield in banana var. Nendran. Carbofuran (0.5 g a.i / plant) was also recommended along with above treatment combination in endemic areas (Sheela *et al.*, 1996 and Shrinivasa *et al.,* 2005).

Integration of neem cake with AMF, *G. mosseae* was the most effective in reducing burrowing nematode population in soils and roots, while karanj cake with *G. mosseae* gave maximum increase in plant growth characters and fruit yield of banana. Maximum mycorrhizal root colonization and number of chlamydospores of *G.mosseae* were in neem cake amended soil (Reddy *et al.,* 1997). They also reported that integration of neem cake at 400 g/plant with *Pasteuria penetrans* at 100 g soil (300 spores) was effective in reducing *R.similis* population both in soil and roots of banana by more than 50 % and increased plant growth parameters. Integration of karanj cake with *G. mosseae* gave maximum banana fruit yield, while neem cake + *G. mosseae* was the most effective in reducing *R. similis* population both in soil and roots (Reddy *et al.,* 1997). Management methods viz., paring and parlinage of suckers with *Trichoderma harzianum* + *Paecilomyces lilacinus* and application of farm yard manure @ 2 kg or vermicompost enriched with *T. harzianum* + *P. lilacinus* (2 kg) at the time of planting and afterwards three times at same dose at three months interval was recommended by Rao, 2005.

Root lesion nematode - *Pratylenchus coffeae* (Zimmerman, 1898) Goodey, 1951

The root lesion nematode, *P. coffeae* is an important nematode pest next only to burrowing nematode. Eight species of *Pratylenchus* root lesion nematodes are reported attacking *Musa* spp. throughout the world. Among these species, two are widespread and recognized as damaging pathogens. These are *P. coffeae* and *P. goodeyi. P. coffeae* causes decline of banana plantations.

In India, yield loss due to root lesion nematode in banana cv. Nendran was reported as 44.4 % (Sundararaju and Cannayane, 2003)

Plant resistance

Banana Cultivars / hybrids *viz.* Kunnan, Vennettu – Kunnan and Hybrid – 74 (Mathi x Pisang Lilin) were tolerant / resistant to *P.coffeae* (Vadivelu *et al*; 1987). Cultivars Nendran and Kalyan Bale were highly susceptible to *P. coffeae* (Sundararaju, 2002). Multiplication of nematodes was more favoured by Nendran than Poovan. This suggests that cv. Nendran is more suscelptible to both nematodes when compared to Poovan (Sundararaju *et al.*, 2002).

The biochemical and molecular changes associated with resistance reaction of banana against root lesion and root knot nematodes on five banana varieties *viz.* Nendran (AAB), Robusta (AAA), Pisang Javi Buaya (AA), Karthobiumthan (ABB) and *Musa balbisiana* (BBB) were reported by Sundararaju and Pantisuba (2006). They reported that the highest protein concentration and increased peroxidase activity were observed in cvs. Nendran and Robusta whereas minimum were observed in cvs. *Musa balbisiana*, Karthobiumtham and Pisang Jari Buaya. The activity of phenol oxidase was reduced in infected plants of cvs. Nendran and Robusta. The activity increased in cvs. Pisang Jari Buaya and Musa balbisiana. The Phenyl Alanine Lyase (PAL) activity was significantly lower in cvs. Nendran and Robusta compared to other three varieties. The phenolic accumulation was increased by 56% in cv. Nendran after nematode infection whereas there was only 2% increase in Karthobiumtham. The molecular analysis indicated a higher rate of mRNA synthesis soon after nematode infection (Sundararaju and Pantisuba, 2006).

Cultural methods

Significant reduction in root-lesion nematode population was noticed in banana field, where *Tagetes* spp. was grown as an intercrop, whereas maximum nematode population was observed in untreated plants. The yield of banana was also significantly increased (12 kg / plant) when intercropped with *Tagetes* spp. compared to untreated control plant (9 kg / plant) (Sundararaju, 2002).

Botanicals

Sundararaju and Cannayane (2003) found that among different neem formulations tested against banana nematodes, Econeem and Nimbecedine show maximum efficacy in reducing nematode population with increased plant growth and yield of Nendran banana.

Among ten plant extracts tested, *Azadirachta indica* exhibited maximum mortality of *P. coffeae* when exposed to 20 h in the 80 % concentration of plant extract and it was followed by *Vitex negundo* and *Crotolaria juncea* (Sundararaju and Cannayane, 2002a).

Biological methods

An attempt was made to mass-produce nematode egg parasitic fungal bio agent, *P. lilacinus* on banana wastes (leaf, petiole and pseudostem), castor and pungam leaves and their combinations and a standard check with pre-soaked sorghum grains. The highest spore load of 8 x 10^6 *P. lilacinus*/g of substrate was harvested in sorghum grains, banana petiole and castor leaf mixture. These mixtures were very effective in nematode management (Sundararaju and Cannayane, 2002b).

Biocontrol potential of endophytic bacteria *Bacillus subtilis* was tested against nematodes of banana viz. *P. coffeae, R. similis, Helicotylenchus multicinctus* and it enhanced activity of defense enzymes responsible for induction of systemic resistance like peroxidase, polyphenol oxidase, phenyl alanine and ammonia lyase (Jonathan and Umamaheswari, 2006).

Integrated method

Similar management practices as in *R.similis*.

Rice cyst nematode (*Heterodera oryzicola* Rao and Jayaprakash, 1978)

The cyst nematode, *H.oryzicola* was reported by Rao and Jayaprakash (1978) from rice and it was reported from banana for the first time in India from Banana Research Station, Kannara, Kerala during 1983 (Charles and Venkitesan, 1984). It was later reported on banana cultivars Saldathi and Njalipoovan from Goa (Koshy *et al.,* 1987).

H. oryzicola causes considerable damage to banana in certain pockets of Kerala and an initial inoculum level of 100 and 1000 viable cysts/plant could reduce bunch weight by 20.5 to 56.6%. The nematode infection affected quality of fruits by increasing acidity and reducing total sugars (Charles, 1989).

Cultural methods

Fallowing for a period of three months after banana, effectively suppressed nematode population (Rajendran *et al.,* 1979).

Chemical methods

The nematode can be controlled by treatment of suckers (by hot water or with chemicals) before planting. Application of phorate or carbofuran @ 2 g a.i./plant three times, planting, 75 and 165 days after planting was effective (Sheela *et al.,* 1994).

The spiral nematode - *Helicotylechus multicinctus* (Cobb 1983), Golden, 1956

The spiral nematode, *H. multicinctus* is probably the most widespread and abundantly distributed nematode on all bananas next to *R. similis*. It is observed infesting plantains. It is often regarded as the main parasitic nematode on bananas where environmental conditions are suboptimal for the crop in relation to latitude, temperature and rainfall.

This nematode is responsible for 20% loss in fruit yield in India (Reddy and Rao, 2003). According to Vadivelu *et al.,* 1987, it is responsible for 33.83% loss in yield, 55.88% loss in fruits / bunch and delayed fruiting by 134 days. The pathogenicity of *H. multicinctus* on banana, studied in pot culture experiment, revealed that plant growth was significantly reduced at 1250 nematodes / plant (Rajendran and Sivakumar, 1996).

Plant resistance

Banana CVS./hybrids *viz.,* Neyvannan, Rasthali, Wather, Peyan, Kullan, Krishna Vazhai, Karpuravalli, Sirumalai, Co-1, Kunnan, Robusta, Gross Michel, Vennettu Kunnan, Hybrid-94, 100, 106 and 109 (all Matti x Tongat) and Hybrid-74 (Matti x Pisang Lilin) were tolerant / resistant to *H. multicinctus* (Vadivelu *et al.,* 1987)

Root knot nematodes - *Meloidogyne* spp.

Five identified species of *Meloidogyne* are reported from *Musa* spp. in warm tropical and sub tropical areas. In India, *M.incognita, M. arenaria* and *M. javanica* readily infest banana (Jonathan *et al.,* 1999). Patel *et al.* (1998) reported 10.9% yield loss in banana due to root knot nematode. Patel *et al.* (2004) reported that *Meloidogyne* spp. and *Rotylenchulus* spp. were the key nematodes prevailing in most of the banana fields of Gujarat while *Helicotylenchus* spp. and *Radopholus* spp. were in negligible number.

Botanicals

There was no larval emergence from egg masses previously treated with argemone, lantana leaf extracts and neem seed kernel suspension (NSKS) treatment (Patel *et al.,* 2004).

Biological methods

Effect of fungal egg parasite *P.lilacinus* for better management of this nematode was reported by Devarajan and Rajendran (2002). Rao (2005) reported management methods *viz.*, paring and parlinage of suckers with *Trichoderma harzianum* + *Paecilomyces lilacinus* and application of farm yard manure @ 2 kg or vermicompost enriched with *T. harzianum* + *P.lilacinus* (2 kg) at the time of planting and afterwards three times at same dose at three months interval.

Reniform Nematode - *Rotylenchus reniformis* Linford and Oliveira, 1940

Yield loss of 19.3 tonnes was observed in fields where *R. reniformis* was the predominant parasitic nematode.

An integrated management strategy for banana nematodes in different cropping situations can be as follows.

Rotation with alternative crops for two to three years.

- Flooding for eight weeks after having destroyed previously infested banana crop depending on the situation.
- Fallow in absence of banana 'volunteers' for 10-12 months
- Selection of disease free suckers
- Use of *in vitro* produced healthy plants
- Paring diseased tissue from corm / sucker
- Paring and leaving large corms under sun for 14 days or parlinage with cow dung slurry/ nematicides/ neem products/ biocontrol agents like *T. harzianum*
- Immersing corms (suckers) in hot water at 55^0 C for 20 minutes
- Coating corms with nematicides in mud
- Applying nematicide in pits (0.5 to 1 g a.i. / plant)
- Spot applications with granular nematicide formulations (cabofuran or phorate) (0.5 g to 1 g a.i / plant) at 45 days interval in endemic areas up to four months of growth.

Citrus

Out of more than 189 species, belonging to 59 nematode genera, associated with citrus in world, approximately 122 species belonging to 57 genera are reported from India. Although more than ten species of phytonematodes are pathogenic on citrus, the citrus nematode *Tylenchulus semipenetrans* was widely distributed and recognized as economically important. *T. semipenetrans* is undoubtedly the most important nematode infesting citrus in the world. In India, it occurs in all citrus growing areas ranging from sub-tropical hilly tracts to tropical areas. The adult female of this nematode is a sedentary semi-endoparasite causing "spreading decline" disease of citrus. *T. semipenetrans* occurrence on citrus trees was first reported from Aligarh (Uttar Pradesh, India) by Siddiqi (1961).

Physical methods

The bare root treatment at 45° C for 25 min. or 46.7° C for 10 min. can disinfect citrus seedlings from citrus nematode (Bindra *et al.*, 1967). This treatment leaves no adverse effect on seedlings. During summer, exposing soil to hot sun controls nematodes substantially (Seghal *et al.*, 1990).

Cultural methods

Populations of citrus nematode in soil and roots can be reduced by removing old feeder roots before growth flush. This practice is followed by application of farm yard manure. This is a general cultural practice adopted by farmers of Haryana and Punjab. Irrigation of citrus orchards with sewage water reduces citrus nematode population to a very low level (Baghel, 1995).

Soil amendments

Use of castor cake can effectively reduce 50% of citrus nematode population (Singh and Sitaramaiah, 1973). Mobin and Khan (1969) concluded that use of neem, mahua, groundnut and mustard oil cakes can effectively reduce citrus nematode population in citrus rhizosphere.

Botanicals

Application of certain plant parts and their extracts affect nematode population either by killing or paralyzing them. *In vitro* studies, aqueous and methanol garlic (*Allium sativum*) extracts (0.5% concentration) cause complete kill of *T. semipenetans* (Nath *et al*., 1982). The extracts of leaves, stem and buds of *Datura stramonium, Ipomea carnea, Tagetus patula* and *Lawsonia alba* (4 mg/ ml dilution) cause 75-100 % larval mortality of *T. semipenetans* (Kumari *et al.,* 1986). Leaf extract of *Calotropis procera, Cannabis sativa, Chenopodium album, Digera arvensis, Oxalis corniculatus* and *Portulaca suffriticosa* at 1: 80 dilutions, under laboratory condition, gave effective mortality (66.9 to 91.9%) of *T. semipenetans*. Water extracts of *Calotropis procera* and *Nerium oleander* and essential oil of Cymbopogon grasses are toxic to *T. semipenetrans* (Sangwan *et al.,* 1985; Verma *et al.,* 1989). Castor (*Ricinus communis*),Oak (*Calotropis procera*) and Datura (*D. stramonium*) chopped leaves applied @ 20 and 40 g / kg soil in 9 m^2 area, around citrus tree trunk, results in reduction of nematode population and increase in yield (Bhagel, 1995).

Intercropping

Growing intercrops at early stages of citrus crop is very common, which influence nematode population. After the citrus plants attain maturity and develop enough canopies, the possibility of growing intercrops close to feeder root zone becomes negligible. Growing of intercrop does not leave much effect.

Plant resistance

Use of resistant citrus root stock offers cost effective method of control of citrus nematode. Varietal screening studies against citrus nematode revealed that trifoliate orange (*Poncirus trifoliatus*), Carrizo and Trover are highly resistant. Cleopatra mandarin Australia, Cleopatra mandarin Morrocco, Smooth lemon Assam and Saoner were moderately resistant. Rangpur lime, used as root stock for Blood Red, Campbell Valencia and Hamlim scions, proved resistant to citrus nematode. Hybrid citrus rootstocks developed by crossing Rangpur lime (*Citrulimon* with *P.trifoliatus* viz, CRH-3, CRH-5 and CRH-41) were highly resistant (Reddy *et al.,* 1987). *C. senensis* x *P.trifoliatus* var.Savage citrange was rated as moderately resistant to citrus nematode (Bhagel, 1995).

Arbuscular mycorrhizal fungus, symbiotic to citrus nematode on citrus, proved beneficial to crop. Varying degree of infection of *G. mosseae* was recorded from six citrus root stocks having different scion combinations. Higher number of mycorrhizal vesicles was recorded from root stocks budded with grapefruit scions. This is an indication that this symbiotic fungus can readily multiply on citrus. *G. mosseae* stimulates growth on *Citrus jambhiri* seedlings and its presence suppresses growth retarding effect of citrus nematode (Baghel *et al.*, 1990).

Biological methods

Predaceous fungi, *Arthrobotrys* spp., *Dactylaria* spp. and *Dactylella* spp. reproduce on citrus nematode and control its population build up on citrus. Application of *P. lilacinus* proved effective in reducing citrus nematode population and increasing fruit size (Jatala, 1986). *P. lilacinus* is a potential biocontrol agent for controlling citrus nematode (Mani *et al.*, 1989) . *P. lilacinus* @ 4g in combination with carbofuran 1 kg ai / ha gave maximum growth parameters and effectively reduced citrus nematode population in acid lime (Reddy *et al.*, 1990). Arbuscular mycorrhizal fungus, symbiotic to citrus nematode on citrus, proved beneficial to crop. Varying degree of infection of *G. mosseae* in orchard was recorded from six citrus root stocks having eleven scion combinations. *G. mosseae* is well known for producing stimulatory effect on growth of *C. jambhiri* seedlings and its presence in roots suppresses growth retarding effect of citrus nematode (Baghel *et al.*, 1990).

Chemical methods

Pre-planting chemical application is not common in India. Post-plant application of chemicals is the major mode of approach for control of nematodes associated with citrus orchard. In early seventies, DBCP, DD, dimethoate, metham sodium and dichlorofenthion were tested and used for controlling citrus nematode population in orchards. The results of DBCP were the best; the doses used varied from 10 to 25 ppm concentration. This chemical gave 73.0 to 98.0 % nematode kill and consequent increase of 41 to 100% yield. In some cases, the nematode population build up was held up for 15 months after the treatment. Temperature and moisture at the time of application of these treatments were considered very important. The efficacy of carbofuran (4 to 6 kg a.i./ha), benfurocarb (4 to 6 kg a.i./ ha) and combination of neem cake (1 kg/ plant + carbofuran @ 2 kg a.i./ha) was tested in the field experiments at different locations of the country under ICAR Coordinated Research Project. Treatment with carbofuran reduced 50.3 to 67.5% nematode population and increased the yield (42.0-60.6%); benfurocarb reduced 37.2 to 60.3% nematode population and also increased yield (31.8 to 57.2%) and neem cake carbofuran combination reduced 33.0% nematode population and increased yield (33.8%). The cost benefit ratios for these treatments varied from 1:4 to 1:7 (AICRP, Nematodes) Report, 1991).

Integrated methods

Different oil cakes like neem, castor, karanj when integrated with parasitic fungi *P. lilacinus* (Reddy *et al.*, 1991), *Verticillium chlamydosporium*, *V. lecani* (Reddy *et al.,* 1996a), antagonistic fungus, *Trichoderma harzianum* (Reddy *et al.,* 1996b) and VA mycorrhiza, *Glomus fasciculatum* (Reddy *et al.,* 1995) gave better plant growth and more nematode reduction in acid lime under glass house conditions. Neem cake (1 kg / plant) combined with Carbofuran (0.5 g.a.i / plant) reduced 33% nematode population and increased the yield by 33.8% (Baghel, 1995). Integration of *P. lilacinus* with carbofuran was effective in increasing plant growth parameters of acid lime and in reducing citrus nematode population both in soil and roots (Reddy *et al.,* 1996c)

Following integrated management strategy (s) against major nematodes like

T. semipenetrans and *Meloidogyne* spp. can be recommended in citrus depending upon conditions of crop (new planting / existing crop).

- Selection of nematode free planting materials through management of nematodes at nursery level.
- Denematization of soil mixture while producing seedlings of sweet orange or lemon with carbofuran or phorate and addition of neem cake and bioagents like *T. viride* or *P. lilacinus* for sustaining soil condition.
- For managing nematodes in the main field, apply neem cake @ 250 g / plant + carbofuran @ 0.5 g a.i. / plant while planting and afterwards neem cake @ 250 g / plant once in six months for a period of four years.
- For existing plantations, apply neem cake (250 g / plant) once in six months.
- If the bioagent formulations are available, apply farm yard manure or vermicompost enriched with bioagent formulations *P. fluorescens* or *P. lilacinus* at the time of planting + four months interval for four years.
- In infested existing plantations, apply carbofuran or phorate (0.5 g a.i / plant + *P. fluorescens* or *P. lilacinus)* two months after chemical application and apply the bioagents at four months interval regularly for two years.

Grape

Several nematode species are associated with grapes but the major ones belong to genera *Meloidogyne*, *Pratylenchus* and *Xiphinema*. Other nematodes associated with this crop are *T. semipenetrans, R. reniformis, Paratrichodorus minor* and *Longidorus attentuatus.* Besides, several new species of nematodes are recorded on grapes from different parts of India.

Physical methods

Hot water treatment is used to disinfect planting material but accurate timings and suitable uniform temperature (125^0 F) are very important for obtaining good results.

Plant resistance

Use of nematode resistant root stocks has an added advantage over methods of control. The cost factor is reduced to minimum. In grapevine, Dogridge (*Vitis champini*), Salt creek (*V. champini),* 1613 (*V. solanis* x Othello), Harmony (1613x Dogridge), St. George (*V. rupestris*), AxR1 (Aroman x Ganzin 1) are well known nematode resistant root stocks. Banquabad Cardinal, Early Muscat, Jasbeli, Loose Perlette and Reisling (all commercial varieties) are resistant to *M. javanica*.

Chemical methods

Application of carbofuran, benfurocarb or phorate at (6 kg a.i./ha), in *M. javanica* infested vineyard gave 63-68% nematode reduction and improved yield (120-133%). Carbofuran at (13 g/sq m) at bud breaking stage is recommended for controlling *M. javanica.*

Integrated management

Concomitant infection of *M. arenaria* and *G.fasciculatum* on grapevine cv. Thompson Seedless negated the growth stimulatory effect of AMF fungus and reduced growth suppressive effect of nematode on host. The fungus stimulated multiplication of nematodes whereas nematodes had no effect on fungus spore production. In combined inoculation of *G. mosseae* and *M. javanica* on grapevine cv. Perlette, Mycorrhizal fungus adversely affected nematode multiplication. Mycorrhizal colonization was not affected by nematodes. Growth suppressive effect of nematode was neutralized by fungus.

Papaya

Root knot nematode, *M. incognita* and reniform nematode, *R. reniformis*, are the major nematode species associated with papaya. Mohanty *et al.* (2000) reported that neem cake and carbofuran treated plants supported minimum nematode population and root knot index in papaya plants. Rao (2005) suggested a package for management of nematodes in papaya starting from preparation of soil mixture for producing seedlings. It includes preparation of soil mixture for producing seedlings by mixing neem cake (50 kg + carbofuran or phorate @ 5 kg + *P. fluorescens* @ 2 kg + P. *lilacinus* @ 2 kg/tonne). This soil mixture can be used for producing seedlings of citrus, applying neem cake (20 g/plant + carbofuran @ 20 g/plant) while planting. Afterwards apply neem cake (250 g/plant) once in four months for a period of two years, applying neem cake @ 250 g/plant once in four months for standing crop. Application of 2 kg farm yard manure or vermi -compost enriched with IIHR bioagent formulations (*P. chlamydosporia* + *P. fluorescens*) at the time of planting and afterwards applying same dosage at four months interval are recommended.

2. Spices

Among large number of plant parasitic nematodes reported in association with spice crops, very a few like *Meloidogyne* spp., *R. similis*, *R. reniformis*, *Helicotylenchus* spp.

are of common occurrence and economic importance in spice crops. The yield losses varied from 38.5 to 64.6% (Mohandas and Ramana, 1991)

Pepper

Plant parasitic nematodes are extremely important limiting factors in black pepper cultivation. The damage due to nematode is more severe in extensive production systems like black pepper nurseries. Although a sizeable number of plant parasitic nematodes are recorded on black pepper, only burrowing nematodes (*R. similis*) and root knot nematodes (*Meloidogyne* spp.) are regarded economically important.

Being a perennial crop trailed on live standards, any management measure needs to be repeated every year under Indian conditions. The perennial multi-cropping systems are ideal situations where burrowing nematode multiplies and causes heavy damage to all the susceptible crops. The high density multi- species cropping pattern does not permit use of nematicides, as most of the crops are export - oriented and tender berries are consumed without any processing or cooking or as pickles. The situation is further complicated where arecanut and coconut are used as live standards as they are very good hosts of *R.similis.*

Cultural methods

Mulching and bio-fumigation ameliorate symptoms of slow-decline/pepper yellows. Addition of chopped leaves of *Glyricidia maculata* (10g / kg of soil) as green manure reduced populations of *R.similis* and increased plant growth (Jasy and Koshy, 1992). Application of botanicals like neem oil cake reduced root-knot nematodes (Ramana *et al.,* 1982).

Among the large number of live standards used in black pepper gardens, coconut and arecanut are good hosts of *R.similis* while several of these are highly susceptible to root knot nematodes too. However, *Garuga pinnata* Roxb. and *Macaranga indica* Wight are not susceptible to root knot nematodes and popular live standards, *Erythrina indica* Lank. and *G. sepium* Walp are less susceptible. Several intercrops like banana, ginger and turmeric and a large number of weeds are also susceptible to these nematodes (Koshy *et al*., 1977; Ramana, 1986). Large numbers of weeds found in pepper gardens are recorded as hosts of root-knot nematodes.

Plant resistance and tolerance

A number of black pepper germplasm accessions, including wild types, were screened against *M.incognita* and *R. similis* by several workers (Venkitesan and Sethi, 1978; Koshy and Sundararaju 1979; Jacob and Kuriyan 1979; Mohandas and Ramana, 1983; Ramana and Mohandas, 1986, 1987b; Koshy, 1987b). Wild collection, Vittal No.430, *Piper hymenophyllum* and *P. attenuatum* recorded the least (less than 30%) root reduction and minimum reproduction of *R.similis*. No resistance or tolerance was found to nematode on screening a number of cultivated and wild germ plasm, intercultivar hybrids and open pollinated seedlings, except for *P. colubrinum*, which is now widely used as a root-stock to graft cultivated pepper plants (Ramana *et al.*, 1987b; Ramana, 1992). A total of 101 cultivars, 74 accessions of wild *Piper* sp. and 140 intercultivar hybrids, were screened against *M.incognita* of which one cultivar, CLT-P-812 was resistant (Ramana and Mohandas, 1986, 1987b). This accession was later released as 'Pournami' for cultivation in root knot infested areas (Ravindran *et al.*, 1992). Some of the wild related species of *Piper* are also resistant to root knot nematodes (Ramana, 1992). The biochemical changes in this variety consequent to nematode infection are studied (Eapen *et al.*, 1999).

Chemical methods

Root-knot infestation in black pepper nurseries is a serious problem in several nurseries. Fumigation of nursery potting mixture was effective in checking infestation (CPCRI, 1974, Koshy, 1986; Mohandas and Ramana, 1987). A number of pesticides are effective in reducing nematode populations on black pepper. Aldicarb / carbofuran / phorate at (3g a.i / vine) applied in May / June and again in September / October resulted in the remission of foliar yellowing and reduction in nematode populations. Among the above three nematicides, phorate was superior (Ramana, 1986; Mohandas and Ramana, 1987; Lokesh and Gangadharappa, 1995; Sundararaju and Sudha, 1998).

Biological methods

By fortifying potting mixtures with biological control agents, nematode- free rooted cuttings can be raised. A number of organisms are tested and found effective in suppressing root knot nematodes. Promising among these are *P. lilacinus* (Ramana, 1994; Sosamma and Koshy, 1997), *Pochonia chlamydosporia (Verticillium chlamydosporium)* (Sreeja *et al.*, 1996), *Pasteuria penetrans* (Sosamma and Koshy, 1997), *Bacillus* spp. (Sheela et *al.*, 1993; Sheela and Venkitesan, 1992; 1994 a and b; 1995 a and b) and *Pseudomonas fluorescens* (Eapen *et al.*, 1997). A number of rhizobacteria, antagonistic to root knot nematodes are isolated recently (Beena *et al.*, 2001). Black pepper plants pre inoculated with vesicular arbuscular mycorrhizal fungi like *G. fasciculatum, G. etunicatum, G. mossae, Gigaspora margarita* recorded a significant increase in growth even in presence of root knot nematodes (Anandaraj *et al.*, 1991; Sivaprasad *et al.*, 1990; 1992). Endophytic bacteria that suppress root knot nematodes also were isolated from black pepper (Aravind *et al.*, 2004).

There are only very a few successful attempts to control *R. similis* by use of any fungal bioagents, probably due to migratory endoparasitic nature of this nematode (Geetha, 1991; Ramana, 1994). The mycorrhizal fungus, *G. fasciculatum* suppressed burrowing nematode infestation (Anandaraj *et al.*, 1996c). Recently, rhizobacteria that suppressed *R. similis* infesting pepper were identified in green house studies (Beena *et al.*, 2003). Endophytic bacteria that suppress root knot nematodes also were isolated from black pepper (Aravind *et al.*, 2004).

The following integrated method can be suggested for management of nematodes in pepper:

- Use nematode free rooted cuttings for raising new plantations
- For the nematode infested vines, apply phorate/carbofuran (1 g a.i/vine) twice a year. The first dose is to be given in May-June at onset of monsoon and second dose in last week of October or early November. The chemicals are to be applied around the vines in shallow basins and incorporated into soil
- Root knot nematode can be effectively managed by application of talc based formulation of *Bacillus macerans* (10 g/vine) in basins at time of planting of vines or just before monsoon period in established plantations.

Cardamom

The major nematode problem is caused by the root- knot nematodes,

Meloidogyne spp., although the lesion nematode, P. *coffeae* and the burrowing nematode, *R. similis* cause root rotting. (D'Souza *et al.*, 1970; Kumar *et al.*, 1971; Khan and Nanjappa, 1972; Viswanathan *et al.*, 1974; Sundararaju *et al.*, 1979b). Reniform nematode, *R. reniformis*, is also recorded on cardamom. (Eapen, 1995a).

Even though there is general awareness about nematodes infesting cardamom, the recommended control measures are not being followed by many growers.

Cultural methods

It is advisable to change nursery sites every year, but this is not always practicable in view of difficulties involved in getting suitable sites having facilities for irrigation. Disinfestations of the nursery beds need to be carried out every year. Disinfestation of nursery beds with methyl bromide (500 g/10m^2) is effective in controlling root-knot infestation in both primary and secondary nurseries (Ali and Koshy, 1982).

A large number of annual weeds present in cardamom plantations and the common shade trees, *Erythrina indica* and *E.lithosperma*, are hosts of root-knot and help in the build up of nematode populations (Muniappan, 1993). No resistance to root-knot nematodes is found and the popular cardamom cultivars, Malabar, Mysore and Vazhukal are all susceptible (Hegde *et al.,* 1993; Eapen, 1995).

Chemical methods

Nursery beds drenched with fenamiphos significantly reduced root-knot nematodes (Ali, 1986). Aldicarb, carbofuran, phorate at 5, 10 or 15 kg a.i. /ha respectively, are applied in primary nurseries of cardamom for control of *M. incognita*. None of the nematicides treatment totally prevented nematode infestation but there was significant reduction in root-knot densities. Aldicarb at the very high level of 15 kg a.i./ha reduced nematode numbers by 90% (Ali, 1987). Application of aldicarb/carbofuran/phorate at 5g and 10g a.i. /plant and neem oil cake at 500g and 1000g/plant twice a year increased yield of cardamom plants infested with *M. incognita* from 47 to 88%. Maximum yield was obtained by the addition of neem oil cake at a rate of 1000g/plant followed by 500 g/plant (Ali, 1984). In another study, application of phorate (2.5-5.0 g a.i./plant) reduced nematode population and increased the yield by more than 40% (Eapen, 1995).

Biological methods

Cardamom nurseries are ideal for practicing biological control. There are reports that *G. margarita* and *G. fasciculatum* suppressed *M. incognita* infestation and enhanced growth and vigour of seedlings (Thomas *et al.,* 1989). *P.lilacinus* reduced root-knot nematodes by 48.5 to 57% in pot culture studies and by 19.7% in field studies (Eapen, 1995b; Eapen and Venugopal, 1995). Some native isolates of *T. harzianum* and other *Trichoderma* spp. are potential antagonists of root-knot nematodes in laboratory, greenhouse and also in cardamom nurseries. (Eapen *et al.,* 2000).

Integrated method

Eapen and Venugopal (1995) reported that integration of soil solarization and application of *P.lilacinus* and *Trichoderma* spp. significantly suppressed *M.incognita* population by 58.30 to 86.90% resulting in improvement in growth and quality improvement. Application of *G. fasciculatum* was very effective in spices like ginger, turmeric, cardamom and pepper. (Sivaprasad and Sheela, 1998).

Nursery management with biological control agents or chemicals is the only option for controlling nematodes in cardamom.

Ginger

Plant parasitic nematodes belonging to 17 genera were reported on ginger (Sundararaju *et al.,* 1979b, Rama and Dasgupta, 1985; Kaur, 1987; Ramana and Eapen, 1998). The most important parasites are *Meloidogyne* spp., *R.similis* and *P.coffeae.* In Kerala, *M.incognita, R.similis* and *Pratylenchus* spp. were the major nematode species found in rhizhosphere of ginger (Mammen, 1973; Charles, 1978; Charles and Kuryan, 1982;Sheela *et al.,* 1995). Several species of root knot nematodes viz. *M. arenaria, M.hapla, M.incognita* and *M. javanica* are reported as parasites of ginger in various countries. Other nematodes observed *include R.reniformis, H. multicinctus, Tylenchorynchus* spp., *Hoplolaimus indicus, Criconemoides* spp. and *Xiphenema* sp. (Sheela *et al.,* 1995).

Integrated method

Neem cake followed by carbofuran treated plots showed minimum gall index in ginger (Mohanty *et al*., 1992). Sheela *et al.* (1995) reported that application of neem cake (2.50 t ha^{-1}) at the time of planting and carbofuran (1 kg ai ha^{-1}) forty five days after planting were effective in reducing *M. incognita* population in soil and root, root-knot count and increasing the yield of ginger. *G. fasciculatum* was very effective in controlling root-knot nematode in ginger, cardamom and pepper (Sivaprasad and Sheela, 1998).

Mulching with green leaves (neem or eupatorium or panal or glyricidia) and application of organic cakes (neem cake or mustard cake) at the time of planting is the best option for management of nematodes in ginger. In endemic areas, application of carbofuran granules (1kg ai/ha) can be recommended at planting.

Turmeric

A number of species of plant parasitic nematodes are reported in association with turmeric in India (Nirula and Kumar, 1963; Sundararaju *et al.*, 1979b; Dasgupta and Rama, 1987; Gunasekharan *et al.*, 1987; Haider *et al.*, 1998a; Rama, 1987; Mani and Prakash, 1992; Routaray *et al.*, 1987b; Bai *et al.*, 1995) of which *Meloidogyne* spp., *R. similis* and *P. coffeae* are of economic importance. Ayyar (1926) first reported root knot nematode infestation in turmeric. Two species of root knot nematodes, *M. incognita* and *M. javanica* are reported on turmeric, but most investigations are concerned with *M. incognita.*

Cultural method

Immersing turmeric rhizomes in hot water at 55° C for 10 min. or 45° C for 50 min. kills *M. incognita* inside the rhizomes (Chen *et al.*, 1986) and this could be used for establishing nematode-free multiplication plots but is unlikely to be economic for large scale field use.

Plant resistance

The cultivars and breeding lines 5379-1-2, 5363-6-3, Kodur, Cheyapuspa 5335-1-7, 5335-27, Ca-17/1, Cli-124/6, Cli-339, Armoor, Duggirala, Guntur-1, Guntur-9, Rajampet, Sugandham and Appalapadu are resistant to *M. incognita* (Mani *et al.*, 1987; Gunasekharan *et al.*, 1987). The species *C. zedoaria* is more resistant to *M. incognita* than *C. domestica* in China (Chen *et al.*, 1986). In Andra Pradesh, India, the high yielding varieties like PCT 8, PCT 10, Suguna, and Sudharshana were free from root knot nematode infestation (Rao *et al.*, 1994). Eight turmeric accessions (Acc. Nos. 31, 82, 84, 142, 178, 182, 198 and 200) were resistant to root knot nematode (Eapen *et al.*, 1999).

Chemical methods

Application of DBCP (15 l a.i./ha) 15 days prior to planting results in yield increase of 253-270% compared with 59-187% increase in yield with application of phenamiphos at (2.5 kg a.i./ha) one day before planting (Patel *et al.*, 1982). Aldicarb and carbofuran applied at (1kg a.i./ha) increase yield by 71% and 68% respectively over control, with a cost benefit ratio of 1:6 in aldicarb and 1:2 in carbofuran treatments (Gunasekharan *et al.*, 1987). Carbofuran (4 kg a.i./ha) applied in rows to 4-months-old turmeric crop resulted in a 81.6% reduction in root knot nematode population against 45% increase in untreated plots (Mani *et al.*, 1987). Similarly, application of carbofuran or phorate (@1kg a.i./ha) reduced root knot nematodes (Haidar *et al.*, 1998b).

Biological methods

Biocontrol agents like *V. chlamydosporium, P. lilacinus, Fusarium* sp., *Aspergillus* sp., and *Scopuloriopsis* sp. controlled root knot nematodes in field trials (Ramana *et al.,* 2002).

3. Masticatory Crops

Betelvine

About 38 species of plant parasitic nematodes are associated with betelvine in India. The most common and serious nematode problem is the root-knot nematode, *M. incognita,* which prevails almost in all the states followed closely by *R. reniformis* and *H. indicus. H. microcephalus, H. incisus, H. dihystera, M. javanica, Tylenchorhynchus indicus, Pratylenchus coffeae, P. zeae and Radopholus similis.* are more localized. According to Koshy and Bridge (1990) *M. incognita, R. reniformis and R. similis* are the principal nematode pests of betelvine in India.

Nematode management in betelvine garden, as a whole is a difficult proposition because of the fact that the produce is a green leaf (chewable) harvested at very short intervals (a week to 10 days) which does not permit use of toxic chemicals.

Cultural method

In new plantations, cultivation of non-host crops like Tagetes, sesamum and groundnut, before planting suppress root-knot nematode populations (Ray, 1979).

Organic amendments

Oilcakes of neem, castor, karanj or groundnut (1tone/ha), and poultry manure, saw dust or green leaves of calotropis at (2 tones/ha), when incorporated into old betelvine gardens suppress populations of root-knot and other nematodes and increase yield. (Jayadil *et al.,* 1985; Siva Kumar and Marimuthur 1986; Acharya and Padhi 1988; Sitharamaiah *et al.,* 1993).

Botanicals

Large number of products mostly of organic origin like azadirachtin, limonoid, nimbidine, thionamone, terthienyls, cucurbitacins and L-chaconine are toxic to nematodes. Avermectins produced by *Streptomyces avermitilis* are also very effective against plant nematodes (Sasser *et al.,* 1982).

Antimetabolites like maleic hydrazide, growth regulators like gibberelic acid and antimitotic agents like morphactins are also effective against endoparasitic nematodes. Aminoacids like DL-alanine and DL-seriene (Krishna Prasad and Shetty, 1974) and many homeopathic drugs (Ray and Pradhan, 1985) also inhibit growth of root-knot nematodes. These products need to be further investigated for commercial utilization.

Cultural methods

Nema (2001) reported that in betel vine cv. Bangla Budagar, application of neem cake (2 t / ha) resulted in the highest reduction in *M. incognita* population (62.11%) and number of galls (57%).

Biological methods

Large number of potential biological control agents belonging to diverse groups are listed acting against plant parasitic nematodes. Among this, the potential one is AMF, *G .fasciculatum* associated with betelvine reported by Sitaramaiah and Maithi (1993). But large scale culture and release of natural biological control agents have practical and economic limitations but once established, it will be the best medium of natural environmental resistance.

Hazarika *et al.* (2000) found that at 4 g Kg^{-1} soil inoculum level of *P. lilacinus*, the plant growth and yield of betel vine increased and root knot infestation decreased. Application of *P. lilacinus* inoculated neem cake (10 g / vine) at 60 days interval significantly reduced root knot nematode infestation and increased vine growth and leaf yield in betel vine (Jonathan *et al.,* 2000).

Integrated method

Integrated nematode management strategy is suggested by (Ray, 1995) which includes application of locally available oil cakes and other organic products, application of biological control agents like *P. lilacinus*, AMF and strictly enforcing statutory phytosanitary regulation in case of *R. similis*.

4. Vegetables

Investigations on parasitic nematodes on vegetables in India date back to 1933 when Ayyer reported root knot nematode *Meloidogyne incognita (Kofoid and White) Chitwood* on vegetable from Kerala, the Southern part of India. Jones (1961) detected presence of potato cyst nematode, (*Globodera rostochinensis)* in Nigeris as a pest of great economic importance.

Plant parasitic nematodes cause direct and indirect damage on vegetable crops. The avoidable yield losses due to plant parasitic nematodes in vegetables comparing the protected with unprotected crop, the magnitude of these losses range 28 to 48% in tomato, 26 to 50% in brinjal, 18 to 33% in chillies, 6 to 90% in okra and 38 to 47% in bitter gourd (Reddy 1989 and 1999). Effect of different inoculum levels of *M. incognita* Race 2 on onion (*Allium cepa*) revealed that 10 juveniles / 500 cc soil could cause significant reduction in weight of onion bulbs (Khan, 2003b).

Indirectly many plant parasitic nematodes increase severity of fungal and bacterial diseases. Some of the plant parasitic nematodes act as a vector for carrying plant diseases. Only species of Dorylaimoid genera, *Xiphinema, Longidorus, Paralongidorus, Paratrichodorus* and *Trichodorus* have thus far been implicated as vectors.

Tomato

Physical methods

Soil solarization

Solarization at 44°C at 10 cm soil depth resulted 100% control of *M. incognita* in tomato nursery (Herrera *et al.*, 1999). Reddy *et al.* (2001) reported that soil solarization with clear transparent sheet for six weeks during hot summer showed an increase in soil temperature (8°C) resulting in significant reduction in population density of *M. incognita* (85.80%) and *Pythium aphanidermatum* (Edson) Fitz. (85.40%) in tomato. Soil solarization using 100 gauge Low Density Poly Ethylene (LDPE) sheet for 15 days was very effective in reducing the nematode population (35%) in tomato nursery (Sheela *et al* 1999; 2002; 2006).

Cultural methods

Organic amendments

Tomato fields amended with oil cakes of neem, groundnut, mustard or mahua showed reduced population of nematodes like *Hoplolaimus indicus* (Sher, 1963), *Tylenchorhynchus brassicae* (Siddique, 1961), *Helicotylenchus* sp. and *M. incognita* (Siddiqui *et al.*, 1976). Vijayalakshmi (1986b) reported that soil amended with cakes of *Shrea robusta* Gaertn f. and *Calophyllum inophyllum* L. resulted in slow egg hatching of *M. incognita*.

Alam (1989) observed that soil amendments like hornmeal, bonemeal and oil seed cakes of mahua, castor, mustard, neem and peanut effectively inhibited *T. brassicae* on their respective hosts (egg plant, chilli, okra, cabbage and cauliflower) and also improved plant growth. In tomato, Goswami and Meshram (1991) observed 50% reduction in root penetration of juveniles of *M. incognita* in soils amended with cakes of mustard and karanj.

Botanicals

Soil mulching with flax, lucerne or orchard grass residue resulted in 77 to 99% reduction of *M. incognita* induced root galling of tomato (Johnson and Curl,1972). Goswami and Vijayalakshmi (1986) reported that extracts of *Andrographis paniculata* (Burm. F.), *Calendula officianlis* L., *Enhydra fluctuam* and *Solonum khasianum* C.B. Clarke reduced root galling by *M. incognita* on tomato transplants.

Pradhan *et al.* (1989) showed that oils of chalmagra, neem and karonj at one half to one eighth dilution given through seedling root dip method were effective against nematodes of tomato. Extracts of leaves of citronella (*Cymbopogan winterianus*) reduced root galls and populations of *M.incognita* in soil and roots at concentrations of S/20, S/10 and S/5 on tomato (Mahapatra and Swain, 1993). Leaf extracts of citronella (*Cymbopogon winterianus*) were effective against *M. incognita* in soil and roots at concentrations of s/20, s/10 and s/ 5 on tomato (Mahapatra and Swain, 1993). Bare root dip of tomato seedlings in extracts of undecomposed castor cake and leaves significantly reduced root knot formation (Akhtar and Alam, 1990; Akhtar, 1994). Sundarababu *et al.* (1993) reported that chopped leaves of bougainvillea, ocimum, onion, prosopis, calotropis and subabul suppressed final population of root knot nematode in tomato and enhanced plant growth. *G. fasciculatum* against *M. incognita* on tomato proved superior among five species of Mycorrhizae. Suppression of nematode population and yield increase (Sundarababu *et al.*, 1993b) were noted. Colonisation by *G. fasciculatum* in tomato resulted in reduced root knot infestation in tomato (Sharma *et al.*, 1994).

Vats and Nandal (1994) found that fresh leaf extracts of Eucalyptus and neem plants (40% w/w) given as root dip suppressed *M. javanica* in tomato effectively. Pannu and Paruthi (1995) reported that the neem formulations (Achook, Margoside, Neemguard and Neemark at 0.0., 1.0, 2.0, 4.0, 6.0, 8.0 and 10.0% concentrations) given against *M. javanica* via root dip showed phytotoxic effects on tomato seedlings.Higher concentrations of organic oils of karanj, neem, mahua and castor proved effective in preventing larval penetration and gall production in roots of tomato (Poornima and Vadivelu, 1997). Naik *et al.* (1998) reported that the extracts of neem products had no adverse effect on growth of tomato plants, but significantly decreased nematode development and reproduction. Maximum reduction in galls, egg masses and egg production was recorded in Nimbecidine treated plants and found superior over seed kernel and cake extracts. Sharma *et al.* (2000) reported that significant reduction in number of galls and improvement in plant growth were observed when okra seeds were soaked in Neemark or Nimbecidine (5%) for six hours

Application of leaves of *A. indica* and *T. patula* improved plant growth of tomato and reduced nematode count and gall index and the effect was on par with that of nematicides, phorate and carbofuran (Khanna and Sharma, 1998). In tomato, plant materials like lemongrass and neem leaves significantly improved plant growth and reduced root-knot disease, producing significantly more number of transplantable and total seedlings (Patel *et al.*, 2000).

Biological methods

P. lilacinus reduced gall index and significantly increased fruit yield by controlling *M. incognita* in tomato (Ekanayake and Jayasundaram, 1994; Goswami *et al.*, 1998). Oka *et al.* (1993) opined that exposure of *Meloidogyne* juveniles to *B. circus* inhibited penetration of nematodes to tomato roots. VAM inhibited growth of *M. incognita* in tomato (Suresh *et al.,* 1985).

The biocontrol potential of *P. penetrans* is established against *Meloidogyne* spp. by a number of workers (Walia and Dalal, 1994; Melki *et al.*, 1995; Sosamma and Koshy, 1997). In tomato, the symbiont, *G. fasciculatum* caused a reduction of 30% of galls and egg masses of root knot nematodes / plant (Sharma *et al.*, 1995).

Chand *et al.* (2001) reported that application of *P. penetrans* both in nursery and transplanting stage of tomato significantly improved plant growth parameters with corresponding reduction in galls and *M. incognita* multiplication. In tomato cv. Pusa Ruby, inoculation of *P. lilacinus* in the second season reduced *M. incognita* in soil by 68.00%

(Khan and Goswami, 2002). Cannayane *et al*. (2004) reported that *P. lilacinus* introduced through seed treatment and seedling bare root dip in cowpea, tomato and chilli protected root surfaces from invasion of *M. incognita* juveniles by forming hyphal network through profuse colonization. Khan and Goswami (2000a) reported that 8 g (57.92 x 10^8 spores) fungus (*P. lilacinus*) infested rice/kg soil was optimum for suppression of *M. incognita* in tomato. *P. lilacinus* isolate 6 showed the highest (75%) of egg parasitism of *M. incognita* (Khan and Goswami, 2000b). Sankaranarayanan *et al.* (2000) reported that *Verticillium chlamydosporium* resulted in 53.70% parasitization of egg masses of *M. incognita* in tomato.

Chand *et al.* (2001) reported that application of *P. penetrans* both in nursery and transplanting stage of tomato significantly improved the plant growth parameters with corresponding reduction in galls and *M. incognita* multiplication. Nagesh *et al.* (2001a) found that supplementation of castor and neem oil cake with nitrogen, phosphorus and potassium in the form of inorganic fertilizers had an additive effect on mycelial growth and sporulation of *P. lilacinus* enhancing the antogonistic potential of *P. lilacinus* against root knot nematode in tomato nursery. Mukhtar *et al.* (2002) reported that in tomato, combined application of *V. chlamydosporium* and *P. penetrans* eaused 48.74, 67.25 and 38.30% reduction in number of galls, egg masses and population of *M. javanica* respectively. Goswami and Sharma (2001) reported that combined application of *Aspergillus terreus* Thom and *P. lilacinus* significantly reduced *M. incognita* population, number of galls and promoted plant growth in tomato.

Chemical method

Application of neem cake in nursery followed by carbofuran (one kg a.i./ha^{-1}) in the main field significantly arrested the rate of multiplication of *M. incognita* and increased the crop yield in tomato (Anitha and Subramanian, 1998 ; Reddy 2003).

Integrated method

Treatment of tomato plants with a combination of either *Trichoderma* or Vydate (Oxamil) or *Trichoderma* and Nemacur (fenamiphos) significantly decreased disease and root-gall index and improved plant height and shoot dry weight (Stephan *et al*., 1996). Application of *P. fluorescens* and carbofuran in combination enhanced yield and improved vigour of plants in tomato by suppressing root-knot nematode (Khan, 2000) and cyst nematode, *H. cajani* in soil and root system (Sujatha *et al*., 2000). Carbofuran applied with *P. penetrans* was quite compatible and was the most effective treatment in controlling the population of *H. cajani* (Gogoi and Gill, 2001).

Aldicarb in combination with *P. lilacinus* resulted the lowest gall index, number of galls and *M. javanica* population in tomato root zone (Owino, 2003).

Deep ploughing decreased nematode population prior to transplanting and when the seedling raised from nematicide treated nursery were transplanted, there was less chance of increase in nematode population (Sheela *et al,* 1998).

In tomato, simultaneous application of *G. fasciculatum* and *P. lilacinus* inocula in nursery beds amended with neem cake or use of mycorrhizal seedlings for root dip in *P. lilacinus* spore suspension revealed synergestic interaction of *P. lilacinus* and *G. fasciculatum.* Significant increase in plant growth parameters of nursery seedlings and yield of tomato were obtained consequent to rational integration of these compounds. Integration of soil solarization, straw mulch 2 to 3cm thick and formalin 4.00% @ 11.1 ml m^{-2} showed significant increase in germination of tomato seeds, plant growth parameters and yield, while the weed growth, *M. incognita* population and root-knot count were reduced (Rao, 2005). Transplanting tomato seedling in nursery beds treated with neem cake and P. *penetrans* in pits incorporated with *P. lilacinus* effectively managed *M. incognita* and increased tomato fruit yield (Reddy *et al.*, 1997). Increased concentrations and dip durations of plant calices of *Calotropics procera* (Wild.), *Euphorbia nesiifolia* L. and *E. tirucalli* were effective against *Rotylenchus reniformis* on tomato (Siddique and Alam, 1995). Transplanting of tomato seedlings in nursery beds treated with neem cake and *P. penetrans* in pits incorporated with *P. lilacinus* effectively managed *M. incognita* and increased tomato fruit yield (Reddy *et al.*, 1997). Integration of calotropics leaf and *G. fasciculatum* (Rao *et al.*, 1996) and oil cakes, bone and horn meals with *P. lilacinus (*Khan and Saxena, 1997a*)* resulted in increased plant growth and reduced population build up of nematodes and root galling in tomato.

Integration of *P. penetrans*, *T. viride* and neem in nursery and rotation of tomato with marigold in field gave significant reduction in root galling, egg mass production and final *M. incognita* population in soil which resulted in increased tomato yield (Reddy *et al*, 2000). Nursery solarisation or application of biological control agents or neem cake can be recommended for practical control of nematodes in tomato.

Brinjal (Egg Plant)

Physical methods

Nursery treatments

Soil Solarization

Kamra and Gaur (1993) reported that in egg plant seedlings, root galling by *M. incognita* was decreased by 66.70% and fresh weight of seedlings increased by 96.70% with nursery soil solarization.

Cultural methods

Organic Amendments

Rao (2005) reported that application of organic amendments *viz.,* neem cake, castor cake and mahua cake each at 500 kg / ha were effective in increasing plant growth parameters and reducing root knot and reniform nematode population in brinjal

Botanicals

Husain *et al.,* (1984) opined that extracts of margosa and marigold leaf given by root dip were more effective than piperazine citrate, chenopodium oil and ground nut cake in reducing root knot development of egg plant.

Biological methods

Nursery treatment with *P. penetrans* resulted in significantly less number of galls (51.52 galls / 20 seedlings) compared to untreated control (178.23 galls / 20 seedlings) in brinjal cv. Pusa Purple Long. (Karund *et al.*, 2001). Goswami and Mittal (2002) reported effect of fungal bioagents for biocontrol of root knot nematode. In Brinjal, *Bacillus macerans* nursery application (25 g m^{-2}) + (drenching seven days after sowing with 2% solution of *B. macerans*) significantly reduced population of *M. incognita* resulting higher yield (30%) (Sheela and Nisha, 2004; Sheela *et al.,* 2006).

P. lilacinus increased growth of brinjal plants and reduced root galling, egg masses and eggs / egg mass of *M. incognita* (Sharma and Trivedi, 1989). Root dip treatment of egg plant seedlings in formulation of *P. lilacinus* resulted in drastic reduction of *M. incognita* population (Rao *et al.*, 1998a). Khanna (2000) reported that an inoculum level of 3.6g fungus / 600 g soil was the optimum level for keeping root-knot nematode under check and enhancing growth of brinjal plants. Application of *P. lilacinus* (10 g / pot) improved plant growth of brinjal and tomato plant by reducing multiplication of *M. incognita* (Verma *et al.*, 2004). In brinjal, *B. macerans* nursery application (25 g m^{-2}) + (drench seven days after sowing) significantly reduced population of *M.incognita* resulting in the highest yield (Sheela and Nisha, 2004). Goswami and Mittal (2002) reported that *Aspergillus niger* and *P. lilacinus* along with neem cake resulted in maximum reduction in root knot nematode population both in soil and root in brinjal.

Chemical methods

Root dip treatment with oxamyl (5000 ppm) in 60 minutes and dimethoate (7500 ppm) reduced galling by *M. incognita* on brinjal (Ahuja, 1978). Root dip in aldicarb or carbofuran or turbofos (500 to 1000 ppm) for thirty minutes was effective against brinjal reniform nematode (Krishnaprasad and Krishnappa, 1981).

Integrated method

Kuriyan and Sheela (1981) reported that deep ploughing plus nursery treatment of methan sodium (25 ml/m^{-2}) plus spot application of aldicarb (one kg a.i. ha^{-1}) had the minimum number of root-knots and population of *M. incognita* in brinjal root. Integration of *P. lilacinus* with carbofuran (one kg a.i./ha) was effective in the management of reniform nematode, *R. reniformis* infesting brinjal (Reddy and Khan, 1989). Solarization for 15 days and nursery treatment with Carbofuran @ 0.30 g a.i. m^{-2} + neem cake treatment @ 200 kg ha^{-1} were effective in increasing yield of brinjal by reducing population of *M. incognita* (Sheela *et al.*, 1999).

The nursery solarisation together with application of farm yard manure or poultry manure (2000 kg ha^{-1}) or carbofuran (1 kg a.i./ ha^{-1}) significantly reduced population of *M. incognita* and increased yield of brinjal (Sheela *et al.*, 2006).

Nursery application of either neem cake or *P.lilacinus, B. macerans* or *P. fluorescens* is the practical solution for managing nematodes in brinjal.

Okra (Bhindi)

Cultural methods

Organic Amendments

Kumar and Nair (1976) observed that soil application of coconut oil cake reduced infestation of root knot nematode in okra and it further promoted plant growth. The extract of *Argemone mexicana* L. acted as a nematicide to *M. javanica* in okra raised in microplots (Nath *et al.*, 1982). Minimum galling and increase in growth of okra after the application of *Clerodendron inerme* (1.5% w/w) was observed by Patel *et al.* (1985). Green leaves of neem and eupatorium for management of nematodes was reported by Ajith and Sheela (1996). Organic amendments with powdered leaves of plants like curry leaf (*Murraya koenigii)*; Jasmine (*Jasminum sambac*); Sour orange (*Citrus aurantifolia*); Patal Garuda (*Rauwolfia serpentina*); Ber (*Zizyphus jujuba*); China rose (*Hibiscus rosachinensis)* and Justicia (*Justicia gandurusa)* were effective in reducing population of *M.incognita*.(Padhi and Behera, 2000). Amendment of soil with mustard cake was more superior and effective than poultry manure with regard to shoot length and root weight in okra (Dahiya *et al.,* 1998).

Biological methods

P. penetrans application significantly reduced population of root knot nematode and gall index in okra (Walia and Mehta, 2000). Seed treatment with biological control agent *B.macerans* @ 3% weight by weight of seeds gave satisfactory control of nematodes and improving fruit yield and also improved vigour of seedlings (Sheela and Nisha, 2006).

Chillies

Plant resistance

Varieties BC 40-1 , BC - 28, BC-24, BC-30, Utkal Ragini, G-2, G-4, Pusa Sadabhar, B.R. Red, Agnirekha, Lamx-235 and Berhampur Local were moderately resistant (Patnaik *et al.,* 2004) to Meloidogyne nemotodes.

Hyoscyamus albus

Effectiveness of organic amendments (oil seed cakes of castor, mustard and linseed) for management of root knot nematode in *Hyoscyamus albus* was reported by Haseeb and Butool (1994).

Carrot

Plant resistance

Arya and Tiagi (1982) reported that among three carrot cultivars cv. Black was moderately resistant to *M. incognita* with root knot index of 1.5.

Other vegetables

Alan (1989) reported that soil amendment with horn meal, bone meal and oil cakes of mahua, castor, mustard, neem and pea nut were effective in chilli, cabbage and cauliflower and improving plant growth. Soil amended with mustard cake was superior and effective than poultry manure with regard to shoot length and root weight in bottle gourd (Dahiya *et al.*, 1998). Neem cake, mustard cake, press mud (25g kg^{-1}) soil and carbofuran (20 and 10 mg a.i. kg^{-1} soil) reduced root-gall formation and increased yield (root weight) and shoot length of red beet.

Solarization alone or combination with low rates of nematicides viz., 1, 3-D (50 and 100 l ha^{-1}), dazomet (125 and 250 kg ha^{-1}) and methyl bromide (10 and 20 g m^{-2}) effectively controlled *M. incognita* and significantly increased marketable carrot yield.

5. Plantation Crops

Coconut

Different nematodes are found in diverse forms of association with living coconut palm (Govindankutty and Koshy 1979; Koshy and Banu 2002). But the major nematode disease affecting the crop is red ring caused by *Bursaphelenchus cocophilus*. Fortunately it was not found in India. The only nematode known to cause severe damage leading to malfunction in coconut is *R. similis* (Koshy *et al.,* 1975). The burrowing nematode, *R.similis*, occurs in most tropical and subtropical areas of the world and is reported from coconut palms in Florida, Jamaica, Sri Lanka and India (Van Weerdt *et al.,* 1959, 1959b; Ekanayake, 1964; Latta, 1966; Weischer, 1967; Koshy *et al.*, 1975).

Management of burrowing nematode on a perennial palm like coconut with a massive root system is difficult, especially under high density multispecies cropping system that exists along the West Coast of South India involving susceptible crops like arecanut, banana, black pepper, betel vine, ginger and turmeric. Unlimited use of nematicides for control of burrowing nematode causes problems of residual toxicity in coconut water and copra (Habeebullah *et al.*, 1983). Apart from this, it may also lead to residual toxicity in the products of intercrops. Control of nematodes by field application of nematicides alone is not a practical proposition.

Cultural methods

Organic Amendments

The cultural practices existing in Kerala and Karnataka (India) are the application of neem / marotti (*Hydrocarpus*) oil cakes at 2-4 kg/palm/year, farm-yard manure @ 50kg / palm / year, and green foliage and tender stem of *G. maculata* to the basins @ 50kg / palm / year.

Intercropping

Growing of green manure crops like cowpea, *Crotolaria* or *Sesbania* in the basins and interfaces during June to August and ploughing in of the entire crop at flowering help in reducing burrowing nematode population and enriching nutritional status of soil. In addition, growing of intercrops like cocoa that enrich soil with sizeable quantities of shed foliage helps in build-up of beneficial organisms and antagonistic micro organisms that may inhibit nematode multiplication (Koshy *et al.,* 1991 a, b; Koshy *et al.*, 2002).

Biological methods

Significant increase in width and leaf area is recorded in coconut seedlings which received AMF alone. Increase in shoot weight, root weight and decrease in lesion indices occur in seedlings inoculated with mycorrhizae prior to *R. similis*. Prior inoculation of mycorrhizae is more effective against *R. similis* than simultaneous inoculation of mycorrhizae and nematode even though root infection and spore production are slightly reduced in nematode inoculated plants. The lesion index and root nematode population can be successfully lowered with mycorrhizae infection. A mixture of mycorrhizae consisting of multiple endophytes, viz. *Acaulospora bireticulata, G. fasciculatum, G. macrocarpum, G. mosseae, G. versiforme, Sclerocystis rubiformis* and *Scutellospora nigra* was effective in improving plant growth and reducing *R. similis* infestation of coconut seedlings (Sosamma, 1994).

Introduction of *P. lilacinus, P. penetrans* and Arbuscular Mycorrhizae into potting mixture contained in polybags of coconut nurseries and again in planting pit at the time of transplantation of coconut seedlings in the field help in better establishment of plants and imparts better growth by offering protection against *R. similis* (Koshy 1998; Sosamma *et al.*, 1998).

A new isolate of *Pasteuria* parasitising *R. similis* in Kerala, India has great potential for use in IPM. The infective propagules of *Pasteuria* adhere to cuticle of adult males, females and juveniles of *R.similis* (Sosamma, 1999, 2000, 2002) Introduction of *P. ilacinus, P. penetrans* and mycorrhizae into potting mixture contained *i*n plastic bags of coconut nurseries and again in planting pit at the time of transplantation of coconut seedlings in the field helps in better establishment of plants and imparts better growth by offering protection against *R. similis* (Koshy, 1998; Koshy *et al.*, 1998; Sosamma *et al.*, 1998). *Catenaria vermicola* was also found parasitising R. *similis* in Kerala (Sosamma, 2000).

Plant resistance and tolerance

All the coconut cultivars (29 exotic, 15 indigenous and 15 hybrids) screened for resistance to *R. similis* in India were susceptible in varying intensities. The dwarf cultivars; Kenthali and Klappawangi, recorded the least nematode multiplication and lesion indices. Similar reactions were noticed in hybrids like Java Giant x Kulasekharam Dwarf Yellow, Kulasekharam Dwarf Yellow x Java Giant, Java Tall x Malayan Yellow Dwarf and San Ramon x Gangabondam (Sosamma *et al.*, 1980, 1988).

Chemical methods

Dipping seedlings in DBCP (1000 ppm) for fifteen min was effective in controlling nematodes in seedlings in coconut nurseries (Koshy and Sosamma, 1979). Effective control of *R. similis* can be obtained with soil application of phenamiphos or phorate at 2.5 kg a.i./ ha during September, December and May in infested coconut nurseries (Koshy and Nair, 1979; Koshy *et al.*, 1985).

The following measures are suggested towards developing an integrated management schedule for *R. similis* infestation on coconut palms (Koshy 2002).

- Application of cow dung (50 kg), oil cakes (2-4 kg) and green manuring with *Glyricidia maculata* (50kg)/palm/year to the basins.
- Growing *Crotolaria juncea*, cow pea or *Sesbania* in the basins and interspaces and incorporating into the soil by ploughing in at flowering stage.
- Application of phorate at (10 g a.i./palm) twice yearly (in June-July and in October-November in India).
- Avoid growing of bananas as a shade crop in coconut nurseries.
- Use of nematode free planting material of coconut and other intercrops.
- Use of tolerant or less susceptible cultivars or their hybrids in infested areas.
- Cut and remove all roots external to the husk of seedlings raised in the field before planting.
- Raise coconut seedlings in potting mixture enriched with bioagents like *P. lilacinus*, *Pasteuria penetrans* and mycorrhyzae in plastic bags.
- Introduce bio agents into the planting pits while planting in the main field.
- Apply phorate (3g a.i./plant) to inter crops like banana, black pepper and arecanut in June-July and October- November.

Arecanut

A number of nematodes are reported from rhizosphere of arecanut (Nair, 1964; Weischer, 1967; Pizarro, 1969; Koshy *et al.,* 1975, 1979; Reddy, 1978; Koshy *et al.*, 1981; Sundararaju & Koshy, 1982a&b; Sundararaju *et al.,* 1984; Rama, 1987; Dasgupta and

Rama, 1987; Subramaniyan *et al.,* 1988) but only *R. similis* is known to be an important parasite of palm. A number of other palms are reported as hosts of *R. similis* and it would not be unexpected if nematode problems with some of these other palms become apparent in years ahead.

Biological method

Sudha *et al.* (2000) reported that *P. lilacinus* significantly reduced population of *R. similis* in soil and root of arecanut.

Plant resistance/tolerance

The cultivars Indonesia 6 (VTL-11), Mahuva 8 and Andaman-5 (VTL-29e) are tolerant to *R. similis* (Koshy *et al.*, 1979; Sundararaju and Koshy, 1982b). The cultivar Indonesia-6 (VTL-11) and Singapore (VTL-17) yield 15% more nuts over local South Canara cultivar. These cultivars could profitably be recommended for *R. similis* infested areas. The hybrid VTL-11 x VTL-17 is highly resistant to *R. similis*.

Chemical methods

As arecanut is chewed directly by many consumers, dosage, frequency and time of application of nematicides and their persistence on arecanut have to be done carefully to avoid residues in nut.

Experiments under field conditions revealed that fensulfothion and aldicarb at (1 g a.i./seedling) applied thrice a year for three consecutive years gave control of *R. similis* both in soil and roots. Increase in plant growth with regard to shoot length, shoot weight, root length, root weight, number of leaves and collar girth with fensulfothion were 46, 168, 33, 173, 25 and 41% respectively over control plants after three years (Sundararaju and Koshy, 1986).

Field experiments were carried out in arecanut monocrop, arecanut + banana and arecanut + banana + pepper to evaluate efficacy of neem oil cake and phorate singly and in combination for control of *R.similis* in the cropping system. Even though, all the treatments were significantly superior over untreated control, the best treatment in these experiments was 1.5 g of phorate in combination with 1 kg neem oil cake which controlled *R. similis* population in arecanut and subsidiary crops very well (Sudha and Sundraraju, 1998).

Integrated methods

Combination of organic amendments and biocontrol agents

The combined effect of organic amendments and biocontrol agents *viz. P. lilacinus*, *P. penetrans* and Arbuscular Mycorrhizal Fungi (AMF) against *R. similis* infecting arecanut (*Areca catechu* var. Mangala) was conducted using sandy loam soil amended with various organic matters. The organic amendments used were neem and marotti oil cakes, leaves and tender shoots of sun hemp and *Glyricidia*, vermicompost, cow dung and coir pith.

Maximum nematode control (95%) was recorded in soil amended with *Glyricidia* leaves and bio agents. Significant reduction in root lesion index and maximum leaf area was recorded in these plants. Increase in height and root growth was maximum in plants grown in coir pith amended soil with bio agents which were at par with plants grown in soil enriched with *Glyricidia* leaves and bio agents.

Control of *R. similis* on arecanut is difficult under high density, multispecies, subsistence farming systems involving perennial crops like coconut, banana, black pepper, betel vine, cardamom and cocoa. Use of nematicides for control of burrowing nematode on coconut or arecanut causes problems of residual toxicity. The following control measures are suggested:- 1) use of nematode-free planting material of arecanut and other intercrops; 2) avoiding *R. similis* susceptible intercrops like black pepper and banana in infested areas; 3) use of resistant/tolerant cultivars of arecanut, when available, and other crops in farming systems; and 4) apply 5-10 kg green manure preferably *Glyricidia* or *Crotolaria* 5) apply 1 kg of neem oil cake/palm/year, and 6) apply phorate (1.5g a.i./plant) to the root zone of arecanut, banana and black pepper in June-July and October-November in arecanut based farming systems (Koshy, 1994; Sudha and Sundaraju, 1998).

Tea

Root-knot nematodes are the most frequently encountered in tea growing regions of India and most of the species of *Meloidogyne* except *M.brevicanda* are associated with nursery plants. Twenty one species of nematode belonging to 14 genera are reported in association with tea roots from India.

Chemical method

Disinfestation of nursery soil and pre planting application of chemicals (fumigants) for control of root-knot nematode are recommented in infected area.

Cultural method

Incorporation of organic matter into nematode infested soil to suppress incidence of parasitic nematodes is advised. Cultivation of marigold (*T. erecta* and *T. patula*) as a pre plant crop considerably reduces population of *Meloidogyne* spp . Intercropping of marigold also is recommended.

Coffee

Quite a few nematodes attack coffee (*Coffea* spp.) throughout the world. The important ones are *Meloidogyne* spp., *Pratylenchus* spp., *Hemicriconemoides* spp., *R. similis* and *R. reniformis*. Of these, only the species of *Pratylenchus* and *Hemicriconemoides* infect Arabica coffee (*C.arabica*) in India (Kumar *et al.*, 1971). Robusta coffee (*C.canephora*) exhibits tolerance against the nematode attack.

Cultural methods

Attempts to check nematodes through addition of soil ameliorants like oil cakes, coffee cherry husk or organic matter or using the chemical nematicides have a limited value use.

Plant resistance

Robusta is fairly tolerant to nematode attack and hence the planters are advised to replace the nematode infected arabica with either robusta or arabica-robusta grafted plants. Wedge-cleft grafting of Arabica scion on robust a root- stock at 'topee' stage is adopted with a fair degree of success (Kumar, 1988).

6. Tuber Crops

Nematodes on tropical tuber crops are almost a neglected group of pests due to various reasons. Occurrence of high degree of resistance to nematodes is a unique feature in tuber crops (Mohandas and Palaniswami, 1996).

Potato

Potato is affected by about 150 species belonging to 50 genera throughout the world. Among these, 92 species of nematodes belonging to about 39 genera are reported in India of which the potato cyst nematodes and root-knot nematodes are the major nematode problems not only in India but also in the world. Sathyarajan *et al.* (1966) reported root knot nematode, (*M. incognita*) infestation in *Solenostemon rotundifolius* from Kerala. Due to attack of *M. incognita* in *S. rotundifolius* conspicuous gall like swellings are formed in roots resulting in malformation of tubers, rendering them unfit for consumption as well as storage. In potato, *M. incognita* population density (I J2 / g of soil) was considered as the damage threshold level (Nagesh, 1996).

Biological methods

P. lilacinus is successfully used to manage root-knot nematode in potato (Jatala *et al.*, 1989 and 1990) and carrot (Sivakumar, 1998). A study conducted by Racke and Sikoro (1992) revealed that plant growth promoting rhizobacteria, *Agrobacterium radiobacter*. Corin and *B. sphaeriacus* increased tuber yield of potato by suppressing population of *Globodera pallida.*

Cassava

A number of plant parasitic nematodes are associated with cassava though detailed report on yield loss and management of the nematodes in cassava are not available. *M. incognita* and *M. javanica* are the most important and widerspread among all the nematodes (Mohandas, 1994). Other species of *Meloidogyne* reported on cassava are *M. arenaria* and *M. hapla*. Among *Pratylenchus species - P. brachyurus, P. safaenis* are from cassava. Other nematodes on cassava are R. *reniformis, H. erythrinae, H. dihystera* and *Scutellonema bradys* which are however, of lesser importance to the crop. Lists of nematode species associated with cassava are reported by Hogger (1971) and Caveness (1980).

Chemical methods

Control of nematodes using nematcides is not desirable in cassava especially because of the residues it may leave behind in tubers and also in environment. As this is a long duration crop, more than one application of the chemical will also be required. Medina *et al.* (1992) tried Sincocin, aldicarb, carbofuran and ethoprop and found Aldicarb (3 kg ai/ha) gave the best result for control of root-knot nematode followed by aldicarb (2 kg ai/ha) and ethoprop.

Plant resistance

Use of resistant varieties is the most desirable for management of root-knot nematode problem. A number of authors reported resistance in cassava cultivars and germplasm accessions to root-knot nematode. Crozzoli and Hidalgo (1992) reported that all ten cultivars evaluated were susceptible. Their rate of reproduction varied from 4.1 to 21.9. They also reported tolerance to nematode infection in two cultivars. On the other hand, Ponte *et al.* (1979) reported that out of 25 cultivars tested for resistance to *M. incognita* and *M. javanica*, 19 cultivars were immune, three were highly resistant and only three cultivars were susceptible.

Sweet Potato

Nematodes on sweet potato and the damage they cause are reviewed by various authors (Jatala and Bridge, 1990; Anguiz and Carto-saenz, 1991; Mohandas, 1994; Sharma *et al.*, 1997). Species of *Meloidogyne* and *R. reniformis* cause reduction in yield and quality of tubers, the world over (Clark and Moyer, 1988). *Pratylenchus* spp. are very serious in Japan whereas *Ditylenchus destructor* and *D. dipsaci* are quite serious in China (Sharma *et al.,* 1997).

R. reniformis is reported in large numbers by Verma and Singh(1969) but nothing is known on yield loss and control (Mohandas, 1994). Gapasin and Valdez (1979) observed 60.6% yield reduction in pot culture experiments. Gapasin (1981) reported that *R. reniformis* reduced yield from 13.4-60.6 % in Philippines. Walters and Barker (1994) studied effect of two populations on sweet potato and found that both populations restricted storage root growth but enhanced shoot growth. Infestation by nematode may also cause cracking of storage roots (Clark and Wright, 1983).

Rajendran *et al.* (1972) reported high population of *Pratylenchus* sp. and *Hoplolaimus* sp. in soil and roots from Coimbatore, Tamil Nadu. *P.coffeae* is very serious in Japan in volcanic acid soils (Suzuki, 1989). Other species reported feeding on sweet potato were *P. flakkeniss, P. brachyurus, P. penetrans, P. vulnus* and *P. zeae.* Auguiz and Canto-Saenz (1991) observed resistance to *P. flakkeniss* among 20 sweet potato cultivars.

In China, *Ditylenchus* spp is a serious problem and programmes on management and breeding for resistance are in full swing. *Ditylenchus* spp. induces “brown ring” in storage root. The species involved are *D. destructor and D. dipsaci* in China. A number of resistant accessions and varieties were reported from China (Wu and Zhang, 1990; Wang *et al*., 1995). Lin *et al*. (1993) observed that exudate secreted by oesophageal glands of *D. destructor* when inoculated with tubers of susceptible sweet potato, the tissue around the inoculated area became brown and the cells were destroyed.

Plant resistance

Resistance to *M. incognita* and *M. javanica* are reported in sweet potato from various countries and active breeding programmes are being conducted in US, Japan and China (Sun and Chen, 1994) which had released high yielding nematode resistant varieties namely ‘Satsumahikari’, ‘Excel’, ‘Yushu 3’, Fusabeni Topa 3’, ‘Hi-Starch’, ‘Red glow’ and ‘J-red’ which showed resistance to nematodes (Kukimura, *et al.,* 1989). A number of germplasm accessions, cultivars/varieties are resistant to the root-knot nematode in Peru (Jatala, 1989; Jatala and Bridge, 1990) and also in India (Mohandas and Palaniswami, 1990a; Mohandas and Rajendran, 1993; Ramakrishnan *et al*., 1997). Differences in susceptibility of certain promising accessions of sweet potato to *M. incognita* were reported by Ramakrishnan *et al*. (1999).

Very high degree of resistance to *M. incognita* and *M. javanica* is reported from the world over and high yielding released varieties are available in many countries including India, USA, China and Japan.

‘Sree Bhadra’, a high yielding released variety of sweet potato was identified as a resistant trap crop to root knot nematode. When this variety was planted in a root-knot nematode infested field, the nematode population declined to below detectible level over single cropping duration of 90-95 days. Subsequently susceptible plants viz. coleus and African yam also escaped nematode damage when planted in such fields. Different centres of AICRP on Nematodes viz KAU, Vellayani; JNKVV, Jabalpur and TNAU, Coimbatore confirmed that root knot nematode population declined in heavily infested fields when Sree Bhadra was planted. The GAU, Anand isolate of RKN infested the Sree Bhadra variety. Sweet Potato varieties Khajangod, Sree Vardhini and Sree Nandhini were resistant to root-knot nematode.

Though many released varieties and major germplasm accessions are either immune or resistant to root knot nematode, a few varieties are susceptible. In such susceptible varieties reduction in shoot and root weight were significant even at a low initial population level of 20 IJ/pot.

Chemical methods

Control of nematodes in sweet potato using chemicals in California was restricted to preplanting fumigation with Methyl bromide (Roberts and Schueurman, 1984). Preplant nematicidal treatment of *M. incognita* infested field doubled yield of marketable sweet potato roots and also reduced proportion of cracked tubers by over 40% (Hall *et al.*, 1988). Nemacur and aldicarb were effective in controlling *Meloidogyne* species (Clark *et al.*, 1980; Gapasin 1981). Application of DS 38697 significantly reduced *M. incognita* in Sweet Potato (Averre *et al.*, 1985).

Yams

Three most important nematode pests of yams are root-knot nematode, yam nematode and lesion nematode. Root-knot nematode infestation produces deformed tubers with uneven surface, whereas the other two nematodes produce typical 'dry rot' of tubers. Steiner (1937), reported *Rotylenchus blaberus* causing a disease in yams originating from Nigeria. Nematode pests on yams and the work carried out were reviewed by Bridge (1982); Degras (1993) and Mohandas (1994). *Scutellonema bradys* is one of the most important pests infesting yams. This nematode is recorded on yams from India, West Africa, the Caribbean and Brazil (Bridge, 1982). The nematode produces 'dry rot' on yams (West, 1934). In field the damage is restricted to outer thin layer of tuber (Adesiyan, 1977). In a survey of nearly 500 samples collected from French West Indies all *Dioscorea* cultivars were infected by *S. bradys*. In tubers, *S. bradys* may be found alone or associated with *Pratylenchus coffeae* (Castaynone-Sereno, *et al.,* 1988). In store, the nematode infection spreads and infection sites coalesce which encircle tuber and could result in 80 to 100% loss of stored yams in Mid-West State of Nigeria (Adesiyan and Odihirin, 1975,1978). Cadet and Quencherve (1994) also reported serious damage of yams during storage and weight loss reached upto 30%, whereas in healthy tubers, it was only 10%. *M. incognita* and *M. javanica* were reported on *Colocasia* and *Xanthosoma* from Southern Florida (McSorley, 1980).

Plant resistance

As is the case with cassava and sweet potato, resistance is very common in yams in India. Very high degree of resistance is identified in *D. esculenta* and *D. alata* (CTCRI, 1997). High yielding and released varieties from the CTCRI, viz. 'Sree Latha' of *D. esculenta* and 'Sree Keerthi' of *D. alata* were highly resistant to the root-knot nematode. Out of 33 *D. alata* accessions tested against *M. incognita,* 3 were immune (Da 66, 120, 228). Da 199 was highly resistant and a further 9 were resistant (Mohandas *et al.*, 1996). Rajendran and Sivagamivadivelu (1990) screened cultivars and reported no resistance in both *D. alata* and *D. esculenta* from Coimbatore. Out of several yam cultivars grown in Southern Nigeria, all except *D. dumetorum* and *D. opbunenyi* were susceptible to *M. incognita* race 2 (Atu et *al.,* 1984).

Though chemical treatments for control of these nematodes are effective, this may not be practical because of long duration. Selection of healthy seed material will be ideal. Hot water treatment of setts used for planting is effective in checking the nematode. Soaking the setts in oxamyl 1200 to 2400 ppm for 40 minutes was the most effective (Hutton *et al.*, 1978 and Roman *et al.*, 1984). Carbofuran, fensulphothion or aldicarb were also effective in controlling nematodes (Roman *et al.,* 1984).

Resistant varieties may be planted in infested fields. There was no damage to susceptible *D.rotundata,* if this was preceeded by resistant Sree Bhadra variety of sweet potato (CTCRI, 1997). Ayala and Acosta (1971) reported reduction of nematode fauna following rotation with *Zea mays*, *Capsicum annuum* and *Nicotiana tabacum*.

Scutellonema bradys is one of the most important pests infesting yams. This nematode is recorded on yams from India, West Africa, the Caribbean and Brazil (Bridge, 1982). The nematode produces 'dry rot' on yams (West, 1943). In field, the damage is restricted to outer thin layer of tuber (Adesiyan, 1977). In a survey of nearly 500 samples collected from French West Indies, all *Dioscorea* cultivars were found infected by *S. bradys*. In tubers, *S. bradys* may be found alone or associated with *P. coffeae* (Castaynone-Sereno, *et al.,* 1988). In store the nematode infection spreads and infection sites coalesce which encircle the tuber and could result in 80 to 100% loss of stored yams in the Mid-West State of Nigeria (Adesiyan and Odihirin, 1975). Cadet and Quencherve (1994) also reported serious damage of yams during storage and the weight loss reached upto 30%, whereas in healthy tubers it was only 10%.

Amorphophallus (Elephant Foot Yam)

Mohandas (1994) reported damage due to infestation by root-knot and root lesion nematodes. The root knot infestation produces typical galls on roots. In corms and cormels, infestation leads to irregular projections, which are relatively more on the latter. Under severe infestation, the tuber tissue gets discoloured and in corms, the infested area dries up a situation resembling dry rot of yams. Farmers in Kerala keep the pest under control by planting nematode free healthy corms. They also practice crop rotation. In Andhra Pradesh the nematode is a very serious problem and severe infestation leads to rotting of tuber before harvest. The root knot nematode multiplies in tuber after harvest during transport and in store, causing more damage. Premature spouting was also recorded, when infestation was severe. Infestation by lesion nematode changes tuber colour to black with cracks on surface. Infestations by both nematodes were very high in irrigated fields as compared to rain fed tracts.

Other nematodes infesting Amorphophallus are *Radopholus similis* and *Radopholoides* sp. Infestation by *R. similis* caused rooting of tuber tissue. The infested roots were devoid of cortical layer. However, infestation is rare. In endemic areas of infestation, seed material treatment of *B.macerans* (0.3 %) weight by weight can be recommended as a precautionary measure (Nisha and Sheela, 2004)

Colocasia and Xanthosoma

Nirula (1959) reported root-knot nematode on colocasia in India. Nadakal and Thomas (1964) reported nematode on colocasia from Kerala. Srivastava *et al.* (1969) reported damage caused by *M. javanica* and discussed its control. Mohandas and Palaniswami (1990b) reported reduction in yield upto 21.4% in *Colocasia* var. Sree Pallavi. Infected cormels were deformed and smaller in size with branches which affected marketability of tubers. Resistance also was reported by them in Sree Reshmi, a high yielding released variety and C9, a popular cultivar.

M. incognita and *M. javanica* were reported on *Colocasia* and *Xanthosoma* from Southern Florida (McSorley, 1980).

Coleus

Among minor tuber crops, coleus (*Solenostemon rotundifolius*) is extensively grown in Kerala and Tamil Nadu and is highly susceptible to root-knot nematode (Sathyarajan *et al*., 1966;). Patnaik and Das (1986) studied pathogenicity of *M. incognita* on coleus. Sheela *et al.* (2004) reported potential of *B. macerans* (2% solution) as dipping of cuttings + drenching in soil seven days after planting in managing *M. incognita* in coleus.

The resistance screening of varieties / lines of coleus against *M. incognita* revealed that the var. Sree Dhara and Nithi were showing resistance and recorded the minimum root-knot index of 1.00. (Nisha,2005). Sheela *et al.* (2006) reported the potential of *B. macerans* (2% solution) as dipping of cuttings + drenching in soil seven days after planting in managing *M. incognita* in coleus nursery.

7. Medicinal Plants

Kacholam

Cultural methods

Organic Amendments

Rajani (1998) reported effectiveness of neem cake (200g m^{-2}) for managing root-knot nematode in Kacholam. Neem cake (200 g/m^2) and AMF (300g inoculum/m^2) were on par in improving leaf production in nematode infested kacholam plant. Neem cake established its superiority in improving shoot weight also. Neem cake, AMF, sawdust and coir pith were statistically on par in improving fresh weight of nematode infested kacholam plants. It was on par with carbofuran. Carbofuran was significantly better in reduing nematode population (number of females and larvae) in the root than neem cake (200g/m^2), AMF (300g inoculum/m^2) and neem leaf (750 g/m^2). Soil mulching with green leaves of neem and chromolaena (5 kg/m^2) improved the fresh plant weight and yield of infested kacholam (Rajani 1998). The effect of neem and chromolaena @ (5 kg/m^2) leaves were on par in reducing root knot count and number of larvae of *M. incognita* kacholam. No significant variation was observed in *R. similis* (Nisha 2001). According to Nisha and Sheela (2003) neem cake (200 g/m^2) was very effective in reducing nematode population in kacholam rhizosphere.

Biological methods

In kacholam, significant improvement in biometric characters like fresh plant weight and root weight was observed when given with rhizome treatment of *Pseudomonas fluoresescens* (3% w/w).The effects of Arbuscular mycorrhizal fungi (AMF) (3% w/w) and hot water treatment (55°C) on *Meloidogyne incognita* and *Radopholus similis* were significantly better than the chemical, dimethoate (0.1%). An yield increase to the tune of 87% was observed as a result of these 3 treatments. *P. flurescens* and AMF were significantly effective in nematode population reduction in kacholam (root knot count and number of females). Hot water treatment and rhizome treatment with *T. viridae* were significantly better than the chemical dimethoate (0.1%) in nematode population reduction. Application of AMF @ (10 g/m^2) was very effective in reducing nematode population in Kacholam and increasing the yield significantly (Nisha and Sheela, 2003). Green mulching with neem, glyricidia and clerodendron also gave good control of nematodes (Nisha and Sheela, 2002).

Chethikoduveli

Survey conducted in the herbal gardens of Kerala revealed association of *M.incognita, R.similis, R. reniformis, Tylenchorynchus* sp., *Helicotylenchus multicinctus, Belanolaimus* sp., *Hoplolaimus* sp., *Criconemoides* sp. and *Hemicriconemoides* sp. in *P.rosea* (Kumar *et al*., 2001). *M. incognita* and *R.similis* were the major nematodes damaging the crop (KAU 2004).

Integrated methods

Management trials were conducted in nematode infested microplots using *Glomus fasciculatum* @ (10 g/m^2) with 10^6 spores/g, neem and groundnut cake @ (0.5 kg/ m^2), carbosulfan (1 kg ai/ha), wood ash (5 kg/m^2), *Paecilomyces lilacinus* and *P.fluorescens* (200 spores/g) @ (10 g/m^2 with 10^6 spores). Among various treatments, *G.fasciculatum* effectively controlled nematodes and produced maximum yield.This was closely followed by *P.fluorescens*. Root knot count was minimum in carbosulfan treatment followed by neem cake and *G.fasciculatum*. Number of larvae emerged / egg mass was the lowest in carbosulfan treatment followed by *G. fasciculatum* and *P. fluorescens*.

The field experiment on management of *R.similis* in *P.rosea* revealed that *G.fasciculatum* @ (10 g/m^2) was the best treatment followed by *P. fluorescens* @ (10 g/m^2). The effect of neem cake and groundnut cake @ (10 g/m^2) and the green leaves of *Glyricidia maculate* and *C.infortunatum* (5 kg/m^2) were effective in managing nematodes and improving biometric characters and yield of plant but the treatments were inferior to above bioagents (KAU,2004).

Thippali

The root knot nematode was pathogenic to *P. longum* at an initial inoculum level of 10 nematodes / plant while *R.similis* was pathogenic at 100 nematodes/plant onwards. Crop loss in term of reduction in plant height (31%) leaf area (9%), fresh weight of root (15%) and fresh weight of shoot (26%) were observed at 100J2 level for *M.incognita*. Significant reduction in yield (31%) was obtained when the plants were inoculated with 1000 second stage juveniles.

Management trials conducted on *R. similis* showed that *G. fasciculatum, P.fluorescens* and carbosulfan improved biometric characters like plant height, number of branches, leaf area, fresh and dry weight of shoot and yield in terms of number and weight of fruits.

8. Ornamental Plants

The nematode pests associated with ornamental plants are root-knot nematodes, lesion nematodes, reniform nematodes, spiral nematodes, needle nematodes, burrowing nematode, bud and leaf nematode and dagger nematode. According to Kumar (2000) plant growth and flower yield of tube rose were significantly reduced at an initial inoculum level of 10 J2/g of soil. Khan (2003a) reported that plant growth of balsam (*Impatiens balsamina* L.) was significantly reduced at 1000J2 of *M. incognita*/kg of soil.

Cultural methods

Intercropping of crossandra with enemy plants like marigold significantly reduced the root-knot and lesion nematode population both in soil and roots. Application of farm yard manure reduced the lesion nematode population in crossandra. Rotation of crossandra with spearmint which is a poor host of *Longidorus africanus* would reduce the needle nematode population (Kolodge *et al.*, 1987).

Biological methods

The parasitic fungus *Paecilomyces marguandi* grown on paddy seeds (2g/kg soil) gave effective control of root-knot nematode infecting crossandra (Khan and Reddy, 1992).

Plant resistance

Ohkawa and Saigusa (1981) found that *Rosa indica* (major) and *R. multiflora* (60-5) were resistant to *P. penetrans* and *P.vulnus* respectively. They also reported that rose cv.Mavetti to be resistant to *M.hapla.*

Chemical methods

Nursery bed treatment

Nursery bed treatment with phorate, aldicarb, fensulfothion and carbofuran each at 5g/m^2 gave good and healthy Crossandra seedlings free from *P. delattrei* (Vadivelu and Muthukrishnan, 1979).

Bare root dip treatment

Dipping of gladiolus corms in thionazin or fensulfothion solution (0.5g a.i. /litre) gave reduced root-knot nematode infestation (Overman, 1970). Dipping of bare roots of rose in 0.1% solution of fenamiphos for 30 minute gave considerable reduction in root-knot infection (Dale, 1973). Root dip treatment for 8 hr in 0.05 to 0.1% solution of aldicarb, carbofuran, fensulfothion, fenamiphos or phorate protected crossandra seedlings from *P. delattrei* for 30 days after transplanting (Vadivelu and Muthukrishnan, 1979).

Foliar application

Folliar application of methyl parathion, chlorpyriphos and quinalphos each at 0.05% were effective in the control of *Aphelenchoides ritzemabosi* infecting zinnia and chrysanthemum (Gill and Walia, 1980; Gill 1981).

Integrated methods

Integrated management of root- knot nematodes infecting crossandra was achieved by rational combination of a biocontrol agent (*T. harzianum or V. chlamidosporium*) with a nematicide (aldicarb) or oil cake (neem cake) (Khan and Reddy, 1992b). Integration of Arbuscular Mycorrhizae like *Glomus mossae* or *G. fasciculatum* with neem cake or aldicarb gave effective management of root knot nematodes infecting tuberose (Khan and Reddy 1992).

9. Future Prospects

Nematodes continues to be a major production constraint, as long as adequate attention is not given for their effective management. Developing cropping systems, avoiding susceptible intercrops, incorporating an Integrated Nematode Management system with minimum or no nematicide application, should be the main thrust in research to increase yield in areas infested with damaging species of nematodes. Exploitation of the wide variability available in horticultural crops to nematodes, particularly *R.similis* and *H.oryzicola, M. incognita, T. semipenetrans, H. multicinctus, R. reniformis* and *P. coffeae* should be a major priority. In addition, the general soil health should be sustained by addition of suitable soil amendments. The procedure for studying population ecology of nematode communities feeding in or on crop roots should be examined in greater detail. Importance of some ectoparasitic species and sedentary parasites, *R. reniformis* and *Meloidogyne* spp. is not well understood. Such studies are to be taken up separately from established procedures. Research on production of nematicides with novel mode of action will continue for foreseeable future, particularly for those growers supplying horticultural produce to high value markets and exports.

Exploitation of resistance of above nematodes on various crops including wild types with good agricultural traits should be a major priority although plant characters like root vigour that confers tolerance to nematodes should also be considered, particularly in programmes for improvement of cultivars.

Development of micro propagation enables mass production of plants for new commercial plantings. This has considerable advantages over conventional techniques as it ensures that plantations are free (at least initially) from nematode parasites and borers. The incidence of undesirable somaclonal variants (Vyulsteke *et al.,* 1988) may become a cause for some concern. Micro propagation should be of benefit to Nematologists who should be able to devise critical tests for pathogenicity of different nematode species on breeding lines and new cultivars.

Nematicides are the only available means of controlling nematodes in established plantations. Cost and high mammalian toxicities discourage nematicide use in most growing systems other than for international export. Further, research into suitability of nematicides in certain soil types and environment will be necessary and greater attention will have to be given to possible development of microbially active soils. With increased application efficiency, selection of soil floras with enhanced ability to degrade nematicides may become a major technical problem requiring careful planning of nematicide rotations.

Effective awareness programmes have to be launched among farming community for imparting a clear understanding of nematode damage. Considering non-availability of safe nematicides, cost-effective biological formulations coupled with seed/seed material treatment have to be promoted for early root protection against nematode attack. Chemical management of nematodes is ruled out in many crops due to mass drive towards organic produce. Therefore, alternate eco-friendly measures to be developed and this cost-benefit relationship is to be worked out. Studies on nematode damage to horticultural crops and their economic importance need to receive greater attention in future.

References

Adesiyan, S.O. (1977) Penetration and multiplication of *Scutellonema bradys* in yams (*Dioscorea* spp.) *Nematol. Medit.* 5: 313-317.

Adesiyan, S.O. and Odihirin, R.A. (1975) Distribution of nematode parasites of yam tubers in Mid-West State, Nigeria. *Occ. Publs. Niger. Soc. Pl. Prot.* 1: 21.

Adesiyan, S.O. and Odihirin, R.A. (1978) Root-knot nematodes as pests of yam (*Dioscorea* spp.) in South Nigeria. *Nematologica*. 24:132-134.

Ajith, K. and Sheela, M.S. (1996) Utilizations of green leaves of Neem and Eupatorium for the management of soil organisms in bhindi and cowpea. *Indian J. Nematol.* 139-141.

Alam, M.M. (1989) Control of root knot and stunt nematodes with horn meal, bone meal and oil seed crops. *Indian J. Nematol.* 19: 166-170.

Ali, S.S. (1984) Occurrence of root-knot nematodes in cardamom plantations of Tamil Nadu. (In) *Proc. PLACROSYM- VI 1982*. (Eds) Bavappa, K.V.A., Iyer, R.D., Das, P.K., Sethuraj, M.R.,Ranganathan, V., Nambiar, K.K.N. and Chacko, M.J. Indian Society for Plantation Crops, Kasaragod, Kerala, India, pp. 615-620.

Ali, S.S. (1986) Occurrence of root-knot nematodes in cardamom plantations of Karnataka.*Indian J. Nematol*, 16: 269-270.

Ali, S.S. (1987) Effect of three systemic nematicides against root-knot nematodes in cardamom nursery. *Nematologia mediterranea,* 15:155-158.

Ali, S.S. and Koshy, P.K. (1982) Occurrence and distribution of root - knot nematodes in cardamom plantations of Kerala. *Nematologia mediterranea,* 10: 107-110.

Ananadaraj, M., Ramana, K.V. and Sharma, Y.R. (1991) Interaction between vesicular arbuscular mycorrhizal fungi and *Meloidogyne incognita* in black pepper. (In): Mycorrhizal Symbiosis and Plant Growth. (eds.) Bagyaraj, D.J. and Manjunath, A. University of Agricultural Sciences, Bangalore, India , pp.110-112.

Ananadaraj, M., Ramana, K.V. and Sharma, Y.R. (1996) Suppressive effect of VAM on root damage caused by *Phytophthora capsici, Radopholus similis* and *Meloidogyne incognita* in black pepper. (In) Impact of Diseases and Pests in Tropical Forests, (Eds.) Nair, K.S.S., Sharma, J.K. and Varma, R.V., Kerala Forest Research Institute, Peechi, India, pp. 232-238.

Anguiz, R. and Canto-Saenz, M. (1991) Reaction of sweet potato cultivars (*Ipomoea batatas* (L.) Lam. *Pratylenchus flakkensis* Seinhorst. *Nematropica*. 21:197-201.

Anitha, B. and Subramanian, S. (1998) Management of the reniform nematode, *Rotylenchulus reniformis* in tomato. *Proc. Third Int. Symp. Afro-Asian Soc. Nematologists.* (ed) Mehta, U.K. Sugarcane Breeding Institute, Coimbatore, pp. 249-250.

Aravind, R., Antony, D., Santhosh,J.E., Kumar, A. and Ramana, R.V. (2004) Isolation and *in vitro* evaluation of endophytic bacteria against root-knot nematodes infesting black pepper (*Piper nigrum* L.). *National Nematology Symposium, 17-19 November 2004, University of Agricultural Sciences, Bangalore Absract p*.

Arya, M. and Tiagi, B. (1982) Reaction of some carrot cultivars to root - knot nematode, *Meloidogyne incognita*. *Indian J. Nematol*. 12: 397.

Atu, U.G., Okoli, O.O. and Oghuji, R.O. (1984) Reactions of yams (*Dioscorea* spp.) to *Meloidogyne incognita* race 2. *International Nematology Network Newsletter*. 1: 10-12.

Averre, C.W., Barker, K.R. and Jones, T.L. (1985) Nematicides for the control of root-knot nematodes on sweet potato, 1984. *Fungicide and Nematicide Tests*. 41: 80-81.

Ayala, A. and Acosta, N. (1971) Observations on yam *Dioscorea alata* nematodes. *Nematropica*. 1:39-40.

Ayyar, P.N.K. 1926. A preliminary note on the root gall nematode, *Heterodera oryzicola* and its economic importance in South India. *Madras agric. J.* 14: 113-118.

Bai, H., Sheela, M.S. and Jiji, T. (1995) Nemic association and avoidable yield loss in turmeric, *Cucurma longa* L. *Pest Mgmt. Hort. Ecosystems* 1: 105-110.

Beena, B., Eapen, S.J. and Ramana, K.V. (2001) Antogonistic rhizobacteria of root knot nematodes infesting black pepper (*Piper nigrum* L.). (In) Abstracts- National Congress on Centenary of Nematology in India. Indian Agricultural Research Institute, New Delhi, India, pp. 143.

Beena, B., Eapen, S.J. and Ramana, K.V. (2003) Native rhizobacteria for the biological suppression of *Radopholus similis* infesting black pepper (*Piper nigrum* L.). (In) Abstracts and Short Papers - Sixth International PGPR Workshop, Calicut, India. pp145.

Blake, C.D. (1961) Root rot of bananas caused by *Radopholus similis* (Cobb) and its control in New South Wales. *Nematologica*, 6: 295-310.

Blake, C.D. (1966) The histological changes in banana roots caused by *Radopholus similis* and *Helocotylenchus multicinctus*. *Nematologica*, 12: 129-137.

Bridge, J. (1986) Control strategies in subsistence agriculture. (In) *Principles and Practices of Nematode control in crops* (eds) Brown, R.H. and Kerry, B.R Academic Press, Australia. p.447.

Cadet, P. and Quencherve, P. (1994) Natural fluctuations of *Scutellonema bradys* (*Nematoda Hoplolaimidae*) during yam growth (*Dioscorea alata*) in Martinique (West Indies). *Nematologica*. 40: 587-600.

Cannayane, I., Sivakumar, C.V. and Jonathan, E.I. (2004) Colonization of black gram, cowpea, tomato and chilli roots by *Paecilomyces lilacinus* protects *Meloidogyne incognita* invasion. *Nat. Symp. Paradigms Nematological Res. Biodynamic Farming, November* 17th to 19th, *2004.* Nematological Society of India, University of Agricultural Sciences, Bangalore. Abstract: 97.

Castagnone-Sereno, P. (1988) Disinfection of yam tubers by hot water and chemical treatment: Nematicide efficacy and agronomical consequences. *Turrialba*.38:337-340.

Caveness, F.E. (1980) Plant parasitic Nematode on cassava. *Proceedings of the workshop on cassava production and extension in Central Africa*. (ed.) Ezumah, H.C. 4: 83-106.

Central Plantation Crop Research Institute (CPCRI) (1974) Annual Report, 1973, Kasargod, 62 pp. 207.

Chand, R., Gill, J.S. and Yadav, B.D. (2001) Management of root knot nematode using *Pasturia penetrans* on the first and second subsequent tomato crop. Nat. Congr. Centenary Nematology India Appraisal Future Plans, 5-7 December (2001) Nematological Society of India, Indian Council of Agricultural Research, National Academy of Agricultural Sciences, Department of Sciences and Technology, Society of Plant Protection Sciences, Indian Agricultural Research Institute, New Delhi. *Abstract:* 141.

Channabasappa, B.S, Krishnappa, K and Reddy, B.M.R. (1995) Utilization of eco-friendly biological agents and bio components in the integrated management of *Radopholous similis* on banana. Paper presented in the *National Symposium on Nematode Management with Eco-friendly Approaches and Bio components,* New Delhi, Abstr. pp. 42.

Charles, J.S.K. (1978) Studies on the nematode diseases of ginger. M.Sc (Ag) Thesis, Kerala Agricultural University, Thrissur, 72 p.

Charles, J,S.K. (1989) Bioecology of the cyst nematode infecting banana. Ph.D. Thesis, Kerala Agricultural University, Trichur, 122 p.

Charles, J.S. and Kuriyan, K.J. (1982) Studies on nematode incidence in ginger. (In) Proc. Placrosym -II 1979. Venkatram (ed.) Indian Society for Plantation Crops, Kasargod, Kerala, India, pp. 50-57.

Charles, J.S.K. and Venketesan, T. S. (1993) Pathogenicity of *Heterodera oryzicola* (Nemata: Tylenchina) towards banana (Musa AAB cv. Nendran). *Fundamentals of Appl. Nematol.* 16: 359-385.

Charles, J.S.K., Venkitesan, T.S. and Thomas, Y. (1983) Field screening of banana against the burrowing nematode, *Radopholus similis* (Cobb) Thorne. *Third Nematol. Symp. Solan* p. 45.

Charles, J.S.K, Venkitesan, T.S and Thomas, Y. (1985) Comparative efficiency of antagonistic intercrops with carbofuran in control of burrowing nematodes in the banana cultivar Nendran. *Indian J. Nematol.* 15: 241-242.

Charles, J.S.K, Rani, T.G and Krishnan, S. (1996) Management of banana nematodes using cultural and chemical strategies. *Pest Mangmt. Hortl. Ecosystems*. 2: 19-22.

Chen, C.M., Li, H.Y. and Lii, O.Y. (1986) The study on root knot nematodes of common turmeric, *Cucurma domestica* (Valet.). *Herald. agric.Sci.* 1: 16-22.

Clark, C.A. and Moyer, J.W. (1988) *Compendium of Sweet Potato Diseases*. APS Press, American Phytopathological Society. p. 74.

Crozzoli, P.R. and Hidalgo, S.O. (1992) Response of ten cassava cultivars to the nematode *Meloidogyne incognita. Fitopatologia Venezolana.* 5:20-22.

CTCRI, 1997. Annual report for (1996-97) Central Tuber Crops Research Institute, Trivandrum, India. pp.99.

D'Souza, G.I., Kumar, A.C., Viswanathan, P.R.K. and Shamanna, H.V. (1970) Relative distribution and prevalence of plant parasitic nematodes in coffee tracts of South Western India. *Indian Coffee* 34: 329-330.

Dahiya, R.S., Bansal, R.K. and Paruthi, I.J. (1998) Efficacy of mustard cake and poultry manure for the management of *Meloidogyne javanica* in bottle gourd and okra. *Proc. National Symposium on rational Approaches in Nematode Management for Sustainable Agriculture,* November 23rd - 25th, 1998 Nematological Society of India, IARI, New Delhi, pp.100-102.

Dasgupta, N.K. and Rama, K. (1987) Plant parasitic nematode associated with plantation crops. *Review of Tropical Plant Pathology,* 4: 289-304.

Davide, R.G. and Marasigan, L.Q. (1985) Yield loss assessment and evaluation of resistance of banana cultivars to the nematodes *Radopholus similis* Thorne and *Meloidogyne incognita* Chitwood. *Philippine Agriculturist* 68:335-349.

Deepthi, N.P. (1993) Control of collar rot and root-knot of cowpea with VA mycorhizae. M.Sc. (Ag.) Thesis, Kerala Agricultural University, Vellanikkara, 109 p.

Degras, L. (1993) The yam, A tropical root crop (Ed.) Rene Coste. MacMillan Press Ltd. 408 p.

Devarajan, K and Rajendran, G, (2001) Effect of fungus *Paeciliomyces lilacinus* (Thom.) Samson on the burrowing nematode *Radopholis similis* (Cobb.) Thorn. in banana. *Pest Mgmt Hort. Ecosystems* 7: 171-173.

Devarajan, K and Rajendran, G. (2002) Effect of fungal egg parasite, *Paecilomyces lilacinus* (Thom.) Samson on *Meloidogyne incognita* in Banana. *Indian J. Nematol*. 32: 78-101.

Dhangar, D.S., Gupta, D.C.and Jain, R.K. (1996) Studies on aqueous extract of different parts of *Tagetus erecta* on hatching, larval mortality and larval penetration of *Meloidogyne javanica. Indian J. Nematol.* 26: 46-51.

Eapen, S.J. (1995a) A note on the incidence of *Rotylenchulus reniformis* in cardamom plants. *Indian J. Nematol.* 25: 213.

Eapen, S.J. (1995b) Investigations on Plant Parasitic Nematodes Associated with Cardamom. Final Report, Indian Institute of Spices Research, Calicut, Kerala, India.

Eapen, S.J. and Venugopal, M.N. (1995) Field evaluation of *Trichoderma* spp. and *Paecilomyces lilacinus* for control of root knot nematodes and fungal diseases in cardamom nurseries. *Indian J. Nematol.* 25: 15-16.

Eapen, S.J., Ramana, K.V. and Sarma, Y.R. (1997) Evaluation of *Pseudomonas fluorescens* isolates for control of *Meloidogyne incognita* in black pepper (*Piper nigrum* L.). (In) *Biotechnology of Spices, Medicinal and Aromatic Plant* - Proceedings of the National Seminar on Biotechnology of Spices and Aromatic Plants, (eds.) Edison, S., Ramana, K.V., Sasikumar, B., Nirmal, B.K. and Eapen, S.J. Indian Society for Spices, Calicut, India, pp. 129-133.

Eapen, S.J., Chempakam, B. and Ramana, K.V. (1999) Role of defense related enzymes in the resistance of black pepper to *Meloidogyne incognita*. (In) Abstract of Papers - National Seminar on Nematological Research in India, C.S. Azad University of Agriculture and Technology, Kanpur.pp72.

Eapen, S.J., Beena, B. and Ramana, K.V. (2000) Preliminary screening of some hyphomycetous fungi for the biological control of root-knot nematodes (*Meloidogyne incognita*) infesting spice crops. (In) Abstract of Papers - National Nematology Symposium on Integrated Nematode Management, Orissa University of Agriculture and Technology, Bhubaneshwar, Odisha. p. 18.

Gapasin, R.M. (1981) Control of *Meloidogyne incognita* and *Rotylenchulus reniformis* and its effect on the yield of sweet potato and cassava. *Ann. Tropical Res.* 3:92-100.

Geetha, S.M. (1991) Studies on the Biology, pathogenicity and biocontrol of different populations of *Radopholus similis*. Ph. D. Thesis, Kerala University, Kerala. pp. 216.

Generalo, L and Davide, R.G. (1986) Effects of *Paecilomyces lilacinus, Penicilliunm anatolicum* and *Arthrohotrys cladodes* on the control of *Radopholous similis* on banana. Paper presented at *Second International Workshop on Biological Control of Nematodes,* Feb 6th-9th, 1986, Los Bonos, Phillipinies, Abstract.p29.

Gogoi, B.B. and Gill, J.S. (2001) Compatability of *Pasteuria penetrans* with carbofuran and organic amendments, its effect on *Heterodera cajani. Ann. Pl. Protection Sci.* 9: 254-257.

Goswamy, B.K. and Vijayalakshmi, K. (1986a) Nematicidal properties of some indigenous plant materials against root-knot nematode, *Meloidogyne incognita* on tomato. *Indian J. Nematol.* 16: 65-68.

Goswamy, B.K. and Vijayalakshmi, K. (1986b) Efficacy of some indigenous plant materials and non-edible seed cakes against *Meloidogyne incognita* on tomato. *Indian J. Nematol.* 16: 280-281.

Goswami, B.K. and Sharma, S.B. (2001) Application of *Aspergillus terreus* and *Paeciliomyces lilacinus* for the management of *Meloidogyne incognita* on tomato. *Int. J. Nematol.*11: 270-273.

Goswami, B.K. and Mittal, A. (2002) Effect of some fungal bioagents on root-knot nematode, *Meloidogyne incognita* infecting brinjal. *Pakist. J. Nematol.* 20: 55-59.

Govindankutty, M.P. and Koshy, P.K. (1979) Nematodes associated with the coconut palm (*Cocos nucifera* L.) (In) *Technical Bulletin 2 - Nematodes, fungi, insects and mites associated with the coconut palm.* Central Plantation Crops Research Institute, Kasaragod, India, 1-50.

Gunasekharan, C.R., Vadivelu, S. and Jayaraj, S. (1987) Experiments on Nematodes of Turmeric - A Review. Proc. *Third Group Discussions on the Nematological Problems of Plantation Crops.* October 29-30, 1987. Sugarcane Breeding Institute, Coimbatore, Tamil Nadu, India. Abstract: 45.

Haider, M.G., Jha, R.N. and Nath, R.P. (1998a) Studies on the nematodes of spices. Pathogenic effect of root knot (*Meloidogyne incognita*) and reniform nematodes (*Rotylenchulus reniformis*) alone and in combination on turmeric (*Curcurma longa* L.). *Indian J. Nematol.* 28: 52-55.

Haider, M.G., Jha, R.N. and Nath, R.P. (1998b) Efficacy of some nematicides and organic soil amendments on nematode population and growth characters of turmeric (*Curcurma longa* L.). *J. Res. Birsa Agric. Univ.* 10: 211-213.

Hall, M.R., Johnson, A.W. and Smittle, D.A. (1988) Nematode population dynamics and yield of sweet potato and onion as affected by nematicide and time of application. *Ann.Appl. Nematol.* 2:15-21.

Haseeb, A. and Butool, F. (1994) Evaluation of different pesticides and oil cakes for the control of *Meloidogyne incognita* infesting *Hyoscyamus albus. Shebin El-Kom,* Egypt: 117-120.

Hazharika, K. Dutta, P.K., Saikia, L. and Dewka, S.C. (2000) Biological control of root knot nematode in betelvine using parasitic fungus, *Paecilomyces lilacinus. Crop Res.* 19: 338-342.

Hedge, L., Krishnappa, K. and Bojappa, K.M. (1993) Reaction of different cultivars of cardamom (*Elettaria cardamomum* Maton) to root knot nematode (*Meloidogyne incognita*). *Indian J. Nematol.* 23: 127-128.

Hogger, C.H. (1971) Plant parasitic nematodes associated with cassava. *Tropical Root Horticulture Society* 94: 321-323.

Hrishi, N. and Mohankumar, C.R. (1976) Coleus for homestead gardens. *Indian Farming.* 26(3): 33-35.

Husain, I.S., Kumar, R. Khan, T.A. and Titov, A. (1984) Effect of root dip treatment of egg plant seedlings with plant extracts, nematicides, oil cake extracts and antihelminthic drugs on plant growth and root-knot development. *Pakist. J. Nematol.* 2: 79-83.

Hussaini, S.S., Bharatha, K.S. and Rao, R.V.V.P. (2001) Soil solarization - comparison of LDPE gauge and periods for control of root knot nematode and weeds in FCV tobacco nursery. *Curr. Nematol.* 12: 39-43.

Hutton, D.G., Coates-Beckford, P.L., Jones, A.G. and Brathwaite, C.W.D. (1978) Effect of various treatments for nematode control on early growth of yams. *Nematropica* 8:6.

Jacob, A. and Kuriyan, J. 1979a. Screening of pepper varieties for resistance against root-knot nematode (*Meloidogyne incognita*). *Agric. Res. J. Kerala* 17: 90.

Jacob, A. and Kuriyan, J. (1979b) Survey of nematodes associated with pepper in Kerala. *Agric. Res. J. Kerala* 17: 270-271.

Jasy, T. and Koshy, P.K. (1992) Effect of certain leaf extracts and leaves of *Glyricidia maculata*, (H.B and K.) Steud. as green manure on *Radopholous similis. Indian J. Nematol.* 22:117-121.

Jatala, P.R., Kaltenbach, A.J., Devaux and Campos, R. (1980) Field application of *Paecilomyces lilacinus* for controlling *Meloidogyne incognita* on potatoes. *J. Nematol.* 12: 226-227.

Jatala, P. (1989) Important nematode parasites of sweet potatoes and their management. (In) *Improvement of sweet potato (Ipomoea batatas) in* Asia International Potato Centre, Lima, Peru. pp.213-218.

Jatala, P. and Bridge, J. (1990) Nematode parasites of root and tuber crops. (In) *Plant Parasitic Nematodes in Subtropical and Tropical Agriculture.* (eds) Luc, M., Sikora, R.A. and Bridge, J. Pub. CAB International. pp 137-180.

Jhala, A.J. and Patel, R.H. (2004) Management of root knot nematode, *Meloidogyne incognita* through soil solarization and intercropping. Nat. Symp. Paradigms Nematological Res. Biodynamic Farming, 17-19 November 2004. Nematological Society of India, University of Agricultural Sciences, GKVK, Bangalore. *Abstract*: 87.

Jiji, T., Sheela, M.S. and Premila, K.S. (1999) Nematodes associated with Banana rhizosphere in different cropping systems in Kerala. *Proc. National Seminar on Nematological Research in India. Challenges and Preparedness for the New Millenium, KAU, 1983,* Abstr:10.

Johnson, L.F. and Curl, E.A. (1972) *Methods for Research in the Ecology of Soil Borne Plant pathogens.* Burgess Publishing Co., Minneapolis, 247pp.

Jonathan, E.I and Rajendran, G. (2000) Biocontrol potential of the parasitic fungus *Paecilomyces lilacinus* against the root-knot nematode, in banana *J. Biol.control* 14(2):67-69.

Jonanthan, E.I. and Umamaheswari, R. (2006) Biomanagement of nematodes infesting banana by bacterial endophytes (*B. subtilis*). *Indian J. Nematol.* 36: 230-233.

Jonathan, E.I., Barker, K.R. and Abd- El – Aleem, F.F. (1999) Host status of banana for four major species and host races of *Meloidogyne. Nematol. Medit.* 27: 123-125.

Jonathan, E.I., Arulmozhiyan, R., Muthusamy, S. and Mannuel, K.W. (2000) Field application of *Paecilomyces lilacinus* for the control of *Meloidogyne incognita* on betel vine, *Piper betle. Nematol. Medit.* 28: 131-133.

Joseph, P.J., Sivaprasad,P., Sheela, M.S. and Jacob, A. (2001) Management of root-knot nematode *Meloidogyne incognita* (Kofoid &White) Chitwood in ginger using arbuscular mycorrhizal fungi (AMF). *Symposium on Biocontrol Based Pest Management for Quality Crop Protection in the current millennium*. July 18th -19th, 2001. PAU, Ludhiana. 102 p.

Kamra, D.A. and Gaur, H.S. (1993) Efficiency of soil solarization and formaldehyde application alone and in combination on plant parasitic nematodes and growth of egg plant seedlings. Nat. Symp. Recent Approaches Integrated Nematode Mgmt. Agric. Crops, 6-7 August 1993. Haryana Agricultural University. *Abstract*: 80.

Karuna, K., Krishnappa, K. and Reddy, B.M.R. (2001) Utilization of *Pasteuria penetrans* in the management of root knot nematode on brinjal. *Nat. Congr. Centenary Nematology India Appraisal Future Plans, December* 5th- 7th 2001. Nemaological Society of India, Indian Council of Agricultural Research, National Academy of Agricultural Sciences, Department of Science and Technology, Society of Plant Protection Sciences, Indian Agricultural Research Institute, New Delhi. Abstract: 142.

KAU. 1999. Final report (1999). ICAR Adhoc Scheme on Development of a suitable natural bio-control agent against the cyst nematode, *Heterodera oryzicola* infesting banana, Banana Research Station, Kannara, Kerala Agricultural University, p. 15.

Kaur, K.J. (1987) Studies on nematodes associated with ginger (*Zingiber officinale* Rosc.) in Himachal Pradesh. Thesis submitted to Himachal Pradesh University, Shimla, India, 198pp.

Khan, E. and Nanjappa, C.K. (1972) Four new species of Criconematoidea (Nematoda) from India. *Indian J. Nematol*. 2: 59-68.

Khan, H.U., Ahmad, R., Ahmad, W., Khan, S.M. and Akhtar, A.S. (2000) Evaluation of chemical vs biological control treatments against root knot nematode, *Meloidogyne incognita* on tomato. *Pakist.J. Phytopath*. 12: 118-120.

Khan, M.R. and Goswami, B.K. (2000a) Effect of different doses of *Paecilomyces lilacinus* isolate 6 on *Meloidogyne incognita* infecting tomato. *Indian J. Nematol.* 30: 5-7.

Khan, M.R. and Goswami, B.K. (2000b) *In vitro* evaluation of Indian isolates of *Paecilomyces lilacinus* against *Meloidogyne incognita* eggs. *Indian J. Nematol.* 30: 86-110 .

Khan, M.R., Goswami, B.K. (2002) Evaluation of *Paecilomyces lilacinus* isolate 6 against *Meloidogyne incognita* infecting tomato. *Int. J. Nematol*. 30: 86-110.

Khan, M.R. (2003a) Life cycle, Pathogenicity and management of root knot nematode, *Meloidogyne incognita* Races infecting Balsam (*Impatiens balsamina* L.). *Indian J. Nematol.* 33: 61-68.

Khan, M.R. (2003b) Effect of different inoculum levels of root-knot nematode, *Meloidogyne incognita* Race 2 on onion (*Allium cepa* L.). *Indian J. Nematol.* 33: 61-66.

Khanna, S.K. and Sharma, K.N. (1998) Phytotherapeutic effect of some indigenous plants / nematicides on *Meloidogyne incognita* infesting tomato. *Proceedings of National Symposium on Rational Approaches in Nematode Management for Sustainable Agriculture,* November 23rd – 25th 1998 (eds. Dhawan, S.C. and Kaushal, K.K.) Nematological Society of India, Indian Agricultural Research Institute, New Delhi, pp. 4-6.

Khanna, T.A. (2000) Biomanagement of *Meloidogyne incognita* (Kofoid and White) Chitwood by *Paecilomyces lilacinus* (Thom.) Samson on egg plant. *Pest Mgmt. Economic Zool.* 8: 133-136.

Koshy, P. K. (1986a) Research on plant parasitic nematodes at the Central Plantation Crops Research Institute, Kayamkulam, India, October 1972 to June 1986, Report, 45p.

Koshy, P.K. (1987) Integrated Nematode Management in Plantation Crops. *Proc. of the Third Group Discussion on the Nematological Problems of Plantation Crops. October, 29th -30th, 1987.* Sugarcane Breeding Institute, Coimbatore, Tamil Nadu, India: 49-51.

Koshy, P.K and Bridge, J. (1990) Nematode parasites of spices. (In) *Plant parasitic nematodes in subtropical Agriculture* (eds) Luc, M., Sikora, R.A and Bridge, J. Pub. CAB International. pp557-582.

Koshy, P. K. (1998) Biological control of burrowing nematode on palms and black pepper. Final Tech. Report, USIF Project No. 1 N –AES-779 C.P.C.R.I., Kayangulam, Kerala, India, 690533.pp359.

Koshy, P.K. and Nair, C.P.R. (1979) Control of *Radopholous similis* (Cobb 1893) Throne, 1949 in coconut nursery. *Indian J. Nematol.* 9:15-19.

Koshy, P. K. and Sosamma, V. K. (1979) Control of the burrowing nematode *Radopholus similis* on coconut seedlings with DBCP. *Indian J. of Nematol*, 7:32-33.

Koshy, P. K., Sosamma, V. K. and Sundararaju, P. (1979) Reactions of thirty one *Areca* germplasm collections to the burrowing nematode *Radopholus similis. Plant Disease Reporter*, 63: 433-435.

Koshy, P.K. and Sundararaju, P. (1979) Response of seven black pepper cultivars to *Meloidogyne incognita. Nematologia Mediterranea* 7: 123-135.

Koshy, P. K., Sosamma, V. K. and Nair, C. P. R. (1975) Preliminary studies on *Radopholus similis* (Cobb, 1893) Thorne, 1949) infesting coconut and arecanut palms in South India. *Indian J. Nematol.* 5: 26-35.

Koshy, P.K., Sosamma, V.K. and Sundararaju, P. (1977) Screening of plants used as pepper standards against root-knot nematode. *Indian Phytopathol.* 30: 128-129.

Koshy, P. K., Raski, D. J. and Sosamma, V. K (1981) *Brachydorus swarupi.* (Nemata Dolichodorinae) from soil about roots of arecanut palms in Kerala state. *Indian Journal of Nematology* 13: 401-404.

Koshy, P.K., Sosamma, V.K. and Faleiro, J.R. (1987) Occurrence of *Heterodera oryzicola* on banana in Goa. *Indian J. Nematol.* 17: 334.

Koshy, P.K., Sosamma, V. K. and Sundararaju, P. (1991a) Nematode Management in coconut and coconut based cropping system. Invited paper Session I on Pest Management, *International Symposium on Coconut Research and Development – II,* Kasaragod, Kerala, 26-29 November, 1991.

Koshy, P.K., Sosamma, V. K. and Sundararaju, P. (1991b) *Radopholus similis,* the burrowing nematode of Coconut. *Journal of Plantation Crops.* 19 : 139-152.

Koshy, P.K., Sosamma, V.K. and Banu, J.G. (2002) *Milestones of Nematological Research in Kerala.* Released during the seminar on Centenary on Nematological Research in Kerala during February 21st -22nd 2002 at CPCRI, Kayangulam, Kerala.

Kukimura, J., Yoshida, T. and Komaki, K. (1989) Satsumahikari; a new sweet potato cultivar. *Bull. Kyushu Natl. Agric. Exp. Station.* 25: 225-250.

Kumar, A.C., Viswanathan, P.R.K. and D'Souza, G.I. (1971) A study of plant parasitic nematodes of certain commercial crops in coffee tracts of South India. *Indian Coffee* 35: 222-224.

Kumar, S. and Vadivelu, S. (1996). Evaluation of organic amendments for the management of root knot and reniform nematodes infesting Brinjal, as compared with carbofuran. *Pest Mgmt Hort. Ecosystems* 2: 71-74.

Kuriyan, K.J. and Sheela, M.S. (1981) Integrated control of root-knot nematode *Meloidogyne incognita* in brinjal. *Indian J. Nematol.* 11:129.

Lokesh, M.S. and Gangadharappa, P.M. (1995) Management of Phytophthora foot rot and nematode diseases in black pepper (*Piper nigrum L.*). *J. Spices and Aromatic Crops* 4: 61-63.

Loos, C.A. 1961. Eradication of the burrowing nematode, *Radopholus similis* from bananas. *Plant Disease Reporter*. 45: 457-461.

Mahapatra, S.N. and Swain, P.K. (1993) Nematicidal efficacy of citronella leaf extract against *Meloiogyne incognita* in tomato. *Indian J. Nematol.* 23: 9.

Mammen, K.V. (1973) Root gall nematodes as a serious pest of ginger in Kerala. *Curr. Sci.* 42: 549.

Mani, A. and Prakash, K.S.(1992). Plant parasitic nematodes associated with turmeric in Andhra Pradesh, India. *Int. Nematol. Network Newsl.* 4: 13-18.

McSorley, R. (1980) Nematodes associated with sweet potato and edible aroids in Southern Florida. *Proc. Flor. State Hort. Soc.* 93: 283-85.

Medina, F., Crozzoli, P.R. and Rivas, D. (1992) Control of the nematode *Meloidogyne incognita* in cassava with different biocides. *Fitopatologiia Venezolana.* 5:26-29.

Mehta, U.K. (1992) Nematode pests of sugarcane. (In) Nematode Pests of crops, D.S Bhatti and R.K. Walia (Eds.) CBS Publishers, New Delhi. pp 159-176.

Melin, P. and Vilardebo, A. (1973) Efficacete de quelques nematicides en bananeraie dans sols volcaniques de la region du Mungo (Cameroun) *Fruits* 28 : 3-17.

Mohandas, C. (1994) Nematodes in Tuber Crops. (In) *Advances in Horticulture Vol. 8 - Tuber Crops*. (Eds.) Chadha, K.L. and Nayar, G.G. Malhotra Publishing House, New Delhi, India. p. 523-533.

Mohandas, C. and Ramana, K.V. (1983) Effect of different levels of *Meloidogyne incognita* on plant growth of two cultivars of black pepper (*Piper nigrum* L.) Abstracts of papers. *Third Nematology Symposium Organised by Nematological Society of India*, Himachal Pradesh, India: 9-10.

Mohandas, C. and Palaniswami, M.S. (1990a) Resistance in sweet potato (*Ipomoea batatas* L.) to *Meloidgyyne incognita* (Kofoid and white) Chitwood in India. *J.Root Crops* 148-149.

Mohandas, C. and Palaniswami, M.S. (1990b) Effect of *Meloidogyne incognita* on three popular cultivars of *Colocasia esculenta*. *National Symp. on Recent Advances in the Production and Utilization of Tropical Tuber Crops*. November 7th - 9th 1990. ISRC, CTCRI, Thiruvananthapuram, India.

Mohandas, C. and Rajendran, P.G. (1993) Resistance in sweet potato germplasm to root - knot *Meloidgyyne incognita* (Kofoid and white, 1919). *J.Root Crops* 19: 134-135.

Mohandas, C. and Palaniswami, M.S. (1996) Resistance in tuber crops to *Meloidogyne incognita* (Kofoid and White, 1919). In *Tropical Tuber Crops - Problems Prospects and future strategies* (eds.) Kurup, G.T., Palaniswami, M.S., Potty, V.P., Padmaja, G., Kabeerathumma, S. and Pillai, S.V. Oxford and IBH Publishing Co. Pvt. Ltd., New Delhi and Calcutta. pp. 366-369.

Mohanty, K.C., Ray, S., Mahapatra, S.N., Patnaik, P.R. and Ray, P. (1995) Integrated management of root knot nematode in ginger (*Zingiber officinale* Rosc.). *J. Spices & Aromatic Crops* 4: 70-73.

Mojumder, V. and Basu, R. (1999). Seed coating of chickpea with neem based pesticidal formulations for the management of *Meloidogyne incognita*. *Indian J. Nematol*. 29: 28-32.

Mukhtar, T., Ahmad, R. and Javed, N. (2002) Efficacy of *Pasteuria penetrans* and *Verticillium chlamydosporium* in the biological control of *Meloidogyne javanica*. *Pakist. J. Phytopath*. 14:79-83.

Nadakal, A.M. and Thomas, A.N. (1964). Studies on plant parasitic nematodes of Kerala. *Curr.Sci.* 33:247-48.

Nagesh, M. (1996) Relationship between the population densities of root knot nematode, *Meloidogyne incognita* (Kofoid and White,1919), Chitwood, 1949 and growth of potato, *Solanum tuberosum* L. *Pest Mgmt. Hort. Ecosystems* 2: 9-14.

Naik, B.G., Nanjegowda and Ravi, K. (1998) Effect of neem products on the growth of tomato and development of root - knot nematode *Meloidogyne incognita*. *Proceedings of First National Symposium on Pest Management on Horticultural Crops, 15-17 October 1997* (eds. Reddy, P.P. Kumar, N.K.K. and Verghese, A.). Association for Advancement of Pest Management in Horticultural Ecosystems, Indian Institute of Horticultural Research, Bangalore, pp. 321-323.

Nair, K.K.R. (1979) Studies on the chemical control of banana nematodes. *Agri.Res. J. Kerala*. 17: 232-235.

Nair, R. B. (1964) Nematodes from arecanut soils. *Arecanut Journal*, 15: 87-88.

Nema, A.G. (2001) Control of nematodes in betelvine in standing gardens infested with root-knot nematodes. *Adv. Pl. Sci.* 14: 589-590.

Nirula, K.K. (1959) Root-knot nematode on colocasia. *Curr. Sci*., 28:125-26.

Nirula, K.K. and Kumar, R. (1963) Collateral host plant of root knot nematodes. *Curr. Sci.* 32: 221-222.

Nisha, M.S. (2001) Ecofriendly management of root-knot and burrowing nematodes associated with Kacholam (*Kaempferia galangal* Linn.) M.Sc (Ag) Thesis, KAU, Thrissur , p 84.

Nisha, M.S. (2005) Integrated management of root-knot nematode, *Meloidgyyne incognita* (Kofoid and white) Chitwood in Coleus. Ph.D.T thesis, Kerala Agricultural University, Vellanikkara, p.132.

Nisha, M.S. and Sheela, M.S. (2002) Effect of green leaf mulching for the management of root-knot nematode in kacholam. *Indian J. Nematol.* 32: 183-233.

Nisha, M.S. and Sheela, M.S. (2002) Rhizome treatment – a low cost technology for the management of Nematodes in kacholam, an important medicinal plant. Paper presented in the *Natl Symp. On Paradigms in Nematological Research for Biodynamic Farming Abstract-14.*

Nisha, M.S. and Sheela, M.S. (2003) Preliminary study of the effect of bioagents and organic amendments for the management of root-knot nematode associated with kacholam *Kaempferia galanga* Linn. *Pest Mgmt. Hort. Ecosystems* 9: 165-168.

Owino, P.O. (2003) Integrated management of root-knot nematodes using agrochemicals, organic matter and the antagonistic fungus, *Paecilomyces lilacinus* in natural field soil. *Nematol. Medit.* 31: 121-123.

Padhi, N. N. and Behera, G. (2000) Evaluation of nematicidal potential in ten indigenous plant species against *Meloidogyne incognita*. *Indian Phytopathol.* 53: 28-31.

Pant, H. and Pandey, G. (2002) Use of *Trichoderma* harzianum and neem cake alone and in combination on *Meloidogyne incognita* galls in chickpea. *Ann. P l. Protection Sci.* 10: 175.

Patel, A.D., Patel, D.J. and Patel, N.B. (2004) Effect of aqueous leaf extracts of botanicals on egg hatching and larval penetration of *Meloidogyne incognita* in banana. *Indian J. Nematol.* 34: 33-36.

Patel, D.J., Makadia, B.M. and Shah, H.M. (1982) Occurrence of root-knot on turmeric (*Cucurma longa* L.) and its chemical control. *Indian J. Nematol.* 12: 168-171.

Patel, R.G., Patel, B.A., Vyas, R.V., Patel, H.R. and Patel, D.J. (1998) *Proc. Natl. Symp. Rational Approaches in Nematode Management for Sustainable Agriculture,* November 23rd - 25th , GAU, Anand . pp. 89-90.

Patel, R.G., Patel, B.A., Patel, H.R., Vyas, R.V. and Patel, D.J. (2004) Distribution of plant parasitic nematodes in banana growing areas of Central Gujarat. *Indian J. Nematol.* 34: 231-232.

Patel, S.K., Patel, D.J. and Patel, H.V. and Patel, B.N. (2000) Management of root-knot nematodes in tomato nursery through botanical plant materials. *Proc. National Nematology Symposium on Integrated Nematode Management,* November 23rd – 27th 2000. Orissa University of Agricultural Science and Technology, Bhubaneswar. Nematological Society of India, Indian Agricultural Research Institute, New Delhi, Abstract: 21.

Patnaik, P.R. and Das, S.N. (1986) Pathogenicity of *Meloidogyne incognita* on edible coleus. *Indian J. Nematol.* 16: 271-2 72.

Ponte, J.J.DA., Torres, J. and Simplicio, M.E. (1979) Behaviour of cassava cultivars in relation to root-knot nematodes. (In) *Trabalhos apresentados a IV Reuniao Brasileirade Nematologia.* pp. 16-20.

Poornima, K. and Vadivelu, S. (1997) Effect of some organic oils and their cakes in the infectivity of root knot nematode, *Meloidogyne incognita* affecting tomato. *Indian J. Nematol.* 27: 165-172.

Rajagopalan, P. and Naganathan, T.G. (1977) Scheme for studies on the nematode parasite of banana. *Tamil Nadu Agric. Univ. Ann. Rep.* 6:129.

Rajagopalan, P, Chinnarajan, A.N and Vadivelu, S. (1977) Control of banana nematodes by rhizome treatments. *South Horticulture* 25:133-137.

Rajani, T.S. (1998) Bio- ecology and management of root-knot nematode, *Meloidogyne incognita* (Kofoid and White) Chitwood in Kacholam, *Kaempferia galanga* Linn. MSc. (Ag) thesis. p.72.

Rajendran, G., Naganathan, T.G. and Sivagami, V. (1979) Studies on the banana nematodes. *Indian J. Nematol.* 9: 54.

Rajendran, P.G. and Vadivelu, S (1991) Varietal screening of Dioscorea spp. to the root-knot nematode. *J.Root Crops.* 17: 223.

Rajendran, G and Sivakumar, C.V. (1996) Pathogenicity of the spiral nematode *Helicotylenchus multicinctus* and effect on yield and nutrient status of banana. *Pest Managmt. Hortil. Ecosystems.* 2: 23-27.

Ramana, K.V. (1986) Slow-wilt disease of black pepper and the role of plant parasitic nematodes in it's etiology. Proc. Second Group Discussion on Nematological Problems of Plantation Crops, April 24-25, 1985. Central Coffee Research Station, Balehonnur, Karnataka, India: 17-21.

Ramana, K.V. (1992) Role of nematodes in the incidence of slow decline (slow wilt disease) of black pepper (*Piper nigrum* L.) in Calicut district , Kerala. Abstracts of Papers, Third Nematology Symposium Organised by Nematological Society of India. Himachal Pradesh Krishi Viswa Vidyalaya, Solan, Himachal Pradesh, India: 9.

Ramana, K.V. (1994) Efficacy of *Paecilomyces lilacinus* (Thom.) Samson in suppressing nematode infestations in black pepper *(Piper nigrum L.). Journal of Spices and Aromatic Crops*, 3: 130-34.

Ramana, K.V. and Mohandas, C. (1986) Reaction of black pepper germplasm to root-knot nematode, *Meloidogyne incognita*. *Indian J. Nematol.* 16: 138-139.

Ramana, K.V. and Mohandas, C. (1987) Nematological Problems in Black pepper (*Piper nigrum* L.) Proceedings of the Third Group Discusion on Nematological Problems of Plantation Crops. October 29-30, 1987. Sugarcane Breeding Institute, Coimbatore, TamilNadu. India. pp.43-44.

Ramana, K.V. and Mohandas, C. and Ravindran, P.N. (1987) Reaction of black pepper germplasm to the burrowing nematode (*Radopholus similis*). *J. Plantation Crops* 15: 65-66.

Ramana, K.V. and Eapen, S.J. (1998) Plant parasitic nematodes associated with spices and condiments. In: *Nematode Diseases in Plants*. (ed.) Trivedi, P.C. CBS Publishers and Distributors, New Delhi, pp. 217-251.

Ramana, K.V., Eapen, S.J. and Beena, B. (2002) Biological Control of Plant Parasitic Nematodes of Major Spices, Final Report of ICAR Ad- hoc Project, Indian Institute of Spices Research, Calicut, Kerala, India.

Rao, M.S. (2005) Novel approaches for the management of nematodes in horticultural crops and standardization of production of bio-nematicides. In: *Biotechnological Management of Nematode Pests and Scope of Entomopathogenic Nematodes.* (Eds. Sithanantham, S., Vasantharaj David, B. and Selvaraj, P.) 2005. Sun Agro Biotech Research Centre Publication Sun Ray No. 1.

Ravichandra, N.G. and Krishnappa, K. (1985) Reaction of banana cultivars popularly grown in Karnataka to the burrowing nematode, *Radopholus similis. Indian J. Nematol.* 15: 240.

Ravindran, P.N., Ramana, K.V., Nair, M.K., Babu, K.N. and Mohandas, C. (1992) Pournami- a high yielding black pepper selection tolerant to root-knot nematode (*Meloidogyne incognita*). *J. Spices and Aromatic Crops* 4: 67-69.

Reddy, D. B. (1978) Insect pests, diseases and nematodes of *Areca catechu* and betel vine (*Piper betle*) recorded in Asia and the Pacific. *Technical Document No. 114. Plant Protection Committee for the South East Asia, Asia and Pacific region, FAO, Bangkok*, 6p.

Reddy, P.P. (1986) Analysis of crop loss in certain vegetables due to *Meloidogyne incognit*a. *Int. Nematol. Network Newsl.* 3: 3-5.

Reddy, P.P. and Khan, R.M. (1987) First report on the occurrence of *Radopholus similis in* Andhra pradesh, India. *Inter.Natl.Nematol.Network Newsletter.* 4:11.

Reddy, P.P. and Khan, R.M. (1989) Evaluation of biocontrol agent *Paecilomyces lilacinus* and carbofuran for the management of *Rotylenchulus reniformis* infecting brinjal. *Pakist. J. Nematol.* 7: 55-59.

Reddy, P.P., Rao, M.S. and Nagesh, M. (1996) Crop loss estimation in banana due to the burrowing nematode, *Radopholus similis*. *Pest Magmt. Hortl. Ecosystems* 2:85-86.

Reddy, P.P. Khan, R.M. and Rao, M.S. (1991) Evaluation of chemicals for the management of the burrowing nematode, *Radopholus similis*. *Curr. Nematol.* 2: 1-2.

Reddy, B.M.R., Krishnappa, K. and Karuna, K. (1997) Seedling bare root dip with chemicals for the manangement of root knot nematode in brinjal. *Indian J. Nematol.* 27: 258-259.

Reddy, P.P., Nagesh, M. and Rao, M.S. and Devappa, V. (1997) Integrated management of the burrowing nematode, *Radopholus similis* using endomycorrhiza, *Glomus mosseae* and oil cakes. *Pest Magmt. Hortl. Ecosystems* 3: 25-30.

Reddy, P.P. and Rao, M.S. (2003) Current trends in integrated nematode management in fruit crops. In : *Nematode management in plants*. (ed) Scientific Publishers India, pp.269-291.

Roberts, P.A. and Schueurman, R.W. (1984) Managing nematodes in sweet potatoes with resistance and nematicides. *California Afric.* 38: 26-27.

Roman, J., Oramas, D. and Green, J. (1984) Nematicide nematodes of yam (*Dioscorea rotundata* Poir). *J. Agriculture of the University of Pureto Rico*, 68:157-162.

Routary, B.N., Sahoo, H. and Das, S.N. (1987b) Nemic associations of ginger and turmeric in Orissa. *Indian J. Nematol.* 17: 122-123.

Sankaranarayanan, C., Hussaini, S.S., Sreeramakumar, P. and Rangeshwaran, R. (2000) Granular application of antagonistic fungi for the biological control of *Meloidogyne incognita* on tomato. *Indian J. Nematol.* 30: 157-161.

Sankaranarayanan, C., Hussaini, S.S., Kumar, P.S. and Rangeshwaran, R. (2001) Evaluation of substrates for the multiplication of *Verticillium chlamydosporium* Goddarad and its biocontrol efficacy against *Heterodera cajani* Koshy on pigeonpea. *Ann. Pl. Prot. Sci.* 9: 73-76.

Sathyarajan, P.K., Das, N.M. and Nair, M.R.G.K. (1966) Root-knot nematode as a pest of *Coleus parviflorus* in Kerala. *Kerala Agric. J.* 2:144-45.

Sharma, A. and Trivedi, P.C. (1989) Influence of inoculum levels of fungus, *Paecoilomyces lilacinus* (Thom.) Samson on the biocontrol of root-knot nematode, *Meloidogyne incognita* (Chitwood). *Int. Nematol. Network Newsl.* 6: 27-29.

Sheela, M.S. (1991) Control of root-knot nematode *Meilodogyne incognita* (Kofoid and White) Chitwood, 1949 infesting black pepper *(Piper nigrum L.)* by bacterial pathogens. Ph.D. Thesis, Kerala Agricultural University, Vellanikkara, 139 p.

Sheela, M.S. and Venkitesan, T.S. (1992) Bacterial pathogens associated with root-knot nematode in Kerala - their occurrence and status *Proc. 4th Kerala Science Congress,* (ed.) Ramachandran Nair, R. Kerala State Committee on Science, Technology & Environment, Government of Kerala, pp.231-233.

Sheela, M.S. and Venkitesan, T.S. (1994) Effect of Pesticides on the nematode pathogenic bacteria, *Bacillus macerans in vitro. J. Biol. Control*. 8: 102-104.

Sheela, M.S. and Venkitesan, T.S. and Mohandas, N. (1993) Status of *Bacillus spp.* as bio control agents of root-knot nematode (*Meloidgyne incognita*) infesting black pepper *(Piper nigrum L.). J. Plant. Crops*, 21: 218-222.

Sheela, M.S., Kuriyan, K.J., Jiji,T., Hebsybai and Mercey, J.N. (1994) Control of nematode complex in banana *Proc. Sixth. Kerala Sci. Cong*. pp. 192-193.

Sheela, M.S. and Venkitesan, T.S. (1994) Potential of the bacterial pathogen *Bacillus macerans* on root-knot nematode infesting black pepper. *Proc. Acad.Environ. Biol*. 6: 97-99.

Sheela, M.S. and Venkitesan, T.S. (1995a) Effect of pesticides on the nematode pathogenic bacterial *Bacillus macerans* Schardinger. *in vitro. J.Biol. Control* 8: 102-104.

Sheela, M.S. and Venkitesan, T.S. (1995b) Selection of media for the mass production of the bioagents (*Bacillus* spp) of root-knot nematode. Paper.

Sheela, M.S., Bai, H., Jiji, T. and Kuriyan, K.J. (1995c) Nematode associated with ginger rhizosphere and their management in Kerala. *Pest Mgmt. Hort. Ecosystems* 1: 43-48.

Sheela, M.S., Hebsybai and Anitha, N. (1996) Ecofriendly management strategies for burrowing nematode of banana in Kerala. *Proc. Int Natl. Symp. Tech. Advt. banana / plantain prodn. Processing India / International*. pp. 136-137.

Sheela, M.S., Jiji, T. and Hebsybai. (1999). Integrated nematode management strategy- A case study in brinjal. *Proc.XI Kerala Sci.Congr.* p. 227-229.

Sheela, M.S., Jiji, T. and Premila, K.S. (1999) Management of root-knot nematode through the plant rhizosphere bacterium *Pseudomonas fluorescens*. Migula. *Nat.Sem. Nematol. Res. India.,* Abstr., pp.33-34.

Sheela, M.S., Jiji, T. and Nisha, M.S. (2002) Evaluation of different control strategies for the management of nematodes associated with vegetables (brinjal). *Proc. International Conference on Vegetables.* p. 268–269.

Sheela, M.S., Hebsybai and Sivaprasad, P. (2003) Management of phytonematodes by Bioagents. (In): Nematode management in plants. Trivedi, P.S. (Ed) Pub. Scientific publishers, India, 185-201.

Sheela, M.S. and Nisha, M.S. (2004) Impact of biological control agents for the management of root-knot nematode in brinjal. *Nat. Symp. Paradigms Nematological Res. Biodynamic Fmg,* Nematological Society of India, University of Agricultural Sciences, Bangalore. Abstract: 63.

Sheela, M.S. and Mohandas, C. (2004) Non-chemical methods of nematode management in sugarcane. (In): Sugarcane Pathology Vol.III: Bacterial and nematode diseases. (eds) G.P. Rao, A. Salem Sanumtally and Philippe Rott. pp 327-336.

Sheela, M.S., Nisha, M.S. and Mohandas, C. (2004) Eco-friendly management of nematodes associated with Chinese potato (Coleus), *Solenostemnon rotundifolius* (Poir) Mortan. Paper presente in *Natl. Symp. On Paradigms in Nematological Research for Biodynamic Farming.* Abstract. p. 90.

Sheela, M.S., Hebsy Bai and Girija, V.K. (ed.) (2005) Nematode pests of crops in Kerala – an overview pp126.

Sheela, M.S. and Nisha, M.S. (2005) Eco-friendly management of nematodes associated with vegetables. Paper presented in the *International Seminar Prithvi* Abstr.p.80.

Sheela, M.S., Nisha, M.S and Manu, C.R. (2006) *Fungi – A Tool for Nematode Management in Plant Diseases – Biological management* (ed.) Sampat Nehra Pub: Aavishkar Publishers, Distributers, Jaipur pp-220-35.

Shreenivasa, K.R., Krishnappa, K., Reddy, B.M.R., Ravichandra, N.G., Karuna, K. and Kantharaju, V. (2005) Integrated management of nematode complex of banana. *Indian J. Nematol.* 35: 37- 40.

Sivakumar, M. (1998) Non-chemical management of *Meloidogyne hapla* Chitwood on carrot (*Dacus carote*). *Proc. Third int. Symp. Afro-Asian Soc. Nematologists,* (ed. Mehta, U.K.), Sugarcane Breeding Institute,Coimbatore, pp. 209-212.

Sivaprasad, P. Jacob, A. Nair, S.K. and George, B. (1990) Influence of VA mycorrhizal colonization on root knot nematode infestation in *Piper nigrum* L. Trends in mycorrhizal research. *Proc. National Conference on Mycorrhiza*, Haryana Agricultural University, Hisar, India: 100-101.

Sivaprasad, P., Jacob, A., Sulochana, K.K., Visalakshi, A. and George, B. (1992) Growth and root-knot nematode infestation and phosphorus nutrition in *Piper nigrum* (L.) as influenced by vesicular arbuscular mycorrhizae. *Proc. Third International Conference on Plant Protection in the Tropics*, Genting Highlands, Malaysia. 6: 34-37.

Sosamma, V. K., Koshy, P. K. and Bhaskara Rao, E. V. V. (1980) Susceptibility of coconut cultivars and hybrids to *Radopholus similis* in field. *Indian Journal of Nematol.* 10: 250-252.

Sosamma, V. K., Koshy, P. K. and Bhaskara Rao, E. V. V. (1988) Response of coconut cultivars to the burrowing nematode, *Radopholus similis*. *Indian Journal of Nematol.* 18: 136-137.

Sosamma, V. K. (1994) Interactions between Vesicular Arbuscular Mycorrhizae(VAM) and burrowing nematode on Coconut. Final report ICAR AP cess fund scheme 1988-93. C.P.C.R.I, Kayangulam, Krishnapuram 690533, Kerala, India. pp.121.

Sosamma, V.K., Geetha, S.M. and Koshy, P.K. (1994) Effect of the fungus *Paecilomyces lilacinus* on the burrowing nematode, *Radopholus similis* infesting betel vine. *Indian J. Nematol.* 24:50-53.

Sosamma, V.K. (1996) Integrated nematode management- Principles and Practices. In: *Recent Advances in Integrated Management of Pests and Diseases of Coconut*. (CPCRI, Kayangulam) pp.215-221.

Sosamma, V.K.and Koshy, P.K. (1997) Biological control of *Meloidogyne incognita* on black pepper by *Pasteuria penetrans* and *Paecilomyces lilacinus*. *J. Plantation Crops* 25: 72-76.

Sosamma, V.K, Koshy, P.K and Samuel, R.(1998) Effect of VAM, *Glomus mosseae* on nematode infesting banana under coconut based high density multispecies cropping system. *Proc. III Int .Symp. AASN*, p.94-97.

Sosamma, V, K. (1998) Nematode parasites of coconut and their management. *Indian Coconut J.* 29: 107-112.

Sosamma, V. K. (1999) Occurrence of *Pasteuria* sp. a hyperparasite of *Radopholus similis* and *Heterodera oryzicola* from Kerala. Annual meeting and Symposium on Plant disease Management for sustainable Agriculture.14-16th Dec. 1999 at CPCRI, Kayangulam. Abstr p.50.

Sosamma, V. K. (2002) Nematodes of coconut, arecanut and oil palm. Paper presented in the seminar on Centenary of Nematological Research in Kerala. during 21-22nd Feb.2002 at CPCRI, Kayangulam, Kerala.

Sreeja, T.P., Eapen, S.J. and Ramana, K.V. (1996) Occurrence of *Verticillium chlamydosporium* Goddard in a black pepper garden in Kerala, *Indian J. Spices & Aromatic Crops* 5:143-147.

Sreeja, P and Charles, J.S.K. (1998) Screening of botanicals against the burrowing nematode, *Radopholous similis*. *Pest Mgmt. Hortl. Ecosystems*. 4: 36-39.

Srivastava, A.S., Gupta, R.L., Singh, B. and Ram, S. (1969) Damage and control of *Meloidogyne javanica* (Treub) in Colocasia (*C.antiquorum*). Labdev. *J. Sci. Tech.* 7B:306-307.

Stephan, Z.A., El-Behaldi, A.H., Al-Zahroo, H.H., Antoon, B.G. and Georgees, M.S. (1996) Control of root-knot wilt disease complex on tomato plants. *Dirasat-Series-B-Pure Appl. Sci.* 23: 13-16.

Subramaniyan, S., Rajendran, G. and Vadivelu, S. (1990) Estimation of loss in tomato due to *Meloidogyne incognita* and *Rotylenchulus reniformis*. *Indian J.Nematol.* 28: 239.

Sun, F.R. and Chen, H.C. (1994) New Sweet potato variety with high resistance to stem nematode-Lushu 78066. *Crop Genetic Resources.* 4: 58.

Sundararju, P., Koshy, P.K. and Sosamma, V.K. (1979b) Survey of plant parasitic nematodes associated with spices in Kerala and Karnataka. *Proc. Second Symposium on Plantation Crops (Plant Protection)*, Ootacamund. Tamilnadu, India:39-44.

Sundararaju, P. and Koshy, P.K. (1982b) Response of *Areca* collections against *Radopholus similis*. *Indian J. of Nematol.* 12: 404-405.

Sundararaju, P. and Koshy, P.K. (1986) Control of *Radopholus similis* on arecanut seedlings with aldicarb, aldicarb sulfone, carbofuran and fensulfothion. *Indian J. Nematol.* 16: 4-7.

Sundararaju, P and Sudha, S. (1993) Nematode management in arecanut and arecanut based farming system. *Proc. Natl. Symp. On Recent Approaches in Integrated Nematode Management of Agricultural Crops,* HAU, Hissar, p.78.

Sundararaju, P. (2002) Seasonal fluctuations of *Radopholus similis* and *Pratylenchus coffeae* in certain cultivars of banana. *Infomusa* 11: 16-18.

Sundararaju, P., Thangavelu, R. and Cannayane, I. (2002) Management of *Pratylenchus coffeae* and *Meloidogyne incognita* on banana by using endophytic fungi. *Current Nematol.* 13: 77-81.

Sundararaju, P and Cannayane, I. 2002. Antinemic activity of plant extracts against Pratylenchus *coffeae* infecting banana. *Indian J. Nematol.* 32:121-124.

Sundararaju, P. and Jeyabhaskaran, K.J. (2002) Evaluation of different soil types on multiplication of Pratylenchus coffeae and growth of banana seedlings var. Nendran. *Nematol. Medit.* 31: 151- 153.

Sundararju, P. and Cannayane, I. (2002a) Antinemic activity of plant extracts against Pratylenchus *coffeae* infecting banana. *Indian J. Nematol.* 32: 121-124.

Sundararju, P. and Cannayane, I. (2002b) Production of nematode egg parasitic fungus, *Paecilomyces lilacinus* on banana wastes and certain plant leaves. *IndianJ. Nematol.* 32:188-189.

Sundararaju, P. and Cannayane, I. (2003) Evaluation of different biopesticides against major nematode pathogens infesting banana cv. Nendran. In: *National Symposium on Organic Farming in Horticulture for Sustainable Production. August 29th – 30th, 2003, CISH, Lucknow, India,* pp. 66-67.

Sundararaju, P. and Sathiamoorthy, S. (2005) Present status of banana nematodes and their management. In: *Biotechnological Management of Nematode Pests and Scope of Entomopathogenic Nematodes.* (eds.) Sithananatham, S., David, B.V. and Selvaraj, P. Sun Agro Biotech Research Centre Publication, Sun Ray No.1. pp. 43-47.

Sundararju, P. and Pantisuba, K. (2006) Biochemical changes in banana plants induced by *Pratylenchus coffeae* and *M. incognita*. *Indian J. Nematol.* 36: 256-259.

Sundarababu, R., Sankaranarayanan, C. and Vadivelu, S. (1993) Nematode management with plant products. *Indian J. Nematol.* 23: 177-178.

Sundarababu,R., Nageswari,S.,Poomima,K. and Suguna,N.(1998) Biological control potential of *Glomus fasciculatum* against *Meloidogyne incognita* on tomato and okra. *Proc. of the First National Symposium on Pest Management in Horticultltral Crops*. Bangalore, Abstract. p.312.

Sudha, S. and Sudararaju, P. (1998) Effect of neem oil cake and nematicides for the control of *Radopholus similis* in arecanut based cropping system. (In) Third international symposium of Afro Asian Society of Nematologists held at Coimbatore from April 16-19, 1998, Abstract p.

Sujatha, B., Lakshman, P.L. and Kumari, N.S. (2000) Evaluation of rhizobacteria, *Pseudomonas fluorescens* against the cyst nematode, *Heterodera cajani*. *Nat. Nematology Symp. Integrated Nematode Mgmt Sustainable Agric. Changing Agro-ecol. Econ. Scenario New Millenium, November 23rd – 24th, 2000*. Orissa University of Agricultural Sciences and Technology, Bhubaneswar, Nematological Society of India, Indian Agricultural Research Institute, New Delhi. Abstract: 27.

Suresh, C. K., Bagyaraj, D. J. and Reddy, D. D. R.(1985) Effect of vesicular arbuscular mycorrhizae on the survival, penetration and development of root-knot nematode in tomato. *Plant and soil*.87: 305-308.

Swarup. G, Dasguptha, D.R and Gill, J.S (eds) (1995) *Nematode Pest Management- An Appraisal of Eco-friendly Approaches* Published by Nematological Society of India, pp 300.

Thomas, G.V., Sundararaju, P., Ali, S.S. and Ghai, S.K. (1989) Individual and interactive effects of VA mycorrhizal fungi and root-knot nematodes, *Meloidogyne incognita* on cardamom. *Tropic. Agric.* 66: 21-24.

Tiwari, S.P. and Dave, G.S. (1985) Burrowing nematode on banana in Madhya Pradesh. *Curr.Sci.* 54:1286.

Tiyagi, S. and Ajaz, S. (2004) Biological control of plant parasitic nematodes associated with chickpea using oil cakes and *Paecilomyces lilacinus. Indian J. Nematol.* 34: 44-48.

Vadhera, I., Tiwari, S.P. and Dave, G.S. (1998b) Integrated management of root-knot nematode, *Meloidogyne incognita* in ginger. *Indian Phytopathol.* 51:161-163.

Vadivelu,R.S., Rajendran, G., Naganathan, T.G. and Jayaraj, S. (1987) Studies on the nematode pests of banana and their management – A review. *Proc. Third Group Discussions on the Nematological problems of Plantation crops.* Sugarcane Breeding Inst. Coimbatore. pp. 32-33.

Venkitesan, T.S. (1976) *Studies on the burrowing nematode Radopholus similis on pepper and its role in slow wilt disease.* Ph.D thesis. Univ. Agrl.Sci. pp 231.

Venkitesan, T.S. and Setty, K.G.H. (1979) Control of the burrowing nematode *Radopholous similis* (Cobb) on black pepper. *Pesticides*, 13: 40-43.

Venkitesan, T.S. and Charles, J.S.K. (1983) Sucker dip treatment for banana nematodes control. *Third Nematol. Symp*, May 24th – 26th, 1983, Solan, Abstract. p.45.

Verma, S.K. and Singh, S.P. (1969) Bio-ecological studies of reniform nematode, *Potylenchulus reniformis* Linford and Oliveira. *Indian J. Ent.* 31: 36-47.

Viswanathan, P.R.K., Kumar, A. C. and D'Souza, G.I. (1974) Nematode association of cardamom in South India. *Cardamom News* 6: 5-19.

Walia, R.K. and Mehta, S.K. (2000) Efficacy of *Pasteuria penetrans* against *Meloidogyne incognita* on okra under different adaphic factors. *Nat. Nematology Symp. Integrated Nematode Mgmt Sustainable Agric. Changing Agro-ecol Econ. Scenario New Millenium,* 23-24 November 2000. Orissa University of Agricultural Science and Technology, Bhubaneshwar. Nematological Society of India, Indian Agricultural Research Institute, New Delhi. Abstract: 56.

Walters, S.A. and Barker, K.R. (1994) Reproduction and damage potential of two populations of *Rotylenchulus reniformis* on sweet potato and related comparisons with *Meloidogyne javanica* on tomato. *J. Nematol* 25.:830-35.

West, J. (1943) Dry rot of yams. *Bull. Imp. Inst.* London, 32: 449-50.

Chapter – 14

Biological Control in Vegetables

T. Jiji, G. Suja and K.S. Meenakumari

Biological control can be defined as use of natural enemies to reduce damage caused by a pest population. Biological control is, generally, the human use of a specially chosen living organism to control a particular pest. This chosen organism may be a predator, parasite or disease causing organism which attacks the harmful insect. Predators, like lady bird beetles and lacewings, are mainly free-living species that consume a large number of preys during their lifetime. Parasitoids are species whose immature stage develops on or within a single insect host, ultimately killing the host. Many species of wasps and some flies are parasitoids. A complete Biological Control program ranges from choosing a pesticide which will be the least harmful to beneficial insects, to raising and releasing one insect to attack another, almost like a "living insecticide" (Hoffmann and Frodsham, 1993).

Biocontrol of disease is the biological destruction of harmful microorganisms causing plant diseases without disturbing the ecological balance (Mukhopadhyay *et al.,* 1992). Biological control agents of plant diseases are the most often referred to as antagonists. It is the reduction in pathogen inoculum or its disease producing capacity by action of one or more organisms accomplished naturally or through manipulation of environment, host or antagonist or by mass production of one or more antagonists (Cook, 1990).

Biological control methods can be used as part of an overall Integrated Pest and Disease Management (IPDM) program to reduce the legal, environmental and public safety hazards of chemicals. In addition, it may be a more economical alternative to insecticides. A few biological control measures prevent economic damage to agricultural crops. Unlike most insecticides, biological control is often specific for a particular pest. Other helpful insects, animals or people can go unaffected or undisturbed by their use and there is less danger of impact on environment and water quality.

1. Biological Control of Insect Pests

Biological control of insect pests with parasites and predators forms a major area of eco-friendly pest management. History of biological control in India can be chronicled through two major approaches – classical biological control for invasive pests or weeds through introduction of natural enemies from other countries or from India into other countries and augmentative biological control for native species through appropriate inoculative or inundative releases (Manjunath and Ramani, 2006).

The 99% of species of pests are kept at level, not much of economic importance by various biotic and abiotic factors. Some of these species also play important role in maintaining natural enemy-guild which adds to these insect fauna diversity and provide large stable equilibrium of pests. This phenomenon is called natural biological control (NCIPM, 2003).

The components of modern intensive agriculture like high yielding varieties, sumptuous fertilization and irrigation alter the crop physiology, morphology and phenology to render the crop attractive for pests and diseases. Promoting natural enemies of key pests of important crops is one of the most viable and indispensable components of integrated pest management. It offers available, sustainable, ecological and economic solution to the pest problem.

Three main approaches to biological control

Biological control programme is a three-pronged approach, *viz.* importation and colonization of exotic natural enemies; conservation of existing natural enemies by modifying the conditions which favour and promote natural enemies and by making periodic release of natural enemies for inoculation and inundation (NCIPM, 2003)

Importation is use of introduced natural enemies to control an exotic, invasive pest species. This is also known as classical biocontrol. Conservation involves actions to protect and maintain established populations of natural enemies in a given area. It is done by identifying any factor that limits effectiveness of a particular natural enemy and changing them to help beneficial species. Augmentation involves supplemental release of natural enemies. Relatively a few natural enemies may be released at a critical time of the season (inoculative release) or literally millions may be released (inundative release). Additionallly, the cropping system may be modified to favour or augment the natural enemies, termed as habitat manipulation biocontrol (Hoffmann and Frodsham, 1993).

Biological control of insect pests – An overview

Biological control with the help of parasites and predators is an ecologically sustainable component of IPM and continues to play important role. Several organizations have played significant role in promoting this technology.

Successes in classical biological control world over

No. of pest species permanently controlled up to 1988	165
Completely controlled	75
Substantially controlled	74
Partially controlled	15
Worldwide successful control project	384

(NCIPM, 2003)

India is credited with the earliest recorded instance of biological control of a pest by employing an imported natural enemy in the world when in 1762 the mynah bird, *Acridotheres tristis*, was successfully introduced from India Mauritius as a predator to control the red locust, *Nomadacris septemfasciata*. Some of the outstanding successes recorded in the history of biological control were achieved with natural enemies of Indian origin like control of the Rhodes grass scale, *Antonina graminis* by *Neodusmetia sangawani* in the USA, Israel and other countries (1959-60); the California red scale, *Aonidiella aurantti* by *Aphytis melinus* in California (1956-57), mango mealybug, *Rastrococcus invadens* by *Gyranusoidea tebygi* and *Anagyrus mangicola* in Central and West Africa (1987).

First successful classical biological control was achieved in India when the cochineal insect, *Dactylopius ceylonicus*, was introduced from Brazil in 1795 and it became readily established on drooping prickly pear, *Opuntia vulgaris* in North and Central India resulting in spectacular and sustained suppression of this aggressive weed. In 1926, another beneficial mealybug, *D. opuntiae*, a North American species, was obtained from Sri Lanka and released in Southern India where *O. dilleni* was predominant and it proved very successful. Several other weeds like *Lantana camera* (1809), *Eichhornia crassipes* (1895), *Chromolaena odorata* (1914), *Salvinia molesta* (1950) and *Parthenium hysterophorus* (1955) invaded India and attempts were made to control these with introduced natural enemies. Results reported on the control of *E. crassipes* by host specific weevils, *Neochetina eichhorniae* and *N. bruchi*, native of Argentina, introduced from Florida (USA) in 1982, control of *S. molesta* by *Cyrtobagous salviniae*, native of Brazil, from Australia in 1982, and that of *Parthenium hysterophorus* by the Mexican beetle, *Zygogramma bicolorata*, from Mexico in 1983 raised new hopes of managing these weeds.

The first deliberate introduction of an exotic natural enemy for biological control of an insect pest in India was the Australian ladybird beetle, *Cryptolaemus montrouzieri*, from Australia in 1898 to control the green scale, *Coccus viridis*, in Southern India. The predator became quickly established and is now widely distributed in India, providing appreciable control of several species of soft scales and mealybugs infesting guava, coffee, citrus and grapevine. A few of the other successful cases include control of cottony cushion scale, *Icerya purchai,* by *Rodolia cardinalis*, from the USA in 1929 and from Egypt in 1930, the apple woolly aphid, *Eriosoma lanigerum*, by the North American parasitoid, *Aphelinus mali*, from England in 1937; the San Jose scale, *Quadraspidiotus perniciosus*, by the eulophid parasitoid, *Encarsia perniciosi*, from the USA in 1958-60; and control of citrus mealybug, *Planococcus citri*, by the encyrtid parasitoid, *Leptomastic dactylopii*, introduced in 1982 from Trinidad.

Several insect pests got accidentally introduced into India recently which include the subabul psyllid, *Heteropsylla cubana* (1988), the American serpentine leaf miner, *Liriomyza trifolii* (1990-91), the coffee berry borer, *Hypothenemus hampei* (1990) and the spiraling whitefly, *Aleurodicus disperses* for which natural enemies were introduced from other parts of the world.

The augmentative biological control is practiced in India from early 1900s. Inundative releases of egg parasitoid, *Trichogramma chilonis* and other species, are adopted on a large scale in several parts of the country for control of sugarcane borers, cotton bollworms and other lepidopteron pests. Release of larval parasitoid, *Perisierola nephantidis*, for control of coconut black-headed caterpillar is *in vogue* in several Southern states for the past several decades. Successful use of *Epiricania melonoleuca* against the sugarcane pyrilla, *Pyrilla perpusilla*, in the Northern parts of India is a sort of classical biological control through redistribution. Various pathogenic micro-organisms are also studied and utilized in biological control. These include the Nuclear Polyhedrosis Viruses (NPV); the fungi *Metarrhizium anisopliae*, *Verticillium lecanii* and *Beauveria bassiana*; and the bacterium *Bacillus thuringiensis*. These insect pathogens and also other biocontrol agents like *Chrysoperla carnea*, several species of predacious ladybird beetles and parasitoids are commercially produced and utilized for control of several important pests in India.

Biological control agents of insect pests used commercially in India

Bio-control agents	Target pests
Predators	
1. *Chrysoperla carnea*	General predator against soft bodied insects.
2. *Crytolaemus montrouzieri*	Mealy bugs
Parasitoids	
3. *Trichogramma* spp.	Lepidopterous insect pests
4. *Cheilomenes blackburni*	Pink ball worm
5. *Leptomastrix dactylopii*	Mealy bug
6. *Copidosoma* spp.	Potato tuber moth
Insect pathogens	
7. HaNPV	*Helicoverpa armigera*
8. SINPV	*Sodoptera litura*
9. *Beauveria bassiana*	Lepidopterous and Coleopterous pests
10. *Metarhizium* spp.	Hemipterous pests
Fungal Antagonist	
11.*Trichoderma* spp	Fungal diseases

(NCIPM, 2003)

The concept of integrated pest management gained foothold consequent to the adverse effects of pesticide use, in the form of secondary pest outbreak, pest resurgence and high content of toxic residues in food products.

Classical biological control (Importation)

Classical biological control is the introduction of natural enemies to a new locale where they did not originate or do not occur naturally. This involves travelling to the country or area from where a newly introduced pest originated and returned with some of the natural enemies that attacked it and kept it from being a pest there. New pests are arriving accidentally or intentionally. If they become a pest, introducing some of their natural enemies can be an important way to reduce the amount of harm they do. This is usually done by government authorities. The first step in the process is to determine the origin of introduced pest and then collect appropriate natural enemies associated with the pest or closely related species. The natural enemy is then passed through a rigorous quarantine process, to ensure that no unwanted organisms (such as hyperparasitoids) are introduced, then they are mass produced and released. Follow-up studies are conducted to determine if the natural enemy become successfully established at the site of release and to assess the long-term benefit of its abilty to manage the pest (Mahr and. Ridgway, 1993). Exotic pests are obviously suitable subject but native pests are also controlled by imported enemies and should receive serious consideration. Effective natural enemies should be obtained either from target pest or related species. Effectiveness of natural enemies cannot be predicted in advance but it should have some desirable attributes *viz.* high searching ability, high degree of host specificity or preference, good reproductive potential and adaptation to a wide range of ecological conditions.

Conservation of natural enemies

Conservation of natural enemies is an important part in any biological control effort. This involves identifying factors which limit effectiveness of a particular natural enemy and changing them to help beneficial species. Conservation of natural enemies is possible by manipulating the environment to favour natural enemies, by removing or mitigating adverse factors for natural enemies by providing the lacking requisites. Lacewings, lady beetles, hover fly larvae and parasitized aphid mummies are almost always present naturally. Fungus infected insects are often common following periods of high humidity. These naturally occurring biological control agents are often susceptible to the same pesticides used to target their hosts. Preventing the accidental eradication of natural enemies is termed simple conservation.

Indiscriminate use of pesticides precludes the general possibility of obtaining satisfactory biological control. If their use is a must, they should be used in a manner so as to have their least adverse effects on natural enemy population and there by protecting effectiveness of natural enemies. This is possible by employing selective pesticides at the least injurious time and in the least disruptive formulation at the right time.

Cultural practices like burning the stubble destroy parasitized larvae while the healthy ones escape by pupating deep in the soil. Each aspect is required to be investigated and evaluated on its ecological implications before execution. It is very essential to examine the *pros* and *cons* critically before adapting any control intervention. Conservation can also be achieved by manipulating microhabitat, by irrigation, pruning and planting trap crops. Source of food, like pollen, honeydew or nectar is essential for many adult parasites and predators. Spraying of artificial honeydew (yeast hydrolysate, sugar, water and little tryptophan added as an attractant) may attract adult green lacewings and also increase their fecundity (van den Bosch and Telford, 1964; Mahr and Ridgeway, 1993).

Nesting site for polistes wasp and insectivorous birds and providing drinking water for insectivorous birds attract these natural enemies and help reduce the pest load. Straw bundles are used in China to provide shelter for spiders in rice field, and this practice alone has reportedly reduced 50-60 % insecticides use (Biliotti, 1977; Sparks *et al.*, 1982).

Augmentation of natural enemies

This is a method of increasing population of a natural enemy which attacks a pest. This can be done by mass producing a natural enemy in a laboratory and releasing it into the field at proper time. Augmentation involves supplemental release of natural enemies. Relatively a few natural enemies may be released at a critical time of the season (inoculative release) or literally millions may be released (inundative release). Additionally, the cropping system may be modified to favour or augment the natural enemies. This latter practice is frequently referred to as habitat manipulation (Mahr and Ridgway, 1993).

Another method of augmentation is breeding a better natural enemy which can attack or find its prey more effectively. Mass rearings can be released at special times when the pest is the most susceptible and natural enemies are not yet present, or they can be released in such large numbers that a few pests go untouched by their enemies. The augmentation method relies upon continual human management and does not provide a permanent solution unlike the importation or conservation approaches.

Augmentation of natural enemies to increase their effectiveness involves direct manipulation either by mass production and periodic colonization or employing chemicals which affect their behaviour. Augmentation in general should be given the lowest priority in biological control research endeavour and should not be resorted to until it is determined that the solution does not exist in foreign exploration and importation of new natural enemies or some type of conservation. The importation and establishment or conservation can solve the problem permanently, and often very cheaply.

Augmentation should be restricted to those natural enemies which have proven effectiveness to decimate the pest population but are prevented from doing so because they are not adequately adapted to weather extremes; are not synchronized with the necessary stages of the host or are otherwise rendered ineffective because of the unfavourable environmental conditions. Augmentation is done in two types of situations. Periodic colonization may involve either inoculative release or inundative release (Biliotti, 1977).

Inoculative release

Release of natural enemies is done occasionally in small numbers for re establishing species of natural enemies whose population is decimated due to adverse conditions. Control of the pest is achieved from progenies of released insects. A small number released at the onset of favourable weather may increase in number, sufficient to suppress the pest population.

Inundative release

In inundative release, natural enemies are mass cultured and released into fields in large numbers. The released insects themselves control the pest. Lady bird beetles, lacewings, or parasitoids like *Trichogramma* are frequently released in large numbers (inundative release). Recommended release rates for *Trichogramma* in vegetables range from 5,000 to 200,000

/ acre / week depending on level of pest infestation. Similarly, entomopathogenic nematodes are released at rates of millions and even billions / acre for control of certain soil-dwelling insect pests.

Inundative release may be regarded as application of biotic insecticides. They are more expensive than importation or inoculative releases. They are the most suitable for univoltine pest species or against pests, which cause economic damage only during limited period of the year. The natural enemies selected for inundative releases need not possess all the attributes desirable for a species imported for permanent establishment or even for a species used for inoculative releases. Their ability to reproduce, disperse or overwinter in the field may be relatively unimportant. Their ability to cause high host/prey mortality and the rearing and application cost are of outstanding importance in an inundative release programme.

The belief that *Trochogramma* spp. is very polyphagous has proved untrue. Some of the species are rather highly host specific. In US 3.5 billion *Trichogramma* are produced annually, with more than 2 billion of them reared for commercial release on 2,00,000 ha of crops (Ridgeway *et al*., 1981). In general, reduction in pest population was about 80 to 85 % and there was subsequent yield increase in cabbages.

Taxonomic studies *i.e.* ecotype of the species and then their host preference are essential prerequisites before launching the inundative release programme. The species of *Trichogramma* is therefore selected based on zone, region, crop and pest rather the earlier empirical use.

Success in the *Trichogramma* release programme rekindled the interest in its use in India and Pakistan. It is now immensely being used in tomato. Likewise Nuclear Polyhedrosis Virus of *Helicoverpa armigera* showed promising results in tomato and many other crops. Release of *C. carnea* for a variety of soft bodied insects is promising.

2. Biocontrol Agents in Vegetables

Predators

Insect predators like lady bird beetles and lacewings, are mainly free-living species that consume a large number of prey during their lifetime (Lefroy, 1909; Agarwala *et al*.,1988; Annie Bernice, 2000; Suja, 2003).

Coccinellids

Around the world about 420 genera and 5500 species of coccinellids are recorded, while in India, 401 species belonging to 79 genera are recorded so far (Poorani, 2002).

World wide, there were 155 attempts to control aphids and 613 to control coccids by introducing ladybird beetles. The outcome of these attempts indicates that coccidophagous ladybirds are more effective than aphidophagous ladybirds (Dixon and Kindlmann, 1998). In India, 22 species were introduced, most of these for controlling sugarcane scale, *Melanaspis glomerata* (Green). Three introduced species, *viz*., *Cryptolaemus montrouzieri,*

Curinus coeruleus and *R.cardinalis* are established and are providing substantial natural control (Rao *et al.*, 1971).

Lady bird beetles (Coccinellidae : Coleoptera) are the most important group of predatory insects. Majority of them are predaceous on injurious insects like aphids, mealy bugs, scales, thrips, leafhoppers, mites and other soft bodied insect pests of horitcultural field crops. Of these, *Rodolia ordinalis* (Mulsant), *Cryptolaemus montrouzieri* Mulsant and *Curinus coeruleus* Mulsant are successfully used in classical biological control. Indigenous coccinellids like *Chilocorus nigrita* (Fabricius), *Cheilomenes sexmaculata* (Fabricius), *Coccinella septempunctata* Linnaeus, *Scymnus coccivora* Ayyar and *Pharoscymnus horni* (Weise) have potential for exploitation in augmentative biological control.

Coccinellids are oval to hemispherical in shape and strongly convex. They are often brightly coloured with red, orange or yellow elytra frequently spotted black or striped. Body length varies from 0.8 to 16 mm. The larvae of these beetles are elongate campodeiform with elongate legs, somewhat flattened and covered with minute tubercles or spines.

The efficiency of predation by coccinellids is reported by many workers (Sharma, 1973; Anand, 1983; Lokhande and Mohan, 1990).

Based on their host range, coccinellids are classified as aphidophagous (feeding on aphids) or coccidophagous (feeding on scales and mealy bugs). Members of these groups differ in their behaviour and general characteristics, (PDBC, 2003).

In vegetable cowpea, the coccinellid, *Cheilomenes sexmaculatus* (Fabr.) was recorded as a predator of the aphid *A.craccivora* (Lefroy, 1909; Bagas and Trahan, 1949). *Micraspis discolour* F. (Agarwala *et al.*, 1988) and *Brumus* and *coccinella sp*. (Falerio *et al.*, 1990) were the other aphid predators reported. The coccinellids, *Sthethorus* spp. identified as important predators of tetranichid mites are distributed throughout the world (Singh and Ray, 1977). *M.sexmaculatus* was the most abundant and persistent predator of *A.craccivora* because of its short life cycle, larger population and fairly high feeding potential (Saharia, 1980). Dooria (1982) observed *Stethorus pauperculus* Weise to prey on *Eutetranychus orientalis* under field situation.

Ofuya (1995) observed that the population of *A.craccivora* can be considerably reduced by action of coccinellid predator *Cheilomenes lunata* (F.). Rani (1995) and Bindu (1997) indicated presence of coccinellid predators *viz. M.sexmaculatus, C.septempunctata, Scymnus* spp. *and Micrapsis crocea* on *A. craccivora* infesting cowpea. Nandakumar and Sheela (1996) reported presence of nine species of arthropods preying on cowpea pests *A.craccivora*, of which *M. sexmaculatus* was the most abundant and efficient one. The most abundant and efficient predator of *A. craccivora* included the coccinellids *viz. C.septempunctata*, *M. sexmaculatus* and the others were *Coccinella* sp. and *M.discolor* (Sharma, 1991and Agarwala and Bardhanroy, 1999). *M. sexmaculatus* and *M.croccea* were the most efficient predator of aphids in the chilli ecosystem. They did not feed on thrips (Santhosh Kumar, 1999).

Regarding efficiency of potential predators, coccinellids ranked first. Among them *C.transversalis, H.octmaculata* and *M.sexmaculatus* were the predominant species in cowpea fields. *H.octomaculata* had the shortest lifecycle of 13.27 ± 0.31 days followed by *C.transversalis* with 13.78 ± 0.33 days and *M.sexmaculatus* with 18.70 ± 0.91 days. Among the major predators, the grubs of *C. transversalis* consumed the maximum of 251.8 ± 6.74 *A.craccivora* followed by *H.octomaculata* with 198.22 ± 12.90 aphids and *M.sexmaculatus* with 127.60 ± 3.35 aphids. A similar trend in the consumption of *A.crasuivora* was observed in adults also.The grubs and adults of coccinellids ate only live aphids. Adult females were more efficient than the male in recognizing, attacking and consuming aphids (Suja, 2003).

Syrphids

Syrphid flies, also known as hover flies for their ability to hover in flight, are common predators of aphids and other soft bodied insects. Three of the more common species are *Syrphus rectus* Osten Sacken, *Allograpta obliqua* (Say) and *Eupeodes americanus* (Weidemann). Adults feed on pollen, nectar and aphid honeydew. Females must feed on pollen for proper egg maturation. Females oviposit in the midst of aphid colonies, using honeydew as an ovipositional stimulant, and laying hundreds of eggs each. Eggs are often the first sign of aphid predators seen in the spring. Eggs are white, elliptical, and about 4/100 inch (1 mm) in length. There are three larval instars. Larvae are elongate, tapering gradually toward the head end. Larvae search for prey by casting the forward end from side to side; prey must be contacted directly to be detected. Larvae pierce aphids with their mouth hooks, sucking fluids from the bodies before moving to other prey. Each larva consumes hundreds of aphids during development. Larval feeding success influences future fecundity. Third instar larvae have the greatest impact on aphid numbers. Larvae of some species are relatively plump. There may be 5-7 generations/year. Multivoltine species overwinter as adults and univoltine species as third instar larvae. These are among the most voracious of aphid predators.

Variation in feeding potential of syrphids was reported by several workers. Lefroy (1909) reported about 67 species of syrphids in India, fed exclusively on aphids. A single syrphid larva destroyed about 484 aphids in 4 hours (Deoras, 1942) where as a single larva of the predator *Xanthogramma scutellare* required an average 123 *A. craccivora* / day (Sitaraman, 1966). The maggot of *X.scutellare* consumed 386.86 aphids during its larval period of 5.07 days (Devi, 1967). Zeren and Duzguner (1983) reported that syphid predator *Ishiodon aegyptius* is effective against *A.gossypii* and *M. persicae*.

Ishiodon scutellare consumed an average of 471 *A.gossypii* during their larval period of 6 days. Another syrphid predator, *Paragus serratus* (Fabrius) consumed an average of 429 *A.gossypi* during a larval period of 6 days (Dahiya *et al.*, 1988).The total consumption of *A. craccivora* by the maggots of *P. serratus* was 258.3 with 9.2, 33 and 34 aphids during the first, second and third instars, respectively (Patro and Behera, 1993).

The mean number of aphids consumed by the first, second and third instar larvae of *X.scutellare* were 49.45,139.11 and 259.90 respectively. The total consumption of entire larval duration being 448.46 aphids and durations of active feeding of three instars were 2,

3 and 4 days, respectively (Rani, 1995). *I. scutellare* completed its life cycle in 13.30 days and the maggot consumed an average of 406 aphids (Nandakumar, 1999). A single maggot of the syrphid predator *I.scutellare* consumed a mean number of 459.5 ± 7.71 aphids during its larval period. The searching speed and turning rate increased with increase in prey density but decreased with increase in hunger level. The total life cycle of the syrphid, *I. scutellare* was worked as 16.43 ± 0.45 (Suja, 2003).

Chrysopids and Hemerobids

Chrysopids (green lace wings) and hemerobids (brown lace wings) form another major group of promising predators for biological control.

In India, *C.carnea* is reported from Western and Northern parts of the country feeding on *Heliothes* species and aphids (Manjunath *et al.*, 1976). Maximum consumption of aphids was by the third instar larvae of *C.carnea* followed by second and first instars (El-dakroury *et al.*, 1979). Grubs of Chrysopids having mandibulo-suctorial mouth parts are well adapted to predatory life. The general tolerance of Chrysipids to many insecticides commonly used against the pests led to considerable interest in the use of Chrysopids in Intergrated Pest Management Programme (Canard *et al.*, 1984; Pree *et al.*, 1989).

Periodical release of *Chrysoperla carnea* in 1:1 predator prey ratio gave up to 98% control of aphids (Gurbanov, 1982). Singh and Kumar (2000) observed that an average of 11.48, 19.82 and 83.3 aphids were consumed by first, second and third instar of *C. carnea* in mustard. In vegetable cowpea, the mean number of aphids consumed by the first, second and third instars larvae were 73.00 ± 15.8, 184 ± 18.27 and 161.4 ± 23.5 respectively and the total number of nymphs and adults of *A. craccivora* consumed by the predator larva during the development period was 419 ± 9.14 (Abhilash, 2001). The total life cycle of the *C. carnea* was worked out as 25.80 ± 0.77 days, and the total consumption of the adult *A. craccivora* by a singe larvae of *C. carnea* was 421.1 ± 1.23 (Suja, 2003).

Acarine predators

Amblyseius finlandicus (Oudemans) occurred in the ecosystem of many phytophagous mites and was a promising predator on eggs and larvae of *Eutetranychus orientalis*, (Gupta *et al.*, 1971). Acarine predators and insect predators were important natural enemies of phytophagous mites in India and abroad (Singh and Ray, 1977; Dhooria, 1982; Sudharma, 1996). Gupta and Gupta (1992) gave an account of the predatory mites in India, their hosts and importance in biological control. The predatory mites *A. finlandicus, A.ovalis, A.longispinosus, A.largoensis, A.multidentatus* and *A. tertanychivorus* were highly effective in checking population of *E. orientalis, Oligonychus coffeae, O. mangiferous, T. neocaledonicus* and *T. ludeni* respectively. Fan and Petitti (1994) reported *Neosciulus barkeri* Hughes as an important predator of *Polyphagotarsonemus latus*.

Predatory mites belonging to genus Amblyseius of the family Phytoseiidae were observed the dominant predators of *T. ludeni* in cowpea and *P. latus* under field situations. In the lab, a macrocheyletid mite, *Macrochelus merdarius* found in rice bran was also observed as an efficient predator of *P. latus.*

A steady decline in the population of *T.ludeni* and a progressive increase in predatory mites were also observed when the bioagents *C.carnea* and the predatory mite *A.longispinosus* were released together (Abhilash, 2001).

Spiders

Spiders in the ecosystem bring about good amount of natural control. Nandakumar and Saradamma (1996) reported occurrence of spiders *Tetragnatha* sp, and *Oxyopes* sp from bittergourd fields of Thiruvananthapuram, Kerala.

Survey conducted in Assam by Debnath and Borah (2002) on natural enemies of potato tuber moth revealed that the spiders *Oxypes* sp. and *Plexippus* sp. were present in the field throughout the year. Thamilvel (2004) observed activity of spider predators *Oxyopes* sp. and *Tetragnatha* sp. in bhindi in Kerala.

Other predators

In brinjal, the reduvid predator, *Rhinocoris fuscipes* Fab. was recorded on Epilachna and fed on all the stages of the beetle in South India (Cherian and Brahmachari, 1941). *Dicyphus tamaninii* preyed on the larvae of agromyzid, *L. trifolii* (Salamero *et al.*, 1987).

Patel and Yadav (1992) reported that *H.armigera* was attacked by a predatory mirid *Nesidiocoris tenuis*.

Parasites

Parasitoids are species whose immature stage develops on or within a single insect host, ultimately killing the host. Many species of wasps and some flies are parasitoids.

Trichogramma are extremely tiny wasps in the family *Trichogrammatidae*. *Trichogramma* wasps occur naturally in almost every terrestrial habitat, and some aquatic habitats as well. They parasitize insect eggs, especially eggs of moths and butterflies. Some of the most important caterpillar pests of field crops, forests and fruit and nut trees are attacked by *Trichogramma* wasps. World wide use of *Trichogramma* for biological control was already reported (Li, Li-Ying, 1994). *Trichogramma* was released to control 28 different caterpillar pests attacking corn, rice, sugarcane, cotton, vegetables, sugar beets, fruit trees and pine and spruce trees (Hassan, 1994). *Trichogramma* wasps primarily parasitize eggs of moths and butterflies (Lepidoptera). Certain species of *Trichogramma* also parasitize eggs of beetles (Coleoptera), flies (Diptera), true bugs (Heteroptera), other wasps (Hymenoptera), and lacewings and their relatives (Neuroptera). Assessment of of quality four commercially available species of *Trichogramma* (Hymenoptera: Trichogrammatidae) was already reported (Losey and Calvin, 1995). Systematics of the Trichogrammatidae with emphasis on *Trichogramma* was studied (Pinto and Stouthamer, 1994). Laboratory evaluation of commercial Trichogrammatid products for potential use against *Plutella xylostella* (L.) was done. (Vasquez, *et al.*, 1997).

Maruca testulalis larvae in cowpea were studied by Don and Pedro (1983). Hanumanna *et al.* (1984) observed that the parasitisim by *Trichogramma* on *H. armigera* ranged 88.48 to 98.49 %. Pandey and Rajendra Singh (1984) reported that the braconid *Trioxys indicus* was an important parasitoid of *A. craccivora*. Selim *et al.* (1987) reported *Aphidius colemani and Lysiphlebus fabarum* as important parasitoids in aphid colonies. *L. trifolii* was frequently attacked by hymenopteran parasitoids like *Diglyphus begini* and the population was kept well below economic damage (Schuster, and Wharton. 1993). The natural enemy complex of *H. rmigera* consists of *Brachymeria wittei* (Setmitz), *Voria ruralis* (Fallen), *Charops bicolor* (Szepligets) and *Copoletis chlorida* (Uchida) (Joginder *et al.*, 1990).

Okeyo *et al.* (1991) revealed presence of at least seven parasitoids attacking *M. testilalis* on cowpea crop. A pupal endoparasitoid *Antrocephalus* sp. was the most predominant. Kamel *et.al* (1994) observed that the aphids were parasitized by *L.confusus, A.metricariae, Praon volucre and T.angelicae*. Richter and Zhumanov (1994) observed a tachinid parasite *Goniophthalmus* on *H. armigera*. According to Dover *et al.* (1995), *Micropolitis demolitor* was a braconid wasp which parasitized the larval stage of *H. armigera.*

In chillies,many parasites are reported. Machauer (1968) and Van Emden *et al.*, (1969) listed parasites of green peach aphid and their control potential. Stary and Gosh (1975) recorded three parasites viz. *Aphidius matricariae*, *Toxares shigai* and *Trioxys indicus* from India. Easwara Moorthy *et al.* (1976) recorded *Aphidius platensis* parasitizing upto 93.37% of *M. persicae* from Tamil Nadu. Tian and Young (1981) also collected *Lysiphlebia japonica* and *T.indicus* from *A.gossypii.*

In brinjal, the egg parasite, *Tetrastichus ovulorum* Ferr. (Eulophidae) was recorded on Epilachna from Karnataka (Krishnamurthy1932; Krishnamurthi and Usman, 1954).

Pleurotropis foveolatus is a common larval parasite and 15 to 20 are recorded from one to fourth instar parasitized larva. Eighteen generations of this parasite were observed during a year, but there were a few during December to January and March to June (Lal, 1946). This parasite had a wide host range and appears to flourish under fairly cold and humid conditions being useful in hilly and other areas (Lal and Gupta, 1947). Efficient parasitization, especially on third and fourth instar larvae was observed, the % of parasitism varied from 35 to 70 in the field (Lal 1961; Mathur and Srivastava, 1964). Kunjamma and Abraham (1973) recorded *Chrysocharis johnsoni* as an important larval parasite on Epilachna. Trivedi (1988) reported parasitization by the eulophid, *Pediobius foveolatus* in Bangalore with upto 9 % parasitism of grubs and pupae of epilachna beetle.

In amaranths, Mueseveck and Rao (1958) reported *Apantles delhiensi* from *Hymenia recurvalis.* Many parasitoids were recorded on the leaf webber, *H.recurvalis*. Field collected *Hymenia. recurvalis* larvae exhibited upto (11.43%) natural parasitism by the parasitioid *A. delhiensis* (Bhattarherjee and Menon (1964). In Hawaii, *Cremastus hymeniae* Viereck and *Casinaria infesta* (Gressom) are very effective on the pest. In USA, *Trichogramma minutum* Reley, attacks the eggs and Apanteles *marginiventris* (Cresson), *Eupteromalus viridescens* Walsh and *Sagaritis provencheri* attack the larvae. *Apanteles hymeniae* Wilkinson breeds on the larvae in Fiji. According to Nair (1975) *Apanteles ruidus* Wlk and *Chardiochilies lymeniae*. also attack the pest larvae. Nair (1975) reported that parasites like *Telenomus javensis* Dodd., *Pareuderus torymoides* Ferr and *Erytoma curculinum* Mayr attack different stages of weevil *H.truncatulus*. A solitary braconid endoparasitoid, *Apanteles opacus* (Ashmead) infested the second instar larvae of *H.recurvalis*. Population of pupae of the parasitoid was maximum during first fortnight of July (Asha, 2005).

In tomato, when *Trichogramma brasiliensis* was released @ 50,000 / ha / week, the tomato fruit borer incidence came down from 42 to 19% and egg parasitism increased from 12.3 to 78.4% (PDBC,1994). Kakar *et al.* (1990) released 5 species of *Trichogramma* against *H.armigera* on tomato and reported egg parasitism up to 100%. Divakar and Pawar (1987) could achieve 62.9 % egg parasitism of *H.armigera* in tomato by inundative releases. The differences in egg parasitism may be due to different dosages of parasitoid release.

Seven parasitoids viz, *C.rexia, Asecodes* sp., *H.indica, Agathidine* sp., *C.agromyzae, H.brevipedicellus* and *Entomacis sp* were recorded from the larvae of the serpentine leaf miner. The parasitoids were recorded for the first time in Kerala. Among these *C.rexia* and *Asecodes* sp. were seen abundant in the ecosystem. Possibility of utilizing these parasitoids for *L.trifoli* which develops resistance to insecticides easily is to be explored (Reji, 2002). *Cotesia plutellae* was described as a promising parasitoid of *Plutella xylostella* (Hemachandra and Singh, 2007).

3. Mass Multiplication of Predators and Parasites

Coccinellids

Mass production of coccinellids is perfected by many workers. Scale feeding cocinellids can be multiplied on diaspine scales *viz, A.aurantii, Coccus* sp., *H. lataniae, M. glomerata* and others. Mealybug feeding cocinellids can be multiplied on mealy bugs grown on ripe pumpkin.

Aphidophagous coccinellids are multiplied on their natural host *viz*., aphids. Adult aphidophagous coccinellids feed and produce viable eggs on Ferrisia. *virgata.* To economise use of aphid hosts which are relatively more difficult to multiply than *F. virgata*, adults of coccinellids are reared on *F. virgata* and their larval stages multiplied on *A.craccivora*. For better egg production, freshly emerged beetles are pre-fed with aphids for about ten days before releasing on *F. virgata* infested pumpkins (PDBC, 2003).

Mass production of Trichogrammatids and syrphids are standardized and perfected by many workers.

Rearing *Trichogramma* requires first rearing an insect, typically a species of moth, to produce eggs in which the wasps develop. The Angoumois grain moth, *Sitotroga cerealella*, and the Mediterranean flour moth, *Ephestia kuehniella*, are easily and inexpensively reared on wheat or other grains and are commonly used to rear *Trichogramma* (Morrison *et al.* 1976). *Corcyra cephalonica* can be used for mass production of Trichogrammatids (PDBC, 2005).

Mass production of *Corcyra cephalonica*

Corcyra cephalonica, a laboratory host for mass multiplication of trichogrammatids and chrysopids was multiplied on rearing medium of sorghum or bajra. Different quantities of rearing medium were mixed with 0.5, 1.0, 1.5 and 2.0 ml of Corcyra eggs and the emerging moths counted. For collecting the emerging moths, aspirator, 0.25 HP oil sealed vacuum pump and industrial room cleaner along with modified containers were used to test their efficiency and safety to adult moths.

Mass production of Trichogrammatids

Different species of trichogrammatids were also mass produced by modifying methodology. Corcyra eggs were gummed using a semi-automatic method which included an ink pad to which sponge pieces were glued to use as gumming pad on 'tricho' card. The 'tricho' cards prepared in this way were exposed to adult parasitoids in polythene bags instead of glass/plastic containers to test feasibility of using polythene bags for en masse parasitisation.

Mass production of Chrysopnids

The chrysophid larvae are cannibalistic and pose a great problem in mass multiplication. Methods were developed by using locally available plastic fittings to enclose these individual larvae in cells/cubes for rearing. Plastic louvers were used along with acrylic sheets to enclose the larvae. Quantity of food (eggs of Corcyra) was standardized for rearing 100 larvae (PDBC, 2004).

Storage, packing, transportation and field delivery of biocontrol agents

The 'tricho' cards after exposing to adult parasitoids were kept in the polythene bags till parasitoids eggs turned black indicating complete development of the pupae. Such cards were stored in temperature of 5 to 10 ºC in BOD incubators or refrigerators. Similar studies were made with eggs of chrysopid to find out suitable temperature at which the 'tricho' cards or Chrysopid eggs can be stored for augmenting the field releases. Locally available materials were utilized to pack the 'tricho' cards or Chrysopid eggs and for transportation ((PDBC, 2004).

Parasitism of *Trichogramma* on host insects was also studied. Studies were made by raising castor, cabbage, cauliflower, knoll-khol, beetroot, tobacco, cotton and radish in pots. A bouquet of these plants was kept in insect rearing cages of 30 cm^2 and *S. litura* eggs affixed on leaves of each plant. *T. chilonis* adults were released in the cage and enclosed for 24 hours. Observations on parasitism egg mass-wise (%) were recorded and each egg separated out later to observe egg-wise parasitism.

Field release of *Trichogramma brasiliensis* on tomato against *Helicoverpa armigera* Release of Trichogrammatids @ 50,000/ha on tomato was standardized by (PDBC, 2006).

Studies were carried out through release of *T. brasiliensis* @ 50.000/ha from 30th day after sowing onwards for six weeks at weekly intervals. An observation on egg parasitism was recorded by collecting *H.armigera* eggs from 25 plants of four different places in a one ha. field. Bored tomato fruits (%) were also, recorded randomly.

4. Microbial Control

Biological control of insects using microbial agents viz, fungi, bacteria, virus, protozoa and actinomycetes are of considerable importance in pest managment. Important entomopathogenic fungi belonging to the genera *Beauveria*, *Entomophthora, Fusarium Metarhizium, Nomuraea, Paecilomyces* and *Verticillium,* bacteria, *Bacillus and thurigiensis*, virus NPV are well known biocontrol agents.

Fungal pathogens

A few insect species, including many pests, are particularly susceptible to infection by naturally occurring, insect-pathogenic fungi. These fungi are very specific to insects, often to particular species, and do not infect animals or plants. Fungal growth is favoured by moist conditions but fungi also have resistant stages that maintain infection potential under dry conditions. Fungi have considerable epizootic potential and can spread quickly through an insect population and cause its collapse. Because fungi penetrate the insect body, they infect sucking insects like aphids and whiteflies that are not susceptible to bacteria and viruses.

Several fungal species have potential as microbial insecticides and in a few countries, are commercially available in formulations that can be applied using conventional spray equipment. Entomopathogenic fungi causing mycosis in insects are identified from most of the insect orders *viz*. Orthoptera, Isoptera, Hemiptera, Diptera, Coleoptera, and Lepidoptera. Fungal pathogens are generally effective against chewing and sucking insects. Enzymes produced by fungi help in the degradation and penetration of the cuticle. High humidity and temperature are essential for sporulation and germination of the fungus. The infectivity depends upon virulence of the fungi and environmental factors. Once infection occurs, the fungal pathogens overcome the internal defence response by development of an outer coat, neutral to circulating haemocytes or are effectively masked by host proteins or by producing immuno modulating substances which suppress the cellular defence system (Narayanan and Subramaniam, 1975).

• *Beauveria* spp.

Beauveria bassiana is a fungus that occurs naturally in soils throughout the world. It causes a fatal disease in various insects and belongs to entomopathogenic fungi. The species is named after the Italian entomologist Agostino Bassi who discovered it in 1835 as the cause of muscardine disease of domesticated silkworms. *Beauveria bassiana* (formerly also known as *Tritirachium shiotae*) is the anamorph (asexually reproducing form) of *Cordyceps bassiana*. The latter teleomorph (the sexually reproducing form) has been collected only in Eastern Asia. In culture, *B. bassiana* grows as a white mold. On most common cultural media, it produces many dry, powdery conidia in distinctive white spore balls. Each spore ball is composed of a cluster of conidiogenous cells. The conidiogenous cells of *B. bassiana* are short and ovoid, and terminate in a narrow apical extension called a rachis. The rachis elongates after each conidium is produced, resulting in a long zig-zag extension. The conidia are single-celled, haploid and hydrophobic.

The fungus widely known as white muscardine fungus infects a wide spectrum of insects numbering more than 700 species. The effectiveness of *B. bassiana* was proven against *Aphis craccivora* (Zaki,1998), *Leptinotarsa decemlineata* (Todorova *et al.*, 2000), *Spodoptera litura* (Gloriana *et al.*, 2001), *Spilosoma obliqua* (Mondal and Bhattacharya, 2003) *Sylepta derogata* (Ramesh *et al.*, 1999).

Beauveria bassiana produces the toxin Beauvericin, bassionolide and beauverolides which cause cellular alterations. Hafez *et al.* (1994) conducted bioassay using *B. bassiana* against *Pthorimoea operculella* and obtained LC_{50} value of 4.7×10^8 conidia / ml. In brinjal, it was also observed that direct spraying of *Beauveria bassiana* killed 58.1% first instar larvae and 35.2% pre pupal stage larvae. The adults were not susceptible to *B.bassiana* (Rajendran and Gopalan, 1999) According to Masuda (2000) 95% mortality of *Plutella xylostella* was caused by *B. bassiana* at 10^7 conidia / ml. A potent local isolate of *B. bassiana* (1TCC 6063) was reported by Jiji *et al.*(2005). The efficacy of this isolate on the management of vegetable crops was also reported (Jiji *et al.*, 2006 a). Molecular characterisation of the same isolate was also done (Jiji *et al.*, 2006 d).

• *Metarhizum anisopliae* (Metsch.) Sorok.

Metarhizum anisopliae, the green muscardine fungus is pathogenic to grass hoppers, bugs, beetles, fruit flies and lepidopteran caterpillars. A few of the hosts are *Ceratatis capitata* (Ekesi *et al.*, 2002), *Helicoverpa armigera* (Gundannavar and Lingappa, 2003), *Plutella xylostella* (Ma, 2000) and *Thrips tabai* (Maniani *et al.*, 2003). Roberts (1981) stated that the importance of toxins to virulence of an entomopathogenic fungus is difficult to evaluate, because toxin production is preceded by several activities of the fungus. *Metarhizium anisopliae* infects insects through cuticle, spiracles and pores of sense organs. The insects are killed mainly by histolytic action of the fungus.

Growth of mycelium within the body creates mechanical blockage of the gut and other internal organs. Toxins viz. Dextruxin A and B are produced by *M. anisopliae* and they cause pathological changes in haemolymph and other vital sites. Of the two varieties of Metarhizium, *M. anisopliae* var. *anisopilae* (spore size 5-8 µm) and *M. anisopliae var. major* (spore size 10-14 µm), Lin and Roberts (1986) found that *M. anisopliae var. anisopliae* is the major source of mycotoxin, dextruxin A compared *to M. anisopliae* var. *major*. The biological effects of dextruxin were investigated by Samuels *et al.* (1988).

Pathogenicity of *M. anisopliae* to many insect species was reported by many workers. Loss of appetite and sluggishness are the major symptoms of infection. Initially white fungal colonies appeared at the joints, the integument and it later turned greenish following sporulation of the fungus. Finally the cadaver became black. Spore concentrations varying from 1.5 x 10^5 spores/ ml. to 1.5 x 10^{10} spores /ml. were reported effective against various pests (Wickramatileke *et al*., 2000; Rajendran, 2002 and Pandey, 2003). The infectivity of *M.anisopliae* on vegetable pest was reported by Jiji *et al*., (2006d).

• *Paecilomyces* spp.

Among fungal pathogens, the genera *Paecilomyces* also have a role in microbial control of insects. *Paecilomyces* species occur naturally in populations of infected and dead insects and randomly in soils around the world. The active ingredient can be cultured readily on various culture media and exists in several morphological stages, depending upon the culture medium. On solid medium, conidiophores, conidia and mycelia are predominant while submerged culture produces blastospores and mycelia. They are infective to lepidopteran larvae, plant hoppers, white flies and aphids. Economically important pests *viz Plutella xylostella* (Gopalakrishnan *et al*., 2002) and *Trialeurodes vaporarium* (Feng *et al*., 2004) were susceptible to infection by *P. fumosoroseus*. The fungal active ingredient, *Paecilomyces fumosoroseus* Apopka strain 97, is originally isolated in 1986 from a mealy bug in a greenhouse in Apopka, FL. The pure culture was identified in 1988 and deposited at the American Type Culture Collection (ATCC 20874). A local isolate of Paecilomyces was identified and deposited in the Indian type culture collection (ITCC 6064) from Kerala by Jiji *et al.* (2006 b). The pathogenicity of the same was proven among different insect hosts infesting vegetables (Jiji *et al*., 2006 a). The pathogen was characterized by using RAPD markers (Jiji *et al*., 2006 c).

• *Verticillium lecanii* (Zimmerman) Vieges

Verticillium lecanii is an entomopathogen against homopteran bugs though they are infective to other groups of insects. Reports on host range of insects include infection against many insects. In chillies, Easwaramoorthy and Jayaraj (1977) found *Verticillium lecanii* to be effective in controlling *M. persicae, Bemisia argentifolii* (Gindin *et al*., 2000) and *Thrips palmi* (Visalakshy *et al*., 2004). Pathogenicity of entomopathogenic fungi, *Verticillium lecanii* to *Scirtothrips dorsalis* was reported by Ganga Visalakshy *et al*. (2006). In a study conducted by IIHR, Bangalore, *Verticillium lecanii* sprayed (1.0 x 10 9 spores/ ha) against *Scirtothrips dorsalis* caused high mortality to all stages (PDBC, 2005).

• *Fusarium* spp.

Different species of *Fusarium* viz. *Fusarium equisetti* , *Fusarium moniliformae var. subglutinans* , *Fusarium semitectum* and *Fusarium pallidoroseum* are known insect pathogens. The infectivity of *F. equisetti* on *Coccidohystrix insolita* (Gopinathan *et al*., 1982) and *F. pallidoroseum* on *Aphis craccivora* (Hareendranath, 1989) was already reported. Hareendranath *et al*. (1987) in the bioassay using *F. pallidoroseum* on *A. craccivora* in cowpea revealed that LC_{50} of fungus was 3.408 x 10^6 spores / ml and it was not found pathogenic to *M. sexmaculatus*. Importance of *Fusarium pallidoroseum* as a biocontrol agent of *Aphis craccivora* was elucidated (Hareendranath, 1987, Faizal, 1996, Sunitha,

1997, Mathai, 1999). Molecular characterisation of the fungus was also done (Jiji *et al.*, 2001). Rejirani (2001) reported that the pathogen caused death of *A. craccivora* by invasion and multiplication of spore in the haemocoel and by production of toxin, fusaric acid. The pathogen @ 7 x 10 ^{6} spores / ml. was effective in controlling *A. craccivora*.

• *Rhizopus* spp.

Occurrence of *Rhizopus oryzae* is on several hemipteran insects from India. The hosts include *Aphis craccivora*, *Anchon pilosum, Coptosoma cribraria*, *Nezara viridula*, *Pseudococcus lilacinus* and *Riptortus pedestris* (Narayanasamy, 1998; Hebsy Bai *et al.*, 1999; Mathai, 1999). Pathogenicity of the *Rhizopus oryzae* on *Riptortus pedestris* on field condition was reported by Jiji *et al.* (2006 d).

Other pathogens

Strawiinski (1930) reported frequent epizootics caused by entomogenous fungus, *Entomophthoral sphaerosperma* on *Pieris brassicae*. Dickson and Laird (1962) observed that population of *M.persicae* was kept down by entomophagus fungi in wet spring. Ramaseshiah (1967) recorded *E. coronata* on *M.persicae* in India. The pathogen *Mucor spp.* is infective to *Aphis craccivora*. In brinjal, *Aspergillus flavus* Link parasitized and controlled the population in all stages. Spores of the fungus when sprayed on the insects killed them in three days. The crude toxin filtrate and the spore of *Aspergillus flavus* have a quick knock down effect. The crude extract killed the insects and was used in dispersal of fungus in the fields, where it proved effective that it may be used in integrated control (Krishnamoorthy and Shankara Naidu, 1971). *Aspergillus flavus, A. niger, Cladosporium* sp. *Entomophthora delphacis, Nomuraea rileyi, Pandora delphacis* and *Zoophthora radicans* are a few of the entomopathogenic fungi which cause epizootics in insects (Ambethgar, 2002).

Pathogens for biological suppression of phytophagus mites

Hirsutella thompsonii causes infectivity on mites. Exudates of *H. thompsonii* caused 77.8 % mortality of *Tetranychus urticae* after seven days and mortality decreased as exudate concentration was reduced. The metabolites *H. thomosonii* collected from one month old shake culture were toxic to *T. urticae* (68.9% mortality after 7 days), but the mortality decreased with dilution. A wettable powder formulation of *H. thomosonii* when sprayed twice at the rate of 50 g/l, reduced *T. urticae* population by up to 62.7% on okra in farmers' field in Coimbatore (PDBC, 2005).

Growth of the fungus in artificial condition and method of multiplication of the fungus were also studied. *Hirustella thomosonii* growth was excellent at 25^{0}C. Conidial production of *H. thomosonii* was hindered at 37^{0}C and above. When oatmeal agar (OMA) discs were used on the culture, the inoculum load, the CFU, pellet number and wet and dry weights increased proportionately. A combination of OMA plugs and SDB medium was the best with maximum CFU, wet and dry conditions. A simple magnetic stirrer technique for faster mass production of *H. thompsoni* was developed (PDBC, 2005). The feasibility of wide application of this entomopathogen against vegetable pests is yet to be exploited.

Mass multiplication

Cost effective production technology for production of bioagents through simple procedures was developed by many workers. Cheaply available substrates like coconut water, molasses, yeast broth, rice bran and wheat bran. were identified as suitable substrates for mass multiplication of fungi (Beevi, 1979; Danger *et al.*, 1991; Faizal and Mathai, 1998; Haraprasad *et al.*, 2001 and Verghese *et al.*, 2003). Technique for mass production of *B. bassiana* was standardized by Ibrahim and Low (1993). According to Haraprasad *et al.* (2001), *B. bassiana* grew well on rice bran, rice bran coarse and wheat bran yielding a maximum sporulation. Techniques for mass multiplication of *M. anisopliae* using coconut water and cassava chips are also developed (CPCRI, 1999). Technique for large scale production of *F. pallidoroseum* in rice bran in poly propylene bags was also developed. Waste materials after rearing rice moth in biocontrol laboratories were also suitable for growth of fungus (Faizal and Mathai, 1996; Sudharma *et al.*, 2001). Beneficial effects of coconut water for mass production of *F. pallidoroseum* were reported by Rejirani (2001). Technology(s) for mass production of *B.bassiana, M. anisopliae* and *R. oryzae* and *Paecilomyces* were also developed (Jiji *et al.*, 2006 b).

Formulations

Formulations were developed by appropriate combination of ingredients for large scale use, easiness of transportation and storage. Dust, granule, bait and oil based formulations are also developed.

Long shelf life of a biopesticide is an important factor in gaining market and further its use in the field. Fifty fold increase in viability of spores of *Verticillium lecani* was found in wettable powder formulation of the fungus compared to its blastospore suspension in water (Ravensburg *et al.* 1990). Andersch (1992) patented the process of formulating *Metarhizium anisopliae* as pellets.

Moore *et al.* (1995) evaluated oil formulations of *M. flavoviridae* and found that formulations retained their virulence for 30 months. Moore (1996) suggested that addition of sachets of silica gel could improve conidial survival by 33 %. A long shelf for *M. anisopliae* was identified by Alves *et al.* (1988) who evaluated 64 formulations of *M. anisopliae*. He found that fungus could be stored up to 660 days with 33% viability. *B. bassiana* and *V. lecanii* in powder and liquid formulations were effective against brinjal jassids, *Amrasca bigutula biguttula. V. lecanii* is compatible with natural enemies of cabbage aphid and *T. chilonis* and is harmless to silk worm and therefore can be used for integrated pest management of cabbage aphid (Ali Derakhshansh *et al.*, 2006). Jiji *et al.* (2006 c) developed wettable powder formulations of *B.bassiana* and *P.lilacinus* for field use.

Bacteria

Bacterial pathogens of insects generally belong to families Bacillaceae, Enterobacteriaceae, Pseudomonadaceae, Streptococcaceae and Micrococcacae of the order Rickettsiales and class Mollicutes (Tanada and Kaya, 1993). They include spore formers and non spore formers. The endospore producing spore formers can tide over unfavourable conditions. Toxic crystal producing crystalliferous and non crystalliferous bacteria are present among the spore formers. Many species of *Bacillus* are crystalliferous and produce endotoxins. Spore forming crystalliferous bacteria are very important as microbial control agents, of which the most exploited one is *Bacillus thuringiensis* Berliner.

Bacillus thuringiensis (Bt) (Eubacteriales): Bacillaceae first isolated from diseased silk worm larvae in 1901, was recognized for its utility under field condition in late 1920s. Bt is an aerobic, gram positive, rod shaped and spore forming bacterium, widely present in nature. It produces a number of insect toxins, the most distinctive of which are protein crystal during sporulation. These crystalline protein inclusions or endotoxins are the principal active ingredients in Bt formulation currently in use. Bt is classified into many subspecies, strains and serotypes. A new nomenclature based on amino acid identity was suggested by Crickmore *et al.* (1996). According to Singh (2001), sub sp. Kenyae (ISPC – 1, ISPC–7), sub sp. *tolworthi* (HD–2), sub sp. *thuringensis* (HD- 263), sub sp *kurstaki* (HD–709, ISPC –4, H3a, 3b, 3c), sub sp. *isralensis* (VCRC B –17, B-113, IPS –80, H –14) and sub sp. *tenebrionis* infect over 68 species of lepidopteran, dipteran and coleopteran insects.

Spores and crystalline insecticidal proteins produced by *B. thuringiensis* are used as specific insecticides under trade names , Dipel and Thuricide. Because of their specificity, these pesticides are regarded as environmentally friendly, with little or no effect on humans, wildlife, pollinators and most other beneficial insects. *B. thurigiensis*-based insecticisdes are often applied as liquid sprays on crop plants, where the insecticide must be ingested to be effective. The solubilized toxins form pores in midgut epithelium of susceptible larvae. Recent research suggested that the midgut bacteria of susceptible larvae are required for *B. thuringiensis* insecticidal activity.

In brinjal, *Bacillus thuringiensis* sub sp. *kurstaki* suspension (0.25%) caused a maximum of 68.9 % in the first instar larvae and a minimum of 13 % in the prepupal stage. *Bacillus thuringiensis* gave complete mortality within 48 hrs (Venkataraman *et al.*, 1962).

Evaluation of commercial formulation of Bt var. kurstaki by Singh *et al.* (2000) indicated that the pathogen was highly effective against *H. armigera, P. xylostella* and *C. binotalis* and non effective against *S. litura*. The effectiveness of Bt. in the management of insect pests was reported by Mote and Sivubhavikatti (2001). They found that Bt (0. 04%) (Delfin) protected brinjal fruits against pests and gave the highest yield and net profit.

The bacteria enter the body of insect through food and is capable of producing septicemia and toxaemia. The other route of infection is injured integument of the insect. Symptoms vary with the strain of *Bacillus* and the host. The diversity and specificity of insecticidal proteins were thoroughly studied during the last decades, over 140 genes code for more than 30 types of Cry proteins. The molecular mode of action of delta endotoxin consists of solubilisation of the crystal and release of the active Cry protein, binding of the activated toxin to receptors and opening of ion channels or pores. Specificity of Bt strains may be caused by presence or absence of specific proteases in the insect midgut (Panda, 2004). The Bt receptor protein was identified as a member of the aminopeptidase N family.

5. Transgenic Plants

Genetically modified plants having Bt crystal genes coded for pest control are developed with development of molecular biology. These plants provide continuous protection from the injurious stages of pests for the entire crop period irrespective of the weather conditions. As defence is built in the plant, protection is ensured to all parts inaccessible to pesticide spray. Bt potato (New leaf) was produced by Monsanto. Transgenic tomato plant carrying CP gene of TMV was reported (Nelson *et al.*, 1988, Byoung-Cheorl Kang, 2007). The US Environmental Protection Agency(EPA) mandated resistance management plan for all commercialized transgenic crops expressing Bt. toxins in USA. Efforts are made to counter the problem of resistance development. According to Neema and Bhatnagar (2004), protein pyramiding with multiple Cry genes, tissue specific protein synthesis and field tactics like refuge crop and Cry gene crop rotation can be employed effectively for resistance management.

Genetic improvement of Bt to increase effectiveness and spectrum of activity involves manipulation of Bt strains insecticidal crystal proteins (ICP) – plasmid content adopting recombinant and non recombinant methodologies through plasmid curing and conjugal transfer. It is possible to increase the amount of desired Cry proteins in strains crystals which often increases potency and spectrum. EG 2424 is produced through conjugation. Synthetic genes are designed for optimal expression.

Mass production of *Bacillus thuringiensis*

Mass production techniques of *B. thuringiensis* are developed by Singh (1995) and Balaraman (2005). Details on nutrient selection, contaminants and their elimination and other requirements are provided by these workers.

Bacillus cereus is infective to coleopterans, lepidopterans and hymenopterans. Vegetative insecticidal proteins from the supernatants of vegetative *Bacillus cereus* culture have two compounds ; VIP 1 and VIP 2 and they exhibit toxicity to insects like delta endotoxin from Bt.

B. magaterium, B. subtilis, B sphaericus (H-5), *Pseudomonas aeruginosa, Serratia marcescence* and *Streptococcus faecalis* are the other bacteria associated with insect pests (Singh, 2001).

Pseudomonas flourescens

There are many reports on the efficacy of *Pseudomonas flouresence* against insect pests viz. *Leptinotarsa decemlineata* , *Epilachna vigintioctopunctata* (Consanzo *et al.*,1998;

Casreillo *et al.*, 2000 ; Commare *et al.*, 2002; Otsu *et al.*, 2004). It was reported that the population of aphids and leaf hoppers were reduced by about 79 and 81%, respectively due to spraying with B 25 strain of *Pseudomonas fluorescens*.

Viral pathogens

Viruses are successfully used for biological control of insect pests especially belonging to lepidopteran, coleopteran and hymenopteran orders. Viruses are specific in their action, infecting a single genus or a single species. There are 11 families of entomopathogenic viruses of which members of the family Baculoviridae are the best exploited on account of its environmental safety. They are safe to non target organisms including human beings The genera, Nucleopolyhedrovirus and Granulovirus which come under Baculoviridae are of wide application for biological control. The infective units ie., the virions of Nucleopolyhedrovirus (NPV's) are occluded in a crystalline polyhedral body within insect cell nuclei while that of Granuloviruses (GV's) are occluded in capsular proteinaceous occlusion bodies (Dhaliwal and Arora, 2001).

There are many reports on occurrence of virus in host insects. Singh (2001) enlisted a few of the host insects from which NPV was reported included *Agrotis ipsilon, Amsacta albistriga, Anadevidia peponis, Anomis flava, H. armigera, Hellula undalis, Mythimna separata, Pericallia ricini, Pieris brassicae, Plutella xylostella, Spilosoma obliqua, Spodoptera exigua* and *S. litura*. Similarly *Grnulosis* viruses were recorded from, *A. ipsilon, P. ricini, P. operculella, P. xylostella, P. bassicae, S. oblique* and *S. litura.*

The infection due to viruses occurs through ingestion. On ingestion, they replicate and cause cell and tissue damage. Internal organs are liquefied. Rupturing of body cuticle occurs and polyhedra are released. The occluded bodies persist in water, soil and carrier materials.

American boll worm (*Helicoverpa armigera*) and Tobacco caterpillar (*Spodoptera litura*) are controlled by c. HaNPV multiplied on larvae of *Helicoverpa armigera*. The production procedure was developed by PDBC, Bangalore, renamed ICAR-NISAIR, Bangalore (Singh, 1995). Procedure for multiplication and release of *Oryctes baculovirus* (OBV) were outlined by ICAR-CPCRI (1999). Retention of biological activity of the ingredient and avoidance of replication of contaminant microorganisms in the formulations are a few of the points to be considered. Baculovirus replication in insect cell lines was recognized for several Nucleo polyhedrosis viruses (Lynn, 1999). Media like TNM-FH, TC 100 and IPL 41 required for cell culture are available commercially. Cell lines can be established from various tissues like embryos, haemocytes, fat bodies, imaginal disc, pupal and adult ovaries and larval gonads. National Institute of Virology, Pune have come out successfully in establishing cell lines useful in the propagation of baculoviruses of *Spodoptera litura* and *Heliothes armigera* with high virus yield.

Application of Ha NPV as spray was effective in regulating population of *H. armigera* (Butani, *et al.,* 1997). Sucrose (0.5%), Jaggery (0.5%), egg white (3%), and chickpea flour increased the activity of Ha NPV (Sonalkar *et al.*, 1997). Identification of improved strains of NPV with increased virulence to *H. armigera* and *S. litura* was done (Subramanian, 1998; Geetha and Rabeendra, 1999). Activity of Ha NPV was enhanced by inserting a gene coding for an insect specific neurotoxin from *Pyemotes tritici* into the gene coding ecdysteroid UDP glucosyl transferase in Ha NPV (Popham *et al.*, 1997). Sun *et al.* (2003) developed engineered HaNPV virus with 40 fold increased speed of kill. Four releases of *T.chilonis* @ 50000/ha and two sprays of *Ha* NPV at 1.5 x 10^{12} POB/ ha applied after second and fourth release of the parasitoids checked *H. armigera* on tomato as effectively as insecticides. Fruit damage due to *H. armigera* was significantly lower when *Ha* NPV was applied at full dose 1.5 x 10^{12} POB/ha than at lower doses on tomato (PDBC, 2005).

Protozoa

Protozoans generally are debilitating agents and their action is slow. Virulent forms are also available. Microsporidians *Nosema locustae* and *Varimorpha neeatrix* have a much better place as biocontrol agents. Infection occurs by ingestion of spores. Spores germinate and penetrate gut wall. On reaching haemocoel, the sporoplasm is inoculated into cytoplasm of host cell in the fat body and haemocytes. The virulent, microsporidian viz. *Vairimorpha necatrix* is promising as a short term biocontrol agent because of its virulence against noctuids like *Helicoverpa armigera*. Studies on *Vairimorpha* sp. (Ha strain) undertaken in Tamil Nadu Agricultural University resulted in developing a wettable powder formulation of the microsporidian, effective for control of *H. armigera* in chick pea.

Nematode

During the last decade, there are hightened interest world over in the augmentative biological control of insects using entomopathogenic nematodes (EPN) belonging to families, Steinernematidae and Heterorhabditidae. These nematodes are mutually associated with insect pathogenic bacteria in the genus, *Photorhabdus / Xenorhabdus*. It is a nematode bacterium complex that works together as biological control unit to kill an insect host. These nematode bacterium association may meet many criteria for augmentative control of insect through inundative release including broad host range, ability to kill host rapidity, long duration storage of infective juveniles (IJ), distribution and persistence, available inexpensive mass production technologies and safety to plant, invertebrates and non target organisms and feasibility of application with existing spray equipments. There is no known negative effect on environment and is amenable for genetic selection. These nematodes are exempted from registration in many countries. The intense interest in EPN as biological control agent of insect pest resulted in a large number of research efforts and subsequent publications.

A major difference between the steinernematids and heterorhabditids is that all but one species in the former group are amphimictic, whereas species in the later group are hermaphrodites in the first generation. Thus steinernematids require a male and female infective juvenile to invade an insect host to produce progeny, whereas heterorhabditids need only one infective juvenile to penetrate and into a host as the resulting hermaphrodite adult is self-fertile.

The bacteria mutually associated with EPN are *Xenorhabdus* and *Photorhabdus* which are motile, gram negative, facultative, non spore forming and anaerobicrods in the family Enterobacteriaceae. In the genus *Xenorhabdus,* five species are associated with *Steinernema* whereas in the genus *Photorhabdus* three species are associated with *Heterorhabditis* with one species *P. lumininescens*, divided into five sub species. The subspecies of *P.luminescens* are subsp. *luminescens, laumondii, akhursti, kayaii and thraciaensis*. One species, *P. asymbiotica* is isolated from human clinical cases and is not associated with nematodes.

Effect of two native isolates of *Rhabditis was* assessed in terms of dipteran and coleopteran pests of bitter gourd namely fruit fly, *Bacterocera cucurbitae* Coq. and epilachna beetle, *Epilachna septima* Fab. The native isolates from the rhizosphere of vegetables using wax moth, *Galleria melonella* as trap were used for assessing the mortality of maggots of *B.cucurbitae* and grubs of *E.septima* of bitter gourd. Active third stage juvenile (dauer juveniles) stored for seven days in sterile tap water were used for inoculation following standard methods. Three levels of native isolates of Rhabditis were inoculated (50,100 and 200 DI/maggot or grub). The results reveal that the mortality of *B.cucurbitae* ranged from 6-47 % at an inoculum range of 50, 100 and 200 DI/insect larvae. At the highest dose, 87% mortality was recorded at 48 h after treatment and it became 100% at 72h after treatment. The mortality of grubs of epilachna beetle due to infestation by isolates of *Rhabditis* ranged 67-87% at 48 h after treatment. At 72 h after inoculation, all the isolates showed 100% mortality. The nematode multiplication was seen in insect cadaver 72 h after treatment.

Effect of two isolates of endomopathogenic nematodes, *Heterorhabditis* sp. was assessed in terms of larval mortality of leaf eating caterpillars of vegetables *viz*., *Sylepta derogata* L. (Pyralidae:Lepidoptera), *Diaphania indica* Saund. (Pyralidae: Lepidoptera) and *Spilosoma oblique* W. (Arctiidae: Lepidoptera). All the three pests showed high mortality after 48 h of inoculation (40-100 %). Based on result of study, mortality of the above three larvae was comparatively higher in *Heterorhabditis* isolate (2) at 24 and 48 h after inoculation while the effect of *Heterorhabditis* isolate (1) was superior in all the above pests at 72 inoculation. The mortality at 24, 48 and 72 h varied depending upon the isolate and insect pest studied revealing the potential of nematodes as biological agents against lepidopteran pests (Sheela *et al.*, 2002). The efficiency of Rhabitis in managing pests of bittergourd was also reported (Sheela *et a*l., 2002a).

Results indicated that *S. carpocapsae* is suitable as a biological control agent for both *H. undalis* and *C. binotalis*. Steinernematids and Heterorhabditids infect over 200 species of insects from several orders. It was reported that *S.carpocapsae* infected *S.liture* in both in lab and field conditions. *Heterorhabditis* sp; *S. carpocapsae, S. glasseri* and *S. feltiae* infected *P. zylostelle. S. feltiae* was effective against *H. armigera* under field conditions. *S. abbasi, S.carpocapsae, S. glaseri* and *H. indica* were infective to *A. ipsilon. H. indica* was effective against *H.armigera* in field condition (Gulsar Banu and Sheela, 2005).

Novel microbial insecticides

Spinosyns, avermectins, milbemycins, thiangazole, cyclophosphate, chlorfenapyr and ulosantoin are novel insecticides of microbial origin. (Dureja and Neera, 2000). Dow Agro

Sciences, USA commercialized spinosad, a mixture of Spinosyn A and Spinosyn D occurred together in cultures of *Saccharopolyspora spinosa*. It was first isolated in 1989 from soil samples of Caribbean islands.

Spinosad is highly active against lepidopteran insects by contact and ingestion. It is insecticidal to thrips, beetles, wasps and flies. It gets absorbed locally on the leaves that extends its activity against leaf feeders like leaf rollers which account for its effect on leaf miners. It is neurotoxin acting on nicotinic receptors but in a unique way without interfering with the action of acetylcholine itself or with the action of neonicotinyl compounds. Trade names include Tracer, Success and Spinosad 48SC. Tracer includes fermentation products Spinosyn A (85%) and Spinosyn D (15 %). D Spinosad acts both as a stomach and contact poison and has low mammalian toxicity and is selectively active on insect pests belonging to lepidopteran pests at LC_{50} values ranging from 0.02 to 5.8 ppm. (Adan *et al.*, 1996). Spinosad has contact and stomach activity and is effective against insects belonging to different groups (Dhaliwal and Arora, 2001). Spinosad 2.8 SC @ 15 g ai / ha was effective in protecting *Plutella xylostella* (Walunj and Pawar, 2001). Spinosad provided effective control of lepidopteran pests of cabbage *Plutella xylostella* (the diamond black moth), *Hellula undalis* (the cabbage head borer) and *Spodoptera littoralis*. (the cabbage leaf caterpillar)) at a dosage level of 15-25 g ai / ha. Efficacy persisted for seven days. Spinosad at this rate showed very little adverse effect on the most important parasitoids of DBM infesting cabbage. (Dey and Somchoudhury, 2001).

Avermectin

The avermectins were isolated from fermentation broth produced by the culture of *Streptomyces avermectlis* in Japan. The metabolites are active against certain nematodes, arthropods like caterpillars and leaf miners and other plant parasitic mites at low dosages. Avermectin is also called abamectin. The trade names are Affirm, Agrimek, Avid, Vertimec and Zephyr.

Abamectins are a mixture of avermectins containing about 80% avermectin B 1a and 20% avermectin B 1b having similar biological and toxicological properties (Dureja and Neera, 2000).

6. Biological Control of Diseases

Biocontrol is the biological destruction of harmful microorganisms causing plant diseases without disturbing the ecological balance. It is the reduction of pathogen inoculum or its disease producing capacity by action of one or more organisms accomplished naturally or through manipulation of environment, host or antagonist or by mass production of one or more antagonists. (Cook, 1990). This strategy aims at suppressing inoculum load of target pathogen below the level that causes economically significant outbreak of disease (Mukerji *et al.*, 1999). Biological agents are microorganisms with ability to interfere with growth or survival of plant pathogens. Potential agents for biocontrol include rhizosphere competent fungi and bacteria which in addition to their antagonistic properties are capable of inducing growth responses either by controlling pathogens or producing growth stimulating factors (Yedidia *et al*., 1999).

Wide range of organisms is being exploited in this technology. Highly efficient native microbial strains are identified and developed as biocontrol agents after through screening and field trials. Fungal antagonists like species of *Trichoderma* and bacterial antagonists such as fluorescent *Pseudomonas*, *Bacillus* and *Agrobacterium*. are widely exploited and are available in the international market. *Trichoderma*, a fungi that are present in all agricultural soils and other diverse habitat are potential biocontrol agents. The antagonistic nature of *Trichoderma* was demonstrated more than seven decades ago (Weindling, 1934), although the first report on biological control experiment using *Trichoderma* under natural field field condition came much later (Chet *et al*., 1987). The antifungal ability of these beneficial microbes is known since 1930s.They grow tropically towards hyphae of other fungi, coil around them in a lectin- mediated reaction and degrade cell wall of the target fungi by secretion of different lytic enzymes. Species of these genus like *T.harzianum, T.viride, T.koningi, T.asperdium, T.hamatum* and *T.atroviride* were highly successful for the suppression of a broad spectrum of phytopathogens including *Pythium* spp., *Armillaria mellea, Phytophthora* spp. and *Aphanomyces* spp. (Pal and Jalali, 1998). *Trichoderma* is frequently used for biocontrol of aerial environment as well (Tronsomo, 1986; Dubos, 1987). It can directly increase root growth and increase uptake and concentration of variety of nutrients viz., Cu, Mn, P, Fe and Na. Recently several new mechanisms for both biocontrol and enhanced plant growth are demonstrated based on mycoparasitism, antibiosis, competion for nutrients or space, tolerance to stress through enhanced root and plant development, solubilisation and sequestrations of inorganic nutrients, induced resistance and inactivation of pathogen enzymes (Elad and Freeman, 2002., Howel, 2003, Harman *et al*., 2004). Strains of *T.harzianum* produced two volatile metabolites, 6-n-pentenyl-2H-pyran-2-one and 6n- pentyl-2H-pyran-2-one, lactones, alcohols and terpene derivatives of pyran. These volatile substances inhibit mycelia of a spectrum of pathogens. Extra cellular enzymes like chitinase and glucanase produced by *Trichoderma* spp. degrade the cell wall of the pathogen.

Fluorescent pseudomonads emerged as the largest and potentially the most promising group of PGPR involved in biocontrol of plant diseases (Kloepper *et al*., 1980, Suslow, 1982). Fluorescent pseudomonads (*P.cepacia. P.aeruginosa, P.fluorescens, P. putida)* and closely related species are important in soil. Disease suppression by fluorescent pseudomonads is attributed to antibiosis, siderphore production, production of HCN and competition. Fluorescent pseudomonads compete with pathogen for iron. They produce non-sequesterin siderophores, which decrease iron and inhibit growth of pathogens. Siderophores like pyochelin, pyoverdine, pseudobactin, ferribactin, ferrichrome and ferroxamine are secreted by pseudomonads. Studies proved the ability of pseudomonads in biocontrol of wilt disease by *Fusarium oxysporum* (Scher and Baker, 1980; Scher and Baker, 1987) and damping off disease of cotton by *Pythium ultimum* (Kloepper *et al*., 1980). Pseudobactin–siderophores was effective against pathogens like *Erwinia carotovora, Pythium aphanidermatum* and *Thielauopsis basicola* (Misaghi *et al*., 1982, Lazzaretti *et al.*, 1995). Secondary metabolites produced by fluorescent *Pseudomonas* effective in biocontrol of pathogens include antibiotics like Oomycin A, acetyl phloroglucinols, phenazines (for iron competition) like oxychlororaphin, pyocyanin, pyrroles like pyoluteorin, ptrrolnitrin, tropolones like pseudomonic acid, B-lactum compounds like bactobolins, thiotropocin, salicylic acid and IAA. (Johri *et al.*, 1997). Fluorescent pseudomonads are successfully used in the control of *Phytophthora* root rot of soyabean (Lifshistz *et al.*, 1986), tobacco black root rot (Keel *et al.*, 1989), potato seed decay (Xu and Gross 1986), several wilt diseases due to *Fusarium* spp. (Kloepper *et al.*, 1980a and b; Scher and Baker, 1982; Sneh, 1981), and fungal diseases of orange and lemon citrus roots (Gardner *et al* 1984) and ornamental plants (Yuen and Schroth, 1986), bacterial wilt of vegetables (Meenakumari *et al.*, 2003; Sonia, 2005) spices and paddy (Sivaprasad,1998; Sivaprasad *et al.*, 2003; Sulochana *et al.*, 2003).

Bacillus is yet another promising group of microorganisms involved in effective biocontrol of most of the phytopathogenic bacteria. *B.subtilis* produces antibiotics iturin, bacilycia and fengacin which inhibit *Rhizoctonia, Botrytis* and *Xanthomonas*. It also produces antagonistic proteins strongly inhibitory to seed-borne fungi and bacteria (Lazzaretti *et al.*, 1995). Importance of *Bacillus* spp. in controlling bacterial wilt under glass house and field conditions was demonstrated. (Shekhawat *et al.*, 1992; 1993).

Arbuscular mycorrhizae are now well recognized as a biofertilizer cum deterrent to root pathogens and widely used for biocontrol of root diseases of crop plants. Host plants pre-inoculated with a mycorrhizal symbiont exhibit increased tolerance to a wide range of root diseases This includes a spectrum of diseases caused by species of *Fusarium, Phytophthora, Pythium, Rhizoctonia, Thielaviopsis, Macrophomina, Verticillium, Sclerotium, Cylindrocladium* and *Olpidium* (Joseph and Sivaprasad, 2000; Kavitha, 2003). The AM symbiont induces resistance against foot rot of black pepper, rhizome rot of ginger, azhukal disease of cardamon and root-knot nematode infestation in most of the crop plants. Mycorrhizal symbiosis induces plant defense mechanisms like accumulation of aminoacids, higher activity of chitinase, glucanase, peroxidase, phenolics and phytoalexins (Joseph and Sivaprasad, 2000).

7. Conclusion

Biological control is popular as an ecofriendly method for the management of crop pests and diseases. Identification and conservation of natural enemies and antagonists are very much relevant in this context. Technology for mass multiplication of natural enemies has to be popularised. Farmers should be given proper training for identification and conservation of natural enemies. Microbial pathogens should be made available to farmers for large scale use. It is important to collect the naturally occurring isolates of entomopathogens and study their potential. Commercial production of good quality formulations, having long shelf life is necessary. Biotechnological interventions may help increase the efficacy of promising pathogens.

References

Abhilash, B. (2001) Biocontrol of mites on yard long bean (*Vigna unguiculata* ssp.sesquipedalis (L.) Verdcourt) and chilli (*Capsicum annuum* L.). M. Sc. (Ag.) thesis, Kerala Agricultural University, Thrissur p.8-14.

Adan, A., Del , E.P. and Vinuela, E. (1996) Spinosad: an overview of the insecticidal activity of this novel biologically derived compound. Paper presented in XX International Congress of Entomology held at Firenze, Italy.

Agarwala, B.K. and Bardhanroy, P. (1999) Numerical response of lady bird beetle (Coleoptera : Coccinellidae) to aphid prey (Homoptera : Aphididae) in field bean in North East. *Indian Journal of Applied Entomology.* 123: 401-405.

Agarwala, B.K. and Gosh, A. K. (1988) Prey records of aphidophagous coccinellidae in India. A review and bibliography. *Tropical Pest Management* 34: 1-14.

Ali Derakhshan S H, Rabindra, R.J. and Ramanujam, B. (2006) Impact of the entomopathogenic fungus, *Verticillium lecanii* of natural enemies of cabbage aphid and beneficial insects. *National Symposium on Biological Control of Sucking Pests in India*, Bangalore, p. 11.

Alves, S.B., Neto, S., Pereira, R. M. and Macedop, N. (1988) Study of formulations of *Metarhizium anisopliae* (Metsch.) Sor. in different storage conditions. *Ecossistema* 12: 78- 87.

Ambethgar. (2002) Record of entomopathogenic fungi from Tamil Nadu. *Journal of Entomological Research* 26 (2): 161-167.

Anand, R.K. (1983) Predation by *Coccinella septumpunctata* Linn., and *Menochilus sexmaculata* Fab., on five species of aphids. *Pranikee*, 4: 234-237.

Andersch, W. (1992) Production of fungi as crop protection agents. *(In) Pflanzenschutz. – Nachrichtn, Bayer* 45: 129-142.

Annie Bernice, T.S. (2000) Eco-friendly pest management in brinjal (*Solanum melongena.* L). M.Sc. (Ag.) thesis, Kerala Agricultural University, Thrissur.

Asha, J. S. (2005) Seasonal Occurrence and eco friendly management of pests of amaranthus (*Amaranthus tricolor* L.) MSc. (Ag.) thesis, Kerala Agricultural University, Thrissur.

Bagas, S. R. and Trehan, K. N. (1949) Life history and bionomics of two predacious and one mycophagous species of coccinellidae. *J. Bombay Nat. Hist. Soc.* 45: 566-575.

Balaraman, K. (2005) Production of Microbial insecticides (*In*) *Microbial Fermentations* Gunasekharan, S. Marimuthu, P. and Natarajan, T. (Ed.), Centre for Advanced studies in Agricultural Microbiology, Department of Agricultural Microbiology, Tamil Nadu Agricultural University, Coimbatore. pp 127-132..

Beevi, N.S. (1979) Studies on entomogenous fungi associated with lady bird beetle *Henosepilachna vigintioctopuncta* (Fabr.) (Coccinellidae: Coleoptera). M.Sc. (Ag.) Thesis, Kerala Agricultural University, Thrissur, p.72.

Bhattacherjee, N.S. and Menon. M.G.R. (1964) Biology and control of *Hymenia recurvalis* (Fab). *Indian Journal of Entomolgy.* 26: 179-183.

Biliotti, E. (1977) Augmentation of natural enemies in Western Europe. (*In*) *Biological Control by. Augmentation of Natural Enemies* pp. 341-7. R, L. Ridgeway and S.B. Vinson (Eds)., Plenum Press, New York.

Bindu, S.S. (1997) Biological Control of Pests of Vegetable Cowpea (*Vigna unguiculata* sub sp. *sesquipedalis* L.) Verdcourt). M.Sc. (Ag.) Thesis, Kerala University, Trivandrum, p 84.

Butani, P. G. Kapadia, M. N. and Parsana , G. J. (1997) Comparative efficacy and economics of Nuclear polyhedrosis virus (NPV) for the control of *Helicoverpa armigera* (Hubner) on groundnut. *Journal of Oilseeds Research.* 14: 85-87.

Byoung-Cheorl Kang, Inhwa Yeam, Hanxia Li, Kari W. Perez and Molly M. John (2007) Ectopic expression of a recessive resistance gene generates dominant potyvirus resistance in plants. *Plant Biotechnology Journal* 5 (4), 526–536.

Canard, M., Semeria Y. and New, T. R. (1984) *Biology of Chrysopidae*, Dr. W. Junk Publishers *The Hague*, p. 287.

Casreillo, L.A., Lee., Jr R.E., Lee M.R and Rutherford (2000) Indentification of ice nucleating active *Pseudomonas fluorescens* strains for biological control of over wintering Colorado potato beetles (Coleoptera : Chrysomelidae). *Journal of Economic. Entomolog* 93(2): 226-233.

Cherian, M.C. and Brahmachari, K. (1941) Notes on three predatory Hemiptera from South India. *Indian Journal of Entomology.*3 : 115-119.

Chet, I. (1987) Trichoderma applications, mode of action and potential as a biocontrol agent of soil borne plant pathogenic fungi.. (*In*) *Innovative approach to plant disease control*, pp.137-160,. Chet.I (Ed.). John Wiley and Sons, New York.

Commare, R. R. Nandakumar, A., Kandan, S., Suresh, S, Bharathi, M.. Ragu chander, T., and Swamiayppan, R. R. (2002) *Pseudomonas fluorescens* based formulation for management of sheath blight disease and leaf folder insect in rice. *Crop Protection.* 21: 671-677.

Consanzo, J.P., Humphreys T.L., Lee., Jr R.E., Moore J.B., Lee M.R and Wyman J.A. (1998) Long term reduction of cold hardiness following ingestion of ice nucleating bacteria in the Colorado potato beetle, *Leptinotarsa decemlineata. Journal of Insect Physiology* 44: 1173-1180.

Cook, R. J. (1990) Current Status and future prospects in research and practices of Biological Control (Abst.) *Inernational Conference on biological control of plant diseases and virus vectors*. Set. 17-21. 1990, Tsukuba, Japan. pp. 11-15.

CPCRI. (1999) *Biological suppression of coconut pests. Technical Bulletin no. 37*. Central Plantation Crops Research Institute, Regional Station, Kayamkulam. p.13.

Crickmore, N., Zeigler, D.R., Feitelson, J. Schnepf, E., Lanrie, J., Lambert, B., Lereclus, D., Bawn., J. and Dean, D. H. (1996) Revision of the nomenclature for *Bacillus thuringiensis* pesticidal crystal proteins. *Microbiol. Mol. Biol. Rev.* 62: 807-813.

Dahiya, A.S., Lakra, R.K. and Zile Singh. (1988) Predatory behaviour and development of two species of syrphid aphidophagous to *Aphis gossypii* glover. *Proceedings of the National Symposium on Integrated Pest Control, Progress and Perspectives* October 15-17. 1987 Mohandas, N. and Koshy, G. (Eds.). Kerala University. pp 200-207.

Danger, J.K., Geetha, L., Jayapal, S.P. and Pillai, G.B. (1991) Mass production of entomopathogen *Metarhizium anisopliae* in coconut water wasted from copra making industry. *Journal of Plantation Crops* 19:54-69.

Debnath, M.C. and Borah, B.K. (2002) Record of natural enemies of potato tuber moth, *Phthorimaea operculella* (Zeller.) in Assam. *Insect Enviroment* 8 (4): 161-162.

Deoras, P. J. (1942) Description and biological notes on a few species of Syrphidae from India. *Indian Journal of Entomology* 4 : 217-219.

Devi, B. (1967) Feeding potential and food requirement of some aphidivorous insects. M.Sc. (Ag) Thesis, Kerala Agricultural University, Thrissur.

Dey, P.K. and Somchoudhury (2001) Evalution of Spinosyns A+D (Spinosad 48 SC) against Lepidopterian pest complex of cabbage and its effect on natural enemies DBM under field condition of West Bengal. *Pestology* 25 (1):54-58.

Dhaliwal, G. S. and Arora R. (2001) Integrated pest management - Concepts and Approaches. Kalyani Publishers, New Delhi. 427p.

Dickson, R.C. and Laird, E.F. (1962) Green peach aphid population on deseart sugar beet. *J. Econ. Entomol.,* 55 : 501-504.

Divakar,B.J. and Pawar, A.D. (1987) Biocontrol of tomato fruit borer, *Heliothis armigera* (Hb.) in Karnataka. *Indian Journal of Plant Protection,*15 :57-61.

Dixon, A.F.G. and Kindlmann, P. (1998) Generation time ratio and the effectiveness of ladybirds as classical biological control agents. *Proceedings of the Australian Applied Entomological Research Conference*, Vol.M.P.Zalucki.R.A.I., Dran & G.G. White. (Eds.), pp. 314-320.I.

Don-Pedro, K.N. and Pedro, K.N.D. (1983) Level of parasitisation of *Maruca testulalis* Geyer (Lepidoptera : Pyralidae) larvae in early and at cowpea (*Vigna unguiculata*) cultivar in Nigeria. *Revue de*

Zoologie Africaine. 97: 677-683.

Dooria, M. S. (1982) Natural enemy complex of Citrus mite, *Eutetranychus orientalis* in Delhi, India, *Acarology Newsletter*. 11: 6.

Dover, B. A., Menon, A., Brown, R.C. and Strand, M.R. (1995) Suppression of juvenile hormone esterase in *Heliothis virescens* by *Microplitis demolitor* calyx fluid. *Journal of Insect Physiology*. 41: 809-817.

Dubos, B. (1987) Fungal antagonism in aerial agrobiocenoses. Innovative Approaches to plant diseases control, pp.107-35. Chet, I. (Ed.) , New York: Wiley.

Dureja, P. and Neera, S. (2000) Potential chemicals of microbial and animal origin for pest control in future. *Proc. The recent developments in agrochemical research to meet the challenges of the 21 st century.* Division of Agricultural Chemicals, Indian Agricultural Research Institute, New Delhi. Chapter 7.

Ekesi, S., Maniania N.K. and. Lux. S.A. (2002) Mortality in three African fruit fly puparia and adults caused by the entomopathogenic fungi *Metarhizium anisopliae* and *Beauveria bassiana. Biocontrol Sci. Technol.*, 12 (1&2) : 7-17.

Elad.Y. and Freeman, S (2002) Biological control of Fungal plant pathogens. The Mycota A. comprehensive Treatise on Fungi as experimental systems for basic and applied research. XI. Agricultural Applications. Springer, Heidelberg, Germany. pp 93-109.

El-dakroury, M.S.I., Abbas, M.S.I.. El-heneidy, A.I.T and Awadallah, K.T. (1979) The efficiency of *Chrysopa carnea* (Steph) on eggs and larvae of *Heliothes armigera* (Hb.) *Agricultural Res. Rev.* 55 :151-156.

Eswaramoorhty, S., Chelliah, S. and Jayaraj, S. (1976) *Aphidius platensis* Brethes- A potential parasite on *Myzus persicae* (Sulz). *Madras Agric. J.*, 63: 182-183.

Eswaramoorhty, S. and Jayaraj, S. (1977) Control of Guava scale *Pulvinaria psidii* Mask and Chilli aphid *Myzus persicae* (Sulz), with *Cephalosporium lecanii* Zimm. and insecticides. *Indian Journal of Agricultural Science.*, 47 : 136-139.

Faizal, M. H. and Mathai, S. (1996) Evaluation of different substrates for mass production of *Fusarium pallidoroseum* , an entomogenous fungus. *Indian Journal of Entomology,* 58 : 99-102.

Falerio, J. R., Singh, K. M. and Singh, B.N. (1990) Influence of abiotic factors on the population build up of important pest of cowpea, *Vigna unguiculata* and their biotic agents, recorded at IARI, Delhi. *Indian Journal of Entomology* 52:675-680.

Fan, Y and Petitt, L.F. (1994) Biological control of Broad mite, *Polyphagotarsonemus latus* (Banks) by *Neoseiulus barkeri* Hughes on pepper. *Biological Control* 4: 390-395.

Feng, M.G., Chen B.and Ying S.H. (2004) Trials of *Beauveria bassiana, Paecilomyces fumosoroseus* and imidacloprid for management of *Trialeurodes vaporarirum* (Homoptera: Aleyrodidae) on greenhouse grown lettuce. *Biocontrol Sci. Technol.,* 14(6): 531-534.

Ganga Visalakshy P.N, Manojkumar, A., Krishnamoorthy, A. and Mani, M. (2006) Pathogenicity of the entomopathogenic fungi, *Verticillium lecanii* to *Scirtothrips dorsalis* Hood. *National Symosium.on Biological. Control.of Sucking Pests in India*. pp43.

Gardner, J. M., J. L. Chandler, and A.W. Feldman. (1984) Growth promotion and inhibition by antibiotic-producing fluorescent pseudomonads on citrus roots. *Plant soil*. 77:103-113.

Geetha, N. and Rabeendra, R. J. (1999) Homology and comparative virulence of *in vivo* clones of nuclear polyhedrosis virus of cotton boll worm, *Helicoverpa armigera* (Hubner) Abstract, *National Symposium. on role of Biochem. and Biotech. in 21st Century* UAS, Bangalore PS III 54.

Gindin, G., Geschtovt N. U., Raccah B. and Barash I. (2000) Pathogenicity of *Verticillium lecanii* to different developmental stages of the silver leaf whitefly *Bemisia argentifolii* Genn. *Phytoparasitica*, 28 (3):229-239.

Gloriana, A.S., Raja N., Seshadri, S., Janarthanan S. and Ignacimuthu, S. (2001) Pathogenicity of entomopathogen *Bacillus thuringiensis* sub sp. *kurstaki* and *Beauvera bassiana* to the larvae of *Spodoptera litura* (F) and *Pericallia ricini* (F.). *Biol. Agric. Hortic.*, 18(3): 235-242.

Gopalakrishnan, C. K., Narayanan and Anusuya. (2002) Determination of LD 50 and LT 50 of *Paecilomyces farinosus* (Holmskiold) Brown and Smith on larval instars of *Plutella xylostella* L. *Entomon* 27 (3) : 249-254.

Gopinathan P. B., Beevi, S, N. and Nair, M. R. G. K. (1982) Occurrence of *Fusarium equsetti* (Corda.) Sacc. as a fungal pathogen of brinjal mealy bug, *Coccidohystrix insolita* (Green) *Entomon* 7: 120.

Gulsar Banu, J. and Sheela, M.S. (2005) Entomopathogenic Nematodes. Nematode Pests of Crops of Kerala-an overview, Sheela, M.S.; Hebsy Bai and Girija, V.K , (Eds.) pp.109-118.

Gundannavar, K. P. and Lingappa, S. (2003) Study of dose mortality response between instars of *Helicoverpa armigera* and *Metarhizium anisopliae* (Metscnikoff) Sorokin. National Symposium on Biomanagement of Insect Pests , 29 to 31, March, 200. Annamalai University, Annamalainagar, Abstract : 64.

Gupta, S. K., Dhooria. M. S. and Sidhu, A.S. (1971) A note on predators of citrus mite in Punjab. *Sci. Cul.* 37 (10): 484.

Gupta, S.K., and Gupta, A. (1992) Predatory plant mites of India and their importance in biological control *(In)* Man mites and Environment , pp.146-156. Haq, M. A. and Ramani, N. (Eds) Anjengo publications, Calicut, Kerala.

Gurbanov, G.G. (1982) Effectiveness of the use of common lacewing (*Chrysopa carnea* Steph) in the control of sucking pests and the cotton moth on cotton. *Investiga Akademics Nauk Azerbaidzhanskoi* SSR, 2 : 92-96.

Hariprasad, N., Niranjana S.R., and Shetty, H.S. (2001) Mass production on solid substrates and enhancement of conidiation of *Beauveria bassiana* for control of coffee berry borer. *Proc. Symp. Biocontrol based Pest Mgmt Quality Crop Prot. Curr. Millenium* July 18-19 , 2001 Singh, D., Mahal, M.S. Sohi, A.S. , Dilawari, V.K., Brar, K.S. and Singh, S.P. (Eds.) Punjab Agricultural University pp. 132-133.

Hareendranath, V. (1989) Control of *Aphis craccivora* Koch. with fungal pathogens and their impact on the natural enemies of the pest. M.Sc. (Ag.) thesis, Kerala Agricultural University, Thrissur, p. 87.

Hareendranath, V., Nair, K. P. V. and Paulose, S. (1987) *Fusarium pallidoroseum* (cooke) Sacc as a fungal pathogen of *A.craccivora*. Koch. *Entomon*. 12 : 392-394.

Hafez, M. Zaki, F. N, Moursy, A. and Sabour, M. (1994) Biological effect of the entomopathogenic fungus *Beauveria bassiana* on the potato tuber moth *Pthorimoea operculella* (Sneller). *J. Islamic Acad. Sci*. 7 : 132-136.

Hanummana, M., Thotadarya, T.S. and Rao, K.J. (1984) A note on the acceptance of the eggs of cotton bollworms by different *Trichogramma parasitoids*. *Curr. Res*. 13 :37.

Hassan, S. A. (1994) Strategies to select *Trichogramma* species for use in biological control. *Biological Control With Egg Parasitoids*, pp. 55-71. E. Wajnberg and S. A. Hassan (Eds).,: CAB International. Oxon, U.K.

Harman. GOE, Howel CR, Viterbo A, Chet I and Lorito M. (2004) *Trichoderma* spp – opportunistic, avirulent plant symbiots. *Nat Rev. Microbiol.* 2 (1): 43-56.

Hebsy Bai, Mathai S and Gokulapalan, C. (1999) *Rhizopus oryzae* Went and Gerlings, a new fungal pathogen of *Dysdercus cingulatus* (F). *Insect Environment*. 4 : 129-130.

Hemachandra and Singh, T.K. (2007) *Cotesia plutellae* (Kurdj.), a promising larvae parasitoid of *Plutella xylostella* (Linnaeus) in Manipur *J. Biol. Control*. 21(1) : 153-156.

Hoffmann, M.P. and Frodsham, A.C. (1993) Natural Enemies of Vegetable Insect Pests. Cooperative Extension, Cornell University, Ithaca, NY. 63 pp.

Howel, C.R. (2003) Mechanisms employed by Trichoderma species in the biocontrol of plant disease. The history and evolution of current concepts. *P. lilacinus DIS* 87 (1): 4-10.

Howel, C.R. and Stipanovic, R.D. (1980) Suppression of *Pythium ultimum*-induced damping off of cotton seedlings by *Pseudomonas fluorescens* and its antibiotic pyoluterin. *Phytopathology* 70:712-715.

Ibrahim, Y.B. and Low, W. (1993) Potential of mass production and field efficacy of isolates of the entomopathogenic fungi *Beauveria bassiana* and *Paecilomyces fumosoroseus* against *Plutella xylostella*. *Int. J. Pest. Mgmt*. 39:288-292.

Jiji, T., Praveena, R., Anitha, N. and Naseema, A (2005) *Beauveria bassiana* (Bals) Vuill.on *Sylepta derogata* Fab. and its cross infectivity on other major pests of vegetable. *Insect Environment* 11(4):144.

Jiji, T., Praveena, R., Kavitha Babu and Naseema, A. (2006 a) Evaluation of two fungi for biological control of fruit flies. *Insect Environment*. 12(1):25-26.

Jiji, T., Praveena, R., Kavitha Babu, Naseema, A. and Anitha, N. (2006 b) Three promising fungal strains pathogenic to fruit flies. *Seventh International Symposium on Fruit flies of Economic Importance*, 10th-15th September, Salvador, Bahia, Brazil.

Jiji, T., Praveena, R., Kavitha Babu, Naseema, A. and Vijayaraghavakumar (2006 d) Some promising fungal strains for the biological control of sucking pests infesting vegetables. *National Symposium on Biological Control held at Bangalore*, 26th-27th May 2006.

Jiji, T. Praveena, R., Babu, K. and Naseema, A. (2006 e) Occurrence of *Paecilomyces lilacinus* on Melon fly *Bactrocera cucurbitae* Coq. and its cross infectivity on *B. dorsalis* (Hendel) and Bhindi leaf roller *Sylepta derogata* Fb. *Abst. Colloquium on Nanoscale Science and Arthropod Bioresources*, Trivandrum, India p. 18.

Jiji, T., K. B. Soni, P. Prajeesh, Kavitha Babu and Rajmohan. K. (2006 c) Molecular characterisation of promising fungal strains of *Beauveria bassiana* and *Paecilomyces lilacinus* using RAPD markers. *Insect Environment.*, 12 (1) : 33-34.

Jiji, T., Susamma Mathai, K.B. Soni, S. Simi and Rajmohan. K. (2001) Molecular characterisation of *Fusarium pallidoroseaum* (Cooke) Sacc., a biocontrol agent against cowpea aphid. *Insect Environment* 7 (3): 132-133.

Joginder, S., Sandhu, S. S. and Singla, M. L. (1990) Ecology of *Heliothis armigera* (Hub) on chick pea in Punjab. *Journal Insect Science*. 3 :47-52.

Johri B.N, Roa, V..S. and Reeta Goel. (1997) Fluorescent pseudomonads in plant disease management. Biotechnological approaches in soil micro-organisms for sustainable crop production K.R. Dadarwal (Ed.). Scientific publishers, Jodhpur-ISBN: 81-7233-147-9.

Joseph, P.J and Sivaprasad, P. (2000) The potential Arbuscular Mycorrhizal associations for biocontrol of soil borne diseases. (In) Biocontrol potential and its exploitation in sustainable Agriculture pp.139-153. Upadhyay, R.K., Mukherji, K.G. and Chamola, B.P. (Eds) Kluwer Academic /Plenum Publishers, New York.

Kakar, K.L., Sharma, J.P. and Dogra, G.S. (1990) Feasibility of using *Trichogramma* spp. against *Heliothis armigera* Hubner on tomato. *Indian Journal of Plant Protection* 18: 237-239.

Kamel, B. H. M., Rabasse, J.M. and Benthamonda, M.H. (1994) The aphids of Citrus fruits and their natural enemies in Turnisia. *Tropiculture* 12: 145-147.

Kapadia, M.N. (1995) Population of parasitoids of *Liriomyza trifoli*. (Burgess on summer host plants and its record as a disease carrier. *Int.J. Trop.Agric*. 13 : 273-275.

Kavitha, K., Meenakumari, K.S and Sivaprasad, P. (2003) Effect of dual inoculation of native arbuscular mycorrhizal fungi and Azospirillum on suppression of damping off in chilli. *Indian Phytopathology.* 56:112-113.

Keel, C., C. Voisard, C.P.H. Berling, G. Kadr, and Defage. G. (1989) Iron sufficiency, a prerequisite for the suppression of tobacco root rot by *Pseudomonas fluorescens* strain CHAO under gnotobiotiic conditions. *Phytopathology.* 79:584-589.

Kloepper, J.W., D.J. Leong, M. Teontze, and Schroth. M.N. (1980a) *Pseudomonas siderophores*: a mechanism explaining disease suppressive soils. *Curr. Microbiol*. 4: 317-320.

Kloepper, J.W., M.N. Schroth and Miller. T.D. (1980b) Effects of rhizosphere colonization by plant growth-promoting rhizobacteria on potato plant development and yield. *Phytopathology.* 70: 1078-1082.

Krishnamoorthy, C. and Shankara Naidu. (1971) *Aspergillus flavus*, a fungal parasite of *Epilachna vigintioctopunctata*. *Current Science*. 24: 666-667.

Krishnamurthi, B. and Usman, S. (1954) Some insect parasites of economic importance noted in Mysore state. *Indian J. Entomol.* 16 : 327-344.

Krishnamurthy Rao, B. (1932) The potato epilachna beetle, *E.vigintioctopunctata.* Bulletin of the Department of Agriculture, Mysore (Entomology Series) 9: 1-16.

Kunjamma, P. M. and Abraham, C.C. (1973) The biology of *Chrysocharis johnsoni,* a larval parasite of *Epilachna vigintioctopunctata* Fab. *Agri. Res. J. Kerala* 11 (2) : 148-149.

Lal, B. (1946) Biological notes on *Pleurotropis foveolatus*, a larval parasite of *E. vigintioctopunctata. Current Science.* 15: 138-139.

Lal, B.S. (1961) On the biology of *Pediobuis faveolatus* (Eulophidae : Hymenoptera). *Indian J. Entomol.* 23: 268-273.

Lal, K.B. and Gupta, K. M. (1947) An unusual effective control of a pest by a parasite. *Current Science.* 16 : 158-159.

Lazzaretti. E, Menten. J.O and Bettiol, W. (1995) Treatment of wheat seed with *Bacillus subtilis* for the control of *Pyricularia oryzae, Bipolaris sorokiniana* and *Alterman atenuis summa. Phytopathologica* 21:163-167.

Lefroy, H. M. (1909) *Indian Insect Life.* Today and Tomorrows Printers and Publishers, New Delhi 764 p.

Lifshitz, R., C. Simonson, F. M. Scher. J. W. Kloepper, C. Rodrick-Semple, and Zaleska. I. (1986). Effect of rhizobacteria on the severity of phytophthora root rot of soyabean. *Can. J. Plant Pathol.* 8:102-106.

Li, Li-Ying. (1994) *Worldwide use of Trichogramma for biological control on different crops: A survey.* CAB International, Oxon, U.K. pp. 37-53.

Lin, K.J. and Roberts, D.W. (1986) *Metarhizium anisopliae* major, investigation of production of dextruxin by culture. *Journal Invertebrate Pathology.* 53:307-314.

Lokhande, R.K. and Mohan, P. (1990) Study on biocontrol of aphid , *Aphis craccivora* Koch by lady bird beetle , *Menochilus sexmacuala* F. in chillies *Advance in Plant Science,* 3:281-286.

Losey, J.E. and Calvin D.D. (1995) Quality assessment of four commercially available species of *Trichogramma* (Hymenoptera: Trichogrammatidae). *J. Econ. Entomol.* 88:1243-1250.

Lynn, D.E. (1999) Comparison of cell line maintenance procedures on insect cells used for producing baculoviruses. *In Vitro Cellular & Developmental Biology - Animal* 35:5:248-251.

Ma, J. (2000) Laboratory susceptibility of *Plutella xylostella* to *Metarhizium anisopliae* and *Nomuraea rileyi. Acta Entomologica Sinica*, 7(1):53-57.

Machauer, M. (1968) Insect parasites of green peach apid *Mysus persicae* and their control potential. *Entomophaga*, 13:91-106.

Mahr, D.L. and Ridgeway. N.M. (1993) Biological control of insects and mites: An introduction to beneficial natural enemies and their use in pest management. North Centr. Reg. Ext. Publ. No. 481.

Maniani, N. K., Sithanantham S., Ekesi, S. Among-Nyarko, K., Baumgartner, J. Lobr, B. and Motoka, C.M. (2003) A Field trial of the entomogenos fungus *Metarhizium anisopliae* for control of onion thrips, *Thrips tabaci Crop Protection.*, 22 (3&4) 553- 559.

Manjunath, T.M, Patel, R.C. and Yadev, D.N. (1976) Observations on *Heliothes peltigera* (Schiff)(Lepidoptera: Noctuidae) and its natural enemies in Anand. (Gujarat state, India). *Proc. Indian Acad.Sci.* 93:55-65.

Manjunath, T.M and Ramani, S. (2006) History of Biological Control in India: An overview and prospects. *National Symposium on Biological Control* of sucking pest, Bangalore pp10.

Masuda, J. (2000) Microbial control of the diamond back moth, *Plutela xylostella* by an entomopathogenic fungus, *Beauveria bassiana.* Effects of temperature on mycoses and conidial invasion time. *J. Appl. Ent.* 44: 177-182.

Mathai, S. (1999) Host range of a potent entomopathogenic fungus *Rhizopus oryzae* Went and Gerlings. *Insect Environment.* 4: 129-130.

Mathai, S., Geetha, D. and Mohandas, N. (1998) Use of different locally available and cheaper substrates for mass multiplication of *Fusarium pallidoroseum* (Cooke) Sacc. an effective fungal pathogen against the pea aphid *Aphis craccivora* Koch. *New Trends in Biotechnology*, 1992, Subbarao, N. S. Balagopalan, C and Ramakrishnan, S.V) (Eds). pp. 249-251.

Mathai, S., Peethambaran, C.K., Sunitha, V.S. and Sudharma, K. (1999) *Fusarium pallidoroseum* (Cooke) Sacc a natural insecticide for the management of pea aphid *Aphis craccivora* Koch. Abstr. *Int. Sem. Integrated Pest Management*, Hyderabad, pp 41-42.

Mathur, A.C. and Srivastava, J.B. (1964) *E. vigintioctopunctata* as a defoliator of some solanaceous medicinal plants. *Indian J. Entomol* 26 : 246.

Meenakumari, K.S., P. Sivaparasad., K.K. Sulochana and Lisa E. (2003) Suppression of Bacterial wilt of chilli and tomato using native isolates of Fluorescent Pseudomonads. *Proc-6th International PGPR Workshop*, 5-10 Oct 2003 Calicut, India p.1 69-175.

Menn, J.J. (1996) Biopesticide: Has their time come? *Journal of Enviornmental science and Health, Contaminents and agricultural wastes* (Special issue in pesticide chemistry for sustainable agriculture) 31: 383-389.

Moore , D., Batemann, R.P. Carey, M. and Prior , C. (1995) Long term storage *of Metarhizium flavoviride* conidia in oil formulations for the control of locusts and grass hoppers. *Biocontrol Sci. Technol.*, 5:193-199.

Moore, D., Douro-kpindou, O.K., Jenkins, N. E. and Lomer, C. J. (1996) Effect of moisture content and temperature on storage of *Metarhizium flavoviridae* conidia. *Biocontrol Sci. Technol.* 6:51-61.

Mondal, P. and Bhattacharya. A.K. (2003) Differential sensitivity of *Spilosoma oblique* to two isolates of *Beauveria bassiana. Indian J. Ent.*, 65(3): 324-329.

Morrison, R. K., Stinner, R. E. and Ridgway, R. L. (1976) Mass production of *Trichogramma pretiosum* on eggs of the Angoumois grain moth. *Southwestern Entomol.* 2:74-80.

Mote and Sivubhavikatti (2001) Comparative efficacy of chemical pesticides with biopesticides against major pests of brinjal in *kharif* season *Proc. Second National Symposium on Integrated Pest Management (IPM) in Horticultural Crops. New molecules, Biopesticides and Environment*, Abraham Varghese and Parvatha Reddy (Eds.) October 17-19, 2001, Bangalore 56 pp.

Mueseveck, C.F.W. and Rao, B.R.S. (1958) A new parasite of *H.recurvalis* (F). *Indian J. Ent.* 20 (1):27.

Mukerji, K.G. (1999) Mycorrhizae in control of plant pathogens *Molecular Approachs. Biotechnological Approaches in Biocontrol of Plant Pathogens,* Mukerji. K.G., Chamola, B. P. and Upadhyay, R. K. (Eds.). Kluwar Academic, New York, USA.

Mukhopadhyay, A.N. Shrestha, S.M. and Mukherjee, P.K. (1992) Biological seed treatment for control of soil borne plant pathogens. *FAO Plant. Prot. Bull.* 40:21-30.

Nair, M.R.G.K. (1975) Insects and Mites of Crops in India. Second edition. Indian Council of Agricultural Research, New Delhi, 408 p.

Nandakumar, C. (1999) Monitoring and management of the pest complex of bitter gourd (*Momordica charantia* L.). Ph. D. Thesis, Kerala Agricultural University, p. 210.

Nandakumar, C. and Saradamma, K (1996) Botanical pesticides are benign to the natural enemies of epilachna beetle. *Insect Environment*. 2(1): 27.

Nandakumar, C. and Sheela, K.R. (1996) Predator fauna control of cowpea aphid. *Insect Environment* 2:111-112.

Narayanaswamy, P. (1998) Mycoparasites check rice brown plant hopper (BPH) in India. *Insect. Environment* pp. 104-106.

NCIPM, (2003) Lecture Compendium on National Training course on IPM cotton, rice and vegetable crops. National centre for Integrated Pest Management, LBS Building, Pusa Campus, New Delhi. p.215.

Neema Agarwal and Bhatnagar R.K. (2004) *Bacillus thuringiensis* versus insects: face to face *Proc. Nat. Symp. on Frontier Areas of Entomol.* Res. Nov. 5-7 2003. IARI, New Delhi, India pp. 25-54.

Nelson, R.S. Mc. Cormick S.M., X Delanny, P., Dube, J., Layten, E.J., Anderson, M., Kaniewskam, R.K., Prooksch, R.B., Horsch, S.G., Rogers, R.T., Fralley and Beachy R.N. (1988) *Biotechnology* pp 403-409.

Ofuya, T.I. (1995a) Colonization and control of *Aphis craccivora* (Koch.) (Homoptera : Aphididae) by coccinellid predators in some resistant and susceptible cowpea varieties in Nigeria. *Crop Protection* 14: 47-50.

Okeyo-Ownor, J. B., Oloo, G. W. and Agwaro, P. O. (1991) Natural enemies of legume pod borer, *Maruca testulalis* Geyer (Lepidoptera : Pyralidae) in small scale farming systems of western Kenya. *Insect Sci.Applic.* 12 : 35-42.

Otsu, Y., Y. Matsuda, H. Mori, H. Ueki, T. Nakajima, K. Fujiwara, W. Matsumoto, N. Azuma, K. Kakutani, T. Nonomura, Y. Sakuratani, T. Shinogi, Y. Tosa, S. Mayama and Toyeda. H. (2004) Stable phylloplane colonization by entomopathogenic bacterium *Pseudomonas fluorescens* KPM-018P and biological control of phytophagous ladybird beetles *Epilachna vigintioctopunctata* (Coleoptera : Coccinellidae). *Biocontrol Sci. Technol.*, 14(5):427-439.

Pal, V. and Jalali, I. (1998) Rhizosphere bacteria for Biocontrol of plant diseases. *Indian J. Microbiol.*, 38: 187-204.

Panda, N. (2004) Host plant resistance and insect pest management. *Proc. Nat. Symp. on Frontier Areas of Entomol. Res. Nov.* 5-7. 2003 IARI New Delhi. 377-395.

Pandey, A.K. (2003) Susceptibility of egg and pupae of lepidoptera to *Beauveria bassiana* (Balsamo) Vuillemin and *Metarhizium anisopliae* (Metschnikoff) Sorokin. *Insect Environ* 9:123-124

Pandey, R. K. and Rajendra Singh, T.B. (1984) Bionomics of *Trioxys indicus* an aphidid parasitoid of *Aphis craccivora*. *Entomon* 99:239-245.

Patel, R.K. and Yadev, D.N. (1992) Impact of Marigold on *Heliothis armigera* Hubner and its natural enemies in seed crop of tobacco. *Tobacco Res.* 18: 65-72.

Patro, B and Behera, M.K. (1993) Bionomics of *Paragus serrates* (Fabricius), Diptera, Syrphidae, a predator of bean aphid *A.craccivora* Koch. *Trop.Sci.*London Whurr Publishers Ltd. 33(2):131-135.

Pawar U.M., Thombre, U.T. and Borikar, P.S. (2001) Advances in Biological Control. Punjab Agricultural University, Ludhiana. pp. 75-82.

Pinto, J.D. and Stouthamer. R. (1994) Systematics of the Trichogrammatidae with emphasis on *Trichogramma*. In: *Biological Control with Egg Parasitoids*, E. Wajnberg and S. A. Hassan (Eds)., pp. 1-36. CAB International, Oxon, U.K.

Poorani, J. (2002) An annotated checklist of the Coccinellidae (excluding Epilachinae) of the Indian subregion. *Oriental Insects* 36: 307-383.

Popham, H.J. R., Li, Y.H. and Miller, L.K. (1997) Genetic improvement of *Helicoverpa zea* nuclear polyhedrosis virus as a biopesticide. *Biological Control* 10:83-91.

Pree, D.J., Archibald, D.E. and Morrison, R.K. (1989) Resistance to insecticides in the common green lace wing *Chrysoperla carnea* (Neuroptera : Chrysopidae) in Southern Ontario. *J. Econ Entomol.* 82:29-34.

Project Directorate on Biological Control Renamed ICAR-NBAIR Bangalore (1994) *Chrysopids and Trichogrammatids,* Strain selection and utilization. Singh, S.P., Bhumannavar, B.S., Jalali S.K., Bakthavalsalam, N. Pushpalatha N.A. (Eds.) p 45.

Project Directorate on Biological Control (2003) Production and use of cocinellid predators, Technical Bulletin 32 Sunil Joshi, Prasanth Mohanraj, Rabindra, R. J. and Rao, N. S. (Eds)., Bangalore. p 25.

Project Directorate on Biological Control (2004) Annual Report, 2004, Bangalore. pp. 7-11.

Project Directorate on Biological Control (2005) Annual Report, 2005, Bangalore. pp. 12.

Project Directorate on Biological Control (2006) Annual Report, 2006, Bangalore. pp 13.

Rajendran, B. and Gopalan, N. (1999) Effect of biopathogens (*Bacillus thuringiensis* Berlines and *Beauveria bassiana* Vaillene) on different stages of egg plant spotted beetle, *Heno sepilachna vigintiotopunitata* (Fab) (Coleoptera : Coccinellidae). *J.Entomol.Res.* 23(1) : 9-14.

Rajendran, P. (2002) Preliminary studies on the effect of green muscardine fungus *Metarhizium anisopliae* (Metsch.) Sorokin on eggplant spotted beetle, *Henosepilachna vigintioctopunctata* Fab. (Coleoptera : Coccinellidae) *Pest Mgmt. Hort. Ecosyst* 8: 11-14.

Ramaseshiah, G. (1967) The fungus *Entomophthora coronata* parasitic on the three species of aphids infesting cruicifers in India. *J. Invert. Path.*, 9:128-130.

Rani, O.P.R. (1995) Bioecology and life table of the pea aphid *Aphis craccivora* Koch. and its natural enemies. M.Sc. (Ag) Thesis. Kerala Agricultural University. Thrissur, p.145.

Rao, V.P., Ghani, M.A., Sankaran, T. and Mathur, K.C. (1971) A review of the biological control of insects and other pests in south east Asia and the pacific region. Technical Communication No. 6, CIBC,West Indies, pp.149.

Ravensberg, W. J., Nalais, M. and Van der Schaaf, D. A. (1990) *Verticillium lecanii* as a microbial insecticide against glass house whitefly. *Brighton Crop Potection Conference – Pests and Diseases* Vol. 3. British Crop Protection Council, Farnham, pp 265-268.

Reji, G. V. (2002) Management of the American Serpentine leaf mines, *Liromyza trifoli* (Burgess) Dietars on Cowpea, *Vigna anguiculata* (L.) Walp. M.Sc. Thesis. Kerala Agricultural University, Thrissur.

Rejirani, O. P. (2001) Production and evaluation of the fungus, *Fusarium pallidoroseum* (Cooke)Sacc. as a biopesticide against the pea aphid, *Aphis craccivora* Koch. Ph D Thesis Kerala Agricultural University, Thrissur pp 136.

Ramesh, K. Mohan, H. Lakshmi, K.A. Padmavathi and Devi. K.U. (1999) *Beauveria bassiana* (Bals) Vuill (Hyphomycetes: Moniliales) in cotton pest management. Afield trial on the cotton leaf roller, *Sylepta derogate* Fabricious (Coleopter : Pyraustidae) *J. Ent. Res.* 23(3):267-271.

Richter, V. A. and Zhumanov, B. (1994) The trachinid genus *Goniophthalmus* new to the fauna of middle Asia (Dipthera Zoosystematica Rossica) 3:146.

Ridgway, R.L., J.R. Ables, C. Goodpasture and Hartstack A.W. (1981) Trichogramma and its utilization for crop protection in the USA. *Proceedings of the joint American Soviet Conference on use of Beneficial Organisms in the control of Crop Pests* Washington, D.C., 1979. Coulson. J.R. (Ed.), Entomological Society of America., pp. 41-8.

Roberts, D. W. (1981) Toxins of entomopathogenic fungi. *Microbial control of Pests and Plant Diseases* (ed. H. D. Burges), Academic Press London. pp. 441-464.

Salamero, A., Gabarra, R. and Albajer, R. (1987) Observation on the predatory and phytphagous habits of *Dicyphus tamaninii* Waguer (Heteroptera : Miridae). *Bulletin-SROP* 10:165-169.

Saharia, D. (1980) Some aspects of biology of Coccinellids predators associated with *A.craccivora* Koch on cowpea. *J. Res.Assam Agric.Uni.* 1:82-89.

Samuels, R. I. Reynolds, S.E. and Charnley, A. K. (1988) Calcium channel activation of insect muscle by dextruxins, insecticidal compounds produced by entomopathogenic fungus *Metarhizium anisopliae. Comparative Biochemistry and Physiology* 90 :403- 412.

Sandhu.S.S. Rajak. R. C. and Agarwal , G.P. (1993) Studies on prolonged storage of *B. basssiana* conidia. Effects of temperature and relative humidity on conidia viability and virulence against chickpea borer, *Helicoverpa armigra Biocontrol Sci. Technol.* 3: 47-53.

Santhosh Kumar, T. (1999) Eco-friendly management of important sucking pests of Chilli (*Capsicum annuum* L.) M. Sc. (Ag) Thesis. Kerala Agricultural University.

Scher, F.M., and Baker. R. (1982) Effect of *Pseudomonas putida* and a synthetic iron chelator on induction of soil suppressiveness to Fusarium wilt pathogens. *Phytopathology* 72: 1567-1573.

Schuster D.J., and Wharton. R.A. (1993) *Hymenopterous parasitoids* of leaf-mining *Liriomyza* spp. (Diptera: Agromyzidae) on tomato in Florida. Environ. *Entomol.* 22:1188–1191.

Selim, A. A. ,El-Rafai, S.A., El-Gantiry, A., Refai, S.G. El-Gantiry, A., Refai, S. A. and El-Gantiry, A.El. (1987) Seasonal fluctuation in the population of *Aphis craccivora* Koch. *Ann.Agric.Sci.* 32 :1837-1848.

Sharma, J.C. (1973) Predataion of *Menochilus sexmaculata* Fab. on different species of aphids. *Indian Journal of Entomology* 35: 343 -344.

Sharma, R.P. (1991) Relative efficary of some insecticides against field population of bean aphid *A.craccivora* Koch. and safety to the coccinellids complex on Lathyrus, lentil and chickpea. *J. Entomol.Res.* 15 : 251-259.

Sheela, M.S., Mohandas, C., Arunrajkumar, K. R. and Nisha, M.S. (2002) Potential of endomopathogenic nematodes for the management of lepidopteran pests. *Proc. XII Swadeshi Sci. Cong.* p. 85-87.

Sheela, M. S., Mohandas, C., Nisha, M. S. and Manu, C. R. (2002) Potential of *Rhabditis* sp. for the management of pests of bitter gourd, *Momordica charantia* L. Green pesticides for insect pest management pp. 203-205.

Shekhawat.G.S., S.K. Chakrabarti and Gadewar.A.V. (1992) Potato bacterial wilt in India *Tec. Bull No. 38*, CPCRI, Shimla 52.77.

Shekhawat.G.S., S.K. Chakrabarti and V.Kishore., V. Sunaina and Gadewar. A.V. (1993) (Eds.). GC. Harman and A.C. Hayward, *Bacterial wilt ACIAR, Proc.* 45:327-330.

Singh, A.P., Arora, Ramesh, Battu, G.S. and Arora, R. (2000) Laboratory evaluation of three *Bacillus thuringiensis* Berliner based biopesticides against the diamondback moth *Plutella xylostella* (Linnaeus). *Pesticide Research Journal*. 12: (1), 54-62.

Singh, J. and Ray, R. (1977) *Stethorus* sp. (Coleoptera.Cocinellidae) – a predator of *Tetranychus neocaledonicus* Andre on okra at Varanasi. *Acarol. Newsl.* 4:5-6.

Singh, S, P. (2001) Prospects of biological control. *Entomon* 26 (special issue): 13-28 (2001) Pub. Association for Advancement of Entomology, Thiruvananthapuram, Kerala. 455-463.

Singh, S. P. (1995) Technology for Production of Natural Enemies. Project Directorate of Biological Control, Bangalore. p. 221.

Sitaraman, R. (1966) Studies on the biology of aphidophagous flies *Xanthogramma scutellare* Fb., (Syrphidae) and *Leucopis* sp. (Ochthiphilidae) M.Sc. (Ag) Thesis, Kerala Agriculture University, Thrissur.

Sivaprasad, P. (1998) Mangement of root disease of important spice crops of Kerala with VA mycorrhiza. DBT Project Final Report, Kerala Agricultural University, Trichur.

Sivaprasad. P., Joseph P. J. and Sulochana K. K, (1999) Foot rot of black pepper and rhizome rot of ginger as influenced by AMF in the field. *National Conference on Mycorrhiza*, 1999, Bhopal. p. 61.

Sivaprasad. P., Sulochana, K.K, Kavitha, M.S., Joseph, P.J. and Meenakumari, K.S. (2003) Effect of Fluorescent pseudomonads isolates of Foot Rot diseases and Growth of Black pepper (*Piper Nigrum* L) *Proc. 6th international PGPR workshop,* 5-10 Oct 2003, Calicut, India-68-74.

Sneh, B. (1981) Use of rhizosphere chitinolytic bacteria for biological control of *Fusarium oxysporum* F. sp. *dianthi* in carnation. *Phytopathol.*. 100:251-256.

Sonalkar, V.U. , Deshmukh, S.D., and Satpute, U.S. (1997) Infuence of feeding stimulants on the incubation period of nuclear polyhedrosis virus of *Helicoverpa armigera* (Hubner). *J. Biol. Control*. 11: 85-87.

Sonia, (2005) Management of bacterial wilt of chilli caused by *Ralstonia solanacearum* (E.F.Smith) Yabuuchi *et al.* using AMF and Fluorescent pseudomonads. MSc (Ag) thesis, Kerala Agricultural University. Vellanikkara, Thrissur, 61.p.

Sparks, A.N., Ables, J.R. and Jones. R.L. (1982) Notes on Biological control of Stem borers in corn, sugarcane and rice in the Peoples Republic of China. *Biological Control of Pests in China*, pp.193-215. US Department of Agriculture, Washington D.C.

Stary, P. and Ghosh, A.K. (1975) Aphid parasites from Meghalaya, India. *Oriental Insects*, 9 :343-349.

Strawiinski, K. (1930) Cabbage butterfly: biology and control. *Poisk. Primo. Ent.* 8: 227-248.

Subramanian, S. (1998) Studies on Nuclear Polyhedrosis virus of *Spodoptera litura* (*Fabricius*) M.Sc. (Ag) Thesis Tamil Nadu Agric. Univ. Coimbatore 112p.

Sudharma, K. (1996) Distribution and bioecology of phytophagous mites of vegetables, medicinal plants and ornamental plants in Thiruvananthapuram district Ph.D thesis. Kerala Agricultural University, Thrissur, p. 171.

Sudharma , K., Reji Nair and Peethambaran C, K. (2001) Mass production of *Fusarium pallodoroseum* (Cooke) Sacc. utilizing waste materials from biocontrol laboratories. Symp. On Biocontrol based Pest Mgmt. for Crop Prot. In the Current Millennium. July 18-19 2001 (Eds.). Darshan Singh, Mahal, M. S, Sohi, A. S, Dilawari, V.K., Brar, K. S and Singh, S. P.. PAU. Ludhiana. p.117.

Suja, G. (2003) Intergrated pest management in grain and vegetable cowpea (*Vigna unguiculata* (L.) Walp Ph. D. Thesis. Kerala Agricultural University, Thrissur p. 13-19.

Sulochana. K.K., P. Sivaprasad, V.I. Soumya and Kamala Nayar (2003) Biological control of sheath Blight of Rice using Fluorescent pseudomonas and AMP in the field. Proc. 6th International PGPR Workshop, 5-10 Oct 2003 Calicut, India P.331-33.

Sun, J., Fuxa, J.R. and Henderson G. (2003) Effects of virulence, sporulation and temperature on *Metarhizium anisopliae* and *Beauveria bassiana* laboratory transmission in *Coptotermes formosanus. J. Inertebr. Pathol.*, 84(1):38-46.

Sunitha, (1997) *Fusrium pallidoroseum* (Cooke) Sacc. as a biocontrol agent for the pea aphid *Aphis craccivora* Koch. Msc. Ag. Thesis. Kerala Agricultural University.

Suslow TV (1982) Role of root colonizing bacteria on plant growth. *Phytopathogenic prokaryotes*. (eds. M.S. Mount and G.H. Lacy). Academic Press. London, U.K. pp 187-223.

Tanada, Y. and Kaya, H. K.(1993*) Insect Pathology*. Academic Press Inc. Harcourt Brace Jovanovich Publishers, San Diego.

Thamilvel, D. (2004) Management of insect pest complex in okra (Abelmoschus esculentus Linn.). M.Sc. (Ag.) thesis, Kerala Agricultural University, Thrissur, 107 p.

Tian, L.X. and Young, L.F. (1981) Apreliminary study on aphid parasites in cotton field (Hymen optera : Aphididae). *Insect Knowledge Kunchong Zhishi*, 18:158-160.

Todorova, S.I., Coderre, D. and Cote, J.C. (2000) Pathogenicity of *Beauveria bassiana* isolates toward *Leptinotarsa decemlineata* (Coleoptera: Chrysomelidae), Myzus persicae (Homoptera: Aphididae) and their predator *Coleomegilla maculate lengi* (Coleoptera: Coccinellidae). *Phytoprotection*, 81(1):15-22.

Trivedi, T. P. (1988) Natural enemies of potato pest in Karnataka. JIPA 15.

Tronsoma, A. (1986) Use of *Trichoderma* spp in biological control of necrotropic pathogen. 348-62.

Van den Bosch, R. and Telford. A. D. (1964) Environmental modification and biological control. Biological Control of Pests and Weeds.(ed. *P. DeBach)* Chapman Hall, London. 459-88.

Van Emden, H.F., Eastop, V.F., Huges, R.D. and Way, M.J. (1969) Ecology of *Myzus persicae. A.Rev. Ent.,* 14:197-270.

Vasquez, L. A., A. M. Shelton, M.P. Hoffmann and Roush. R.T. (1997) Laboratory evaulation of commercial Trichogrammatid products for potential use against *Plutella xylostella* (L.) (Lepidoptera: Plutellidae). *Biological Control* 9:143-148.

Verghese, A., Nagaraju, D.K., Jayanthi, P.D.K. and Gopalakrishnan, C. (2003) Report of Entomopathogenic fungus *Beauveria bassiana* (Balsamo) *Vuillemin* on mango stone weevil. *Insect Environ* 8:146-147.

Venkataraman, T.V., Mathur, V.K and Chander, R. (1962) Effects on the possible use of *Bacillus thuringienses* in the control of crop pests. *Indian J. Entomol*. 24: 274-277.

Visalakshy, P.N. G. Kumar, A.M. and Krishnamoorthy, A.(2004) Epizootics of fungal pathogen *Verticillium lecanii* Zimmerman on Thrips palmi Karny. *Insects Environ*. 10(3):134-135V.

Walunj, A.R. and Pawar, (2001) Evaluation of new molecule, Spinosad 2.8 SC for the control of DBM (*Plutella xylostella* (L.) on cabbage. Proc. Second National Symposium on Integrated Pest Management in Horticultural Crops: New Molecules, Biopesticides and Environment Abraham Verghese and Parvatha Reddy (Eds.) IIHR, Bangalore 253.

Weindling, R. (1934) Studies on a lethal principle effective in the parasitic action of *Trichoderma lignorum* on *Rhizoctonia solani* and other soil fungi. *Phytopathology*. 24: 1173-1179.

Wickramatileke W.S.N. Ahangama, D. and Banda, D.M.A. (2000) Evaluation of the effectiveness of entomopathogenic fungus *Metahizium anisopliae* var. Major on cabbage semilooper (*Chrysodeixis erisoma* Doubt.) *Trop. Agric. Res*. 12:177-185.

Xu, G.W. and Gross. D.C (1986) Selection of fluorescent pseudomonads antagonistic to *Erwinia carotovora* and suppressive of potato seed decay. *Phytopathology* 76:414-422.

Yedidia I., Benhamou, N. and Chet, I. (1999) Induction of defense responses in cucumber plants (*Cucumis sativus* L.) by the biocontrol agent *Trichoderma harzianum*. *Appl. Environ. Microbiol*. 65:1061-1070.

Yuen,G.Y. and Schroth M.N. (1986) Interaction of *Pseudomonas fluorescens* E6 with ornamental plants and its effect on the composition of root colonising microflora. *Phytopathology* 76:176-180.

Zaki, F.N. (1998) Efficiency of the entomopathogenic fungus, *Beauveria bassiana* (Bals) against *Aphis craccivora* Koch and *Bemisia tabaci* Gennandius. *J. Appl. Ent*., 122:397-399.

Zeren, O. and Duzguner, Z. (1983) Investigations on the natural enemies of aphidoid species damaging vegetables in the Cukuroa region. *Turkiye Bitki Koruma Dergizi*. 7:199-221.

Chapter – 15

Marketing of Fresh Fruits and Vegetables

Binoi K. Menon and T.R. Gopalakrishnan

The horticulture sector has emerged as a core sector in agriculture providing the best option for diversification in agriculture for nutritional security, employment generation and enhancing farm income. Investments made in horticulture during the ninth and tenth 5 year plans are highly productive in transforming the agrarian economy. The economic importance of fruits and vegetables is also emphasised in recent years owing to an increased demand in the domestic and export markets. The increasing domestic demand could be attributed to increase in income, population growth, changing consumption pattern and growing awareness for better nutrition while the increase in international demand is due to changing global perspective on trade related agriculture (Singh, 2007).

With the economic developments taking place, agriculture is becoming more and more market oriented. Marketing involves finding out what your customers want and supplying to them at a profit. This description stresses two crucial points that govern marketing. *Firstly*, that the whole marketing process has to be customer oriented. Production must supply customers with what they want or need. This is the only reason people spend their money. *Secondly*, that marketing is a commercial process and is only sustainable if it provide all the participants with a profit (Patnaik, 2006). Horticultural products are mainly sold fresh; some are eaten raw while others are cooked. Some horticultural products are traditionally processed when no other form of storage was available, e.g. dried fruit and jams. As society develops and becomes more affluent, the markets for processed and prepared horticultural products develop.

Horticultural crop prices fluctuate widely from year to year, from the start of the season to the main supply period, from day to day and from market to market. This extreme variance in price makes horticultural production potentially both very profitable and very risky. Often, success depends on marketing skills and on obtaining good prices rather than on production expertise.

1. Marketing Scenario of Fruits and Vegetables

Fruit and vegetable marketing include the movement of produce from farms where it is produced to the consumers or manufacturers. This covers physical handling and transport, preliminary processing and packing to simplified handling and reduction in wastage, grading and quality control to simplify sales transactions and to meet different consumers' requirements, and holding over time to match concentrated harvest seasons with the continuing demands of consumers. For the farmer, the strategic function of the marketing system is to offer him a convenient outlet for his produce at a remunerative price. To consumers and manufacturers of agricultural raw materials, assurance of a steady supply at a reasonable price is the vital service. Prices are determined through free market process by negotiations at rural purchase, wholesale and retail stages, and represent a balance between consumers' ability to pay and farmers' need for incentive to produce. An effective marketing system will be geared towards expanding the range and types of consumer service and thus offer procurement outlets.

An efficient marketing system is vital to provide an incentive to farmers to produce more, convey the changing production needs of economy to producers to enable production planning and foster true competition among traders and eliminate exploitation of farmers, particularly the small and marginal ones, who predominate agrarian sector in India (Gandhi and Namboodiri, 2005).

The fruit and vegetable marketing sector is dominated by middle men. Practices of these middle men are often exploitative and the cultivators are often the worst hit. Whenever there is rise in prices of agricultural commodities, the farmers are the least benefited and when there is fall in prices, they are the worst affected. Cultivation of fruits and vegetables is at the mercy of middle men more than other crops because of the highly perishable nature of produce coupled with imperfect market structure ruled by unscrupulous intermediaries. Natural characteristics of fruit and vegetable crops like high moisture content, soft texture and unit size make the marketing quite complex and risky (Bibin, 2007). Because of their high perishability, seasonality and bulkiness, horticultural produces require special care and attention in providing time, form and place utilities which in turn add to their marketing costs. An imperfect market structure is prevalent for these crops mainly due to existence of private traders and other intermediaries in channels of distribution. These factors have great bearing on the marketing system influencing the ultimate prices realized by growers. The production centers are often localized and are far remote from concentrated urban areas of consumption. This calls for quick and efficient methods of transportation and proper packing which do not often happen in practice. This results in a colossal amount of post harvest loses in fruits and vegetables rising up to 25-30% (Khan, 2005).

Most farmers or producers perform one or more marketing functions. They sell their produce either to the pre harvest contractors in the flowering stage itself or to the commission agents in the area. The farmer believes that they are relieved from bearing the risk of damage in transportation and storage and extra expenses related to the transfer of produce from the farm gate to the market. Some large farmers transport the produce to nearby markets and sell either to market agent or to wholesaler. There is an increase in marketing cost and margins with increase in length of marketing channel, while the producers share of consumer rupee decreased with increase in length of marketing channels.

Studies conducted on marketing of fruits and vegetables indicate that the grave problem that farmers face are unstable prices and trader curtails, due to lack of bargaining power. Reason was mainly due to the fact that the farmers "as individuals" are dealing with an established group of traders in the market. Other marketing problems include high marketing costs incurred, ignorance on prevailing market data, time and effort involved and lack of adequate product range (Binoi, 2007).

The fruit and vegetable marketing sector in Kerala has huge potential of around Rs. 1000 crores. Vegetable cultivation in Kerala is undertaken predominantly by marginal farmers and landless agricultural workers either in lease land or own land. The agricultural revolution in Kerala is characterised by rise of mono-culture cash crops to a prominent position, relegating multiple cropping to background. A major casuality in the process was vegetable cultivation. Growing import bill of the state is only a symbol of Kerala's food dependency. The farmers undertaking fruit and vegetable cultivation are at mercy of traders and middle men who reap the success. There exists a chain of intermediaries between the producer and the final consumer. They include village merchant, itinerant trader, wholesaler, commission agent, pre-harvest contractor and retailer. They take their own margins for the services, they render. But these margins are generally ex-orbitant, making commodities costly for consumers and reducing the producer's share in consumer's price. Apart from this, several malpractices prevail in marketing of agricultural produce like arbitrary deductions from the produce, manipulation of weights and measures and cheating the farmers, expenses by way of service charges in market for entry, incorrect, weighing, wastage of time and energy, Transportation expenses to the wholesale market, collusion between middle man and buyer while fixing prices, delay in payment due to farmers etc. (Binoi, 2007). The result is that the farmers are indebted to trader or a moneylender. The main marketing channels operating in the state are:

Farmer - Village merchant-Wholesaler-Retailer-Consumer

Farmer - Commission agent-Wholesaler-Retailer-Consumer

Farmer - Wholesaler-Retailer-Consumer

Farmer - Pre-harvest Contractor-Retailer-Consumer

Farmer - Farmer Market-Consumer

The fruit and vegetable sector has witnessed remarkable changes in the last decade with activities of Vegetable and Fruit Promotion Council, Keralam. Apart from promoting fruit and vegetable cultivation in the state, the organization has set up an alternate marketing system for perishable fruits and vegetables.

2. Marketing Systems of Fruits and Vegetables

Kerala is pre-dominantly a consumer state, depending on the neighbouring states for major food commodities. Fruit and vegetable cultivation have substantially increased in the state in recent years as a result of developmental schemes launched. The state undertakes cultivation of major tropical vegetables like bitter gourd, snake gourd, cowpea, little gourd, ash gourd, tomato, bhindi, amaranths, cucumber, pumpkin, yams, brinjal, colocasia and cool season vegetables like carrot, potato, beans, cabbage, cauliflower, radish and garlic.

There is a three tier system prevalent for marketing of fruits and vegetables (Patnaik, 2006).

Rural primary markets are the traditional and the oldest outlets for trading of produce. The farmers sell their produce to consumers or to the small rural retailers. The trading volumes are generally low and lack of infrastructure facilities shortens the shelf life of these perishable commodities. Primary markets cater to local demand.

Secondary markets cater to distant demand. These markets attract potential buyers or traders who assemble the produce and consolidate a truck load for sales in city wholesale market. These operations are also primarily carried out in an informal manner.

Wholesale markets provide a convenient point for gathering a large amount of produce from different sources and for its division into small assortments to meet needs of retailers in the country. Procurement of agri-produces by various government agencies also takes place through these markets. The volumes handled in these markets are much larger. These markets, therefore, require not only an elaborate physical infrastructure but also some kind of regulation to protect interest of both the producer and the consumer. Most wholesale markets are covered under the Agriculture Produce Marketing Committee Acts and are also called regulated markets. In major cities like Delhi, Kolkata and Bangalore, these markets perform dual function; transit market for supplies to the hinterland and distant markets and terminal market for supplies to the retailers for local consumption.

The primary wholesale markets are located in important towns near centres of production. The producer-farmers bring major part of produce for sale themselves in these markets. The secondary wholesale markets are generally located in district headquarters or important trade centres. The bulks of the arrivals in these markets are from other markets. Major transactions take place between village level commission agents and wholesalers. In the terminal markets, the produce is either finally disposed of to consumers or processors or is assembled for dispatch to distant markets and also for exports. In view of their importance in the marketing of agri-produces in the country, the Government has taken the mantle upon itself to provide the requisite infrastructure and other facilities in these markets.

In Kerala, there are transit markets at the lowest stage which function only during the specified time and a majority of farmers depends on these markets. In the next stage, there are rural markets, fed by the adjoining major production belts. These markets operate during specific timings on all days and feed the small wholesalers and majority retailers. Local produces are marketed in the transit and rural markets. Wholesale markets operate in all districts where marketing of both local and out of state produce are undertaken and they form the major feeding points for wholesalers, processors, hoteleers and vegetable vendors.

Regulated Markets are results of Government's initiative to promote organized marketing of agricultural commodities in the country through a network. The basic objective of setting up a network of physical markets is to ensure reasonable gain to the farmers by creating environment in markets for fair play of supply and demand forces, regulate market practices and attain transparency in transactions.

The Government of Kerala established wholesale markets in the state with the economic assistance of the European Union. These markets have regulations of activities to a large extent. There are six wholesale markets established, 3 in rural areas and 3 in urban areas in Trivandrum, Ernakulam and Calicut.

The basic objective of market regulation is to regulate the trade practices, increase market efficiency through reduction in market charges, elimination of intermediaries and to protect interest of producer-seller (Anon, 2006). Although regulation of markets has improved their functioning and has helped in reducing multiple trade charges and levies on producer-seller, verification of accurate weights and scales, establishment of market committees in which the agricultural producer is given due representation, judicious utilization of market funds, fair settlement of disputes, arrangements for better storage facilities and market intelligence. But the existing machinery has failed to check trading malpractices and has made the agricultural marketing system highly restrictive and inefficient. Monopolistic practices and procedures have taken root and prevented development of free and competitive trade in agriculture. Processing industries cannot buy directly from the farmers, except through notified markets where intermediaries take away a sizeable of the price of the produce. There is no direct link up between farm and factory. Hence, the agri-production is oriented towards direct consumption, to the detriment of processing and value addition.

Direct and Alternative Marketing: In the highly regulated and monopolistic wholesale markets, the farmers are unable to bargain effectively in the market. Intermediaries and system inefficiencies consume a disproportionate share of consumer price. There is no incentive for the farmer to improve quality and productivity. Farmer's dependency on the trader is high due to which he loses the freedom to find the best buyer. Even the technical services to the farmers are inadequate and ineffective. The only option left to modernize marketing system is to set up an alternative marketing system that may operate parallel to existing system. These markets would be outside the purview of the APMC Acts and would ensure transparency, quality control, efficiency and fair play.

Direct Marketing by Farmers enables farmers to meet specific requirements of wholesalers from their inventory of graded produce, as also of retail consumers, thereby improving their price realization considerably. This also ensures grading of farm produces at the farm gate. It also obviates the need for hauling the produce to the regulated markets, saving both time and money on transportation. Direct marketing pre-supposes prior knowledge of the buyers' needs among the farmers. It also enables transactions through purchase orders for long-term deliveries and payments. This commercialization of agriculture based on resources business principles is a very healthy sign.

Micro-level farmer marketing entities, as an instrument of direct marketing, have tremendous potential to address the desiderata of the traditional marketing system. These instruments of direct marketing can not only bring economies of scale by ensuring optimum utilization of local resources, but also can go a long way towards reducing costs and margins in marketing of produce.

Farmer marketing entities, as an instrument of direct marketing with producer-sellers has the potential to bring substantial reduction in expenses incurred on different aforesaid heads. Needless to say that direct marketing by a group pf producer-sellers will help them in avoiding commission of commission agents constituting a very high portion of all charges and taxes. Homogeneous farmers form themselves into self-help groups at district levels for undertaking marketing of their produces collectively and reap the benefits of group-marketing through different activities like joint market intelligence, joint hiring of trucks, group bargaining, joint storing, grading, preliminary processing and packing. This will not only ensure remunerative prices for their produces by putting them in a better bargaining position *vis-a-vis* the buyers but also will go a long way towards bringing saving in different costs both in terms of time and money.

Direct marketing (retailing) by farmers was experimented through *Apni Mandis* in Punjab and Haryana. A modified concept was introduced in Andhra Pradesh through *Rythu Bazars* and in Tamil Nadu through *Uzhavar Santhaigal*. These markets, having a catchment area of 60-80 km, have become popular for retailing fruits and vegetables in major towns. These markets are being run by state governments, as a promotional measure, to introduce the principle of marketing without middlemen for the benefit of small and marginal farmer(Ray, 2005). More such markets need to come up in the organized sector with private investments so that appropriate backward and forward linkages could also be developed.

Farmer markets of Vegetable and Fruit Promotion Council, Keralam (VFPCK): VFPCK working in the fruit and vegetable sector of Kerala developed a novel concept of group marketing of produces by the farmers themselves. The concept of group marketing, in simple terms, is marketing managed by farmer groups. As the name implies, the main focus is to empower and facilitate the farmers to take more effective decisions with regard to the marketing of their produce. The concept of group marketing thereby provides the member farmers better access to markets and therefore a greater share in the consumer's rupee. This system has been acting as a better and efficient alternative for the fruit and vegetable farmers of Kerala.

Under this concept, 150-200 farmers in the production area bulk their produce at a common point on prefixed days, where the traders are invited. The group concept increases the bargaining power of farmers and location of bulking point in the production area provides the farmers a home turf advantage. Under group marketing, 10-15 Self Help Groups (SHGs) numbering 150-200 farmers come together under the banner of Swasraya Karshaka Samithi (SKS), bulk and trade their produces collectively. This system helps the member farmers to have a good volume thereby being in a better position to negotiate with wholesalers to 'optimise their returns'. Large volumes induce traders to buy at the SKS. The produce is sold by auction or bargaining to ensure competitive prices. In addition to better prices, farmers enjoy benefits of reduced transportation and handling costs, proper grading and handling, accurate weighing and timely payment. More importantly, farmers save the time that they would be spending in doing these activities. As the traders are attracted to the SKS, farmers have a home turf advantage. This is of great benefit to farmers considering the difficulties they experience in the existing public markets, which normally leads to distress sales. Weighing at the Samithi is done by farmers and is transparent and accurate. The Samithi is free of loading /unloading workers since the farmers do it on their own and this ensures careful handling of produce. Prompt payment is guaranteed within the prescribed period, as there is a collective effort in recovery from among debtors. The SKS is not envisaged to be an organisation, involved in pricing and buying/selling of produces but facilitates trading between farmers and traders. The SKS is therefore a means for farmers to improve their bargaining power through better access to markets and traders.

There are about 210 farmer markets operating in Kerala under the brand name *Swasraya Karshaka Samithi (SKS)*. The concept of linking the farmers directly to the traders has resulted in a situation of mutual coexistence. Prices are fixed based on market information on daily prices and arrivals provided by VFPCK. The samithis are registered bodies with systematic and uniform accounting and transparency in all activities. The SKS has now evolved to form self sustaining social institutions providing a single window for catering to the different needs of farmers including inputs, seeds, farm machinery and campaigns with social cause etc. The SKS is able to enhance income of farmers by reducing costs and obtaining reasonable prices. Apart from this, the SKS is also able to boost confidence of farmers in undertaking fruits and vegetables cultivation. Price spread analysis conducted reveals that farmer share of consumer price has increased from below 40 to 60% in VFPCK farmer markets (Binoi, 2007). The farmers are motivated to consider farming as a commercial venture by undertaking production in accordance to market demand and promoting diversification in produces with market changes.

The sector is now facing the entry of multinationals into retailing of fruits and vegetables. The cohesive farmer groups of VFPCK are better placed to counter possible exploitation by retail networks since group force delivers better bargaining power and enhanced returns.

Forward and Future Markets: Direct marketing can be effective only if farmers have an opportunity to access reliable reference prices for a range of grades, qualities and delivery times. Further, with the coming into force of WTO provisions on agriculture, need for effective price risk management has become imperative to protect the agriculture sector- primary and processed- from price volatility. Forward and futures contracts enable this price discovery and price stability. They also afford an opportunity for the farmer to hedge his risk by deciding in advance what to produce/sell/store etc. Forward contract is an agreement between two parties to buy and sell a commodity at a pre-determined price on a future date. Future markets can help in stabilizing prices in times of good and poor harvests.

Co-operatives: Co-operatives are established in different states with the success of the Anand (Gujarat) pattern in milk production and marketing. Generally all cooperatives follow a three-tier system with the apex body coordinating processing and marketing activities.

The most successful story is that of Mahagrapes, one of the largest brands of fresh grapes (mainly seedless grapes) exported from India. The farmers are given a Minimum Support Price for their produce and they are provided consultancy on pesticide usage.

Their main plank for inculcating brand equity in their customers is excellence in quality. They even have customers who give them advances. They also help promote their products through keeping personal relations and sending across samples and photographs. Their future plans are to sell their products through e-commerce and enter the market of frozen juices.

Other success stories in the sector include HOPCOMS and Safal market of NDDB. The most significant benefit arises from the fact that HOPCOMS purchases its produce directly from the farmers. This eliminates the intermediaries, and consequently, a remunerative price is paid to farmers - usually 10-15% higher than the open market prices. Furthermore, during periods when there is an excess supply of certain produce in the market, the open market price drops. HOPCOMS, however, assures a minimum price for produces during times of poor sales. Another major benefit is that cash is paid to farmers on the day of transaction, thereby eliminating need for extending credit, which is prevalent in private business (Smitha, 2005). HOPCOMS has weighbridges at each procurement centre, hence assuring farmers of correct weights. HOPCOMS also has infrastructure facilities like cold storage and godowns to store produces, at procurement centres where farmers can sell directly to consumers. At the procurement centres, HOPCOMS also sells seeds and fertilisers at subsidised rates, as well as crates for transport. In addition, because farmers travel long distances with large quantities of produces, HOPCOMS provides lodging and boarding facilities.

The Safal fruits and vegetables auction markets have introduced transparent, efficient and computerized marketing of fruits and vegetables for the first time in India. Safal gives importance to physical infrastructure and backward and forward linkages. It has established associations with horticulture produce farmer's associations in the major production belts of the country and with major wholesale traders and exporters.

Partnership with Farmers – Private Sector Initiatives: Precision Farming: The days of public distribution system are fast changing. The concepts of local food security, neighbourhood production and processing to meet market demand, universalization of production and marketing information, both for producers and consumers, call for precision in micro-level planning and operations. A new producer-consumer relation is emerging. There is gradual shift in the agriculture scene from being supply driven to becoming demand led. Certain private sector organizations in India have sought to surmount challenges, involved in the management of our agricultural supply chain. The fragmentation, lack of infrastructure, presence of middlemen and the plethora of rules and regulations have not proved to be too daunting for certain inventive and resourceful organizations. What is different is that partnerships become the prime criteria in supply chain management. These partnerships are in Processing, Research and Development, Extension services and Marketing with the farmers directly. The focus of these private sector initiatives is more towards elimination of waste, increase in farm incomes and productivity enhancements. A good example of successful backward linkage in horticulture and agriculture crop procurement is the one by Pepsi foods which is in operation in Punjab for more than 10 years now. Pepsi pioneered the concept of contract farming for the complete bulk procurement of a variety of crops like potato, tomato, groundnut, chilli and paddy (Khan, 2005).

The e-Choupal model promoted by ITC is specifically designed for the farmer's benefit. It is construed to overcome the challenges owing to the unique character of the Indian agriculture. For instance, the problems associated with small and fragmented farm holdings are weak infrastructure, supply chain intermediaries and lack of quality and real time information. To face the rising challenges related to agriculture sector, ITC has set up small Internet kiosks at village level to provide farmers real time market and pricing related information and highlighting arbitrage opportunity in sales between various markets.

Tata Rallis, by marketing the key farm inputs including agri-chemicals, pesticides, seeds and fertilizers has succeeded in leveraging their strength to the benefit of farmers and other partners. For instance, processors interested in raw material procurement for processing are ropedin to partner Tata Rallis. Also filling in key gaps are the financial service institutional partners who offer credit on inputs to farmers as well as funding working capital requirements.

Terminal market: This works on the following principles.

It has a hub and spoke format, that is the Terminal Market (the hub) to be linked to a number of collection centres (the spokes). The collection centres

(Spokes) are to be conveniently located at key production centres to allow easy access to farmers(Anon, 2006). The following state of facilities will be provided like electronic auction, grading, washing and packing lines, processing and exports, banking and transport including reefer vans .

e-marketing: Modern information and communications technologies (ICTs) and web based marketing of agricultural produces hold great promise for socio-economic development of rural hinterlands in India. The 'anytime-anywhere' advantage of e marketing leads to efficient price discovery and offers economy of transaction for agricultural trading. This attracts many rural developmental agencies to deploy websites for marketing agricultural produce. For instance, rural entrepreneurs can benefit because ICTs help to improve access to markets or supply chains and provide a broader base for decision-making, thus making risk more calculable. Many local communities have experienced that ICTs have increased bottom-up participation in the governance processes and may expand the reach and accessibility of government services and public infrastructure. The electronic village project of M.S.Swaminathan Research Foundation (MSSRF) in Pondicherry received the Stockholm award for its promise.

3. Vertical Coordination in Food Chain

The dimension of commoditisation to commercialization cycle calls for tight coordination of links between production, processing and marketing to accomodate manufacture of products based on consumer driven demand. Vertical coordination will shift part of the marketing effort towards discovery of consumer preferences rather than manipulation of same by retailers. Balancing demand expansion with production increases implies integration of demand side research and production side research. Peterson (1997) conceptualized coordination as a continum running from open markets to complete vertical integration (multiple market stages under single ownership). Although this idea of a continum is intuitively appealing, most of the prior theoretical work has focused on two ends of the continuum ie., spot markets and vertical integration, while the middle of the continum is largely unexplored.

Strategic options for vertical coordination –vertical coordination continum

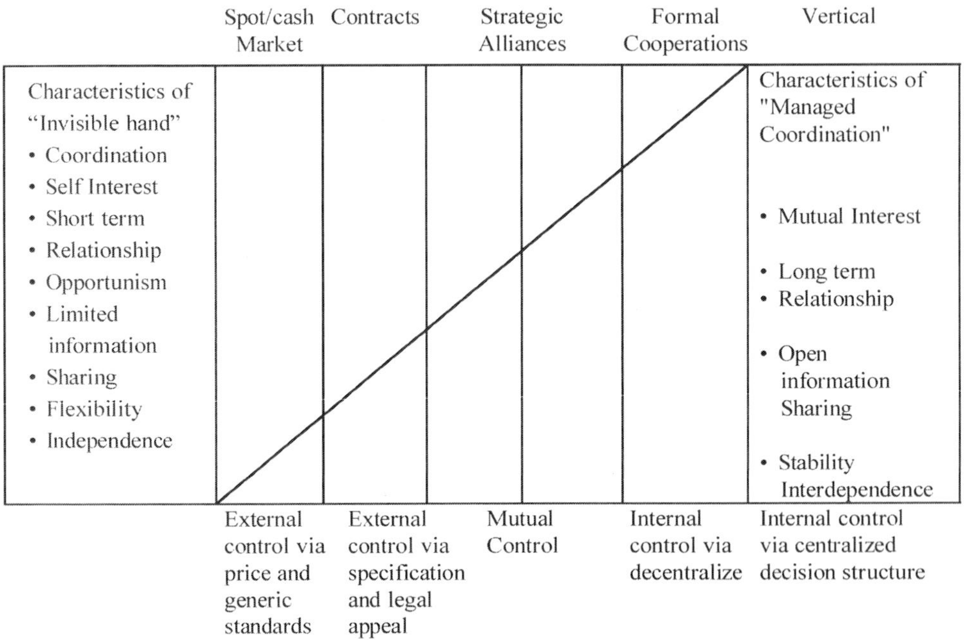

The five major categories of coordination are suggested from spot markets to vertical integration. The continuum suggests that as strategies are considered from left to right, coordination moves from being dominated by invisible hand characteristics (self interest) through a changing mix of invisible hand/managed characteristics to being dominated by managed characteristics (mutual interest). In between the continuum is *contracting*- the legally enforceable establishment of specific and detailed conditions of exchange; *strategic alliance*- as an exchange relationship in which the firms involved share risks and benefits emanating from mutually identified objectives and *formal cooperation* through a formal organization that has an identity distinct from the exchange actors and that is designed to be their joint agent in the conduct of a cooperative exchange. The agricultural cooperatives lie at this point on the continuum.

The vertical integration in this version of continuum is a mechanism that relies upon centralized control to achieve coordination. It can be observed that coordinating mechanisms move from completely external in form(spot markets) to completely internal in form(vertical integration) while passing through several transitional stages of mixed form(contracts, strategic alliances and formal cooperation).

Vertical integration: Here each individual farm loses its identity and becomes a company owned farm. The key to vertical integration is asset ownership. It is assumed that the transaction costs are reduced by internalizing production and processing assets. Peterson(1997) observed that vertical integration is a mechanism that relies upon centralized control to achieve coordination.

The growing phenomena of globalisation and its gradual liberalization have triggered the process of vertical coordination in Indian agriculture. Resultantly, different steps in production, processing and marketing become interdependent and farmer also becomes a part of the large food production system in due course. A review of past and ongoing initiatives in vertical coordination seems to be moving in three streams- evolution of commodity and location specific linkages between producers and agroindustry under the umbrella of contract farming, evolution of supply chain models integrating technological advancements and promotion and regulation of these mechanisms (Ray, 2005). Three core processes are critical in establishing effective agriculture market linkages which include building effective backward linkages, conduct and performance of markets and marketing institutions and demand generation through value addition.

Vertical coordination means synchronization of the successive stages of production and marketing regarding the quantity, quality and timing of product flows. Methods of vertical coordination include open production or spot market, contract production and vertical integration. In *open production* no commitment to sell produce is made before completing production and spot process coordinates the sale across the different stages of production. In *contract production*, production is done aimed at future delivery, there is more interaction between buyer and seller and delivery schedule, pricing method and product characteristics are negotiated. In *vertical integration*, a single firm controls two or more successive stages of vertical coordination. In vertically integrated firms, the management decision dictates the transfer of resources across stages (Ray, 2005). Movement along the continuum of vertical coordination from open market production to vertical integration represents the degree to which the control of production has shifted to the contractor or integrator as more functions are transferred from the producer.

Vertical integration reduces costs of contracting and spot market trading but also introduces new types of transaction costs including costs related to communicating information within a firm. In Indian agriculture, vertical integration is difficult to achieve because of the land ceiling acts which prohibit the corporate acquisition of cultivable land. Vertical integration is a more viable option for poultry and dairy in India (Ray, 2005). There are questions to ponder, which include whether the larger farms will generate new efficiencies in food production and whether the contractual arrangements will insulate the producer from market risks or will it introduce new risks. It is yet to be ascertained whether the new system will provide opportunities for new farmers or the existing small farmers. The system will have to cater to needs of small farmers if it has to generate results and hence clustering of small farms will be the viable option.

In case of successful cooperatives, the bondage between cooperatives and membership revolved around three major factors - strong backward integration through provision of services, benefits of collective marketing and transparency in the process of market transaction-. To retain the trust of their membership, cooperatives need to take two important steps – firstly to continue to improve the quality of services needed by majority of the membership and secondly to introduce organizing element to further align production system with the needs of the market and generate economies of scale to improve farmers bargaining power. On the marketing front, the cooperatives need to put greater effort in improving cooperative market interface through upgrading professional managerial input and value addition to its products. Unlike cooperatives, private firms do not attempt to build broad based relationships with the contract farmers. Farmer managed organizations have greater farmer/producer orientation in building backward linkages. Private firms have greater procurement orientation and their relationship with the farmers is based on this. In each of the above cases, the common theme is that each firm has attempted to hedge against the vagaries of agribusiness by adopting innovative formats for vertical coordination and supply chain management. Agribusiness firms are now directly interacting with the producer farmers, and not relying completely on the intermediaries in the supply chain. Value addition and quality consciousness are becoming the driving forces for the firms and rural farmers entering into collaborative partnerships. Firms are leveraging modern technology to overcome the infrastructure bottlenecks. Use of IT for vertical coordination in agribusiness is amply demonstrated by ITC(www.mckinsey.com). Also firms are investing resources in organising farmers at the backend of the supply chain and establishing procurement bases. The dairy cooperatives in the country and the fruit cooperatives in Western India are good examples of vertical coordination in the cooperative sector.

4. Supply Chain Management

Globalization has a dramatic impact on both agriculture and processing, worldwide. It has not only opened up markets but led to increased production and also heightened competitiveness and driven commodity prices down. At the same time, increasing consumer awareness has raised concerns of food safety and introduced new concepts like tracing and tracking. This led to a trend towards vertical integration of production and processing, not only to achieve greater efficiencies of scale to survive in the fiercely competitive market, but also to satisfy food safety and other consumer norms. In future, successful ventures in the agri-processing sector are likely to be vertically integrated where production, post-harvest management, storage, processing and distribution shall be linked in an integrated supply chain. Yet, the horticulture marketing practices lack systems approach. Trading and marketing structures are traditional consisting of a long chain of intermediaries. As many as 75% of farmers sell their produce at the farm level. They cannot afford to go to distant mandis on account of lack of facilities, expensive transportation and malpractices in the assembling markets. The long marketing channel is detrimental to quality and safety of those perishable products. There long chain of intermediaries between the farmer and the retailer, up to 6-7 add cost but no value to the product. It also blocks the flow of market related information to the farmers resulting in his exploitation. Further, the farmer is not aware of the market demand and continues to produce as per his own judgement. This has led to one of the biggest handicaps in Indian agriculture – *it is supply driven rather than being demand led.*

The uncertainty of quality and regular supply have adversely affected the processing of agri-produces in the country. The quality issues can be mitigated to a large extent by putting the post-harvest management infrastructure in place and having backward linkages with R&D institutions and input suppliers, including seeds and seedlings. The supply chain can be shortened by the concept of "contract farming". This is a system by which the right of the farmer to take his own decisions is respected. Under the contract system, the farmer has the option to enter into a contract to sell his produces to the processor at a pre-determined price or formula, which assures him a minimum return. Risks, which are inherent in any agricultural operation, are shared; the farmer carrying the production risks and the processor carrying the market risks. Even the production risks are sometimes mitigated to an extent by supply of inputs and services by the processor. In the process, the farmer and the processor become joint stakeholders in the venture to enhance production and productivity. The imperatives to improve yields and to drive down costs often facilitate transfer of technology from the processor to the farmer. Despite the obvious benefits, the system of contract farming has not become very popular except in a few cases, like the Punjab Agro–Pepsi experience in growing tomato in Punjab. The primary reason being the fear in the minds of either party of the contracts not being honoured, since there is no firm legal backing to the system.

5. Post Harvest Management

With globalization of the Indian economy and the impact of WTO on Indian agriculture, importance of post harvest management cannot be over emphasized. The Indian agricultural commodities will have to compete with the world commodities not only in foreign markets but also in domestic markets(Pushpasurendra, 2007). An integrated supply chain, from farm to market, would require the following infrastructure for all types of perishable horticultural produce.

- **Cleaning, Grading and Packaging:** Facilities for cleaning, washing, grading and packaging. at the farm level are again negligible. Only limited grading and sorting facilities exist in markets for removing unwanted biodegradable biomass and for proper packaging. Our effort in the past has been primarily focused on increasing production, without adequate attention being paid to development of post-harvest management infrastructure. The need for this infrastructure is imperative in our country, having a tropical climate which accelerates perishability of fruits and vegetables.

  There is need to create facilities for cleaning, grading and packaging not only at the primary level but also at the village level. This would not only remove the unwanted biodegradable biomass, which in turn can be used as cattle feed, but could also reduce the volumes of the market arrivals by as much as 25%. This would help in reducing congestion in the markets. Post-grading packaging by the farmers would also reduce the chances of adulteration. Scientific packaging at the farm level needs to be promoted. Considering the export potential of fruits and vegetables and with a view to cater to the needs of bulk buyers in respect of well-graded and packed produce, it would be necessary to provide the facilities for grading and packing in the growing regions by setting up mechanized Grading and Packing houses.

- **Cold Storages/Cool Chain:** Development of adequate cold storage capacity/cool chain would help not only in increasing the shelf life and minimizing post-harvest losses through proper preservation but would also help the farmer in taking timely marketing decisions. In view of the future requirements of fresh/pre-cooked/frozen food items and the anticipated change in food habits in favour of processed foods, the capacity requirement for post harvest management of perishables needs to be augmented by another million tons over the next 10 years. The private sector should be tapped for this purpose by giving appropriate incentives. Further Commodity specific post harvest practices are required to be developed, both for short and long distance marketing.

Promoting Value Addition: Despite being the largest producer of fruits and vegetables, less than 2% of the fruits and vegetables are commercially processed. Value addition through food processing generates demand for agricultural raw materials and also has a multiplier effect on the rural economy (Baisya, 2006).

Agricultural development in most of the agriculturally advanced countries has taken place by streamlining the marketing of produce. Agricultural marketing is considered an important part of the general process of agricultural development. Essentially the marketing channels should operate in a manner that the producer is able to sell his produce at a price that its intensive value merits. For this purpose, the system should feed back the farmer the ruling prices and price trends, the size and dimensions of the market and the buying and selling activities at the market at which he generally sells his produce.This knowledge would gradually induce the farmer to respond to market needs and develop a cropping pattern that would lead to specialisation or maximisation of available resources. At the same time, the marketing system should create a congenial environment and provide flexible institutional arrangements. It should also be operationally efficient and progressive, attend to produce requirements, and be capable of adopting marketing development and innovations over time

The importance of good marketing can be conveniently considered from the four different perspectives of the national economy, the farmer, the product, the processor and the consumer. At the national level as societies and countries develop, there is a movement of people from the countryside into the towns and cities. Populations in developing countries are expanding, normally at around three % a year. Urban populations, however, are expanding on an average at about four % a year. This means that the number of people needing to be fed by the rural communities will double in 16 years. In addition, since the amount of food eaten by each individual normally increases as people become wealthier, the supply of food for the towns and cities will need to double approximately every 10 to 14 years. This change in population distribution will create new or improved opportunities for both farmers and rural employment, particularly if new roads and improved transport are provided.

Subsistence farming (that is, just providing enough food for the farmer and the immediate family) will become less important. Although proportionately, there will be fewer farmers, their role will become ever more important because their task will be to feed the growing urban populations. To do so will require the farmer to become more specialized and more skilled so that more food can be produced. In developed agriculture, successful small farms can and do survive. Viable small farms tend to specialize in high output enterprises. These are crops or livestock systems, capable of generating high incomes / unit of land. There are four ways to increase profitability: by lowering costs, reducing losses, obtaining higher prices and selling more produce

Lowering costs and reducing wastage are high priorities because they provide the chance of cheaper produce. In all developed societies, cost of food has fallen in real terms. Cheaper food, in turn, means fruit and vegetables will be affordable by the poorer sectors of society. Not only will the market for horticultural produces increase but higher consumption levels can also lead to improvements in health and nutrition. Lower unit costs can be achieved by higher yields and lower transport, packaging and distribution costs. Lower wastage mainly depends on improved post-harvest handling but also on better production and harvesting techniques. *Higher prices* can be obtained by producing right crops, at the right time and the correct quality. They may also depend on negotiating skills and targeting high price buyers. *Increased sales* may depend on lower costs but also on finding and developing new markets, new business ventures, new crops and new periods of supply.

Traditionally extension officers provided technical production advice. Development brings with it the increased commercialization of agriculture. Farmers need assistance in this transition. Marketing and business advice are a new dimensions for extension officers, which require technical, financial, marketing and management skills. It is an exciting area with scope for innovation and the opportunity to significantly improve rural incomes. It is better to place a marketing organisation at the facilitating end if the objective is to reach poor. The basic precept is for facilitators and promoters to do as little as possible by way of provision of services, and as much as possible by way of capacity-building, networking, linking and mainstreaming. Ultimately, the most important investment that can be made is to consult with clients. An effort to build people's organisations must be made and the people themselves must be aware of the functioning of their institutions.

A trend means that somebody has identified a market opportunity and begun to exploit it. All marketing opportunities are limited in scope; the gap in the market gets filled, and then overfilled in the course of time. The people who make money out of those trends are those who start them, not those who follow them. Those who try to jump on a bandwagon when it is moving fairly fast are liable to fall off and get hurt. Starting a trend requires analysis, insight and, finally, courage.

References

Anon, (2006) Concept note on Modern terminal market for fruits, vegetables and other perishables, Department of Agriculture and Cooperation, Ministry of Agriculture, Government of India.

Anon, (2006) Trends in agricultural marketing in India,India Infoline.

Baisya, R.K., (2006) Indian processed food industry- present status and future challenges, *Indian Food Packer*, pp 55-62.

Bibin, M., (2007) Group Marketing system for fruits and vegetables in Kerala, M.Sc thesis submitted to Kerala Agricultural University.

Binoi, K. (2007) Management of marketing system for perishable agriculture commodities – Problems and future prospects in Kerala, Ph.D thesis submitted to School of Gandhian thought and Development Studies, Mahatma Gandhi University, Kerala.

Gandhi Vasant, P. and Namboodiri, N.V. (2004) Marketing of Fruits and Vegetables in India: A Study Covering the Ahmedabad, Chennai and Kolkata Markets, IIMA working paper.

Khan, Amir Ullah, (2005) Trade liberalization and the Agriculture sector in India . pp 5-32. Seminar on the food processing industry in India: Challenges and opportunities for investment and trade.

Patnaik, G., (2006) Marketing, Storage and Extension - Agriculture in India, pp 4-21.

Pushpa Surendra, (2006) Bitter Fruits of harvest, India Together.

Ray, Sandipan, (2005). The structure of rural markets for achieving vertical coordination in the Indian Food value chain. pp 119-124, Seminar on the food processing industry in India Challenges and opportunities for investment and trade.

Singh, H.P., (2007) Technological interventions in Indian Horticulture – a way ahead, *Indian Horticulture*, pp 3-9.

Smitha, P. (2005) Hopcoms a success story of horticulture cooperative marketing.

www.mahagrapes.net

www.mckinsey.com

www.wto.org

Chapter – 16

Floriculture Global and Domestic Scenario

V.L. Sheela

Floriculture is increasingly regarded as a viable diversification from traditional field crops due to increased per unit returns and increasing habit of "saying it with flowers" during all the occasions. All over the world, the floricultural sector is experiencing rapid changes. Opening up to the world, market in the WTO regime paving way to free movement of floricultural products worldwide, globalization and its effect on income generation have all contributed to increase in per capita consumption of flowers in most countries. Noticeable feature of global floriculture is the development and expansion of floriculture in non-traditional areas. Netherlands, Italy, Germany and Japan had strong tradition for growing and consumption of flowers. The concept of commercial floriculture was perpetuated across the world from those regions. New production centres are developing in Latin America, Africa and also in Asia to meet the demand of consuming countries and also to expand domestic market, commensuration with improved economic conditions these countries provide highly qualitative floricultural products with low cost price which is finding a favourable place in markets of Western Europe, America and Japan. Emergence of new production centres has made floriculture more competitive and this in turn is benefiting the ultimate consumers.

The flower industry comprises the cultivation of and trade in cut flower, cut foliage, potted plants and bedding plants. The main representatives of cut flowers are rose, chrysanthemum carnation and lily. Potted plants and cut flowers have an almost 80% share of world trade in ornamental plant products. World cut flower markets are growing at an approximate rate 6-9 % per year. In international terms, the consumption of cut flowers is concentrated in three regions Western Europe, North America and Japan. The highest growth is expected in Japan and USA. Consumption is rising not only in Japan but also in other Asian countries with rise in expendable income and a flower minded culture. At present, USA with about one third of consumption has the largest share of total world consumption, followed by Germany (about 20%), Italy and France.

Top Ten cut Flowers and Pot plants

No.	Top Ten Cut Flowers	Top Ten Pot Plants
1.	Rose	Ficus
2.	Dendranthema (Chrysanthemum)	Dracaena
3.	Tulip	Kalanchoe
4.	Lily	Dendranthema
5.	Carnation	Begonia
6.	Gerbera	Hedera
7.	Freesia	Azalea
8.	Cymbidium	Saintpaulia
9.	Alstroemeria	Hydrangea
10.	Gypsophilia	Spathiphyllum

1. World Export/Import of Ornamental Plants

The value of world import/export is increasing every year. Europe is the biggest import market for ornamental floricultural products. For cut flowers, the North American Market is the biggest import market.

The world's largest exporter of floricultural products is by far The Netherlands, responsible for about 60% of international export. The international trade in cut flowers and potted plants is growing every year. From the total production of cut flowers and potted plants, only a small part is exported to the International market. About 75% of international trade comes from within Europe. German alone accounts for 30% of world import of cut flowers. Besides Germany, the United States, France and United Kingdom are big importers. The Netherlands is also a big importer but re-exports most of its imports. The total value of the world export of cut flowers rose from 1.25 billion dollars in 1982 to 3.6 billion dollars in 1995. Next to Netherlands, Columbia, Israel, Kenya and Ecuador are the biggest exporters.

In terms of purpose, European consumption of floricultural products is very heterogeneous. Consumption pattern differs from country to country in Europe. Flowers are generally purchased for gifts (45%), special occasions, such as birth days and weddings (25%) and home decoration. Quality of flowers, their colour, beauty, scent, vase life etc. are factors that matter. The definition of floriculture differs from country to country making it difficult to compare statistics. If regional factor is considered, Latin America is the biggest export region followed by Europe. Asia playa a rather small role in international trade. Most Asian trade is within Asia itself. Only for special products like orchids (Thailand) and Bonsai products (China/Japan), there is a marginal export position.

2. Recent Trends

Consumers are becoming more aware of what they want to buy. They are becoming professional buyers. They tend to ask for higher choices in product quality levels depending on purpose of purchase, as well as for higher levels of service and a wider and deeper assortment. The Association of Speciality Cut Flower Growers (ASFGC) established in 1988 is a body that addresses production and marketing needs of speciality cut flower growers. Importance of protected cultivation is also increasing. The trend has shifted from seasonal to year round production. The aim is make available flowers when they are most wanted. (Like Valentine's Day, mother's day).

3. Flower Auctions

Flower auctions in Holland were set up 75 years by Dutch growers who had become tired of cut throat competition. Today there are three major auction centres in the Netherlands, Aalsmeer BVH and Tele flower. Through these auctions, several billion dollars' worth of Dutch and imported floricultural products from Africa, South America, and Asia are sold throughout Europe. These auctions provide a huge audience of buyers and payment is guaranteed.

Demand for Floricultural products is expected to touch 160-240 billion from the present 60 billion US $. Due to stringent environmental regulations, increased energy costs, ever increasing wages, shrinking agricultural lands, worsening climates during winter, the floricultural developed nations would not be in a position to expand their production and hence the demand has to be met from developing nations. India holds a fair chance in such a scenario. Due to rich diversity in agro climatic conditions, we have the distinction of producing an array of floricultural crops both tropical and temperate.

4. Domestic Trade of Flowers

Area and production of floricultural crops have increased from 21000 ha with 1.48 tonnes of loose flowers and 6806 lakh number of cut flower/spikes during nineties. Karnataka ranks first in area (22%) followed by Tamil Nadu (20%) Andhra Pradesh,Telengana (20%), West Bengal (15%) and Maharashtra (7.0%). These five states together account for about 87% of the area under flowers and can be called the "flower growing belt" of the country. Export of Horticultural products has increased from a mere Rs. 150 million in 1991-92 to Rs. 1650 million in 2002-03 (Gajana, 2004). Domestic flower trade was about Rs. 5000 million which is indicative of the relative significance of traditional flowers for the Indian Economy, despite the growing market for modern flowers.

The demand for flowers can be categorized based on taste and preferences of consumers namely

a) Those used for adornment, garlands and bouquets. The flowers used are jasmine, tuberose, chrysanthemum, crossandra and rose.

b) Those used for house decorations, ceremonies, social gettogethers and parties. The long stalked flowers like dahlia, roses, orchids, asters and gladiolus are used.

c) Those needed by the perfumery industry to process the flowers to extract essential oils. These include flowers like jasmine, champaka and roses. India is the largest exporter of jasmine oil, used in perfumeries.

The major traditional flowers traded in the market are rose, crossandra, marigold, jasmine, chrysanthemum, aster and tube rose.

5. Marketing Channels and Price Spread in Traditional Flowers

The main marketing, channels by which the traditional flowers move from point of production to the ultimate consumers are as under (Jiyaulla, 1993, Vendini, 1994, Ramesh, 1997).

A producer – whole saler cum-commission agent-retailer-consumers and B Pre-harvest contractors- whole saler- Retailer- consumers

Flower being perishable and number of intermediaries involved being more with limited scope for direct marketing, the producers receive relatively less hare in consumer rupee. Traditional flower market is faced with a number of setbacks like dominance of commission of agents (forced sales) and unchecked activities of pre harvest contractors, risk in terms of price fluctuations associated with high commission of agents, high transportation costs, lack of organized market ensuring competitive trading practices, absence of processing facilities to convert to other forms, and lack of government support as given to export oriented cut flower industry. Further demand for traditional flowers is limited to domestic consumers with virtually no scope for export. The export rejects of commercially cultivated hi-tech flowers often flood the market resulting in lower prices for domestic flowers. However the domestic demand for traditional flowers is quite strong since they are part of our culture. There is need to streamline the trading practices by establishing specialized flower markets managed by professionals.

6. E-commerce in Floriculture

Information technology has brought in miraculous changes in the life of modern man. Among several utilities of information technology internet the information super high way has become an imperative tool for common man to exchange information at a greater speed, without manual intervention to boost industrial business.

Floriculture industry is one among the several areas wherein such technology is used to a great extend for quick shipment of produce and maintenance of the same is of paramount importance due to highly perishable nature of product. E-commerce has changed the way florists do business. This is achieved by developing an online information system using which growers/bidders can participate in the online auction of their products and without any manual intervention, products can be sold and purchased. In fact in the next four years, more than 300% increase is expected in online flower sales. It is also noteworthy that flowers are currently the third most profitable sellers on the web. The internet allows each customer to see the product they want they buy. It is estimated that sixty seven % of all fresh cut flower purchases are made by women who are also currently the fastest growing segment of internet users.

A grower engaged in commercial floriculture business can approach an online auction center with samples of different products of flowers. After acquiring user name and password the grower enters the online auction center. Samples of the products will be serially numbered and full description of each of the product with respect to plant characteristics and grower's field coupled with package of practices followed will also be displaced. Auction center in their website specifies the time of auction of such products well in advance. At the same time, bidders sitting across the globe would also log on to the auction center by specifying their respective user name and password. Once the auction starts, computers will display the scanned images of the products with their identification number along with the growers' price. A successful bidder at the end of the process will make use of the e-banking facility to arrange for payments. Such a procedure not only saves time, but also fetches the growers' international market price for the produce.

7. On-line Information Systems Dealing with e-floriculture

World commerce online

World commerce on line-Inc. is a leader in global business to business electronic commerce solutions for the perishable products industries, having the acquisition of flower. Auction.com. Flower Auction.com is an e-commerce company targeting the US floriculture industry providing interactive on line e-commerce solutions for flower growers. It enables businesses in the perishable products industry with e-commerce technology solutions. Further details about this online information system can be browsed at www.flower Auction.com.

Rosebazaar.Com

It is a global market place for flowers. It is an information and learning centre on flower trade. It is also an auction centre for flowers from around the globe. Flower shop is a unique retail shop wherein user can buy boquets, plants and flowers. Flora shop offers regional prices depending on the destination. Further details about this information system can be browsed at www.rosebazar.com

The main intention of e-commerce in floriculture is to reduce arbitrage costs by creating a virtual trade platform for the global flower business. It the fastest Real Time Auction with response speeds of 1/22nd of a second. It also ensures interactivity. It promotes the global trade and encourages bulk order which results in foreign exchange. E Commerce promotes buying flowers at our convenience. It is also an information centre. It provides a newsletter giving highlights of floriculture business around the world. The news item would provide links to authentic online sources for detailed news whenever they are available and permitted. Some of the regular features would be on foreign Exchange Rates, Ruling Market Prices for flowers, important stock market information and current trade and technical events. It also provides detailed trade articles on various subjects of interest like climatic conditions, kinds of flowers cultivated, trade statistics, market conditions government policies, exports, imports and trade directory. There are "green pages" which is world Trade Directory for floriculture which provides a list of registered companies under several sections such as growers, exporters trade houses etc.

Websites dealing with e-commerce in floriculture

- www.auctionbiz.com
- www.auctusdev.com
- www.bloomauction.com
- www.buh.nl
- www.faj.cojp
- www.florabuy.com
- www.floraholland.nl
- www.iflora.com
- www.interflora.com
- www.justflowers.com
- www.rosebazar.com
- www.stockmantra.com
- www.ufgca.com
- www.vba.com
- www.rosebazar.com

8. Flower Auctions

There are seven flower auction centres in the Netherlands. The two largest are Aalsmeer (VBA) and Naaldwjk (BVH). The remaining auctions sell speciality flowe3rs or have more local distribution. Aalsmeer, owned by a cooperative of about 5000 growers, is the world's biggest flower auction. The impressive Aalsmeer flower Auction grounds with all its facilities are the property of Co-Operative association of about 500 growers. According to Guinness Book of Records, it is the largest commercial building in the world for a single purpose. It is located near Schipel Airport, one of the Europe's largest air cargo transportation centres.

The core of Aalsmeer flower auctions is essentially found in two elements, the clock and the auction hall. The clock displays the prices by way of an electronic light; the buyers sit behind their desks where they can press a button to stop the light at the price they are willing to pay for a certain lot. The buyers are mostly whole sellers and exporters. The VBA enables them to rent space under the same roof so that they can prepare their recently purchased flowers and plants for further despatch without transportation costs and without risking loss of quality. Accommodation is also provided for parties that may be important to buyers in terms of service, such as banks, administrators etc. Originally only the products of the members were sold but in recent years imported products from other countries have also been added so that all types of floricultural products are available throughout the year for buyers. One an average, 15 million flowers and potted plants are traded daily in 50,000 transactions.

Tele VBA

Tele VBA is an electronic information system with which suppliers can call up personal information (daily price information), personal statistics (per-day and period), container information and VBA information (Statistics per-day and period) on the whole VBA assortment. It is user friendly being based on internet technology.

Flower Auction Holland

Flower Auction Holland is one of the largest co-operative flower auctions in the world. With one auction center in Naaldwijk and another in Bleiswijk, it is also right in the heart of the world's largest glass house horticultural district. The growers come from a number of countries. Quality of service and the product is of utmost importance. The co-operative includes 5600 growers and 2,500 buyers.

United Flower Growers Auction USA

The auction at the united flower growers employs state of the art computers to facilities its sales process. Large auction digital 'clocks' and reader boards indicate the starting price and sales information for each item offered for sale. The auction process starts early morning with the arrival of floricultural products which can be visually inspected for quality. As auction progresses, customers make purchase by using computerized keypads located on every disk. This sales method allows transactions to occur at amazing speed. In fact, over 4000 individual purchases, initiated by hundreds of customers, can be transacted every hour.

Bloom auction.com (Canada)

Bloom auction.com was formed to enhance the market reach of physical flower auctions by increasing participation of buyers who are far from the auctions. This is achieved by remote bidding which may or may not be synchronized with the flower auctions.

9. Floriculture and WTO

India has so far been governed by the patent act 1970, which did not permit the patenting or any form of intellectual property protection for live forms. However after becoming the signatory of the GATT agreement, it became mandatory for the Indian patent Act to be modified. India thus adapted the *sui generis* system acceptable to the rules of the land in question. According to the provisions under the article 27.3 (b) of GATT prescriptions, after a long and fierce debate across the country, India adopted the plant variety protection and farmers' Right Bill 2001, as its IPR (intellectual property Rights) for the plants and other living organizations. This feature follow the plant breeders Right (PBR) type of IPRS. The significant features include.

An exclusive right for the farmers and breeders to save, conserve utilize, use and exchange the plant mat3erials and seed without any legal obligations.

The legislation explicitly defines the variety, extended variety and derived variety along with the legal angles and implications of their use in the breeding programmes.

The legislation also provides scope for registration of varieties by farmer if they fulfil the DUS criteria. The act has been adopted w.e.f. 2001. Besides the Biodiversity Act 2001 and the geographic Indication act 1999 adopted by the Indian Government are expected to effectively cover the native flora and fauna under the *sui generis* system of IPRS. The Indian Acts have also taken appropriate care to safeguard the interests of the farmers and researchers who are the custodians of all forms of knowledge on plants and varieties.

10. Documentation Indexing

The genetic pool of flowers (both traditional and modern) shows wide variability. Much of this has not been documented or exploited. It is important that our traditional varieties and plant types need to be registered and indexed. India still has a long way to go to establish clarity in the Government policy on PRS, give IPR protection to new types and also avoid risks of selection and defamation in export of floricultural products.

Further Reading

Sheela,V.L. 2007. Peter, K V (Ed) Flowers for Trade. New India Publishing Agency, New Delhi. 369 p.

Chapter – 17

Features of A Garden

S.S. Sindhu and Sapna Panwar

Landscape architecture involves the perfect planning of the proper choice of the ornamental plants and landscape design elements. The landscape plan should be visualized as a complete plan instead of visualizing different units of the plan. The landscape models focuses on both the integrated master landscape planning and specific garden design. The role of landscape gardening in human welfare cannot be ignored as landscape designing is directly or indirectly involved in increasing the aesthetic value the city, changing microclimate, social behavioural of the peoples, checks pollution etc. Even in an developing country as India, people do not live by bread alone they also need some finer things of life like clean and green environment for staying healthy. Hence in this context, people should realise the importance of gardening through nature like parks and gardens so urban population find a place to relax and enjoy the beauty of nature. The chapter describes the very basic features of gardens.

1. Entrance Gate

The purpose of gate is to extend warm welcome to the visitors. It should be attractive, aesthetic and at same time should provide protection to the garden. It is generally made of iron in form of designer grills etc.

2. Garden Walls

Traditionally walls were meant for the purpose of giving protection to the place. But today the concept of putting walls is to give the protection as well as aesthetic appeal to the garden. Moreover, the walls should not block the view of garden but it should also present the view of the garden from outside. Sometimes walls are used to screen out the ugly

Fig. 1: Garden wall

places which are interfering with the beauty of the garden. The size of garden wall varies depending upon the style and size of the garden. Generally walls from 1.80 to 5 m may be raised upon the size of the garden. The materials used for making garden walls are stones, bricks etc. Sometimes, designer grills are erected over brick base so that inside garden view from outside is not obstructed.

3. Dry Wall

Dry walls refer to the wall or walls planted with different plants so as to give aesthetic view. The hanging plants in the dry wall give the pleasant look and introduce the softscape element. Nowadays, it is a common feature of the most of English gardens. The dry wall can be constructed independently or can be laid against vertical or near vertical surface between the two different levels of the place. The construction of wall should finish just before the onset of rainy season so that planting can be done in rainy season. The slope of land supporting the wall should be taken proper care before laying out the foundation. The wall can be raised vertically if the total wall height does not exceed 75 cm but slanting position with a slope of 1:6 is recommended in case when the height exceeds more than 75 cm.

Fig. 2: Dry wall

Plants suitable for the dry wall are: Sweet alyssum, *Gazania splendens*, *Gypsophila elegans*, *Portulaca spp.*, *Zebrina pendula* *Viola* spp., *Verbena* spp., *Phlox* spp., etc.

Some ferns are very useful as they give natural effect to the wall e.g. *Adiantum caudatum*, *Adiantum venustum*, *Polystichum* spp., *Pteris cretica* etc.

4. Green Wall

Green walls are found most often in metropolitan cities where the plants reduce overall temperatures of the building. Green wall is a wall, either free-standing or part of a building that is partially or completely covered with vegetation.

Benefits:

- Extra layer of insulation to the building.
- Keeps places cool in summer.
- Aesthetic benefits.
- Urban air quality improvement.
- Reduction in noise pollution.

Fig. 3: Green wall at the Universidad del Claustro de Sor Juana, Mexico

There are two main categories of green walls:

- **Green façades**
- **Living walls**

Green façades: Green façades are made up of climbing plants either growing directly on a wall or, more recently, specially designed supporting structures. The plant shoot system grows up the side of the building while being rooted in the ground.

Living wall: Living wall the modular panels are often made of stainless steel containers, geo-textiles, irrigation systems, a growing medium and vegetation.

Most green walls are implemented to create an element of aesthetic diversity that adds to the palette of building materials and the surfaces created. Designers of green walls can work at a great range of scales, from the creation of small private space interactions like intimate garden settings, to the monumental scale of a multi-storey project. Green walls can also be integrated into the entire building site design and utilize multiple systems and forms. Green wall technologies provide a wide range of options for designers who are interested in using the building envelope to accomplish multiple objectives and to provide new free standing design features on the interior and exterior of buildings. Living walls may also be a means for water reuse. The plants may purify slightly polluted water by absorbing the dissolved nutrients. Bacteria mineralize the organic components to make them available to the plants. Green walls are a key component of living architecture and they will become increasingly important fixtures in our cities in the years to come.

Plants suitable for green walls: Climbers and wall shrubs are the simplest way of adding interest to walls. It includes range of herbaceous perennials, grasses, herbs etc. Plants for summer interest like annuals, patio plants or bedding perform well e.g. *Begonia* 'Million Kisses Series', *Bidens ferulifolia*, ivy-leaved pelargoniums, petunias and verbena.

5. Fences

Fences are boundary elements essentially needed for protection purpose. Several materials are utilized for making fences depending upon the style and type of gardens. Most commonly used materials are wood, bamboo, wire, wire netting and chain fences. Wooden and bamboo fences mostly suited to the traditional type of landscapes. Some carving in wooden provides artistic look to the fence. The major drawback with wooden fences is that they are less durable than wire fences. To secure aesthetic view of the garden, creepers like *Pyrostegia venusta, Lonicera japonica* etc are trained over it.

Fig. 4: A wooden fence

6. Steps

Steps are the means by which different levels of the gardens can be connected. Steps are utility item and made of various materials like concrete, stone, wood etc. The steps in the garden should aim to make the climb easy and comfortable.

Fig. 5: A stone step Garden Drives

7. Garden Drives

The garden must have a proper carriage drive leading to important parts of the garden like office, nursery area, garage etc. in order to provide pleasant passage mainly to the persons or vehicles. For long drives a slope 1: 100 may be recommended. The width of road of the garden varies from 3-5m. Mostly gravel drive and concrete drive are used in the garden.

Fig. 6: A Garden drive at Lodhi Garden, Delhi

8. Garden Paths

It is the utility features of a garden which provide an access to different parts of the garden. The difference between garden drives and paths is that paths are narrower than drives and not meant for heavy traffic use. They are meant for the passage of mainly for public. The curvaceous path with gentle curves breaks the monotonous view of straight path. Paths are made of gravel, paving stones, crazy paving, bricks, or grass. A garden path should never be less than 60 cm wide but should be preferably between 90 and 120 cm, if sufficient space available.

Fig. 7: A zigzag garden path

Types of paths

- Gravel path
- Brick path
- Stone paving
- Crazy paving
- Grass paths

9.Hedge

Shrubs or trees when planted at regular intervals to form a continuous screen is called a hedge. Sometimes hedge is also made of sub-shrubs, succulents and cactus.

Benefits

- Provide protection of an area
- Demarcation of the place
- Screen unwanted and ugly sites
- Acts as windbreak and anti-pollutants

Types of hedges:

1. Protective hedge

Characteristics:

- Hardy in nature
- Quick growing

Thorny nature

- Drought resistance
- Withstand frequent pruning

2. Ornamental hedges

Characteristics:

- Attractive foliage/flowers
- Dense growth habit
- Withstand frequent pruning

Fig. 8: Hedge

A) Trees as hedge: *Acacia farnesiana, Diospyros embryopteris, Erythrina spp., Grevillea robusta, Putranjiva roxburghii, Thevetia nerifolia* etc.

B) Shrubs as hedge: *Bougainvillea spectabilis, Bougainvillea glabra, Caesalpinia pulcherrima, Clerodendron inerme, Dodonea viscosa, Hamelia patens* etc.

10. Edging

Lining of borders of paths, drives, border, lawn and shrubbery with live plants or brick, concrete, etc. is known as edging. The main purpose of edging is to provide a lining for the purpose of demarcation.

Fig. 9: Edging

Types of edging.

a) **Formal edging:** It is made of stone, bricks, tiles etc.

b) **Informal edging:** It is made of living plants.

Foliage plants for edging: *Alternanthera* spp., *Cineraria maritima, Coleus* spp., *Pilea muscosa* , *Iresene* spp., *etc.*

Flowering plants for edging: *Allysum spp., Amaryllis spp.*, *Brachycome* spp., *Gerbera* spp *Zephyranthes* spp, *Torenia* spp. etc.

11. Shrubbery

A specific area meant for growing of only shrubs is known as shrubbery. Shrubs may be planted in groups or as accent in the form of a specimen plant. Several kinds of shrubs including tall and dwarf growing, deciduous and evergreen and those flowering in different seasons and having different coloured flowers are planted in a shrubbery to give a natural, pleasing and colourful effect almost throughout the year. The common width of a shrubbery is between 3-6 m and length varies according to the space available for this purpose. The shrubbery was arranged as a walk ideally a winding from one corner to the other corner of the garden.

Fig. 10: Shrubbery

12. Herbaceous Border

It is an informal arrangement of annuals or biennials and perennials or combinations to create a diverse effect through colour, shape. The term herbaceous border is mostly in use in the United Kingdom and the Commonwealth. In North America, the term perennial border is normally used. Herbaceous borders as they are known today were first popularly used in gardens in the Victorian era. Hybridization and new imported plant species revolutionized the form of British gardens in the 18th and 19th centuries. In addition, the works of Gertrude Jekyll, a British 20th century garden designer and prolific writer, popularized the use of the herbaceous border through a revival of the British cottage garden. The world's longest herbaceous border at 215-metres (705 ft), is according to the Guinness Book of Records at Dirleton Castle, East Lothian, Scotland.

Fig. 11: Herbaceous border at East Lothian, Scotland

There are two types of herbaceous border

a) **Annual herbaceous border:** Annuals of various kinds are planted in a border which produce large variations through plant height, growth habit and colour and form.

b) **Perennial herbaceous border:** Herbaceous perennial plants are ideal for making herbaceous border because of their performance and less care required in cultivation.

Examples: Tall plants: *Solidago* spp., *Rudbeckia* spp, *Tithonia* spp.

Medium plants: Perennial forms of *Coreopsis* spp, *Phlox* spp, *Gaillardia* spp

Dwarf plants: *Portulaca* spp, *Gerbera* spp ,*Oxalis* spp *and Verbena* spp

13. Flower Beds

A flower bed is a garden area especially prepared for growing flowers. The most essential criterion of a flower bed is to display the flowers in its best form. The simple designs as circles, squares, rectangles and ellipses are easier to maintain.

Fig. 12: Flower beds

The different kinds of annuals are:

Winter season annuals: Sown during September/October, e.g. Petunia, Pansy, Phlox, Salvia, Cineraria, Corn flower, Stock, Sweet alyssum, Sweet pea, Sweet Sultan, sweet William etc.

Summer season annuals: Sown during February /March, e.g. Kochia, Gaillardia, Cosmos, Sunflower , Tithonia, Zinnia, Marigold etc.

Rainy season annuals: Sown during June/July e.g. Amaranthus, Balsam, Celosia, Gomphrena etc.

14. Carpet Bedding

Carpet bedding employs covering an area according to a set design with contrasting ornamental plants having neat, dwarf habit, texture and distinct colour so as to create geometric displays. It is often used to create letters, logos or trademarks, floral clocks etc. For contrast and designed effects, plants of one colour are grown as background and other coloured groups are utilized for foreground in form of letters, lines etc.

Fig. 13: Carpet bedding

Plants suitable for carpet bedding are *Alternenthera* spp., *Cineraria maritima, Coleus* spp., *Iresene* spp., *Portulaca* spp., *Sedum rupestre* etc.

15. Ground Covers

Ground covers refer to those plants which form a low canopy of foliage or spreading type in nature and cover the ground very densely so that weeds cannot grow under them. The low growing and trailing perennial and /or annual plants, both foliage and flowering are ideal as ground covers. Dwarf shrubs are suitable for this purpose which responds to hard pruning and trimming to keep them low growing. It is used to provide protection from erosion and drought, and to improve its aesthetic appearance of the landscape.

Fig. 14. Ground cover of *Zebrina pendula*

a) **Upright and spreading type:** These grow upright about 15-30 cm with spreading type of growth which cover the larger area in short period.

   Examples: *Duranta goldiana, Alternenthera versicolour, Iresene lindinii, Cuphea gerlonica, Opheophogon japonicas, Gazania rigens* etc.

b) **Trailing type:** *Ipomoea charyteruse*, *Ipomoea tricolour, Wadelia trilobata, Zebrina pendula* etc.

16. Topiary

It is the art of clipping and shearing shrubs and small tress and sometimes even herbaceous perennials into various decorative geometrical shapes, birds, animals, sculptures and called topiary. The term is derived from the Roman words *ars topiaria* meaning ornamental gardening. It is a common feature in English, French, Portuguese and Japanese gardens.

Fig. 15: Elephants shape topiary at Bang Pa-In Royal Palace, Thailand

Characteristics of plants used in topiary should be

- Evergreen, mostly woody, have small leaves or needles,

- Produce dense foliage,
- Compact and/or columnar shape
- Tolerant to heavily training and pruning.

Common species commonly used for topiary include *Buxus sempervirens*, *Thuja* spp., *Clerodendron inerme*, *Duranta* spp., *Polyalthia longifolia*, *Putranjiva roxburghiii* etc.

17. Creating Shapes

Wreath: Creating a living wreath isn't difficult. Start with a young plant (ivy or a pliable stemmed plant) and a wire support/frame. Plant two small plants in the container and gently bend each to follow the lines of the frame, securing it with stretch tape or twist tie. If possible, wind the stem around the wire to give it a firmer grasp and camouflage the support. Pinch the growing tip off. As the plant fills in, keep the side shoots clipped to encourage branching.

Stuffed Topiary (animal shapes): These are easier than one-dimensional forms. Acquire chicken wire, make a frame and fill with damp coarse sphagnum moss. Plant directly into the moss with small plants (ivy, baby tears or creeping fig), spacing evenly around the topiary. Keep turning the frame every few days to expose all sides to light. Keep the inner mosses evenly moist not allowing it to dry out.

Standard Topiary (lollipop): Start out with a plant that has a single stem (fuchsia, coleus, rosemary etc) stick a bamboo stake as close to the stem as possible into the soil. Tie the stem to the stake. As it grows, remove all side branches on the bottom 2 /3^{rds} of the stem leaving a tuft of growth on top. When it has reached the height you want, pinch the top and let the upper branches grow out to form a crown. All the while, the standard in-training will need constant rotating to make certain that the stem stays straight and the side branches grow out symmetrically.

Spirals and Cones: Usually, the same shaping as the standard, but not removing the bottom branches and pruning to shape into a spiral or a cone.

18. Lawn

It can be defined as piece of land thickly covered with uniformly green, soft grass. The lawn is the soul of garden or landscape. It helps to unify all the landscape elements in aesthetic way. It adds beauty and softness against walls and other concrete structures and

Fig. 16. A beautiful lawn around India Gate, New Delhi

harmonizes with the surrounding and provides a natural green colour. The lawn provides a perfect setting for a flower bed, a shrubbery or a specimen tree or a shrub. A lawn also has its spiritual value. A lawn is the source of charm and pride and reduces tension of the mind after a day's hard work in the materialistic world. It helps in reduction of air pollution, noise pollution. It provides cooling effect during summers. The lawn grass should be able to withstand regular mowing and wear tear.

Species suitable for lawn:

For temperate region: *Agrostis tenuis, Festuca rubra, Lolium perenne, Poa pratensis* etc.

For tropical and sub-tropical region: *Cynodon dactylon , Zoysia japonica , Zoysia matrella, Zoysia tenuifolia* etc.

19. Arches

The arches are mostly constructed near the gate of the garden and climbers and ramblers are trained over it. They are usually made of angle iron and are about 2 m high and 1m broad. Galvanized wire netting may be fixed on the sides as to help the creepers to climb up. The arches can be angular roofed, flat roofed and arch roofed.

Fig. 17: Arches

20. Pergolas

A series of arches joined in combined manner forming a covered walk is called pergola. It is ideal for the display of ornamental climbers and also for screening and partitioning. It is generally constructed over pathways, which add beauty to a garden. Like arches the support can be made of wooden or stone or brick pillars, angel iron and G.I. pipes. The roof may be made-up of angel iron of different sections with longitudinal and cross rails in shape of arch or flat. The width of pergolas is kept within 2-2.5 m and height is also same but preferably on the higher side and length depends upon area to be covered for varied purposes. In large size pergola, sometimes the tender plants can be kept so as to avoid hot sun.

Fig. 18: Pergola

21. Bridges

It is a necessary item in a garden to cross a stream. The size, shape, and material of bridge depend upon the type of water barrier it is to transverse. They are made of stone, wood or concrete etc. To create a magnificent view, some bridges are built much taller than necessary. This type, often found in East-Asian style gardens, is called a Moon bridge, evoking a rising full moon. A moon bridge is a highly arched pedestrian bridge associated with gardens in China and Japan. The moon bridge originated in China and was later introduced to Japan.

Fig. 19: A bridge in the Japanese Tea Garden at Golden Gate Park in San Francisco, California

22. Gazebo

A gazebo is a pavilion structure, sometimes octagonal, that may be built, in parks, gardens and spacious public areas. Gazebos are freestanding or attached to a garden wall, roofed, and open on all sides; they provide shade, shelter, ornamental features in a landscape, and a place to rest. Some gazebos in public parks are large enough to serve as bandstands or rain shelters.

Fig. 20: Gazebo at Will Rogers Gardens, Oklahoma City, UK

23. Vertical Garden

This new concept of gardening has developed in Switzerland and is now gaining popularity in India. This is very suitable for those areas where no space available for establishment of a garden and people want to enjoy the beauty of nature inside the house. It consists of a wooden/iron frame of thickness varying from 15 cm to 30 cm depending upon convenience and the height and breadth of the frame according to available space. The border paces are enclosed with wire frame which serves as the growing medium. It is possible to grow many dwarfs flowering and foliage plants in such frames provided the medium is supplied with nutrient and watered regularly. Shallow rooted plants needing very little anchorage will grow well.

Fig. 21: Vertical Garden

Many vertical gardens have been made in Delhi during Commonwealth games 2010 and the concept has become very popular and being spread in other part of the country.

Plants suitable for vertical gardens should be dwarf in nature and herbaceous like *Syngonium,* ivy, ferns, *Zebrina, Epipremnum, Tradescantia* etc. However petunia, pansy phlox can also be grown.

24. Roof Gardening

A roof garden is any garden constructed on the roof of a building. In modern times, homes with a compound and lawn especially in cities and towns are becoming rare and skyscrapers are replacing such homes. As a result the private home gardens are vanishing and the only places left for gardening are the roofs of houses and the balcony. A spacious and well planned roof garden can be a place of joy and recreation. Besides the decorative benefits, roof plantings may provide additional advantages viz. food, temperature control, hydrological benefits, architectural enhancement, habitats or corridors for wildlife, and recreational opportunities. Roof gardens are most common in urban areas. Plants have the ability to reduce the overall heat absorption of the building which then reduces energy consumption.

Fig. 22: Roof garden near island of Sentosa adjacent to Singapore

25. Planning

Thing to check before starting a roof garden

- Whether the roof surface can bear the weight of the soil especially wet soil,
- Roof should be made waterproof to prevent seepage of water into the room below.

The concept of roof gardening itself is artificial; therefore while planning and designing, some points should be kept in mind as for example it is not possible to create natural scenery with hills and valleys or natural forest scenery. On a roof the space available for gardening is limited, therefore it will be good idea to maximum encourage vertical growth, so this is the reason why a roof garden lover should use more climbers and trailers in the garden. Hanging baskets also play a major role in a roof gardening. Small lawn can be established.

26. Plant Materials Suitable for Roof Gardening

Flowering Annuals: Stock, Pansy, Dog flower, Dahlia, Marigold, Sweet alyssum etc.

Herbaceous Perennials: *Pelargonium* spp., *Canna* spp., *Portulaca* spp., *Vinca rosea*, *Verbena* spp etc

Shrubs: Many of dwarf and medium shrubs can be grown.

Trees: *Plumeria* spp., *Callistemon* spp., can be grown as a specimen plant.

Creepers: *Cobaea scandens*, *Ficus repens*, *Vernonia elaegnifolia*, *Thunbergia alata* etc.

27. Terrace Gardening

A terrace is a raised space of ground constructed around a dwelling house or at the corner of a garden or on the sides of hill. When the terraces are used for some sort of gardening this is known as terrace gardening. Generally, this type of garden is generally constructed in front of the house from where a view of whole garden can be observed. however this can also be constructed in any other places. A terrace garden is very often referred to as the outdoor living or drawing room.

Fig. 23: Baha'i Shrine and Gardens, Haifa, Israel

28. Planting

Planting in the terrace should be done after much thorough planning. Terrace gardening can be naturally undertaken in hilly areas where natural slope exits however in plains also it is possible to create artificial terraces. All the traditional Mughal gardens were laid in terraces. In home gardens, first, it is to be decided if the whole area will be paved garden or some place will be left for lawns, beds, specimen trees and creepers. In the paved areas, as a general principle, there should be little plant growth and here wider joints of the paving allow more growth of grass which is difficult to clip. Only a few low growing creeping plants such as *Portulaca* spp., *Lantana sellowiana* for the tropics and *Thymus serpylllum*, *Veronica repens* for the temperate region can be recommended for the planting in the wider joints. Small specimen dwarf trees or shrubs can be planted amidst grass in the circular, rectangular or square beds left out. Container gardening, hanging baskets can also be displayed in the terrace. Garden ornaments like sundial or bird bath, stone sculpture also add charm. Chairs and tables may be arranged for resting.

29. Water Garden

Planned planting of aquatic and semi- aquatic plants in and around a pool is known as water gardening. Water possesses natural qualities that are both tangible and intangible. Water is a strong attraction for people. The reflections of the sky and plants on the water surface, the movement of fishes in water, all gives us a very pleasant and happy feeling. Water garden may be natural or artificial. Natural water garden fit well in an informal design and are often modeled from an existing water body. An artificial water garden is usually a concrete structure, its form corresponding to style of the garden .The water garden may be formal or informal .In the formal water garden, the pool is situated in a central position and it is in a regular shape like circle, square, rectangle, oblong, oval or any other geometric form with distinct outline. The informal pool is irregular in shape with its outer edges hidden by plants or stones. It blends with any natural landscape. There are three types of pools used in a garden namely concrete pool, container or pre –fabricated pool and polythene lined excavated pool. The concrete pool is durable and can be constructed in any shape and size. Its depth can be variable being deep in the centre and shallow on the margins.

Fig. 24: Plants in water garden

30. Plants for Water Garden

a. **Surface flowering aquatics:** *Euryale ferox, Limnanthemum indicum, Nelumbo lutea, Nymphea, Victoria regia* etc.

b. **Oxygenator Plants:** *Callitriche autumnalis*, *Elodea Canadensis*, *Ranunculus aquatilis, Vallisneria spiralis*

c. **Floating plants:** *Azolla caroliniana*, *Eichhornia crassipes, Lemma gibba, Trapa natans*.

d. **Marginal Plants:** *Acorus calamus, Cyperus alternifolius, Iris* spp. *Sagittaria sagittifolia, Typha.*

31. Rock Garden

The Japanese were the first to recognize the enriching quality of stones in the landscape. Generally rock gardens are made with large stones depicting mountains etc. Most of features are made by cutting the rock. The site of rockery can be at the end of garden or path, under a tree, in the corner of the garden, against a wall etc. The sloppy area is ideal for rockery, but it can also be constructed on a flat ground. A real rock or alpine garden cannot be created at the elevations less than 2,000-2,500 m because most of the true rock or alpine plants do not grow below this altitude.

Fig. 25: A view of Rock Garden, Chandigarh

32. Plants Suitable for a Rock Garden

Cactus: *Cereus* spp., *Cephalocereus* spp ., *Mammilaria* spp., *Opuntia* spp. etc

Succulents: *Adenium* spp., *Agave* spp., *Aloe* spp . , *Euphorbia milli, Furcraea* spp. etc.

Flowering shrubs: *Cuphea* spp., *Lantana depressa* , *Malpighia coccigera* , *Plumbago auriculata Russelia juncea* etc .

Flowering annuals: *Dianthus* spp ., *Allysum* spp., *Celosia* spp., *Verbena* spp .,*Portulaca* spp. etc.

33. Healing Garden

The "healing garden" is an evolving concept gaining popularity today. What does the word heal mean in the context of a garden? If we look in the Webster dictionary the word heal is defined as "vt to make whole and healthy; to cure; to remedy, repair. - vi to grow sound." Healing is defined as "the act or process by which anything is healed or cured; the power to heal. – adj. Tending to cure or heal." (Webster, 1990, p 257).

The garden specifically designed to meet the physical, psychological, social and spiritual needs of the people using the garden. The use of the word healing in the case of a 'healing garden' encompasses these definitions to a certain extent, but instead of stressing the idea that they can cure a person, the benefits are related more to the alleviation of stress and the abilities of the space to soothe, to calm, to rejuvenate or to restore one's mental and emotional health. A main role of the space is to provide sanctuary, to allow for meditation or to evoke other qualities desired by the garden user. The World Health Organization defines health as "a state of complete physical, mental and social well-being and not merely the absence of disease or infirmity." (www.who.int/aboutwho/en/definition.html)

Fig. 26: Healing Garden at Swedish Covenant Hospital, United States

The elements of a healing garden

Natural elements: Plants that attract birds and butterflies, ornamental trees, shrubs, perennials, and ornamental grasses, annual plants, sunlight and natural shade, wind precipitation etc.

Constructed garden elements: Patio, courtyard paved walking pathways, tables, chairs and benches, landscape lighting, gazebos, pergolas or umbrellas with tables fountain, water garden, or other water features, electrical outdoor outlet for music and related activities etc.

34. Clock Garden

The floral clock garden originated with Carolus Linnaeus, an 18th century Swedish botanist. He hypothesized that flowers could accurately predict time based on when they opened and when they closed. He called it specifically the Horologium Florae (lit. "flower clock"), and proposed the concept in the 1751 publication *Philosophia Botanica*. In fact, many such gardens were planted in the early 19th century using his designs. Many plants exhibit a strong circadian rhythm, and a few have been observed to open at quite a regular time, but the accuracy of such a clock is diminished because flowering time is affected by weather and seasonal effects. The flowering times recorded by Linnaeus are also subject to differences in daylight due to latitude: his measurements are based on flowering times in Uppsala, where he taught and had received his university education.

Fig. 27: Floral Clock at International Peace Garden, North Dakota, US

The plants suggested for use by Linnaeus are given in a table below, ordered by recorded time. A "-" signifies that data are missing.

Botanical name	Common name	Opening time	Closing time
Tragopogon pratensis	Goat's-Beard	3 a.m.	-
Leontodon hispidum L.	Rough Hawkbit	by 4 a.m.	-
Helminthotheca echioides L.	Bristly ox-tongue	4-5 a.m.	
Cichorium intybus L.	Chicory	4-5 a.m.	-
Crepis tectorum L.	Hawk's Beard	4-5 a.m.	-
Reichardia tingitana L.	False Sow thistle	by 6 a.m.	10 a.m.
Sonchus oleraceus L.	Sow thistle	5 a.m.	12 p.m.
Taraxacum officinale Weber	Dandelion	5 a.m.	8-9 a.m.
Crepis alpina L.	Hawk's beard	5 a.m.	11 a.m.
Tragopogon hybridus L.	Goat's beard	5 a.m.	11 a.m.
Rhagadiolus edulis Gaertner	-	5 a.m.	10 a.m.
Lapsana chondrilloides L.	-	5 a.m.	-
Convolvulus tricolor L.	Morning Glory	5 a.m.	-
Hypochaeris maculata L.	Spotted cat's ear	6 a.m.	4-5 p.m.
Hieracium umbellatum L.	Hawkweed	6 a.m.	5 p.m.
Hieracium murorum L.	Hawkweed	6 a.m.	2 p.m.

Crepis rubra L.	-	6 a.m.	1-2 p.m.
Sonchus arvensis L.	Field milk-thistle	6 a.m.	-
Sonchus palustris L.	Marsh Sow-Thistle	by 7 a.m.	2 p.m.
Leontodon autumnale L.	Hawkweed	7 a.m.	3 p.m.
Hieracium sabaudum L.	Hawkweed	7 a.m.	1-2 p.m.
Cicerbita alpina (L.) Wallr.	Blue sow-thistle	7 a.m.	12 p.m.
Lactuca sativa L.	Garden Lettuce	7 a.m.	10 a.m
Calendula pluvialis L.	-	7 a.m.	3-4 p.m.
Nymphaea alba L.	White Waterlily	7 a.m.	5 p.m.
Anthericum ramosum L.	St. Bernard's Lily	7 a.m.	-
Hypochaeris achyrophorus L.	-	7-8 a.m.	2 p.m.
Hedypnois rhagadioloides L.	-	7-8 a.m.	2 p.m.
Trichodiadema babrata L.	-	7-8 a.m.	2 p.m.
Hieracium pilosella L.	Mouse-ear Hawkweed	8 a.m.	-
Anagallis arvensis L.	Scarlet pimpernell	8 a.m.	-
Petrorhagia prolifera L.	Proliferous Pink	8 a.m.	1 p.m.
Hypochaeris glabra L.	Smooth cat's-ear	9 a.m.	1 p.m.
Malva caroliniana L.	-	9-10 a.m.	1 p.m.
Spergularia rubra L.	Sand spurrey	9-10 a.m.	2-3 p.m.
Mesembryanthemum crystallinum	Ice-Plant	9-10 a.m.	3-4 p.m.
Cryophytum nodiflorum L.	Ice-Plant	10-11 a.m	3 p.m.
Calendula officinalis L.	Pot marigold	-	3 p.m.
Hieracium aurantiacum	Hawkweed	-	3-4 p.m.
Anthericium ramosum L.	-	-	3-4 p.m.
Alyssum alyssoides L.	-	-	4 p.m.
Papaver nudicaule L.	Iceland poppy	-	7 p.m.
Hemerocallis lilioasphodelus L.	Day-lily	-	7-8 p.m.

Conservatory

It is used for keeping shade or partial shade loving plants. Examples are ferns, orchids, palms, cacti and succulents, *Alocasia* spp., *Anthurium* spp., Foliage plants like *Dieffenbachia* spp., *Philodendron* spp., *Dracaena* spp., *Maranta* spp. etc.

35. Greenhouse

It is a covered structure for growing plants inside with complete or partial climate control. The modern greenhouse is climate controlled for cultivation of crops under protected environment, particularly when it is not possible to grow the crop in the open due to unfavourable climatic conditions. High value low volume crops are most suited for greenhouse cultivation.

36. Garden Adornments

In a garden several adornments should be provided to make it look more ornamental and useful. For easy understanding, you can see these things in any.

Garden seats: These are necessity in any types of garden. The seats should be comfortable, durable, artistic looking, and should not look out of place. Wooden and fabric seats are more comfortable to sit as compared to those built with stone or iron.

Fig. 28: Garden seat

Ornamental tubs, urns and vases: Plants displayed in ornamental tubs or urns at suitable places looks beautiful. It can be made of timber, concrete, stone etc. It can be fixed permanently in a place or sometimes it can be movable also.

Fig. 29: Vases

Bird baths: These are nothing but large bowl-shaped containers, generally made of concrete, fixed over a pillar or column which is about one meter tall. Water is stored in the bowl for the birds to come and drink or bath in it. Bird baths should be constructed at a quiet corner of the garden. A birdbath is an attraction for many different species of birds to visit gardens, especially during the summer and drought periods. Birdbaths that provide a reliable source of water year round add to the popularity and "micro-habitat" support.

Fig. 30: Bird bath

Sundials: A sundial can be used as focal point in a garden, can be placed at the end of lawn and at the junction or termination of a path. In India it can be made with the digits. It should be positioned in a place where the shadow from a tree or building does not fall for a long duration, otherwise the feature becomes meaningless. Before fixing the dial, the movement of the sun should be observed and then the dial is screwed to the base. Sundial is used in the garden to know the time.

Fig. 31: Sundial in Singapore Botanic Gardens

Fountains: Fountain can also be used as focal point. Unless there is water under pressure one cannot have any good fountain. Fountains are made to work by circulating the same water contained in a pool. Today various designs of fountain are available i.e. straight upright water jet or finer water jets converging in the form of an umbrella. Now day's musical fountains are gaining popularity.

Fig. 32: Musical fountains at Akshardham Temple, New Delhi

Fig. 33: The Four Godesses of the Seasons Statues

Statues: A statue is a sculpture of animate or inanimate objects in the gardens. The statues may be made up of stone, copper, bronze, concrete etc. Artistically accenting a garden with extensive selection of statues improves the aesthetic value of the garden. Like in Japanese type garden we generally found the statue of Lord Buddha. For example Buddha jayanti parks, New Delhi have a big statue of Lord Buddha.

Fig. 34: Ornamental stones

Ornamental stones: These stones of various colours and designs can be placed near polls, along or in the midst of various streams and waterfalls. It improves the look of the area to a great extend. Large ornamental piece can also be played at the corner of the doorway and other suitable places. The ornamental stones are most common in the Japanese gardens.

References

Arora, J.S. 2004. *Introductory Ornamental Horticulture,* Kalyani Publishers

Bose, T.K., Maiti, R.G, Dhua, R.S. and Das, P. 1999. Floriculture and Landscaping, Kolkata, India, Naya Prakash.

Dubey, K.N. and Singh, S.B. 1985. Ornamental gardens and their layout, *In: Olericulture and Floriculture*, Bharati Bhandar, Baraut, Meerut,pp: 99-112

Furuta, T. 1983. Interior Landscaping, Reston, Virginia, Reston Publishing Company, Inc.

Lancaster, S.P., Bose, T.K. and Mukherjee, D. 1997. *Gardening in India*, Oxford & I BH Publishing Co. Pvt. Ltd., New Delhi.pp 1-15.

Randhawa, G.S. and Mukhopadhyay, A. 1996. *Floriculture In India*, Allied Publishers Ltd, New Delhi, pp 1-631

Sindhu, S.S. 2016 (Ed) Ornamental Horticulture.New India Publishing Agency,New Delhi 485p.

Swarup, V. 1997. *Ornamental Horticulture*, Macmillan India Ltd, Daryaganj, New Delhi, pp 1-449.

Chapter – 18

House Plants

S.S. Sindhu and Sapna Panwar

House plants are valuable ornamental that provides decorative value and aesthetic sense to the living place. Today with the increase in urbanization and modern life style, the house plants occupies most distinguished position. They compliment interior settings and also bring life to indoor environment. Besides interior decoration, they also contribute their role in environmental security. Air quality is also not free from air pollutants and generates number of serious health problems. They can be very beneficial for our living environment by purifying and renewing our polluted air. They acts as natural air filters by absorbing various toxic compounds, pollutants etc and also contribute to balance the humidity. They are important component of Interiorscaping and provides softscape element to the hardscape elements of the concrete structures.

1. Placement

These attractive house plants rightly deserve their place as interior decorations. "Shade loving" is an adjective often used to describe indoor plants, but this is a misnomer, Plants may be shade tolerant but they do not actually seek poor light. Diffused light through the glasses or window's can be effective. Supplement the sun's rays with artificial light in the form of a "warm white" fluorescent tube. Some plants may be shifted at frequent intervals in the morning sun to fulfil their light requirements.

2. Growing Media

The growing media must be of best quality. The media should provide be porous so that it gives good root aeration. At the same time, it should be well drained and should have good retention of both water and nutrients. More commonly used media should be locally available like soils but nowadays artificial media containing inert material are also being used for growing house plants. The growing media should possess the following properties:

- It should be easily available
- It should be cheap
- It should provide good aeration
- It should be inert in nature
- It should be well drained
- It should be free from disease and pests
- It should retain appropriate moisture content
- It should posses good physical and chemical properties

Most common components of growing media are

(i) **Vermiculite:** Vermiculite is a light weight, sterile, micaceous product having good amount of K and Mg. The ph is usually 7.0 to 7.5. The main benefit of using is its tendency to compress and loss drainage enhancing quality with time.

(ii) **Perlite:** Perlite is sterile and light weight, porous and light rock material of volcanic origin. It is white in colour. The ph usually ranges from 7.0 to 7.5 It does not hold water and nutrients. It is essentially high expanded aluminium silicate rock and improves aeration and neutral in reaction.

(iii) **Coco peat:** It is prepared from the coconut husk. It has very high porosity and very good drainage quality. It is free from pathogen and provides good aeration.

(iv) **Sand:** Sand should be sufficiently gritty and with a good particle size. The main disadvantage of sand is that it is very heavy and evaporates moisture too quickly. It provides aeration.

(v) **Soil:** Soil is commonly used growing media. It should be rich in nutrients. Soils have generally higher water retention and cation exchange capacity. The major disadvantage is that it is too heavy and contain various harmful microorganism that causes diseases. So, soil sterilization is needed prior to planting.

3. Media Preparation

A good medium for house plants should contain most commonly locally available media components. Recommendations that are generally followed are garden soil-2 parts, organic matter-1 part, sand-1 part and leaf mould-1 part. The ratio of media components can be changed depending upon plant type and environmental conditions. Nowadays most commonly used media is 3:1:1 of coco peat, vermiculite and perlite. Some of the media that can be prepared especially suitable for particular crop

House plant mix-1 (1:1:1): 1 part peat moss, 1 part vermiculite and 1 part perlite. This media is suitable for growing succulents that don't require extreme dryness.

House plant mix-2 (2:1:1): 2 part peat moss, 1 part vermiculite and 1 part perlite. This is good for non-woody foliage plants.

House plant mix-3 (3:2:1): 3 part sphagnum moss, 2 part vermiculite and 1 part perlite. This is best for shrubbery, woody plants and most vines.

4. Containers

Clay or terracotta pots as are commonly known as are the best kind to grow house plants. Earthenware pots are most commonly containers used in form of dishes, pots, shallow bowls and troughs. A variety of potting containers available in varied designs and colours. Metal receptacles of brass and copper are also used to enhance the aesthetic look.

5. Potting

- Potting mixture suitable for these plants should consist of 2 parts of garden soil, 1 part of well decomposed organic manure which may be vermicompost l/4 of sand
- Mixture thoroughly and fill each pot, some additional requirement for pots of 25 to 30 cm size is about 150 gms of coco peat or leaf mould powder 100 gms of Neem cake powder 100 gms of insecticides may be anti termite powder or any other branded insecticide to prevent soil borne infections.
- While filling the pot, cover the drainage hole in the base with some crocks (broken pieces of earthen pot).
- Take out the plant along with the root ball of earth and put in the soil mixture, making a hole to accommodate the plant and press properly.
- Fill the edges with growing mix and tap the pot to shake it down well, and firm it with your fingers.
- Finally water the plant carefully.
- Keep the newly potted plant in shade and semi shade for about a week.

6. Propagation

The method of propagation varies from plant to plant. The house plants can be propagated by sexual or asexual method of propagation. The following are the examples of the plants propagated by various methods

I. Seeds: Asparagus, Cacti, Azalea, *Impatiens, Kalanchoe, Calceolaria, Cyclamen, Gloxinia, Coleus, Schefflera, Araucaria, Begonia, Aralia*, African violet, Palm, Philodendron, *Cineraria,* etc.

II. Stem cuttings: They are further divided into subcategories depending upon texture of plant tissues

(i) **Herbaceous cuttings:** *Portulaca, Kulfa, Pothos, Coleus.*

(ii) **Soft wood cuttings:** *Euphorbia splendens, Pentas lanceolata, Pedilanthus*, *Eranthemum, Tradescantia, Alternanthra.*

(iii) **Semi hardwood cuttings**: Acalypha, Crotons, *Hamelia patens*, Jatropha, *Dieffenbachia, Crinum asiaticum, Dracaena.*

(iv) **Hardwood cuttings**: *Ficus lyrata, Ficus elastica, Dracaena, Bougainvillea, Hibiscus, Duranta.*

III. Through leaves

(i) **Leaf cuttings:** Kalanchoe, African violet, Rex begonia, Begonia, Peperomia. Sansevieria, Bryophyllum etc.

(ii) **Leaf petiole:** Begonia, Peperomia, Scindapsus, Gloxinia and African violet

(iii) **Leaf lamina:** Begonia, Peperomia.

(iv) **Leaf bud:** Kalanchoe, Poinsettia, Hydrangea etc.

IV. Cane or nodal cuttings: Dieffenbachia, Aglaonema, Monstera , Dracaena, etc.

V. Air layering: Rubber plant, Hedera helix, crotons.

7. Watering

More house plants are killed by kindness then by neglect. And the most common sort of kindness is to give them too much to drink. The amount of water a house plant needs depends on the conditions you keep it in and on the weather. The role of plant roots is to forage for soil water and the nutrients that are dissolved in it. But the growing roots also need air. As the soil becomes wetter and wetter, the water drives air from the space between soil particles and the soil eventually becomes waterlogged. When this happens, roots will first stop growing, then rot.

Ways of watering

There are generally two ways of watering

1. Top watering
2. Bottom watering

Top watering

In this method, the pot is watered thoroughly with a hose from top. Here the excess water flows out of the container through the drainage hole. The main advantage of this method is that salts that are trapped in the soil are leached out with the excess water.

Bottom watering

In this method soil absorbs water from below surface of the pot. The main advantage is that medium absorbs appropriate quantity of water thus reduces wastage of water. Generally followed by professionals. The main drawback is that salt build up is inevitable in this method. The examples of bottom watering are:

- Capillary mats
- Wick watering
- Pots with attached saucers

8. Feeding

A regular feeding at least once in a month with a handful of Neem cake powder or vermicompost or coco peat in rotation provides regular diet to the plants. Feeding a dosage of about 250 gms. Mixture of all the three organic manure with a tablespoon ful of Bone-meal added to it that during spring and monsoon keeps the house plants in their full health.

9. Diseases

The most common fungus which attack the pot plants are *Alternaria, Cercospora Collectotrichum Fusarium, Phytophthora, Rhizoctonia Sclerotinia* etc. The most common bacteria that attack house plants are Erwinia, Xanthomonas and Pseudomonas.

Insect pests: Aphids, Mealy Bugs, Scales, Thrips, Mites, Slugs and snails, Nematodes

10. Precautions

Success with your house plants depends on how you choose them and on how you treat them from the first moment on. Make sure when selecting plants from the nursery that they are "hardened off" to cope with the ups and downs of the prevailing climatic requirement and specific house hold environment. Look for plants that are sturdy and make sure there are no signs of damage or disease. Most expensive plants are not necessarily the best specimen. Ensure that in the process of shifting or transportation, they are not subjected to shock or suffer from draughts and subjected to direct sunlight and take care not to over water them.

11. Indoor Plants

1. **Aglaonema:** are decorative plants with mottled leaves. It can grow in poor light conditions as well. It needs regular watering.
2. **Aspidistra**: a hardy house plant which can grow with ease in shade.

3. **Areca palm**: A very decorative house plant popular and graceful which canwith stand indoor conditions reasonably well.
4. **Spider plant**: (chlorophytum): This is one of the easiest and most popular of all house plants. It is practically indestructible, and reproduces itself freely, a ready source of gifts for friends. It will tolerate a wide range of temperatures
5. **Dracaena varieties**: A number of exciting foliage plants come under this name, different species of the same genus. With wide range of selection this can give you much colourful and green foliage.
6. **Philodendrons**: The most philodendrons make excellent house plant. It can be grown in a pot of any size and can be trained on a moss stick and hydro-ponic Another similar looking but with dark green leaves is sweet heart Vine also being known as oxycadium Grows best in shade.
7. **Ferns**: An excellent and easy plant to grow indoors. This is very attractive with different kinds of green colour and foliage.
8. **Umbrella tree** (*Scheffelera actinophylla*): Can grow tall with broad leaves but very hardy indoor normally called Brassia. But quite a few species of schaffelera makes lovely and hardy house plants including the one with variegated leaves.
9. **Mother in law's tongue** (*Sansevieria)* despite its name this is a very pleasing house plant. It has long narrow leaves. Very tolerant of most environments A very good house plants for cleaning of indoor air pollution
10. ***Hedera helix*:** Common Ivy as is known can grow indoors. A climbing by habit, it can be trained and supported over the frame or even grown in pots. To fill the pots. An ideal plant for hanging baskets or cylindrical containers for being arranged as hanging house plants.
11. ***Begonia rex:*** The plant is very attractive in its look**.** Leaves are marbled, mottles and blotched in various patterns in green, purple, red, brown or silver colours.
12. ***Scindapsus aureus***: Known commercially as "pothos" ,. It is viny and climbs tall by rootlets; leaves heart shaped, waxy dark green , yellow variegated.
13. ***Euphorbia pulcherrima***: It is a branching shrub grows up to 3 m height, The decorative value of the plants is the terminal shoots forming dark red , velvet, lanceolate bract leaves surrounding the tiny yellowish flowers.
14. ***Spathiphyllum***: It posses glossy green , thin leathery, oblong lanceolate leaves, inflorescence on reed like stems. The ovate spathe, white turning green with age , 8-10 cm long; maize like white spadix also looks attractive
15. ***Monstera:*** It is woody stemmed, climber with hanging aerial roots having hanging thick leathery leaves which are glossy green, pinnately cut and perforated with holes. The leaves of a young plant are heart shaped and without holes. They often overlap and clinig closely to a tree trunk.

16. ***Codiaeum variegatum:*** Crotons are beautiful tropical ornamental shrubs. The ornamental value is of leaves. Leaves are thick leathery, glabrous, ovate to linear, entire or lobed or spirally twisted, 10-30 cm long, variegated in beautiful colours . The green and yellow colour of leaves often change to shades of red.

Chapter – 19

Care of Trees

Sabeena Jose

Trees are the largest plant components of gardens and landscapes. However much architectural features dwarf them, for human beings, they are the largest living entities in the environment. Profoundly influencing our habits, styles of habitation, culture and daily life, trees have been revered, sanctified and given symbolic importance and above all, depended upon as means of livelihood by entire communities through the ages. Through folklore, traditional knowledge and rituals, the usefulness of trees has been handed down from generation to generation-in societies all over the world. In India, cave paintings and carvings in addition to ancient literary works document the aesthetic and medicinal knowledge of our forefathers about trees found in the sub-continent.

Trees differ in stature, branching habit, girth of trunk and branches, canopy structure, outline and crown shape. The colour of the foliage, its texture, glaze, shape and size, foliage colour, floral features and season of flowering also differ among trees. These characteristics in addition to giving distinctiveness, also determine their ornamental use in gardens and landscapes. In tropical gardens evergreen dicotyledons and palms dominate among the trees, whereas in temperate situations evergreen conifers are selected over the deciduous dicots for tree planting. Palms among trees are universally grown as avenue trees or as clumps in areas where they can be easily grown while the other trees are grown as such or pruned and trained into various forms or as bonsai in gardens.

1. Classification of Trees

Classification Based on Stature

The stature attained by trees at maturity enables them to be classified as Large trees, Medium-large trees and Small trees. Large trees are those, which attain heights of over 15m. Medium large are those, which attain heights of 8 to 15m and Small trees are those that attain less than 8m height.

Large trees: *Azadirachta indica, Ficus religiosa, Ficus bengalensis, Delonix regia, Enterolobium saman, Couropita guianensis, Casuarina equisetifolia, Millingtonia hortensis, Grevillea robusta.*

Medium large trees: *Nyctanthes arbor-tristis, Jacaranda mimosaefolia, Lagestroemia flos-reginae, Filicium decipiens, Ficus retusa, Crataeva nurvala. Citharexylum quadrangulare, Cassia fistula.*

Small trees: *Muntingia calabura, Saraca indica, Melia azaderach, Caesalpinea coriaria, Pisonia alba.*

Classification Based on Canopy Outline and Form

The arrangement of branches and the size, shape, texture and orientation of leaves give trees a particular form and canopy outline. Trees that attain their maximum stature unencumbered, are able to develop their distinctive characteristics fully. Many coniferous trees exhibit an architectural form, which enable them to be classed as having pyramidal or conical (Araucarias, Cypresses), fastigiated, or columnar (*Polyalthia longifolia* var. *pendula*, Junipers) outlines. In most of the dicotyledonous trees, canopies are spreading (*Samanea*, *Delonix*, Divi-divi), open-headed (*Ficus religiosa, Jaccaranda,* the Cassias, *Tabebuia*), or rounded or dome-shaped (Palms, Cannon-ball tree, the Fiddle-wood tree). A weeping crown with drooping branches and pendant leaves are unique to contain species of pines, willows and cypresses and to the Bottlebrush and the Weeping fig.

Classification Based on the Decorative Nature of Leaves and Flowers

Trees with inconspicuous flowers but having ornamental foliage are called **Foliage trees**. They are primarily grown for shade and the decorative nature of their foliage cover. Trees with conspicuously ornamental flowers are called **Flowering trees**. Their nature of flowering and the size, colour and number of blooms per group or bunch make them spectacular or be noticed for their flowers.

Classification Based on Floral Characteristics

Though trees are large entities, the flowers of very few trees are more than 3cm in diameter. Relatively large sized flowers are seen in the Tulip tree, the Portia tree, the Cannon Ball tree, the Gul Mohr and in the jacarandas. Most trees have small flowers in fairly small sized clusters produced all over the canopy. The size of the clusters and their colour contrasts with the foliage or the absence of the foliage at flowering make them conspicuous and more or less attractive. Spectacular flowering trees of the tropics, which flower in a decidous or semi-decidous state, include the cassias, tecomas, the jacarandas and the sacred Barna.

i) Classification based on the season of flowering

The months of March to May which heralds Spring in Northern India and in the hills of Central and Southern India and at the same time denotes the onset of summer in the rest of India is the main flowering season of trees in the sub-continent. The cassias, *Bauhinia*, *Lagestroemia*, *Michelia*, Gul mohr, Asokam, and *Citharexylum* are among the spring/summer flowering trees. In response to the lowering of temperatures in December, flowering takes place by January in *Amherstia*, *Bombax* and *Mimusops*. The rainy season from June to August sets off flowering in *Couropita* and *Nyctanthes* and a second burst of flowers in *Peltophorum* and *Michelia*.

ii) Classification based on floral scent

Intoxicating fragrances to pungent aromas are associated with flowers of trees. They may be delicately scented as in the honey-scented flowers of Neem and Persian lilac or lightly scented as in the flowers of cassias. Strongly scented are the flowers of the Cannon – Ball tree, the Golden Champa, the Coral Jasmine, the Ylang-ylang and the Plumerias. Among fragrant flowered trees unique with a spicy scent are the flowers of *Alstonia scholaris*.

Classification Based on Foliage Characteristics

Based on foliage size, surface texture and leaf orientation, trees may have densely filled crowns as in the Copper-pod tree, the Noble Amherstia and the Cannon – Ball tree while the crown may be sparsely filled as in the Bottle brush or the *Casuarina*. The surface texture presented by the foliage may be coarse as in the Cannon-Ball tree, the Cassias and the *Citharexylum* or fine textured as in *Casuarina*, the Divi-divi, the Junipers and the Thujas.

Classification Based on Suitability for Gardening and Landscaping

a) For roadsides and highway avenues

Large flowering trees: *Peltophorum*, Pink Cassias, Indian Laburnum, *Lagestroemia*, *Spathodea*.

Small flowering trees : *Bauhinia*, *Michelia*, *Tabebuia*, *Saraca*

Large foliage trees: *Enterolobium* , *Azadirachta*, *Adenanthera*, *Grevillea*, *Mimusops*, Tamarind

Small foliage trees: *Muntingia*, Divi-divi, *Pisonia alba* and small leaved Ficuses.

b) For central avenue planting

Polyalthia longifolia var. *pendula*, Traveller's palm, *Araucaria* spp. *Cupressus* spp.

c) Group planting for Grove-effect

Cycads, suckering Palms, Pines.

d) As specimen plants

The Baobab tree, the Bodhi trees, the Cannon-ball tree, the Pagoda trees.

e) For Bonsai

Cupressus spp., *Ficus* spp., *Pinus* spp.

2. Flowering and Foliage Trees of Indian Gardens

Ornamental flowering trees and foliage trees popularly grown in gardens and used for landscaping in India are listed below:

Adansonia digitata L. Bombacaceae

African Calabash tree, Baobab tree, Monkey bread tree (Eng.), Gorak Amli, Gorak Lichora (Hindi/Guj.), Gorak Chinch (Mar.), Anaipuli (Tam.).

Trees are deciduous and native to the western coast of Africa. They are considered to be very long living. The trunk attains disproportionate stoutness-of up to 10m – and tapers upwards abruptly. The bark is smooth and gray. Leaves are palmate. The large white flowers are about 15 cm in diameter and have purple stamens.

Fruits resemble long gourds and are eaten by monkeys. Trees were introduced into Botanic gardens in India as specimen plants. But in cultivation the trunk does not attain the spectacular proportions observed in its native habitat.

Adenanthera pavonina L. Fabaceae

Bead tree, Coral-wood tree (Eng.), Rakhta kambal (Beng.), Moti Chanothi (Guj.), Ratan Gunj, Thorli Gunj (Mar.), Bandiguruvenda (Tel.),Anilkundumani (Tam.), Manjadi (Mal.).

Trees have spreading branches and have a rough grayish coloured bark. They are native to China and Malaya. The heartwood is coral red in colour. The leaves are dark green and bipinnate. The yellow fragrant flowers are produced in axillary panicles from March to May. The long curved pods contain scarlet lenticular seeds. The pods dehisce on the tree and the two halves contort and curl exposing the bright red seeds adhering to the silvery white inner surface of the pods. These are popular avenue trees of public parks and gardens.

Alstonia scholaris (L.) R. Br. Apocynaceae

Devils tree, Dita bark (Eng.), Saptaparni (Sans.), Chatian (Hindi), Chattim (Beng), Satvin (Mar.), Palaigh (Tel., Tam.), Eyzhilam Paala (Mal.).

A tall elegant tree attaining a height of up to 15 m with horizontal branches in whorls making it an ideal shade tree. The basal portion of the trunk becomes buttressed with age. The flowers are villous, greenish white opening at night and have a pungent spicy scent. Leaves are usually seven in a whorl. All plant parts contain bitter alkaloids. Not preferred for planting in gardens due to its legendary association with evil spirits. Hence, confined to roadsides and untended lands in the humid tropical belt of South India.

Amherstia nobilis Wall. Fabaceae

Noble Amherstia, Queen of Flowering trees

Trees are evergreen and attain heights of up to 15m and carry a dense crown of elongated pinnate leaves which when young are coppery pink and contrast beautifully with the mature dark green leaves. The orange-red flowers are tipped yellow and hang in pendant racemes. These flaming and drooping inflorescences make *Amherstia* one of the most beautiful trees of gardens. Though plants thrive in the high rainfall belt of South India, the fail to flower as profusely as in the North.

Amoora rohituka (Roxb.) Wight & Arn. Meliaceae

Sohaga (Hindi), Amoor (Beng.).

A tall evergreen adapted to moist shady locations along the western coast of peninsular India. The leaves are oblong, lanceolate, imparipinnate and in 4 to 8 pairs along with the odd leaflet. The flowers are small, white, borne on stiff axillary spikes. The fruits are globular or egg shaped. They split open when ripe, exposing the deep orange coloured aril.

Anthocephalus cadamba Miq. Rubiaceae

The Kadamba tree (Eng.), Kadamba (Hindi / Guj.), Kadambamu (Tel.), Attutek (Mal.).

A large deciduous tree attaining a height of up to 20m with a spreading crown and drooping branches. The leaves are long stalked and cordate-ovate in shape. The globose yellow flower heads are delicately scented and emerge in June-July. They are considered to be harbingers of the monsoons. The fruits are creamy yellow and globose, shaped like an orange. In India, trees are revered due to their association with Lord Krishna. They are highly susceptible to frost.

Araucaria cookii R. Br. ex D. Don Araucariaceae syn. *Araucaria columnaris*

Araucaria, Cook Pine

An elegant conifer with horizontally oriented symmetric branches carrying dark green closely spaced claw like needles. Plants thrive in the tropical belt of South India and attain a height of up to 60m but produce cones at high altitudes only. The architectural features of this tree make it a prized plant of home gardens. In India, trees are majestically displayed in avenues and as specimen plants in public parks and gardens. Other species introduced into India having similar canopy structure include the Bunya-bunya Pine *Araucaria bidwillii* Hook. and the Norfolk Island Pine *Araucaria heterophylla*

Azadirachta indica A. Juss. Meliaceae

Margosa (Eng.), Neem (Hindi / Beg.), Limdo (Guj.), Arya Veppu (Tel./ Tam./ Mal.).

Trees are the best-known among the medicinal trees of India. They are considered to be sacred and are grown throughout the country. They are medium to large sized attaining heights of up to 15 m. Leaves are imparipinnate, lanceolate and oblique with a broadly toothed margin. White honey scented flowers are produced in small axillary panicles. The oval fruits are single seeded. They are relished by birds when ripe. All plant parts except the ripe pericarp is bitter in taste. Trees are ideal for avenue planting along highways.

Bauhinia purpurea DC. Ex Walp. Fabaceae

Purple Bauhinia, Camel hoof tree (Eng.), Gulabi Kachnar (Hindi / Guj.), Dev Kanchan, Rakhta Kanchan (Beng.), Karath (Punj.), Boroda (Oriya), Atmatti (Mar.), Kanchanam (Tel.) Basavanapada (Kann.), Mandari (Tam.), Chuvanna Mandaram (Mal.).

A beautiful deciduous flowering tree with upright slightly crooked branches. The leaves are rounded, two lobed with an overall resemblance to camel hoofs. In northern India, leaf shedding occurs before March. In the high rainfall belt of southern India, plants are seldom leafless at flowering. Flowering starts in March and continues up to November. The fruits are flattened and carry five to six seeds. Trees are ideal for planting in residential avenues and small gardens. *Bauhinia variegata* with pale pink flowers is also grown. *Bauhinia racemosa* having white flowers occur in deciduous forests in India are referred to as Mountain Ebony.

Bignonia megapotamica Spreg. Bignoniaceae

Rio-Grande Trumpet flower

Plants are attractive evergreens with shiny leaves growing to a height of 12m. Native to Brazil, they were introduced into India as shade trees for gardens and avenues. The flowers are pale pink, typically trumpet shaped and are produced in small clusters during March-April.

Caesalpinia coriaria Willd. Fabaceae

Divi-divi

A small sized evergreen tree with an umbrella like canopy formed by low spreading branches. The leaves are dark green, pinnate and feathery in appearance. The flowers are greenish white, and the fruits are characteristically twisted. In its native habitat in central America, the fruits are used for tanning and dyeing. A popular shade tree planted around the borders of playgrounds and stadiums.

Callistemon lanceolatus (Sm.) Sweet Myrtaceae

The Bottle Brush tree

A slow growing medium sized tree attaining a height of 8m or more with characteristically drooping branches. The leaves are narrow and needle like and provide sparse shade beneath trees. The flowers with their showy red stamens are borne terminally on older branches giving them the appearance of bottlebrushes. Flowering extends from February to November. Ideal for home gardens as it can be trained to the stature of a small tree.

Cananga odorata Hook. Annonaceae
Ylang-ylang, Lang-Ilang

Trees are slender evergreens native to Burma and Malaya having gracefully arching branches. The scented greenish yellow flowers are produced in axillary clusters. The ovoid fruits are green and turn yellow on ripening. Flowers are used for extraction of perfumery oil.

Cassia fistula L. Fabaceae

Indian Laburnum (Eng.), Amaltas (Hindi), Bander lathi, Sonali, Sondal (Beng.), Sonaru (Assa.), Sunari (Oriya), Alash (Punj.), Garmalo (Guj.), Bhava (Mar.), Rela (Tel), Kakke (Kann.), Konnai (Tam.), Kani-konnai (Mal.).

Trees are the most spectacular among the yellow cassias grown in India. They are naturalized in deciduous forests up to an elevation of 1500m above M.S.L. They attain heights of up to 10m and are deciduous to semi deciduous in different parts of the country. The slender branches have a drooping habit and carry long pinnate leaves with 4 to 8 pairs of leaflets. The bright yellow rounded blooms are produced in pendant panicles when trees are leafless in March – April or along with the new leaves from late April to May. The pods are long and cylindrical turning black on drying. The tree has the status of being the State Flower of Kerala.

Cassia javanica Linn. Fabaceae

Java Cassia

A medium sized tree attaining a height of 15m. The spreading branches and branchlets give trees an umbrella shaped canopy. The pinnate leaves have smooth and silky textured leaflets with rounded tips. Trees are completely deciduous in the dry tracts and semi-deciduous in humid regions of India. The pink flowers appear with the new leaves in May. The pods are cylindrical and contain flattened seeds.

A native of Java and Sumatra, trees were introduced into Indian gardens for their spectacular flowers.

Cassia nodosa Buch.- Ham. ex Roxb. Fabaceae

Flowering Cassia, Pink Cassia

Trees are relatively large deciduous evergreens and are similar in branching habit to *Cassia javanica*. The pink flower clusters arise from nodes when the trees are leafless from May to June. It is a particularly beautiful species suited for planting in large gardens and parks.

Cassia renigera Wall. Fabaceae

Burmese Pink Cassia

Trees are deciduous, attaining heights of up to 9m in Indian gardens. Upright branches bearing drooping branchlets give them an attractive canopy. The pinnate leaves bear feathery and downy leaflets. The clusters of rose-pink flowers arising from the nodes are lightly scented. Older flowers gradually fade to a white colour resulting in varying shades of pink and white flowers in the same cluster. Trees are ideal avenue trees for roadsides.

Casuarina equisetifolia L. Casuarinaceae

Beef wood tree, the Casuarina (Eng.), Jangli Saru (Hindi), Jhav (Beng.), Vilayati Saru (Guj.), Suru (Mar.), Sarugudu, Saruku, Chowka (Tel.), Chavukku (Kann.), Savukku (Tam.), Kattadi (Mal.).

A tall tree attaining heights of up to 30m with longitudinally exfoliating bark. The leaves are 6 to 8 jointed needles. Plants produce cones profusely from February to October. Trees are ideal for training as barrier hedges along garden boundaries. They have proven to be suitable for afforestation of sandy tracts and as salt resistant windbreaks along the seacoasts of South India. In *Casuarina* groves, the passage of wind through the branches and leaves produce a characteristic audible murmur.

Citharexylum quadrangulare Jacq. Verbenaceae

Fiddle-wood tree (Eng.), Paarijatham (Mal.).

Trees are upright, producing slender branches all around and attaining heights of up to 10m. Drooping racemes of small fragrant white flowers are produced in leaf axils towards the tips of branches. Plants remain evergreen in the humid tracts of peninsular India.

Couropita guianensis Hook. Lecythidaceae

Cannon-ball tree (Eng.), Topi gola (Hindi), Naga linga (Beng.), Kailaspathi (Mar.), Nagaalingam (Tam. / Mal.)

A tall erect tree attaining heights of up to 20m.The branches are pendant and leaves are crowded at the tips. The leaves are alternate, simple, obovate, glabrous and are shed once or twice a year. The flushes are light green in colour. The scented flowers are borne on long woody cauliflorous racemes and have a fascinating structure. The flower buds are rounded in shape. The sepals and petals are cupped and coloured in tints of pink and yellow. They surround the essential organs consisting of the gynoecium which resembles a "Shiva Lingam" over hung on one side by a snake's hood like structure formed by fused stamens. The divine significance attributed to these flowers makes this tree a specimen tree of gardens. The fruits are brown and cannon ball like and hang pendant close to the trunk. A native of tropical America, the tree was introduced for providing shade in public parks and gardens.

Crataeva nurvala Buch.-Ham. Capparidaceae

Caper tree, the Sacred Barna (Eng.), Barna (Hind / Punj.), Barun (Ass.), Barum (Beng.), Baruna (Oriya), Varno (Guj.), Haravarna, Vayavarna (Mar.), Mavilinga (Kann.) Maralingam (Tam.), Neeer mathalam (Mal.).

Native to deciduous forests, trees are distributed in Northern Kerala and attain heights up to 10m. They are planted near sacred places. The trifoliate leaves with ovate-lanceolate leaflets are shed in December- January. Small white flowers in dense clusters produced when leafless, make trees singularly beautiful. The flowers turn yellow with age and are shed before refoliation. Fruits are globular, woody and contain angular seeds with hard seed coats.

Cupressus spp. Cupressaceae

Cypress trees

Several species of evergreens with architectural canopy outline are grown as avenue trees in gardens located in the temperate belt of India. They include the mourning or weeping cypress (*Cupressus pendula*), the Kashmiri cypress (*C. cashmeriana*), the Monterey cypress (*C. macrocarpa*), the Mediterranean cypress (*C. sempervirens*), the Italian cypress (*C. fastigata*) and the Bhutan cypress (*C. nepalensis*).

Delonix regia Rafin. Caesalpinaceae

The Flamboyant tree (Eng.), Gul Mohr (Hindi / Mar. / Beng. / Guj.), Krishnachura, Radhachura, Rakhthachura (Ass. / Oriya), Bhima Sankesula (Tel.), Mayooram (Tam.), Alasipoomaram (Mal.).

Trees are large with stout trunks and a spreading canopy and attain heights of up to 15m.The bipinnate leaves give the canopy a fine feathery texture. The flowers are orange red to scarlet coloured. The trees are truly flamboyant when in flower giving the impression of the crown being ablaze. Leaf shedding occurs just before flowering. Occasionally flowers emerge with the new leaves. The fruits are long, flat and sword like. Trees are native to Madagascar.

Enterolobium saman Prain. Mimosaceae

Syn. *Samanea saman* Merrill., *Pithecolobium saman* Benth.

The Rain Tree (Eng.), Vilayati siris (Hindi, Guj., Mar.), Bilati siris, Koroi (Beng.), Nidraganneru (Tel.) Thoonga moonji (Tam.), Urakkam thoongi (Mal.).

Trees are large with thick and heavy spreading branches. The extensive canopy provides dense shade except during the short deciduous spell of December-January. Refoliation follows immediately in the humid tracts of India. Leaves are bipinnate and with a relatively large terminal pinna. Flowers are produced in small clusters and are pink or rarely white. Trees are native to Brazil and are popular along roadsides and playgrounds.

Erythrina variegata Fabaceae

Variegated Coral tree

Leaves are attractively variegated with patches of creamy white, pink and light green.

Ficus benghalensis Linn. Moraceae

The Banyan (Eng.), Bar (Hindi), Bat, Bar (Beng.), Borh (Punj.), Wad (Mar.), Aal (Tam.), Peyraal (Mal.).

Trees are large spreading and attain a grove like structure by virtue of their thick prop roots growing like supporting columns around the main stem. Birds relish fruits and the groves resound with their noisy chirping when fruits ripen. Suited for planting in large parks and gardens and for making Bonsai. However, the trees are revered as they are considered to be one of the Bodhi trees of other Bodhisattwas (previous incarnations of Buddha) and are cared for wherever they are grown.

Ficus comosa Roxb. Moraceae

Syn. *Ficus benjamina* L. var. *comosa*

Java Fig, Golden fig, Weeping fig (Eng.), Marri (Tel.), Aalmaram (Tam.), Ithi (Mal.)

A very ornamental evergreen tree that is native to and distributed in tropical Asia. The drooping branches carry ovate leaves that are glossy and leathery. The aerial roots hang down the branches and are thin and wiry. After flowering, small globose fruits shower the branches and ripen to a bright red colour. Birds relish them. A small to medium sized tree that can be pruned to suit small gardens.

Ficus elastica Roxb. Moraceae

India rubber plant (Eng.), Atha Bor (Beng.)

Trees are tall and attain heights of over 15m. The leaves are leathery and emerge covered in showy bracts which are coloured light pink to dark maroon. The mid rib may also be coloured similarly. Important cultivars include "Decora" and "Rubra". Leaves having variegations of pink, yellow and white are grouped under *Ficus elastica* var. *vareigata*. "Decora Schijvareana" is a popular cultivar. This species is more popular as ornamental foliage plants.

Ficus glomerata Roxb. Moraceae

Syn. *Ficus racemosa* Linn.

Gular (Hindi), Damur (Beng.), Umbar, (Mar.), Atti (Tel.), Athi (Kann. /Mal.)

Trees are tall, attaining heights of over about 30m.and have a smooth trunk. The leaves are ovate, oblong to lanceolate and characteristically get infested with galls. Cauliflorous, the fruits are produced in profusion. They ripen to a dark pinkish red colour and are shed like flowers, littering the canopy-basin. The young leaves and bark are used in indigenous medicine.

Ficus hispida Linn. Moraceae

Rough leaved fig, Devils fig (Eng.), Daduri, Katgulkar (Hindi), Dhed umbar (Guj.), Karoti (Mar.), Peyathi (Tam.), Kattathi, Peyathi (Mal.)

Trees are small sized and attain heights of about 10m. The trunk and branches are covered with small stiff hairs. The leaves are opposite, oblong, with an entire or toothed margin and a rough lower surface. Cauliflorous like most *Ficus* spp., the fruits become yellow in colour on ripening.

Ficus infectoria Roxb. Moraceae

Paku, Pilkhan (Hindi)

Trees are small in size. Leaves resemble those of *Ficus comosa* in shape but are bigger with well-marked venation. Tropical America is the center of origin of this tree.

Ficus religiosa Linn. Moraceae

The Peepal tree (Eng.), Peepal (Hindi), Asvatha (Beng.), Ravi, Ragi, Rai, Asvathamu (Tel), Drasi, Arasi (Tam.), Arasu, Arayaal (Mal.).

This is the Bodhi tree under which Gautama Buddha attained enlightenment. It is perhaps the most revered among the indigenous *Ficus* trees of India. Trees are large, attaining more than 30m height with tall straight trunks that branch higher up, creating a compact broadly rounded crown. The glossy heart shaped leaves have elongated tapering tips and they flutter in the slightest breeze. The entire crown resounds musically when windy. Trees are planted in temple compounds and large gardens for shade and resting places are constructed around them. *Ficus religiosa* var. *krishnae* called Makhan katori or Krishna Bar is a relatively small sized tree having cupped leaves that are cone-like at the basal end of the lamina. They are believed to have been used by Lord Krishna for storing stolen butter.

Ficus triangularis Moraceae

Angular-leaved ficus

Plants are relatively short in stature and are ideal for home gardens. The leaves are attractively angular. Shoots can be pruned and maintained like shrubs. They produce aerial roots sparsely.

Filicium decipiens Thw. Sapindaceae

The fern-leaved tree

Trees are medium sized, evergreen and attain heights of about 12m. The leaves are finely cut with winged petioles. This attractive ornamental foliage tree is native to South India and Sri Lanka.

Grevillea robusta A. Cunn. Proteaceae

Silver oak, Silk oak, Australian silk.

Trees are evergreen and grow up to 30m in height. The leaves are fern like and their under surfaces have a silvery sheen. The flowers are golden yellow and appear in April-May. The seedlings are grown as indoor plants.

Jaccaranda ovalifolia R. Ber. Bignoniaceae

Syn. *Jacaranda mimosaefolia* D. Don.

Jaccaranda (Eng.), Neeli Gul mohr (Hindi)

Trees are deciduous and attain a height of 8 to 9m. The open crown carries dark green leaves that are bipinnate. The flowers are blue, bell-shaped and are produced in clusters from February onwards in different parts of India. Trees are popular in gardens and are planted along avenues.

Lagestroemia flos-reginae Retz. Lythraceae

Queen's flower, Pride of India (Eng.), Jarul (Hindi), Jarool (Beng.), Ajhar (Assa.), Jarulo (Oriya), Taman (Mar.), Varagogu (Tel.), Maruvachall (Kann.), Kadali (Tam.), Manimaruthu (Mal.)

Trees are evergreen and grow up to a height of 20m. They are one of the most attractive trees of Indian gardens. The branches are short and sturdy and the relatively large leaves provide a dense canopy cover. The foliage is bright green and contrast beautifully with the pink to mauve coloured flowers. In peninsular India, the trees flower in April-May and August- September.

Melia azaderach Linn. Meliaceae

Bead tree, China tree, China berry, The Persian Lilac (Eng.), Dake, Drek, Bakain (Hindi), Ghora neem, Maha neem (Beng.), Pejri (Mar.), Mahanimbo (Oriya),Taraka vepa (Tel.), Malai vembu, Poo vembu (Tam.), Malaveppu (Mal.)

Trees are medium sized and attain heights of up to 8m. Seedlings are hardier than those of Neem but are damaged by frost. The leaves are dark green with finely cut margins. The canopy provides light shade and is ideal for residential avenues. The flowers are scented and lilac coloured. The fruits ripen to a yellow colour.

Michelia champaka Linn. Magnoliaceae

Golden Champa (Eng.), Champa (Hindi, Punj.), Phul chompa (Assa.), Sonar champa (Beng.), Pito champo (Guj.), Son champa (Mar.), Champakamu (Tel.), Sampige (Kann.), Shambaga (Tam.), Chempakam (Mal.)

These elegant evergreens are indigenous to India and attain heights of about 10 to 12m under cultivation. In the Himalayan foothills, trees attain heights of up to 30m. The trunk grows straight and slender branches are spread all around giving the canopy a broadly pyramidal shape. The leaves are smooth surfaced and ovate-lanceolate in shape. The young shoot buds are protected by scaly stipules. The flowers are light orange-yellow to pale yellow or creamy white in colour and have an intoxicating fragrance. They are produced singly in leaf axils. For earliness in flowering, trees are propagated by layering. The fruits are woody, ovoid in shape and have angular seeds.

Millingtonia hortensis Linn. Bignoniaceae

Indian cork tree, the Jasmine tree (Eng.), Akash Neem, Neem chameli (Hindi) Akas Neem Phulari (Beng.), Bakeni, Reali, Mach- mach, Sita-hara, (Oriya), Akas Nimb, Kavla Nimb (Mar.), Kavuki (Tel.), Beratu (Kann.), Mara Malli, Kattu Malli (Tam.), Mara Mulla (Mal.)

Trees are tall stately and evergreen, attaining heights of up to 20m. The gracefully drooping branches carry small and multifid leaves. The pure white trumpet shaped flowers have long corolla tubes and are delicately scented. They are produced in profusion during monsoons. Flowers open at night and are shed by morning, carpeting the ground below the trees. These were favored flowering trees of Indian gardens of the previous century.

Mimusops elengi Linn. Sapotaceae

Mulsari (Hindi), Bakul (Beng.), Bokolo (Oriya), Ovalli (Mar.), Pagada (Tel./ Kann.), Vagulam (Tam.), Ilenji (Mal.)

Trees are tall growing evergreens indigenous to India and attain heights of more than 15m. They have attractive glabrous leaves. The creamy white flowers are small and fragrant. They fade to a light brown colour and are shed quickly. Seedlings are slow growing. Layered plants have a bushy stature and are used as foliage plants.

Muntingia calabura Linn. Verbenaceae

Bird cherry

Trees are small sized evergreens having horizontally spreading branches that give the canopy an umbrella shape. The leaves are ovate oblique, with a rough surface texture. The flowers are white and the green berries ripen to a light yellow colour. Birds disseminate the seeds widely. They germinate readily like the seeds of the Peepal in crevices of cemented rooftops.

Nyctanthes arbor-tristis Linn. Oleaceae

Har singar (Hindi), Shephalika (Beng.), Parijat (Guj. / Mar.), Paarijaatamu (Tel.), Pavizha-Malli (Mal.)

Indigenous to India, these are small to medium sized trees that have a grayish green bark. The young branches droop downwards and are quadrangular in outline. The leaves are ovate and have widely serrate margins. Flowers are borne in terminal leaf axils from August to December. They are scented and have a bright orange coloured corolla tube terminating in free lobes that are white. Flowers open at night and are shed by morning. They were formerly used for dyeing silk thread and fabric. The fruit is a flat capsule with two seeds.

Peltophorum ferrugineum Benth. Fabaceae

Syn. *Peltophorum pterocarpum* Becker

Copper pod tree (Eng.), Konda chinta (Tel.), Ivalvagai (Tam.), Chara konna (Mal.)

Trees are large, with a spreading crown carrying dark green pinnate leaves. The flowers are golden yellow and borne terminally in pyramidal clusters during March to May. Mature fruits are coppery red in colour. One of the most popularly grown avenue trees of Indian highways, they are widely distributed in South East Asia.

Pisonia alba Nyctaginaceae

Cabbage tree, Lettuce tree, Pisonia

These are small evergreen trees native to the Andaman Islands. They are grown as foliage trees in tropical gardens for their attractive light green to yellowish green leaves. The leaves are broadly ovate and have serrate margins. They have medicinal uses. The young leaves are used as a vegetable in India.

Pongamia glabra Fabaceae

Pongam, Ungu (Mal.)

These are evergreens with a dense cover of drooping branches bearing glabrous ovate leaves. The flowers small, pinkish white and are produced in drooping inflorescences. The seeds are the source of Pongam oil are also being used for preparing bio-insecticides.

Plumeria acutifolia Poir. Apocynaceae

Syn. *Plumeria acuminata*

The Pagoda tree, The Frangipani (Eng.), Chameli, Gul-i-chin (Hindi), Dalanaphula (Beng.), Golachi (Oriya), Khair champa (Mar.), Arhattaganneru (Tel.), Kadusampage (Kann.), Ilattarali (Tam.), Arali, Ezha chembakam (Mal.)

Trees are small to medium sized with an evenly spreading crown. The bark is silvery gray in colour. Leaves are broadly lanceolate, tapering at both ends. The flowers are fragrant and coloured in varying shades of pink and yellow. They are produced from March- April to October. Trees are native to Mexico and Guatemala. They are conventionally planted near pagodas and mandapams.

Plumeria alba Linn. Apocynaceae

The Pagoda tree, White Frangipani (Eng.), Veyyivarahalu (Tel.), Peru (Tam.)

Trees are short in stature with dark green leaves. They are native to India. The scented white flowers are larger than the flowers of other species and have reflexed petals.

Plumeria rubra Linn. Apocynaceae

Lal Champa

Trees resemble *Plumeria alba* in stature. The flowers are attractive and pinkish-red in colour. They are produced in loosely hanging clusters during March April and July-October.

Polyalthia longifolia Benth. & Hook. Annonaceae

The mast tree (Eng.), Asokam, Deodar (Hindi), Debdaru (Beng.), Aranamaram (Mal.)

Trees are tall and evergreen. They have evenly spreading branches giving the crown a pyramidal outline. The leaves are glossy and have wavy margins. Greenish flowers are produced in March- April. *Polyalthia longifilia* var. *pendula* has pendant branches that hang close to the trunk, covering it and is more popular for planting in gardens and along central avenues of double-lane roads and highways.

Ravenala madagascariensis J. F. Gmel. Musaceae

Travellers Palm, Travellers tree (Eng.), Khajur Pankhi (Hindi), Panthapadup (Beng.)

Trees have fan like leaves held on either sides of the main shoot tip in a single plane. The leaves resemble the leaves of bananas. However the trees have a woody trunk resembling that of palms. They are planted as specimen plants in gardens for these unique features.

Saraca asoka (Roxb.) De Wilde Fabaceae

Syn. *Saraca indica* Linn.

The Asoka tree (Eng.), Ashok, Sita Asoka (Hindi / Beng.), Ashopalava (Guj.), Osoko (Oriya), Asoka, Jasundi (Mar.), Asokamu (Tel.), Asoka (Kann.), Asogam (Tam.) Asokam, Hemapushpam (Mal.)

Trees are evergreen, short in stature and the trunk has a warty surface. Leaves are pari-pinnate, with oblong-lanceolate leaflets. Flowers are orange red in colour and are produced in dense clusters. The fruits are compressed and contain 2 to 3 seeds. Trees are adapted to the shady conditions of evergreen forests and occur in groups near streams.

Spathodea campanulata Beauv. Bignoniaceae

The African tulip tree, Fountain tree, Scarlet bell tree (Eng.), Patadi (Tel.), Nirukavi (Kann.)

Trees are large, evergreen and attain heights of 25-30m. The branches are short, giving trees a compact crown. Leaves are pinnate, ovate-lanceolate and consist of 9-19 pairs of leaflets in addition to the odd terminal one. Flowers are orange-red coloured and are produced in terminal erect panicles from October to December. They carry nectar and birds largely effect pollination. Fruits are erect pointed pods. Trees are native to central Africa and were introduced into India for planting in gardens and along roadsides. At present it is widely distributed in the lower slopes of the Western Ghats.

Swietenia mahagonii Jacq. Meliaceae

Mahogany (Eng.), Abnus (Hindi), Mehagoni (Beng.) Mahagony (Mal.)

Trees are large, deciduous, native to central and South America. They were introduced into India for their silvicultural importance. The glossy leaves and extensive canopy provide shade almost throughout the year except during the brief period of leaf fall in December-January. The fruits are woody, ovate to rounded in shape and more conspicuous than the small white flowers. They dehisce on drying releasing winged seeds in large numbers that are dispersed by wind.

Tabebuia spectabilis (Planch. & Linden) G. Nicholson Bignoniaceae

Syn *Tabebuia chrysantha* (Jacq.) Nichols.

Golden Trumpet tree (Eng.), Basant Rani (Hindi)

Trees are native to South America. They are small in size and have a rough textured and fissured bark. The older branches have a crooked appearance, which enhance the aged and attractive look of the trees. The mature leaves have a grayish green colour. Golden yellow flowers are produced in profusion during March-April. *Tabebuia rosea* D.C., native to Mexico is the Pink Trumpet tree and is grown in the less humid tracts of India. *Tabebuia aurea* Benth. & Hook. f. ex S. Moore is the Silver Tabebuia or Carribean Trumpet tree.

Tamarindus indica Linn. Caesalpinaceae

Amli, Ambli (Hindi / Guj. / Beng.), Tentul (Beng.), Chincha (Mar.), Chintachettu (Tel.), Puli (Tam.), Vaalan puli (Mal.)

A large and long-lived tree native to India and Africa, it is one of the more popular shade trees of roadsides in South India. Leaves are compound with a chanelled axis and carry 10 –20 pairs of leaflets. The mature leaves are jade green in colour. The flowers are small and yellowish in colour with pink stripes on petals. They are produced in May-June. The fruits are slightly curved and greenish brown in colour. The seeds are shiny coffee brown in colour, square in outline and slightly flattened. The fruits are harvested when tree-ripe and are the source of an acidic pulp having culinary and medicinal uses.

Terminalia arjuna Wight & Arn. Combretaceae

Arjun (Hindi / Beng.), Yermaddi (Tel.), Kulamaruthu, Vellamaruthu (Tam.), Neermaruthu (Mal.)

Plants are semi-deciduous foliage trees grown for shade in gardens and along road sides. the leaves hang gracefully on drooping branches.The flowers are creamy white and inconspicuous.

Terminalia catappa Linn. Combretaceae

Syn. *Terminalia badamia*

Tropical almond, Indian almond, Bengal almond, Country badam (Eng.), Badam, Deshi badam, (Hindi), Bangla badam, Baksa badam (Beng.), Badamo (Oriya), Jungli badam (Mar.), Ingudi, Nadu badamu (Tel.), Badamalili, Nadu badami (Kann.), Amandi, Nattuvadum (Tam.), Thalli thenga (Mal.)

Trees are tall, deciduous and have a straight trunk and whorled branches that spread horizontally. Leaves are large and found tufted at the end of branches. They acquire a red colour before shedding in December- January. Flowers are small, white and are produced in slender spikes. The fruits are egg-shaped and flattened along the edges like a ridge. They turn yellowish red on ripening. Bats relish the outer pulp of the fibrous fruits. Trees are grown for shade and for their canopy architecture in gardens and roadsides.

Thespesia populnea Soland. Malvaceae

Portia tree (Eng.), Paras pipal (Hindi), Bendi, Parasapiplo (Guj.), Parash, Porassh (Beng.), Bhendi (Mar.), Gangaravi (Tel.), Kallal, Poovarasu (Tam.), Cheelanthi, Poovaras (Mal.)

Trees are small to medium sized, evergreen and have closely spaced branches and a rounded canopy. The leaves are heart shaped with acuminate tips. The flowers are large, yellow, cupped and have a maroon-red colour on the inner surface of the petals at the basal end. They are salt tolerant and also withstand waterlogged situations are hence planted in coastal regions and in frequently inundated areas. Trees were popular shade tree of roadsides during the previous millennium in South India.

Platycladus orientalis (L.) Franco Cupressaceae

Chinese Arbor vitae / Biota

Plants are similar to *Thuja occidentalis* (Arbor Vitae) and were previously named *Thuja orientalis* L. The finding of a distant relationship only with the former has resulted in revision of its systematic position and renaming. Small sized trees of this conifer are grown in Indian gardens. They are also planted along avenues and in pots for their architectural shape and curiously flattened and angular branches carrying scale like leaves that are not aromatic unlike the 'Arbor vitae'. With ageing, the trunk becomes twisted and the bark sheds in longitudinal strips.

3. Palms and Cycads

Palms belonging to Arecaceae and Cycads belonging to Cycadaceae having ornamental value are grown in gardens against the backdrop of an expansive lawn or along avenues and as indoor plants. They attain the stature of trees and have a compact crown of evergreen leaves. Adapted to warm humid tropical situations, they are hardy and are propagated by seeds. The suckering types can be propagated by separation of suckers from clumps.

Palms

Ornamental palms are grouped into two, based on the extent to which their leaflets are united. Those in which the individual pinnae are free are called the Feather leaved Palms and those in which the pinnae are partially to completely united are called the Fan leaved Palms. The popularly grown ornamental palms of our gardens belonging to both these groups are listed below: -

Feather leaved palms

Caryota mitis Lour.

Clustering Fish-tail palm

These are quick growing and attain heights of up to 7m. The stems are stout and ringed. The leaves are 1-3m long and have wedge shaped leaflets that resemble tail fins of fish. They produce suckers and are indigenous to Burma and Indonesia.

Chrysalidocarpus lutescens H. Wendl.

Golden Cane palm, Parlour palm

These are native to the Bourbon Islands. The stems are slender and golden yellow in colour. These palms are grown in large clumps. The leaves become beautifully arched as palms mature. They are excellent for planting in lawns and indoors.

Cyrtostachys renda

Syn. *Cyrtostachys lakka*

Ornamental Sealing-wax palm

These are tall graceful palms, native to Borneo. The sheathing petiole and midrib of the leaves are bright red in colour. They are very ornamental when young. Tall stems are cut back to maintain clumps attractive.

Phoenix canariensis

Canary Date palm

These are tall stately palms having spiny leaves that are held straight without an arch.

Phoenix roebelenii

Pygmy Date palm

Plants have finely divided arching pinnae. They are short, slow growing and are suited for sunny indoor areas.

Phoenix sylvestris Roxb.

Wild Date palm

This is a slow growing palm with feathery dark green leaflets. It is suited for growing indoors in the juvenile stage.

Roystonia regia C.F.Cook

The Royal palm, the Bottle palm

Palms are majestic and stately attaining heights of up to 20m. The trunk swells gracefully at the middle as plants age. Slow growing and suited for avenue planting. Leaves are feathery and 2-4m long. The sheathing petiole fully encloses the trunk. They are native to the West Indies

Fan Leaved Palms

Licuala grandis H. Wendl.

The Table palm

These are short in stature and are very popular in gardens. The leaves have ruffled or pleated lobes and the margins of leaflets are united and have toothed tips. The petiole is slender and spiny. Grown as an indoor plant too.

Livistona chinensis R. Br.

Chinese fan palm

These are hardy palms that attain heights of up to 9m. They are extensively grown in gardens. The leaves are broad and deeply forked and have leaflets that are partially free. The crown is compact and globular.

Rhapis excelsa

Syn. *Rhapis flabelliformis*

Ground Rattan palm, Rathan cane palm Large Lady palm (Eng.), Kannonochiku (Jap.)

These have thin bamboo like stems that attain heights of up to 3.5m. They grow in clusters. Persistent fibrous leaf sheaths cover the stem. The leaves are glossy, dark green and cut into several segments. Petioles are toothed. They are native to China.

Cycads

Cycas circinalis Linn.

Fern Palm, Palm Cycad

These have cylindrical stems that attain heights of up to 4.5m. The stems are initially erect and tend to recline with age and are covered by leaf scars. Branching is frequently observed in old plants.The crown carries a rosette of glossy pinnate leaves. Plants may be male or female. They are distributed in India, Madagascar and New Guinea.

Cycas revoluta Thunb.

Ornamental Sago palms, Dwarf Cycas

These cycads are dwarf in stature, attaining heights of up to 3m. The main stem branches occasionally above the ground level. The leaves are dark green and have short and stiff leaflets. Plants are native to Java and are found naturalized up to Southern Japan.

Zamia furfuracea

Cone-leaved Cycas

Plants are similar to *Cycas circinalis* in habit. They are slow growing and thick-stemmed and the main stem remains subterranean for a long time. The leaves have stout mid ribs and the leaflets are thick, leathery, dull green, angular and broadly pointed towards the tip. Suckers are produced from the base in its natural habitat. In gardens, plants thrive near pool sides.

4. Coniferous Trees and Deciduous Trees for Temperate Dlimatic Situations

In the hill zones of the tropics and in regions characterized by warm to cool temperate climatic conditions, evergreen and deciduous trees are grown in gardens and landscapes. Conifers are mainly valued for their architectural features and evergreen nature while deciduous trees that thrive in these situations transform the landscape from late summer through autumn with the vivid colours attained by their foliage.

Coniferous Trees

These are mostly large evergreens having columnar to pyramidal structure and needle like or scale like leaves and usually produce seed bearing cones. They are mostly confined to the northern hemisphere and are among the more primitive of seed producing plants.

Coniferous trees grown in gardens include the following:

Abies sp. Mill. Pinaceae

Fir trees

Thisis a genus of evergreen trees having large to small conical crowns and are distributed widely in the temperate regions. Firs are very popular as Christmas trees. *Abies grandis* the Giant Fir, *Abies cephalonica* the Grecian Fir and *Abies pinsapo* the Spanish Fir are large in size and *Abies koreana* the Korean Fir is a small tree and *Abies balsamea* the Balsam Fir and *Abies concolor* the white Fir are compact to spreading shrubs.

Agathis australis Araucariaceae

Dammar Pine, Kauri Pine, New Zealand Kauri

These are native to the southern hemisphere and have conical crowns when young and develop large trunks and rounded canopies when mature.

Araucaria araucana Araucariaceae

Monkey's Puzzle tree

Araucaria Juss. is an ancient genus of evergreens native to the southern hemisphere. They are large trees with erect stems and horizontal branches carrying condensed and curved needle like leaves. The elongated conical to pyramidal shape of the trees and dark green foliage make them popular in formal plantings. The Monkey's Puzzle is the national tree of Chile. *Araucaria angustifolia* the Parana Pine and *Araucaria cunninghami* the Hoop Pine are also popular in South American gardens. A few species are also grown in Indian gardens – for details see under 'Flowering and Foliage Trees'.

Casuarina equisetifolia (See under 'Flowering and Foliage Trees')

Cedrus deodara (Roxb.) G. Don Pinaceae

Deodar, Himalayan Cedar (Eng.) Devadaru (Mal.), Deva Tharu (San.)

Native to the Himalayas, stands of these tall growing large trees with broadly pyramidal crowns and slightly pendulous branches are a majestic sight of the slopes and valleys and are called the 'Trees of Gods'. The young leaves are blue-green and have a fine texture.

Cryptomeria japonica (L.f.) D. Don Cupressaceae

Japanese Cedar, Sugi (Jap.)

This is the national tree of Japan. Plants have large trunks and spirally arranged needle like leaves. The wood is scented and extensively used for buildings. A traditional tree of Japanese gardens, plants with shrub like habit and having juvenile leaves alone are also grown.

Cupressus L. Cupressaceae

Cypress trees

This genus has evergreen trees or arboreal shrubs native to the warm temperate regions of the northern hemisphere. They are decorative plants of traditional English and Mughal gardens. Several species having an architectural canopy outline are grown as avenue trees in gardens located in the temperate belt of India. They include the mourning or weeping cypress (*Cupressus pendula*) and the Kashmiri cypress (*Cupressus cashmeriana*). Others having ornamental value are the Monterey cypress (*Cupressus macrocarpa*), the Mediterranean cypress (*Cupressus sempervirens*), the Italian cypress (*Cupressus fastigata*) and the Bhutan cypress (*Cupressus nepalensis*). The leaves are scale like and arranged in pairs.

Ginkgo biloba L. Ginkgoaceae

Maidenhair tree

Plants are long-lived deciduous tall growing trees distributed in Eastern China. Considered to be living fossils related to gymnosperms among present day plants, they have fan shaped leaves with notched edges. The veins are radiating and branch dichotomously without forming a network.

Juniperus L. Cupressaceae

Junipers

This is a genus of coniferous trees and arboreal shrubs distributed in the northern hemisphere. *Juniperus chinensis* L. the Chinese Juniper and *Juniperus communis* the Common Juniper are popular garden plants of Indian gardens too. The berries are traditionally used for flavouring food in their centers of origin.

Larix decidua Mill. Pinaceae

European Larch

These are deciduous trees with open crowns and a silvery-gray to grayish-brown bark. The leaves are flattened and sessile.

Metasequoia glyptostroboides Hu & Cheng Cupressaceae

Dawn Redwood

Plants are large, fast growing deciduous conifers that thrive in waterlogged soils. They are endangered in the wild and are protected in national parks in N. America.

Picea abies Pinaceae

European Spruce

These belong to a genus of coniferous evergreens distributed in the northern temperate and artic regions. They are long-lived and have a conical structure with whorled branches and are extensively used as Christmas trees.

Pinus L. Pinaceae

Pine

This is a large genus of coniferous trees native to the northern hemisphere. They are the chief timber trees of the temperate belt. The mature leaves are found in whorl like spirals and are called needles. The seed bearing cones are used ornamentally. Pines are considered to be invasive and large pine forests tend to reduce the natural plant diversity of regions. Species diversity of pines is high in Central America.

Podocarpus Labill. Podocarpaceae

Fern Pine, Pine

This is a genus of monoecious or dioecious evergreen trees or arboreal shrubs of ancient origin. They are found in almost all the continents including Antarctica. The leaves have a distinct midrib and are spirally arranged on the shoots. Foliage is used in fresh floral arrangements. The female cones at maturity become fleshy and reddish purple in colour and are edible. *Podocarpus macrophyllus* the Buddhist Pine is an arboreal shrub. *Podocarpus nerifolia* is found to thrive in the humid belt of South India. The leaves are used decoratively.

Sequoia sempervirens Cupressaceae

Coast Redwood, California Redwood

Plants are evergreens and are the tallest among the trees of the world. Together with two other species they belong to the only living genus of redwoods. The freshly cut wood of these plants is reddish brown in colour. Plant specimens are maintained in bio reserves and national parks in N. America.

Sequoiadendron giganteum (Lindl.) J. Buchh. Cupressaceae

Giant Sequoia, Sierra Redwood, Wellingtonia

These are the world's largest trees and occur naturally in California in scattered groves that are protected. They are evergreen with awl shaped leaves that are arranged spirally.

Metasequoia glyptostroboides Hu & Cheng Cupressaceae

Dawn Redwood

Plants are fast growing deciduous conifers that can thrive in waterlogged soils. Trees attain a height of 30m and a diameter of 1.3m. They are critically endangered in the wild.

Taxodium Rich. Cupressaceae

Cypress trees

This is a genus of deciduous to evergreen trees endemic to the eastern coast of North America. A number of species are distributed in swampy areas and highlands. They include the Pond Cypress *Taxodium ascendens* and the Montezuma Cypress *Taxodium mucronatum*.

Taxus baccata Taxaceae

European Yew

These are slow growing long-lived coniferous trees that have spirally arranged lanceolate leaves held distichously. They are the conventional hedge plants of European gardens. Seed cones have a modified scale that develops into an aril enclosing the seeds. These are bright red and are called yew berries and are eaten by birds. All plant parts except the aril are poisonous. The Himalayan Yew is *Taxus wallichiana*. The only tropical species is the Sumatran Yew *Taxus sumatrana*.

Thuja occidentalis Cupressaceae

Arbor vitae (Tree of Life), Eastern *Arbor vitae*, Northern White Cedar, Thuja, Thuya

These are small evergreen trees having fan like flattened branches held on a single plane.

The foliage consists of scale like leaves and is characteristically scented. They are garden favourites in Europe and North America.

Tsuga Carr. Pinaceae

Hemlock

This is a genus of evergreens adapted to cool and moist areas of temperate regions. Many species are very cold hardy. The plants have irregular to conical crowns carrying flattened branched with drooping ends. The leaves are aromatic and their aroma resembles that of a herb poison called hemlock obtained from an unrelated plant. *Tsuga dumosa* the Himalayan Hemlock has pendulous pubescent branches.

Deciduous Trees

These are the trees most commonly found in large gardens of the temperate regions. They are grown for shade, for the beauty of their foliage and flowers and for the bright colours the leaves attain in the autumn season or fall. A large number of these are distributed in the northern hemisphere. Most of them are cold hardy but those that cannot tolerate spring frost that harms new buds and leaves are usually seen confined to warmer areas.

Popularly grown ornamental deciduous trees include the following: -

Acer L. Aceraceae

Maple

This is a genus of trees and shrubs native to the northern hemisphere. The species distributed in Asia and the Mediterranean regions are mostly evergreen and those distributed further north are deciduous. The fruit is a samara and is referred to as 'maple key'. Around 15 species including *Acer caesium* are found in the Himalayas. The Sugar Maple is *Acer saccharum*, the source of maple syrup, which is the sap tapped from the bark of mature trees. The foliage of Maple trees attains brilliant yellow and red colours before leaf fall.

Betula L. Betulaceae

Birch (Eng.), Bhoorj (Hindi)

This is a genus of trees having bark coloured red, black, white, yellow or silver and marked with horizontal lenticels that separate in thin papery plates. These, referred to as 'Bhoorj Patra' were used in ancient times in India for writing. Birch is the national tree of Russia. The Silver Birch is the national tree of Finland.

Carya glabra Juglandaceae

Pecan, Pignut Hickory, Sweet Pignut

These have ornamental foliage. The leaves are yellowish green and turn reddish in Autumn. The flowers are drooping catkins. The seeds are edible.

Cercis canadensis Fabaceae

Eastern Redbud / Judas Tree

These are flowering trees native to N. America and are popular in gardens. A number of species are grown in C. America, Europe and the Mediterranean region.

Cornus florida Cornaceae

Flowering Dogwood

These are popular shade adapted garden trees of N. America and are grown for their brilliant red autumn foliage and clusters of white to pink floriferous bracts that cover plants during the flowering season.

Cotoneaster Medik. Rosaceae

Cotoneaster

The genus is native to temperate regions and is grown in gardens for their white to rosy pink flowers and showy bright red berries. The Himalayan Tree Cotoneaster is grown in parks and large gardens of N. India. Other species grown are *Cotoneaster khasiensis* having rose-pink flowers and *Cotoneaster acuminata,* which is used for making walking sticks.

Fagus sylvatica Fagaceae

European Beech

These are native to the mountainous regions of Europe. They occur in dense stands crowding out other vegetation and allowing very little light into their interiors. The leaves are simple with crenulate margins. The flowers are produced in small catkins and the fruits called beechnuts are edible.

Fraxinus excelsior Oleaceae

Common Ash, European Ash

Plants are ornamental foliage trees with drooping branches that are popularly grown in gardens. The flowering Ash is *Fraxinus ornus*.

Laburnum anagyroides Fabaceae

Laburnum, Golden Chain tree

Plants are small trees native to Europe. The flowers are golden yellow and densely packed in pendulous racemes of up to 20cm length. The resemblance of these flowers to those of *Cassia fistula* of the same family has resulted in the common name 'Indian Laburnum' of the latter. All the plant parts are poisonous.

Liquidambar orientalis Altingiaceae

Oriental Sweet Gum (Eng.), Silaras (Hin.)

These were introduced into Indian gardens during the colonial period. The leaves are palmately lobed and yellow flowers are produced in globose heads. A resin named storax produced from the seeds has perfumery use.

Magnolia grandiflora Magnoliaceae

Southern Magnolia, Bull Bay

These are small ornamental trees native to N. America and are widely grown in temperate zone gardens. The tepals are showy, fragrant and white to pink in colour. *Magnolia campbelli* is native to E. Himalayas and Manipur. *Magnolia obovata* is the Japanese Big leaf Magnolia and *Magnolia liliflora* is the Lily Magnolia or the Tulip Magnolia native to China.

Platanus orientalis Linn. Platanaceae

Chinar, Oriental Plane tree, Sycamore, Buttonwood

These are trees that are often avenue planted along highways. The leaves are lobed and have serrate margins. The flowers are globose balls.

Populus Linn. Salicaceae

Aspen, Cottonwood, Poplar

This is a genus that has more than 25 species distributed in the northern hemisphere. They are differentiated into three groups and referred to by three common names- aspens, cottonwoods and poplars. Aspens are native to Europe and N. America. They have leaves with flat petioles that tremble in the slightest breeze. The Common Aspen or Trembling Aspen of Europe is *Populus tremula* and the Quaking Aspen native to N. America is *Populus tremuloides*. Cottonwood trees are native to the wetlands of N. America. The Swamp Cottonwood is *Populus heterophylla*. Poplars are native to S. Europe and Asia. The White Poplar is *Populus alba* and the Black Poplar is *Populus nigra*.

Prunus Linn. Rosaceae

Flowering Cherry

These are cherry trees having multi-species parentage. They are popularly avenue planted as flowering trees.

Quercus L. Fagaceae

Oak

This is a large genus with more than 300 species of deciduous to evergreen trees attaining majestic proportions in size. They are native to the northern hemisphere. The leaves are characteristically lobed. The fruit is an edible nut called an acorn. Plants are components of large gardens and important sources of timber. Oak trees are the national trees of England, Wales, France, Germany, Estonia, Lithuania, Poland, Serbia and the United States of America. The English Oak is *Quercus robur*.

Salix babylonica Salicaceae

The Willow

Plants are important foliage trees of gardens of the temperate regions. The spreading branches have drooping tips. The Weeping Willow is *Salix chrysocoma*.

Ulmus hollandica Ulmaceae

Elm, European Elm

These are important foliage trees of European gardens. The branches have a weeping habit. *Ulmus wallichiana* is the Himalayan Elm.

5. Propagation and Planting of Trees

In the tropics, trees are to be selected based on their suitability to growing conditions. Seed propagation is resorted to for ensuring long life and the characteristic structure and branching habit of species. In dicotyledonous trees with dimorphic branching, vegetative propagation of orthotropic shoots will maintain plant architecture as in seed propagated plants while plageotropic shoots will not. Most of the leguminous tree seeds have dormancy and require scarification for quicker germination. Unscarified seeds take 2-3 months to germinate as their dormancy is overcome in nature over a period of one to three months.

Hardened one to three years old nursery plants are the most suitable for planting. Planting should preferably coincide with the monsoons for quicker acclimatization of plants. Pits of 1 cubic meter size are to be taken and if possible, left to weather for 2 to 3 weeks after partially filling them with 10-15 kilograms of organic manure. For row planting, small sized trees are spaced 5m apart while medium to large sized trees are spaced 9-10m apart. For grove planting, depending upon the species selected, a closer spacing is to be adopted.

6. After Care of Trees

After planting, trees are to be staked and those planted along roadsides are to be protected with a tree guard. Except for avenue planted conifers and pines, tree trunks are generally left bare of branches up to a height of at least 2.5m. Lopping of shoot suckers (orthotropic shoots arising from near the base) and pruning of excessive lateral growth is also to be done during the initial years. Pruning of heavy branches is to be done with a suitable implement that can make a clean cut without unnecessary wounding of tree bark.

The life span of trees varies. Trees like the Redwoods, the immortal Banyans and several others have survived climatic and environmental variations over several millennia. However most of the ornamental trees of gardens show decline in vigour and re-growth of branches with ageing. Old trees are to be examined for symptoms of progressive decay. Those in which the trunks have become hollow and partially diseased are to be removed quickly. Under- planting can be done well in advance of felling old trees.

Further Reading

Sabeens Jose. 2008. Peter, K.V. (Ed) Ornamental Plants in a garden.New India Publishing Agency, New Delhi 324p.

Chapter – 20

Principles and Elements of Landscape Design

S.S. Sindhu and Sapna Panwar

Landscaping is an art and science to imitate nature and create a functional, aesthetically pleasing outdoor environment. The main focus of landscape design is to blend both hardscape and softscape elements in an aesthetic way. The basic knowledge of art elements and design principles is desirable in order to create the pleasing landscape design. The landscape designers must visualize the design in critical manner and should have vision that how the landscape design will look after 10-15 years when the plants will acquire the full growth so he should plan in that way taking care of future needs. Landscape designer should have knowledge of both landscape architecture and botany. Landscape designers use a design process that systematically considers all aspects of the land, the environment, the growing plants, and the needs of the user to ensure a visually pleasing, functional, and ecologically healthy design. The design process should begin with proper planning taking care the needs and desires of the user and the conditions of the site.

1. Basic Principles

The principles are the fundamental concepts of composition—proportion, order, repetition, and unity—that serve as guidelines to arrange or organize the features to create an aesthetically pleasing or beautiful landscape. Landscape principles are important in order to organize the elements in a way that create a visually pleasing landscape. Physical and psychological comforts are two important concepts in design that are achieved through use of these principles. People feel more comfortable in a landscape design which provides soothing and harmonious effect of overall landscape. A harmonious composition can be achieved through the principles of proportion, order, repetition, and unity. All landscape principles are related with each other in one or other way and all together produce a perfect landscape.

The basic principles of landscape design are as under:

2. Scale and Proportion

It usually bears reference to the size of a thing or object that appears to have a pleasing relationship to other things or to the design as a whole or it is the size relationship of different parts of the landscape design and also depicts relation to each other and to the design as a whole.

Two types:

Relative proportion: It is the size of an object in relation to other objects.

Absolute proportion: It is the scale or size of an object.

An important absolute scale in any landscape design is the human scale i.e. according to size of the human body. The designs are created by keeping in view that the size of objects is relative to size to humans. All the features should be placed in well proportionate manner so that they present a united look. Plant material, garden structures, and ornaments should be considered relative to human scale. Other important relative proportions include the size of the house, yard, and the area to be planted. Proportion in landscape design usually relates to people and their activities. For example a small pool in a large open lawn would not be proportionate but at same time it fits beautifully into a small private area. Similarly a large gate in small garden or small gate in large garden looks disproportionate.

Perfect proportion of garden elements to its setting

3. Balance

Balance is the visual equilibrium in landscape design around a central axis or it is concept of equal visual attraction and weight, usually around a real or imaginary central axis. Form, color, size, and texture all affect balance.

Balance can be following types

- Symmetrical Balance
- Asymmetrical Balance
- Perspective Balance

Symmetrical balance: In formal balance, the mass or weight or numbers of objects on either side of a central axis should be exactly the same. Plants are frequently clipped, lines tend to be straight, and edges are clearly defined.

Asymmetrical balance: In informal balance, plants should be irregularly placed on either side of an imaginary axis so that the mass or weight on either side of the axis appears to be balanced. Curved lines, obscure and merging edges and natural contours identify asymmetry in the garden.

Perspective balance: It is concerned with the balance of the foreground, midground, and background. When looking at a composition, the objects in front usually have greater visual weight because they are closer to the viewer. This can be balanced, if desired, by using larger objects, brighter colors, or coarse texture in the background. In most cases, either the foreground or background should be dominant.

Formal balance of the garden

4. Rhythm

Rhythm refers to repetition of same object at equidistance so as to give continuous effect or it refers to the repeated use of features like plants with identical shape, line, form, texture and/ or color. Too much repetition creates monotony but when used effectively can lead to rhythm, focalization or emphasis. Rhythm is achieved when the elements of a design create a feeling of motion which leads the viewer's eye through or even beyond the designed area. Tools like color schemes, line and form can be repeated to attain rhythm in landscape design. Rhythm is expressed through the placement of plants, park furniture, etc., either individually or as group.

Repetition of fountains creates continuous movement in garden

5. Simplicity

The design presents a comfortable look if design is simple. Too much inclusion of elements or features in design makes the things complex and moreover, it creates confusion. Simplicity goes hand-in-hand with repetition and can be achieved by elimination of unnecessary detail. Too much variety or detail creates confusion of perception. Simplicity is the reduction of a design to its simplest, functional form, which avoids unnecessary cost and maintenance.

6. Unity

Unity means that people should visualize the landscape design as a complete unit instead of visualizing the different parts of the composition or landscape. Unity gives the artistic look to the design. The different components of garden should unite with each other so as to present harmonious effect. A natural feeling evolves when each activity area belongs to and blends with the entire landscape. Everything selected for a landscape must complement the central scheme and must, above all, serve some functional purpose. The simplest way to create unity is through the use of a design theme or a design style.

7. Transition

Transition refers to the gradual change in the landscape design and can be obtained by the arrangement of objects with varying textures, forms, or sizes in a logical sequential order. It breaks the monotony of the design and appeals pleasing. For example, coarse to medium to fine textures, round to oval to linear structural forms, or cylindrical to globular to prostrate plants. Transition from shorter to taller plants and from fine to coarse textures would enframe the scene and makes it appear closer, like a painting on a wall. Generally, transition assists in the gradual movement of a viewer's eye to the design and within it.

8. Focal point or Emphasis

It is the centre of attraction of any landscape design. Through the use of emphasis, eye movement is directed towards a center of interest that takes a position of prominence in the landscape. Generally an architectural feature like fountain, statue etc are used for creation of focal point. This could be a single tree, a beautifully designed water feature, a piece of sculpture, or a collection of ericaceous plants that automatically draw the eye to this point of interest. Open lawn areas, paths and strategically placed plants can lead the eye to the principal feature without distraction. Plantings should be placed to easily lead the eye to this center of heightened interest.

Tree sculpture create emphasis in the landscape

9. Elements

The elements of landscape design are as under:

1. Line

Lines are a powerful tool for the designer because they can be used to create an infinite variety of shapes and forms, and they control movement of the eye and the body. Line plays the following functions:

- Create patterns
- Develop spaces
- Create forms
- Control movement
- Establish dominance
- Create a cohesive theme in a landscape

Types

- Curvilinear lines: They have weak, scallopy edges will not be visually interesting or pleasing to the eye. Curvilinear, meandering lines suggest a naturalistic look that invites the user to casually stroll through and experience the landscape.
- Linear lines: Such as those found in a straight hedge or the edges of paving materials suggest quick, direct movement.
- Angled lines can create opportunities for creating the "bones or the framework of the landscape". Lines that interconnect at right angles create an opportunity for reflection, stopping or sitting.

2. Form

Form relates to the natural shape of the plant. Form is found in both hardscape and plants, and it is typically the dominant visual element that spatially organizes the landscape and often determines the style of the garden. The form of structures, plant beds, and garden ornaments also determines the overall form theme of the garden.

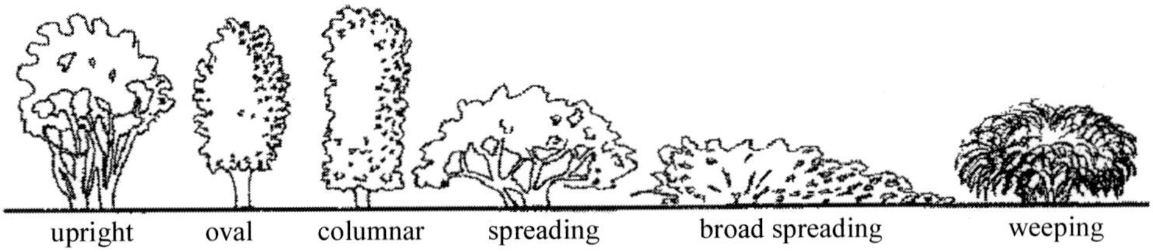

Fig. 1. Form refers to the shape and structure of a plant or mass of plants.

- Upright: *Ginkgo biloba* "Princeton Sentry"- Princeton Sentry Ginkgo
- Horizontal or spreading: *Taxus* x *media* "Hillii"-Hill's Yew- and Quercus palustris- Pin Oak.
- Weeping, drooping of pendulous forms: *Fagus sylvatica* "Purple Fountain" – Purple fountain Beech.

3. Texture

Texture refers to surface character of any garden feature. It depicts how coarse or fine the surface of the plant or hardscape material feels and/or looks. Texture is used to provide variety, interest, and contrast. Texture can be depicted by the plant's foliage, flowers, bark, and overall branching pattern all have texture. The size and shape of the leaves often determines the perceived texture of the plant. A plant can generally be described as having a coarse, medium, or fine texture.

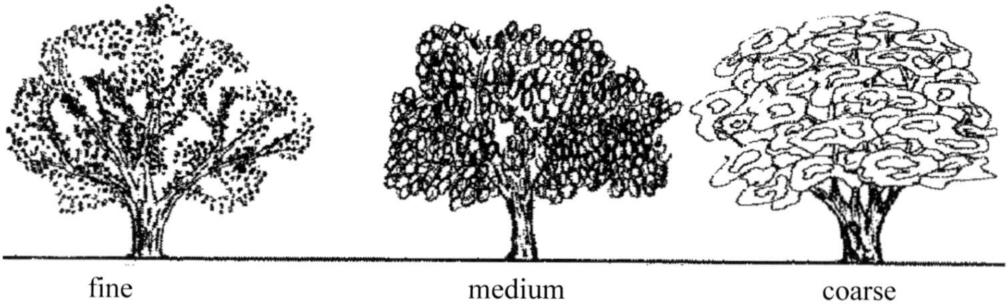

Fig. 2: The texture of plants differs as the relationships between the leaves, the twigs and the branches differ.

When using ornamental grasses for example, a gradation of textures from fine to medium to coarse could be as follows. Texture in the landscape depends upon the distance from which the plant is viewed by the observer. In distant views, the overall mass of the plant is the dominating feature and the fineness or softness of a leaf or branching pattern is lost. In terms of the overall planting plan, texture must balance in relationship to the axis. Weight on one side should equal the mass on the other side of the axis. For example, much fine texture- as the case would be in using Buxus- is required to balance relatively little coarse texture, as the case would be in the use of *Viburnum rhytidophyllum*, the *Leather leaf Viburnum*. Intermediate plants are recommended to provide the necessary transition from one textural extreme to the other.

4. Colour

Colour in plant material and hardscape adds interest and variety to the landscape. Colour is the most conspicuous element in the landscape and is usually the focus of most people; however, it is also the most temporary element, usually lasting only a few weeks a year for individual plants.

Colour wheel

The use of color is guided by colour theory by the use of the colour wheel. It is used to create color schemes. A simple description of the colour wheel includes the three primary colors of red, blue, and yellow; the three secondary colors (a mix of two primaries) of green, orange, and violet; and six tertiary colors (a mix of one adjacent primary and secondary color), such as red-orange. Color theory explains the relationship of colors to each other and how they should be used in a composition. The basic color schemes are monochromatic, analogous, and complementary. Colour theory is a very complex and very personal matter that expresses individual taste and feelings.

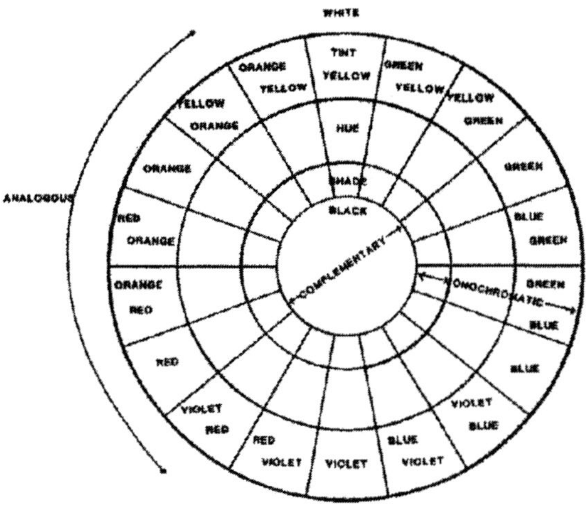

Fig. 3: A color wheel can be used to explain color variations.

Warm colours: Such as reds, oranges and yellows tend to advance towards to viewer Cool colours: blues, violets and greens tend to recede into the landscape.

Warm colours read well and affect the eye more quickly than do cool colours. When using warm colours, they should be used in sequence which must be smooth and gradual. For example, red to scarlet to orange scarlet to orange to bronze to orange yellow to yellow to pale yellow to cream to white. Consideration of the use of colour in plantings requires a thorough, practical understanding of the personality of the plants. To vigorously use colour and effective colour combinations requires a thorough knowledge of plants, their colours and seasonal changes with detail of twig, leaf, flower and fruit as well as principles of colour.

10. Variety

Variety is the spice of life. It breaks the monotony and brings diversity in landscape design. Variety is depicted through lines, forms, textures and colours are required in order to achieve an interesting landscape. Without variety in both the use of 'hard' and 'soft' landscape materials, this can lead to loss in interest in landscaping of that area.

A variety of forms creates significant landscape interest

11. Grouping

Grouping creates a strong feeling of mass and a bold landscape statement. Much greater appeal is achieved when odd numbers of plants are used in the landscape like groupings of three, five, seven, nine plants etc. Plants should be irregularly spaced and every effort should be made to avoid placement of plants in an equilateral triangle. While grouping, a designer usually starts with a specimen that establishes the scale of the landscape. Around it are grouped slightly less important plants which complement the specimen in colour, texture and habit of growth.

12. Mass

Made up of plants that cannot be seen in their entirety from any one vantage point. Seasonal stability and variety in plant mass is accomplished through a mix of evergreens and deciduous plants. Only rarely should a design consist exclusively of evergreens or deciduous material instead of a mixture of both.To create a harmonious effect in any group, a designer should strive to properly fit together plant forms, textures and colours into a harmonious whole or mass. Size of any mass or composition depends upon its location in relationship to other factors such as the need for screening, proximity to other groups, etc. Mass can be any size, but smaller masses or clumps are not normally as effective as larger, bolder mass plantings.

8. Sequence

The effective use of sequence is oftentimes employed to create visual movement in the landscape. It is an important consideration to take into account in the development of the overall planting pattern. For example, sequence could be an orderly natural combination of plant material. In this case, low objects would appear in the foreground, intermediate objects in the middle ground, and tall objects in the background.

Further Reading

Sindhu, S.S. 2015 (Ed). Ornamental Horticulture. New India Publishing Agency, New Delhi. 485 p.

Chapter – 21

Glossary of Selected Horticultural Terms

P.G. Sadhan Kumar and S. Nirmala Devi

A

Abscission - Shedding of a leaf, flower or fruit as a result of formation of a layer of loosely adhering cells at its base which break apart easily.

Abscission layer - A layer of thin-walled cells formed at the base of the organ.

Abscission zone- Region with abscission layer which is located near the base of petiole of leaves.

Absorption - The process of taking up water and holding it in pores or open spaces.

Acaulis - Without stem.

Achene - A dry, indehiscent, one-seeded fruit.

Acicular - Slender or needle-shaped.

Acid soil - A soil with pH lower than 7.0.

Actinomorphic - With radial symmetry, the parts similar in size and shape.

Acuminate - Coming to a point with concave, curved sides.

Adaptation - Ability of plants to perform in a particular environment.

Adnate - Fusion of unlike parts eg. fusion of stamens with corolla.

Adsorption - The adhesion of molecules of gases or dissolved substance to the surface of solid particles. It can be distinguished from absorption, which is not a surface phenomenon.

Adventitious embryony - This is a type of apomixis where the embryos rise from a cell or group of cells either in the nucellus or in the integuments. This is also known as nucellar embryony or nucellar budding.

Adventitious plants - Plant developiung asexually on the leaves or stems of mother plant.

Adventitious roots - Roots arising from any part of the plant other than the radicle.

Aerate - Loosening or puncturing the soil to increase water penetration.

Aerial - Stem which remains above the ground in air.

Aerial roots - Roots produced above ground, often used for climbing.

Aeroponics - The process of growing plants in an air or mist environment without the use of soil or any other medium.

Afforestation - Planting of trees and grasses.

After-ripening - Completion of maturation processes within the seeds to enable germination.

Agamospermy - All type of apomixes in which seeds are formed by asexual means without fertilization.

Aggregate fruit - The product of fusion of several ovaries in the same flower. It is different from a collective fruit which is formed from a complete inflorescence.

Aging - Chemical or structural changes a cell,tissue or organ undergoes during its life cycle which result in death/failure of the plant/plant part.

Agroforestry - A land use system of growing trees,crops and animals simultaneously or sequentially on the same piece of land.

Aitionomic - Ability to develop parthenocarpic fruits in response to some stimulus external to the ovary.

Albinism - Complete absence of pigment from the leaves.

Aleurone - A layer of protein surrounding the endosperm.

Alkaline soil- A soil with a pH higher than 7.0.

Allele or Allelomorph - An allele is an alternate form of gene which is located at specific position on a specific chromosome.

Allogamy - When the egg in a flower gets fertilized by not its own pollen, it is referred to as allogamy.

Allopolyploid (Alloploid) - A polyploid with more than two sets of chromosomes, each being derived from a different species.

Alternate bearing (Biennial bearing) - Bearing heavy and lean crops in alternate years.

Alternate host - A plant other than the main crop.

Amendments - Organic or mineral materials such as peat moss, compost, perlite

Amphidiploid - A polyploidy whose chromosome complement is made up of the entire somatic complements of two species.

Androecious - Production of only staminate flowers in a plant.

Androecium - Group of stamens wherein pollen is produced.

Androdioecious - Group of plants consisting of male and hermaphrodite plants.

Androgynous - Having staminate flowers above and pistillate flowers below in the same spike.

Andromonoecious - Production of staminate and perfect flowers separately on the same plant.

Androphore - A support or column on which stamens are raised above the perianth.

Anemophily - Pollination by wind.

Aneuploid - An organism whose somatic chromosome number is not an even multiple of the haploid number.

Angular stem - Stem which shows many angles in a transverse section.

Annual - A plant which completes its entire life cycle, from seed to seed, and dies in a single season.

Anther - The pollen containing part of stamen, usually consisting of two sacs.

Anthesis - The time of flower opening.

Anthocyanin - Pigments in plant giving scarlet, purple or blue colour to leaves, stems and flowers in plants.

Anti-oxidant - Material used to prevent oxidation or discolouration of food.
Apetalous - Without petals.
Apical dominance - Inhibition of growth of lateral buds by the growing apical buds.
Apomixis - Production of seed from unfertilized egg, or nucellar cells or embryosac cells.
Apiculate - Ending in an abrupt pointed tip.
Apogamy - Development of embryo from male gametophytic cells i. e., synergids or antipodal cells without fertilization.
Apomixis - Reproduction in which sexual organs or related structures take part but fertilization does not occur and the resulting seed is vegetatively produced.
Apospory - Apomixis.
Aquaponics - Land based production of fish in tanks combined with recirculation of the water from the fish tanks through hydroponic systems to produce high value vegetable/ horticultural crops. Waste products from the fish tank are converted into soluble nutrients for the plants and clean water is circulated back.
Aquatic - Plants growing in water.
Arboreal - Tree like or pertaining to trees.
Arboretum - A garden with a large collection of trees and shrubs cultivated for research or educational purposes.
Arching - Curving.
Aromatic - Fragrant, spicy.
Ascorbic acid - Vitamin C. Water soluble vitamin found in fruits and vegetables.
Asepalous - Without sepals.
Asexual - Not involving union of gametes.
Asexual propagation - Vegetative propagation or vegetative reproduction in plants by methods other than seed propagation.
Assimilation - Conversion of substances taken in from the outside into living tissue of plants or animals.
Autogamy - Self-fertilization ie. the fertilization of a flower by its own pollen
Autopolyploid (Autoploid) - Polyploid with more than two sets of similar chromosomes derived from the same species.
Awl-shaped - Narrow and gradually tapering to a sharp point.
Axillary - Arising from the axil.

B

Backcross - A cross of a hybrid to either of its parents.
Backcross breeding - A breeding method where recurrent backcrosses are made to one of the parents of a hybrid, accompanied by selection for a specific character or characters.
Bare root - Plants without soil adhering to the roots.
Basic number - The number of chromosomes in ancestral diploid ancestors of polyploids, represented by *x*.
Basipetal - From the growing tip of shoot to or root towards the base.
Beaked - Ending in a point

Beared - Having long hairs.

Berry - A juicy indehiscent fruit with the seeds immersed in the pulp; usually several seeded without a strong layer surrounding the seeds.

Biennial - A plant which completes its life-cycle in two seasons.

Bilocular - Fruits with two locules.

Bioremediation - Use of living organisms toremove or detoxify unwanted chemicals from the environment.

Biotype - A group of individuals with the same genotype. Biotypes may be homozygous or heterozygous.

Blossom end rot - A deficiency problem in tomato caused by calcium deficiency.

Bolting - Premature flowering of plants before attaining enough vegetative growth.Usually due to late planting and too warm temperatures.

Bonsai - Art of growing trained, dwarf plants in containers.

Bract - Modified leaves growing just below a flower.

Breeder's seed - Breeder's seed is directly controlled by the originating or sponsoring plant breeding institution, firm or individual. It is the source for the production of seed of the certified classes. It has the maximum genetic purity.

Broad elliptic - Wider than elliptic.

Broad ovate -Wider than ovate.

Bulb - An enlarged, fleshy, thick, underground part of a stem surrounded by a mass of leafy scales. Scales of a bulb are actually thickened and shortened leaves. Roots develop from the base of a bulb.

Bulbil - Small bulbs arising around the parent bulb.

Bulblet - Small bulbs which arise in leaf axils.

Buttoning - Production of small sized curd in cauliflower.

C

Caducuous - Falling off early.

Callus - It is an irregular mass of parenchyma cells in various stages of lignification.

Cambium - Thin mucilaginous, cellular layer between the wood of a tree and its inner bark which is converted into xylem on the inner side and the phloem on the outer side.

Campanulate - Bell shaped.

Capsule - A simple dry fruit formed from a compound gynoecium splitting along two or more lines of suture in various ways.

Carpel - One of the foliar units of a compound pistil or ovary. A simple pistil has only one carpel.

Catkin - A slender spikelike or drooping flower cluster.

Cauliflorous - Flowers borne on the stem from the old wood.

Central leader - Training system of tree in which it is trained to form a trunk extending from the soil to the top of the tree.

Certified seed - Seed used for commercial crop production produced from foundation, registered, or certified seed by maintaining satisfactory genetic purity and identity under the regulation of a legally constituted agency.

Chance seedling - A fruit tree which grows from seed scattered by wind, bird etc.,

Character - An attribute of an organism resulting from the interaction of a gene or genes with environment.

Chelates - Metal-containing organic compounds used to supply minerals to plants.

Chimeras - Chimeras are plants composed of two or more genetically different tissues growing separately but adjacent to each other in one plant.

Chip budding - Method of budding where a single bud with a large piece of wood attached to it is removed from the scion and fitted firmly by tying to similar cut in the stock

Chlorosis - Yellowing of leaves due to loss or breakdown of chlorophyll.

Chromosome - A structural unit in the nucleus which carries the genes in a linear constant order; the number being typically constant in any species.

Citrange - A hybrid (*Poncirus trifoliate* x *Citrus sinensis*) bearing very acidic fruits, used as a good stock for citrus.

Cleistogamy - Condition where flowers remain unopened leading to self pollination.

Climacteric peak - The maximum point of respiration rate of a mature fruit.

Climber - A plant with thin stem and special organs of attachment by which they cling to neighbouring objects for the support of their body.

Clone - A population of cells or organisms derived from a single cell or common ancestor by mitosis or a group of plants originated by vegetative propagation from a single plant.

Colchicin - A drug obtained from *Colchicum autumnale* having the ability to cause chromosome doubling resulting in the production of polyploids.

Collective fruit - Fruit formed from a complete inflorescence

Colloid - A state of a substance where the units are very large molecules or molecule complexes that do not form true solution and diffuse slowly / not at all through plant or animal membrane, cell membrane or cell wall.

Combining ability - Ability of a strain to produce superior progeny upon hybridization with another strain.

Compatibility - It is the ability to combine. In the case of sex cells, it is the ability to unite and form a fertilized egg that can grow to maturity. In case of grafting,it is determined by the degree of success of the union between stock and scion.

Complete fertilizer - A plant food which contains all the three primary elements Viz., nitrogen, phosphorus and potassium.

Compost - Decomposed organic matter.

Concentric cracking - Cracking which develop concentrically around the shoulder of tomato fruits.

Conifer - A cone bearing tree having tiny needlelike leaves.

Connate - United or joined.

Cordate - Heart-shaped, with the notch basal.

Cordon - A system of training where the trunk is not developed straight but is trained in a horizontal direction along the lower wire of trellis at a height of 60 cm.

Corm - A solid, fleshy bulb-like part of the stem, usually subterranean and spherical in shape.

Cormel - A corm arising vegetatively from a parent corm.

Corymb - A short, broad, more or less flat-topped, indeterminate inflorescence in which all the flowers reach one level because of differential growth of their stalks; the outer flowers opening first.

Cover crop - A crop planted in place of normal crop to control weeds which add humus to the soil when ploughed before regular planting regular planting.

Creeper - A trailing shoot that roots throughout most of its length.

Crossing over - The exchange of corresponding segments between chromatids of homologous chromosomes during meiotic prophase resulting in recombination of linked genes.

Cross-Pollination - Pollination of a flower with pollen from another flower.

Crown - Plant part at the surface of ground from where new shoots are produced.

Cultivar - An agricultural or horticultural variety, which has originated and persisted under cultivation, as distinct from a botanical variety.

Cuneate - Wedge-shaped, with the broad end apical.

Cybrids - The product of fusion of protoplasts of two different genetic backgrounds with mixed cytoplasm.

Cylindrical - Elongated with a circular cross-section.

Cymose - A broad, more or less flat-topped, determinate inflorescence, with central flower opening first.

Cytoplasm - The protoplasm of a cell excluding the nucleus.

Cytoplasmic inheritance - Transmission of hereditary characters through the cytoplasm as detected by differing contribution of male and female parents in reciprocal crosses.

D

Damping off - This is a term used to describe the death of small seedlings resulting from attacks by certain fungi like *Pythium ultimum* , *Rhizoctonia solani, Botrytis cinerea* and *Phytophthora* spp.

Day-neutral plant - Plants in which flowering is not influenced by day-length.

Deblossoming - Removing blossoms from a plant.

Deciduous - Falling off soon after maturity.

Decumbent - Reclining or lying on the ground.

Dedifferentiation - It is the capability of mature cells to return to a meristematic condition and develop a new growing point.

Dehiscent - Opening spontaneously when ripe

Deltoid - Triangular; delta shaped.

Dentate - Having marginal teeth where apices are perpendicular to the margin and do not point forward.

Designated diseases - Refers to the diseases specified for certification of seeds and in whose regard certification standards must be met.

Desuckering - Removal of excessive sucker from banana clump.

Determinate - Growth habit where the terminal or central flower of an inflorescence opens first and the prolongation of the axis is arrested.

Diallel cross - Crossing of a number of genotypes in all possible combinations.

Dichogamy - Condition in which male and female parts of a flower mature at different times, ruling out the possibility of self-pollination.

Diclinous - Having male and female germ cells in separate flowers either on the same plant or on different plants.

Dichotomous - Ending with two divisions.

Differentiation - The process of changes in composition, structure, and functions of cells and tissues during growth.

Dihybrid - Heterozygous with respect to two genes.

Dioecious - Plants where staminate and pistillate flowers occur on different individuals.

Diploid - An organism with two chromosomes of each kind.

Disinfectants - Disinfectants are chemicals used to eliminate pathogenic organisms within the seed.

Disinfectants - Disinfectants are chemicals used to eliminate pathogenic organisms present on surface of the seeds.

Disk florets - The tubular flowers in the centre of heads of most composites.

Dominant - A gene which expresses itself in the hybrid to the exclusion of its contrasting recessive allele

Dormancy - An internal condition or stage of development, of a viable seed that prevents its germination even under favourable conditions of germination which can be broken by time or providing special conditions

Dormant - Resting, inactive, quiescent, inactive stage of development in an organism.

Double cross - A cross between two F_1 hybrids.

Double fertilization - Union of one male gamete with the egg producing zygote and the other with the two already fused polar nuclei to produce the endosperm.

Doubled haploid - A plant produced by doubling the chromosomenumber ofa haploid plant.

Drepaniform - Sickle-shaped.

Drift - Changes in gene and genotypic frequencies in small populations due to random processes.

Drought - A period of dryness coupled with low atmospheric humidity and soil moisture deficiency in the root zone of plant.

Drupe - A fleshy, one-seeded, indehiscent fruit with the seed enclosed in a stony endocarp.

E

Ecology - Interaction of organism and environment.

Ecophenes - Plants with similar genetics exhibiting differences due to varied natural environment.

Ecosystem - Communities of plants, animals and microbes together with their physic-chemicalenvironment.

Ecotype - Individuals which are interfertile with each other and members of other ecotypes of the same species, but maintain their individuality through environmental isolation and selection.

Ellipsoid - Elliptical in section, like a football

Elliptical - Oval in outline, being narrowed to rounded at ends and widest at or about the middle.

Emarginate - Apex markedly notched.

Emasculate - To remove the anthers from a bud or flower before the pollen is shed.

Embryo - The rudimentary plant still enclosed in the seed which arises from the zygote.

Embryo culture - Excising embryo from seed and germinating it in a aseptic culture.

Embryo rescue - Using embryo culture to restore interspecific hybrids that would be lost otherwise during embryo development.

Endocarp - The innermost layer of the pericarp.

Endophyte - A plant living within another plant.

Endosperm - Nutritive tissue surrounding and nourishing the embryo in seed plants.

Ensiform - Sword-shaped, gladiate.

Entomophily - Pollination by insects.

Environment - The sum total of the external conditions which affect growth and development of an organism.

Ephemeral - A plant with very short life-cycle, having several generations in one year.

Epicalyx - A whorl of bracts below the calyx.

Epicarp - The outer layer of fruit wall.

Epicotyl - The growing axis above the cotyledon.

Epigeal - Plants in which the cotyledons appear above the surface of the soil during germination.

Epigynous - The floral organs are attached to the upper part of the ovary, the ovary then appearing inferior in position.

Epiphyte - A plant which grows on another plant but gets its nourishment from air and rainfall. It is not harmful to the host plant.

Epinasty - More growth in upper surface than in the lower surface which causes the leaf to curl downward.

Epistasis - Interallelic interaction whereby manifestation at any locus is affected by genetic phase at any or all other loci or dominance of one gene over a nonallelic gene .

Erect - A rigid and strong stem holding itself in an upright position.

Erosion - The removal of soil by wind, water or man.

Etiolation - Development of plants or plant parts in the absence of light.

Etiology - The study of causal organism of diseases and its relation with the plant.

Evaporation - This is the process by which water in soil, earth and plants returns to the air. Higher temperatures speed the process of evaporation.

Evergreen - Not deciduous; persistent.

Exocarp - The outermost of the three layers forming the pericarp of a fruit.

Exotic - A plant introduced from a foreign country.

Explant - Original tissue used for growing plants in an artificial medium.

Exserted - Projecting beyond, as stamens from a perianth.

Eye - An undeveloped bud growth which will later produce new growth.

F

F_1, F_2, etc. - Symbols used to designate the first generation, second generation, etc., after a cross.

Falcate - Sickle-or scythe-shaped

Fallow - A period during which the land is not put under cultivation.

False fruit - A fruit that develops from ovary and other parts of the flower viz., receptacle or complete inflorescence

Family - A group of individuals directly related by descent from a common ancestor.

Fasciated - Abnormally flattened and seemingly several units fused together.

Fertigation - Applying fertilizers along with irrigation water .

Fertilization - Union of the gametes (egg and sperm) to form a zygote.

Fertilizer - Organic or inorganic plant foods in liquid or granular form which is used to amend the soil for improving the quality or quantity of plant growth.

Filament - The stalk of a stamen.

Floret - Individual small flower.

Foliar feeding - Applying liquid fertilizers on the leaves.

Forcing - The process of hastening plant growth to maturity or flowering.

Foundation seed - Seed stock produced from breeder seed by or under the direct control of an agricultural experiment station. Foundation seed is the source of certified seed, either directly or through registered seed.

Fresh ungerminated seed - Seeds, other than hard seeds, which remain firm and apparently viable even after the appropriate treatment for breaking dormancy.

Fruit - The ripened ovary with adnate parts.

Functional male sterility - Pollengrain is normal and functional. But anther does not dehisce to release pollen.

G

Gamopetalous - With united petals either throughout their length or at the base.

Gamosepalous - With a calyx whose sepals are marginally connate, in whorl or in part.

General combining ability - Average performance of a line in a series of crosses.

Genetic gain - Average improvement in a progeny over the mean of the parents.

Genome - A set of chromosomes as contained within the gamete and corresponds to the haploid chromosome number of the species.

Genotype - The genetic make-up of an organism comprising the sum total of its genes, both dominant and recessive; a group of organisms with the same genetic makeup.

Germination - The resumption of growth by the embryo resulting in the development of a young plant from the seed.

Germplasm - The sum total of hereditary materials in a species.

Girdling - Destroying an area of bark around the periphery of a stem of the living plant.

Glabrous - Without hairs, smooth.

Grading - Separating or sorting of produce according to size or quality.

Graftage - Method of inserting buds, twigs, or shoots on other stems or roots for fusion of tissues.

Grafting - Grafting is the process of connecting two pieces of living plant tissue together in such a manner that they unite and grow as one plant.

Green house - A structure built using glass,plastic or fibre glass with controlled temperature and humidity.

Ground cover - A group of plants usually used to cover bare earth and create a uniform appearance.

Grow-out test - Performed to determine the genuineness of seed as to species or variety, or freedom from seed-borne infection.

Growth habit - Mode of growth.

Growth regulator - Any substance, either external or internal, that influences plant growth..

Gummosis - A disorder, where copious exudations or deposits of gum is seen, particularly in stone fruits and citrus as a result of mechanical injury, pathogenic fungi, insects etc.

Gynodioecious - Group of plants consisting of pistillate and hermaphrodite plants.

Gynoecious - A condition where plant produces only pistillate flowers.

Gynophore - A stalk on which an ovary is elevated above the floral axis.

H

Habit - Characteristic form or bodily appearance of an organism.

Habitat - Natural home of a plant.

Hairy - Covered with hairs.

Halophyte - A plant which can tolerate high soil salinity.

Haploid - A cell or organism with the gametic chromosome number(n).

Hard seeds - Seeds which remain hard at the end of prescribed test period because they have not absorbed water owing to an impermeable seed coat .

Hardening - Subjecting plants to adverse conditions to hasten tissue maturation for increasing hardiness.

Hardiness - The ability of a plant to withstand low temperatures or frost.

Hardwood cutting - Cutting from mature hardwood for propagation

Hastate - Arrow-shaped with the basal lobes turned outward.

Heading back - Removal of only a portion, leaving another portion from which new growth can take place.

Herb - Any non-woody plant without persistent aerial parts having green colour and leafy texture.

Herbaceous cuttings - These are cuttings taken from succulent herbaceous plants.

Heritability - The proportion of observed variability due to heredity (additive effect of genes)

Hermaphrodite - Flowers with stamens and pistil in the same flower.

Hesperidium - A berry with a tough, leathery rind

Heterogamous - With two or more kinds or forms of flowers.

Heterogeneous - Lacking in uniformity; exhibiting variability.

Heteromorphism - Dissimilarity in the position and placement of floral parts.

Heterosis - Increase in vigour, growth or yield of an F_1 hybrid over its parents .

Heterostyly - Condition in which length of style differs in flowers of different plants of a species; a device for ensuring cross-pollination.

Heterozygous - Having unlike alleles at corresponding loci of homologous chromosomes; an organism can be heterozygous for one or several genes.

Hexaploid - Having six sets of chromosomes (6n).

Hilum - Scar on seed coat, marking point of former attachment of seed to funicle.

Homogamy - Simultaneous maturing of pollen and stigmas in a perfect flower.

Homologous chromosomes - Chromosomes which pair at the first division in meiosis and each member of a pair has a corresponding sequence of gene loci and is derived from a different parent.

Homozygous - Having like alleles at corresponding loci on homologous chromosomes. An organism can be homozygous at one, several, or all loci.

Hormone - A substance synthesized by an animal or organism that is able to influence growth at some point other than the point of synthesis; comparable to an auxin in plants. Sometimes the term is used in broad sense to include auxins.

Horticulture - Art and science of growing fruits, vegetables and ornamental plants.

Hybrid - The first generation offspring of a cross between two individuals differing in one or more genes.

Hybridization - The crossing of individuals of unlike genetic constitution.

Hydrophily - Pollination by water.

Hydroponics- The science of growing plants in mineral solutions or liquid.

Hypocotyl - The part of the axis below the cotyledons in a seedling.

Hypogeal - Below ground; in hypogeal germination the cotyledons remain below ground within the testa.

Hypogynous - A flower in which all other floral organs are situated beneath the ovary.

I

Illegitimate seed - Resulting from natural cross pollination between plants or clones where the male parent cannot be ascertained.

Ideotype - An ideal type of a plant which performs predictably well.

Immune - Not liable to attack by by pests/diseases.

Inarching - Approach grafting where scion remains a part of tree till its union with stock is complete. The scion is severed off 2-3 months after the union.

Inbred line - Product of continued inbreeding developed by self-pollination and selection.

Inbreeding - The mating of individuals more closely related than individuals mating at random.

Incised - Cut rather deeply.

Incompatibility - Failure to obtain fertilization and seed formation after self pollination, within or between clones.

Incomplete flower - A flower lacking one or more sets of floral organs.

Indehiscent - Not opening when ripe.

Indeterminate - An inflorescence in which the terminal flowers are the last to open, so that the floral axis may be prolonged indefinitely by a terminal bud.

Indigenous - Native to a particular area or region.

Integument -The envelope of an ovule

Internode -The portion of a stem between two adjacent nodes.

Interstock (Intermediate stock, Interstem)-This is a piece of stem inserted by means of two graft unions between the *scion* and the *rootstock.*

Irradiation - Exposing seeds,pollen or other plant parts to x-rays or other short wavelength(gamma0 radiations to increase mutation rates.

Isogenic lines - Two or more lines differing from each other genetically at one locus only.

Isolation - Separation of one group from another so that mating between groups is prevented.

Isolation distance - Distance to be maintained between the seed crop and the contaminant.

J

June drop - The abscission of partly developed fruits (often occurring in June).

Juvenile stage - Early or vegetative phase of growth characterized by carbohydrate utilization.

K

Kniffin - System of training where the arms of the vine are tied to horizontal wire on the same level above the ground.

L

Lanceolate - Lance shaped; much longer than broad; widening above the base and tapering to the apex.

Layering - Layering is a propagation method in which adventitious roots are initiated on a stem while it is still attached to the parent plant.

Leaf bud cuttings - Here the leaf blade, petiole and a small piece of stem with attached axillary bud is the propagating material.

Leaf cuttings - Here the leaf blade or leaf blade and petiole is used as the propagating material.

Leaf spines - Leaves modified into sharp, pointed structure, the spines.

Leaf tendrils - Leaves modified into slender, wiry, often closely coiled structures, the tendrils.

Legume - A dry, dehiscent fruit, formed from single carpel that splits along both sutures.

Line - A group of individuals from a common ancestry.

Linear - Long and narrow, the sides parallel or nearly so.

Linkage - Association of characters in inheritance due to location of genes in proximity on the same chromosome.

Locule - Compartment or cell of an ovary, fruit or anther.

Locus - The position occupied by a gene in a chromosome.
Long day - A plant which requires a day critical than its day length in order to flower.
Lorate - Strap-shaped, ligulate.
Lycopene - A carotenoid responsible for the red colour of tomato fruit.
Lyrate - Lyre-shaped, pinnatifid with an enlarged terminal lobe and smaller lower lobes.

M

Macro element - Elements present and also required in relatively large amounts by plants, viz., N, P, K, Ca, H, etc.
Male sterility - Absence or non-functional.nature of pollen in flowering plants
Marker gene - A gene of known function and known location on the chromosome
Mass selection - A form of selection in which individual plants are selected and the next generation propagated from the aggregate of their seeds.
Mass-pedigree method - A system of breeding in which a population is propagated in mass until conditions favourable for selection occur, after which pedigree selection is practiced.
Meiosis - Nuclear divisions in which the diploid chromosomes number is reduced to half that of the parent cell to give the haploid number, as in gametes.
Meristem - Region of actively growing cells.
Mesocarp - The middle layer of the pericarp or fruit wall which is often fleshy or succulent.
Mesophyte - A plant that grows normally under conditions of moderate humidity, that neither requires nor survives culture in water or extreme drought.
Metaxenia - Influence of pollen on maternal tissue of the fruit.
Microelement - Element essentially needed in small quantity, for the nutrition which also termed as minor element viz., Cu, Zn, Mn, Mg, B.
Micro irrigation - Irrigation method that provides water in and around root zone of plant with the help of emitters.
Micropropagation - Production of plants from very small plant parts,tissues or cells grown aseptically in a test tube or other container where the environment and nutrition can be rigidly controlled.
Mitosis - The normal process of the division of somatic cells in which the chromosomes are duplicated longitudinally to give two daughter nuclei each having a chromosome complement equal to that of the original nucleus.
Monocapic - Fruiting once and then dying.
Monoecious - Both staminate and pistillate flowers borne separately on the same individual plant.
Monoembryonic - A seed with a single embryo which is usually of gametic origin.
Monohybrid - An individual heterozygous for one chromosome.
Monopodial - A primary axis which continues its original line of growth from the same apical meristem to produce successive lateral branches.
Mulch - Plant residue or other materials placed on the surface of soil to conserve moisture, control erosion, runoff, check temperature and weed growth.
Mutant - A plant that arises as a result of a mutation.

Multiple alleles - A series of alleles or alternative forms of a gene.

Multiple fruit - A cluster of fruits traceable to the gynoecia of separate flowers.

Multiple genes - Two or more independent pairs of genes which produce complementary or cumulative effects upon a single character of the phenotype.

Mutant - Individual which has suddenly acquired a heritable variation which is not present in the parent form.

Mutation - A sudden variation in the hereditary material of a cell.

Mycorrhiza - A symbiotic association between roots of a plant with a fungus which may form a layer outside the root (ectotrophic) or within the outer tissues (endotrophic).

N

Necrosis - Death of plant part/tissue.

Node - A point on a stem where leaves or branches are attached.

Nonrecurrent apomixis - Here an embryo arises directly from the egg nucleus without fertilization.

Nucellar embryo - An embryo developed from the tissue of the nucellus and consequently of the same genotype as the female parent.

Nucellar embryony - A form of apomixis in which the embryo arises directly from the nucellus

Nut - A simple, dry, indehiscent fruit with a bony shell, characteristically derived from a compound gynoecium, but one-seeded by abortion.

Nutrients - Elements available through soil, air and water which the plant utilizes for growth, development and reproduction.

O

Obcordate - Heart-shaped, with the notch apical.

Obdeltoid - Deltoid but the attachment of the petiole at one of the angles instead of on a side of the triangle.

Oblanceolate - The reverse of lanceolate, a leaf broader at the distal third than at the middle and tapering toward the base.

Oblique - Slanting; unequal-sided.

Oblong - Longer than broad, with the margins nearly parallel.

Obovate - The reverse of ovate, broader toward the apex.

Offset(Offshoot) - This is a type of lateral shoot or branch that develops from the base of the main stem in certain plants.

Off-type - Plant or seed deviating significantly from the characteristics of a variety as described by the plant breeder in any observable respect.

Orbicular - Cir7cular or disc-shaped.

Ornithophily - Pollination by birds.

Osmosis - Passage of material through a membrane from areas of higher concentration to areas of lower concentration.

Oval - Elliptical with the width greater than half the length.

Ovary - Part of the pistil which contains the ovules. This is a swollen part which later becomes the fruit.

Ovate - Oval, but broader toward the base.

Ovule - A structure containing the egg which later becomes the seed after fertilization.

P

Pandurate - Fiddle-shaped; obovate with a marked concavity along each basal side.

Panicle - A branched cluster of stalked flowers.

Parthenocarpic - Development of fruit without fertilization of ovule

Parthenogenesis - Development of an organism from a sex cell but without fertilization of ovule.

Pedate - Palmately divided or parted with the lateral divisions clefted.

Pedicel - Stalk of individual flower of an inflorescence.

Pedigree - A record of the ancestry of an individual, family or strain.

Peduncle - The stalk of an inflorescence or partial inflorescence.

Pellucid - Translucent.

Peltate - Shield shaped with the petiole attached to the underside.

Pendulous - Drooping or hanging down.

Pepo - A one-celled, many-seeded, inferior type of fruit with parietal placenta with a pulpy interior.

Perennial - A plant that continues its growth from many years.

Perianth - A collective term for corolla and calyx.

Pericarp -The wall of the ripened ovary or fruit wall of which the layers may be fused into one, or more or less divisible into exocarp, mesocarp and endocarp.

Perigynous - The floral organs borne and arising from around the ovary and not beneath it.

Persistent - Remaining attached; not falling off.

Petal - One of the separate parts of a corolla which is usually coloured and showy.

Petiolule - The stalk of a leaflet.

Phenotype - The external appearance of an organism which is the result of interaction of genotype with the environment in which the plant is growing.

Phyllotaxy - The arrangement of leaves on the stem.

Physiological races - Pathogens of the same species and variety, which arestructurally similar but differ in physiological and pathological characteristics.

Phytochrome - A pigment in plant associated with daylenth response.

Pinching back - Using the thumb and forefinger tip of a branch or stem is nipped.

Pistil - The female part of a flower (gynoecium) consisting of ovary, styles and stigmas of one or more carpels.

Pistillode - A rudimentary pistil present in some staminate flowers.

Placentation - The position of the placentae in the ovary.

Plicate - Folded as in a fan or approaching this condition.

Pod - A dry dehiscent fruit.

Polar nuclei - Two of the nuclei of the embryo sac which unite with the second sperm from the pollen in a triple fusion, giving rise to the endosperm, which is thus triploid.

Pollarding- This is the process of heading back old,tall and ill shaped branches to produce good shape and also to promote new growth on which fruits are produced in plenty. This is a common operation in drumstick.

Pollination - The transfer of pollen from an anther to the receptive stigma of a flower.

Polycarpic - Flowering and fruiting many times.

Polycross - Open pollination of a group of genotypes (generally selected) in isolation from other compatible genotypes in such a way as to promote random mating *inter se.*

Polyembryony - The production of two or more embryos within an ovule.

Polygamodioecious - A condition where two sorts of plants occur; one have some perfect and some staminate flowers while the other have some perfect and some pistillate flowers.

Polygamomonoecious - A condition in which some perfect and some staminate flowers are produced on the same plant.

Polygamous - The same or different individual plants producing both bisexual and unisexual flowers.

Polygenes - Genes whose effects are too slight to be identified individually but which, through similar and supplementary effects, can have important effects on total variability.

Polymorphism - The occurrence together in the same population of two or more distinct forms at frequencies too great to be explained by recurrent mutation.

Polypetalous - Having a corolla of separate petals.

Polyploid - An organism with more than two sets (genomes) of chromosomes in its somatic cells.

Polysepalous - Having a calyx of separate sepals.

Pome - A fruit in which the seeds are surrounded by a tough but not woody layer derived from the inner part of the fruit wall, the whole fused with the deeply cup-shaped fleshy receptacle.

Pomology - The study of fruits.

Precision farming - Precision farming often referred as hi-tech farming, is a farming management concept based on observing and responding to intra-field variations. The objective is optimisation of inputs use to facilitate optimal output resulting in saving of valuable resources like water and energy.

Precocious - Appearing or developing early.

Primary meristems - They are group of cells directly descended from embryonic cells that have never involved in meristematic activity.These are also known as performed meristems.

Protandrous - Stamens shedding pollen before the stigma is receptive.

Protandry - A condition in which the anthers mature before the pistil in the same flower, the pollen being dispersed before the stigma is receptive. This is a condition favouring cross pollination.

Protogyny - A condition in which the gynoecium mature before the anthers. This condition also favours cross pollination.

Pruning - The judicious removal of leaves, shoots, twigs, branches or roots of a plant to increase its fruitfulness.

Pseudobulb - This is a specialized storage structure produced by many orchid species consisting of an enlarged, fleshy section of the stem made up of one to several nodes.

Pubescent - Covered with soft short hairs.

Pure line - A strain in which all members have descended by self-fertilization from a single homozygous individual.

R

Raceme - A simple, elongated, indeterminate inflorescence with stalked flowers.

Radial cracking - Cracking which radiates from the stem end of tomato fruits.

Ratoon - New shoots arising from perennial crops after the first crop which is used as propagating material of subsequent crops.

Ray florets - Outer modified floret of some composites with an extended or strap-like part to the corolla.

Recurrent selection - Used in breeding to increase the frequency of favourable genes for yield and other characters by repeated cycles of selection.

Reniform - Kidney shaped.

Rhizome - An underground stem which is distinguished from a root by the presence of nodes, buds and leaves or scales.

Rhomboidal - Rhombic shaped.

Ringing - Removing a narrow strip of bark round the periphery of the stem to prevent downward translocation of food beyond that point.

Roguing - The removal of individual plants which deviate in a significant manner from the normal type in an established variety or during seed production.

Root cuttings - Here roots are the propagating material.

Rootlet - A very slender root or the branch of a root.

Rootstock - The seedling or plant over which a shoot is grafted

Runcinate - Coarsely serrate to sharply incised with the teeth pointing toward the base.

Runner - A slender elongated lateral shoot with long internodes and rooting at the nodes

S

S_1, S_2, etc. - Symbols denoting the first selfed generation, second selfed generation etc.

Sagittate - Arrow shaped or triangular with the basal lobes pointing downward.

Scale leaves - Leaves thin, dry, sessile, membranous and usually brownish in colour.

Scarification - Scarification is any process which soften the seed coat making them permeable to water and gases.

Scion - A detached shoot or bud used in vegetative propagation for grafting or budding.

Secondary meristem - These are group of cells that have differentiated and functioned in some tissue system which again resume meristematic activity. Also known as wound meristem.

Seed - Seed is a mature ovule which consists of embryo , seed-coat and in some cases an endosperm.

Segregation - The separation of homologous chromosomes and genes from different parents at meiosis.

Selection - Any process which results in an increase in the proportion of genotypes of desired characteristics in succeeding generations.

Self fertile - Capable of fertilization and setting seed after self-pollination.

Self incompatibility - The inability of a plant producing functional gametes to set seed when self-pollinated

Self-pollination (Autogamy) - Pollination by its own pollen.

Self-sterile - Failure to complete fertilization and produce seed after self- pollination.

Semihardwood cuttings - These are cuttings taken from woody, broad leaved evergreen species. Cuttings from partially matured wood of deciduous plants are also considered as semihardwood cuttings.

Senescence - A physiological aging process in which tissues in an organism deteriorate and finally die.

Sepal - A member of the outer series of perianth segments or one of the separate parts of a calyx, usually green and foliaceous.

Short day plants - Plants which flower only when the day length is less than the critical length.

Shrub - A woody plant of limited height, typically branching at or near the base and not developing much of a trunk.

Siblings - A group of individuals within a family.

Sibs - Progeny of the same parents derived from different gametes.

Silicula - A short and broad siliqua containing only a few seeds.

Siliqua - A dry, dehiscent fruit formed from two fused carpels separated by a false septum from which carpel walls diverge when ripe.

Simple fruit - Fruits derived from a single carpel or from a number of fused carpels.

Single cross - A cross between two genotypes, usually two inbred lines.

Smudging - Induction of flowering in mango through building of slow fires emitting smoke.

Softwood cuttings - Cuttings taken from soft, succulent, new spring growth of deciduous or evergreen plants are called as softwood cuttings.

Somaclone - Plant developed from cell culture.

Spadix - A thick or fleshy spike surrounded or subtended by a spathe.

Spathe - A large bract enclosing a spadix or two or more bracts enclosing a flower cluster.

Spatulate - Spoon shaped.

Species - The unit of taxonomic classification into which genera are subdivided. A group of similar individuals different from other similar arrays of individuals.

Spike - An indeterminate inflorescence of sessile or subsessile flowers.

Stacking - Practice of driving a stake into the ground next to the plant for supporting it.

Stamen - Male reproductive organs of a flower; a unit of the androecium. typically consisting of anther and filament.

Staminate - A unisexual flower with stamens.

Staminode (Staminodium) - A sterile stamen.

Steckling - A root (such as beetroot, carrot, turnip) which is stored to over winter for replanting in the following spring for seed production.

Stigma - The portion of the pistil which receives the pollen.

Stock (Rootstock, Understock) - This is the lower portion of the graft which develops into the root system of the grafted plant.

Stolon - A slender, horizontal creeping stem which at or below the surface of the ground gives rise to a new plant at its tip.

Strain - A group of similar individuals within a variety.

Stratification - This is a method of breaking dormancy where the imbibed seeds are subjected to a period of chilling to after ripen the embryo.

Structural male sterility - The male part (stamen) is absent.

Subulate - Awl shaped or tapering from base to apex.

Succulent - Fleshy.

Sucker - This is a shoot that arises on a plant from below ground.

Superior - An ovary with the perianth inserted below or around its base, the ovary being free.

Syncarp - A fleshy aggregate fruit.

Synthetic variety - A variety produced by crossing *inter se* a number of genotypes selected for good combining ability in all possible hybrid combinations, with subsequent maintenance of the variety by open pollination.

T

Tap root - The primary descending root, forming a direct continuation from the radicle.

Tendril - A thread like climbing organ formed from the whole or part of a stem, leaf or petiole.

Tepal - Individual member of perianth, not clearly differentiated into corolla and calyx.

Test cross - A cross of a double or multiple heterozygote to the corresponding multiple recessive to test for homozygosity or linkage.

Testa - The outer coat of the seed which is usually hard and brittle.

Tetraploid - An organism with four basic (x) sets of chromosomes.

Thinning out - Pruning of entire stems or branches to give remaining stems or branches more space and more reserve food for improved growth.

Three way cross - A cross between a single cross and an inbred line.

Top cross - A cross between an inbred line and an open pollinated variety.

Topiary - A method of pruning and training of plants into formal shapes such as animals.

Top working - Converting an undesirable plant into a suitable variety by shield budding or cleft, crown or tongue grafting by cutting back the main scaffold branches and inserting one or more scions of desirable varieties in each cut end.

Totipotency - Totipotency is the ability of living plant cell to regenerate into a new plant.

Trailing - A weak stem which creeps on the ground without rooting at the nodes.

Transgenic plants - Genetically transformed plants owing to combination of different kinds of DNA.

Transpiration - The giving off of water vapour by plants through cuticle,stomata and lenticels.

Trap crop - Crops attractive to pests more than the main crop.

Tree - A woody plant which produces one main trunk and a more or less distinct and elevated head.

Trichome - A bristle or hair.

Trimonoecious - Male flowers, female flowers and hermaphrodite flowers are separate and seen in one plant.

Triploid - An organism with three basic (x) sets of chromosomes.

True fruits - Fruits which are derived from the carpel(s) alone.

Truss - A cluster of flowers.

Tuber - A fleshy, underground swollen portion of stem or root bearing scale-like leaves with buds (eyes) in their axils and acts as a storage organ and propagule

Tuberous root - Thick and fleshy root without any particular shape.

Tuberous stems - They are produced by the enlargement of the hypocotyl section of the seedling plant, but may include the first nodes of the epicotyl and the upper section of the primary root.

U

Umbel - An indeterminate, often flat-topped inflorescence consisting of several pedicellate flowers having a common point of attachment.

Unilocular - Having a single cavity.

Unisexual - A flower having only one sex.

Univalent - An unpaired chromosome in meiosis.

V

Valvate - Jointed by edges without overlapping.

Variability - The amount of variation present among members of a population.

Variation - Difference among individuals of same or of different species.

Variegated - Leaves which are marked with multiple colours.

Variety - A group of strains or single strain of closely related plants of common origin which have similar characteristics and can be differentiated on the basis of structural and functional characters from another group of plants. Vase life- Longevity of flower in vase after cutting from the stem.

Vegetable - The edible product of a herbaceous garden plant.

Vegetative apomixis- Here vegetative buds or bulbils are produced in the inflorescence in place of flowers.

Vernalization - Treatment of seeds or bulbs before planting to hasten flowering.

Vernation - The arrangement of leaves within a bud.

Vigour - Sum total of all seed attributes which favour rapid and uniform crop stand in the field.

Vine - A slender stemmed climbing or trailing plant.

Virulence - Capacity of a pathogen to incite a disease.

Vitamins - A group of food substances other than fats, carbohydrates, proteins and minerals/salts which are essential to nutrition and serve to prevent diseases.

Viticulture - The science of grpe cultivation.

Vivipary - Germination of seeds while fruits are still attached to the parent plant.

Volunteer Plants - Unwanted plants growing from seed that remains in the field from a previous crop.

W

Waxing - Coating fruit surface with wax to reduce water loss.

Wilting - The drooping and loss of freshness of plant foliage.

Wilting coefficient - (Wilting point, Wilting range). The percentage of moisture held by the soil within the feeding zone of root when permanent wilting of plants takes place.

Witche's broom - A disorder of the stem or its branch in which there is extreme amount of branching, resulting in a very bushy condition.

Wither tip - A form of dieback reported to be due to copper deficiency.

X

Xenia - Influence of pollen upon phenotype.

Xylem - A complex tissue of tough,fibrous, elongated cells forming vessels and woody tissue, responsible for upward movement of water and minerals from roots to leaves.

Z

Zoophily - Pollination by animals.

Zygomorphic - Capable of division by only one plane of symmetry; having the parts of the same kind of different sizes or shapes, so as to be bilaterally symmetrical.

Zygote - The cell formed from the fusion of two gametes; a fertilized egg.

Appendices

Appendix I-X

Appendix I: Released Varieties / Hybrids in Aromatic Crops from KAU

P. Skaria Baby, P.P. Joy, Samuel Mathew, Ancy Joseph and Ragina Joseph

Name of the crop	Lemongrass	Palmarosa	Vettiver	Kacholam	Kacholam	Cinnamon
Name of the variety	OD-19 (Sugandhi)	ODP-1	ODV-3	Kasthuri	Rajani	Sugandhini (ODC-130)
Source/parent/pedigree	Local collection from Nenmara	Germplasm collection at Aromatic and Medicinal Plants Research Station, Odakkali	Germplasm collection at Aromatic and Medicinal Plants Research Station Odakkali	Collection from Koothattukulam in Ernakulam district.	Collection from Chelavu in Thodupuzha of Idukki district.	Germplasm collection at Aromatic and Medicinal Plants Research Station, Odakkali
Method of Breeding Year of release and authority	Selection Kerala State Seed Committee meeting held during 1988.	Selection	Selection	Clonal selection Released in the 21st Kerala State Seed Sub Committee meeting held on 18-09-2002.	Clonal Selection Released in the 21st Kerala State Seed Sub Committee Meeting held on 18-09-2002.	Selection Released in the Kerala State Seed Committee meeting during 2000.
Institution/Station responsible for the development	Aromatic and Medicinal Plants Research Station, Odakkali	Aromatic and Medicinal Plants Research Station, Odakkali	Aromatic and Medicinal Plants Research Station, Odakkali	College of Horticulture, Vellanikkara	College of Horticulture, Vellanikkara	Aromatic and Medicinal Plants Research Station, Odakkali
Specific advantages/features	Better yield with good quality oil	0.4-0.45% oil 95-98% geraniol	0.50% oil	Duration 210-240 days. Tolerant to shade and can be grown as intercrop in coconut gardens	High oleoresin content (7.68%), duration 210-240 days	Oil recovery 1.2-1.6%, Eugenol content of leaf oil 93.67%, No serious pests and diseases
Key characters for identification	Compact plants with reddish purple culm and	Perennial grass with lanceolate leaves. Panicles	Glabrous perennial grass with thicker	Robust plants with more number of leaves which	Leaves fewer and small. Rhizomes medium	Semi-compact canopy, medium to large

Contd.

Name of the crop	Lemongrass	Palmarosa	Vettiver	Kacholam	Kacholam	Cinnamon
	straight, thin, green leaves slightly rolled or folded up	turning reddish, often very bright when mature	stem, with less branching roots and wider leaves	are larger in size. Light brown rhizome	bold, creamy white in colour and more in number, particularly secondary rhizomes	leaves with purple coloured new flush
Productivity	Oil yield 100-125 kg/ha	30 t/ha herb yield FWB 120-130 kg/ha oil yield	5 t/ha root yield (shade dry), FWB 25 kg/ha oil yield	2.27 t/ha of dry rhizome	2.19 t/ha of dry rhizomes	180-200 kg leaf oil/ha
Area of adaptation	Major lemongrass growing areas throughout India	Throughout Kerala	Throughout Kerala	Central zone of Kerala	Central zone of Kerala	Midlands and high ranges of Kerala, both in open and as intercrop in coconut plantation
Tips of cultivation	Propagated by slips and seeds Spacing 15-20 cm Manures compost 2500 kg/ha Wood ash 1875 kg/ha	Propagated by slips and seeds. Spacing 20 x 60 cm. FYM 10t/ha	FYM or compost 5t/ha. Spacing 60 x 30-60 cm (based on soil fertility status climate etc.)	Spacing 20x15 cm. Seed rate 750 kg/ha	Spacing 20 x 15 cm. Seed rate 750 kg/ha	Spacing 4 x 4 m as monocrop Nutrient requirement organic manure 20 kg/plant NPK 200: 180-200 g/plant
Source books	Rajan S and KP Prameela 2006. Crop varieties released from KAU, KAU Directorate	Baby P Skaria, PP Joy, Samuel Mathew, Gracy Mathew, Ancy Joseph and Ragina	Baby P, Skaria, PP Joy, Samuel Mathew, Gracy Mathew, Ancy Joseph and Ragina	S Rajan and KP Prameela 2004. Crop varieties released from KAU, Kerala Agricultural	S, Rajan and KP Prameela 2004. Crop varieties released from	S, Rajan and KP Prameela 2004. Crop varieties released from

Contd.

Name of the crop	Lemongrass	Palmarosa	Vettiver	Kacholam	Kacholam	Cinnamon
	of Research, Vellanikkara, Thrissur, pp 199	Joseph 2007. Aromatic Plants. Horticulture Science Series -01 New India Publishing Agency, pp 118	Joseph 2007. Aromatic Plants. Horticulture Science Series-01, New India Publishing Agency pp 131	University, Directorate of Research, Vellanikkara, pp 197	KAU, Kerala Agricultural University, Directorate of Research, Vellanikkara, pp 198	KAU, Kerala Agricultural University, Directorate of Research, Vellanikkara Thrissur, pp 146
	Baby P Skaria, PP Joy, Samuel Mathew, Gracy Mathew, Ancy Joseph and Ragina Joseph 2007. Aromatic Plants, Horticulture Science Series 01, New India Publishing Agency 270 pp					Baby P Skaria, PP Joy, Samuel Mathew, Gracy Mathew, Ancy Joseph and Ragina Joseph 2007. Aromatic Plants, Horticulture Science Series 01, New India Publishing Agency 270 pp
Authority	D/s. Baby P Skaria, PP Joy, Samuel Mathew, Gracy Mathew, Ancy Joseph (Aromatic and Medicinal Plants Research Station, Odakkali, Asamannoor PO, Ernakulam – 383549, Tel: 0484-2658221, amprs@kau.in)	D/s. Baby P Skaria, PP Joy, Samuel Mathew, Gracy Mathew, Ancy Joseph (Aromatic and Medicinal Plants Research Station, Odakkali, Asamannoor PO, Ernakulam-383 549, Tel: 0484-2658221, amprs@kau.in)	D/s. Baby P Skaria, PP Joy, Samuel Mathew, Gracy Mathew, Ancy Joseph (Aromatic and Medicinal Plants Research Station, Asamannoor PO, Ernakulam-383549, T. 0484-2658221 amprs@kau.in)	D/s. Baby P Skaria, PP Joy, Samuel Mathew, Gracy Mathew, Ancy Joseph (Aromatic and Medicinal Plants Research Station, Asamannoor PO, Ernakulam-383549, T: 0484-2658221, amprs@kau.in)	D/s. Baby P Skaria, PP Joy, Samuel Mathew, Gracy Mathew, Ancy Joseph (Aromatic and Medicinal Plants Research Station, Asamannoor PO, Ernakulam-383549 T: 0484-2658221 amprs@kau.in)	D/s. Baby P Skaria, PP Joy, Samuel Mathew, Gracy Mathew, Ancy Joseph (Aromatic and Medicinal Plants Research Station, Asamannoor PO, Ernakulam-383549 T: 0484- 2658221 amprs@kau.in)

Appendix – II a
Released Varieties of Fruit Crops in India

T. Radha, Professor , KAU, PO KAU, Thrissur – 680 656

Name of the crop	Name of the variety	Parentage of hybrids	Released from	Specific characters
Banana	CO-1(1984)	Evolved through multiple crossing	TNAU, Coimbatore	The crop duration is 14-15 months. The mean bunch weight is 10.57 kg with seven hands per bunch and a total of 83 fruits per bunch. The fruits contain a TSS of 22.6^0 brix and acidity of 0.58 percent.
Lemon	Kagzikalan		ICAR-IARI, New Delhi	Plant is semi-vigrous, low spreading, and prolific bearer.
Guava	TRY(G)1(2005)		TNAU	It is an offseason bearing, shiny greenish yellow fruit with desirable aroma. It is resistant to fruit fly. Withstands drought and is sodicity tolerant.
	Pusa Srijan		ICAR-IARI, New Delhi	Extremely dwarf rootstock, hence recommended for high density planting.
	Arka Kiran		ICAR-IIHR, Bangalore	Pink pulp with soft seeded variety, mean fruit weight of 230g.
Papaya	CO-8	Dioecious derived through improvement of CO2		Red pulped, suitable for dessert, papain and processing. Duration 2 years. Yield 200-300 t/ ha.
	Arka Prabhath		ICAR-IIHR, Bangalore	Gynodioecious hybrid, deep pink pulp colour, 13-14 brix, 90-100 kg yield per plant.
	Coorg Honeydew	Selection from Honey Dew	ICAR-IIHR, Bangalore	Gynodioecious in nature and semi dwarf. Fruit skin is dark in colour with ridging on the surface. Pulp is yellow in colour, Fruits on an average weigh about 1.5- 2 kg.
	Surya	It is the offspring of		Plant is gynodioecious. Skin is smooth, becomes

Contd.

Contd. Appendix II a

Name of the crop	Name of the variety	Parentage of hybrids	Released from	Specific characters
		Sunrise Solo x Pink flesh sweet.		uniformly yellow in colour on ripening. Fruits are medium in size of about 600-800 with a small fruit cavity.
Mango	Pusa Lalima (2011)		ICAR-IARI New Delhi	Regular bearer. Semi vigorous and suitable for closer planting. Fruits are attractive in shape and having red peel and orange pulp.
	Pusa shreshth (2011)		ICAR-IARI New Delhi	Regular bearer, attractive elongated shape, red peel and orange pulp. Plants are suitable for closer planting.
Passion Fruit	Kaveri	Yellow passion fruit x Purple type.	ICAR-IIHR Bangalore	Vines are very vigorous. Produce ovoid to round, purple dotted large size (80-90) fruits with 30-35% juice.
Jack Fruit	PLR (J)2 (2007)	Clonal selection from Pathirakkotai Local	TNAU, Coimbatore	Good quality and bigger sized fruits with highly palatable and edible flakes. Suitable for growing in tropical climates.
Sapota	PKM (sa) 5 (2007)	Selection from open pollinated seedlings maintained in a private orchard		High yielder, High TSS and suitable for preparation of dry flakes, milk shake powder and mixed fruit jam. Oval shaped attractive fruit with smooth light brown skin flesh is crisp and retain coppery brown colour.
	PKM (sa) 7 (2003)	It is a open pollinated clones of PKM 1	TNAU, Coimatore	An open pollinated clone of PKM-1. High yielder with cluster bearing character. Fruits spindle shaped, less seeded and suitable for dry flake production.

Appendix – II b
Released Varieties of Arid Fruits

S.K. Sharma, CIAH, Bikaner, Rajasthan – 334 006

S.No.	Name of Variety	Parentage in case of hybrids	Released from	Specific Characters
1.	**Ber/Jujube**	*Zizipus Mauritiana* Lamk.		F: Rhamnaceae
	Goma Kirti(1997)	Clonal selection of Umran	ICAR-CHES, Godhra	Highly yielder and early maturing variety. Resistant to various diseases and pests by virtue of its earliness.
	Thar Sevika(2006)	Hybrid between Seb x Katha	ICAR-CIAH, Bikaner	An early maturing variety. The fruits are juicy, sweet with a TSS content of 22-24%. Fruits after consumption do not cause throat soaring, which is common in other cultivars. Average fruit yield is 30-32 Kg/tree. Also suitable for staggered picking which can be done up to third week of January.
	Thar Bhubhraj (2007)	A selection from local material of Bhusavar area of Bharatpur	ICAR-CIAH, Bikaner	An early maturing cultivar having an average yield potential of 30-36 kg/tree. The fruits district of Rajasthan are ready for harvesting during last week of December-First week of January. The fruits are very juicy, sweet with a TSS content of 22-23%.
	Narendra Ber -1*	Selection	NDUAT, Faizabad	Trees are spreading type, heavy yielder, fruits are round to oblong, yellowish green in colour.
	Narendra Ber - 2*	Selection	NDUAT, Faizabad	Trees are medium in height, semi spreading, fruits are obviate with yellowish green skin.
2.	**Aonla**	*Emblica officinalis* Gaertn.		F: Euphorbiaceae
	Goma Aishwariya (2006)(CHES Sel.1)	A clonal selection of NA-7	ICAR-CHES, Godhra	An early and drought tolerant variety. The average yield potential is 102.9 kg/tree. It has low fiber content and is suitable for processing and export.
3.	**Bael**	*Aegle marmelos* (L.) Corr. Serr.		F: Rutaceae
	Narendra Bael (NB)-4	Selection	NDUAT, Faizabad	Trees are spreading and fruits are oblong in shape, fruit quality excellent and heavy bearer.

Contd.

Contd. Appendix II b

S.No.	Name of Variety	Parentage in case of hybrids	Released from	Specific Characters
	Narendra Bael (NB)-5	Selection	NDUAT, Faizabad	Prolific bearer and fruits are medium in size, round with thin skull, low fibre and seed content.
	Narendra Bael (NB)-7	Selection	NDUAT, Faizabad	Fruits are very large in size, flattened round, yellowish green in colour.
	Narendra Bael (NB)-9	Selection	NDUAT, Faizabad	Prolific bearing, fuits are medium to large size with oblong in shape, low fibre and seed content.
	Narendra Bael (NB)-16	Selection	NDUAT, Faizabad	Fruits are elliptical round, pulp yellow, average weight 1.3 kg, T.S.S. 31%, and low fibre content.
	Narendra Bael (NB)-17*	Selection	NDUAT, Faizabad	Fruits are attractive, average weight 1.75 kg, fibre content low.
	CISH B-1	Selection	ICAR-CISH, Lucknow	It is a mid season variety which matures during April-May. Trees are tall, vigorous with dense canopy, erect growth habit, precocious and heavy bearer. Fruit shape is oval to oblong. Average fruit weight 1.0 kg. Suitable for canning and slices preparation.
	CISH B-2	Selection	ICAR-CISH, Lucknow	Trees are dwarf with medium spreading habit. Foliage is sparse and almost thornless, precocious with moderate bearing habit. Fruits are oblong to round in shape. Suitable for processing with pleasantly aromatic pulp.
	Goma Yashi	Selection	ICAR-CHES, Godhra	Good quality fruits with large in size. Ovate in shape, greenish yellow in colour. Flesh colour is straw.
	Pant Aparna	Selection	GBPUAT, Pantnagar	Its trees are dwarf with drooping foliage, almost thornless, precocious and heavy bearer. The leaves are large, dark green and pear shaped. Fruits are globose in shape with average weight 1.0 kg.
	Pant Shivani	Selection	GBPUAT, Pantnagar	It is an early mid season variety. Trees are tall, vigorous, dense, upright growth, precocious and heavy bearer. Fruit weight range from 2 to 2.5 kg.

Contd.

Contd. Appendix II b

S.No.	Name of Variety	Parentage in case of hybrids	Released from	Specific Characters
	Pant Sujata	Selection	GBPUAT, Pantnagar	Trees are medium dwarf with drooping and spreading foliage, dense, precocious and heavy bearer. Fruit weight varied from 1 to 1.5 kg.
	Pant Urvashi	Selection	GBPUAT, Pantnagar	It is mid season variety. Trees are tall, vigorous, dense, upright growth, precocious and heavy bearer. Fruits are ovoid, oblong. The fruit weight range from 1.5 to 2.5 kg.
4.	**Custard Apple**	*Annona squamosa* L.		F: Annonaceae
	Arka Sahan	Island gem (*Annona atemoya* Hort.) X Mammoth (*A. squamosa* L.) with 22.8% total sugars.	ICAR-IIHR, Bangalore	Fruits are light green in colour, moderately thick (0.5 cm) with large, flat eyes. The edible pulp is remarkable for its sweetness
	Balanagar	Selection	Selected near Hyderabad	Fruits are very large and heart shaped, very good pulp and heavy bearer.
	APK (Ca) 1(2003)	Selection	TNAU, (ARS), Aruppukottai	It bears higher yield in rainfed vertisol (black soil). It is a drought tolerant variety which bears sweet fruits (TSS of 24.5 Brix with an acidity of 0.2 per cent). The Fruit weigh ranges about 207.5 g with a mean of 72 fruits per tree per year. The first bearing commences in a graft/budded plant at 3-3 ½ years. The Optimum productive life is 25 yrs. The best season for planting is May – June / August – September. It yields about 7300 kg fruits / ha (14.90 kg / tree which is 30.7 per cent higher than Balanagar). It is suitable for cultivation in plains of Tamil Nadu especially in semi arid regions and marginal soils of both vertisol and alfisol in dry tracts. It is suitable for both rainfed and irrigated conditions.

Contd.

Contd. Appendix II b

S.No.	Name of Variety	Parentage in case of hybrids	Released from	Specific Characters
5.	**Fig**	*Ficus carica* L.		F: Moraceae
	YCD.1 Timla Fig	Introduction	HRS, TNAU, Yercaud	The trees are well adapted to the rainfed situations of Shevroys hills and to the poor shallow and rocky soils. The trees showed high drought tolerance; besides, it is exceptionally hardy nature and free from any pest or disease including the common fig rust.
6.	**Karonda**	*Carissa carandas* L.	F: Apocynaceae	
	Pant Suvarna (1986)	Selection	GBPUAT, Pantnagar	Plant upright growing & sparse, fruit size 2.16 × 1.66 cm, colour dark brown at ripening.
	Pant Manohar (1991)	Selection	GBPUAT, Pantnagar	Medium sized, dense bushes, fruit size 2.13 × 1.69 cm, flesh 88.27 %, TSS 3.92%, acidity 1.82%, dark brown colour
	Pant Sudarshan (1991)	Selection	GBPUAT, Pantnagar	Fruit size 2.16 × 1.69 cm, colour dark brown, flesh 88.47%, TSS 3.45%, acidity 1.89 %.
7.	**Tamarind**	*Tamarindus indica* L.		F: Leguminosae
	PKM 1 (1997)	Clonal selection from local type Endapuli	TNAU, Periyakulam	High yielder, bears fruits in clusters (37), fruits are characteristically semi circle in shape, high pulp recovery (39%) with high tartaric acid and ascorbic acid content.
	DTS-1 (2005)	Clonal selection	UAS, Dharwad	High yield potential and starts bearing within 5 to 6 years of planting.
	Akola Smruti (AKT-10) (2010)	Selection	PDKV, Akola	Regular bearing, slightly curving fruits, very negligible incidence of fruit borer.
	Urigam	Selection	Local genotype selected from Dharmpuri district in Tamilnadu	Short in size. Pods are lengthy, fleshy and tasty. Pods are flat and exhibit a typical inward curvature. Average fruit weight is between 150-200g. Seed content varies from 10-12/pod.
	Prathisthan	Selection	FRS, Aurangabad	High Yielder.
	Yogeshwari	Selection	Forest Department, Karnataka	Red pulp.

* Varieties identified in National Workshops/Group Meetings of AICRP's.

Appendix – III
Released Varieties of Ornamental Plants

P.K. Valsala Kumari, College of Horticulture, KAU, P.O. KAU, Thrissur – 680 656, Kerala

Rose

S.No.	IARI (Hybid Teas)
1.	Abhisarika
2.	Anurag
3.	Arjun
4.	Bhim
5.	Charugandha
6.	Chitra
7.	Chitralekha
8.	Chitwan
9.	Dr. B.P. Pal
10.	Dr. Benjamin Pal
11.	Dr. Bharat Ram
12.	Ganga
13.	Gulzar
14.	Hans
15.	Jawahar
16.	Madhosh
17.	Mother Teresa
18.	Mridula
19.	Mohini
20.	Mrinalini
21.	Nehru Centenary
22.	Nurjehan
23.	Preyasi
24.	Priyadarshini
25.	Pusa Bahadur
26.	Pusa Christina
27.	Pusa Garima
28.	Pusa Gaurav
29.	Pusa Priya
30.	Pusa Sonora
31.	Rajkumari
32.	Raktagandha
33.	Raktima
34.	Shreyasi
35.	Soma
36.	Sujata

S.No.	ICAR-IARI (Hybid Teas)
37.	Surabhi
38.	Surekha
39.	Vasant
40.	Arunima
41.	Banjaran
42.	Chandrama
43.	Deepika
44.	Deepshika
45.	Delhi Princess
46.	Kavita
47.	Lahar
48.	Manasi
49.	Navneet
50.	Nav Sadhahar
51.	Neelambari
52.	Prema
53.	Pusa Virangana
54.	Pusa Barahmasi
55.	Pusa Pitambar
56.	Rupali
57.	Shabanam
58.	Shringar
59.	Sindoor
60.	Suchitra
61.	Usha

S.No.	Other Varieties
62.	Pusa Mohit
63.	Pusa Abhishek
64.	Pusa Manhar
65.	Pusa Muskan
66.	Pusa Urmil
67.	Pusa Ranjana

Contd.

S.No.	TNAU
68.	YCD-2
69.	YCD-1

S.No.	NBRI (Hybrid Teas)
1.	Kronenburg
2.	Light Pink Prize
3.	Mrinalini Lighter Mutant
4.	Mrinalini Stripe
5.	Pink Montzuma
6.	Salmon Beauty
7.	Summer Holiday Mutant
8.	Winter Holiday Mutant
9.	Sylvia White
10.	Girija

S.No.	NBRI (Floribundas)
11.	Angara
12.	Curio
13.	Pink Contempo
14.	Pink Imperator
15.	Salmon Beauty Mutant
16.	Sharada
17.	Sukumari
18.	Tangerine Contempo
19.	Twinkle
20.	Yellow Contempo
21.	Zorina Pink Mutant

S.No.	Miniature Roses
22.	Miniature Cultivar Windy City Mutant (Light Pink)

Gladiolous

S.No.	ICAR-IARI
1.	Agnirekha
2.	Mayur
3.	Suchitra
4.	Neelam
5.	Anjali
6.	Archan
7.	Bindiya
8.	Shweta
9.	Chandini
10.	Kamini
11.	Mohini
12.	Vandana
13.	Sukanya
14.	Urmil
15.	Jyotsana
16.	Gulaal
17.	Shabnam
18.	Urvasi
19.	Pusa Suhagin
20.	Chirag
21.	Sarang
22.	Sunayna

S.No.	ICAR-IARI
23.	Dhanvantri
24.	Gunjan
25.	Lohit
26.	Rangmahal
27.	Swapnil
28.	Swarnima

S.N.	NBRI
1.	Archana
2.	Basant Bahar
3.	Gazel
4.	Jwala
5.	Manhar
6.	Manisha
7.	Manmohan
8.	Mohini
9.	Mukta
10.	Pitambar
11.	Smita
12.	Triloki
13.	Aldebaran

Contd.

S.N.	NBRI
14.	Friendship
15.	Hans
16.	Indrani
17.	Kalima
18.	Kohra
19.	Mridula
20.	Priyadarshini
21.	Sada Bahar
22.	Arun
23.	Dhiraj
24.	Sanyukta

S.No.	ICAR-IIHR
1.	Aarti
2.	Apsara
3.	Meera
4.	Poonam
5.	Sapna
6.	Shobha
7.	Grock

S.No.	ICAR-IIHR
8.	Jacksonville Gold
9.	White Goddess
10.	Kumkum
11.	Nazrana
12.	Sindoor
13.	Arka Kesar
14.	Darshan
15.	Sagar
16.	Shakti

S.No.	PAU Ludhiana
1.	Punjab Dawn
2.	Punjab Morning
3.	Shan – e- Punjab

S.No.	MPKV
1.	Shree Ganesh
2.	Prerana
3.	Neelrekha

China Aster

S.No.	NARP (MPKV)
1.	Phule Ganesh White
2.	Phule Ganesh Violet
3.	Phule Ganesh Pink
4.	Phule Ganesh Purple

S.No.	ICAR-IIHR
1.	Kamini
2.	Poornima
3.	Shashank
4.	Violet Cushion

Orchids

S.No.	KAU
1.	Deep Blush
2.	Lemon Glow
3.	Master Delight
4.	Pink Cascade
5.	Velvet Soft

S.No.	ICAR-IIHR
1.	Vanda (IIHR-164, Purple)
2.	Dendrobium (IIHR – 38, Purple Violet)

Contd.

Marigold

S.No.	ICAR-IARI
1.	Pusa Basanti Gainda
2.	Pusa Narangi Gainda
3.	MS 8 × Selection 7

Tuberose

S.No.	ICAR-IIHR	S.No.	NBRI
1.	Sringar	1.	Rajat Rekha
2.	Suvasini	2.	Swarna Rekha

Bougainvillea

S.No.	NBRI	S.No.	ICAR-IIHR	S.No.	IARI
1.	Arjun	1.	Chitrawati	1.	Vishakha
2.	Begam Sikandar	2.	Dr. H.B. Singh		
3.	Chitra	3.	Jawaharlal Nehru		
4.	Mary Palmer Special	4.	Purple Wonder		
5.	Nirmal	5.	Sholay		
6.	Parthasarthy	6.	Usha		
7.	Surekha	7.	Zakariana		
8.	Shweta				
9.	Wazid Ali Shah				

Jasmine

S.No.	TNAU	S.No.	ICAR-IIHR
1.	*J. auriculatum* Parimullai	1.	*J. grandiflorum* Arka Surabhi
2.	CO–1 & CO-2 mullai *J. grandiflorum*		
3.	CO–1 & CO-2 pitchi *J. sambac*		
4.	Gundumalli		
5.	Sujimalli		
6.	Madanabanam		
7.	Ramabanam		

Contd.

Hippeastrum

S.No.	NBRI	S.No.	ICAR-IARI
1.	Kiran	1.	Surya Kiran
2.	Samrat		
3.	Jyoti		

Chrysanthemum

NBRI

1. Agnishikha	42. Taruni	83. Neelima	124. Y2 K
2. Alankar	43. Tulika	84. Niharika	125. Yellow Charm
3. Anamika	44. Ajay	85. Nirmal	126. Yellow Prolific
4. Aruna	45. Appu	86. Pancho	
5. Asha	46. Apsara	87. Peet-singar	
6. Ashankit	47. Apurva	88. Phuhar	
7. Basant	48. Apurva Singar	89. Priya	
8. Basanti	49. Arun Kumar	90. Prof.Harris	
9. Batik	50. Arun Singar	91. Puja	
10. Colchi Bahar	51. Bindiya	92. Ragini	
11. Cosmonaut	52. Birbal Sahni	93. Rangoli	
12. Gairik	53. Dhawal	94. Sadhbhavni	
13. Hemanti	54. Diana	95. Santhi	
14. Himani	55. Gauri	96. Ratna	
15. Jhalar	56. Gulal	97. Sharda	
16. Jugnu	57. Guldasta	98. Sharad Kanti	
17. Kanak	58. Haldighati	99. Sharad Kumar	
18. Kansya	59. Hemant Singar	100. Sharad Mala	
19. Kapish	60. Himanshu	101. Sharad Mukta	
20. Kumkum	61. Jaya	102. Sharad Sandhya	
21. Kunchita	62. Jayanti	103. Sharad Sobha	
22. Lohita	63. Jubilee	104. Sharad Singar	
23. Man Bhawan	64. Jwala	105. Shizuka	
24. Navneet	65. Jyothi	106. Shyamal	
25. Navneet Yellow	66. Jyotsna	107. Suhag Singar	
26. Nirbhaya	67. Kargil-99	108. Sujata	
27. Nirbhik	68. Kaumadi	109. Suneel	
28. Pingal	69. Kiran	110. Sunayana	
29. Pitika	70. Kirti	111. Suparna	
30. Pitambe	71. Kundan	112. Surekha Yellow	

Contd.

Contd. Appendix - III

NBRI		
31. Purnima	72. Lal Kila	113. Surya
32. Rohit	73. Lalima	114. Swarn Singar
33. Shabnam	74. Lalpari	115. Sweta Singar
34. Shafali	75. Lilith	116. Tushar
35. Sharadhar	76. Maghi	117. Usha
36. Sheela	77. May-Day	118. Vandana
37. Sheveta	78. Mayur	119. Vasantika
38. Surekha Yellow	79. Meghdoot	120. Vijay
39. Sonali	80. Mini-Queen	121. Vinaya
40. Subarna	81. Mohini	122. White Charm
41. Tamra	82. Mother Teresa	123. White Profilic

S. No.	ICAR-IIHR	S.No.	TNAU
1.	Chandrakant	1.	CO-1
2.	Chandrika	2.	CO-2
3.	Indira	3.	MDU
4.	Kirti		
5.	Nilima		
6.	Pankaj		
7.	Rakhee		
8.	Ravi Kiran		
9.	Red Gold		
10.	Yellow Star		

Appendix IV a
Released Varieties of Coconut
P. Maheswarappa
Co-ordinator AICRP on Palms, ICAR-CPCRI, Kudlu P.O., Kasargod - 671 124

Dwarf Cultivars
Kalpa Raksha

Research institute: ICAR-CPCRI. This is a semi tall variety with sweet tender nut water and with higher resistance to root (wilt) disease of coconut. It comes to flowering by 54 months from planting. Its annual yield is 87 nuts/palm, 16.38 kg copra/ palm and 10.65 kg oil/palm. In root (wilt) disease affected tracts, it gives an annual yield 65 nuts/palm. The quantity of tender nut water is 290ml. This is released as a variety for tender nut and for cultivation in root (wilt) prevalent areas of Kerala

Chowghat Orange Dwarf (COD)

Research institute: ICAR-CPCRI, This is an early flowering cultivar and takes about 3-4 years for initial flowering. The average annual yield is 63 nuts/palm/ year. It is best suited for tender nut water. This cultivar was released by CPCRI in 1991 for large scale cultivation as tender nut variety.

Kalpa Sree

Research institute: ICAR-CPCRI. Kalpasree has superior quality of coconut oil, very sweet tender nut water and meat and is resistant to root (wilt) disease. This is the earliest flowering cultivar and takes about 2.5 to 3 years for flowering. It is found to be resistant to root (wilt) disease. The mean annual yield is 90 nuts/palm with a copra content of 96.3g. This variety is released for root (wilt) prevalent areas of Kerala.

Kalpa Jyothi

Research institute: ICAR-CPCRI. Dwarf variety with yellow fruits, higher average yield of 114 nuts per palm per year under rainfed conditions with estimated copra yield of over 16 kg per palm per year. Recommended for cultivation in Kerala and Karnataka for tender nut purpose.

Kalpa Surya
Research institute: ICAR-CPCRI Dwarf with Orange fruits recommended for cultivation in Kerala, Karnataka and Tamil Nadu for tender nut purpose. The average yield is 123 nuts per palm per year under irrigated conditions with estimated copra out turn of 23 kg per palm per year.

Gauthami Ganga
Research institute: AICRP Centre, Ambajipeta (APAU)Recommended region for cultivationFor cultivation in coastal zone of Andhra Pradesh. Salient features:It is a selection from the local early bearing germplasm accession, Gangabondam, with an average nut yield of 80 to 90 nuts/palm/year. Dwarf palm with semi circular canopy with oblong shaped green colour fruits. Starts yielding at the age of 36 months. Excellent tender coconut variety. Yields 12813 nuts/ha per annum with copra content of 156.68g/nut and an oil content of 68%. Quantity of tender nut water is 467 ml/ nut with very sweet taste. Total sugar content is 6.4 g/ 100 ml, potassium content: 2035 ppm, sodium: 23.00 ppm and amino acid content is 1.7 mg /100 ml.

Hybrid Cultivars

Chandra Sankara (COD x WCT)
Research Institute: ICAR-CPCRI
The palms come to bearing early when compared to tall palms. It is a heavy yielder and produces 116 nuts/palm with a range of 100-150 nuts. The copra content in nut is 160-230 g. It is susceptible to drought and hence irrigation is required during summer months. Chandrasankara was released by CPCRI in 1985 for cultivation in Kerala and Karnataka.

Kera Sankara (WCT x COD)
Research institute: ICAR-CPCRI

The palm comes to bearing by the fourth year of planting. The mean annual yield of nuts is 108 with a range of 70-130 nuts. The copra content is 187g/ nut. This hybrid was released by CPCRI in 1991, for large scale cultivation in Kerala, coastal Andhra Pradesh and coastal Maharashtra.

Chandra Laksha (LCT x COD)
Research Institute: ICAR-CPCRI

The hybrid palm comes to bearing in about 4-5 years after planting. The annual yield is 109 nuts/ palm with a copra content of 150-210g/nut. The hybrid was released by CPCRI.

Kalpa Samrudhi (MYD x WCT)
Research Institute: ICAR-CPCRI

The mean annual yield is 117 nuts per palm. The copra yield is 4.38 t/ha and oil is 3.04 t/ha. The hybrid is suitable for tender nut purpose. This hybrid was recommended for cultivation in Kerala, and Assam.

Kalpa Sankara (CGD x WCT)
Research Institute: ICAR-CPCRI

The mean annual yield is 85 nuts per palm. The copra yield is 2.5 t/ha and oil is 1.69 t/ha. This hybrid was recommended for cultivation in root (wilt) disease prevalent tracts of Kerala.

Konkan Bhatye Coconut hybrid 1: GBGD x ECT
Research institute: AICRP Palms Centre, Bhatye:

BSKKV, Dapoli Recommended region for cultivationFor cultivation in Konkan region of Maharashtra.**Salient features:**Tall palm with semi-circular canopy bearing at 66 months with green colour oval shaped fruit. It is a high yielder with better copra quality than West Coast Tall and Pratap and with better oil content than ECT (67.10%). This hybrid yields 122 nuts/palm/year (20300 nuts/ha/ year) with copra yield of 22.08 kg/palm/year i.e. 3.43 tonnes/ha/year.

Godhavari Ganga (ECT x GBGD)
Research institute: AICRP Palms Centre, Ambajipeta, APAU, A.P.

Hybrid released in the year 1991.
First coconut hybrid in Andhra Pradesh.
Cross between ECT as female parent and GBGD as male parent.
The palm comes to bearing in 4 years after planting.
High yielder with an average nut yield of 140 nuts/palm/year.
The copra content is 150 g/nut with 68% oil content.
Recommended for cultivation in Andhra Pradesh and Tamil Nadu.

Tall Cultivars

Kalpa Dhenu
Research Institute: ICAR-CPCRI

It commences flowering in 67 months after planting in the field. This variety has the potential to produce 22,794 nuts per hectare per year in Tamil Nadu state. The quantity of tender nut water is 290 ml. This is recommended for cultivation in Kerala, East Coast region of Tamil Nadu, Andhra Pradesh and Andaman & Nicobar Islands.

Kalpa Pratibha
Research Institute: ICAR-CPCRI

This variety has the potential to produce 23,275 nuts per hectare per year in West Bengal state. The quantity of tender nut water is 448 ml. This is recommended for cultivation in Kerala, Maharashtra, interior zone of Tamil Nadu and coastal zone of Andhra Pradesh

Kalpa Mitra
Research institute: ICAR-CPCRI

This variety has the potential to produce 80 nuts/palm/year. The quantity of tender nut water is 495 ml. This is recommended for cultivation in Kerala and West Bengal.

Chandra Kalpa
Research institute: ICAR-CPCRI

The average annual yield is 100 nuts/palm. The palm grows in all types of soil and it can withstand moisture stress. This cultivar is recommended for cultivation in the states of Kerala, Karnataka, Andhra Pradesh and Maharashtra.

Kera Chandra
Research institute: ICAR-CPCRI

The annual yield of nuts varies from 90 to 200 with an average of 110 nuts/palm. This is recommended for the west coast including Konkan Region, coastal Andhra Pradesh and West Bengal.

Kalpatharu
Research institute: ICAR-CPCRI

This variety is recommended for ball copra production. It yields around 116 nuts per palm per year with copra content of 176g, under rainfed situations. This is recommended for cultivation in Karnataka, Kerala and Tamil Nadu.

Kalpa Haritha
Research institute: ICAR-CPCRI

A superior high yielding tall selection with lesser incidence of eriophyid mite infestation. The average nut yield is 118 nuts per palm per year under rainfed conditions with estimated copra yield of 25.5 kg per palm per year. Recommended for Kerala and Karnataka suitable for copra and tender nut purpose.

Banawali Green Round (Pratap)
Research institute: ICAR-AICRP on Palms, Bhatye centre : BSKKV, Dapoli

Recommended for Konkan region of Maharashtra. Tall palm with semicircular canopy and green colour round shaped nuts. Yield about 140 to 145 nuts/palm/year, Oil content: 68 %

Kera Bastar
Research institute: ICAR-AICRP on Palms, Jagdalpur centre: IGKV, Raipur

Recommended region for cultivation: For cultivation in coastal zone of Andhra Pradesh,Tamil Nadu, Konkan region of Maharashtra and Bastar region of Chhattisgarh.

Salient Features

Excellent coconut variety with wide adaptability. The nut yield ranges from 110 to 117 nuts/palm/year with a copra yield of 2.5 to 3.1 tonnes /hectare. Total sugar content in tender nut is 6.2 g/ 100 ml.

At the Jagadalpur centre (Chattisgarh state) located in the non-traditional coconut belt, the variety could yield 14,438 nuts/ha with 16.3 kg of copra/palm. Quantity of tender nut water is 332 ml.

Kamrupa

Research institute: ICAR-AICRP on Palms, Kahikuchi Centre: AAU, Assam

Recommended for the Region: Assam

Gives an average yield of 17600 nuts/ha/year 101 nuts/palm/year. This cultivar yields 16.3 kg copra/palm/year leading to a copra yield of 2.86 tonnes/ha and tender nut water of 253ml. Nut water contains 5.16g/100 ml of total sugar and 2294 ppm of potassium. The high yielding capacity, better nut quality, greater tolerance to pest and disease and ability to withstand abiotic constraints like low temperature and semi water logging conditions are the positive attributes of this variety.

Aliyarnagar Tall

Research institute: ICAR-AICRP on Palms, Aliyarnagar Centre, TNAU, Coimbatore

- Pedigree : Selection from Arasampatti local (Tall)
- Time taken for first flowering : 48 months after planting
- Nut characters : Small to medium sized , Oblong shaped , Green coloured
- Average nut yield : 126 nuts / palm / year

- Copra content : 131 g / nut
- Copra yield : 16.5 kg / palm / year
- Oil content : 66.5 %
- Special attributes : Tall palms with high nut yield , Early bearing

This cultivar records an yield level of 22015 nuts/ha/year (126 nuts/palm/year), with copra yield of 2.88t/ha. The tender nut of this cultivar contains 310 ml of nut water. The nut water contains 110 ppm of potassium and 4.3 ppm of sodium. This cultivar is highly acceptable to the farming community because of its early bearing tendency (four years for first flowering), high yielding nature and ability to withstand moisture stress.

Kalyani Coconut -1
Research institute: ICAR-AICRP on Palms, Mondouri, BCKVV, West Bengal

Recommended region for cultivation: For cultivation in rain fed and irrigated areas of West Bengal.

Salient Features

Suitable for rain fed cropping also. Low rainfall condition does not affect the yield while the higher rainfall has positive influence. It gives 14066 nuts / ha / year with an average yield 80 nuts/palm/ year. Oil content is higher than the present varieties; The copra content is 154 g/nut with a copra yield of 12.3kg/palm/year (2.17 tones of copra/ha). Moderately tolerant to drought.

Tender nut water quality of the variety is good with 350 ml of water, 4.9% of total sugar content and high potassium (2347 ppm) content. It also contains higher quantities of sodium (27 ppm) and amino acids (1.8%).

Kera Keralam
Research Institute: ICAR-AICRP on Palms, Veppankulam Centre, TNAU, Coimbatore

Recommended region for cultivation: For cultivation in Kerala, Tamil Nadu and West Bengal States

Salient Features

It is a selection from the popular local accession of west coast region of India, viz., West Coast

Tall. Highly promising and adaptive to wide ranging cultivation zones; Yields 147 nuts/palm/year with irrigation and other better management conditions and 109 nuts/palm/year under rain fed situations respectively in North Kerala region.

In East Coast region of Tamil Nadu, it recorded higher nut yield by 19 to 48% over the previously released varieties with a copra yield of 3.58 tonnes/ha. The oil content is 67.8%. It yields 2.43 tones/ha of oil i.e., 21.5% higher than ECT.

Moderately tolerant to drought. It comes up well in varied types of soil including sandy, sandy loam and red sandy loam with pH ranging from 6.0 to 8.0.

Appendix IV b
Released Varieties of Cocoa

P. Maheswarappa, Co-ordinator, ICAR-AICRP on Palms. CPCRI, Kudlu, P.O., Kasargod – 671124

VTLCC-1

Vittal Cocoa Clone 1 Early, heavy bearer and both self and cross compatible clone.Green to yellow pods: 75/tree/year. No. of beans/pod: 36. Single dry bean weight: 1.05 g. Shelling: 12%, Fat content: 52.5%.Dry bean yield: 1.33 kg/tree/year, 911 kg/ha.

VTLCS-1

Vittal Cocoa Selection 1 Precocious, stable and high yielder both under arecanut and coconut with medium canopy, withstands biotic and abiotic stress. Red to Orange pods: 54.5 / tree/year No. of beans/ pod: 42Single dry bean wt: 1.13 gShelling: 11%, Fat-52.1%, Dry bean yield/ tree/ year: 2.52 kg/tree/year, 1700 kg/ha

VTLCS-2

Vittal Cocoa Selection 2 Early, stable and high yielder both under arecanut and coconut, have bold and bigger beans, less incidence of pests and diseases. Green to Yellow pods: 55 /tree/year No. of beans/ pod: 42Single dry bean wt: 1.21 g. Shelling: 15% Fat: 53% Dry bean yield: 2.7 kg/ tree/ year, 1850 kg/ha

VTLCH-1

Vittal Cocoa Hybrid 1 Vigorous, early and heavy bearer. Green to yellow pods: 50/tree/year. No. of beans/pod: 42. Single dry bean weight: 1.00 g. Shelling: 13%, Fat content: 53.6%. Dry bean yield: 1.48 kg/tree/year, 1014 kg/ha.

VTLCH-2

Vittal Cocoa Hybrid 2 Heavy bearer, medium canopy and tolerant to black pod rot.Green to yellow pods: 70/tree/year. No. of beans/pod: 40. Single dry bean weight: 1.15 g. Shelling: 11%, Fat content: 54%. Dry bean yield: 1.15 kg/tree/year, 800 kg/ha.

VTLCH-3

Vittal Cocoa Hybrid 3 Suitable for rainfed & irrigated areca and coconut gardens of Karnataka, Kerala, Tamil Nadu and Andhra Pradesh.Green to yellow pods: 45/tree/year. No. of beans/pod: 43. Single dry bean weight: 1.07 g. Shelling: 13%, Fat content: 52. Stomatal resistance (s/cm): 2.42. Dry bean yield: 1.45 kg/tree/year, 993 kg/ha.

VTLCH-4

Vittal Cocoa Hybrid 4 Suitable for rainfed & irrigated areca and coconut gardens of Karnataka, Kerala, Tamil Nadu and Andhra Pradesh. Red to orange pods: 40/tree/year. No. of beans/pod: 43. Single dry bean weight: 1.01 g. Shelling: 12%, Fat content: 53%. Stomatal resistance (s/cm): 2.41. Dry bean yield: 1.25 kg/tree/year, 856 kg/ha.

Appendix IVb. Continued

Arecanut

Varieties	No.of pods /tree/year	No.of beans/pod	Single dry bean wt	Dry bean yield (kg/tree/yr)	Yield /Ha (kg)	Shell (%)	Nib recovery (%)	Fat (%)	Remarks
VTLCH-1	50	42	1.00	1.48	1014	13	87	54	Early, high yielder
VTLCH-2	50	40	1.15	1.15	800	11	89	54	High yielder with medium canopy, pod rot resistant
VTLCH-3	45	45	1.07	1.45	993	13	87	50	High yielder & drought tolerant
VTLCH-4	40	43	1.01	1.25	856	12	88	50	High yielder & drought tolerant
VTLCC-1	70	37	1.05	1.33	911	12	88	53	Both self and cross compatible, high yielder in multi locations
VTLCS-1	55	42	1.13	2.52	1700	11	89	52	Suitable both under arecanut and coconutcanopy, withstands both biotic and abiotic stress
VTLCS-2	55	40	1.21	2.70	1850	15	85	53	Suitable both under arecanut and coconut canopy, bolder and big sized beans, tolerant to biotic and abiotic stress

Appendix-IV C
Released Varieties of Plantation Crops

P. Maheswarappa, Co-ordinator, ICAR-AICRP on Palms, ICAR-CPCRI, Kudlu P.O., Kasargod – 671 124

Arecanut

Mangala

It is a selection from VTL 3, an introduction from China. It is semi tall and has partially drooping crown with well spread leaves and having more number of leaflets. Leaflets are dark green with characteristic crinkling at tip. It is semi tall, early bearing with quicker stabilisation of yield and producing nuts with good chewing and market quality. Nuts are small and round in shape. It produces 10 kg ripe nuts/ palm/ year which gives 2 kg chali. This variety is released from ICAR-CPCRI Regional Station, Vittal.

Sumangala

It is an introduction from Indonesia. Palm is tall with partially drooping habit. Under ideal conditions, palm flowers in 4–5 years. Nuts are medium sized, deep yellow to orange in colour and oblong to round in shape. It gives an average yield of 17.25kg ripe nuts/palm/ year at 10th year. This variety is released from ICAR-CPCRI Regional Station, Vittal.

Sreemangala

This variety was selected from an introduction from Singapore. Habit, flowering and fruit characters are similar to Sumangala. Nuts are round and bold. It gives an average yield of 15.63 kg ripe nuts/palm/year. This variety is released from ICAR-CPCRI Regional Station, Vittal.

Swarnamangala

Evaluation of exotic germplasm and selection for economic traits resulted in identifying a high yielding variety VTL-12 (Saigon). It is a tall variety with homogeneous population. Bunches are well spaced, nuts are bold and heavier with high recovery of chali/ dry kernel (26.40 %). Average yield is 3.88kg chali/ palm/year. This variety is released from ICAR-CPCRI Regional Station, Vittal.

Mohitnagar

This is an indigenous cultivar from West Bengal. Important feature of this variety is its greater uniformity. Bunches are well spaced and nuts are loosely arranged on spikes which help in uniform development enabling efficient plant protection measures. Nuts are round and medium in size. Early stabilization of yield and high annual yield potential of 3.7 kg chali/ palm (15.08-kg ripe nuts) were also noticed.

Samrudhi

This variety recommended for Andaman and Nicobar Islands is tall with longer internodes and crown. Striking feature of this variety is its consistent and high yield potential (18.89 kg ripe nuts/ palm/ year with a chali yield of 4.34 kg/ palm) having well placed bunches with round and bold nuts. This variety is released from ICAR-Central Island Agricultural Research Institute, Port Blair, Andaman and Nicobar Islands

SAS I (Sirsi Arecanut Selection I)

A variety recommended for hill zone of Karnataka, it is tall with compact canopy and is a regular bearing variety. Nuts are round and even sized and closely arranged on compact bunches. It is suitable for both tender and ripe nut processing. It has high curing(%) and yields 4.60 kg chali/ palm /year.

Kahikuchi local

Kahikuchi is a selection variety from VTL-64 (Kahikuchi) released in the year 2009 by ICAR-CPCRI, Kasaragod. A high yielding variety recommended for release for Assam and NEH Region. It is a tall cultivar with medium thick stem possessing comparatively shorter internodes. Nuts are bigger and heavier with high recovery of chali/ dry kernel (25.16%) from fresh fruits. The average yield of this cultivar is 3.70 kg dry kernel/palm/year. It is recommended for the arecanut cultivated areas in Assam and North Eastern regions of India.

Kahikuchi local

Maduramangala

Selection from Sreevardhan

The variety produces superior quality of nuts with marble appearance of the split nut, which fetches higher price in the market. The variety possesses high yielding nature, medium tall palms, medium thick stem, shorter internodes, partially drooping crown, homogeneous population, regular bearer, consistent in yield, bunches are well placed on the stem, orange to yellow colour, oval and round shaped nuts, high recovery (25.90%) of dry kernel from fresh nuts, higher content of polysaccharides and polyphenols, medium maturity group, comes to bearing by 4th year. The average yield is 3.54 kg dry kernel palm^{-1} year^{-1} and about 4500-5000 kg dry kernel ha^{-1} year^{-1}and 2.95 kg dry tender processed nuts palm^{-1} year^{-1} and about 3800-4500 kg dry tender processed nuts ha^{-1} year^{-1}.

Nalbari: Selection from a genotype collected from Assam. It was proposed based on the yield performance in Mohitnagar (West Bengal) and Vittal (Karnataka). The yield performance of VTL 75 is higher as compared all the earlier released varieties and found suitable for ripe nut processing. The variety possesses high yielding nature, tall type with medium thick stem, longer internodes, partially drooping crown, homogeneous population, regular bearer, consistent in yield, bunches are well placed on the stem, round shaped yellow colour nuts, high recovery (25.18 %) of dry kernel from fresh nuts, comes to bearing by 5th year and economic yield can be realized up to 40-45 years. The average yield is 4.15 kg Chali/dry kernel palm^{-1} year^{-1} and 5600 kg dry kernel ha^{-1} year^{-1}.

Hybrids

VTLAH 1 (Hirehalli Dwarf x Sumangala)

A hybrid variety which is dwarf in nature with superimposed nodes, reduced canopy, partial drooping crown, medium sized, oval to round and yellow orange coloured nuts.

Early stabilization of yield and high recovery of chali (26.45 %) are striking features of this hybrid. Average chali yield is 2.24 kg/palm/year.

VTLAH 2 (Hirehalli Dwarf x Mohitnagar)

For Karnataka and Kerala
Dwarf, reduced canopy
Early yield stabilisation
Oval, medium, orange nuts
Average yield (chali/ palm) : 2.64 Kg

Hybrids

VTLAH 1 (Hirehalli Dwarf x Sumangala)

A hybrid variety which is dwarf in nature with superimposed nodes, reduced canopy, partial drooping crown, medium sized, oval to round and yellow orange coloured nuts. Early stabilization of yield and high recovery of chali (26.45 %) are striking features of this hybrid. Average chali yield is 2.24 kg/palm/year.

VTLAH 2 (Hirehalli Dwarf x Mohitnagar)

For Karnataka and Kerala Dwarf, reduced canopy Early yield stabilisation Oval, medium, orange nuts Average yield (chali/ palm) : 2.64 Kg

Varieties	Habit	Nut yield (dry, kg/palm)	Recommended for States
Mangala	Semi tall	3.00	Karnataka & Kerala
Sumangala	Tall	3.20	Karnataka & Kerala
Sreemangala	Tall	3.18	Karnataka & Kerala
Mohitnagar	Tall	3.67	W. B., Kerala, Karnataka
Swarnamangala	Tall	3.78	Karnataka & Kerala
Kahikuchi	Tall	3.70	Assam & NEH Region
Hybrids			
VTLAH1	Dwarf	2.54	Karnataka & Kerala
VTLAH2	Dwarf	2.64	Karnataka & Kerala

Appendix- V a

Released Varieties of Spices from Kerala Agricultural University (KAU)

Asha Sankar

Department of Plantation Crops and Spices, College of Horticulture, P.O. KAU – 680 656, Thrissur, Kerala

Black Pepper

Name of variety	Panniyur -8
Parents	Panniyur-2xNeelamundi
Agronomic features	Steady yield, tolerance to drought and foot rot. Performs well in open and shade. Early spiking of vines. More number of laterals and sib laterals and berries. Compact setting of berries. Bold berries with glossy green texture.
Average fresh yield	5760 kg/ha
Average dry yield	2130 kg/ha
Dry recovery	37.00 %
Potential fresh yield	8125 kg/ha
Potential dry yield	3000 kg/ha
Released by	Panniyur Research Station, Kerala Agricultural University

Name of variety	Vijay Hybrid
Parents	Panniyur-2xNeelamundi
Agronomic features	Vigorous vine. Large number of laterals and spikes per unit area. Bold berries, high yield and good quality. Field tolerant to foot rot.
Average fresh yield	6.00 kg/plant
Spike length	13.4 cm
Dry recovery	39.8 %
Volatile oil	3.33 %
Oleoresin	10.19 %
Released by	Department of Plantation Crops & Spices, Kerala Agricultural University

Ginger

Name of variety	Athira
Breeding method	Selection from somaclones of Maran
Agronomic features	Suited as pure and intercrop and for fresh and dry ginger. Tolerant to soft rot and bacterial wilt diseases
Average fresh yield	21.00 t/ha
Potential fresh yield	30.00 t/ha
Rhizomes	Bold
Dry recovery	22.6%
Volatile oil	3.1%
Oleoresin	6.8%
Crude fibre	3.4%
Released by	Department of Plantation Crops & Spices, Kerala Agricultural University

Contd.

Name of variety	Karthika
Breeding method	Selection from somaclones of Maran
Agronomic features	Suited as pure and intercrop and for fresh and dry ginger. Tolerant to soft rot and bacterial wilt diseases. Low infestation of shoot borer
Average fresh yield	19.00 t/ha
Potential fresh yield	27.00 t/ha
Rhizomes	Bold
Dry recovery	21.6%
Volatile oil	3.2%
Oleoresin	7.2%
Crude fibre	3.7%
Released by	Department of Plantation Crops & Spices, Kerala Agricultural University

Name of variety	Aswathy
Breeding method	Selection from somaclones of Rio- de Janeiro
Agronomic features	Suited as pure and intercrop and for fresh ginger. Highly suited to homesteads
Average fresh yield	23.00 t/ha
Potential fresh yield	30.00 t/ha
Rhizomes	Bold
Dry recovery	19.7%
Volatile oil	3.32%
Oleoresin	7.45%
Crude fibre	3.5%
Released by	Department of Plantation Crops & Spices, Kerala Agricultural University

Spice varieties newly released by UHS, Bagalkot

Cardamom

The cardamom variety Mudigere-3 was released by Mudigere Research Station under ICAR-AICRP on Spices.

Appendix V a Continued

Released Varieties of Spice Crops Through All India Coordinated Research Project on Spices

K. Nirmal Babu, Co-ordinator ICAR-AICRP on Spices, ICAR-IISR, Kozhikode, Kerala – 673 012

Turmeric

Sl No.	Variety	Pedigree/Parentage	*Av. yield (kg/ha)	Dry recovery %	Curcumin %	Oleoresin %	Essential oil %	Crop duration	Salient features
1	IISR Prabha	Open pollinated progeny selection	37.0	19.5	6.5	15	6.5	205	High yielding varitey
2	IISR Prathibha	Open pollinated progeny selection	39.1	18.5	6.2	16.2	6.2	225	High yielding varitey
3	IISR Alleppey Supreme	Selection from Alleppey Finger turmeric	35.4	19	5.55	16	-	210	Tolerant to leaf blotch.

* Yield tones/ha (fresh)

Appendix V a Continued

Cinnamon

Sl No.	Variety	Pedigree/Parentage	*Av. yield (kg/ha)	Bark recovery (%)	Bark oil (%)	Leaf oil (%)	Salient features
1.	YCD.1	Clonal selection from op seedings progenies of Sri Lankan type	360	35.3	2.8	3	Good bark recovery adopted to wide rangeof soil and rainfed conditions. Recommended for high ranges at 500-1000m above MSL.
2.	PPI (C) -1	Selection from OP seedlings progeny introduced from Sri Lanka	980	34.22	2.9	3.3	Suitable for cultivation in high rainfall zones and hill regions of tamil Nadu at an altitude range of 100-500 m MSL.
3.	Konkan Tej	Seedling selection from progenies of Sri Lankan accessions	334	29.1651.78	3.2	2.28	Superior qualities with 3.2% bark oil with bark recovery 29.16%,cinnamaldehyde in bark oil 70.23,eugenol in bark oil 6.93%, eugenol in leaf oil 75.5%,yields 4.10kg fresh bark.
4.	Sugandhini (ODC-130)	Single tree selection from Wayanadu local collection.A Sri Lankan type	640	51.0	0.94	1.6	Recommended for cultivation for leaf oil production,cinnamaldehyde in bark oil 45%, eugenol in leaf oil 93.7%;released mainly for leaf oil purpose.densly foliage.
5.	RRL(B) C-6	Selection from germplasm collection-OP seedling progenies	250				High quality,sweet and pungent bark with 83% cinnaldehyde content in bark oil, 94.0% eugenol in leaf oil,leaf oil 0.8%. Spreading,branching,nature.
6.	IISR Nithyashree	Clonal selection from OP seedling progeny	200		2.7	3	Good regeneration capacity,bark and leaf oleoresin contents are high.Good bark recovery with good aroma and taste.Bark oleoresin-10.0%.
7.	IISR Navashree	Selection from Op seedling progeny of Sri Lankan collection	200	40.6	2.7	2.8	High quality line,good bark recovery with good aroma and taste,grow well in plains and high elevations.High cinnamaldehyde content (73%) in bark oil,medium quality High shoot regeneration.Bark oleoresin 8%, cinnamaldehyde in leaf oil 15%,eugenol in bark oil 6%,eugenol in leaf oil 62%.

Contd.

Appendix V a Continued

Nutmeg

Sl No.	Variety	Pedigree/Parentage	*Av. yield	Salient features
1.	Konkan Sugandha	Single plant selection from local seedling population	200- 526 fruits/ tree	Adaptable in Konkan region. Tree canopy is conical and compact. No incidence of major pests / disease
2.	Konkan Swad	Selection from nutmeg seedling from Ratnagiri district	761.38 fruits/tree	Adapted in Konkan region with warm,humid condition as well as shade provision. canopy erect, conical shape. contain 39.85 essential oil in seed and 10.9 % in mace. No incidence of major pests / disease are noticed
3.	IISR Viswasree	Clonal selection from elite germplasm	1000 fruits/tree (1.33Kg mace, 9kg dry,3122Kg/ha)	Bushy and compact canopy, low incidence of fruit rot. Nut recovery 70%,mace recovery 35%,oil, 7.14% maceoil, oleoresin 2.48% and mace 13.85 respectively,butter 30.9%,myristicin 12.48% and mace 20.03% respectively. Especially suitable under mixed cropping system.

Coriander

Sl No.	Variety	Pedigree/Parentage	*Av. yield (kg/ha)	Essential oil %	Duration (days)	Salient features
1	Sindhu	Selection germplasm, Warangal local	1000	0.40	100-110	Oval medium breakable grains, suitable for rainfed areas, tolerant to wilt, powdery mildew as well as drought condition,medium duration.
2	Hisar Anand	Mass selection from Haryana collection	1400	0.35	-	A medium tall dual purpose variety,oval medium sized seeds, wider

Contd.

Appendix V a Continued

Sl No.	Variety	Pedigree/Parentage	*Av. yield	Essential oil (%)	Duration	Salient features
						adaptability to different soil conditions.Resistant to lodging due to spreading habit.
3	Hisar Sugandh	Mass selection from indigenous germplasm	1400			Suitable for irrigated conditions. Resistant to stem gall diseases.
4	Hisar Surabhi	Mass selection from local germplasm	1800	0.4-0.5	130-140	Bushy erect plant type, seed medium, oblong; tolerant to frost, less susceptabile to aphids, medium duration
5	Azad Dhania-1	Mass selection from Kalyanpur germplasm collection	1000	0.29	120-125	Erect, early branching, number of umbellates per umbel 5, tolerant to moisture stress, powdery mildew and aphids.
6	Pant haritima	Selection from local type Pant Dhania	1200	0.4	150-160	Tall erect plant, a dual purpose type, good yielder of leaves, smaller seeds with high oil. Resistant to stem gall..
7	DWA 3*	Purelineselection from Karnataka collection	400	0.27	-	A dual purpose variety and for seed production in rabi crop, moderately tolerant to powdery mildew, black clay soils are best suited
8	CIMPOS-33	Selection from germplasm introduced from Bulgaria	2100	1.3	-	Tall erect, compact, profusely branching and flowering, grains small and bold. Mainly recommended for oil production.
9	ACR- 01-256 (NRCSS ACR-1)	Reselection from EC-467683 from Russia	1100	0.35-5	-	Dual purpose variety, long duration, resistant to stem gall and wilt.
10	RCr 20	Recurrent half sib election from Jaipur local	900	0.25	100-110	Medium sized bushy plant suitable for rainfed crop or limited moisture condition and heavy soils of south

Contd.

Appendix V a Continued

Sl No.	Variety	Pedigree/Parentage	*Av. yield	Essential oil (%)	Duration	Salient features
						Rajastan. Moderately resistant to stem gall, bold grains,early maturity.
11	RCr.41	Recurrent half sib selection from local type from "Kotta"	909	0.25	130-140	A tall erect plant with thick stem.Grows well under irrigated conditions, resistant to stem gall, wilt and moderately resistant to powdery mildew; small seeds (9.3g/1000 seed), long duration varitey
12	RCr 435	Recurrent selection from local germplasm from Jalore	1000	0.33	110-130	Plants are bushy, erect, bold seeds, medium sized, medium maturing variety, adapted for irrigated condition moderately resistant to root knot and powdery mildew.
13	RCr 436	Recurrent half sib selection from local germplasm from Kotta	1100	0.33	90-100	Plants semi dwarf, bushy type with quick early growth and bold seeds. Resistant to root rot and root knot nematodes most suitable for limited moisture condition and heavy soils of south Rajasthan
14	RCr446	Half sib selection from local type from Jaipur local	1200	0.33		Plants tall, are leafy erect with higher number of seeds per umbel.Seeds are medium in size and moderately resistant to stem gall.
15	RCr 684	Mutation breeding of gamma rays. Induced mutant of Rcr-20	990	0.32	110-120	A variety, resistant to stemgall and less susceptible the powdery mildew Adapted to medium heavy textured soil and sandy loam soil under irrigation. Seeds of the variety are bold.Plants are toleran and erect with higher number of seeds per umbel, medium maturity.

Contd.

Appendix V a Continued

Cumin

Sl No.	Variety	Pedigree/Parentage	*Av. yield (kg/ha)	Essential oil (%)	Crude fibre	Duration (days)	Salient features
1	5-404	Selection from local germplasm	350	2.2	7.7		An erect plant, medium sized fruit, moderately tolerant to powdery mildew.
2	RZ-209	Recurrent single plant progeny selection from Jore	650			120-130	A variety shown some resistance with blight and wilt disease
3	RZ-223	Mutation breeding in UC-216	600	3.0-3.5		120-130	Wider adaptability, resistant to wilt, superior in yield and seed quality over RZ-19.Plants bushy, semi-erect, long bold attractive seeds, medium duration.
4	Ac-01-167	Reselection from EC-243373	515	3			Bold seeds resistant to wilt.

* Yield Kg/ha (Dry)

Fennel

Sl No.	Variety	Pedigree/Parentage	*Av. yield (kg/ha)	Essential oil (%)	Crude fibre	Duration (days)	Salient features
1.	PF – 35	Selection from local germplasm	1280				Plant tall and spreading moderately tolerant to leaf spot, leaf blight and sugary diseases
2.	Co.1	Reselection from PF 35	570			220	Medium statured, diffuse branching, Suitable for intercropping and border cropping with chilli and turmeric. Suitable for drought prone, water logged, saline and alkaline conditions

Contd.

Appendix V a Continued

Sl No.	Variety	Pedigree/Parentage	*Av. yield	Essential oil (%)	Crude fibre	Duration	Salient features
3.	Guj Fennel 1	Pure line selection from Vijaypur local	1695				Plant tall and bushy, shattering and lodging, suitable for early sowing and rabi crop, reasonably tolerant to drought, moderately tolerant to sugery disease, oblong, medium bold and dark green seeds.
4.	Guj fennel 2	Pedigree selection from local germplasms	1940	2.4			Plants bushy, bold grains, rich in volatile oil and suitable for mplasm both rain fed and irrigated condition
5.	RF 101	Recurrent half sib selection from local germplasm collection from Jobner	1400			150-160	Errect medium tall nature,medium maturity type with long bold grains, most suitable for loamy and black cotton soil.
6.	S-7-9	Slection	1100	1.2	24	210	A bushy plant with big umbel, moderately tolerant to blight.
7.	Guj Fennel II	Selection based on individual plant progeny performance from local germplasm	2489	1.8		148	A medium maturity type adopted to rabi season un der irrigation; seeds medium bold.
8.	RF 125	Reccurent half sib selection is an exortic collection EC 243380 from Italy	1700			110-130	Plants are short statured with seeds compact umbels and long bold when green presence, denser view of plants. Tollerent to sugery disease.
9.	Hisar Sawrup	Mass selection from indigenous germplasm of Haryana	1600	1.6		175-185	Plants grow up right, spreading, gives a bushy appearance. A late maturity type grain long and bold, resistant to lodging, no shattering of grains.
10.	Azad Sanuf-1	Selection from germplasm	1500	2		160-170	Medium plants, resistant to blight and root rot desases. Escapes attack of aphids due to early maturity, seeds are bold green.

Contd.

Appendix V a Continued

Sl No.	Variety	Pedigree/Parentage	*Av. yield	Essential oil (%)	Crude	Duration	Salient features
11.	Pant Madhurika	Pure line selection from local germplasm			180-185		Tall robust eruct plant with big umbels having bold seeds with green fine ridgs sweet in taste, medium maturity.
12.	Rajendra Sourabha	NA	NA	NA	NA	NA	NA
13.	AF-01-119 NRCSS-AF1	Recurrent selection from individual plant progeny	1950				Medium maturity seed, bold,tolerent to blight
14.	RF 143	Recurrent selection from individual plant progeny	1200	1.87			Medium tall and reccomented for loamy and black cotton soils

* Yield kg/ha (Dry)

Fenugreek

Sl No.	Variety	Pedigree/Parentage (kg/ha)	*Av. yield	Duration (days)	Salient features
1.	CO_2	Selection from CF 390	480	85-90	Short duration dual purpose variety , field tolerant to *Rhizoctonia* root rot disease, suitable for both kharif and rabi season.Early maturity,short duration.
2.	Hisar Sonali	Pure line selection from germpalsm	1700		Tall and bushy vigorous growing variety, dual purpose variety, late maturity (140-145 days), suitable for cultivation under irrigated condition. Moderately resistant to root rot and aphids.

Contd.

Appendix V a Continued

Sl No.	Variety	Pedigree/Parentage (kg/ha)	*Av. yield	Duration (days)	Salient features
3.	Hisar suvarna	Pureline selection from local germplasm	1600		A quick growing, erect and tall, dual purpose, medium maturity (130-140days), moderately resistant to percospora and powdery mildew. Wider adaptability,suitable for cultivation throughout the country.
4.	Hisar Madhavi	Pureline selection from local germplasm of UP	1900		A quick growing, erect and tall, dual purpose, medium maturity (130-140days) , moderately resistant to powdery mildew and to downy mildew. A variety with under adaptability suitable for both irrigated and rainfed condition.
5.	Hisar Muktha	Pureline selection natural green seed coated mutant line from UP	2000		A quick growing seedtype variety, medium maturity (135-140days) , moderately resistant to powdery mildew and to downy mildew.Erect and tall plants.Wide adaptability.Suitable for both irrigated and rainfed condition
6.	RMt 303	Mutation breeding from variety RMt 1	1900		Medium maturity variety (145- 150 days) seeds bold ,with typical yellow colour ,less susceptible to powdery mildew
7.	RMt305	Mutation breeding from v ariety RMt 1	1300		First determinant type, multipodant, early maturing, wider adaptability, resistant to powdery mildew and rootknot nematodes .Seeds bold, attractive and yellow, duration 120-125 days.
8.	Guj Methi 1	Recurrent selection based on pure line selection from J. Fenu102	1864		The first variety from Gujarat released for the state.Plant dwarf.

Contd.

Appendix V a Continued

Sl No.	Variety	Pedigree/Parentage (kg/ha)	*Av. yield	Duration (days)	Salient features
9.	RMt143	RMt143	1600	140-150	Moderately resistant to powdery mildew , seeds bold yellow colour,suitable for heavier soils.
10.	Rajendra Abha (Kasuri Methi)	NA	NA	NA	NA
11.	Pant Ragini	Selection from local germplasm	1200	170-175	A dual purpose tall bushy type resistant to downy mildew and root rots, medium maturity. Seed contain 2-2.5% essential oil
12.	Rajendra Khushba	NA	NA	NA	NA
13.	AM-01-35	Selection from local germplasm	1720		Dual purpose, tolerant to powdery mildew

Varieties of seed spices released by NRCSS Ajmer

Coriander	Ajmer Coriander-1 (ACr-1)	ICAR-NRCSS Ajmer
Ajowain	Ajmer Ajwain-1, Ajmer Ajwain-2 (AA-1,2)	ICAR-NRCSS Ajmer
Fennel	Ajmer Fennel-AF-1 (AF-1,2)	ICAR-NRCSS Ajmer
Fenugreek	Ajmer Fenugreek-1, Ajmer Fenugreek-2 (AGm-1,2)	ICAR-NRCSS Ajmer
Dill	Ajmer Dill-1,2 (AD-1,2)	ICAR-NRCSS Ajmer
Nigella	Ajmer Nigella-1 (AN-1)	ICAR-NRCSS Ajmer
Anise	Ajmer Anise-1 (Ani-1)	ICAR-NRCSS Ajmer
Celery	Ajmer Celery-1 (ACel-1)	ICAR-NRCSS Ajmer

Appendix – V b

Released Varieties of Spice Crops through All India Coordinated Research Project on Spices

K.Nirmal Babu, Co-ordinator AICRP on Spices, ICAR-IISR, Kozhikkode, Kerala – 673012

Black pepper

Sl No	Variety	Centre which developed	Year of release	Pedigree/Parentage	*Av. yield (kg/ha)	Dry recovery (%)	Piperine (%)	Oleoresin (%)	Essential oil (%)	Salient features
1	Panniyur 1	Pepper Research Station, KAU, Panniyur, Kerala	1967	F_1 of Uthirankotta x Cheriyakaniyakadan	1242	35.3	5.3	3.5	3.5	Highest yield potential. Do not tolerate shade, moderately high oleoresin (11.8%) long spikes & bold berries
2	Panniyur 2	Pepper Research Station, KAU, Panniyur, Kerala	1991	Open pollinated progeny of Balankotta	2570	35.7	6.6	10.9	3.4	Shade tolerant, rich in oleoresin high pipline
3	Panniyur 3	Pepper Research Station, KAU, Panniyur, Kerala	1991	F_1 of Uthirankotta x Cheriyankaniyakadan	1953	27.8	5.2	12.7	3.1	Late maturing, suitable for all pepper growing region, performs well under open situation. Long spikes & bold berries
4	Panniyur 4	Pepper Research Station, KAU, Panniyur, Kerala	1991	Clonal selection from Kuthiravally	1277	34.7	4.4	9.2	2.1	Stable yielder, performs well under adverse condition also
5	Panniyur 5	Pepper Research Station, KAU, Panniyur, Kerala	1996	Open pollinated progeny of Perumkodi	1107	35.7	5.3	12.33	3.8	Suitable for both monocropping & mixed crop in coconut/arecanut gardens. Long spikes
6	Panniyur 6	Pepper Research Station, KAU, Panniyur, Kerala	2000	Clonal selection from Karimunda	2127	33	4.9	8.27	1.33	Steady and stable yielder tolerant to drought and adverse climatic conditions. Suitable for open condition as well as partial shade
7	Panniyur 7	Pepper Research Station, KAU, Panniyur, Kerala	2000	Open pollinated progeny of Kalluvally	1410	34	5.6	10.6	1.5	Vigorous, hardy and a regular bearer, long spike, high piperine (5.6%) tolerates adverse climatic condition suitable open and shaded conditions.

8	Panniyur 8	Pepper Research Station, KAU, Panniyur, Kerala	2013	Hybrid of Panniyur 6 x Panniyur 5	5760	37	5	12	3.8	Suited to all pepper growing regions of Kerala. Field tolerant to drought situations and *Phytophthora* foot rot.
9	Sreekara	Indian Institute of Spices Research, Calicut, Kerala	1990	Clonal selection from Karimunda	2677	35	5.1	13.0	7.0	High quality and high volatile oil content with wider adaptability to all pepper growing tracts.
10	Subhakara	Indian Institute of Spices Research, Calicut, Kerala	1990	Clonal selection from Karimunda	2352	35	3.4	12.4	6.0	High quality and high volatile oil content with wider adaptability to all pepper growing tracts.
11	Panchami	Indian Institute of Spices Research, Calicut, Kerala	1991	Clonal selection from Aimpiriyan	2828	34	4.7	12.5	3.4	Late maturing variety with excellent fruit set.
12	Pournami	Indian Institute of Spices Research, Calicut, Kerala	1991	Clonal selection from Ottaplackal	2333	31	4.1	13.8	3.4	High yielding variety, tolerant to root knot nematode.
13	IISR Sakthi	Indian Institute of Spices Research, Calicut, Kerala	2006	Open pollinated progeny of Perambramundi.	2253	43	3.3	10.2	3.7	Tolerant to *Phytophthora capsici*
14	IISR Thevam	Indian Institute of Spices Research, Calicut, Kerala	2006	Clonal selection of Thevamundi,	2481	32.5	1.6	8.15	3.1	Vines grow vigorously,stable yielding field tolerant to *Phytophthora*
15	IISR Girimunda	Indian Institute of Spices Research, Calicut, Kerala	2004	Hybrid between Narayakodi x Neelamundi	2880	32	2.2	9.65	3.4	Recommended for rainfed conditions,suitable for high elevation
16	IISR Malabar Excel	Indian Institute of Spices Research, Calicut, Kerala	2006	Hybrid between Cholamundi x Panniyur-1	1440	32.3	11.7	2.4	2.8	Suitable for high elevation and plains
17	PLD –2	Indian Institute of Spices Research, Calicut, Kerala	1971	Clonal selection from Kottanadan	2475	-	3	15.45	4.8	Late maturity high quality cultivar, recommended for Trivandrum and Quilon districts of Kerala.

18	Arka Coorg Excel	Central Horticultural Experiment Station, Chettalli, IIHR	2012	Seedling Selection	-	-	-	-	-	To develop bold seeded, long spiked, high yielding pepper variety

* Yield Kg/ha (Dry)

Cardamom

Sl No.	Variety	Centre which developed	Year of release	Pedigree/Parentage	*Av. yield (kg/ha)	Dry recovery (%)	Oil (%)	1,8 Cineol (%)	α - terpenyl acetate (%)	Salient features
1	Mudigere 1	Regional Research Station, UAS, Mudigere. Karnataka	1984	Clonal selection from Malabar type	300	20	8	36	42	Erect and compact panicle, suitable for high density planting, moderately tolerant to thrips, hairy caterpillar and white grubs, pubescent leaf. Short panicle, pale green, oval bold capsule
2	Mudigere 2	Regional Research Station, UAS, Mudigere. Karnataka	1994/1996	Clonal selection from open pollination of Malabar type	475		8	45	38	Early maturing variety, suitable for high density planting, round/oval bold capsules.
3	PV 1	Cardamom Research Station, KAU, Pampadumpara, Idukki, Kerala	1991	A selection from Walayar collection, Malabar type	260	19.9	6.8	33	46	An early maturing type,short panicle, elongated slightly ribbed light green capsules, Long, bold capsule
4	PV 2	Cardamom Research Station, KAU, Pampadumpara, Idukki, Kerala	2001	Selection from OP Seedlings of PV-1,a Malabar type	982	23.8	10.45	-	-	Early maturing, lengthy panicle, Long bold capsules, high dry recovery percentage, field tolerant to stem borer and thrips, suitable fro elevation range of 1000-1200 m above MSL.
5	ICRI 1	ICRI (Spices Board), Myladumpara,	1992	Selection from Chakkupalam collection, Malabar	325	22.9	8.7	29	38	An early maturing type globose ,round and extra bold dark green capsules ; medium sized panicle with profusely flowering , early

6	ICRI 2	ICRI (Spices Board), Myladumpara, Kerala	1992	Clonal selection from germplasm, Mysore type	375	22.5	6.67	29	36	Performs well under high altitude and irrigated condition, medium long panicles, oblong bold and parrot green capsules,tolerant to azhukal disease
7	ICRI 3	ICRI (Spices Board), Myladumpara, Kerala	1994	Selection from Malabar type	440	22	6.6	54	24	Early maturing long pubescent leaves, tolerant to rhizome rot disease, oblong, bold parrot green capsules. suitable for hill zone of Karnataka
8	ICRI 4 TDK 4	ICRI (Spices Board), Myladumpara, Kerala	1997	Clonal selection from lower pulleys, a Malabar type	455	22.76	6.4	-	-	Early maturity, medium sized panicles, Globose bold capsules. Suitable for low rainfall areas, relatively tolerant to rhizome rot and capsule borer
9	ICRI 5	ICRI (Spices Board), Myladumpara, Kerala	2006	Hybrid between MCC 260 x MCC 49	1,543	23.15	7.13	-	-	First hybrid variety, Early maturity Moderately tolerant to drought, High yield under intensive management, Capsule size 68% ; more than 70 mm,
10	ICRI 6 (MCC – 73)	ICRI (Spices Board), Myladumpara, Kerala	2006	Selection from the germplasm(acc. MCC-73)	1,200	19.0	7.33	-		High yield, Medium maturity, Relatively tolerant to drought, High percentage of bold capsules Capsule size 71% ; more than 7mm
11	ICRI 7	ICRI (Spices Board), Myladumpara, Kerala	2010	Hybrid	-	22	8.84	-	-	Suitable for Wayanad, Kerala, Semi-erect panicles Angular bold capsules, Oleoresin 7.99%
12	IISR Kodagu Suvasini (CCS-1)	Indian Institute of Spices Research, Calicut, Kerala	1997	Selection from OP progeny of CL-37 from RRS Mudigere,Malabar type	745	22	8.7	42	37	Early maturing, suitable for high density planting, long panicle.tolerant to rhizome, rot,thrips,shoot/panicle/capsule borer.
13	IISR Avinash (RR-1)	Indian Institute of Spices Research, Calicut, Kerala	2001	A selection from OP progeny of CCS-1, a malabar type	847	20.8	6.7	30.4	35.5	Dark green capsules. Tolerant to rhizome rot, and shoot/panicle/capsule borer.

14	IISR Vijetha (NKE-12)	Indian Institute of Spices Research, Calicut, Kerala	2001	Clonal selection from field resistant plants for Katte,a Malabar type	643	22	7.9	45	23.4	Resistant to katte virus, bold capsules. Field tolerant to thrips and borer.

Ginger

Sl No.	Variety	Centre which developed	Year of release	Pedigree/Parentage	*Av. yield (kg/ha)	Dry recovery (%)	Oleoresin (%)	Crude fibre (%)	Essential oil (%)	Crop Duration	Salient features
1	Suprabha	High Altitude Research Station, OUA &T, Pottangi, Orissa	1988	Clonal selection from Kunduli local	16.6	20.5	8.9	4.4	1.9	229	Plumpy rhizome, less fibre, wide adaptability,suitable for both early and late sowing.
2	Suruchi	High Altitude Research Station, OUA &T, Pottangi, Odisha	1990	*Clonal selection from Kunduli local*	11.6	23.5	10.9	3.8	2.0	218	Profuse tillering, bold rhizome, early maturing, suitable for both rainfed and irrigated condition.
3	Suravi	High Altitude Research Station, OUA &T, Pottangi, Odisha	1991	Induced mutant of Rudrapur local	17.5	23.6	10.2	4.0	2.1	225	Plumpy rhizome, dark skinned yellow fleshed, suitable for both irrigated and rainfed conditions.
4	Himgiri	Department of vegetable crops, YSPUJ&F, Solan, Himachal Pradesh	1996	Clonal selection from Himachal collection	13.5	20.2	4.29	1.6	6.05	230	Best for green ginger less susceptible to rhizome rot disease, suitable for rainfed condition.
5	IISR Varada	ICAR- Indian Institute of Spices Research, Calicut, Kerala	1996	Selection from germplasm	22.66	19.5	6.7	3.29-4.50	1.7	200	High yielder, high quality bold rhizome, low fibre content.Wide adaptability and tolerant to diseases.
6	IISR Mahima	ICAR- Indian Institute of Spices Research, Calicut, Kerala	2004	Selection from germplasm	23.2	23	4.5	3.26	1.72	200	High yielder, plumpy extra bold rhizomes, resistant to *M. incognita and M. javanica* pathotype 1
7	IISR Rejatha	ICAR-Indian Institute of Spices Research, Calicut, Kerala	2004	Selection from germplasm	22.4	20.8	6.3	4	2.36	200	High yielder,plumpy and bold rhizome

8	Aswathy (IC No. 0584128)	Department of Spices and Plantation crops, Kerala Agriculture University, Trichur	-	Single plant selection from somaclones of cultivar Rio-de-Janeiro	23	19.7	7.45	3.5	3.32	220-240	Ideal for cultivation both as pure and intercrop. High yielding high quality clone suitable for green with high recovery of volatile oil and oleoresin. Filed tolerant to Phyllosticta leaf spot.
9	Athira (IC No. 0584128)	Department of Spices and Plantation crops, Kerala Agriculture University, Trichur	-	Selection form somaclones of cultivar Maran	21	22.6	6.8	3.4	3.1	220-240	Ideal for cultivation both as pure intercrop. Suitable for fresh and dry ginger. Tolerant to soft rot and bacterial wilt diseases than parent cultivar. High yielding high quality clone with high zingiberence
10	Karthika (IC No. 0584129)	Department of Spices and Plantation crops, Kerala Agriculture University, Trichur	-	Selection form somaclones of cultivar Maran	19	21.6	7.2	3.7	3.2	220-240	Ideal for cultivation both as pure and intercrop. Suitable for fresh and dry ginger. Tolerant to soft and bacterial wilt diseases than parent cultivar. Low infestation of shoot borer under field conditions. High pungency clone with high gingerol.
11	Subhada	High Altitude Research Station, OUA &T, Pottangi, Orissa	2009	Mutagen in EMS (40 PPM) treatment and selection of mutants	18000	22.4	10.4	3.4	2.0	210	Suitable for hills and plains

* yield tones/ha (fresh)

Turmeric

Sl No.	Variety	Centre which developed	Year of release	Pedigree/Parentage	*Av. yield (kg/ha)	Dry recovery (%)	Curcumin (%)	Oleoresin (%)	Essential oil (%)	Crop duration	Salient features
1	CO.1	Dept. of spices and plantation crops, TNAU, Coimbatore, Tamil Nadu	1982	Vegetative mutant by x-ray irradiation of Erode local	30.5	19.5	3.2	6.7	3.7	270 days	Bold and orange yellow rhizomes, suitable for drought prone areas, water logged, hilly areas saline and alkaline areas
2	BSR.1	Dept. of spices and plantation crops, TNAU, Coimbatore, Tamil Nadu	1986	Clonal selection from Erode local irradiated with x rays	30.7	20.5	4.2	4	3.7	285	Bright yellow rhizome suitable for problem soils and drought prone areas of Tamil Nadu.
3	BSR.2	Dept. of spices and plantation crops, TNAU, Coimbatore, Tamil Nadu	1994	Induced mutant from Erode local	32.7	-	-	-	-	245	A high yielding short duration variety with bigger rhizomes, resistant to scale insects
4	Krishna	Maharastra	1983	Clonal selection from Tekurpeta collection	9.2	16.4	2.8	3.8	2	240	Plumpy rhizomes, moderately resistant to pests and diseases
5	Sugandham	Sardarkrushinagar Dantiwada Agricultural University, Jagudan	1984	Clonal selection from germplasm	15.0	23.3	3.1	11	2.7	210	Thick, round rhizomes with short internodes. Moderately tolerant to pest and diseases
6	Roma	High Altitude Research Station, OUA &T, Pottangi, Orissa	1988	Clonal selection from T.Sunder	20.7	31	6.1	13.2	4.2	250	Suitable for both rainfed and irrigated condition. Ideal for hilly areas and late sown season.
7	Suroma	High Altitude Research Station, OUA &T, Pottangi, Orissa	1989	Clonal selection from T. Sunder by x- ray irradiation	20.0	26	6.1	13.1	4.4	253	Round and plumpy rhizome, field tolerance to leaf blotch, leaf spot and rhizome scales.

8	Ranga	High Altitude Research Station, OUA &T, Pottangi, Odisha	1992	Clonal selection from Rajpuri local	29.0	24.8	6.3	13.5	4.4	250	Bold and spindle shaped mother rhizome, suitable for late sown condition and low lying areas. Moderately resistant to leaf blotch and scales
9	Rasmi	High Altitude Research Station, OUA &T, Pottangi, Odisha	1992	Clonal selection from Rajpuri local	32.0	23	6.4	13.4	4.4	240	Bold rhizomes, suitable for both rainfed and irrigated condition, early and late sown season
10	Rajendra Sonia	Dept. of Hort., Tirhut College of agriculture RAU, Dholi, Bihar	1989	Selection from local germplasm	42.0	18	8.4	-	5	225	Bold and plumpy rhizome
11	Megha turmeric 1	ICAR, R.C..NEH Region, Shillong, Meghalaya	1996	Selection form Lakadong type	23.0	16.37	6.8	-	-	300-315	High curcumin content and bold rhizomes,suitable for North east hill region and North west Bengal.
12	Pant Peetabh	G.B Pant University of Agriculture and technology, Pantmagar, Uttaranchal	2001	Clonal selection from local type	29.0	18.5	7.5	-	1	-	Resistant to rhizome rot
13	Suranjana (TCP-2)	Uttar Bangal Krishi Viswa Vidyala, North Bangal, Pundibari	2000	Clonal selection from local types of west Bengal	29.0	21.2	5.7	10.9	4.1	235	Tolerant to rhizome rot and leaf blotch; resistant to rhizome scales and moderately resistant to shoot borer suitable for open and shaded condition
14	Suvarna	ICAR Indian Institute of Spices Research, Calicut, Kerala	1987	Selection from germplasm,collected from Assam	17.4	20	4.3	13.5	7	200	Bright orange coloured rhizome with slender fingers,field tolerant to pest and diseases.
15	Suguna	ICAR Indian Institute of Spices Research, Calicut, Kerala	1991	Selection from germplasm, collected from AP	29.3	20.4	4.9	13.5	6	190	Early maturing, field tolerant to rhizome rot.

16	Sudarsana	ICAR Indian Institute of Spices Research, Calicut, Kerala	1991	Selection from germplasm, collected from Singhat, Manipur	28.8	20.6	5.3	15	7	190	Early maturing, field tolerant to rhizome rot.
17	IISR Prabha	ICAR Indian Institute of Spices Research, Calicut, Kerala	1996	Open pollinated progeny selection	37.0	19.5	6.5	15	6.5	205	High yielding varitey
18	IISR Prathibha	ICAR Indian Institute of Spices Research, Calicut, Kerala	1996	Open pollinated progeny selection	39.1	18.5	6.2	16.2	6.2	225	High yielding varitey
19	IISR Kedaram	ICAR Indian Institute of Spices Research, Calicut, Kerala	2004	Clonal selection from germplasm	34.5	18.9	5.5	13.6	-	210	Resistant to leaf blotch.
20	IISR Alleppey Supreme	Indian Institute of Spices Research, Calicut, Kerala	2004	Selection from Alleppey Finger turmeric	35.4	19	5.55	16	-	210	Tolerant to leaf blotch.
21	Kanthi	Dept. of plantation crops and spices, KAU, Trichur	1996	Clonal selection from Mydukur variety of Andhra Pradesh	37.65	20.15	7.18	8.25	5.15	240-270	Erect leaf with broad lamina, big mother rhizomes with medium bold fingers and closer internodes
22	Sobha	Dept. of plantation crops and spices, KAU, Trichur	1995	Clonal selection from local type	35.88	19.38	7.39	9.65	4.24	240-270	High Yielding Varitey with high curcumin content (7.39%), Erect leaves with narrow lamina. Mother rhizome big with medium bold figures and closer internodes. Inner core of rhizomes is dark orange like Alleppey. More territory rhizomes.
23	Sona	Dept. of plantation crops and spices, KAU, Trichur	2002	Clonal selection from local germplasm	21.29	18.88	7.12	10.25	4.4	240-270	Orange yellow rhizome, medium bold with low territory fingers.Best suited for central zone of Kerala.rhizome medium bold. Field tolerant to leaf blotch.
24	Varna	Dept. of plantation crops and spices, KAU, Thrissur	2002	Clonal selection from local germplasm	21.89	19.05	7.87	10.8	4.56	240-270	Bright orange yellow rhizome, medium bold with closer internodes,territory fingers present.suited to central zone of Kerala.

25	Haldi – 1	Department of Vegetable Science N.D. University of Agriculture & Technology, Kumarganj, Faizabad.		18)							essential oil
26	Duggirala Red	Dr. Y. S. R. Horticultural University Turmeric Research station, Kammarpally	2013	Mass selection	25	23.5	4.1	-	-	240-270	High yielding variety. Rhizomes are long, plumpy, strong and very deep orange in colour.
27	Narendra Haldi – 2	Department of Vegetable Science N.D. University of Agriculture & Technology, Kumarganj, Faizabad	-	-	-	-	-	-	-	-	High yield potential, good size finger
28	Narendra Haldi – 3	Department of Vegetable Science N.D. University of Agriculture & Technology, Kumarganj, Faizabad	2012	Selection from germplasm (NDH 9)	-	-	-	-	-	-	Highyielding,, Root knot resistant. Moderate resistant against leaf spot and leaf blotch
29	Surangi	High Altitude Research Station, OUA &T, Pottangi, Odisha	2009	Clonal selection	23400	28	4.5-6.5	12.7	4.6	180-200	Suitable for hills and plains

* Yield tones/ha (fresh)

Cinnamon

Sl No.	Variety	Centre which developed	Year of release	Pedigree/Parentage	*Av. yield (kg/ha)	Bark recovery (%)	Bark oil (%)	Leaf oil (%)	Salient features
1	YCD.1	Hort. Research station (TNAU), Yercaud , Tamil Nadu	1996	Clonal selection from op seedings progenies of Sri Lankan type	360	35.3	2.8	3	Good bark recovery adopted to wide range of soil and rainfed conditions. Recommended for high ranges at 500-1000 m above MSL.
2	PPI (C) -1	HRS, (TNAU), Pechiparai, Tamil Nadu	2003	Selection from OP seedlings progeny introduced from Sri Lanka	980	34.22	2.9	3.3	Suitable for cultivation in high rainfall zones and hill regions of Tamil Nadu at an altitude of 100-500 m MSL.
3	Konkan Tej	Regional coconut research station, Dr, BSKKV, Vengurle, Maharashtra	1993	Seedling selection from progenies of Sri Lankan accessions	334	29.16 51.78	3.2	2.28	Superior qualities with 3.2% bark oil with bark recovery 29.16%,cinnamaldehyde in bark oil 70.23, eugenol in bark oil 6.93%,eugenol in leaf oil 75.5%,yields 4.10kg fresh bark.
4	Sugandhini (ODC-130)	Aromatic and Medical Plants Research station, Odakkali, Kerala	2000	Single tree selection from Wayanadu local collection.A Sri Lankan type	640	51.0	0.94	1.6	Recommended for cultivation for leaf oil production, cinnamaldehyde in bark oil 45%, eugenol in leaf oil 93.7%; released mainly for leaf oil purpose.densly foliage.
5	RRL(B) C-6	Regional research Laboratory, CSRi, Bhubaneswar, Orrisa	1996	Selection from germplasm collection-OP seedling progenies	250	-	-	-	High quality,sweet and pungent bark with 83% cinnaldehyde in bark oil, 94.0% eugenol in leaf oil,leaf oil 0.8% . Spreading, branching,nature
6	IISR Nithyashree	ICAR-Indian Institute of Spices Research, Calicut Kerala	1996	Clonal selection from OP seedling progeny	200	-	2.7	3	Good regeneration capacity, bark and leaf oleoresin contents are high. Good bark recovery with good aroma and taste. Bark oleoresin-10.0%

7	IISR Navashree	ICAR-Indian Institute of Spices Research. Calicut Kerala	1996	Selection from Op seedling progeny of Sri Lankan collection	200	40.6	2.7	2.8	High quality line, good bark recovery with good aroma and taste, .High cinnamaldehyde (73%) in bark oil, medium quality High shoot regeneration. Bark oleoresin 8%, cinnamaldehyde in leaf oil 15%,eugenol in bark oil 6%,eugenol in leaf oil 62%

Nutmeg

Sl No.	Variety	Centre which developed	Year of release	Pedigree/Parentage	*Av. yield	Myristicin (%)	Elimicin (%)	Salient features
1	Konkan Sugandha	Regional Fruit Research Station (BSKKU), Vengurla, Maharashtra	1998	Single plant selection from local seedling population	200- 526 fruits/ tree	-	-	Adaptable in Konkan region. Tree canopy is conical and compact. No incidence of major pests / disease
2	Konkan Swad	Regional Coconut Research station, Ratnagairi	2003	Selection from nutmeg seedling from Ratnagiri district	761.38 fruits/tree	-	-	Adapted in Konkan region with warm, humid condition as well as shade provision.canopy erect, conical shape.contain 39.85 essential oil in seed and 10.9 % in mace. No incidence of major pests / disease are noticed
3	IISR Viswasree	ICAR-Indian Institute of Spices Research, Calicut Kerala	2001	Clonal selection from elite germplasm	1000 fruits/tree (1.33Kg mace, 9kg dry,3122Kg/ha)	Nut 12.48 and mace 20.03		Low incidence of fruit rot. Nut recovery 70%, mace recovery 35%, 7.14% mace oil, oleoresin nut 2.48% and mace 13.85% respectively, butter 30.9%, Especially suitable under mixed cropping system

4	IISR Keralashree	ICAR-Indian Institute of Spices Research, Calicut Kerala	2012	Seedling selection from elite mother tree from Burliar Farmer's Participatory Breeding)	2000 fruits/tree	9.4	0.7	High yield, high quality and extra bold fruit mace and nut.

Coriander

Sl No.	Variety	Centre which developed	Year of release	Pedigree/Parentage	*Avg yield (kg/ha)	Essential oil (%)	Duration (days)	Salient features
1	Guj. Cor.2	Sardarkrushinagar Dantiwada Agricultural University, Jagudan	1974	Selection from germplasm	1100	0.35	112	Suitable for early sowing, erect plant,r ound bold grains, moderately tolerant to wilt and powdery mildew
2	Co.1	Dept. of spices and Pl.Crops, HC & RI, TNAU, Coimbatore, Tamil Nadu	1997	Selection from Koilpatti local	440	0.27	110	A dual purpose variety with small statured plant, suitable for rainfed areas, small grain
3	Co.2	Dept. of spices and Pl.Crops, HC & RI, TNAU, Coimbatore, Tamil Nadu	1982	Reselection from culture P2 of Gujarat	520	0.40	90-100	A dual purpose variety, suitable for saline, and alkaline and drought prone areas seeds oblong,medium.
4	Co.3	Dept. of spices and Pl.Crops, HC & RI, TNAU, Coimbatore, Tamil Nadu	1991	Reselection from Acc.695 of IARI, New Delhi type	650	0.38-0.41	85-95	A dual purpose variety, good yielder, medium sized grains, suitable for both rainfed and irrigated condition, rabi as well as kharif season. Field tolerant to powdery mildew, wilt and grain mould.
5	Co(CR).4	Dept. of spices and Pl.Crops, HC & RI, TNAU, Coimbatore, Tamil Nadu	2002	Reselection from germplasm ATP77 guntur collection	600	0.4	65-70	Early maturing variety suitable for both rainfed and irrigated condition; grains oblong and medium; field tolerant to wilt and grain mould
6	Guj.Cor.2	Dept. of spices and Pl.Crops, HC & RI, TNAU, Coimbatore, Tamil Nadu	-	Reselection from Co.2	1450	0.40	110	Semi spreading type, suitable for early sowing,moderately tolerant to powdery mildew,grains oblong,lodging and shattering resistant.
7	Rajendra Swathi	Dept. of Hort., Tirhut College of agriculture RAU,	1988	Pureline selection from Muzaffarpur collection	1300	0.65	90-100	Medium sized plant with fine,aromatic round grains,Suitable for intercropping,

		Dholi, Bihar						field tolerance to aphids
8	Sadhana	Regional Agrl. Research Station, APHU, Lam, Guntur, A P	1989	Mass selection from local Alur collection	1025	0.20	95-110	A dual purpose, semi-erect variety, suitable for rainfed condition field tolerance to white fly, mites and aphids. A mid-late variety withstands moisture stress, responded well to input management under optimum moisture level.
9	Swathi	Regional Agrl. Research Station, APHU, Lam, Guntur, A P	1989	Mass selection from Nandyal germplasm	855	0.30	82-85	Plants medium size semi-erect type, early maturing variety, suitable for rainfed condition, and late sown season. Field tolerant to white fly, Suits well to the areas where the soil moisture retentiveness in compariably less, being early maturity. It escapes powdery mildew disease.
10	CS 287	Regional Agrl. Research Station, APHU, Lam, Guntur, A P	-	Reselection from Guntur collection	600	0.40	110	Early maturing variety, suitable for both rainfed and irrigated condition. Field tolerant to wilt and grain mould.
11	Sindhu	Regional Agrl. Research Station, APHU, Lam, Guntur, A P	1991	Mass selection germplasm, Warangal local	1000	0.40	100-110	Oval medium breakable grains, suitable for rainfed areas, tolerant to wilt, powdery mildew as well as drought condition, medium duration.
12	Hisar Anand	Department of Vegetable Crops, HAU, Hisar, Haryana	1994	Mass selection from Haryana collection	1400	0.35	-	A medium tall dual purpose variety,oval medium sized seeds, wider adaptability to different soil conditions.Resistant to lodging due to spreading habit.
13	Hisar Sugandh	Department of Vegetable Crops, HAU, Hisar, Haryana	2001	Mass selection from indigenous germplasm	1400			Suitable for irrigated conditions.Resistant to stem gall diseases.
14	Hisar Surabhi	Department of Vegetable Crops, HAU, Hisar, Haryana	2004	Mass selection from local germplasm	1800	0.4-0.5	130-140	Bushy erect plant type, seed medium, oblong; tolerant to frost, less susceptabile to aphids, medium duration

15	Azad Dhania-1	C.S. Azad University of agriculture and technology, Kanpur, Uttar Pradesh	1996	Mass selection from Kalyanpur germplasm collection	1000	0.29	120-125	Erect, early branching, number of umbellates per umbel 5, tolerant to moisture stress, powdery mildew and aphids.
16	Pant haritima	G.B Pant University of Agriculture and technology, Pantmagar, Uttaranchal	1993	Selection from local type Pant Dhania	1200	0.4	150-160	Tall erect plant, a dual purpose type, good yielder of leaves, smaller seeds with high oil. Resistant to stem gall.
17	DWA 3*	University of Agricultural Science, Dharward, Karnataka	1999	Purelineselection from Karnataka collection	400	0.27	-	A dual purpose variety and for seed production in rabi crop, moderately tolerant to powdery mildew, black clay soils are best suited
18	CIMPOS-33	Central Institute of Medicinal Plants, Regional centre, Mehsana, Gujarat	-	Selection from germplasm introduced from Bulgaria	2100	1.3	-	Tall erect, compact, profusely branching and flowering, grains small and bold. Mainly recommended for oil production.
19	ACR-01-256 (NRCSS ACR-1)	NRC seed spices, Ajmer, Rajasthan	2005	Reselection from EC-467683 from Russia	1100	0.35-5	-	Dual purpose variety, long duration, resistant to stem gall and wilt.
20	RCr 20	SKN College of Agriculture, RAJAU, Jobner, Rajasthan	1997	Recurrent half sib election from Jaipur local	900	0.25	100-110	Medium sized bushy plant suitable for rainfed crop or limited moisture condition and heavy soils of south Rajastan. Moderately resistant to stem gall, bold grains, early maturity.
21	RCr.41	SKN College of Agriculture, RAJAU, Jobner, Rajasthan	1988	Recurrent half sib selection from local type from "Kotta"	909	0.25	130-140	A tall erect plant with thick stem. Grows well under irrigated conditions, resistant to stem gall, wilt and moderately resistant to powdery mildew. Small seeds (9.3g/1000 seed),long duration variety
22	RCr 435	SKN College of Agriculture, RAJAU, Jobner, Rajasthan	2003	Recurrent selection from local germplasm from Jalore	1000	0.33	110-130	Plants are bushy, erect, bold seeds, medium sized, medium maturing variety, adapted for irrigated condition moderately resistant to root knot and powdery mildew.

23	RCr 436	SKN College of Agriculture, RAJAU, Jobner, Rajasthan	2001	Recurrent half sib selection from local germplasm from Kotta	1100	0.33	90-100	Plants semi dwarf, bushy type with quick early growth and bold seeds. Resistant to root rot and root knot nematodes most suitable for limited moisture condition and heavy soils of south Rajasthan
24	RCr446	SKN College of Agriculture, RAJAU, Jobner, Rajasthan	2001	Half sib selection from local type from Jaipur local	1200	0.33	130	Plants tall, are leafy erect with higher number of seeds per umbel. Seeds medium in size and moderately resistant to stem gall.
25	RCr 684	SKN College of Agriculture, RAJAU, Jobner, Rajasthan	-	Mutation breeding of gamma rays. Induced mutant of Rcr-20	990	0.32	110-120	Resistant to stem gall and less susceptible the powdery mildew. Adapted to medium heavy textured soil and sandy loam soil under irrigation. Bold seeds, Medium maturity.
26	RCr 480	SKN College of Agriculture, RAJAU, Jobner, Rajasthan	2009	Recurrent selection based on individual plant progeny (half sib) performance in accession No.480	-	-	-	-
27	Hisar Sugandh (DH-36)	Dept. of vegetable Crops, CCS, HAU, Hisar, Haryana	2001	Mass selection from germplasm collected from farmer's field of Haryana.	-	-	-	
28	Sudha (LCC-128)	Regional Agrl. Research Station, APHU, Lam, Guntur, A P	2006	Mass selection made from the land race collected from Ongole district	-	-	-	
29	DH – 246 (Hisar Surbhi)	Dept. of vegetable Crops, CCS, HAU, Hisar, Haryana	2004	Selection form local germplasm	-	-	-	-
30	LCC - 234	Dr.YSR. Horticultural University Horticultural Research Station, Lam Guntur	2013	Mass selection made from the land race collected from Prakasam district	15-18	0.15	35-55	High yielding leaf variety suitable for off season production in Andhra Pradesh

31	DH 220	Dept. of vegetable Crops, CCS, HAU, Hisar, Haryana	2012	-	1373	0.39		This variety has out yielded other varieties, Hisar Ananad (National check) and Loval checks under coordinated varietal trials of AICRPS.
32	DH 206	Dept. of vegetable Crops, CCS, HAU, Hisar, Haryana	2009	Selection from germplasm	-	-	-	-
33	RCr 728	SKN College of Agriculture, RAJAU, Jobner, Rajasthan	2009	-	-	-	-	-
34	APHU Dhania -1	Regional Agrl. Research Station, APHU, Lam, Guntur, A P	2010	Mass selection from germplasm	10-12	0.4	80-100	Suitable for growing in Bihar, Madhya Pradesh, Uthar Pradesh and Tamil Nadu.
35	Hisar Bhoomit (DH-228)	Dept. of vegetable Crops, CCS, HAU, Hisar, Haryana	2007 (state release)	Selection from germplasm	180-200	-	110	Small seeded with high oil content, suitable for leaf production

*Yield Kg/ha (Dry)

Cumin

Sl No.	Variety	Centre which developed	Year of release	Pedigree/Parentage	*Av. yield (kg/ha)	Essential oil (%)	Crude fibre (%)	Duration (days)	Salient features
1	Mc.43	Spices Research Station, GAU, Jagudan, Gujarat	1952	Selection from germplasm	580	2.7	15.5	-	Plant semi spreading, grains bold lustering withstand lodging and shattering, moderately tolerant resistant to *Fusarium* wilt, *Alternaria* blight & powdery mildew.
2	Guj. Cumin 1	Spices Research Station, GAU, Jagudan, Gujarat	1983	Selection from local germplasm (Vijaypur -5)	550	3.6	14.25	-	Plants bushy and spreading, grains bold, linear oblong; Withstand shattering and lodging, moderately tolerant to wilt, powdery mildew and blight.
3	RZ-19	SKN College of Agriculture, RAJAU, Jobner, Rajasthan	1988	Recurrent selection from UC.19	500	-	-	-	Erect plant, bold, lustrous grain, tolerant to wilt and blight suitable for late sowing season.
4	Guj Cumin 2	Spices Research Station, GAU, Jagudan, Gujarat	-	Pure line selection from M2 irradiated seeds from MC-43	620	4	22.1	-	Bushy plant, good branching habit, grains bold, medium sized, lustrous grain, tolerant to wilt and blight suitable for late sowing season
5	Guj. Cumin 3	Spices Research Station, GAU, Jagudan, Gujarat	-	Recurrent selection derived from W.German entry EC-232689	620	4.4	-	-	Bushy dwarf plant, fruit medium sized, frost wilt resistant variety suitable for winter season in limited irrigation. Higher essential oil content, seed pungent with good aroma
6	RZ-19	Repeat	1988	Recurrent single plant progeny selection from Ajmeer	560	-	-	140-150	Erect plant, pink flowers, bold, lustrous grain, and gray pubescent, tolerant to wilt and blight suitable for late sowing season.
7	5-404	Spices Research Station, GAU, Jagudan, Gujarat	1952	Selection from local germplasm	350	2.2	7.7		An erect plant, medium sized fruit, moderately tolerant to powdery mildew.
8	RZ-209	SKN College of Agriculture, RAJAU, Jobner, Rajasthan	1995	Recurrent single plant progeny selection from Jore	650	-	-	120-130	A variety shown some resistance with blight and wilt disease

9	RZ-223	SKN College of Agriculture, RAJAU, Jobner, Rajasthan	-	Mutation breeding in UC-216	600	3.0-3.5	-	120-130	Wider adaptability, resistant to wilt, superior in yield and seed quality over RZ-19.Plants bushy, semi-erect, long bold attractive seeds, medium duration.
10	Ac-01-167	NRC seed spices, Ajmer, Rajasthan	2005	Reselection from EC-243373	515	3			Bold seeds resistant to wilt.
11	RZ-345	SKN College of Agriculture, RAJAU, Jobner, Rajasthan	2009	recurrent selection based on individual plant progeny (half sib) performance in accession 345	-	-	-	-	-
12	GC-4 (Gujarat Cumin – 4)	Sardarkrushinagar Dantiwada Agricultural University, Jagudan	2006	The Cumin variety JC-2000-72 is the selection from GC-3.	-	-	-	-	-
13	RZ- 223	SKN College of Agriculture, RAJAU, Jobner, Rajasthan	2004	The variety has been developed through mutation breeding in UC-216	-	-	-	-	-
14	RZ-341 (UC – 341)	SKN College of Agriculture, RAJAU, Jobner, Rajasthan	-	The variety has been developed through poly cross between high volatile oil content and low volatile oil content	-	-	-	-	-

*Yield kg/ha (Dry)

Fennel

Sl No.	Variety	Centre which developed	Year of release	Pedigree/Parentage	*Av. yield (kg/ha)	Essential oil (%)	Crude fibre (%)	Duration (days)	Salient features
1	PF – 35	Sardarkrushinagar Dantiwada Agricultural University, Jagudan	1973	Selection from local germplasm	1280	-	-		Plant tall and spreading moderately tolerant to leaf spot, leaf blight and sugary diseases
2	Co.1	Dept. of spices and plantation crops, HC & RI, TNAU, Coimbatore, Tamil Nadu	-	Reselection from PF 35	570	-	-	220	Medium statured, diffuse branching, Suitable for intercropping and border cropping with chilli and turmeric. Suitable for drought prone, water logged, saline and alkaline conditions.
3	Guj Fennel 1	Sardarkrushinagar Dantiwada Agricultural University, Jagudan	1984	Pure line selection from Vijaypur local	1695	2.2	-	158	Plant tall and bushy, shattering and lodging, suitable for early sowing and rabi crop, reasonably tolerant to drought, moderately tolerant to sugery disease, oblong, medium bold and dark green seeds.
4	Guj fennel 2	Sardarkrushinagar Dantiwada Agricultural University, Jagudan	1997	Pedigree selection from local germplasm	1940	2.4	-	159	Plants bushy, bold grains, rich in volatile oil and suitable for both rain fed and irrigated condition
5	RF 101	SKN College of Agriculture, RAJAU, Jobner, Rajasthan	2001	Recurrent half sib selection from local germplasm collection from Jobner	1400	1.9	-	150-160	Errect medium tall nature, medium maturity type with long bold grains, most suitable for loamy and black cotton soil.
6	S-7-9	-	-	Slection	1100	1.2	24	210	A bushy plant with big umbel, moderately tolerant to blight.
7	Guj Fennel II	Sardarkrushinagar Dantiwada Agricultural University, Jagudan	2003	Selection based on individual plant progeny performance from local germplasm	2489	1.98		148	A medium maturity type adopted to rabi season un der irrigation; seeds medium bold.
8	RF 125	SKN College of Agriculture, RAJAU, Jobner, Rajasthan	2003	Reccurent half sib selection is an exortic collection EC 243380 from Italy	1700	2.5-3.0 (2.78)	-	110-130	Plants are short statured with compact umbels and long bold seeds when presence, denser view of plants green.

									Tollerent to sugery disease.
9	Hisar Sawrup	Dept. of vegetable Crops, CCS, HAU, Hisar, Haryana	2004	Mass selection from indigenous germplasm of Haryana	*1600*	1.6	-	175-185	Plants grow up right, spreading, gives a bushy appearance. A late maturity type grain long and bold, resistant to lodging, no shattering of grains.
10	Azad Sanuf-1	C.S. Azad University of agric. And technology, Kanpur, Uttar Pradesh	1996	Selection from germplasm	1500	2	-	160-170	Medium plants, resistant to blight and root rot desases. Escapes attack of aphids due to early maturity, seeds are bold green.
11	Pant Madhurika	G.B Pant University of Agriculture and technology, Pantnagar, Uttaranchal	2001	Pure line selection from local germplasm	-	-	180-185	110	Tall robust eruct plant with big umbels having bold seeds with green fine ridgs sweet in taste, medium maturity.
12	GF- 11 (Gujarat Fennel – 11)	Spices Research Station, GAU, Jagudan, Gujarat	2004	Recurrent selection based on individual plant progeny performance.	-	-	-	-	
13	RF 143	SKN College of Agriculture, RAJAU, Jobner, Rajasthan		Recurrent selection from individual plant progeny	*1200*	1.87	-	-	Medium tall and reccomented for loamy and black cotton soils
14	AF-01-119 NRCSS-AF1	NRC seed spices, Ajmer, Rajasthan	2005	Recurrent selection from individual plant progeny	*1950*	-	-	-	Medium maturity seed, bold, tolerent to blight
15	RF – 205 (UF – 205)	SKN College of Agriculture, RAJAU, Jobner, Rajasthan	2009	Recurrent selection based on individual plant progeny (half –sib) from F2 generation of a cross between JF-25 x RF – 125.	-	-	-	-	-
16	HF -143	Dept. of vegetable Crops, CCS, HAU, Hisar, Haryana	2012	NA	1779	NA	NA	150	This variety has out yielded other varieties, GF-II (National check) and local checks undr coordinated varietal trials of AICRPS.

17	RF – 281	SKN College of Agriculture, RAJAU, Jobner, Rajasthan	2012	-	-	-	-		
18	RF-178	SKN College of Agriculture, RAJAU, Jobner, Rajasthan							
19	Rajendra saurbha	Dept. of Hort., Tirhut College of agriculture RAU, Dholi, Bihar	2000						
20	HF-33	Dept. of vegetable Crops, CCS, HAU, Hisar, Haryana	2004						High yielding
21	JF-444-1	Spices Research Station, GAU, Jagudan, Gujarat (Recommended for National release)	2010	Selection based on individual plant progeny performance form local germplasm	2588	2.05		154	Compact seeds in umbellate, Hard and flat stem, synchronise maturity, Small umbellate at center of umbel.

*Yield kg/ha (Dry)

Fenugreek

Sl No.	Variety	Centre which developed	Year of release	Pedigree/Parentage	*Av. yield (kg/ha)	Seed protein (%)	Duration (days)	Salient features
1	Co.1	Dept. of spices and plantation crops, HC & RI, TNAU, Coimbatore, Tamil Nadu	1982	Reselection from TG-2356 introduced for North India	680	-	80-85	A quick growing, dual purpose , early maturing variety tolerant to root rot disease. Seeds contain 21.7% protein.
2	Co 2	Dept. of spices and plantation crops, HC & RI, TNAU, Coimbatore, Tamil Nadu	1999	Selection from CF 390	480	-	85-90	Short duration dual purpose variety , field tolerant to *Rhizoctonia* root rot disease, suitable for both *kharif* and *rabi* season.Early maturity, short duration.
3	Rajendra kanti	Dept. of Hort., Tirhut College of agriculture RAU, Dholi, Bihar	1988	Pure line selection from Reghunathpur collection	1300	9.5	-	Medium sized bushy plant; early maturity, suitable for intercropping in both kharif and rabi season,, field tolerant to *cercospora* leaf spot, powdery mildew and aphids.
4	RMt.1	SKN College of Agriculture, RAJAU, Jobner, Rajasthan	1990	Pure line selection from Nagpur local	1400	21	-	Vigorous semi erect medium sized,moderately branched growth habit, medium sized,bold and attractive typically yellow coloured grains, moderately resistant to root knot nematode and powdery mildew and aphids
5	Lam sel.1	Dr.YSR. Horticultural University ,Horticultural Research Station, Lam Guntur	1992	Selection from germplasm collection of Uttar Pradesh	740	53	-	Dual purpose varieties, early muturing, bushy type and medium height940cm0, more number of branches and green matter. When cultivated for green leaf purpose it gives an average green yield of 12 tonnes per hectare. Field tolerant to major pests and diseases.
6	Hisar Sonali	Dept. of vegetable Crops, CCS, HAU, Hisar, Haryana	1994	Pure line selection from germpalsm	1700	-	-	Tall and bushy vigorous growing variety, dual purpose variety, late maturity (140-145 days), suitable for cultivation under irrigated condition.Moderately resistant to root rot and aphids.

7	Hisar Suvarna	Dept. of vegetable Crops, CCS, HAU, Hisar, Haryana	2001	Pureline selection from local germplasm	1600	-	-	A quick growing, erect and tall, dual purpose, medium maturity (130-140days) , moderately resistant to percospora and powdery mildew.suitable for cultivation throughout the country.
8	Hisar Madhavi	Dept. of vegetable Crops, CCS, HAU, Hisar, Haryana	2001	Pureline selection from local germplasm of UP	1900	-	-	A quick growing, erect and tall, dual purpose, medium maturity (130-140days), moderately resistant to powdery mildew and to downy mildew.A variety with under adaptability suitable for both irrigated and rainfed condition.
9	Hisar Muktha	Dept. of vegetable Crops, CCS, HAU, Hisar, Haryana	2001	Pureline selection natural green seed coated mutant line from UP	2000	-	-	A quick growing seedtype variety, medium maturity (135-140days), moderately resistant to powdery mildew and to downy mildew.Erect and tall plants.Wide adaptability.Suitable for both irrigated and rainfed condition.
10	RMt 303	SKN College of Agriculture, RAJAU, Jobner, Rajasthan	-	Mutation breeding from variety RMt 1	1900	-	-	Medium maturity variety (145- 150 days) seeds bold ,with typical yellow colour ,less susceptible to powdery mildew
11	RMt 305	SKN College of Agriculture, RAJAU, Jobner, Rajasthan	-	Mutation breeding from variety RMt 1	1300	-	-	First determinant type, multipodant, early maturing, wider adaptability, resistant to powdery mildew and rootknot nematodes.Seeds bold, attractive and yellow, duration 120-125 days.
12	Guj Methi 1	Spices Research Station, GAU, Jagudan, Gujarat	-	Recurrent selection based on pure line selection from J. Fenu 102	1864	-		The first variety from Gujarat released for the state.Plant dwarf.
13	RMt143	SKN College of Agriculture, Jobner, Rajasthan	-	Pureline selection of local collection of Jodhpur	1600	-	140-150	Moderately resistant to powdery mildew, seeds bold yellow colour, suitable for heavier soils.
14	Rajendra Abha (Kasuri Methi)		-	NA	NA	NA	NA	NA
15	Pant Ragini	G.B Pant University of Agriculture and technology,	2001	Selection from local germplasm	1200		170-175	A dual purpose tall bushy type resistant to downy mildew and root rots, medium maturity.

		Pantmagar, Uttaranchal						Seed contain 2-2.5% essential oil
16	NRCS- AM - 1 AM-01-35	ICAR-NRC Seed spices Ajmer, Rajasthan		Selection from local germplasm	1720	-	-	Dual purpose, tolerant to powdery mildew
17	RMt-351 (UM-351)	SKN College of Agriculture, Jobner, Rajasthan	2006	The variety has been developed through irradiation of RMt-1 (a released variety of Rajasthan state) with gamma rays at 20 kr.	-	-	-	-
18	APHU Methi-1 (LFC-84)	Dr.YSR. Horticultural University ,Horticultural Research Station, Lam Guntur	2009 (Recommended for state release)	Selection from a land race collected from Guntur district.	-	-	-	-
19	RMt- 361 (UM- 361)	SKN College of Agriculture, Jobner, Rajasthan	2009	Developed through irradiation of RMt-1 with gamma rays.	-	-	-	-
20	HM – 348	Department of Vegetable Cops, HAU, Hisar, Haryana	2013	Pure line selection form germplasm collected form Haryana	-	-	-	This variety has out yielded other varieties, Hisar Sonali (National check) and local checks under Coordinated varietal trials of AICRPS ar Hisar, Pantnagar & Lam centes.
21	HM – 219	Department of Vegetable Cops, HAU, Hisar, Haryana (Recommended for national release)	2009	Selection form germplasm collected form Haryana	-	-	-	
Ajowain								
1	Gujarat Ajowan – 1	Sardarkrushinagar Dantiwada Agricultural University, Jagudan	-	Selection form germplasm	2269	-	-	Non-shattering, mildly susceptible to powdery mildew and resistant to insects, a late maturity (176 days) variety.
2	Pant Ruchika	G.B Pant University of Agriculture and technology,	2001	pure line selection for local collection	600/800	-	-	Erect, bushy plant, seed light brown, and attaractive; late maturing variety (170-175

		Pantmagar, Uttaranchal						days)
3	RFA – 68	A.R.S. Substation, Udaipur Agricultural University, Chittorgarh. Rajasthan	-	Selection form local germplasm grown in Pratapgarh area	900	-	-	A medium maturity variety, flowers in about 90 days takes about 150 days to mature.
4	Ajmer Ajowan 1	ICAR-NRC Seed Spices, Ajmer, Rajasthan	2004	Selection from Pratapgrah local NRCSS AA-61	1420	-	-	Plant tall, late maturity group (160 days), suitable for early and rabi sowing under irrigated and limited available water conditions, seeds medium size, contains 3.4 % volatile oil.
5	Ajmer Ajowan - 2	ICAR-NRC Seed Spices, Ajmer, Rajasthan	2004	Selection from Gujarat local NRCSS AA-19	1280	-	-	A bushy plant, early matruring group (147 days), moderately tolerant to drough, seeds medium, 3.0 % volatile oil, moderately
6	Lamsel - 1	PRS, ANGRAU, Guntur, Andra Pradesh	-	Mass selection	1000 /1400	-	-	A tall early maturity (140 days) variety.
7	Lamsel – 2	PRS, ANGRAU, Guntur, Andra Pradesh	-	Mass selection	1000/1200	-	-	A spreading bushy trype with more braches, requieing more spacting
8	Rajendra Mani	Dept. of Hort., Tirhut College of agriculture RAU, Dholi, Bihar	-	NA	NA	-	-	NA
Nigella								
1	Azad Kalaunji	C .S. Azad University of Agriculture and technology, Kanpur, Uttar Pradesh	1998	Selection from germplaskm from Kalyanpur	1000/1200	-	-	Erect plant, bold seeds, highly aromatic.maturity 135 to 145 days
2	Rajendra Shyama	Dept. of Hort., Tirhut College of agriculture RAU, Dholi, Bihar	-	-	-	-	-	-
3	Pant Krishna	G.B Pant University of Agriculture and technology, Pantmagar,	-	Selection from indigenous germplasm	500/600	-	-	Plant medium statured, susceptible to damping off, medium maturity, bold seed, suitable for cultivation Uttar Pradesh and Uttaranchal

		Uttaranchal					
4	Ajmer Nigella – 1	ICAR-NRC Seed Spices, Ajmer, Rajasthan -	Selection from Rampura local	720	-	-	Seeds are black, bold, volatile oil content in 0.7%, field resistant to root rot (*Fusarium Oxysporum*). Suitable for growing under irrigation medium duration type (135 days)
Mango gigner							
1	Amba	High Altitude Research Station, OUA &T, Pottangi, Odisha -	Selection from local germplasm	21.9	-	-	Oleoresin 6.48 %, essential oil 0.8 %, dry recovery 18.7%. No major disease and pest problem.

*Yield kg/ha (Dry)

Appendix VI
Released Varieties of Vegetables from Kerala Agricultural University

T.R. Gopalakrishnan, Kerala Agricultural University, P.O. KAU, Thrissur – 680 656, Kerala

SI.No.	Crops	Variety	Description
1.	Chilli	Vellayani Athulya	Medium statured plants with light green fruits at immature stage, tolerant to partial shade, medium pungent , yield 32t/ha
		Vellayani Samrudhi	*Capsicum frutescens* variety with Cream coloured, erect fruits, highly pungent, shade tolerant, yield 20 t/ha
		Vellayani Thejus	*Capsicum chinense* variety with High yield, shade tolerance and good quality, dark green fruits, conical dark red on ripening yield 30t/ha
2.	Tomato	Vellayani Vijay	Determinate, short statured and compact plants, light green fruits at immature stage, tolerant to bacterial wilt, high temperature and partial shade, Yield 37.26 t/ha
		Akshaya	High yielding, indeterminate variety suited for rain shelters. Flat round medium sized fruits. Yield 40t/ha(inside rainshelter)
		Manu Lakshmy	Oval uniform red fruits medium sized, resistant to bacterial wilt. Yield 35t/ha
3.	Snake gourd	Harithasree	Medium long Green coloured fruits with white stripe. High Yield 60t/ha
4.	Ivy gourd (*Koval)*	Sulabha	Pale green,long cylindrical fruits with continuous striations. Yield 60t/ha
5.	Okra	Anjitha	Early maturing, yellow vein mosaic resistant variety, tolerant to shoot and fruit borer; fruits long, smooth and green with prominent ridges, yield 14.6 t/ha
		Manjima	Early maturing, yellow vein mosaic resistant hybrid, tolerant to shoot and fruit borer, long, light green fruits, yield 16 t/ha
6.	Cowpea	Malika	Trailing growth habit, long light green pods with out purple tip, seeds brown coloured with a white speck having irregular shape at one end, purple flowers, yield 9.8 t/ha
		Sharika	Trailing growth habit, long white pods with purple tip, black seeds, purple flowers, 10.62 t/ ha
		Vellayani Jyothika	Vigorous growing plants with attractive long light green pods. Tolerant to *Fusarium* wilt. Yield 19.33t/ha
7.	Dolichos bean	Hima	Photosensitive variety with white flowers, large, succulent, light green, broad, straight pods with brown pods. Yield 14.82 t/ha

Contd.

Sl.No.	Crop	Variety	Description
		Grace	Photosensitive variety with purple stem,lilac flowers, medium broad, greenish purple, slightly curved pods with black seeds. Yield 15.1 t/ha
8.	Winged bean	Revathy	Short day plant, trailing habit, pole type, medium sized four ribbed green pods with dark brown large seeds, violet flowers, yield 2.14 t/ha
9.	Amaranth		
		Arun	Photo insensitive, stem, petiole and leaves maroon red coloured, leaves ovatinate in shape, seeds black coloured, yield 20.1 t/ha/cutting
		Mohini	Multicut green leaved variety, stem, petiole and leaves green coloured, reddish basal stem, seeds black, yield 13.2 t/ha
		Renusree	Highly branching, red fleshy stem and green leaves, multicut type giving 6-8 harvests, yield 15.5t/ha
		Krishnasree	Highly branching, red fleshy stem and coppery red leaves, multicut type giving 6-8harvests, yield 14.8t/ha
10.	Drumstick	Anupama	Early bearing, protracted flowering, more harvests, long fruits, fruit length 65 cm, yield 50 t/ha
11.	Coleus	Nidhi	Characteristic aroma, oblong shaped tubers, early maturity, yield 27.9 t/ha

Appendix VII : Research & Educational Infrastructure in Horticulture in India

P.G. Sadhan Kumar and S. Nirmala Devi, Kerala Agricultural University P.O. KAU, Thrissur – 680 656

Institutes	Address	Website
ICAR Indian Agricultural Research Institute New Delhi	Indian Agricultural Research Institute Pusa- 110012, New Delhi	www.iari.res.in
ICAR-IARI Regional Station, Himachal Pradesh	IARI Regional station Katrain, Kullu Valley-175129, Himachal Pradesh	www.iari.res.in
ICAR-Indian Institute of Horticultural Research Bangalore	Indian Institute of Horticultural Research, Hessaraghatta Lake Post, Bangaluru -560089, Karnataka	www.iihr.res.in
ICAR-Indian Institute of Spices Research Marikunnu	Indian Institute of Spices Research Marikunnu P.O.-673012, Kerala	www.spices.res.in
ICAR-Indian Cardamom Research Institute Myladumpara, Kerala	Indian Cardamom Research Institute Myladumpara, Kailasanadu P.O.Idukki (dt.) 685553, Kerala	www.indianspices.com
ICAR-Indian Institute of Vegetable Research	Indian Instiute of Vegetable Research, Post Bag No 01 PO Jakhin, varanasi -221005, Uttar Pradesh	www.iivr.org.in
ICAR-Indian Institute of Soil Science Bhopal	Indian Institute of Soil Science, Nabi Bagh, Berasia Road Bhopal-462038, Madhya Pradesh	www.iiss.nic.in
Indian Institute of Packaging, Mumbai	Indian Institute of Packaging, E-2,MIDC Area Andheri(E), Mumbai -93	www.iiip-in.com
ICAR-National Bureau of Plant Genetic Resources New Delhi	National Bureau of Plant Genetic Resources, Pusa Campus -110012, New Delhi	www.nbpgr.emet.in
ICAR-National Bureau of Soil Survey & Land Use Planning,Nagpur	National Bureau of Soil Survey & Land Use Planning Sankar Nagar P.O. Amravti Road,Nagpur-440010, Maharashtra	www.nbsslup.org
ICAR-National Centre for Integrated Pest Management New Delhi	National Centre for Integrated Pest Management LBS Bldg. ,Pusa Campus, New Delhi 110012	www.ncipm.org.in
ICAR-National Research Centre for Plant Bio Technology, New Delhi	National Research Centre for Plant Bio Technology LBS Building ,Pusa campus - 110012, New Delhi	www.nrcpb.org
ICAR-National Research Centre on DNA Finger Printing, NBPGR, New Delhi	National Research Centre on DNA Finger Printing NBPGR, Pusa campus -110012 New Delhi	www.nbpgr.ernet.in
ICAR-National Research Centre For Banana Thiruchirapalli	National Research Centre For Banana, Thogamalai road Thayannur Post, Thiruchirapalli-620102, Tamil Nadu	www.nrcb.res.in
ICAR-Directorate of Cashew Research Dakshina Karnataka	Directorate of Cashew Research Station Darbe P.O. Puttur Dakshina Karnataka- 574202, Karnataka	www.cashew.res.in

ICAR- National Research Centre For Citrus, Nagpur	National Research Centre For Citrus, P.B. No.464, Shankar Nagar P.O., Amravati Road Nagpur-440010, Maharashtra	www.nrc.citrus.nic.in
ICAR-National Research Centre For Litchi, Bihar	National Research Centre For Litchi, P.O. MIC Bela Muzhaffarpur-842005, Bihar	www.http//nrclitchi.org
ICAR-Directorate of Floricultural Research Pune, Maharashtra	Directorate of Floricultural Research Agricultural College Shivajinagar, Pune, Maharashtra 411005	www.dfr.icar.gov.in
ICAR-Directorate of Medicinal & Aromatic Plants, Gujarat	Directorate of Medicinal & Aromatic Plants Boriavi Anand - 387310, Gujarat	www.amapr.org.in
ICAR-Directorate of Mushroom Research Himachal Pradesh	Directorate of Mushroom Research Chambaghat Solan - 173213 Himachal Pradesh	www.nrcmushroom.org
ICAR-Indian Institute of Oilpalm Research West Godaveri	Indian Oilpalm Research Institute, Pedavegi West Godavari - 534 450, Andhra Pradesh	www.dopr.org.in
Directorate of Onion and Garlic Research, Pune-Nasik	Directorate of Onion and Garlic Research Pune-Nasik Highway, Rajgurunagar, Pune - 410505 Maharashtra	www.dogr.res.in
ICAR-National Research Centre For Orchids, Gangtok	National Research Centre For Orchids Pakyong Gangtok -737106, Sikkim	www.nrc orchids.nic.in
ICAR-National Research Centre on Pomegranate, Solapur	National Research Centre On Pomegranate C/o Centre on Rabi Sorghum (NRCS), NH-9,Bypass Road Shelgi,Solapur-413006, Maharashtra	www.nrcpomegranate.org
ICAR-NRC on Seed Spices Tabiji Farm, Ajmer	NRC on Seed Spices Tabiji Farm,Ajmer- 3 05206, Rajasthan	www.nrcss.org.in
National Research Centre for Agroforestry, Uttar Pradesh	National Research Centre for Agroforestry Jhansi-284003,Uttar Pradesh	www.nrcaf.ernet.in
ICAR-Directorate of Weed Research, Jabalpur	Directorate of Weed Research Maharajpur, Adhartal Jabalpur - 482 004, Madhya Pradesh	www.nrcws.org.in
National Horticultural Research and Development Foundation, Maharashtra	National Horticultural Research and Development Foundation Chitegaon Phata, Nashik-Aurangabad Highway, Post: Varhe Darna, Taluka: Niphad District: Nashik Maharashtra- 422201	www.nhrdf.com www.nbri-lko.org
National Botanical Research Institute, Lucknow	Rana Pratap Marg, Post Box No. 436, Lucknow-226 001, U.P.	www.ncl-india.org
National Chemical Laboratory, Pune	National Chemical Laboratory Dr. Homi Bhabha, Road, Pune 411 008	
National Institute for Interdisciplinary Science & Technology (NIST)	National Institute for Interdisciplinary Science & Technology (NIST), Industrial Estate P.O. Pappanamcode Trivandrum - 695 019, Kerala, India.	www.niist.res.in

ICAR-Central Arid Zone Research Institute , Jodhpur	Central Arid Zone Research Institute Jodhpur-342003, Rajasthan	www.cazri.res.in
ICAR-Central Island Agricultural Research Institute, Andaman & Nicobar Islands	Central Island Agricultural Research Institute PB No. 181,Portblair-744101, Andaman & Nicobar Islands	www.icar.cari.res.in
ICAR-Central Food Technological Research Institute Mysore	Central Food Technological Research Institute,Mysore-570 020	www.cftri.com
ICAR-Central Institute for Subtropical Horticulture, Lucknow	Central Institute for Subtropical Horticulture, Rehmankhera P.O. Kakori,Lucknow- 226 101, Uttar Pradesh	www.cishres.in
ICAR-Central Institute of Arid Horticulture, Rajasthan	Central Institute of Arid Horticulture, Sri Ganganagar Highway Beechwal, Bikaner- 334006, Rajasthan	www.ciah.ernet.in
ICAR-Central Institute of Temperate Horticulture,J&K	Central Institute of Temperate Horticulture Old Air Field, Rangreth -190005,J&K	www.cith.org.in
ICAR-Central Plantation Crops Research Institute Kasargod	Central Plantation Crops Research Institute Kasargod-671124,Kerala	www.cpcri.gov.in
ICAR-Central Potato Research Institute, Shimla	Central Potato Research Institute, Shimla- 171001 Himachal Pradesh	www.cpri.emet.in
ICAR-Central Tuber Crops Research Institute Thiruvananthapuram	Central Tuber Crops Research Institute, Thiruvananthapuram Sreekariyam, Thiruvananthapuram-695017, Kerala	www.ctcri.org
ICAR-Central Research Institute for Dry Land Agriculture, Hyderabad	Central Research Institute for Dry Land Agriculture Santhoshnagar, Hyderabad- 500059, Andhra Pradesh	www.crida.in
Central Soil & Water Conservation Research & Training Institute, Dehradun	Central Soil & Water Conservation Research & Training Institute 218 ,Kaulagarh Road, Dehradun-248195 Uttarakhand	www.eswertiweb. org
Central Institute of Medicinal and Aromatic Plants	Central Institute of Medicinal and Aromatic Plants P.O. - CIMAP, Near Kukrail Picnic Spot, Lucknow - 226 015, India	www.cimap. res. in
Central Coffee Research Institute, Chickmagalur	Central Coffee Research Institute, Coffee Research Station Chickmagalur - 577 117	www.indiancoffee.org
ICAR-Water Technology Centre For Eastern Region Bhubaneswar	Water Technology Centre For Eastern Region,PO SE Railway Project Complex Chandrasekharpur,Bhubaneswar -751023,Orissa	www.wtcer.emet.in
ICAR-Central Soil Salinity Research Institute. Haryana	Central Soil Salinity Research Institute, Karnal- 132001, Haryana	www.cssri.org
ICAR Research Complex for NEH Region Umrol Road, Meghalaya	ICAR Res. Complex for NEH Region Umrol Road Ri-Bhoi -793103, Meghalaya	www.icameh.emet.in
ICAR Research Complex For Eastern Region ICAR Parisar, Patna	ICAR Research Complex For Eastern Region ICAR Parisar, P.O.B.V. College, Patna- 800014, Bihar	www.icarer.res.in

ICAR Central Coastal Agricultural Research Institute, Goa Ela. Goa	Central Coastal Agricultural Research Institute, Goa Ela Old Goa North Goa -403402, Goa	www.icar.res.in
ICAR-Directorate of Seed Research, Uttar Pradesh	Directorate of Seed Research, Kushmaur, P.O. Kaithauli.Mau Nath Bhanjan-275101,Uttar Pradesh	www.dsr.org.in
ICAR-Project Directorate of Cropping Systems Research, Meerut	Project Directorate of Cropping Systems Research, Modipuram Meerut-250110,Uttar Pradesh	www.pdesr.ernet.in
Acharya NG Ranga Agricultural University, Hyderabad	Acharya NG Ranga Agricultural University, Rajendranagar Hyderabad ,Ranga Reddy 500030	www.angrau.net
Dr. Y.S.R. Horticultural University, West Godavari Dist, Andhra Pradesh	Andhra Pradesh Horticultural University, Venketaremanna Gudam, (Near) Thade Palli gudam West Godavari District, Andhra Pradesh- 534101	www.drysrhu.edu.in
Professor Jayashankar Telangana State Agricultural University	Rajendranagar Hyderabad, Telangana 500030	www.pjtsau.ac.in
Anand Agricultural University, Gujarat	Anand Agricultural University, Anand- 388110 Gujarat	www.aau.in
Assam Agricultural University, Jorhat	Assam agricultural University, Jorhat- 785013 Assam	www.aau.ac.in
Bidhan Chandra Krishi Viswavidyalaya ,West Bengal	Bidhan Chandra Krishi Viswavidyalaya P.O. Krishi Viswavidyalaya, Mohanpur, Nadia - 741252 West Bengal	www.bckv.edu.in
Birsa Agricultural University, Ranchi	Birsa Agricultural University, Kanke Ranchi - 834006, Jharkhand	www.bauranchi.org
Chandra Shekar Azad University of Agriculture & Technology, Kanpur	Chandra Shekar Azad University of Agriculture & Technology Kanpur-208002, Uttar Pradesh	www.csauk.ac.in
Chaudhary Charan Singh Haryana Agricultural University, Haryana	Chaudhary Charan Singh Haryana Agricultural University Hissar-125004, Haryana	www.hau.emet.in
Chaudhary Sarwan Kumar Himachal Pradesh Krishi Vi shvavidy alaya, Palampur	Chaudhary Sarwan Kumar Himachal Pradesh Krishi Vishvavidyalaya,Palampur,Kangra-176062 Himachal Pradesh	www.hillagric.ac.in
Dr. Balasaheb Sawant Konkan Krishi Vidyapeeth	Dr. Balasaheb Sawant Konkan Krishi Vidyapeeth Dapoli Ratnagiri-415712	www.dbskkv.org
Dr.Panjabrao Deshmukh Krishi Vidyapeeth ,Akola	Dr.Panjabrao Deshmukh Krishi Vidyapeeth Krishi Nagar, Akola-444104	www.pdkv.ac.in
Dr Yashwant Singh Parmar Univ of Plorticulture & Forestry, Solan	Dr Yashwant Singh Parmar University of Horticulture & Forestry, Solan, Nauni -173230, Himachal Pradesh	www.yspuniversity.ac.in
Govind Ballabh Pant University of Agriculture & Technology,Uttrakhand	Govind Ballabh Pant University Of Agriculture & Technology Pantnagar,Udhamsingh Nagar - 263145, Uttarakhand	www.gbpuat.ac.in

Indira Gandhi Krishi Vishwavidyalaya, Raipur	Indira Gandhi Krishi vishwavidyalaya, Krishak Nagar Raipur-492006, Chattisgarh	www.igau.edu.in
Jawaharlal Nehru Krishi Viswavidyalaya , Jabalpur	Jawaharlal Nehru Krishi Viswavidyalaya, Krishi Nagar Adhartak Jabalpur-482004, Madhya Pradesh	www.jnkw.org
Junagadh Agricultural University, Gujarat	Junagadh Agricultural University, Moti Baug, Agril. Campus Junagadh-362001, Gujarat	www.jau. In
Kerala Agricultural University, Thrissur	Kerala Agricultural University, K.A.U.P.O., Thrissur-680656	www.kau.in
Maharana Pratap Univ. o f Agriculture & Technology Rajasthan	Maharana Pratap Univ.of Agriculture & Technology Udaipur-313001, Rajasthan	www.mpuat.ac.in
Mahatma Phule Krishi Vidyapeeth, Maharashtra	Mahatma Phule Krishi Vidyapeeth,Rahuri- 413722, Maharashtra	www.mpkvac.in
Vasantrao Naik Marathwada Agricultural University Parbhani	Vasantrao Naik Marathwada Agricultural University Parbhani - 431402, Maharashtra	www.nnmkv.nic.in
Narendra Dev University of Agriculture & Technology Faizabad	Narendra Dev University of Agriculture & Technology Kumarganj, Faizabad-224229, Uttar Pradesh	www.nduat.in
Agriculture University, Jodhpur	Agriculture University, Jodhpur, Jodhpur Mandore Rajasthan-342304	www.aujodhpur.ac.in
Agriculture University, Borkhera	Agriculture University, Borkhera, Baran Road Kota Rajasthan 324001	www.aukota.org
Navsari Agricultural University, Gujarat	Navsari Agricultural University .Vijalpore Navsari-396450, Gujarat	www.nau.in
Orissa Univ. of Agriculture & Technology, Bhubaneswar	Orissa Univ. of Agriculture & Technology Siripur Bhubaneswar, Khurda-751003	www.ouat.ac.in
Punjab Agricultural University, Ludhiana	Punjab Agricultural University, Ludhiana-141004, Punjab	www.pau.edu
Rajendra Agricultural University, Bihar	Rajendra agricultural University, Pusa, Samastipur-848125, Bihar	www.pusavarsity.org.in
Sardar Vallabh Ballabh Bhai Patel Univ. of Agriculture & Technology, Meerut	Sardar Vallabh Ballabh Bhai Patel University of Agriculture & Technology, Roorkee Road, Modipuram Meerut-250110, Uttar Pradesh	www.svbpmeerut.ac.in
Sardarkrushinagar-Dantiwada Agricultural University Gujarat	Sardarkrushinagar - Dantiwada Agricultural University Sardarkrushinagar, Dantiwada.Banaskanha-385506 Gujarat	www.sdau.edu.in
Sher-E- Kashmir Univ of Agricultural Sciences & Technology, Jammu	Sher-E- Kashmir University of Agricultural Sciences & Technology, Railway road ,Jammu-180012 (J&K)	www.skuast.org
Sher-E- Kashmir Univ of Agricultural Sciences & Technology of Kashmir, Srinagar	Sher-E- Kashmir Univ of Agricultural Sciences & Technology of Kashmir, Shalimar, Srinagar- 191121, J&K	www.skuastkashmir.ac.in

Rajasthan Agricutural University, Bikaner	Rajasthan Agricutural University,Bikaner-334006, Rajasthan	www.raubikaner.org
Tamil Nadu Agricutural University, Coimbatore	Tamil Nadu Agricutural University,Coimbatore- 641003 Tamil Nadu	www.tnau.ac.in
University of Agricultural Sciences ,Dharwad	University of Agricultural Sciences, Dharwad , Karnataka	www.uasd.edu
University of Agricultural Sciences Bangalore	University of Agricultural Sciences, Bangalore- 560065, Karnataka	www.uasbangalore.edu.in
University of Agricultural Sciences, Karnataka	University of Agricultural Sciences, PB 329 Raichur - 584102, Karnataka	www.uasraichur.edu.in
Uttar Banga Krishi Viswavidyalaya West Bengal	Uttar Banga Krishi Viswavidyalaya, P.O. Pundibari, Dist. Cooch,Behar-736165 .West Bengal	www.ubkv.ac.in
University of Agricultural and Horticultural Sciences Karnataka	University of Agricultural and Horticultural Sciences No. 126, Shimoga- 577024, Karnataka	www.uahs.in
University of Horticultural Sciences, Karnataka	University of Horticultural Sciences, Bagalkot - 587103 Karnataka	www.uhsbagalkot.edu.in
Uttarakhand University of Horticulture and Forestry Uttarakhand	Uttarakhand University of Horticulture and Forestry Bharasar - 246 123 Uttarakhand	www.uuhf.ac.in
Sri Konda Laxman Telangana State Horticultural University	Sri Konda Laxman Telangana State Horticultural University Rajendranagar, Hyderabad - 500030 Telangana State	www.skltshu.ac.in
Rubber Research Institute of India, Kerala	Rubber Research Institute of India, Rubber Board P.O. Kottayam - 686 009, Kerala, India	www.rubberboard.org. in
Tropical Botanic Garden and Research Institute, Palode	Tropical Botanic Garden and Research Institute Pacha-Palode, Thiruvananthapuram - 695562, Kerala, India	www.bgri.org
United Planters Association of South India	United Planters Association of South India	www.upasi.org
Tea Research F oundation, Coimbatore	Tea Research Foundation, Valparai - 642 127	
ICAR-AICRP on Medicinal Aromatic Plants and Betalvine	AICRP on Medicinal Aromatic Plants and Betalvine ICAR-Directorate of Medicinal and Aromatic Plants Research Boriavi Anand - 387 310	irector.dmapr@gmail.com
ICAR-AICRP on Palms	AICRP on Palms, ICAR-Central Plantation Crops Research Institute, Kasargod - 671 124	aicrppalms@yahoo.com
ICAR-AICRP on Spices	AICRP on species, ICAR-Indian Institute of Spices Research, Marikunnu Post Calicut - 673 012	aicrps@spices.res.in
ICAR-AICRP on Fruits	AICRP on Fruits Project Coordinator (Fruits) ICAR-IIHR Hessaraghatta Lake Post, Bangalore - 560 089	pcfruits@gmail.com

ICAR-AICRP Mushroom	AICRP on Mushroom, DMR Chambaghat, Solan - 173213	www.nrcmushroom.org
ICARP-AINRP on Temperate Fruits	Central Institute of Temperates Horticultures KD Farm, Old AirPort Road Rangreth,Srinagar,Kashmir,India - 190007	www.cith.org.in
ICAR-AICRP on Onion and Garlic	AICRP on Onion and Garlic ICAR-Directorate of Onion and Garlic Research, Rajguru Nagar, Pune - 410 505	www.dogr.res.in
ICAR-AICRP on Floriculture	AICRP on Floriculture, Directorate of Floricultural Research Pune-410 505	pcflori@gmail.com
ICAR-AICRP on Potato	AICRP on Potato, ICAR-Central Potato Research Institute, Shimla - 171 001, Himachal Pradesh	aicrp_potato@hotmail.com
ICAR-AICRP on Tuber Crops	AICRP on Tuber Crops, ICAR-Central Tuber Crops Research Institute, Sreekariyam, Thiruvananthapuram - 695 017	www.aicrptc.in
ICAR-AICRP on Vegetable Crops	AICRP on Vegetable Crops, ICAR-IIVR, PB No. 1 P.O. Jakhani, Varanasi - 221 305	pcvciivr@gmail.com
ICAR-AICRP on Arid Zone Fruits	AICRP on Arid Zone Fruits, ICAR-Central Institute of Arid Horticulture, Sri Ganganagar Highway, Beachwal Bikaner - 334 006	ciah@nic.in
ICAR-AICRP on Cashew	AICRP on Cashew, ICAR-Directorate of Cashew Research Puttur - 574 202. Dakshina Kannada, Karnataka	www.cashew.res.in
Central Agricultural Universities CAU, Imphal, Nagaland	Central Agricultural University, P.B. No, 23, Imphal P.O. Imphal-795004	www.cau.ac.in
Rani Lakshmi Bai Central Agricultural University	Rani Lakshmi Bai Central Agricultural University, Gwalior Road Jhansi, Uttar Pradesh 284003	www.rlbcau.ac.in
Deemed Universities	ICAR-IARI, Pusa New Delhi-110012	www.iari.res.in
Sam Higginbottom Institute of Agriculture, Technology and Sciences	Post Agricultural Institute, Rewa Road, Naini, Allahabad Uttar Pradesh-211007	www.shiats.edu.in
Rajendra Agricultural University, Pusa	Rajendra Agricultural University, Pusa, Smastipur, Bihar - 848125	www.pusavarsity.org.in

Appendix–VIII: Important Websites in Horticulture

T. Pradeepkumar, Kerala Agricultural University, P.O. KAU, Thrrisur – 680 656

Australian Society for Horticultural Science	http://www.aushs.org.au/
Agricultural & Processed Food Products Export Development Authority (APEDA)	http://www.apeda.com/
American Society for Horticultural Science	http://www.ashs.org/
Asian Vegetable Research and Development Centre (AVRDC) World Vegetable Centre	http://www.avrdc.org.tw/
Central Food Technological Research Institute (CFTRI)	http://www.cftri.com/
ICAR-Central Institute of Medicinal & Aromatic Plants (CIMAP)	http://www.cimap.org/
Centre Institute of Post harvest Engineering and Technology	http://www.icar.org.in/ciphet.html
Consultative Group on International Agricultural Research, CGIAR	http://www.cgiar.org/
Coffee Board, India	http://indiacoffee.org/
Department of Agriculture and Co-operation, India	http://agricoop.nic.in/
Department of Bio-technology, India	http://dbtindia.nic.in
Department of Scientific and Industrial Research, India	http://dsir.nic.in/
FAO	http://www.fao.org/
Global Agribusiness Information Network:	http://www.fintrac.com/gain/:
Greenhouse Vegetable Information:	http://www.ghvi.co.nz/
Indo-American Hybrid Seeds	www.indamseeds.com
Institute of Vegetable and Ornamental Crops	http://www.igzev.de/
Institute for Horticultural Development, Victoria, Australia	http://www.nre.vic.gov.au/agvic/ihd/
Iowa State University Department of Horticulture	http://www.hort.iastate.edu/
National Horticulture Board (NHB), India	http://hortibizindia.nic.in/
National Institute of Agricultural Extension Management (MANAGE)	http://www.manage.gov.in/
ICAR-AICRP Directorate of Mushroom Research, Chambaghat, Solan - 173213, Himachal Pradesh	www.nrcmushroom.org
Oregon State University, Department of Horticulture	http://osu.orst.edu/dept/hort
Pineapple News	http://agrss.sherman.hawaii.edu/ pineapple pineappl.htm
Pomology Resources Center	http://www.bsi.fr/pomologie/english/ pomology:
Rubber board, India	http://rubberboard.org.in/
Spice Paprika web site	http://www.paprika.deltav.hu/:
Spices Board, India	http://www.indianspices.com/
Sri Lanka Agribusiness on-line	http://www.agro-lanka.org/
Sustainable Apple Production	http://orchard.uvm.edu/
Tea Board, India	http://tea.nic.in/
The Horticultural Taxonomy Group	http://www.hortax.org.uk/
The International Society of Citriculture	http://www.lal.ufl.edu/isc_ citrus_ homepage.htm

Appendix – VIII a

Horticulture Institutes in India

Kerala Agricultural University, P.O. KAU, Thrissur – 680 656

Indian Council of Agricultural Research (ICAR)
Director General
Indian Council of Agricultural Research, Krishi Bhavan
Dr. Rajendra Prasad Road, New Delhi – 110 114
www.icar.org.in

ICAR-Indian Agricultural Research Institute (IARI)
Director
Indian Agricultural Research Institute, Pusa,
Dr. K. S. Krishnan Marg, New Delhi – 110012
Email: root@iari.ernet.in
Fax: 011-5766420, 5766420
www.iari.res.in

ICAR-Central Tuber Crops Research Institute (CTCRI)
Director
Central Tuber Crops Research Institute
Sreekariyam, Thiruvananthapuram – 695017
Kerala, India Tel: (91)(471) 2598551 to 2598554
Telegram: TUBERSEARCH Fax: 091-471-2590063
E-mail: ctcritvm@yahoo.com
www.ctcri.org

ICAR-Central Potato Research Institute (CPRI)
Director
Central Potato Research Institute
Shimla-171001, Himachal Pradesh
Gram-Potato Research, Shimla
Phone: 91-177-2625182, 2625073 Fax: 91-177-2624460
E-mail: dircpri@icar.org.in; directorcpri@gmail.com
www.cpri.ernet.in

ICAR-Indian Institute of Vegetable Research (IIVR)
Director
Indian Institute of Vegetable Research
Post Bag No. 01, Post Office Jakhani (Shahanshapur)
Varanasi - 221 305, Uttar Pradesh, India
Phone NO. 91-0542-2635247; 2635236 Fax 91-0542-229007
E-mail:directoriivr@gmail.com
www.iivr.org.in

ICAR-Indian Institute of Horticultural Research (IIHR)
Director
Indian Institute of Horticultural Research
Hessaraghatta, Bangalore - 560089
Email: director@iihr.ernet.in, Tel : 080 28466420 Fax: +91 80 28466291
www.iihr.res.in

ICAR-Indian Institute of Spices Research (IISR)
Director
Indian Institute of Spices Research
Marikunnu P.O., Calicut, Kerala, India - 673012
Ph: 0091-495-2731346 Grams : RESEARCH, CALICUT
Email: director@spices.res.in
www.spices.res.in

ICAR-Central Plantation Crops Research Institute (CPCRI)
Director
Central Plantation Crops Research Institute
Kasaragod – 671 124, Kerala
Grams: 'RESEARCH' Kasaragod
Ph : 04994-232 894-5 Fax : 91-4994-232 322
Email: directorcpcri@gmail.com
www.cpcri.gov.in

ICAR-Central Institute for Sub-Tropical Horticulture (CISTH)
Director
Central Institute for Sub-Tropical Horticulture
Rehmankhera (P.O), Kakori, Lucknow – 227 107, Uttar Pradesh
Ph: 0522 – 284 1022/23Fax: 0522 – 284 1025
E-mail: cish.lucknow@gmail.com
www.cishlko.org

ICAR-Central Institute for Arid Horticulture
(earlier National Research Center for Arid Horticulture)
Director, Central Institute for Arid Horticulture
Sri Ganganagar Road, NH - 15, Beechwal, Bikaner 334 006, Rajasthan
Ph - +91 151 2250960, 2250147 Fax - +91 151 2250145
E-mail: ciah@nic.in
http://www.ciah.ernet.in/

ICAR-Central Arid Zone Research Institute
Director
Central Arid Zone Research Institute
Jodhpur - 342003, Rajasthan
Ph : +91 291 2786584, Fax : +91 291 2788706
www.icar.org.in/cazri.htm

ICAR-Central Institute of Temperate Horticulture (CITH)
Director
Central Institute of Temperate Horticulture, P.O. Sanatnagar
Srinagar,Jammu & Kashmir – 190005
Fax: 0194-433178, 433104
E-mail: dircithsgr@icar.org.in

ICAR-Vivekananda Parvatiya Krishi Anusandhan Sansthan
(Vivekananda Institute of Hill Agriculture)
Indian Council of Agricultural Research
Almora – 263 601, Uttarakhand
Tel.: 05962-230208; Fax: 05962-231539
Email: vpkas@nic.in Website:http://vpkas.nic.in

ICAR-National Research Centre for Banana
Director
National Research Centre for Banana
Thogmalai Road, Thayanur Post, Tiruchirapalli – 620 102, Tamil Nadu, India
Tel: 9442553117
Email: nrcbdirector@gmail.com, directornrcb@gmail.com, nrcbanana@gmail.com
Website: www.nrcb.res.in

ICAR-National Research Centre for Grapes
Director
National Research Centre for Grapes
(Indian Council of Agricultural Research)
P.B. No. 3, P.O. Manjri Farm, Solapur Road
Pune – 412307, Maharashtra, India
Phone: 91-20-26956000 (EPABX), Fax: 91-20-26956099
Email: dirnrcg@gmail.com, nrcgrapes@gmail.com
Website: http//nrcgrapes.nic.in/

ICAR-Central Citrus Research Institute
Director
PB No. 464, Shankar Nagar P.O., Nagpur – 440 010
Tel: (Office) 0712-2500249/2500615
Telefax: 0091-0712-2500813
Email: citrus8_ngp@sancharnet.in, dirnrccngp@gmail.com
Website: nrccitrus.nic.in

ICAR-National research Centre for Litchi
Director
Mushahari, P.O. Rama, Muzaffarpur – 842 002, Bihar, India
Tel: Director: 0621-2289475, A.O./C.P.I.O.: 0621-2281160, A.F.&A.O.: 0621-2281161
Fax: 0621-2281162
Website: http://www.nrclitchi.org
Email:www.ccringp.org.in
Feedback: aris@nrclitchi.org, director@nrclitchi.org

ICAR-National Research Centre on Pomegranate
Director
National Research Centre on Pomegranate
NH-9, Solapur-Pune Highway, Kegaon (PO) Solapur – 413 255, Maharashtra
Tel: 0217-2353533 (Office)
Emai: nrcpomegranate@gmail.com, director@nrcpomegranate.org, rkrishnapal@gmail.com
Website: http://www.nrcpomegranate.org

ICAR-National Research Centre on Orchids
Director
NRC for Orchids
P.O. Pakyong, Sikkim – 737 106
Tel: 03592-257954 (O). 257289/257703
Fax: 03592-257289/282
Email: nrcorchids@rediffmail.com
Website: http://nrcorchids.nic.in/Home.html

ICAR-National Research Centre on Seed Spices
Director
National Research Centre on Seed Spices
Tabiji, Ajmer – 305 206 (Rajasthan), India
Tel: +91-145-2680955, 0145-2684400
Telefax: +91-145-2681001, 0145-2684417
Email: nrcss.director@gmail.com, drbsingh2000@yahoo.com
Website: www.nrcss.org

ICAR-National Bureau of Plant Genetics Resources
Director
National Bureau of Plant Genetics Resources
Indian Council of Agricultural Research, Ministry of Agriculture (Govt. of India)
Pusa Campus, New Delhi – 110012, India
Tel: 91-11-25842495
Email: director@nbpgr.ernet.in
Website: http://www.nbpgr.ernet.in

ICAR-Directorate of Seed Research
Director
Directorate of Seed Research
(Indian Council of Agricultural Research)
Post Bag No. 11, Kusmaur
P.O.: Kaithauli, Maunath Bhanjan, UP – 275101
Tel: +91-0547-2530326
Fax: +91-0547-2530325
Email:pd_dsr2005@yahoo.co.in
Website: http://www.dsr.org.in/page.asp?id=173

ICAR-Directorate of Mushroom Research
The Director
Directorate of Mushroom Research
Indian Council of Agriculture Research
Chambaghat – 173213
Solan, Himachal Pradesh
Tel: 91-1792-230767, 230541
Director's Office: 91-1792-230451
Residence: 230401, Fax: 91-1792-231207
Email: directordmr@gmail.com
Website: www.nrcmushroom.org

ICAR-Directorate of Onion and Garlic Research
Director
Director of Onion and Garlic Research
Rajgurunagar – 410 505
Pune, Maharashtra
Tel: 02135-222026, 222697
Fax: 02135-224056
Email: director@dogr.res.in, aris@dogr.res.in
Website: www.dogr.res.in

ICAR-Indian Institute of Oilpalm Research
Director
Indian Institute of Oilpalm Research
West Godavari District, Pedavegi – 534450, Andhra Pradesh
Gram: Palmsearch, Eluru
Tel: 91-8812-259532, 259524
Fax: 91-8812-259531
Email: dopr2009@gmail.com
Website: http://dopr.gov.in

ICAR-Directorate of Medicinal and Aromatic Plants Research
Boriavi – 387310, Anand Gujarat
Tel: 91-2692-271602; Fax: 91-2692-271601
EPABX: 91-2692-271605; 91-2692-271606
Email: director.dmapr@gmail.com
URL: http://www.dmapr.org.in

Appendix - IX
Vital Statistics on Horticultural Crops
National Horticulture Board, *Gurgaon, Haryana*

Table 1. Projected demand of horticultural crops

Commodity	Production (million tons)			
	1991-92	2010-11	2016-17	2020-2021
Fruits	28.63	74.80	90.00	98.00
Vegetables	58.53	141.00	220.00	220.00
Spices	1.90	4.10	6.00	6.50
Coconut	7.49	10.90	15.50	24.00
Cashew nut	0.40	0.67	1.00	2.00
Cocoa, other	-	4.00	7.00	9.50
Total	96.95	240.5	339.50	360.00

Table 2. Global area, production and productivity of Fruits

Country	Area in million hectare	Production in million tones	Productivity in tones/ha
China	11.4	122.18	10.7
India	6.38	74.87	11.7
Brazil	2.44	39.28	16.1
United States of America	1.14	25.38	22.2
Italy	1.27	16.9	13.2
Philippines	1.15	16.18	14
Mexico	1.21	15.25	12.6
Spain	1.56	15.18	9.7
Indonesia	0.66	14.86	22.4
Turkey	1.09	13.94	12.8
Others	26.75	245.22	9.2
Total	55.05	599.24	14.05

Table 3. Changes in Area Production and Productivity of fruits in India

Year	Area in lakh ha	Production in million tones	Productivity in tones/ha
1991-92	2.87	28.63	10.0
2001-02	4.01	43.00	10.7
2002-03	3.78	45.20	11.9
2003-04	4.66	45.94	9.9
2004-05	5.04	50.86	10.1
2005-06	5.32	55.85	10.4
2006-07	5.55	59.56	10.7
2007-08	5.85	65.58	11.2
2008-09	6.10	68.46	11.2
2009-10	6.32	71.51	11.3
2010-11	6.38	74.87	11.7
2011-12	6.70	76.42	11.4
2012.13	6.98	81.29	11.6
2013.14	7.22	88.98	12.3
2014-15	6.36	88.82	14.0
2015-16	6.41	91.44	14.3

Appendix X
Nutritional Values of Horticultural Crops
***Indian Institute of Nutrition,** Hyderabad*

Table 1: Nutritional value of fruits

Fruits	Energy (κCal)	Protein (g)	Carbohydrate (g)	Fat (g)	Calcium (mg)	Fe (mg)
					All values are per 100g	
Amla	58	0.5	13.7	0.1	50	1.2
Apple	59	0.2	13.4	0.5	10	0.66
Apricot (fresh)	53	1	11.6	0.3	20	2.2
Apricot (dry)	306	1.6	73.4	0.7	110	4.6
Bael fruit	137	1.8	31.8	0.3	85	0.6
Banana (ripe)	116	1.2	27.2	0.3	17	0.36
Cherries (red)	64	1.1	13.8	0.5	24	0.57
Currants (black)	316	2.7	75.2	0.5	130	8.5
Dates (dried)	317	2.5	75.8	0.4	120	7.3
Dates (fresh)	144	1.2	33.8	0.4	22	0.96
Figs (Anjeer)	37	1.3	7.6	0.2	80	1
Grapes (blue)	58	0.6	13.1	0.4	20	0.5
Grapes (pale green)	71	0.5	16.5	0.3	20	0.52
Guava	51	0.9	11.2	0.3	10	0.27
Jamun	62	0.7	14	0.3	15	0.43
Lemon	57	1	11.1	0.9	70	0.26
Litchi	61	1.1	13.6	0.2	10	0.70
Loquat	43	0.6	9.6	0.3	30	1.30
Mango (ripe)	74	0.6	16.9	0.4	14	1.30
Musk Melon (Kharbuja)	17	0.3	3.5	0.2	32	1.40
Mulberry (Shahtoot)	49	1.1	10.3	0.4	70	2.3
Orange	48	0.7	10.9	0.2	26	0.32
Orange (juice)	9	0.2	1.9	0.1	5	0.7
Papaya (ripe)	32	0.6	7.2	0.1	17	0.5
Peaches	50	1.2	10.5	0.3	15	2.4
Pears	52	0.6	11.9	0.2	8	0.5
Phalsa	72	1.3	14.7	0.9	129	3.1
Pineapple	46	0.4	10.8	0.1	20	2.4
Plum	52	0.7	11.1	0.5	10	0.6
Pomegranate	65	1.6	14.1	0.1	10	1.79
Raspberry	56	1	11.7	0.6	40	2.3
Sharifa	104	1.6	23.5	0.4	17	4.3
Strawberry	44	0.7	9.8	0.2	30	1.8
Water melon (Tarbuj)	16	0.2	3.3	0.2	11	79

Table 2: Nutritional value of spices and condiments

Spices and Condiments	Energy (κCal)	Protein (g)	Carbohydrate (g)	Fat (g)	Calcium (mg)	Fe (mg)
					All values are per 100g	
Asafoetida	297	4	67.8	1.1	690	39.4
Cardamom	229	10.2	42.1	2.2	130	4.6
Chilli (Dry)	246	15.9	31.6	6.2	160	2.3
Chilli (Green)	29	2.9	3	0.6	30	4.4
Cloves (Dry)	286	5.2	46	8.9	740	11.7
Coriander	288	14.1	21.6	16.1	630	7.1
Cumin Seeds	356	18.7	36.6	15	1080	11.7
Fenugreek Seeds	333	26.2	44.1	5.8	160	6.5
Garlic(Dry)	145	6.3	29.8	0.1	30	1.2
Ginger (Fresh)	67	2.3	12.3	0.9	20	3.5
Mango Powder	337	2.8	64	7.8	180	45.2
Pepper Black (Dry)	304	11.5	49.2	6.8	460	12.4
Pepper Green	98	4.8	13.7	2.7	270	24
Poppy Seeds	408	21.7	36.8	19.3	1584	15.9
Turmeric	349	6.3	69.4	5.1	150	67.8

Table 3: Nutritional value of roots and tubers

Spices and Condiments	Energy (κCal)	Protein (g)	Carbohydrate (g)	Fat (g)	Calcium (mg)	Fe (mg)
					All values are per 100g	
Arrow root	334	0.2	83.1	0.1	10	1
Beet root	43	1.7	8.8	0.1	18.3	1.19
Carrot	48	0.9	10.6	0.2	80	1.03
Colocasia (Arvi)	97	3	21.1	0.1	40	0.42
Onion	50	1.2	11.1	0.1	47	0.6
Potato	97	1.6	22.6	0.1	10	0.48
Radish	17	0.7	3.4	0.1	35	0.40
Sweet potato	1.2	1.2	28.2	0.3	46	0.21
Turnip (Shalgam)	29	0.7	6.2	0.2	30	0.40
Yam (Zimikand)	79	1.2	18.4	0.1	35	1.19

Table 4: Nutritional value of peas, beans and pulses

Peas, beans and pulses	Energy (κCal)	Protein (g)	Carbohydrate (g)	Fat (g)	Calcium (mg)	Fe (mg)
					All values are per 100g	
Channa	360	17.1	60.9	5.3	202	4.6
Channa dal	372	20.8	59.8	5.6	56	5.3
Kala Channa (roasted)	369	22.5	58.1	5.2	58	5.6
Urad dal (whole)	347	24	59.6	1.4	154	3.8
Lobia	323	24.1	54.5	1	77	8.6
Beans (sem,dry)	347	24.9	60.1	0.8	60	2.7
Moong dal (whole)	334	24	56.7	1.3	124	4.4
Moong dal	358	24.5	59.9	1.2	75	3.9
Kala channa	321	22	57.2	0.5	287	6.77
Masoor dal	343	25.1	59	0.7	69	7.58
Moth dal	330	23.6	56.5	1.1	202	9.58
Peas (green)	93	7.2	15.9	0.1	20	1.5
Peas (dry)	315	19.7	56.5	1.1	75	7.05
Peas (roasted)	340	22.9	58.8	1.4	81	6.40
Rajmah	346	22.9	60.6	1.3	260	5.10
Red gram (Arhar)	335	22.3	57.6	1.7	73	2.7
Soyabean	432	43.2	20.9	19.5	240	1075

Table 5: Nutritional value of leaf vegetables

Leaf vegetables	Energy (κCal)	Protein (g)	Carbohydrate (g)	Fat (g)	Calcium (mg)	Fe (mg)
					All values are per 100g	
Bathua	30	3.7	2.9	0.4	150	4.2
Cabbage	27	1.8	4.6	0.1	39	0.8
Celery leaves	37	6.3	1.6	0.6	230	6.3
Colocasia leaves(Arvi leaves)	56	3.9	6.8	1.5	460	0.98
Coriander leaves (Dhania)	44	3.3	6.3	0.6	184	1.42
Fenugreek leaves (Methi)	49	4.4	6	0.9	395	1.93
Lettuce	21	2.1	2.5	0.3	50	2.4
Mustard leaves(Sarso ka saag)	34	4	3.2	0.6	155	16.3
Radish leaves	28	3.8	2.4	0.4	265	0.09
Spinach (Palak)	26	2	2.9	0.7	73	1.14

The demand for high-value commodities (such as horticulture, dairy, livestock and fish) is increasing faster than foodgrains

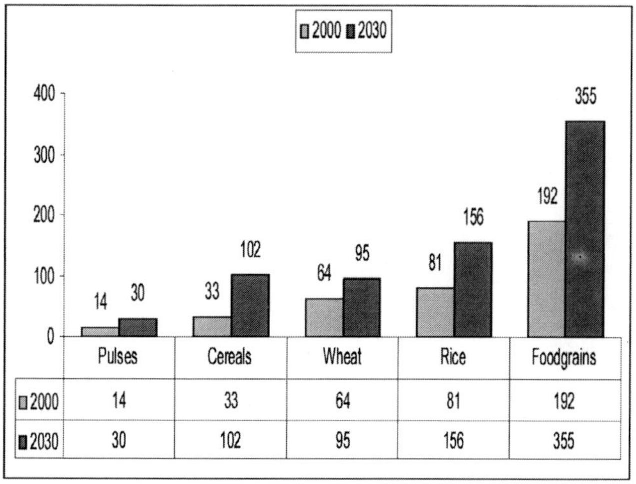

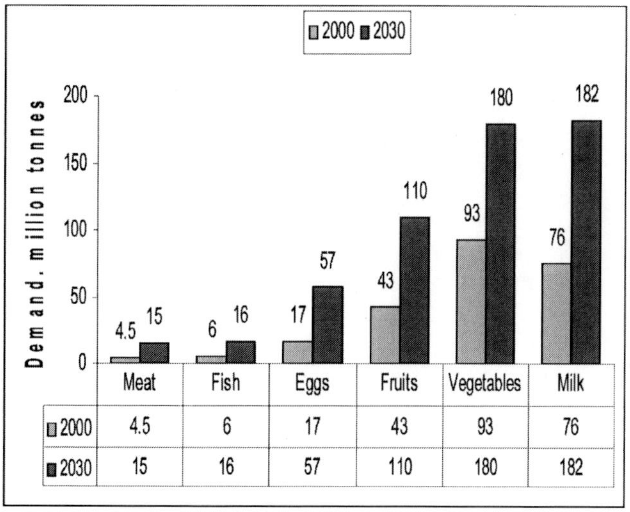

Colour Plates

Chapter 17: Features of A Garden

Fig. 1: Garden wall

Fig. 2: Dry wall

Fig. 3: Green wall at the Universidad del Claustro de Sor Juana, Mexico

Fig. 4: A wooden fence

Fig. 5: A stone step Garden Drives

Fig. 6: A Garden drive at Lodhi Garden, Delhi

Fig. 7: A zigzag garden path

Fig. 8: Hedge

Fig. 9: Edging

Fig. 10: Shrubbery

Fig. 11: Herbaceous border at East Lothian, Scotland

Fig. 12: Flower beds

Fig. 13: Carpet bedding

Fig. 14. Ground cover of *Zebrina pendula*

Fig. 15: Elephants shape topiary at Bang Pa-in Royal Palace, Thailand

Fig. 16. A beautiful lawn around India Gate New Delhi

Fig. 17: Arches

Fig. 18: Pergola

Fig. 19: A bridge in the Japanese Tea Garden at Golden Gate Park in San Francisco, California

Fig. 20: Gazebo at Will Rogers Gardens, Oklahoma City, UK

Fig. 21: Vertical Garden

Fig. 22: Roof garden near island of Sentosa adjacent to Singapore

Fig. 23: Baha'i Shrine and Gardens, Haifa, Israel

Fig. 24: Plants in water garden

Fig. 26: Healing Garden at Swedish Covenant Hospital, United States

Fig. 25: A view of Rock Garden, Chandigarh

Fig. 27: Floral Clock at International Peace Garden, North Dakota, US

Fig. 28: Garden seat

Fig. 29: Vases

Fig. 30: Bird bath

Fig. 31: Sundial in Singapore Botanic Gardens

Fig. 32: Musical fountains at Akshardham Temple, New Delhi

Fig. 33: The Four Godesses of the Seasons Statues

Fig. 34: Ornamental stones

Chapter 20: Principles and Elements of Landscape Design

Perfect proportion of garden elements to its setting

Formal balance of the garden

Repetition of fountains creates continuous movement in garden

Tree sculpture create emphasis in the landscape

A variety of forms creates significant landscape interest

Index

Symbols

A

B

C

D

E

F

G

H

I

J

M

N

O

P

Q

R

S

T

U

V

W

X

Y

Z